Patrick Moore's Practical Astronomy Series

For further volumes:
http://www.springer.com/series/3192

The Amateur Astronomer's Guide to the Deep-Sky Catalogs

Jerry D. Cavin

To
Anoushah Ansari,
 A friend and classmate.
Keep reaching for the
stars,
 Jerry Cavin
 4/14/14

🐴 Springer

Jerry D. Cavin
Austin, TX, USA
jcavin@yahoo.com

ISBN 978-1-4614-0655-6   e-ISBN 978-1-4614-0656-3
DOI 10.1007/978-1-4614-0656-3
Springer New York Dordrecht Heidelberg London

Library of Congress Control Number: 2011936008

© Springer Science+Business Media, LLC 2012
All rights reserved. This work may not be translated or copied in whole or in part without the written permission of the publisher (Springer Science+Business Media, LLC, 233 Spring Street, New York, NY 10013, USA), except for brief excerpts in connection with reviews or scholarly analysis. Use in connection with any form of information storage and retrieval, electronic adaptation, computer software, or by similar or dissimilar methodology now known or hereafter developed is forbidden.
The use in this publication of trade names, trademarks, service marks, and similar terms, even if they are not identified as such, is not to be taken as an expression of opinion as to whether or not they are subject to proprietary rights.

Cover illustrations:

Top left: Johann Elert Bode, German astronomer.
Top right: John Louis Emil Dreyer.
Bottom left: Sir William Herschel, , as depicted in the book Sir William Herschel, His Life and Works by Edward Singleton Holden (W. H. Allen & Co, 13 Waterloo Place, S W, London, (1881), Frontispiece.
Bottom right: Tycho Brahe, illustration from J. L. E. Dreyer biography of his fellow countryman Tycho Brahe – A Picture of Scientific Life and Work in the Sixteenth Century (1890).

Printed on acid-free paper

Springer is part of Springer Science+Business Media (www.springer.com)

Preface

This book provides an introduction to a few of the historical deep space sky catalogs. It is intended to provide amateur astronomers with historical information and the catalog entries with coordinates. In some cases, the information recorded in the original catalog have been omitted to allow the contents to be formatted properly on the page. To this end, I have chosen to keep the data that will allow the amateur the enjoyment of observing the objects of each catalog.

Almost 2,000 years of historical catalogs are covered in this book. Chapter One describes Ptolemy's Almagest created in A. D. 140, and Chapter Twelve describes the Caldwell Catalog created in 1995. In choosing the set of catalogs used in this book, I provided what I thought the most interesting historic catalogs. I apologize if I have omitted to list and describe your favorite. The history of astronomy is rich with many catalogs that equally deserve to be noted, and it is nearly impossible to include all here.

I would like to thank the many people who helped during the preparation of this book. The idea for writing this book came in part from the many research papers I wrote in earning my Master of Science in Astronomy at the Swinburne University of Technology Astronomy Online Program. My sincere thank you goes out to the many Swinburne professors and astronomers who taught me over the years. I would like to the thank the many people who provided the information used to compile the catalogs documented in these chapters, including Joerg Schlimmer, Carol Huston, Neda Mobara, Brenda Branchett and Carol Iorg, of the Astronomy League; Dennis Duke, Professor of Physics at the Florida State University; Dennis J Webb, co-author of the book *The Arp Atlas of Peculiar Galaxies*, and Robert Erdmann of the NGC/IC Project. Finally, I would like to thank my wife Halima, my daughter Sheila, and my son Zachariah for understanding and accepting the private time I needed to compile and write this book.

Jerry D. Cavin

About the Author

Jerry Cavin first turned to amateur astronomy while growing up under the dark skies of Iowa. After leaving the farm, he completed a BS in Computer Science with a minor in Electrical Engineering at the University of Nebraska at Omaha. He worked for 10 years as the Lead Software Engineer at the Control Data Corporation specializing in real time control systems. After getting married he moved to Austin, Texas and spent over 15 years at the University of Texas at Austin working as a Research Scientist Associate on electronic warfare systems and 6 years at Overwatch Systems.

He is currently working as a Quality Engineer at Bridge360 on a wide variety of software projects. He has recently completed his second MSc degree in Astronomy at the Swinburne Astronomy Online, via the Swinburne University of Technology located in Melbourne, Australia.

Contents

Chapter 1

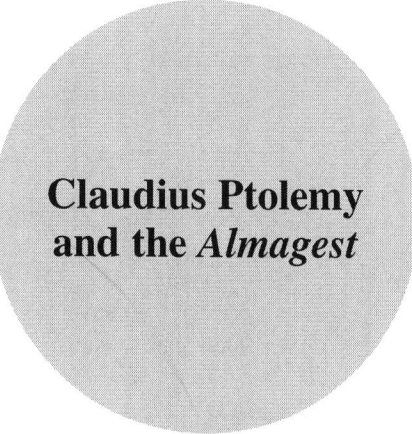

Claudius Ptolemy and the *Almagest*

We know of many ancient astronomers. Aristarchus of Samos (310 BC–c 230 BC) was the first to argue for a heliocentric view of the Solar System putting the Sun at the center of the universe. Eratosthenes of Cyrene (c 276 BC–c 195 BC) created the technique of measuring Earth-based locations by using latitudes and longitudes and also accurately computed the circumference of Earth. The calculations of Thales of Miletus (c 624 BC–c 546 BC) allowed him to accurately predict solar eclipses. Aristyllus (c 280 BC), with the help of Timocharis (c 320 BC–c 260 BC), was the first astronomer to create sky catalogs at the astronomical observatory of the library of Alexandria. The observations of Aristyllus and Timocharis are the earliest known Greek observations and can be dated to the year 290 BC But it was Hipparchus of Nicaea that became known as the founder of astronomy.

Hipparchus of Nicaea (c. 200–126 BC)

What little we know about Hipparchus is found in the writings of Strabo of Amasya and Ptolemy of Alexandria. They recorded that Hipparchus performed his astronomical observations in Bithynia, on the island of Rhodes southwest of Turkey in the Aegean Sea, and in Alexandria. He wrote many astronomy books, including one that documented his accomplishments; unfortunately the only surviving work of Hipparchus is the *Phenomena,* a commentary on a poem about the constellations by Aratus and Eudoxus.

Hipparchus would have been exposed to the Babylonians' knowledge of the stars. Using the historical records he made many improvements to the observational instruments, allowing him to make more precise measurements of the heavens than

J.D. Cavin, *The Amateur Astronomer's Guide to the Deep-Sky Catalogs,*
Patrick Moore's Practical Astronomy Series, DOI 10.1007/978-1-4614-0656-3_1,
© Springer Science+Business Media, LLC 2012

anyone before him. His groundbreaking works in astronomy led him to discover new areas of mathematics.

In 134 BC Hipparchus was witness to a supernova, or "guest star," in the ancient sky. The supernova would have been an extraordinarily bright object in the sky during the day and night. During Hipparchus's time the sky was thought to be unchanging with the exception a few known 'wanderers' (the planets). This brilliant new object would have shaken his belief in an unchanging, static heaven. The event was reported to have encouraged him to measure and compile a catalog of the positions of the stars.

According to Ptolemy, Hipparchus spent 20 years making painstaking measurements of nearly 1,000 stars. After constructing his new star catalog he found another mystery concerning the stars. He compared his measurements to the measurements from the early catalogs of Aristyllus and Timocharis, taken 150 years before him. Although the latitude remained consistent he saw that the longitude measurement showed an increase of 2° from the older measurements. This difference in measurements indicated that the equator was slowly moving towards the ecliptic at a rate of 48 s a year. From this he deduced the "precession of the equinoxes." This is known today to be caused by the gradual shift in the orientation of Earth's axis, which traces out a circle once every 26,000 years.

Claudius Ptolemaeus (AD 90–AD 160)

Claudius Ptolemaeus (Ptolemy) was a mathematician, geographer, astrologer, and astronomer living in Egypt under Roman rule. What is known about Ptolemy is based solely on his three great books that have been passed down through the ages, the *Almagest,* his treatise on Greek astronomy; the *Geography*, a treatise on the cartography of the Persian and Roman world; and the *Tetrabiblos*, a treatise on astrology.

The name Almagest is an Arabic name meaning the "Great Book" or "Great Work." Ptolemy called his work the *"Mathematical Treatise."* The *Almagest* became the most influential work in astronomy for the next 1,500 years. The star catalog of the *Almagest* contained 1,028 stars. Each star is listed with the name of the constellation in which it appeared, its location within the constellation, its ecliptic longitude and latitude, and the magnitude (brightness) of the star, from 1 (brightest) to 6 (dimmest). The method of measuring the brightness of stars is believed to have been created by Hipparchus. With some changes this is still the way we measure the brightness today (Table 1.1).

Ptolemy's *Almagest* has been embroiled in controversy for many years. If the measurements are closely examined it shows that Ptolemy had plagiarized another catalog. Many scholars believe Hipparchus's catalog was the real source of the catalog of the *Almagest*. Ptolemy had simply adjusted the measurements to allow for 150 years of precession. Early Persian and European astronomers such as Ulugh Beg and Tycho Brahe also believed this to be true.

Fig. 1.1 An imagined depiction of Claudius Ptolemy holding a cross staff (some historians have traced the origins of the cross staff to the Chaldeans in 400 BC) used to measure the separation between stars from the book *Great Astronomers* by Sir Robert Stawell Ball, published by the J.B. Lippincott Company in 1895

Table 1.1 The catalog of the *Almagest* listed by constellation. The data in this table is provided with permission of Dennis Duke, professor of physics at the Florida State University

#	HR number	Full star name	Constellation Name	Constellation #	Star	Almagest measurement Long	Almagest measurement Lat	Actual (Epoch 137 AD) Long	Actual (Epoch 137 AD) Lat	V Mag	aV Mag
1	424	1Alp UMi	UMi	1	1	60.17	66	62.63	65.88	3	2
2	6789	23Del UMi	UMi	1	2	62.5	70	65.29	69.73	4	4.4
3	6,322	22Eps UMi	UMi	1	3	70.17	74.33	73.11	73.68	4	4.2
4	5,903	16Zet UMi	UMi	1	4	89.67	75.67	91.07	74.89	4	4.3
5	6,116	21Eta UMi	UMi	1	5	93.67	77.67	94.2	77.73	4	4.9
6	5,563	7Bet UMi	UMi	1	6	107.5	72.83	106.87	72.83	2	2.1
7	5,735	13Gam UMi	UMi	1	7	116.17	74.83	114.98	75.1	2	3
8	5,430	5 UMi	UMi	1	8	103	71.17	102.05	71.25	4	4.3
9	3,323	1Omi UMa	UMa	2	1	85.33	39.83	87.06	40.09	4	3.4
10	3,354	2 UMa	UMa	2	2	85.83	43	85.64	44.38	5	5.5
11	3,403	4Pi 2UMa	UMa	2	3	86.33	43	86.85	43.8	5	4.6
12	3,576	8Rho UMa	UMa	2	4	86.17	47.17	87.96	47.72	5	4.8
13	3,616	13Sig2UMa	UMa	2	5	86.67	47	89.23	47.65	5	4.8
14	3,771	24 UMa	UMa	2	6	88.17	50.5	90.35	51.01	5	4.6
15	3,624	14Tau UMa	UMa	2	7	90.5	43.83	91.48	44.37	4	4.7
16	3,757	23 UMa	UMa	2	8	92.5	44.33	94.71	44.92	4	3.7
17	3,888	29Ups UMa	UMa	2	9	99	42	100.35	42.59	4	3.8
18	3,894	30Phi UMa	UMa	2	10	101	44	103.31	38.06	4.3	4.6
19	3,775	25The UMa	UMa	2	11	100.67	35	101.66	35.15	3	3.2
20	3,569	9Iot UMa	UMa	2	12	95.5	29.33	97.01	29.55	3	3.1
21	3,594	12Kap UMa	UMa	2	13	96.33	28.33	97.95	28.81	3	3.6
22	3,662	18 UMa	UMa	2	14	95.67	36	97.28	35.86	4	4.8
23	3,619	15 UMa	UMa	2	15	95.83	33	97.18	33.29	4	4.5
24	4,301	50Alp UMa	UMa	2	16	107.67	49	109.14	49.57	2	1.8
25	4,295	48Bet UMa	UMa	2	17	112.17	44.5	113.29	44.94	2	2.4
26	4,660	69Del UMa	UMa	2	18	123.17	51	124.81	51.5	3	3.3

27	4,554	64Gam UMa	UMa	2	19	123	46.5	124.27	46.98	2	2.4
28	4,033	33Lam UMa	UMa	2	20	112.67	29.33	113.6	29.78	3	3.5
29	4,069	34Mu UMa	UMa	2	21	114.17	28.25	115.26	28.85	3	3
30	4,335	52Psi UMa	UMa	2	22	121.67	35.25	122.77	35.44	3.7	3
31	4,377	54Nu UMa	UMa	2	23	129.83	25.83	130.64	26.06	3	3.5
32	4,375	53Xi UMa	UMa	2	24	130.33	25	131.39	25	3	3.9
33	4,905	77Eps UMa	UMa	2	25	132.17	53.5	132.62	54.2	2	1.8
34	5,054	79Zet UMa	UMa	2	26	138	55.67	139.32	56.29	2	2.1
35	5,191	85Eta UMa	UMa	2	27	149.83	54	150.75	54.42	2	1.9
36	4,914	12Alp2CVn	UMa	2	28	147.83	39.75	148.59	40.11	3	2.8
37	4,785	8Bet CVn	UMa	2	29	140.17	41.33	142.09	40.56	5	4.3
38	3,705	40Alp Lyn	UMa	2	30	105	17.25	105.99	17.81	4	3.1
39	3,690	38 Lyn	UMa	2	31	103.33	19.17	104.59	19.98	4	3.8
40	3,800	10 LMi	UMa	2	32	106.17	20	107.83	20.58	7	4.6
41	3,809		UMa	2	33	102.17	22.5	106.81	23.64	7	4.8
42	3,612		UMa	2	34	101.17	20.33	101.58	20.71	7	4.6
43	3,275	31 Lyn	UMa	2	35	90	22.25	91.59	22.94	7	4.3
44	6,370	21Mu Dra	Dra	3	1	206.67	76.5	208.46	76.41	4	5.1
45	6,554	25Nu 2Dra	Dra	3	2	221.83	78.5	223.68	78.33	3.7	4.8
46	6,536	23Bet Dra	Dra	3	3	223.17	75.67	225.8	75.51	3	2.8
47	6,688	32Xi Dra	Dra	3	4	237.33	80.33	238.44	80.48	4	3.8
48	6,705	33Gam Dra	Dra	3	5	239.67	75.5	242.04	75.18	3	2.2
49	6,923	39 Dra	Dra	3	6	264.67	82.33	266.99	81.97	4	5
50	7,049	46 Dra	Dra	3	7	272.33	78.25	274.53	78.07	4	5
51	6,978	45 Dra	Dra	3	8	268.83	80.33	270.38	79.98	4	4.8
52	7,125	47Omi Dra	Dra	3	9	289.5	81.5	289.82	80.98	4	4.7
53	7,371	58Pi Dra	Dra	3	10	338	81.67	339.01	81.82	4	4.6
54	7,310	57Del Dra	Dra	3	11	350.5	83	352.74	82.86	4	3.1
55	7,582	63Eps Dra	Dra	3	12	7.67	78.83	7.8	79.4	4	3.8

(continued)

Table 1.1 (continued)

#	HR number	Full star name	Constellation Name	#	Star	Almagest measurement Long	Lat	Actual (Epoch 137 AD) Long	Lat	V Mag	aV Mag
56	7,685	67Rho Dra	Dra	3	13	352.83	77.83	355.4	78.08	4	4.5
57	7,462	61Sig Dra	Dra	3	14	10.67	80.5	11.55	80.93	5	4.7
58	7,180	52Ups Dra	Dra	3	15	21.67	81.33	25.64	83.06	5	4.8
59	7,352	60Tau Dra	Dra	3	16	26.17	80.25	29.44	80.48	5	4.4
60	6,636	31Psi1Dra	Dra	3	17	73.33	84.5	77.03	83.82	4	4.3
61	6,927	44Chi Dra	Dra	3	18	50.33	83.5	53.24	83.24	4	3.6
62	6,920	43Phi Dra	Dra	3	19	41.83	84.83	45.89	84.66	4	4.2
63	6,566	27 Dra	Dra	3	20	118.67	87.5	117.97	86.78	6	5.1
64	6,596	28Ome Dra	Dra	3	21	111.67	86.83	105.47	86.83	6	4.8
65	6,223	18 Dra	Dra	3	22	159	81.25	156.77	81.65	5	4.8
66	6,315	19 Dra	Dra	3	23	159.33	83	156.79	83.22	5	4.9
67	6,396	22Zet Dra	Dra	3	24	158.33	84.83	154.9	84.78	3	3.2
68	6,132	14Eta Dra	Dra	3	25	160	78	167.54	78.49	3	2.7
69	5,986	13The Dra	Dra	3	26	160.33	74.67	170.8	74.51	3.7	4
70	5,744	12Iot Dra	Dra	3	27	162.67	70	158.34	71.11	3	3.3
71	5,226	10 Dra	Dra	3	28	127.33	64.67	128.51	65.27	4	4.7
72	5,291	11Alp Dra	Dra	3	29	131.17	65.5	131.07	66.29	3	3.7
73	4,787	5Kap Dra	Dra	3	30	109.17	61.25	110.05	61.62	3	3.9
74	4,434	1Lam Dra	Dra	3	31	103.17	56.25	104.18	57.08	3	3.8
75	7,750	1Kap Cep	Cep	4	1	35	75.67	37.55	75.28	4	4.4
76	8,974	35Gam Cep	Cep	4	2	33	64.25	34.31	64.41	4	3.2
77	8,238	8Bet Cep	Cep	4	3	7.33	71.17	10.19	71.01	4	3.2
78	8,162	5Alp Cep	Cep	4	4	346.67	69	347.31	68.91	3	2.4
79	7,957	3Eta Cep	Cep	4	5	339.33	72	338.28	71.57	4	3.4
80	7,850	2The Cep	Cep	4	6	340	74	339.77	73.93	4	4.2
81	8,417	17Xi Cep	Cep	4	7	358.5	65.5	358.56	65.71	5	4.3

82	8,694	32Iot Cep	Cep	4	8	7.5	62.5	7.87	62.49	3.7	3.5
83	8,494	23Eps Cep	Cep	4	9	346.33	60.25	347.16	60.04	5	4.2
84	8,465	21Zet Cep	Cep	4	10	347.33	61.25	348.46	61.09	4	3.3
85	8,469	22Lam Cep	Cep	4	11	349	61.33	350.51	61.83	5	5
86	8,316	Mu Cep	Cep	4	12	343.67	64	344.29	64.15	4	4.1
87	8,571	27Del Cep	Cep	4	13	351.33	59.5	352.07	59.47	5	3.8
88	5,328	17Kap2Boo	Boo	5	1	152.33	58.67	153.58	58.88	5	4.4
89	5,350	21Iot Boo	Boo	5	2	154.17	58.33	154.95	58.86	5	4.8
90	5,404	23The Boo	Boo	5	3	155.33	60.17	156.18	60.36	5	4.1
91	5,351	19Lam Boo	Boo	5	4	159.67	54.67	160.92	54.67	5	4.2
92	5,435	27Gam Boo	Boo	5	5	169.67	49	171.61	49.59	3	3
93	5,602	42Bet Boo	Boo	5	6	176.67	53.83	178.04	54.28	3.7	3.5
94	5,681	49Del Boo	Boo	5	7	185.67	48.67	186.91	49.14	3.7	3.5
95	5,733	51Mu 1Boo	Boo	5	8	185.67	53.25	187.13	53.55	4	4.2
96	5,763	52Nu 1Boo	Boo	5	9	185	57.5	186.29	57.21	4	5
97	5,727	2Eta CrB	Boo	5	10	187.67	46.5	190.88	47.01	3.7	5
98	5,709	1Omi CrB	Boo	5	11	188.5	45.5	190.55	46.12	5	5.5
99	5,634	45 Boo	Boo	5	12	188.17	41.33	189.07	40.66	5	4.9
100	5,616	43Psi Boo	Boo	5	13	186.67	41.67	187.52	42.34	5	4.5
101	5,638	46 Boo	Boo	5	14	187	42.5	188.83	42.04	5	5.7
102	5,600	41Ome Boo	Boo	5	15	187.33	40.33	187.71	40.34	5	4.8
103	5,505	36Eps Boo	Boo	5	16	180	40.25	182.04	40.74	3	2.6
104	5,447	28Sig Boo	Boo	5	17	175.67	41.67	177.7	42.14	4	4.5
105	5,429	25Rho Boo	Boo	5	18	175	42.17	176.76	42.52	3.7	3.6
106	5,477	30Zet Boo	Boo	5	19	185.33	28	186.97	28.02	3	3.9
107	5,235	8Eta Boo	Boo	5	20	171.33	28	173.24	28.34	3	2.7
108	5,185	4Tau Boo	Boo	5	21	170.5	26.5	172.16	26.67	4	4.5
109	5,200	5Ups Boo	Boo	5	22	171.33	25	173.23	25.28	4	4.1
110	5,340	16Alp Boo	Boo	5	23	177	31.5	178.32	32.01	1	0

(continued)

Table 1.1 (continued)

#	HR number	Full star name	Constellation		Star	Almagest measurement		Actual (Epoch 137 AD)		V Mag	aV Mag
			Name	#		Long	Lat	Long	Lat		
111	5,793	5Alp CrB	CrB	6	1	194.67	44.5	196.1	44.51	1.7	2.2
112	5,747	3Bet CrB	CrB	6	2	191.67	46.5	193.15	46.2	3.7	3.7
113	5,778	4The CrB	CrB	6	3	191.83	48	193.33	48.72	5	4.1
114	5,855	9Pi CrB	CrB	6	4	193.67	50.5	196.08	50.64	6	5.6
115	5,849	8Gam CrB	CrB	6	5	197.17	44.75	198.87	44.68	4	3.8
116	5,889	10Del CrB	CrB	6	6	199.17	44.83	200.98	44.99	4	4.6
117	5,947	13Eps CrB	CrB	6	7	201.33	46.17	203.07	46.3	4	4.2
118	5,971	14Iot CrB	CrB	6	8	201.67	49.33	202.94	49.36	4	5
119	6,406	64Alp1Her	Her	7	1	227.67	37.5	230.2	37.51	3	3.3
120	6,148	27Bet Her	Her	7	2	213.67	43	215.13	42.94	3	2.8
121	6,095	20Gam Her	Her	7	3	211.67	40.17	213.23	40.2	3	3.8
122	6,008	7Kap Her	Her	7	4	208	37.17	209.72	37.43	4	4.7
123	6,410	65Del Her	Her	7	5	226.67	48	228.78	48.01	3	3.1
124	6,526	76Lam Her	Her	7	6	232	49.5	233.92	49.53	3.7	4.4
125	6,623	86Mu Her	Her	7	7	237.67	52	239.5	51.74	3.7	3.4
126	6,779	103Omi Her	Her	7	8	245.5	52.83	246.8	52.43	3.7	3.8
127	6,707	94Nu Her	Her	7	9	241.67	54	243.54	53.87	3.7	4.4
128	6,703	92Xi Her	Her	7	10	241.5	53	243.21	52.94	4	3.7
129	6,212	40Zet Her	Her	7	11	213.83	53.17	215.84	53.21	3	2.8
130	6,324	58Eps Her	Her	7	12	220.17	53.5	222.33	53.47	5	3.9
131	6,332	59 Her	Her	7	13	220	56.17	221.95	56.13	5	5.3
132	6,377		Her	7	14	221.17	58.5	223.6	58.7	3	5.4
133	6,418	67Pi Her	Her	7	15	224	59.83	226.05	59.79	4	3.2
134	6,436	69 Her	Her	7	16	225.33	60.33	226.97	60.31	4	4.7
135	6,484	75Rho Her	Her	7	17	226.33	61.25	229.39	60.36	3.7	4.1
136	6,695	91The Her	Her	7	18	240.83	61	242.55	60.93	4	3.9

137	6,588	85Iot Her	Her	7	19	232.17	69.33	233.87	69.51	4	3.8
138	6,464	74 Her	Her	7	20	225.33	70.25	224.51	69.23	6	5.6
139	6,509	77 Her	Her	7	21	226.83	71.25	226.49	71.44	6	5.8
140	6,574	82 Her	Her	7	22	229.67	72.25	231.45	71.98	6	5.4
141	6,220	44Eta Her	Her	7	23	210.67	60.25	212.59	60.54	3.7	3.5
142	6,168	35Sig Her	Her	7	24	205.33	63	207.07	63.33	4	4.2
143	6,092	22Tau Her	Her	7	25	195.67	65.5	198.11	65.99	3.7	3.9
144	6,023	11Phi Her	Her	7	26	193.67	63.67	195.39	63.93	4	4.3
145	5,982	6Ups Her	Her	7	27	190.17	64.25	191.97	64.49	4	4.8
146	5,914	1Chi Her	Her	7	28	191.17	60	191.85	60.06	4	4.6
147	5,763	52Nu 1Boo	Her	7	29	185	57.5	186.29	57.21	4	5
148	6,117	24Ome Her	Her	7	30	212.67	38.17	215.54	35.41	5	4.6
149	7,001	3Alp Lyr	Lyr	8	1	257.33	62	259.26	61.84	1	0
150	7,051	4Eps1Lyr	Lyr	8	2	260.33	62.67	262.84	62.6	3.7	3.8
151	7,056	6Zet1Lyr	Lyr	8	3	260.33	61	262.29	60.56	3.7	4.1
152	7,139	12Del2Lyr	Lyr	8	4	263.67	60	265.91	59.54	4	4.3
153	7,298	20Eta Lyr	Lyr	8	5	272	61.33	274.37	60.89	4	4.4
154	7,314	21The Lyr	Lyr	8	6	271.67	60.33	274.82	59.78	4	4.4
155	7,106	10Bet Lyr	Lyr	8	7	261	56.17	263.09	56.22	3	3.5
156	7,102	9Nu 2Lyr	Lyr	8	8	260.83	55	262.8	55.43	4.3	5.3
157	7,178	14Gam Lyr	Lyr	8	9	264.17	55.33	266.14	55.24	3	3.2
158	7,192	15Lam Lyr	Lyr	8	10	264.17	54.75	266.35	54.66	4.3	4.9
159	7,417	6Bet1Cyg	Cyg	9	1	274.5	49	275.49	49.18	3	2.9
160	7,478	12Phi Cyg	Cyg	9	2	279	50.5	279.18	50.81	5	4.7
161	7,615	21Eta Cyg	Cyg	9	3	286.33	54.5	287.28	54.46	3.7	3.9
162	7,796	37Gam Cyg	Cyg	9	4	298.5	57.33	299.24	57.26	3	2.2
163	7,924	50Alp Cyg	Cyg	9	5	309.17	60	309.8	60.01	2	1.3
164	7,528	18Del Cyg	Cyg	9	6	289.33	64.67	290.64	64.57	3	2.9
165	7,469	13The Cyg	Cyg	9	7	292.5	69.67	293.05	69.66	4	4.5

(continued)

Table 1.1 (continued)

#	HR number	Full star name	Constellation Name	#	Star	Almagest measurement Long	Lat	Actual (Epoch 137 AD) Long	Lat	V Mag	aV Mag
166	7,420	10Iot2Cyg	Cyg	9	8	291.17	71.5	292.47	71.56	3.7	3.8
167	7,328	1Kap Cyg	Cyg	9	9	286.67	74	289.4	73.93	3.7	3.8
168	7,949	53Eps Cyg	Cyg	9	10	300.83	49.5	301.71	49.47	3	2.5
169	7,963	54Lam Cyg	Cyg	9	11	303.83	52.17	304.09	51.74	3.7	4.5
170	8,115	64Zet Cyg	Cyg	9	12	306.67	44	307.35	43.84	3	3.2
171	8,028	58Nu Cyg	Cyg	9	13	310	55.17	310.55	55.02	3.7	3.9
172	8,079	62Xi Cyg	Cyg	9	14	314.5	57	315.23	56.66	3.7	3.7
173	7,735	31 Cyg	Cyg	9	15	301.17	64	302.57	63.73	4	3.8
174	7,751	32 Cyg	Cyg	9	16	302.67	64.5	304.32	64.41	4	4
175	7,851	46Ome2Cyg	Cyg	9	17	312.17	64.75	311.31	64.26	5	5.4
176	8,130	65Tau Cyg	Cyg	9	18	310.67	49.67	312.67	50.5	3.7	3.7
177	8,143	67Sig Cyg	Cyg	9	19	313.83	51.67	314.72	51.57	3.7	4.2
178	153	17Zet Cas	Cas	10	1	7.83	45.33	9.34	44.58	3.7	3.7
179	168	18Alp Cas	Cas	10	2	10.83	46.75	12.05	46.5	3	2.2
180	219	24Eta Cas	Cas	10	3	13	47.83	14.1	47.41	4	3.4
181	264	27Gam Cas	Cas	10	4	16.67	49	18.2	48.65	2.7	2.5
182	403	37Del Cas	Cas	10	5	20.67	45.5	22.02	46.32	3	2.7
183	542	45Eps Cas	Cas	10	6	27	47.75	28.98	47.36	4	3.4
184	707	Iot Cas	Cas	10	7	31.67	47.33	36.44	48.74	4	4.5
185	343	33The Cas	Cas	10	8	14.67	44.33	15.92	43.02	4	4.3
186	382	34Phi Cas	Cas	10	9	17.67	45	19.76	44.92	5	5
187	9,071	8Sig Cas	Cas	10	10	2.33	50	4.45	49.29	6	4.9
188	130	15Kap Cas	Cas	10	11	15	52.67	16.91	52.11	4.3	4.2
189	21	11Bet Cas	Cas	10	12	7.83	51.67	9.21	51.32	3	2.3
190	9,045	7Rho Cas	Cas	10	13	3.67	51.67	5.4	51.04	6	4.5
191	10,869		Per	11	1	26.67	40.5	28.4	40.17	8	5.8

192	834	15Eta Per	Per	11	2	31.17	37.5	32.87	37.28	4	3.8
193	915	23Gam Per	Per	11	3	32.67	34.5	34.19	34.32	3.3	2.9
194	799	13The Per	Per	11	4	27.5	32.33	28.66	31.52	4	4.1
195	854	18Tau Per	Per	11	5	30.67	34.5	32.08	34.16	4	4
196	937	Iot Per	Per	11	6	31.5	31.17	32.72	30.67	4	4.1
197	1,017	33Alp Per	Per	11	7	34.83	30	36.22	29.92	2	1.8
198	1,052	35Sig Per	Per	11	8	35.33	27.83	36.74	27.82	4	4.4
199	1,087	37Psi Per	Per	11	9	37	27.67	37.88	27.77	4	4.2
200	1,122	39Del Per	Per	11	10	37.67	27.33	38.93	27.1	3	3
201	941	27Kap Per	Per	11	11	30.5	27	31.77	25.97	4	3.8
202	936	26Bet Per	Per	11	12	29.67	23	30.3	22.22	2	2.1
203	947	28Ome Per	Per	11	13	29.17	21	30.51	20.76	4	4.6
204	921	25Rho Per	Per	11	14	27.67	21	28.99	20.44	4	3.4
205	879	22Pi Per	Per	11	15	26.83	22.25	28.04	21.55	4	4.7
206	1,324		Per	11	16	44.83	28	45.92	28.25	4	4.6
207	1,261	47Lam Per	Per	11	17	43	28.17	43.89	28.67	4	4.3
208	1,273	48 Per	Per	11	18	42.33	25	43.61	26.02	4	4
209	1,303	51Mu Per	Per	11	19	44	26.25	44.91	26.48	4	4.1
210	1,350	53 Per	Per	11	20	44.17	24.5	45.73	24.4	5	4.8
211	1,454	58 Per	Per	11	21	46.33	18.75	47.69	18.78	5	4.3
212	1,135	41Nu Per	Per	11	22	36.83	21.83	37.96	21.93	3.7	3.8
213	1,220	45Eps Per	Per	11	23	38.67	19.25	39.79	18.9	3	2.9
214	1,228	46Xi Per	Per	11	24	38.33	14.75	39.08	14.72	4	4
215	1,131	38Omi Per	Per	11	25	34.17	12	35.25	11.97	3.3	3.8
216	1,203	44Zet Per	Per	11	26	36.33	11	37.23	11.12	2.7	2.8
217	1,306	52 Per	Per	11	27	41.83	18	43.25	18.71	5	4.7
218	1,314		Per	11	28	45	31	46.45	31.51	5	5.2
219	840	16 Per	Per	11	29	24.67	20.67	25.89	20.83	7	4.2
220	2,077	33Del Aur	Aur	12	1	62.5	30	63.95	30.67	4	3.7

(continued)

Table 1.1 (continued)

#	HR number	Full star name	Constellation Name	#	Star	Almagest measurement Long	Lat	Actual (Epoch 137 AD) Long	Lat	V Mag	aV Mag
221	2,029	30Xi Aur	Aur	12	2	62.33	31.83	63.23	32.02	4	5
222	1,708	13Alp Aur	Aur	12	3	55	22.5	55.93	22.84	1	0.1
223	2,088	34Bet Aur	Aur	12	4	62.83	20	64.02	21.26	2	1.9
224	2,012	32Nu Aur	Aur	12	5	61.17	15.25	62.36	15.48	4	4
225	2,095	37The Aur	Aur	12	6	62.83	13.33	64	13.56	3.7	2.6
226	1,605	7Eps Aur	Aur	12	7	52	20.67	52.94	20.7	3.7	3
227	1,641	10Eta Aur	Aur	12	8	52.17	18	53.53	18.08	3.7	3.2
228	1,612	8Zet Aur	Aur	12	9	52	18	52.73	17.97	4	3.8
229	1,577	3Iot Aur	Aur	12	10	49.83	10.17	50.73	10.22	3.3	2.7
230	1,791	112Bet Tau	Aur	12	11	55.67	5	56.65	5.23	2.7	1.6
231	1,843	25Chi Aur	Aur	12	12	56	8.5	58.25	8.65	5	4.8
232	1,805	24Phi Aur	Aur	12	13	56.33	12.17	57.3	11	5	5.1
233	1,706	14 Aur	Aur	12	14	50.67	10.33	54.6	9.36	6	5
234	6,556	55Alp Oph	Oph	13	1	234.83	36	236.43	36.19	2.7	2.1
235	6,603	60Bet Oph	Oph	13	2	238	27.25	239.44	28.1	3.7	2.8
236	6,629	62Gam Oph	Oph	13	3	239	26.5	240.72	26.4	4	3.8
237	6,281	25Iot Oph	Oph	13	4	223.33	33	224.7	32.76	4	4.4
238	6,299	27Kap Oph	Oph	13	5	224.67	31.83	226.03	32.1	4	3.2
239	6,149	10Lam Oph	Oph	13	6	218.33	24.5	219.64	23.82	4	3.8
240	6,056	1Del Oph	Oph	13	7	215	17	216.36	17.54	3	2.7
241	6,075	2Eps Oph	Oph	13	8	216	16.5	217.52	16.63	3	3.2
242	6,567	57Mu Oph	Oph	13	9	236.67	15	238.4	15.47	4	4.6
243	6,698	64Nu Oph	Oph	13	10	242.33	13.67	243.84	13.97	4.3	3.3
244	6,733	69Tau Oph	Oph	13	11	243.33	14.33	244.86	15.52	4	4.8
245	6,378	35Eta Oph	Oph	13	12	231.17	7.5	232.03	7.39	3	2.4
246	6,445	40Xi Oph	Oph	13	13	233.67	2.25	234.84	2.36	3.7	4.4

247	6,401	36 Oph	Oph	13	14	233	−2.25	234.3	−2.7	4	4.3
248	6,453	42The Oph	Oph	13	15	234.33	−1.5	235.48	−1.59	3.7	3.3
249	6,486	44 Oph	Oph	13	16	235	−0.33	236.41	−0.65	4	4.2
250	6,519	51 Oph	Oph	13	17	235.83	−0.25	237.55	−0.43	5	4.8
251	6,595	58 Oph	Oph	13	18	237.17	1	240.28	1.97	5	4.9
252	6,175	13Zet Oph	Oph	13	19	222.17	11.83	223.29	11.61	3	2.6
253	6,147	8Phi Oph	Oph	13	20	221.67	5.33	222.76	5.45	4.7	4.3
254	6,118	7Chi Oph	Oph	13	21	220.67	3.17	222.06	3.47	5	4.4
255	6,104	4Psi Oph	Oph	13	22	219.83	1.67	221.64	1.81	4.7	4.5
256	6,153	9ome Oph	Oph	13	23	222.33	0.67	223.7	0.64	5	4.4
257	6,112	5Rho Oph	Oph	13	24	220.67	−0.75	222.52	−1.51	4	4.6
258	6,712	66 Oph	Oph	13	25	242	28.17	244.16	28.06	4	4.6
259	6,714	67 Oph	Oph	13	26	242.67	26.33	244.26	26.62	4	4
260	6,723	68 Oph	Oph	13	27	243	25	244.56	25	4	4.4
261	6,752	70 Oph	Oph	13	28	243.67	27	245.54	26.68	4	4
262	6,771	72 Oph	Oph	13	29	244.67	33	246.29	33.19	4	3.7
263	5,842	21Iot Ser	Ser	14	1	198.83	38	201.16	38.32	4	4.5
264	5,899	38Rho Ser	Ser	14	2	201.67	40	203.5	40.19	4	4.8
265	5,933	41Gam Ser	Ser	14	3	204.33	36	206.36	35.99	3	3.8
266	5,867	28Bet Ser	Ser	14	4	202	34.25	203.87	34.53	3	3.7
267	5,879	35Kap Ser	Ser	14	5	201.33	37.25	203.75	37.35	4	4.1
268	5,972	44Pi Ser	Ser	14	6	203.17	42.5	206.08	42.64	4	4.8
269	5,788	13Del Ser	Ser	14	7	201.67	29.25	202.38	29.07	3	3.8
270	5,868	27Lam Ser	Ser	14	8	204.83	26.5	206.53	26.8	4	4.4
271	5,854	24Alp Ser	Ser	14	9	204.33	25.33	206.02	25.66	3	2.7
272	5,892	37Eps Ser	Ser	14	10	206.33	24	208.29	24.16	3	3.7
273	5,881	32Mu Ser	Ser	14	11	208.83	16.5	210.03	16.47	4	3.5
274	6,129	3Ups Oph	Ser	14	12	218.17	13.25	220.65	13.45	5	4.6
275	6,446	53Nu Ser	Ser	14	13	233.67	10.5	234.35	10.5	4	4.3

(continued)

Table 1.1 (continued)

#	HR number	Full star name	Constellation Name	#	Star	Almagest measurement Long	Lat	Actual (Epoch 137 AD) Long	Lat	V Mag	aV Mag
276	6,561	55Xi Ser	Ser	14	14	237	8.5	238.64	8.21	3.7	3.5
277	6,581	56Omi Ser	Ser	14	15	237.83	10.83	239.51	10.77	4	4.3
278	6,710	57Zet Ser	Ser	14	16	243.67	20	244.13	20.02	4	4.6
279	6,869	58Eta Ser	Ser	14	17	248.67	21.17	250.09	21.03	3.7	3.3
280	7,141	63The1Ser	Ser	14	18	258.33	27	259.85	27.1	4	4
281	7,635	12Gam Sge	Sge	15	1	280.17	39.33	281.2	39.38	4	3.5
282	7,546	8Zet Sge	Sge	15	2	276.67	39.17	278.23	39.61	6	5
283	7,536	7Del Sge	Sge	15	3	275.83	39.5	277.58	39.11	5	3.8
284	7,479	5Alp Sge	Sge	15	4	274.67	39	275.25	39.01	5	4.4
285	7,488	6Bet Sge	Sge	15	5	273.33	38.67	275.39	38.45	5	4.4
286	7,669	63Tau Aql	Aql	16	1	277.17	26.83	279.18	27.22	4	5.5
287	7,602	60Bet Aql	Aql	16	2	274.83	27.17	276.61	27.11	3	3.7
288	7,557	53Alp Aql	Aql	16	3	273.83	29.17	275.57	29.38	1.7	0.8
289	7,560	54Omi Aql	Aql	16	4	274.67	30	276.26	31.1	3.3	5.1
290	7,525	50Gam Aql	Aql	16	5	273.17	31.5	275.09	31.46	3	2.7
291	7,610	61Phi Aql	Aql	16	6	276	31.5	278.09	31.71	5	5.3
292	7,429	38Mu Aql	Aql	16	7	269.67	28.67	270.85	28.97	5	4.4
293	7,474	44Sig Aql	Aql	16	8	271.17	26.67	271.95	26.69	4.7	5.2
294	7,235	17Zet Aql	Aql	16	9	262.17	36.67	263.95	36.47	3	3
295	7,570	55Eta Aql	Aql	16	10	273.67	21.67	274.56	21.74	3	3.9
296	7,710	65The Aql	Aql	16	11	278.83	19.17	279.02	18.93	3	3.2
297	7,377	30Del Aql	Aql	16	12	266	25	267.62	25.02	3.7	3.4
298	7,447	41Iot Aql	Aql	16	13	268.5	20	269.96	20.24	3	4.4
299	7,446	39Kap Aql	Aql	16	14	269.67	15.5	268.96	14.57	5	4.9
300	7,236	16Lam Aql	Aql	16	15	261.17	18.17	261.45	17.85	3	3.4
301	7,852	2Eps Del	Del	17	1	287.67	29.17	288.24	29.26	3.3	4

302	7,883	5Iot Del	Del	17	2	288.67	29	289.48	29	4.3	5.4
303	7,896	7Kap Del	Del	17	3	288.67	27.75	289.24	27.7	4	5.1
304	7,882	6Bet Del	Del	17	4	288.5	32	290.48	32.13	3.3	3.6
305	7,906	9Alp Del	Del	17	5	290.17	33	291.55	33.19	3.3	3.8
306	7,928	11Del Del	Del	17	6	291.33	32	292.33	32.12	3.3	4.4
307	7,947	12Gam2Del	Del	17	7	293.17	33.17	293.61	32.96	3.3	3.9
308	7,858	3Eta Del	Del	17	8	287.5	30.25	288.97	30.85	6	5.4
309	7,871	4Zet Del	Del	17	9	287.5	31.83	289.92	32.33	6	4.7
310	7,892	8The Del	Del	17	10	289	31.5	290.41	30.78	6	5.7
311	8,131	8Alp Equ	Equ	18	1	296.33	20.5	297.26	20.32	7	3.9
312	8,178	10Bet Equ	Equ	18	2	298	20.67	299.57	21.17	7	5.2
313	8,097	5Gam Equ	Equ	18	3	296.33	25.5	297.61	25.41	7	4.7
314	8,123	7Del Equ	Equ	18	4	297.67	25	298.65	25.03	7	4.5
315	15	21Alp And	Peg	19	1	347.83	26	348.48	25.72	2.3	2.1
316	39	88Gam Peg	Peg	19	2	342.17	12.5	343.29	12.56	2.3	2.8
317	8,775	53Bet Peg	Peg	19	3	332.17	31	333.47	31.12	2.3	2.4
318	8,781	54Alp Peg	Peg	19	4	326.67	19.67	327.63	19.46	2.3	2.5
319	8,880	62Tau Peg	Peg	19	5	334.5	25.5	335.24	25.57	4	4.6
320	8,905	68Ups Peg	Peg	19	6	335	25	336.06	24.81	4	4.4
321	8,650	44Eta Peg	Peg	19	7	329	35	329.97	35.14	3	2.9
322	8,641	43Omi Peg	Peg	19	8	328.5	34.5	329.17	34.45	5	4.8
323	8,667	47Lam Peg	Peg	19	9	326.17	29	327.24	28.84	4	4
324	8,684	48Mu Peg	Peg	19	10	327	29.5	328.54	29.46	4	3.5
325	8,634	42Zet Peg	Peg	19	11	318.83	18	320.27	17.76	3	3.4
326	8,665	46Xi Peg	Peg	19	12	320.5	19	322.1	18.73	4	4.2
327	8,717	50Rho Peg	Peg	19	13	321.33	15	322.67	14.54	5	4.9
328	8,697	49Sig Peg	Peg	19	14	320.5	16	322.21	15.84	5	5.2
329	8,450	26The Peg	Peg	19	15	309.33	16.5	310.83	16.47	3	3.5
330	8,413	22Nu Peg	Peg	19	16	308	16	309.39	15.76	4	4.8

(continued)

Table 1.1 (continued)

#	HR number	Full star name	Constellation		Star	Almagest measurement		Actual (Epoch 137 AD)		V Mag	aV Mag
			Name	#		Long	Lat	Long	Lat		
331	8,308	8Eps Peg	Peg	19	17	305.33	22.5	306.04	22.22	2.7	2.4
332	8,454	29Pi 2Peg	Peg	19	18	323.33	41.17	323.86	41.03	3.7	4.3
333	8,430	24Iot Peg	Peg	19	19	317.33	34.25	318.48	34.37	3.7	3.8
334	8,315	10Kap Peg	Peg	19	20	312.33	36.83	313.16	36.73	3.7	4.1
335	165	31Del And	And	20	1	355.33	24.5	355.96	24.32	3	3.3
336	154	29Pi And	And	20	2	356.33	27	356.87	27.06	4	4.4
337	163	30Eps And	And	20	3	354.33	23	355.28	23	4	4.4
338	68	25Sig And	And	20	4	353.67	32	354.67	31.51	4	4.5
339	63	24The And	And	20	5	354.67	33.5	355.47	33.29	4	4.6
340	82	27Rho And	And	20	6	355	32.33	355.84	32.32	5	5.2
341	8,965	17Iot And	And	20	7	349.67	41	350.36	40.97	4	4.3
342	8,976	19Kap And	And	20	8	350.67	42	351.55	41.67	4	4.1
343	8,961	16Lam And	And	20	9	352.17	44	352.66	43.92	4	3.8
344	215	34Zet And	And	20	10	354.17	17.5	354.8	17.55	4	4.1
345	271	38Eta And	And	20	11	355.67	15.83	356.55	15.85	4	4.4
346	337	43Bet And	And	20	12	3.83	26.33	4.53	25.91	3	2.1
347	269	37Mu And	And	20	13	2	30	3.29	29.55	4	3.9
348	226	35Nu And	And	20	14	1.83	32.5	3.36	32.45	4	4.5
349	603	57Gam1 And	And	20	15	16.83	28	18.39	27.67	3	2.2
350	496	Phi Per	And	20	16	17.17	37.33	18.79	36.69	4.3	4.1
351	464	51 And	And	20	17	15.17	35.67	16.65	35.31	3.7	3.6
352	458	50Ups And	And	20	18	12.33	29	12.92	28.97	4	4.1
353	477	53Tau And	And	20	19	12	28	13.08	27.79	4	4.9
354	335	42Phi And	And	20	20	10.17	35.5	10.65	36.23	5	4.3
355	390	46Xi And	And	20	21	12.67	34.5	12.05	33.68	5	4.9
356	417	48Ome And	And	20	22	14.17	32.5	12.84	33.28	5	4.8

357	8,762	1Omi And	And	20	23	341.67	44	342.08	43.72	3	3.6
358	544	2Alp Tri	Tri	21	1	11	16.5	11.04	16.77	3	3.4
359	622	4Bet Tri	Tri	21	2	16	20.67	16.43	20.46	3	3
360	660	8Del Tri	Tri	21	3	16.33	19.67	17.2	19.48	4	4.9
361	664	9Gam Tri	Tri	21	4	16.83	19	17.65	18.81	3	4
362	545	5Gam2Ari	Ari	22	1	6.67	7.33	7.27	7.09	3.3	4
363	553	6Bet Ari	Ari	22	2	7.67	8.33	8.05	8.42	3	2.6
364	646	17Eta Ari	Ari	22	3	11	7.67	12.15	7.28	5	5.3
365	669	22The Ari	Ari	22	4	11.5	6	12.98	5.6	5	5.6
366	563	8Iot Ari	Ari	22	5	6.83	5.83	7.61	5.34	5	5.1
367	773	32Nu Ari	Ari	22	6	17.67	6	18.24	6	6	5.4
368	887	48Eps Ari	Ari	22	7	21.33	4.83	22.6	3.99	5	4.6
369	951	57Del Ari	Ari	22	8	23.83	1.67	24.86	1.66	4	4.3
370	972	58Zet Ari	Ari	22	9	25.33	2.5	26.06	2.73	4	4.9
371	1,005	61Tau1Ari	Ari	22	10	27	1.83	27.48	2.42	4	5.3
372	869	46Rho3Ari	Ari	22	11	19.67	1.5	20.9	1.15	5	5.6
373	847	43Sig Ari	Ari	22	12	18	-1.5	19.01	-1.45	5	5.5
374	813	87Mu Cet	Ari	22	13	15	-5.25	15.87	-5.67	3.7	4.3
375	617	13Alp Ari	Ari	22	14	10.67	10.5	11.71	9.92	2.7	2
376	838	41 Ari	Ari	22	15	21.67	10.17	22.3	10.33	4	3.6
377	824	39 Ari	Ari	22	16	21.33	12.67	22.43	12.38	5	4.5
378	801	35 Ari	Ari	22	17	19.67	11.17	21.05	11.14	5	4.7
379	782	33 Ari	Ari	22	18	19.17	10.67	20.22	10.73	5	5.3
380	1,066	5 Tau	Tau	23	1	26.33	-6	27.65	-6.11	4	4.1
381	1,061	4 Tau	Tau	23	2	26	-7.25	27.16	-7.63	4	5.1
382	1,038	2Xi Tau	Tau	23	3	24.33	-8.5	25.95	-8.97	4	3.7
383	1,030	1Omi Tau	Tau	23	4	24.33	-9.25	25.27	-9.5	4	3.6
384	1,174	30 Tau	Tau	23	5	29.67	-9.5	31.4	-8.83	5	5.1
385	1,239	35Lam Tau	Tau	23	6	33.67	-8	34.7	-8.17	3	3.5

(continued)

Table 1.1 (continued)

#	HR number	Full star name	Constellation Name	#	Star	Almagest measurement Long	Lat	Actual (Epoch 137 AD) Long	Lat	V Mag	aV Mag
386	1,320	49Mu Tau	Tau	23	7	36.67	−12.67	37.63	−12.39	4	4.3
387		38Nu Tau	Tau	23	8	33	−14.83	33.97	−14.66	4	3.9
388	1,473	90 Tau	Tau	23	9	42.17	−10	43.77	−9.72	4	4.3
389	1,458	88 Tau	Tau	23	10	43	−13	42.84	−11.94	4	4.3
390	1,346	54Gam Tau	Tau	23	11	39	−5.75	39.82	−5.94	3.3	3.7
391	1,373	61Del1Tau	Tau	23	12	40.33	−4.25	40.89	−4.17	3.3	3.8
392	1,411	77The1Tau	Tau	23	13	40.83	−5.83	41.98	−5.96	3.3	3.8
393	1,457	87Alp Tau	Tau	23	14	42.67	−5.17	43.84	−5.6	1	0.9
394	1,409	74Eps Tau	Tau	23	15	41.83	−3	42.49	−2.77	3.3	3.5
395	1,547	97 Tau	Tau	23	16	47.17	−4	47.79	−3.85	4	5.1
396	1,656	104 Tau	Tau	23	17	50.33	−5	51.33	−4.45	5	5
397	1,658	106 Tau	Tau	23	18	50	−3.5	51.88	−2.7	3	5.3
398	1,910	123Zet Tau	Tau	23	19	57.67	−2.5	58.86	−2.43	3	3
399	1,497	94Tau Tau	Tau	23	20	45.67	−0.25	46.24	0.49	4	4.3
400	1,791	112Bet Tau	Tau	23	21	55.67	5	56.65	5.23	3	1.6
401	1,392	69Ups Tau	Tau	23	22	42	0.5	42.53	0.91	5	4.3
402	1,387	65Kap1Tau	Tau	23	23	41.67	0.25	42.24	0.43	5	4.2
403	1,256	37 Tau	Tau	23	24	37	0.67	37.5	1.08	5	4.4
404	1,329	50Ome2Tau	Tau	23	25	39	−1	40.17	−0.95	6	4.9
405	1,287	44 Tau	Tau	23	26	38.5	5	39.77	5.11	5	5.4
406	1,269	42Psi Tau	Tau	23	27	38	7.33	39.44	7.72	5	5.2
407	1,369	59Chi Tau	Tau	23	28	42	3	42.2	3.8	5	5.4
408	1,348	52Phi Tau	Tau	23	29	41.67	5	42.02	5.61	5	4.9
409	1,140	16 Tau	Tau	23	30	32.17	4.5	33.52	4.18	5	5.5
410	1,142	17 Tau	Tau	23	31	32.5	3.67	33.5	4	5	3.7
411	1,165	25Eta Tau	Tau	23	32	33.67	3.33	34.08	3.86	5	2.9

412	1,188		Tau	23	33	33.67	5	35.03	5.2	4	5.3
413	1,101	10 Tau	Tau	23	34	25	−17.5	26.17	−18.42	4	4.3
414	1,620	102Iot Tau	Tau	23	35	50	−2	50.83	−1.42	5	4.6
415	1,739	109 Tau	Tau	23	36	54	−1.75	54.65	−1.21	5	4.9
416	1,810	114 Tau	Tau	23	37	56	−2	56.58	−1.53	5	4.9
417	1,990	130 Tau	Tau	23	38	59	−6.33	61.08	−5.92	5	5.5
418	1,985	129 Tau	Tau	23	39	59	−7.67	60.86	−7.82	5	6
419	1,821	118 Tau	Tau	23	40	57	0.67	57.12	1.67	5	5.5
420	1,928	125 Tau	Tau	23	41	59	1	59.52	2.3	5	5.2
421	2,002	132 Tau	Tau	23	42	61	1.33	61.58	0.91	5	4.9
422	2,034	136 Tau	Tau	23	43	62.33	3.33	62.6	3.94	5	4.6
423	2,084	139 Tau	Tau	23	44	63.33	1.25	63.63	2.27	5	4.8
424	2,890	66Alp Gem	Gem	24	1	83.33	9.5	84.4	9.96	2	1.6
425	2,990	78Bet Gem	Gem	24	2	86.67	6.25	87.6	6.54	2	1.1
426	2,540	34The Gem	Gem	24	3	76.67	10	75.19	10.81	4	3.6
427	2,697	46Tau Gem	Gem	24	4	78.67	7.33	79.53	7.54	4	4.4
428	2,821	60Iot Gem	Gem	24	5	82	5.5	83.09	5.58	4	3.8
429	2,905	69Ups Gem	Gem	24	6	84	4.83	85.43	5.05	4	4.1
430	2,985	77Kap Gem	Gem	24	7	86.67	2.67	87.75	2.88	4	3.6
431	2,808	57 Gem	Gem	24	8	81.67	2.67	82.96	2.75	5	5
432	2,846	63 Gem	Gem	24	9	83.17	0.33	84.44	−0.62	5	5.2
433	2,473	27Eps Gem	Gem	24	10	73	1.5	74.02	1.83	3	3
434	2,650	43Zet Gem	Gem	24	11	78.25	−2.5	79.08	−2.28	3	3.8
435	2,777	55Del Gem	Gem	24	12	81.67	−0.5	82.61	−0.41	3	3.5
436	2,763	54Lam Gem	Gem	24	13	81.67	−6	82.89	−5.85	3	3.6
437	2,216	7Eta Gem	Gem	24	14	66.5	−1.5	67.55	−1.13	3.7	3.3
438	2,286	13Mu Gem	Gem	24	15	68.5	−1.25	69.35	−1.01	3.7	2.9
439	2,343	18Nu Gem	Gem	24	16	70.17	−3.5	70.89	−3.29	3.7	4.2
440	2,421	24Gam Gem	Gem	24	17	72	−7.5	73.19	−6.95	3	1.9

(continued)

Table 1.1 (continued)

#	HR number	Full star name	Constellation Name	#	Star	Almagest measurement Long	Lat	Actual (Epoch 137 AD) Long	Lat	V Mag	aV Mag
441	2,484	31Xi Gem	Gem	24	18	74.67	−10.5	75.35	−10.24	4	3.4
442	2,134	1 Gem	Gem	24	19	64.17	−0.67	65.04	−0.35	4	4.2
443	2,219	44Kap Aur	Gem	24	20	66.5	5.83	67.48	5.99	3.7	4.3
444	2,529	36 Gem	Gem	24	21	75.17	−2.25	76.04	−1.37	5	5.3
445	3,086	85 Gem	Gem	24	22	88.33	−1.33	91.14	−1.08	5	5.3
446	3,003	81 Gem	Gem	24	23	86.33	−3.33	89.21	−2.84	5	4.9
447	2,938	74 Gem	Gem	24	24	86	−4.5	87.76	−3.97	5	5.1
448	3,208	16Zet1Cnc	Gem	24	25	90.67	−2.67	95.4	−2.4	4	4.7
449	12,632		Cnc	25	1	100.33	0.33	101.28	1.37	8	5.8
450	3,366	33Eta Cnc	Cnc	25	2	97.67	1.25	99.5	1.4	4.3	5.3
451	3,357	31The Cnc	Cnc	25	3	98	−1.17	99.84	−0.93	4.3	5.3
452	3,449	43Gam Cnc	Cnc	25	4	100.33	2.67	101.66	3.03	3.7	4.7
453	3,461	47Del Cnc	Cnc	25	5	101.33	−0.17	102.78	0	3.7	3.9
454	3,572	65Alp Cnc	Cnc	25	6	106.5	−5.5	107.71	−5.25	4	4.3
455	3,474	48Iot Cnc	Cnc	25	7	98.33	11.83	100.4	10.26	4	3.9
456	3,176	10Mu 2Cnc	Cnc	25	8	92.67	1	93.54	1.17	5	5.3
457	3,249	17Bet Cnc	Cnc	25	9	97.17	−7.5	98.38	−10.46	3.7	3.5
458	3,669	82Pi 2Cnc	Cnc	25	10	109.67	−2.33	110.75	−1.1	4.3	5.3
459	3,623	76Kap Cnc	Cnc	25	11	111.17	−5.67	110.28	−5.73	4.3	5.2
460	3,595	69Nu Cnc	Cnc	25	12	104	7.25	105.1	7.11	5	5.4
461	3,627	77Xi Cnc	Cnc	25	13	107	4.83	107.28	5.25	5	5.1
462	3,731	1Kap Leo	Leo	26	1	108.33	10	109.35	10.28	4	4.5
463	3,773	4Lam Leo	Leo	26	2	111.17	7.5	111.93	7.75	4	4.3
464	3,905	24Mu Leo	Leo	26	3	114.33	12	115.57	12.26	3	3.9
465	3,873	17Eps Leo	Leo	26	4	114.17	9.5	114.77	9.57	2.7	3
466	4,031	36Zet Leo	Leo	26	5	120.17	11	121.59	11.73	3	3.4

467	4,057	41Gam1Leo	Leo	26	6	122.17	8.5	123.48	8.71	2	2.3
468	3,975	30Eta Leo	Leo	26	7	120.67	4.5	121.97	4.74	3	3.5
469	3,982	32Alp Leo	Leo	26	8	122.5	0.17	124.03	0.39	1	1.4
470	3,980	31 Leo	Leo	26	9	123.5	-1.83	124.53	-1.49	4	4.4
471	3,937	27Nu Leo	Leo	26	10	120	-0.25	121.43	-0.07	5	5.3
472	3,866	16Psi Leo	Leo	26	11	117.33	0	117.56	0.21	5	5.3
473	3,782	5Xi Leo	Leo	26	12	114.17	-3.67	115.77	-3.25	5	5
474	3,852	14Omi Leo	Leo	26	13	117.33	-4.17	118.41	-3.86	4	3.5
475	3,950	29Pi Leo	Leo	26	14	122.5	-4.25	123.42	-4.02	4	4.7
476	4,133	47Rho Leo	Leo	26	15	129.17	-0.17	130.47	0.06	4	3.8
477	4,127	46 Leo	Leo	26	16	127	4	128.55	4.47	6	5.5
478	4,209	52 Leo	Leo	26	17	130.33	5.33	131.76	5.89	6	5.5
479	4,227	53 Leo	Leo	26	18	132.17	2.33	133.75	2.74	6	5.3
480	4,300	60 Leo	Leo	26	19	131.33	12.25	132.89	12.81	6	4.4
481	4,357	68Del Leo	Leo	26	20	134.17	13.67	135.24	14.29	2.3	2.6
482	4,408	81 Leo	Leo	26	21	134.33	11.17	139.66	11.68	5	5.6
483	4,359	70The Leo	Leo	26	22	136.33	9.67	137.48	9.66	3	3.3
484	4,399	78Iot Leo	Leo	26	23	140.33	5.83	141.54	6.06	3	3.9
485	4,386	77Sig Leo	Leo	26	24	141.67	1.25	142.82	1.68	4	4.1
486	4,418	84Tau Leo	Leo	26	25	144.67	-0.83	145.59	-0.59	4	4.9
487	4,471	91Ups Leo	Leo	26	26	147.5	-3.17	149.14	-3.08	5	4.3
488	4,534	94Bet Leo	Leo	26	27	144.5	11.83	145.86	12.39	1.3	2.1
489	4,192	41 LMi	Leo	26	28	126	13.33	127.57	13.88	5	5.1
490	4,259	54 Leo	Leo	26	29	128.17	15.5	129.56	16.42	5	4.3
491	4,310	63Chi Leo	Leo	26	30	137.5	1.17	138.74	1.37	4.3	4.6
492	4,294	59 Leo	Leo	26	31	137.17	-0.5	138.11	-0.25	5	5
493	4,291	58 Leo	Leo	26	32	138	-2.67	139	-2.57	5	4.8
494	4,737	15Gam Com	Leo	26	33	144.83	30	147.86	28.43	7	4.4
495	4,667	7 Com	Leo	26	34	144.33	25	147.62	23.45	7	4.9

(continued)

Table 1.1 (continued)

#	HR number	Full star name	Constellation Name	#	Almagest measurement Long	Lat	Actual (Epoch 137 AD) Long	Lat	V Mag	aV Mag
496	4,789	23 Com	Leo	26	148.5	25.5	152.45	24.11	7	4.8
497	4,517	3Nu Vir	Vir	27	146.33	4.25	148.19	4.65	5	4
498	4,515	2Xi Vir	Vir	27	147	5.67	147.36	6.09	5	4.8
499	4,608	9Omi Vir	Vir	27	150.67	8	151.86	8.54	5	4.1
500	4,589	8Pi Vir	Vir	27	150.5	5.5	151.61	6.16	5	4.7
501	4,540	5Bet Vir	Vir	27	149	0.17	150.84	0.66	3	3.6
502	4,689	15Eta Vir	Vir	27	158.25	1.17	158.93	1.41	3	3.9
503	4,825	29Gam Vir	Vir	27	163.17	2.83	164.52	2.93	3	2.9
504	4,921	44 Vir	Vir	27	167.5	2.83	169.3	2.43	5	5.8
505	4,963	51The Vir	Vir	27	171	1.67	172.32	1.85	4	4.4
506	4,910	43Del Vir	Vir	27	164.33	8.5	165.72	8.79	3	3.4
507	4,828	30Rho Vir	Vir	27	158.17	13.5	159.48	13.59	5	4.9
508	4,847	32 Vir	Vir	27	160.17	11.67	161.51	11.62	6	5.2
509	4,932	47Eps Vir	Vir	27	162.17	15.17	164.09	16.3	2.7	2.8
510	5,056	67Alp Vir	Vir	27	176.67	−2	177.95	−1.93	1	1
511	5,107	79Zet Vir	Vir	27	174.83	8.67	176.33	8.76	3	3.4
512	5,095	74 Vir	Vir	27	176.33	3.33	177.7	3.26	5	4.7
513	5,100	76 Vir	Vir	27	177.25	0.17	179.34	−0.29	6	5.2
514	5,150	82 Vir	Vir	27	180	1.5	180.83	1.82	4.3	5
515	5,064	68 Vir	Vir	27	178	−3	178.91	−3.21	5	5.3
516	5,173	86 Vir	Vir	27	181.67	−1.5	183.13	−1.26	5	5.5
517	5,232	90 Vir	Vir	27	178	8.5	181.27	9.74	5	5.2
518	5,338	99Iot Vir	Vir	27	186.67	7.17	187.79	7.55	4	4.1
519	5,315	98Kap Vir	Vir	27	187.33	2.67	188.59	2.98	4	4.2
520	5,409	105Phi Vir	Vir	27	188.33	11.67	189.56	11.93	4	4.8
521	5,359	100Lam Vir	Vir	27	190	0.5	191.05	0.63	4	4.5

522	5,487	107Mu Vir	Vir	27	26	192.67	9.83	194.08	9.97	4	3.9
523	4,813	26Chi Vir	Vir	27	27	164.67	-3.5	166.28	-3.39	5	4.7
524	4,902	40Psi Vir	Vir	27	28	169	-3.5	170.3	-3.35	5	4.8
525	4,955	49 Vir	Vir	27	29	172.25	-3.33	173.83	-3.17	5	5.2
526	4,981	53 Vir	Vir	27	30	177.17	-7.17	176.79	-7.71	6	5
527	5,019	61 Vir	Vir	27	31	178.17	-8.33	179.43	-8.47	5	4.7
528	5,196	89 Vir	Vir	27	32	185	-7.83	186.1	-6.21	6	5
529	5,531	9Alp2Lib	Lib	28	1	198	0.67	199.2	0.56	2	2.6
530	5,523	7Mu Lib	Lib	28	2	197	2.5	198.27	2.22	5	5.3
531	5,685	27Bet Lib	Lib	28	3	202.17	8.83	203.48	8.71	2	2.6
532	5,586	19Del Lib	Lib	28	4	197.67	8.5	199.37	8.44	5	4.9
533	5,652	24Iot Lib	Lib	28	5	204	-1.67	205.1	-1.64	4	4.5
534	5,622	21Nu Lib	Lib	28	6	201.33	1.25	202.86	1.4	4	5.2
535	5,787	38Gam Lib	Lib	28	7	207.83	4.75	209.18	4.58	4	3.9
536	5,908	46The Lib	Lib	28	8	213	3.5	213.91	3.61	4.3	4.2
537	5,777	37 Lib	Lib	28	9	206.17	9	207.56	9.18	5	4.6
538	5,941	48 Lib	Lib	28	10	213.67	6.67	214.47	6.31	4.3	4.9
539	5,978	Xi Sco	Lib	28	11	214.33	9.25	215.4	9.47	4.3	4.2
540	5,902	45Lam Lib	Lib	28	12	213.5	0.5	214.56	0.31	6	5
541	5,838	43Kap Lib	Lib	28	13	210.67	0.33	211.84	0.25	5	4.7
542	5,810		Lib	28	14	211.17	-1.5	211.27	-1.28	4	5.8
543	5,603	20Sig Lib	Lib	28	15	203	-7.5	204.82	-7.42	3	3.3
544	5,794	39Ups Lib	Lib	28	16	211.17	-8.5	212.72	-8.29	4	3.6
545	5,812	40Tau Lib	Lib	28	17	212	-9.67	213.46	-9.79	4	3.7
546	5,984	8Bet1Sco	Sco	29	1	216.33	1.33	217.27	1.24	3	2.5
547	5,953	7Del Sco	Sco	29	2	215.67	-1.67	216.66	-1.74	3	2.3
548	5,944	6Pi Sco	Sco	29	3	215.67	-5	217.04	-5.24	3	2.9
549	5,928	5Rho Sco	Sco	29	4	216	-7.83	217.25	-8.36	3	3.9
550	6,026	14Nu Sco	Sco	29	5	217	1.67	218.72	1.88	4	3.9

(continued)

Table 1.1 (continued)

#	HR number	Full star name	Constellation Name	#	Star	Almagest measurement Long	Lat	Actual (Epoch 137 AD) Long	Lat	V Mag	aV Mag
551	5,993	9Ome1Sco	Sco	29	6	216.33	0.5	217.75	0.46	4	4
552	6,084	20Sig Sco	Sco	29	7	220.67	-3.75	221.89	-3.8	3	2.9
553	6,134	21Alp Sco	Sco	29	8	222.67	-4	223.85	-4.33	2	1
554	6,165	23Tau Sco	Sco	29	9	224.5	-5.5	225.55	-5.87	3	2.8
555	6,028	13 Sco	Sco	29	10	219.33	-6.5	220.34	-6.44	5	4.6
556	6,070		Sco	29	11	220.67	-6.67	221.78	-6.84	5	4.8
557	6,241	26Eps Sco	Sco	29	12	228.5	-11	229.73	-11.33	3	2.3
558	6,247	Mu 1Sco	Sco	29	13	228.83	-15	230.26	-15.17	3	3.1
559	6,262	Zet1Sco	Sco	29	14	230	-18.67	231.22	-19.42	4	4.7
560	6,271	Zet2Sco	Sco	29	15	230.17	-18.67	231.39	-19.28	4	3.6
561	6,380	Eta Sco	Sco	29	16	233.17	-19.5	234.81	-19.79	3	3.3
562	6,553	The Sco	Sco	29	17	238.17	-18.83	239.68	-19.4	3	1.9
563	6,615	Iot1Sco	Sco	29	18	240.5	-16.67	241.61	-16.46	3	3
564	6,580	Kap Sco	Sco	29	19	239	-15.17	240.56	-15.38	3	2.4
565	6,527	35Lam Sco	Sco	29	20	237.5	-13.33	238.68	-13.53	3	1.6
566	6,508	34Ups Sco	Sco	29	21	237	-13.5	238.1	-13.75	4	2.7
567	16,475	45 Oph	Sco	29	22	241.17	-13.25	242.79	-9.5	8	5.8
568	6,492	3 Sgr	Sco	29	23	235.5	-6.17	236.95	-6.31	5	4.3
569	6,616		Sco	29	24	235.5	-4.17	241.33	-4.17	5	4.5
570	6,746	10Gam2Sgr	Sgr	30	1	244.5	-6.5	245.37	-6.65	3	3
571	6,859	19Del Sgr	Sgr	30	2	247.67	-6.5	248.64	-6.21	3	2.7
572	6,879	20Eps Sgr	Sgr	30	3	248	-10.83	249.18	-10.74	3	1.9
573	6,913	22Lam Sgr	Sgr	30	4	249	-1.5	250.43	-1.8	3	2.8
574	6,812	13Mu Sgr	Sgr	30	5	246.67	2.83	247.29	2.59	4	3.9
575	7,121	34Sig Sgr	Sgr	30	6	255.33	-3.17	256.46	-3.18	3	2
576	7,039	27Phi Sgr	Sgr	30	7	253	-3.5	254.23	-3.71	4	3.2

577	7,116	32Nu 1Sgr	Sgr	30	8	255.17	0.75	256.55	0.35	8	4.8
578	7,150	37Xi 2Sgr	Sgr	30	9	255.67	2.17	257.52	1.91	4	3.5
579	7,217	39Omi Sgr	Sgr	30	10	257.67	1.5	259.04	1.13	4	3.8
580	7,264	41Pi Sgr	Sgr	30	11	259.17	2	260.34	1.69	4	2.9
581	7,304	43 Sgr	Sgr	30	12	261.33	2.83	262.44	3.5	5	5
582	7,340	44Rho1Sgr	Sgr	30	13	262.33	4.5	263.55	4.44	4	3.9
583	7,342	46Ups Sgr	Sgr	30	14	262.83	6.5	263.82	6.33	4	4.6
584	7,489	55 Sgr	Sgr	30	15	265.67	5.5	268.72	5.38	6	5.1
585	7,614	61 Sgr	Sgr	30	16	269.5	5.83	272.53	5.35	5	5
586	7,561	57 Sgr	Sgr	30	17	267.67	2	270.5	2.11	6	5.9
587	7,362	47Chi1Sgr	Sgr	30	18	262.67	-1.83	263.39	-2.23	5	5
588	7,440	52 Sgr	Sgr	30	19	264.83	-2.83	265.89	-3.01	4	4.6
589	7,292	42Psi Sgr	Sgr	30	20	260	-2.5	261.1	-2.67	5	4.8
590	7,234	40Tau Sgr	Sgr	30	21	257.67	-4.5	258.95	-4.73	3.7	3.3
591	7,194	38Zet Sgr	Sgr	30	22	256.33	-6.75	257.72	-6.94	3	2.6
592	7,337	Bet1Sgr	Sgr	30	23	257.67	-23	259.83	-21.9	2	4
593	7,348	Alp Sgr	Sgr	30	24	257	-18	260.69	-18.08	2.3	4
594	6,832	Eta Sgr	Sgr	30	25	246.67	-13	247.78	-13.05	3	3.1
595	7,623	The1Sgr	Sgr	30	26	267.33	-13.5	268.92	-14.15	3	4.4
596	7,581	Iot Sgr	Sgr	30	27	263.83	-20.17	266.58	-20.46	3	4.1
597	7,597	58Ome Sgr	Sgr	30	28	267.67	-4.83	269.81	-5.22	5	4.7
598	7,618	60 Sgr	Sgr	30	29	268.83	-4.83	270.61	-5.25	5	4.8
599	7,604	59 Sgr	Sgr	30	30	268.83	-5.83	269.99	-6.09	5	4.5
600	7,650	62 Sgr	Sgr	30	31	269.67	-6.5	271.12	-6.89	5	4.6
601	7,754	6Alp2Cap	Cap	31	1	277.33	7.33	277.93	7.14	3	3.6
602	7,773	8Nu Cap	Cap	31	2	277.67	6.67	278.53	6.79	6	4.8
603	7,776	9Bet Cap	Cap	31	3	277.33	5	278.11	4.79	3	3
604	7,715	2Xi 2Cap	Cap	31	4	275	8	276.52	7.49	6	5.8
605	7,829	12Omi Cap	Cap	31	5	279	0.75	279.3	0.64	6	5.5

(continued)

Table 1.1 (continued)

#	HR number	Full star name	Constellation #	Name	Star	Almagest measurement Long	Lat	Actual (Epoch 137 AD) Long	Lat	V Mag	aV Mag
606	7,814	10Pi Cap	31	Cap	6	278.67	1.75	278.79	1.11	6	5.3
607	7,822	11Rho Cap	31	Cap	7	278.83	1.5	279.26	1.4	6	4.8
608	7,761	7Sig Cap	31	Cap	8	276.17	0.67	276.76	0.67	5	5.3
609	7,889	14Tau Cap	31	Cap	9	281.67	3.83	282.39	3.55	6	5.2
610	7,900	15Ups Cap	31	Cap	10	281.83	0.83	281.76	0.42	5	5.1
611	7,980	18Ome Cap	31	Cap	11	281.67	-8.67	282.02	-8.77	4	4.1
612	7,936	16Psi Cap	31	Cap	12	280.83	-6.5	281.27	-6.76	4	4.1
613	8,080	24 Cap	31	Cap	13	286.67	-7.67	285.92	-7.9	4	4.5
614	8,204	34Zet Cap	31	Cap	14	290.17	-6.83	290.99	-6.83	4	3.7
615	8,213	36 Cap	31	Cap	15	290.33	-6	291.58	-6.38	5	4.5
616	8,127	28Phi Cap	31	Cap	16	288.5	-4.25	289.09	-4.35	5	5.2
617	8,087	25Chi Cap	31	Cap	17	286.67	-4	287.35	-4.35	5	5.3
618	8,060	22Eta Cap	31	Cap	18	286.67	-2.83	286.83	-2.81	5	4.8
619	8,075	23The Cap	31	Cap	19	286.67	0	287.89	-0.37	4	4.1
620	8,167	32Iot Cap	31	Cap	20	291	-0.83	291.74	-1.2	4	4.3
621	8,260	39Eps Cap	31	Cap	21	293.33	-4.75	294.26	-4.82	4	4.7
622	8,288	43Kap Cap	31	Cap	22	295	-4.5	295.64	-4.66	4	4.7
623	8,278	40Gam Cap	31	Cap	23	294.83	-2.17	295.78	-2.36	3	3.7
624	8,322	49Del Cap	31	Cap	24	296.33	-2	297.54	-2.27	3	2.9
625	8,283	42 Cap	31	Cap	25	296.83	0.33	297.17	0.08	4	5.2
626	8,351	51Mu Cap	31	Cap	26	298.67	0	299.76	-0.5	5	5.1
627	8,319	48Lam Cap	31	Cap	27	297.67	2.83	299.09	2.08	5	5.6
628	8,311	46 Cap	31	Cap	28	298.67	4.33	299.5	4.35	5	5.1
629	8,277	25 Aqr	32	Aqr	1	300.33	15.75	302.12	15.5	3	3
630	8,414	34Alp Aqr	32	Aqr	2	306.33	11	307.47	10.78	3	3
631	8,402	31Omi Aqr	32	Aqr	3	305.17	9.67	306.21	9.29	5	4.7

632	8,232	22Bet Aqr	Aqr	32	4	296.5	8.83	297.5	8.77	3	2.9
633	8,264	23Xi Aqr	Aqr	32	5	297.33	6.25	298.17	6.13	5	4.7
634	8,093	13Nu Aqr	Aqr	32	6	287.67	5.5	290.45	4.95	3	4.5
635	7,990	6Mu Aqr	Aqr	32	7	286.17	8	287.15	8.44	4	4.7
636	7,950	2Eps Aqr	Aqr	32	8	284.67	8.67	285.82	8.28	3	3.8
637	8,518	48Gam Aqr	Aqr	32	9	309.5	8.75	310.76	8.35	3	3.8
638	8,539	52Pi Aqr	Aqr	32	10	311.67	10.75	312.71	10.56	3	4.7
639	8,558	55Zet2Aqr	Aqr	32	11	312	9	312.91	8.97	3	3.7
640	8,597	62Eta Aqr	Aqr	32	12	313.33	8.5	314.49	8.27	3	4
641	8,499	43The Aqr	Aqr	32	13	306.17	3.17	307.3	2.85	4	4.2
642	8,512	46Rho Aqr	Aqr	32	14	307	3	308.11	2.48	5	5.4
643	8,573	57Sig Aqr	Aqr	32	15	308.67	-0.83	309.47	-1.11	4	4.8
644	8,418	33Iot Aqr	Aqr	32	16	301.67	-1.67	302.78	-1.92	4	4.3
645	8,452	38 Aqr	Aqr	32	17	303.17	-0.25	304.55	-0.16	6	5.5
646	8,709	76Del Aqr	Aqr	32	18	311.67	-7.5	312.95	-8.1	3	3.3
647	8,679	71Tau2Aqr	Aqr	32	19	311.33	-5	312.67	-5.56	4	4
648	8,544	53 Aqr	Aqr	32	20	304.67	-5.67	306.16	-6.31	5	5.7
649	8,670	68 Aqr	Aqr	32	21	308.33	-10	309.94	-10.86	5	5.3
650	8,649	66 Aqr	Aqr	32	22	307.83	-9	309.3	-9.85	5	4.7
651	8,610	63Kap Aqr	Aqr	32	23	315	2	313.57	4.24	4	5
652	8,698	73Lam Aqr	Aqr	32	24	314.83	0.17	315.64	-0.32	4	3.7
653	8,782	83 Aqr	Aqr	32	25	317.67	-1.17	318.42	-1.61	4	5.4
654	8,834	90Phi Aqr	Aqr	32	26	320	-0.5	321.23	-0.9	4	4.2
655	8,850	92Chi Aqr	Aqr	32	27	320.5	-1.67	321.14	-2.79	4	5.1
656	8,841	91Psi1Aqr	Aqr	32	28	319	-3.5	320.2	-3.86	4	4.2
657	8,865	95Psi3Aqr	Aqr	32	29	319.83	-4.17	320.84	-4.71	4	5
658	8,866	94 Aqr	Aqr	32	30	317.83	-8.25	319.24	-8.19	5	5.1
659	8,968	102ome1Aqr	Aqr	32	31	322.67	-11	323.69	-10.96	5	5
660	8,988	105ome2Aqr	Aqr	32	32	323.17	-10.83	324.19	-11.53	5	4.5

(continued)

Table 1.1 (continued)

#	HR number	Full star name	Constellation #	Name	Almagest measurement Long	Lat	Actual (Epoch 137 AD) Long	Lat	V Mag	aV Mag
661	8,982	104 Aqr	32	Aqr	321.67	−14	322.62	−14.47	5	4.8
662	8,998	106 Aqr	32	Aqr	322.17	−14.75	322.97	−15.12	5	5.2
663	9,031	108 Aqr	32	Aqr	323.17	−15.67	324.29	−16.4	5	5.2
664	8,892	98 Aqr	32	Aqr	317	−14.17	317.56	−14.7	4	4
665	8,939	101 Aqr	32	Aqr	318.33	−15.75	319.41	−16.46	4	4.7
666	8,906	99 Aqr	32	Aqr	317.5	−15	317.96	−15.49	4	4.4
667	8,789	86 Aqr	32	Aqr	311.83	−14.75	312.32	−16.49	4	4.5
668	8,817	89 Aqr	32	Aqr	312.33	−14.75	313.59	−15.62	4	4.7
669	8,812	88 Aqr	32	Aqr	313.17	−14	314.01	−14.41	4	3.7
670	8,728	24Alp PsA	32	Aqr	307	−20.33	307.73	−20.88	1	1.2
671	9,098	2 Cet	32	Aqr	326.67	−15.5	327.77	−16.2	3.7	4.6
672	33	6 Cet	32	Aqr	329.67	−14.67	330.39	−15.16	3.7	4.9
673	48	7 Cet	32	Aqr	329	−18.25	329.57	−18.72	3.7	4.4
674	8,773	4Bet Psc	33	Psc	321.67	9.25	322.7	9.11	4	4.5
675	8,852	6Gam Psc	33	Psc	324.17	7.5	325.2	7.44	4	3.7
676	8,878	7 Psc	33	Psc	326	9.33	327.12	8.94	4	5.1
677	8,916	10The Psc	33	Psc	328.17	9.5	329.38	9.05	4	4.3
678	8,969	17Iot Psc	33	Psc	330.67	7.5	331.67	7.45	4	4.1
679	8,911	8Kap Psc	33	Psc	326	4.5	326.98	4.52	4	4.9
680	8,984	18Lam Psc	33	Psc	329.67	3.5	330.78	3.48	4	4.5
681	9,072	28Ome Psc	33	Psc	336	6.33	336.65	6.43	4	4
682	80	41 Psc	33	Psc	341	5.75	342.08	5.43	6	5.4
683	132	51 Psc	33	Psc	343	3.75	344.25	3.14	6	5.7
684	224	63Del Psc	33	Psc	347.17	2.25	348.21	2.15	4	4.4
685	294	71Eps Psc	33	Psc	350.5	1.17	351.64	0.99	4	4.3
686	361	86Zet Psc	33	Psc	353	−0.17	353.9	−0.25	4	4.9

687	330	80 Psc	Psc	33	14	352.33	−2	352.14	−1.54	6	5.5
688	378	89 Psc	Psc	33	15	353	−5	353.41	−4.35	6	5.2
689	434	98Mu Psc	Psc	33	16	356.5	−2.33	357.07	−3.08	4	4.8
690	489	106Nu Psc	Psc	33	17	358.67	−4.67	359.58	−4.81	4	4.4
691	549	111Xi Psc	Psc	33	18	0.67	−7.75	1.56	−8.04	4	4.6
692	595	113Alp Psc	Psc	33	19	2.5	−8.5	3.41	−9.17	3	3.8
693	510	110Omi Psc	Psc	33	20	0.5	−1.67	1.78	−1.74	4	4.3
694	463	102Pi Psc	Psc	33	21	0.17	1.83	1.04	1.76	5	5.6
695	437	99Eta Psc	Psc	33	22	0.67	5.33	0.91	5.27	3	3.6
696	413	93Rho Psc	Psc	33	23	0.5	9	1.21	9.27	4	5.4
697	349	82 Psc	Psc	33	24	2	21.75	2.96	21.89	5	5.2
698	352	83Tau Psc	Psc	33	25	1.67	21.67	2.45	20.65	5	4.5
699	274	68 Psc	Psc	33	26	358.67	20	359.07	20.86	6	5.4
700	262	67 Psc	Psc	33	27	357.67	19.83	357.89	19.4	6	6.1
701	230	65 Psc	Psc	33	28	357	20.33	356.76	20.45	6	5.6
702	310	74Psi1Psc	Psc	33	29	355.67	14.33	357.55	13.28	4	4.7
703	328	79Psi2Psc	Psc	33	30	356.33	13.25	357.75	12.44	5	5.6
704	351	84Chi Psc	Psc	33	31	357.67	12	358.64	12.35	4	4.7
705	383	90Ups Psc	Psc	33	32	2.17	17	2.93	17.36	4	4.8
706	360	85Phi Psc	Psc	33	33	359.83	15.33	0.6	15.4	4	4.7
707	339	81Psi3Psc	Psc	33	34	0	11.75	357.75	11.21	4	5.6
708	9,067	27 Psc	Psc	33	35	331.17	−2.67	332.39	−3.1	4	4.9
709	9,087	29 Psc	Psc	33	36	332.25	−2.5	333.28	−2.95	4	5.1
710	9,089	30 Psc	Psc	33	37	330.67	−5.5	332.09	−5.68	4	4.4
711	3	33 Psc	Psc	33	38	332.33	−5.5	332.99	−5.81	4	4.6
712	896	91Lam Cet	Cet	34	1	17.67	−7.75	19.15	−7.95	4	4.7
713	911	92Alp Cet	Cet	34	2	17.67	−12.33	18.38	−12.72	3	2.5
714	804	86Gam Cet	Cet	34	3	12.67	−11.5	13.57	−12.11	3	3.5
715	779	82Del Cet	Cet	34	4	10.5	−14	11.6	−14.61	3	4.1

(continued)

Table 1.1 (continued)

#	HR number	Full star name	Constellation			Almagest measurement		Actual (Epoch 137 AD)		V Mag	aV Mag
			Name	#	Star	Long	Lat	Long	Lat		
716	754	78Nu Cet	Cet	34	5	10.17	−8.17	12.46	−9.33	4	4.9
717	718	73Xi 2Cet	Cet	34	6	12.67	−6.33	11.51	−5.99	4	4.3
718	649	65Xi 1Cet	Cet	34	7	7.67	−4.17	8.12	−4.4	4	4.4
719	708	72Rho Cet	Cet	34	8	3	−24.5	3.71	−25.36	4	4.9
720	740	76Sig Cet	Cet	34	9	3.33	−28	4.15	−28.6	4	4.8
721	781	83Eps Cet	Cet	34	10	6.67	−25.17	7.28	−25.99	4	4.8
722	811	89Pi Cet	Cet	34	11	7	−27.5	7.73	−28.38	3	4.3
723	509	52Tau Cet	Cet	34	12	352	−25.33	352.5	−25.65	3	3.5
724	585	59Ups Cet	Cet	34	13	353	−30.83	353.3	−31.08	4	4
725	539	55Zet Cet	Cet	34	14	355	−20	355.94	−20.4	3	3.7
726	402	45The Cet	Cet	34	15	349.67	−15.67	350.32	−15.76	3	3.6
727	334	31Eta Cet	Cet	34	16	345	−15.67	345.7	−16.06	3	3.5
728	235	19Phi2Cet	Cet	34	17	341	−13.67	341.43	−14.7	5	5.2
729	227		Cet	34	18	340.67	−14.67	339.85	−17.29	5	5.6
730	194	17Phi1Cet	Cet	34	19	339.33	−13	339.94	−14.09	4.7	4.8
731	190		Cet	34	20	339	−14	339.29	−15.29	4.7	6
732	74	8Iot Cet	Cet	34	21	334.33	−9.67	334.97	−10.01	3.3	3.6
733	188	16Bet Cet	Cet	34	22	335.67	−20.33	336.45	−20.76	3	2
734	1,879	39Lam Ori	Ori	35	1	57	−13.5	57.78	−13.61	8	3.4
735	2,061	58Alp Ori	Ori	35	2	62	−17	62.82	−16.28	1.3	0.5
736	1,790	24Gam Ori	Ori	35	3	54	−17.5	55.02	−17.05	2	1.6
737	1,839	32 Ori	Ori	35	4	55	−18	56.46	−17.53	4.3	4.2
738	2,124	61Mu Ori	Ori	35	5	64.33	−14.5	64.68	−14.02	4	4.1
739	2,229	73 Ori	Ori	35	6	66.33	−11.83	68	−11.08	6	5.3
740	2,199	70Xi Ori	Ori	35	7	66.5	−10	67.02	−9.44	4	4.5
741	2,159	67Nu Ori	Ori	35	8	66	−9.75	65.93	−8.89	4	4.4

742	2,223	72 Ori	Ori	35	9	67.33	-8.25	67.82	-7.49	6	5.3
743	2,198	69 Ori	Ori	35	10	66.67	-8.25	67	-7.52	6	4.9
744	2,047	54Chi1Ori	Ori	35	11	61.67	-3.75	62.85	-3.35	5	4.4
745	2,135	62Chi2Ori	Ori	35	12	64.67	-4.25	65	-3.54	5	4.6
746	1,934	47Ome Ori	Ori	35	13	57.83	-19.67	58.58	-19.46	4	4.6
747	1,872	38 Ori	Ori	35	14	56.33	-20	57.28	-19.76	6	5.4
748	1,842	33 Ori	Ori	35	15	55.33	-20.17	56.43	-20.2	6	5.5
749	1,811	30Psi2Ori	Ori	35	16	54.17	-20.67	55.25	-20.33	5	4.6
750	1,676	15 Ori	Ori	35	17	50.5	-8	51.87	-7.54	4	4.8
751	1,638	11 Ori	Ori	35	18	49.33	-8.17	50.6	-7.61	4	4.7
752	1,580	90mi2Ori	Ori	35	19	48	-10.25	48.46	-9.28	4	4.1
753	1,570	7Pi 1Ori	Ori	35	20	46.33	-12.83	47.65	-12.54	4	4.7
754	1,544	2Pi 2Ori	Ori	35	21	45.17	-14.25	46.43	-13.7	4	4.4
755	1,543	1Pi 3Ori	Ori	35	22	44.83	-15.83	45.74	-15.6	3	3.2
756	1,552	3Pi 4Ori	Ori	35	23	44.83	-17.17	46.16	-17.01	3	3.7
757	1,567	8Pi 5Ori	Ori	35	24	45.33	-20.33	46.55	-20.24	3	3.7
758	1,601	10Pi 6Ori	Ori	35	25	46.33	-21.5	47.6	-21.08	3	4.5
759	1,851	34Del Ori	Ori	35	26	55.33	-24.17	56.43	-23.78	2	2.2
760	1,903	46Eps Ori	Ori	35	27	57.33	-24.83	57.53	-24.75	2	1.7
761	1,948	50Zet Ori	Ori	35	28	58.17	-25.67	58.75	-25.54	2	1.9
762	1,788	28Eta Ori	Ori	35	29	53.83	-25.83	54.22	-25.78	3	3.4
763	1,892	42 Ori	Ori	35	30	56.5	-28.33	57.11	-28.37	4	4.6
764	1,893	43The2Ori	Ori	35	31	56.67	-29.17	57.05	-28.92	3.3	4.1
765	1,899	44Iot Ori	Ori	35	32	57	-29.83	57.06	-29.44	3	2.8
766	1,937	49 Ori	Ori	35	33	57.67	-30.67	57.99	-30.76	4	4.8
767	1,855	36Ups Ori	Ori	35	34	56.5	-30.83	55.97	-30.79	4	4.6
768	1,713	19Bet Ori	Ori	35	35	49.83	-31.5	50.88	-31.36	1	0.1
769	1,735	20Tau Ori	Ori	35	36	51	-30.25	51.91	-30.07	3.7	3.6
770	1,784	29 Ori	Ori	35	37	53.33	-31.17	53.64	-31.14	4	4.1

(continued)

Table 1.1 (continued)

#	HR number	Full star name	Constellation Name	#	Star	Almagest measurement Long	Lat	Actual (Epoch 137 AD) Long	Lat	V Mag	aV Mag
771	2,004	53Kap Ori	Ori	35	38	60.17	−33.5	60.47	−33.32	2.7	2.1
772	1,679	69Lam Eri	Eri	36	1	48.33	−31.83	49.26	−31.78	3.7	4.3
773	1,666	67Bet Eri	Eri	36	2	48.5	−28.25	49.38	−28.07	4	2.8
774	1,617	65Psi Eri	Eri	36	3	48	−29.83	47.25	−30	4	4.8
775	1,560	61Ome Eri	Eri	36	4	44.67	−28.25	45.1	−28.04	4	4.4
776	1,520	57Mu Eri	Eri	36	5	43.17	−25.83	43.37	−25.59	4	4
777	1,463	48Nu Eri	Eri	36	6	40.17	−25.33	40.85	−25.35	4	3.9
778	1,383	42Xi Eri	Eri	36	7	36.33	−26	37.37	−25.18	5	5.2
779	1,325	40Omi2Eri	Eri	36	8	35.5	−27	35.9	−27.14	4	4.4
780	1,298	38Omi1Eri	Eri	36	9	32.83	−27.83	33.43	−27.71	4	4
781	1,231	34Gam Eri	Eri	36	10	27	−32.83	27.84	−33.34	3	3
782	1,162	26Pi Eri	Eri	36	11	24.33	−31	24.91	−31.33	4	4.4
783	1,136	23Del Eri	Eri	36	12	24.17	−28.83	24.8	−29.25	3	3.5
784	1,084	18Eps Eri	Eri	36	13	22	−28	22.71	−28.05	3	3.7
785	984	13Zet Eri	Eri	36	14	17.17	−25.5	17.82	−26.12	3	4.8
786	917	9Rho2Eri	Eri	36	15	14.5	−23.83	14.77	−24.03	4	5.3
787	925	10Rho3Eri	Eri	36	16	12.17	−23.5	15.18	−24.07	3	5.3
788	859		Eri	36	17	10.5	−23.25	11.56	−24.94	4	6.3
789	818	1Tau1Eri	Eri	36	18	5.17	−32.17	5.86	−32.84	4	4.5
790	850	2Tau2Eri	Eri	36	19	5.83	−34.83	6.6	−35.65	4	4.8
791	919	11Tau3Eri	Eri	36	20	8.83	−38.5	8.56	−39.05	4	4.1
792	1,003	16Tau4Eri	Eri	36	21	13.83	−38.17	14	−38.67	4	3.7
793	1,088	19Tau5Eri	Eri	36	22	17.5	−39	18.12	−39.59	4	4.3
794	1,173	27Tau6Eri	Eri	36	23	21.33	−41.33	21.51	−41.83	4	4.2
795	1,181	28Tau7Eri	Eri	36	24	21.5	−42.5	21.24	−42.74	5	5.2
796	1,213	33Tau8Eri	Eri	36	25	22.17	−43.25	22.77	−43.83	4	4.7

797	1,240	36Tau9Eri	Eri	36	26	24.67	−43.33	24.9	−43.67	4	4.7
798	1,453	50Ups1Eri	Eri	36	27	34.17	−50.33	33.57	−51.03	4	4.5
799	1,464	52Ups2Eri	Eri	36	28	35	−51.75	33.85	−52.03	4	3.8
800	1,393	43 Eri	Eri	36	29	28.17	−53.83	28.29	−54.76	4	4
801	1,347	41Ups4Eri	Eri	36	30	25.83	−53.17	26.33	−54.15	4	3.6
802	1,214		Eri	36	31	17.83	−53	17.82	−53.41	4	5.1
803	1,195		Eri	36	32	14.83	−53.5	15.64	−54.46	4	4.2
804	1,143		Eri	36	33	11.83	−52.5	12.82	−54.98	4	4.6
805	897	The1Eri	Eri	36	34	0.17	−53.5	357.08	−53.86	1	2.9
806	1,696	3Iot Lep	Lep	37	1	49.67	−35	49.78	−34.94	5	4.4
807	1,705	4Kap Lep	Lep	37	2	49.83	−36.5	49.95	−36.05	5	4.4
808	1,757	7Nu Lep	Lep	37	3	51.33	−35.67	52.05	−35.59	5	5.3
809	1,756	6Lam Lep	Lep	37	4	51.33	−36.67	51.83	−36.43	5	4.3
810	1,702	5Mu Lep	Lep	37	5	49.17	−39.25	49.4	−39.28	3.7	3.3
811	1,654	2Eps Lep	Lep	37	6	46.17	−45.25	46.06	−45.16	3.7	3.2
812	1,865	11Alp Lep	Lep	37	7	55.83	−41.5	55.43	−41.3	3	2.6
813	1,829	9Bet Lep	Lep	37	8	54.83	−44.33	53.72	−44.11	3	2.8
814	2,035	15Del Lep	Lep	37	9	61	−44.17	61.08	−44.21	3.7	3.8
815	1,983	13Gam Lep	Lep	37	10	59	−45.83	59.14	−45.88	3.7	3.5
816	1,998	14Zet Lep	Lep	37	11	60	−38.33	60.07	−38.46	3.7	3.5
817	2,085	16Eta Lep	Lep	37	12	62.67	−38.17	63.01	−37.92	3.7	3.7
818	2,491	9Alp CMa	Cma	38	1	77.67	−39.17	78.5	−39.19	1	−1.5
819	2,574	14The CMa	Cma	38	2	79.67	−35	80.4	−34.94	4	4.1
820	2,593	18Mu CMa	Cma	38	3	81.33	−36.5	81.18	−36.89	5	5
821	2,657	23Gam CMa	Cma	38	4	83.33	−37.75	83.76	−38.22	4	4.1
822	2,596	20Iot CMa	Cma	38	5	85.33	−40	81.67	−39.88	4	4.4
823	2,590	19Pi CMa	Cma	38	6	80.5	−42.67	82.01	−42.97	5	4.7
824	2,443	8Nu 3CMa	Cma	38	7	76.17	−41.25	76.14	−41.52	5	4.4
825	2,429	7Nu 2CMa	Cma	38	8	76	−42.5	75.82	−42.53	5	4

(continued)

Table 1.1 (continued)

#	HR number	Full star name	Constellation Name	#	Star	Almagest measurement Long	Lat	Actual (Epoch 137 AD) Long	Lat	V Mag	aV Mag
826	2,294	2Bet CMa	Cma	38	9	71	−41.33	71.3	−41.5	3	2
827	2,387	4Xi 1CMa	Cma	38	10	74.67	−46.5	74.79	−46.81	5	4.3
828	2,414	5Xi 2CMa	Cma	38	11	76.17	−45.83	75.79	−46.3	5	4.5
829	2,653	24Omi2CMa	Cma	38	12	84.67	−46.17	85.18	−46.36	4	3
830	2,580	16Omi1CMa	Cma	38	13	81.67	−47	82.33	−47.01	5	3.9
831	2,693	25Del CMa	Cma	38	14	86.67	−48.75	87.59	−48.68	3.3	1.8
832	2,618	21Eps CMa	Cma	38	15	83.67	−51.5	84.96	−51.59	3	1.5
833	2,538	13Kap CMa	Cma	38	16	83	−55.17	82.77	−55.38	4	4
834	2,282	1Zet CMa	Cma	38	17	69.67	−53.75	71.5	−53.62	3	3
835	2,827	31Eta CMa	Cma	38	18	92.17	−50.67	93.77	−50.82	3.3	2.5
836	2,648	19 Mon	Cma	38	19	79.5	−25.25	81.74	−26.96	4	5
837	2,177	The Col	Cma	38	20	70	−61.5	67.17	−60.92	4	5
838	2,256	Kap Col	Cma	38	21	71.33	−58.75	70.62	−58.78	4	4.4
839	2,296	Del Col	Cma	38	22	73	−57	72.58	−56.93	4	3.8
840	2,361	Lam CMa	Cma	38	23	74.17	−56	74.75	−56.01	4	4.5
841	1,996	Mu Col	Cma	38	24	58	−55.5	58.8	−55.92	4	5.2
842	2,056	Lam Col	Cma	38	25	60.33	−57.67	61.43	−57.49	4	4.9
843	2,106	Gam Col	Cma	38	26	62.33	−59.83	63.11	−58.97	4	4.4
844	2,040	Bet Col	Cma	38	27	59	−59.67	60.41	−59.63	2	3.1
845	1,956	Alp Col	Cma	38	28	56	−57.67	56.2	−57.61	2	2.6
846	1,862	Eps Col	Cma	38	29	52.17	−59.5	52.68	−58.85	4	3.9
847	2,845	3Bet CMi	Cmi	39	1	85	−14	86.32	−13.69	4	2.9
848	2,943	10Alp CMi	Cmi	39	2	89.17	−16.17	90.18	−15.65	1	0.4
849	3,102	11 Pup	Arg	40	1	100.33	−42.5	101.88	−42.77	5	4.2
850	3,185	15Rho Pup	Arg	40	2	104.33	−43.33	105.69	−43.46	3	2.8
851	3,045	7Xi Pup	Arg	40	3	98.83	−45	100.27	−45.14	4	3.3

852	3,034	Omi Pup	Arg	40	4	98.67	−46.17	100.31	−46.25	4	4.5
853	2,944		Arg	40	5	95.33	−45.5	97	−46.25	4	4.7
854	2,948		Arg	40	6	96.33	−47.25	97.71	−47.63	3	3.8
855	2,922		Arg	40	7	95.33	−49.75	97.21	−49.31	4	4.6
856	2,996	3 Pup	Arg	40	8	99.33	−49.5	100.14	−49.41	4	4
857	2,993	1 Pup	Arg	40	9	98.5	−49.25	99.84	−48.91	4	4.6
858	3,113		Arg	40	10	104	−49.83	105.21	−49.86	4	4.8
859	2,823		Arg	40	11	94	−53	94.7	−53.67	4	5.4
860	2,773	Pi Pup	Arg	40	12	94	−58.67	94.6	−58.74	3	2.7
861	2,937		Arg	40	13	100.17	−55.5	100.78	−55.57	5	4.5
862	2,961		Arg	40	14	102.17	−58.67	103.42	−58.62	5	4.8
863	3,017		Arg	40	15	103.67	−57.25	105.24	−57.92	4	3.6
864	3,084		Arg	40	16	106.5	−57.83	108.39	−58.25	4	4.5
865	3,165	Zet Pup	Arg	40	17	111.17	−58.67	112.97	−58.51	2	2.3
866	3,080		Arg	40	18	108.17	−60	109.48	−59.87	5	3.7
867	3,162		Arg	40	19	111	−59.33	113.69	−59.78	5	5.5
868	3,225		Arg	40	20	113.17	−56.67	115.24	−57.56	5	4.4
869	3,243		Arg	40	21	114.33	−57.67	116.61	−57.99	5	4.4
870	3,535		Arg	40	22	125.67	−51.5	127.32	−53.26	3.7	5.8
871	3,477		Arg	40	23	126.17	−55.67	128.15	−57.47	3.7	4.1
872	3,426		Arg	40	24	124	−57.17	126.42	−58.37	3.7	4.1
873	3,487		Arg	40	25	129.17	−60	131.99	−60.23	3.7	3.9
874	3,445		Arg	40	26	129	−61.25	131.04	−61.24	3.7	3.8
875	3,438	Bet Pyx	Arg	40	27	120.17	−51.83	121.13	−51.28	3	4
876	3,468	Alp Pyx	Arg	40	28	119.33	−49	120.83	−49.06	3	3.7
877	3,518	Gam Pyx	Arg	40	29	118	−43.33	119.85	−43.44	4	4
878	3,556	Del Pyx	Arg	40	30	119	−43.5	121.05	−42.95	4	4.9
879	3,634	Lam Vel	Arg	40	31	134.17	−54.5	135.64	−55.95	2	2.2
880	3,786	Psi Vel	Arg	40	32	137.5	−51.25	139.24	−51.21	2.3	3.6

(continued)

Table 1.1 (continued)

#	HR number	Full star name	Constellation Name	#	Star	Almagest measurement Long	Lat	Actual (Epoch 137 AD) Long	Lat	V Mag	aV Mag
881	2,878	Sig Pup	Arg	40	33	101.17	−63	103.24	−64.05	4	3.3
882	3,055		Arg	40	34	109	−64.5	113.57	−65.39	6	4.1
883	3,207	Gam2Vel	Arg	40	35	120	−63.83	121.88	−64.6	2	1.7
884	3,117	Chi Car	Arg	40	36	128.5	−69.67	125.47	−70.44	2	3.5
885	3,447		Arg	40	37	135.17	−65.67	139.41	−66.34	3	3.6
886	3,485	Del Vel	Arg	40	38	141.33	−65.83	143.48	−67.21	3	2
887	3,498		Arg	40	39	146	−67.33	148.02	−68.5	2	4.5
888	3,734	Kap Vel	Arg	40	40	151	−62.83	153.49	−63.72	3	2.5
889	3,803		Arg	40	41	158	−62.25	158.84	−64.21	3	3.1
890	2,120	Eta Col	Arg	40	42	64	−65.83	63.67	−66.49	3.7	4
891	2,451	Nu Pup	Arg	40	43	80.17	−65.67	81.4	−66.31	2.7	3.2
892	2,326	Alp Car	Arg	40	44	77.17	−75	79.28	−76.08	1	−0.7
893	2,553	Tau Pup	Arg	40	45	89	−71.75	92.1	−73.04	2.7	2.9
894	3,418	5Sig Hya	Hya	41	1	104	−15	105.34	−14.77	4	4.4
895	3,410	4Del Hya	Hya	41	2	103.33	−13.17	104.46	−12.57	4	4.2
896	3,482	11Eps Hya	Hya	41	3	105.33	−11.5	106.57	−11.23	4	3.4
897	3,454	7Eta Hya	Hya	41	4	105.5	−14.25	106.44	−14.43	4	4.3
898	3,547	16Zet Hya	Hya	41	5	107.5	−12.25	108.74	−11.14	4	3.1
899	3,613	18Ome Hya	Hya	41	6	110.33	−11.83	111.52	−11.19	5	5
900	3,665	22The Hya	Hya	41	7	113.33	−13.67	114.31	−13.07	4	3.9
901	3,787	32Tau2Hya	Hya	41	8	118.83	−15.33	119.88	−15.1	4	4.6
902	3,845	35Iot Hya	Hya	41	9	120.67	−14.83	121.74	−14.38	4	3.9
903	3,759	31Tau1Hya	Hya	41	10	118.5	−17.17	119.68	−16.85	4	4.6
904	3,750		Hya	41	11	119.17	−19.75	120.66	−19.99	6	5.4
905	3,748	30Alp Hya	Hya	41	12	120	−20.5	121.46	−22.53	2	2
906	3,849	38Kap Hya	Hya	41	13	126	−26.5	126.88	−26.7	4	5.1

907	3,903	39Ups1Hya	Hya	41	14	128.67	−26	129.87	−26.17	4	4.1
908	3,970	40Ups2Hya	Hya	41	15	131.17	−23.25	132.53	−23.27	4	4.6
909	4,094	42Mu Hya	Hya	41	16	138	−24.67	139.28	−24.67	3	3.8
910	4,171	Phi3Hya	Hya	41	17	140	−23.25	142.29	−23.54	4	4.9
911	4,232	Nu Hya	Hya	41	18	143	−22.17	144.54	−21.95	3	3.1
912	4,343	11Bet Crt	Hya	41	19	151.5	−25.75	152.73	−25.6	3.7	4.5
913	4,314	Chi1Hya	Hya	41	20	152.33	−30.17	153.65	−30.23	4	4.9
914	4,450	Xi Hya	Hya	41	21	162.17	−31.33	162.32	−31.49	4	3.5
915	4,494	Omi Hya	Hya	41	22	164.5	−33.17	165.44	−33.38	4	4.7
916	4,552	Bet Hya	Hya	41	23	166.17	−31.33	167.71	−31.4	3	4.3
917	5,020	46Gam Hya	Hya	41	24	180	−13.67	181.11	−13.62	3.7	3
918	5,287	49Pi Hya	Hya	41	25	193.5	−17.67	192.7	−12.83	3.7	3.3
919	3,314		Hya	41	26	102.5	−23.25	104.03	−22.62	3	3.9
920	4,042	22Eps Sex	Hya	41	27	131	−16.33	133.58	−17.45	3	5.2
921	4,287	7Alp Crt	Crt	42	1	146.33	−23	148.14	−22.7	4	4.1
922	4,405	15Gam Crt	Crt	42	2	152.5	−19.33	153.45	−19.65	4	4.1
923	4,382	12Del Crt	Crt	42	3	150	−18	150.95	−17.65	4	3.6
924	4,514	27Zet Crt	Crt	42	4	157	−18.5	158.21	−18.27	3.7	4.7
925	4,402	14Eps Crt	Crt	42	5	149.33	−13.67	150.41	−13.49	4	4.8
926	4,567	30Eta Crt	Crt	42	6	159.17	−16.17	160.27	−16.05	4.3	5.2
927	4,468	21The Crt	Crt	42	7	151.67	−11.5	152.75	−11.3	4	4.7
928	4,623	1Alp Crv	Crv	43	1	165.33	−21.67	166.36	−21.7	3	4
929	4,630	2Eps Crv	Crv	43	2	164.33	−19.67	165.87	−19.61	3	3
930	4,696	5Zet Crv	Crv	43	3	166.67	−18.17	168.01	−18.2	5	5.2
931	4,662	4Gam Crv	Crv	43	4	163.33	−14.83	164.95	−14.43	3	2.6
932	4,757	7Del Crv	Crv	43	5	166.67	−12.5	167.66	−12.03	3	3
933	4,775	8Eta Crv	Crv	43	6	167	−11.67	168.14	−11.51	4	4.3
934	4,786	9Bet Crv	Crv	43	7	170.5	−18.17	171.51	−17.94	3	2.7
935	5,192	2 Cen	Cen	44	1	190.5	−21.67	192.19	−21.42	4.7	4.2

(continued)

Table 1.1 (continued)

#	HR number	Full star name	Constellation Name	#	Star	Almagest measurement Long	Lat	Actual (Epoch 137AD) Long	Lat	V Mag	aV Mag
936	5,221	4 Cen	Cen	44	2	190	−18.83	191.95	−18.81	4.7	4.7
937	5,168	1 Cen	Cen	44	3	189.17	−20.5	191.04	−20.29	3.7	4.2
938	5,210	3 Cen	Cen	44	4	190	−20	192.09	−19.9	4.7	4.3
939	5,028	Iot Cen	Cen	44	5	186.17	−25.67	187.47	−25.76	3	2.8
940	5,288	5The Cen	Cen	44	6	195.67	−22.5	196.63	−21.56	3	2.1
941	5,089		Cen	44	7	189.17	−27.5	190.67	−27.46	4	3.9
942	5,367	Psi Cen	Cen	44	8	198.17	−22.33	199.87	−22.31	4	4.1
943	5,378		Cen	44	9	199.17	−23.75	200.97	−23.66	4	4.4
944	5,485		Cen	44	10	202	−18.25	203.52	−18.11	4	4.1
945	5,471		Cen	44	11	202.5	−20.83	204.08	−20.78	4	4
946	5,190	Nu Cen	Cen	44	12	193.33	−28.33	195.34	−28.09	3.7	3.4
947	5,193	Mu Cen	Cen	44	13	194	−29.33	195.73	−28.8	3.7	3
948	5,248	Phi Cen	Cen	44	14	195.17	−28	197.22	−27.82	3.7	3.8
949	5,285	Chi Cen	Cen	44	15	196.33	−26.5	198.32	−26.4	3.7	4.4
950	5,440	Eta Cen	Cen	44	16	202.83	−25.25	204.42	−25.3	3	2.3
951	5,576	Kap Cen	Cen	44	17	207.5	−24.25	208.94	−23.81	4	3.1
952	5,231	Zet Cen	Cen	44	18	198	−33.5	199.16	−32.73	2.7	2.5
953	5,260	Ups2Cen	Cen	44	19	197.67	−31	199.46	−30.78	5	4.3
954	5,249	Ups1Cen	Cen	44	20	196.83	−30.33	198.54	−30.26	5	3.9
955	15,139		Cen	44	21	192.17	−34.83	193.44	−34.21	5	5.8
956	4,940		Cen	44	22	189	−37.67	190.8	−37.54	5	4.7
957	4,819	Gam Cen	Cen	44	23	185.83	−40	186.68	−39.98	3	2.2
958	4,802	Tau Cen	Cen	44	24	185	−40.33	185.69	−39.94	4	3.9
959	4,743	Sig Cen	Cen	44	25	182.67	−41	185.01	−42.24	5	3.9
960	4,621	Del Cen	Cen	44	26	182.67	−46.17	181.8	−44.38	3	2.6
961	4,638	Rho Cen	Cen	44	27	183.33	−46.75	183.71	−45.44	4	4

962	5,172		Cen	44	28	198.33	−40.75	199.74	−37.12	4	4.7
963	5,132	Eps Cen	Cen	44	29	196.33	−43	199.78	−39.4	2	2.3
964	5,141		Cen	44	30	197.67	−43.75	200.78	−40.22	3	5
965	4,763	Gam Cru	Cen	44	31	190	−51.17	190.9	−47.58	2	1.6
966	4,853	Bet Cru	Cen	44	32	195.33	−51.67	195.96	−48.45	2	1.3
967	4,656	Del Cru	Cen	44	33	186.33	−55.17	190	−50.26	4	2.8
968	4,730	Alp1Cru	Cen	44	34	191.17	−55.33	196.2	−52.69	2	0.7
969	5,459	Alp1Cen	Cen	44	35	218.33	−41.17	216.16	−41.81	1	−0.3
970	5,267	Bet Cen	Cen	44	36	204.17	−45.33	208.03	−43.92	2	0.6
971	4,898	Mu 1Cru	Cen	44	37	194.67	−49.17	194.9	−45.89	4	3.7
972	5,571	Bet Lup	Lup	45	1	208	−24.83	209.18	−24.82	3	2.7
973	5,469	Alp Lup	Lup	45	2	205.83	−29.17	207.68	−29.81	3	2.3
974	5,695	Del Lup	Lup	45	3	211	−21.25	212.79	−21.2	4	3.2
975	5,776	Gam Lup	Lup	45	4	214.17	−21	215.63	−21.01	4	2.8
976	5,708	Eps Lup	Lup	45	5	213	−25.17	214.27	−25.02	4	3.4
977	5,626	Lam Lup	Lup	45	6	210.17	−27	211.86	−26.29	5	4.1
978	5,605	Pi Lup	Lup	45	7	210.5	−29	211.79	−28.18	5	4
979	5,683	Mu Lup	Lup	45	8	214.67	−28.5	214.53	−28.26	5	4.3
980	5,646	Kap1Lup	Lup	45	9	213.67	−30.17	213.67	−29.4	5	3.7
981	5,649	Zet Lup	Lup	45	10	215.67	−33.17	214.97	−32.56	5	3.4
982	5,453	Rho Lup	Lup	45	11	202	−31.33	207.8	−31.95	5	4.1
983	5,354	Iot Lup	Lup	45	12	201.83	−30.5	202.97	−30	4	3.5
984	5,396	Tau2Lup	Lup	45	13	203	−29.33	203.92	−28.93	3.7	4.3
985	5,948	Eta Lup	Lup	45	14	218.83	−17	219.89	−17.2	4	3.4
986	5,987	The Lup	Lup	45	15	219.33	−15.33	220.86	−15.38	3.7	4.2
987	5,883	5Chi Lup	Lup	45	16	215.67	−13.33	216.95	−12.95	4	4
988	5,925	Xi 1Lup	Lup	45	17	216.67	−11.83	218.24	−13	4	4.6
989	5,660	1 Lup	Lup	45	18	207.17	−11.83	208.81	−12.8	3.7	4.9
990	5,686	2 Lup	Lup	45	19	206.5	−10	209.11	−11.3	3.7	4.3

(continued)

Table 1.1 (continued)

#	HR number	Full star name	Constellation Name	#	Star	Almagest measurement Long	Lat	Actual (Epoch 137 AD) Long	Lat	V Mag	aV Mag
991	6,537	1Sig Ara	Ara	46	1	237.67	−22.67	239.55	−22.9	5	4.6
992	6,743	The Ara	Ara	46	2	243	−25.75	245.27	−26.41	4	3.7
993	6,510	Alp Ara	Ara	46	3	236.33	−26.5	239.04	−26.28	3.7	3
994	6,295	Eps1Ara	Ara	46	4	230.67	−30.33	233.69	−30.03	5	4.1
995	6,462	Gam Ara	Ara	46	5	235.17	−34.17	238.39	−32.86	3.7	3.3
996	6,461	Bet Ara	Ara	46	6	235	−33.33	238.31	−32.01	4	2.8
997	6,285	Zet Ara	Ara	46	7	230.83	−34.25	233.94	−32.83	4	3.1
998	6,897	Alp Tel	CrA	47	1	249.17	−21.5	249.16	−22.38	4	3.5
999	7,062	Eta1CrA	CrA	47	2	251.67	−21	253.46	−20.38	5	5.5
1000	7,122		CrA	47	3	253.17	−20.33	255.03	−19.55	5	5.4
1001	7,188	Zet CrA	CrA	47	4	254.83	−20	256.38	−19.06	4	4.8
1002	7,242	Del CrA	CrA	47	5	256.17	−18.5	257.62	−17.6	5	4.6
1003	7,259	Bet CrA	CrA	47	6	257	−17.17	258.12	−16.49	4	4.1
1004	7,254	Alp CrA	CrA	47	7	256.83	−16	258.16	−15.02	4	4.1
1005	7,226	Gam CrA	CrA	47	8	256.5	−15.17	257.65	−14.02	4	4.2
1006	7,152	Eps CrA	CrA	47	9	255.17	−15.33	256.18	−13.99	6	4.9
1007	7,129		CrA	47	10	254.67	−14.83	255.66	−14.22	6	5.4
1008	7,021	Lam CrA	CrA	47	11	251.83	−14.67	252.95	−14.95	5	5.1
1009	6,942		CrA	47	12	249.67	−15.83	250.54	−16.18	5	5.2
1010	6,951	The CrA	CrA	47	13	249.17	−18.5	250.6	−18.81	5	4.6
1011	8,728	24Alp PsA	PsA	48	1	307	−20.33	307.73	−20.88	1	1.2

1012	8,576	17Bet PsA	PsA	48	2	300.67	−20.33	301.16	−21.21	4	4.3
1013	8,695	22Gam PsA	PsA	48	3	304.17	−22.25	305.34	−23.53	4	4.5
1014	8,720	23Del PsA	PsA	48	4	305.33	−22.5	306.17	−23.54	4	4.2
1015	8,628	18Eps PsA	PsA	48	5	304.33	−16.25	305.33	−17.15	3.7	4.2
1016	8,431	14Mu PsA	PsA	48	6	295.17	−19.5	296.09	−19.91	5	4.5
1017	8,570	Zet PsA	PsA	48	7	301.17	−15.17	303.64	−15.37	5	6.4
1018	8,478	16Lam PsA	PsA	48	8	298.83	−14.67	299.42	−15.57	4	5.4
1019	8,386	12Eta PsA	PsA	48	9	295.17	−15	296.29	−15.12	4	5.4
1020	8,326	10The PsA	PsA	48	10	291.83	−16.5	292.65	−16.4	4	5
1021	8,305	9Iot PsA	PsA	48	11	291	−18.17	291.27	−18.13	4	4.3
1022	8,353	Gam Gru	PsA	48	12	290.17	−22.25	291.38	−22.86	4	3
1023	8,069	Eta Mic	PsA	48	13	278	−22.33	280.54	−23.47	3.3	5.5
1024	8,151	The1Mic	PsA	48	14	281.17	−22.17	283.5	−23.81	3.3	4.8
1025	8,229	Xi Gru	PsA	48	15	284	−21.17	285.59	−24.91	3.3	5.3
1026	8,180	The2Mic	PsA	48	16	282	−20.83	284.16	−24.23	5	5.8
1027	7,965	Alp Mic	PsA	48	17	283.83	−17	279.67	−15.24	4	4.9
1028	8,039	Gam Mic	PsA	48	18	283.83	−14.83	282.47	−14.47	4	4.7

The 'Almagest Measurement' column presents the measurements listed by Ptolemy in the *Almagest*; the 'Actual' column provides a modern value of the object adjusted backward for Epoch AD 137. The 'V mag' column presents the magnitude recorded in the Almagest and the 'aV mag' column presents the actual visual magnitude of the star known today. The data in this table is provided with permission of Dennis Duke, professor of physics at the Florida State University

Chapter 2

The Abd-Al-Rahman Al-Sufi Catalog

Abd-al-Rahman Al-Sufi (also called Azophi or Al-Sufi "The Mystic") was a Persian astronomer and nobleman living at the court of the Emire Adud ad-Daula in Isfahan (Persia). There is not a lot of information to tell us about the life of Al-Sufi, but we know he worked on the translation of early Greek astronomical texts, in particular the *Almagest* of Ptolemy.

Europe was in the Dark Ages during the time of Al-Sufi. Religious fervor had gripped European nations as the bubonic plague devastated the population. Because of the chaotic state of Europe the knowledge of the ancient Greeks had not reached much of the rest of Europe. Meanwhile the scholars of the Middle East were busy translating the ancient Greek texts and improving upon the knowledge. This period is known as the Golden Age of Islam, when Muslim scientists, engineers, and scholars made significant contributions to the world. One of the great Muslim astronomers was Al-Sufi.

About AD 964 Al-Sufi published his greatest work, *Kitab al-Kawatib al-Thabit al-Musawwar* (the "Book of Fixed Stars"), a revision of Ptolemy's *Almagest* based on his own observations. In the book Al-Sufi describes the position, color, and brightness of 1,018 objects. The book became known to the West because of thirteenth century Castilian translations and more recently by the translations of M. Schjellerup. The *Book of Fixed Stars* is the source of many of the modern names of stars coming from Latin translations of the Arabic names given by Al-Sufi. The main body of work in the *Book of Fixed Stars* is a detailed description of each constellation and the stars they contain. Two illustrations of the Arabic constellations are provided, one illustration from inside the celestial globe and one from outside of the celestial globe. It is still considered a masterpiece in the study of the proper motion of stars and for the study of long period variables. Al-Sufi also

J.D. Cavin, *The Amateur Astronomer's Guide to the Deep-Sky Catalogs*,
Patrick Moore's Practical Astronomy Series, DOI 10.1007/978-1-4614-0656-3_2,
© Springer Science+Business Media, LLC 2012

Fig. 2.1 The image of the constellation Sagittarius from Al-Sufi's book *The Depiction of Celestial Constellations*

attempted to match the Greek star names and constellations with the Arabic names and constellations.

Al-Sufi was the first astronomer to report on the Andromeda Galaxy (M31). He called it "a Little Cloud" lying before the mouth of a Big Fish, an Arabic constellation. After receiving reports from Arab navigators he also described a southern group of stars as al-Baqar al-Abyad (the "White Bull"). This group of stars is known today as the Large Magellanic Cloud. In addition to reporting a number of new objects, Al-Sufi's observations improved upon Ptolemy's brightness (magnitude) measurements.

An annual international observing competition honoring Al-Sufi has been created by the amateur section of the Astronomical Society of Iran (ASIAC). The Al-Sufi list is similar to the Messier object list but has a variety of more difficult objects. This is the list currently being used for the Sufi Observing Competition held by the ASIAC (Table 2.1).

Table 2.1 The table lists the 120 objects of the Al-Sufi observing competition. The Al-Sufi observing competition list has been compiled and edited with the help of Neda Mobara

#	NGC/IC/Messier number	Common name	Type	Constellation	RA (2000)	Dec (2000)
1*	NGC 6475/M7	Ptolemy's cluster	Open cluster	Scorpius	17 h 53 m 51.1 s	−34° 47' 34"
2	NGC 6405/M6	Butterfly cluster	Open cluster	Scorpius	17 h 40 m 20.7 s	−32° 15' 15"
3	NGC 6533/M8	Lagoon Nebula	Star cluster with Nebula	Sagittarius	18 h 04 m 04.0 s	−24° 23' 49"
4	NGC 6514/M20	Trifid Nebula	Star cluster with Nebula	Sagittarius	18 h 02 m 20.9 s	−23° 01' 38"
5	NGC 6656/M22		Globular cluster	Sagittarius	18 h 36 m 24.1 s	−23° 54' 12"
6	NGC 6266/M62		Globular cluster	Ophiuchus	17 h 01 m 12.5 s	−30° 06' 44"
7	NGC 6093/M80		Globular cluster	Scorpius	16 h 17 m 02.51 s	−22° 58' 30.4"
8	NGC 6231		Open cluster	Scorpius	16 h 54 m 10.9 s	−41° 49' 27"
9	NGC 6281		Open cluster	Scorpius	17 h 04 m 41.2 s	−37° 59' 07"
10	NGC 6541		Globular cluster	Corona Australis	18 h 08 m 02.3 s	−43° 42' 57"
11	NGC 6441		Globular cluster	Scorpius	17 h 50 m 12.9 s	−37° 03' 04"
12	NGC 4826/M64	Black Eye Galaxy	Galaxy	Coma Berenices	12 h 56 m 44.2 s	+21° 40' 58"
13	NGC 3587/M97	Owl Nebula	Planetary Nebula	Ursa Major	11 h 14 m 47.71 s	+55° 01' 07.7"
14	NGC 3556/M108		Galaxy	Ursa Major	11 h 11 m 31.2 s	+55° 40' 24"
15	NGC 3992/M109		Galaxy	Ursa Major	11 h 57 m 36.0 s	+53° 23' 28"
16	NGC 4258/M106		Galaxy	Canes Venatici	12 h 18 m 57.5 s	+47° 18' 15"
17	NGC 4736/M94		Galaxy	Canes Venatici	12 h 50 m 53.0 s	+41° 07' 12"
18	NGC 5055/M63	Sunflower Galaxy	Galaxy	Canes Venatici	13 h 15 m 49.1 s	+42° 01' 50"
19	NGC5194/M51	Whirlpool Galaxy	Galaxy	Canes Venatici	13 h 29 m 52.1 s	+47° 11' 43"
20	NGC5457/M101	Pinwheel Galaxy	Galaxy	Ursa Major	14 h 03 m 12.4 s	+54° 20' 55"
21	NGC 5272/M3		Globular cluster	Canes Venatici	13 h 42 m 11.2 s	+28° 22' 32"
22	NGC 5904/M5		Globular cluster	Serpens Caput	15 h 18 m 33.7 s	+02° 04' 58"
23	NGC 3031/M81	Bode's Galaxy	Galaxy	Ursa Major	09 h 55 m 32.9 s	+69° 03' 55"
24	NGC 3034/M82		Galaxy	Ursa Major	09 h 55 m 50.7 s	+69° 40' 43"
25	NGC5457/M102		Galaxy	Ursa Major	14 h 03 m 12.4 s	+54° 20' 55"

(continued)

Table 2.1 (continued)

#	NGC/IC/Messier number	Common name	Type	Constellation	RA (2000)	Dec (2000)
26	NGC 5907		Galaxy	Draco	15 h 15 m 54.0 s	+56° 19' 45"
27	NGC 6946	Fireworks Galaxy	Galaxy	Cygnus	20 h 34 m 52.7 s	+60° 09' 11"
28	NGC 7092/M39		Open cluster	Cygnus	21 h 31 m 48.3 s	+48° 26' 55"
29	NGC 7243		Open cluster	Lacerta	22 h 15 m 08.5 s	+49° 53' 51"
30	NGC 7331		Galaxy	Pegasus	22 h 37 m 04.5 s	+34° 25' 01"
31	NGC 6218/M12		Globular cluster	Ophiuchus	16 h 47 m 14.4 s	−01° 56' 52"
32	NGC 6333/M9		Globular cluster	Ophiuchus	17 h 19 m 11.7 s	−18° 30' 59"
33	NGC 6626/M28		Globular cluster	Sagittarius	18 h 24 m 32.8 s	−24° 52' 12"
34	NGC 6715/M54		Globular cluster	Sagittarius	18 h 55 m 03.3 s	−30° 28' 42"
35	NGC 6809/M55		Globular cluster	Sagittarius	19 h 39 m 59.3 s	−30° 57' 44"
36	NGC 6494/M23		Open cluster	Sagittarius	17 h 57 m 04.7 s	−18° 59' 07"
37	NGC 6611/M16	Eagle Nebula	Open cluster	Sagittarius	18 h 18 m 48.1 s	−13° 48' 26"
38	NGC 6618/M17	Horseshoe Nebula	Bright Nebula	Sagittarius	18 h 20 m 47.1 s	−16° 10' 18"
39	IC 4725/M25		Open cluster	Sagittarius	18 h 31 m 36.0 s	−19° 15' 00"
40	*	Asterism of Nu1 and Nu2	asterism	Sagittarius	18 h 20 m 26 s	−16° 10' 36"
41	NGC 6716		Open cluster	Sagittarius	18 h 54 m 34.3 s	−19° 54' 04"
42	NGC 6402/M14		Globular cluster	Ophiuchus	17 h 37 m 36.1 s	−03° 14' 46"
43	IC 4665		Open cluster	Ophiuchus	17 h 46 m 18 s	+05° 43' 00"
44	IC 4756	Graff's cluster	Open cluster	Serpens	18 h 39 m 04.9 s	+05° 27' 09"
45	NGC 6705/M11	Wild Duck cluster	Open cluster	Scutum	18 h 51 m 05.9 s	−06° 16' 12"
46	NGC 6694/M26		Open cluster	Scutum	18 h 45 m 18.6 s	−09° 23' 01"
47	NGC 6712		Globular cluster	Scutum	18 h 53 m 04.3 s	−08° 42' 22"
48	NGC 6864/M75		Globular cluster	Sagittarius	20 h 06 m 04.7 s	−21° 42' 22"
49	NGC 224/M31*	Andromeda Galaxy	Galaxy	Andromeda	00 h 42 m 44.3 s	+41° 16' 06"
50	NGC 205/M110	Part of M31	Globular cluster	Andromeda	00 h 40 m 21.9 s	+41° 41' 26"
51	NGC 185		Galaxy	Cassiopeia	00 h 38 m 58.1 s	+48° 20' 27"

52	NGC 7654/M52		Open cluster	Cassiopeia	23 h 24 m 50.4 s	+61° 36' 24"
53	NGC 40	Bow-Tie Nebula	Planetary Nebula	Cepheus	00 h 13 m 00.9 s	+72° 31' 20"
54	NGC 7789	Caroline's[a] Rose cluster	Open cluster	Cassiopeia	23 h 57 m 24.0 s	+56° 42' 30"
55	NGC 129		Open cluster	Cassiopeia	00 h 29 m 54.1 s	+60° 12' 35"
56	NGC 457	Kachina Doll cluster	Open cluster	Cassiopeia	01 h 19 m 32.6 s	+58° 17' 27"
57	NGC 663		Open cluster	Cassiopeia	01 h 46 m 16.0 s	+61° 13' 06"
58	NGC 6205/M13	The Great cluster in Hercules	Globular cluster	Hercules	16 h 41 m 41.4 s	+36° 27' 36"
59	NGC 6341/M92		Globular cluster	Hercules	17 h 17 m 07.5 s	+43° 08' 11"
60	NGC 6229		Globular cluster	Hercules	16 h 46 m 58.8 s	+47° 31' 40"
61	NGC 6503		Galaxy	Draco	17 h 49 m 26.2 s	+70° 08' 42"
62	NGC 6981/M72		Globular cluster	Aquarius	20 h 53 m 27.9 s	−12° 32' 13"
63	NGC 7009	Saturn Nebula	Planetary Nebula	Aquarius	21 h 04 m 10.7 s	−11° 21' 49"
64	NGC 7099/M30		Globular cluster	Capricornus	21 h 40 m 21.9 s	−23° 10' 45"
65	NGC 7293	Helix Nebula	Planetary Nebula	Aquarius	22 h 29 m 38.35 s	−20° 50' 13.2"
66	NGC 628/M74		Galaxy	Pisces	01 h 36 m 41.6 s	+15° 47' 03"
67	NGC 598/M33	Pinwheel Galaxy	Galaxy	Triangulum	01 h 33 m 50.8 s	+30° 39' 37"
68	NGC 752		Open cluster	Andromeda	01 h 57 m 47.9 s	+37° 51' 00"
69	NGC 650/M76	Part of Little Dumbbell Nebula	Planetary Nebula	Perseus	01 h 42 m 18.1 s	+51° 34' 16"
70	NGC 869/884*	Double cluster	Open cluster	Perseus	02 h 19 m 03.8 s	+57° 08' 06"
71	NGC 891		Galaxy	Andromeda	02 h 22 m 33.5 s	+42° 21' 03"
72	NGC 1039/M34		Open cluster	Perseus	02 h 42 m 07.4 s	+42° 44' 46"
73	NGC 1023		Group of Galaxies	Perseus	02 h 40 m 23.9 s	+39° 03' 48"
74	Collinder 399*	Al-Sufi's cluster[b]	Asterism	Vulpecula	19 h 25 m 24 s	+20° 11' 00"
75	NGC 6838/M71		Globular cluster	Sagittarius	19 h 53 m 46.1 s	+18° 46' 42"
76	NGC 6853/M27	Dumbbell Nebula[c]	Planetary Nebula	Vulpecula	19 h 59 m 36.33 s	+22° 43' 16.0"
77	NGC 6934		Globular cluster	Delphinus	20 h 34 m 11.5 s	+07° 24' 16"
78	NGC 6720/M57	Ring Nebula	Planetary Nebula	Lyra	18 h 53 m 35.09 s	+33° 01' 44.5"

(continued)

Table 2.1 (continued)

#	NGC/IC/Messier number	Common name	Type	Constellation	RA (2000)	Dec (2000)
79	NGC 6913/M29		Open cluster	Cygnus	20 h 23 m 57.7 s	+38° 30' 28"
80	NGC 7582	Brightest member of Grus Quartet	Galaxy	Grus	23 h 18 m 23.7 s	−42° 22' 15"
81	NGC 55	Brightest in Sculptor Group	Galaxy	Sculptor	00 h 15 m 08.4 s	−39° 13' 14"
82	NGC 300		Galaxy	Sculptor	00 h 54 m 53.4 s	−37° 41' 00"
83	NGC 7793	Part of Sculptor Group	Galaxy	Sculptor	23 h 57 m 49.7 s	−32° 35' 30"
84	NGC 613		Galaxy	Sculptor	01 h 34 m 17.5 s	−29° 24' 58"
85	NGC 247		Galaxy	Cetus	00 h 47 m 08.4 s	−20° 45' 36"
86	NGC 253	Sculptor Galaxy	Galaxy	Sculptor	00 h 47 m 33.1 s	−25° 17' 17"
87	NGC 288		Globular cluster	Sculptor	00 h 52 m 47.4 s	−26° 35' 24"
88	Melotte 20/Collinder 39	Alpha Persei Moving cluster	Open cluster	Perseus	03 h 26.9 m	+49° 07'
89	NGC 1342		Open cluster	Perseus	03 h 31 m 40.1 s	+37° 22' 28"
90	NGC 1502		Open cluster	Camelopardalis	04 h 07 m 49.3 s	+62° 19' 54"
91	NGC 1528		Open cluster	Perseus	04 h 15 m 18.9 s	+51° 12' 41"
92	NGC 246	Skull Nebula	Planetary Nebula	Cetus	00 h 47 m 30.2 s	−11° 52' 20"
93	NGC 936	Darth Vader's Galaxy	Galaxy	Cetus	02 h 27 m 37.5 s	−01° 09' 23"
94	NGC 1068/M77	Seyfert Galaxy	Galaxy	Cetus	2 h 42 m 40.7 s	−00° 00' 47"
95	Melotte 22/M45	The Pleiades	Open cluster	Tau	03 h 47 m 28 s	+24° 06' 18"
96	NGC 1746		Open cluster	Taurus	05 h 03 m 50.2 s	+23° 46' 04"
97	NGC 1907		Open cluster	Auriga	05 h 28 m 04.5 s	+35° 19' 32"
98	NGC 1912/M38		Open cluster	Auriga	05 h 28 m 42.5 s	+35° 51' 18"
99	NGC 1960/M36		Open cluster	Auriga	05 h 36 m 17.7 s	+34° 08' 27"
100	NGC 2099/M37		Open cluster	Auriga	05 h 52 m 18.3 s	+32° 33' 11"
101	NGC 1952/M1	Crab Nebula	Bright Nebula	Taurus	05 h 34 m 31.9 s	+22° 00' 52"
102	NGC 2168/M35		Open cluster	Gemini	06 h 08 m 55.9 s	+24° 21' 28"

103	NGC 2158		Open cluster	Gemini	06 h 07 m 25.6 s	+24° 05' 46"
104	NGC 2403	Member of M81 Group	Galaxy	Camelopardalis	07 h 36 m 51.8 s	+65° 36' 13"
105	NGC 7089/M2		Globular cluster	Aquarius	21 h 33 m 29.3 s	-00° 49' 23"
106	NGC 7078/M15		Globular cluster	Pegasus	21 h 29 m 58.4 s	+12° 10' 00"
107	NGC 1976/M42	Orion Nebula	Nebula	Orion	05 h 35 m 17.2 s	-05° 23' 27"
108	NGC 2068/M78		Nebula	Orion	05 h 46 m 45.8 s	+00° 04' 45"
109	NGC 2244	Cluster in the center of the Rose Nebula	Open cluster	Monoceros	06 h 31 m 55.6 s	+04° 56' 35"
110	NGC 2264	Cone Nebula[d]	Cluster with Nebula	Monoceros	06 h 40 m 58.3 s	+09° 53' 44"
111	NGC 1097		Galaxy	Fornax	02 h 46 m 19.0 s	-30° 16' 28"
112	NGC 1360		Galaxy	Fornax	03 h 33 m 14.63 s	-25° 52' 18.6"
113	NGC 1365	Great Barred Spiral Galaxy	Galaxy	Fornax	03 h 33 m 36.3 s	-36° 08' 25"
114	NGC 1316	Fornax A (largest in group)	Galaxy	Fornax	03 h 22 m 41.7 s	-37° 12' 30"
115	NGC 1291		Galaxy	Eridanus	03 h 17 m 18.3 s	-41° 06' 26"
116	NGC 1904/M79		Globular cluster	Lepus	05 h 24 m 10.6 s	-24° 31' 27"
117	NGC 2392	Eskimo Nebula	Planetary Nebula	Gemini	07 h 29 m 10.8 s	+20° 54' 42"
118	NGC 2323/M50		Open cluster	Monoceros	07 h 02 m 42.3 s	-08° 23' 26"
119	NGC 2287/M41		Open cluster	Canis Major	06 h 46 m 00.0 s	-20° 45' 15"
120	NGC 2632 M44	Beehive cluster	Open cluster	Cancer	08 h 40 m 22.2 s	+19° 40' 19"
121	NGC 1912/M38		Open cluster	Cancer	05 h 28 m 42.5 s	+35° 51' 18"
122	NGC 2281*	Asterism of lambda (diamond shaped asterism near the core of the cluster)	Open cluster	Auriga	06 h 48 m 17.8 s	+41° 04' 44"

The "*" indicates the object is contained in Al-Sufi's original catalog. The Al-Sufi Observing Competition list has been compiled and edited with the help of Neda Mobara

[a] The Caroline's Rose cluster (or the White Rose cluster) is named after Caroline Herschel, who discovered the object in 1783

[b] The Al-Sufi's Cluster is also known as the Brocchi's Cluster or the Coat Hanger asterism

[c] The Dumbbell Nebula is also known as the Apple Core Nebula

[d] The Cone Nebula is also known as the Christmas Tree Nebula

Chapter 3

Ulugh Beg

Mīrzā Muhammad Tāriq ibn Shāhrukh (Ulugh Beg) was born in Soltaniyeh, Persia.[1] The translation of his common name, Ulugh Beg, is the "Great Prince," a nickname he received as a child. He was the grandson of the Mongol conqueror Tamerlane and the only son of Shah Rukh. He spent his childhood days with his grandfather and father as they traveled through the lands of the Middle East and India conquering the indigenous people. At the age of 16 Ulugh Beg received an entire province of Mawaraunnahr to govern from his father. The province included the great city of Samarkand. But he was not as interested in ruling as his father and grandfather. Instead, in Samarkand he founded and built a great madrasa (university) where the greatest Islamic mathematicians and astronomers of his time came to study. The Samarkand madrasa was built in Rigestan Square, where it still stands today.

It has been reported that Beg became interested in astronomy when he visited the ruins of Maragheh Observatory used by the astronomer Nasir al-Din Tusi (1201–1274). To improve upon his madrasa he built the greatest observatory of his time, the Samarkand Observatory. It became the first observatory to permanently mount the astronomical instruments directly into the structure of the building. The observatory was staffed with 70 of the greatest scientists and astronomers, making the Samarkand Observatory the most advanced scientific research center on Earth. A quote carved into the stone above Samarkand Observatory entry may provide some insight into Ulugh Beg's character: "Religions disperse, kingdoms fall apart, but works of science remain for all ages."

---

[1]The ancient city of Soltaniyeh is found in modern day Iran, approximately 240 km northwest of Tehran.

J.D. Cavin, *The Amateur Astronomer's Guide to the Deep-Sky Catalogs*,
Patrick Moore's Practical Astronomy Series, DOI 10.1007/978-1-4614-0656-3_3,
© Springer Science+Business Media, LLC 2012

Fig. 3.1 This is a depiction of the mathematician and astronomer Ulugh Beg and his observatory at Samarkand. The picture is a USSR stamp issued on October 8, 1968

The Samarkand Observatory was a three-level drum-shaped building 160 ft in diameter and 120 ft tall. The observatory contained a marble sextant, an armillary sphere, and a triquetram. The most significant instrument of the Samarkand Observatory was a giant marble quadrant known as the Fakhri Quadrant. It ran through the center of the observatory between two high walls in a semicircle from the ceiling on one side of the building sloping downward to below the floor and arcing up to the ceiling of the opposite wall. Along the arc ran two set of marble tiles inscribed with digits marking the degrees. The astronomers would sight the star through a small circular sighting device at the top of one wall and record its position along the marble scale. The enormous size of the quadrant, a radius of 116 ft and an optical separation of 180 s of arc, allowed for measurements to be more accurate than ever before achieved.

Ulugh Beg may have been inspired by the 150 books written by Nasir al-Din Tusi when he began to compile his own observations. In 1437 Beg's *Zij-i Sultani* ("Catalog of Stars") was published, listing about 1,000 stars. The *Zij-i Sultani* was the first star catalog published since Ptolemy's *Almagest*. Unfortunately Beg's work was unknown in Europe until 50 years after Brahe's publication of his catalog. But for his time Ulugh Beg's measurements were far superior to those of Ptolemy (Table 3.1).

Table 3.1 A comparison of the star positions compared with the same stars in Ptolemy-Hipparchus from Ulugh Beg's catalog republished by Edward Ball Knobel in 1917

Constellations	Number of stars	Average error	
		Ulugh Beg	Ptolemy-Hipparchus
The northern constellations			
Ursa Minor "The Little Bear"	7	15.0	17.2
Ursa Major "The Great Bear"	32	10.2	29.8
Draco "The Dragon"	27	17.8	23.8
Cepheus "The King"	9	8.9	14.1
Boötes "The Plowman"	18	16.6	24.5
Corona Borealis "The Northern Crown"	7	12.4	11.1
Hercules	25	11.1	14.2
Lyra "The Lyre"	15	12.0	17.0
Cygnus "The Swan"	13	13.3	11.5
Cassiopeia "The Seated Queen"	35	33.9	38.0
Auriga "The Charioteer"	8	19.5	19.2
Ophiuchus "The Snake Holder"	27	17.1	17.6
Serpens "The Snake"	18	12.8	9.4
Sagitta "The Arrow"	5	11.4	18.4
Aquila "The Equal"	12	20.7	18.2
Delphinus "The Dolphin"	10	10.7	15.3
Equuleus "The Foal"	4	14.5	10.7
Pegasus "The Winged Horse"	17	20.7	9.6
Andromeda "The Woman Chained"	17	25.4	17.0
Triangulum "The Triangle"	4	27.2	14.0
The zodiacal constellations			
Aries "The Ram"	17	22.2	18.2
Taurus "The Bull"	38	20.8	23.2
Gemini "The Twins"	16	17.2	19.0
Cancer "The Crab"	10	14.3	21.8
Leo "The Lion"	31	11.4	26.0
Virgo "The Virgin"	27	7.3	14.2
Libra "The Scale"	13	20.1	9.8
Scorpius "The Scorpion"	21	12.8	14.7
Saggitarius "The Archer"	27	13.7	14.4
Capricornus "The Sea Goat"	23	13.3	12.4
Aquarius "The Water-Bearer"	38	18.6	26.8
Pisces "The Fish"	32	19.1	23.3
The southern constellations			
Cetus "The Whale"	30	26.0	36.0
Orion	34	13.5	17.1
Eridanus "The River"	29	10.7	28.3
Lepus "The Hare"	11	12.9	9.3
Canis Major "The Greater Dog"	19	6.3	16.0
Canis Minor "The Lesser Dog"	2	13.0	23.5

(continued)

Table 3.1 (continued)

Constellations	Number of stars	Average error	
		Ulugh Beg	Ptolemy-Hipparchus
Argo Navis[a] "The Argonaut"	28	14.4	26.8
Hydra "The Sea Serpent"	22	8.9	20.9
Crater "The Cup"	5	10.0	14.0
Corvus "The Crow or Raven"	5	9.4	16.4
Centaurus "The Centaur"	18	13.3	14.0
Lupus "The Wolf"	17	14.6	31.2
Ara "The Altar"	..	....	....
Corona Australis "The Southern Crown"	11	36.4	47.1
Piscis Austrinus "The Southern Fish"	9	20.9	41.2

[a] The Argo Navis was a large constellation representing the *Argo*, the ship sailed by Jason and the Argonauts. In 1752 the French astronomer Nicolas Louis de Lacaille divided the Argo Navis into three smaller constellations: the Carina, Puppis, and Vela

The observatory quadrant was the primary instrument used to compile the *Zij-i-Sultani*.[2] The compilation was a collaborative effort between Ulugh Beg and fellow students and astronomers working under his supervision, such as Jamshīd al-Kāshī, Ali Qushji, and others. The star measurements contained in the *Zij-i-Sultani* were made between 1420 and 1437. It listed 994 stars originally compiled from other Arabic catalogs, which were based on Ptolomy's star catalog in the *Almagest*. The improved star measurements in the *Zij-i-Sultani* provided the most significant improvement in star measurement since the *Almagest* 1,300 years earlier.

Soon after the death of Ulugh Beg's father there was a power struggle for Ulugh Beg's province. Being an only son, Ulugh Beg was easily overthrown. On October 27, 1449, at the age of 56, he was beheaded on an order from his own son, Abd ul-Latif. Ulugh Beg's tomb and remains were found in Samarkand by archaeologists in 1941. When the archeologists examined the body of Ulugh Beg it was discovered he was buried as a *shahid* (wearing the clothes he died in), a sign that he was considered a martyr at the time of his death.

---

[2] A full listing of Ulugh Beg's star catalog can be found in Edward Bell Knobels book *Ulugh Beg's Catalogue of Stars*, published by the Carnegie Institution of Washington in 1917.

Chapter 4

The Tycho Brahe Catalog (1598)

Fig. 4.1 Illustration of Tycho Brahe from J. L. E. Dreyer biography of his fellow countryman *Tycho Brahe – A Picture of Scientific Life and Work in the Sixteenth Century* (1890)

J.D. Cavin, *The Amateur Astronomer's Guide to the Deep-Sky Catalogs*,
Patrick Moore's Practical Astronomy Series, DOI 10.1007/978-1-4614-0656-3_4,
© Springer Science+Business Media, LLC 2012

The greatest observational astronomer of the sixteenth century was Tycho Brahe. The instruments he built to measure the position of the stars were the most accurate instruments on Earth. Brahe is also remembered for supplying the data used by Johannes Kepler (1571–1630) to derive the Laws of Planetary Motion. Kepler's work would then provide the foundation for Isaac Newton's (1643–1727) to develop the laws of gravitation (Table 4.1).

Tycho Brahe was born the only surviving twin of a noble Danish family. His father Otte Brahe was a nobleman and worked in the court of the Danish King while his mother Beate Brahe was from a family of politicians and churchmen. His uncle, Jørgen Brahe was a scholar and a vice-admiral who inherited and acquired a significant amount of properties. Being the first born Tycho appeared to be was destined for a life of aristocratic leisure. But, Tycho's father had made an agreement with Jørgen before the birth of his son, that if Tycho was a boy, he could adopt and raise Tycho. After his birth Otte Brahe did not want to follow through with his agreement. When Tycho was 2 years old he was kidnapped by his uncle.

Table 4.1 The table contains three versions of Tycho Brahe's star catalogs

1598 list	1602 list	1627 list	Const	Star seq #	Ecliptic longitude Degree	Arcmin	Ecliptic latitude Degree	Arcmin	Brahe's V Mag
336	307	1	UMi	1	23	2.5	66	2	2
337	308	2	UMi	2	25	36	69	50.5	4
338	309	3	UMi	3	3	24	73	50	4
339	310	4	UMi	4	21	29	75	0	4
340	311	5	UMi	5	24	52	77	38.5	5
341	312	6	UMi	6	7	16.5	72	51.5	2
342	313	7	UMi	7	14	41	75	23.5	3
343	0	8	UMi	8	2	54	71	23	6
344	0	9	UMi	9	27	20	70	18	6
345	0	10	UMi	10	17	17	35	50	6
346	0	11	UMi	11	17	28	37	20	6
347	0	12	UMi	12	17	45	40	13	6
348	0	13	UMi	13	18	3	42	56	6
349	0	14	UMi	14	21	38	57	55	6
350	0	15	UMi	15	21	55	70	42	6
351	0	16	UMi	16	24	31	69	3	6
352	0	17	UMi	17	15	7	68	4	6
353	0	18	UMi	18	7	22	67	43	6
354	0	19	UMi	19	9	57	67	22	6
355	0	20	UMi	20	26	30	63	55	6
356	314	21	UMa	1	17	36.5	40	2.5	4
357	315	22	UMa	2	17	10	43	55.5	4
358	316	23	UMa	3	16	8	44	22	5
359	317	24	UMa	4	18	25	47	50.5	4

(continued)

Table 4.1 (continued)

1598 list	1602 list	1627 list	Const	Star seq #	Ecliptic longitude Degree	Arcmin	Ecliptic latitude Degree	Arcmin	Brahe's V Mag
360	318	25	UMa	5	19	44.5	47	44.5	4
361	319	26	UMa	6	24	42.5	51	36.5	5
362	320	27	UMa	7	23	50	42	30	5
363	321	28	UMa	8	25	2	45	3	4
364	322	29	UMa	9	28	0	46	21.5	5
365	323	30	UMa	10	0	38	42	36	4
366	324	31	UMa	11	3	38.5	38	15.5	4
367	325	32	UMa	12	0	32.5	34	34.5	3
368	326	33	UMa	13	25	56	29	15.5	3
369	327	34	UMa	14	27	10	28	38	3
370	328	35	UMa	15	27	7	33	30	5
371	329	36	UMa	16	27	26	36	6	5
372	330	37	UMa	17	9	34	49	40	2
373	331	38	UMa	18	13	43.5	45	3.5	2
374	332	39	UMa	19	25	25.5	51	37	2
375	333	40	UMa	20	24	45	47	6.5	2
376	334	41	UMa	21	13	56.5	29	51.5	4
377	335	42	UMa	22	15	4.5	28	45	4
378	336	43	UMa	23	22	33	35	14	4
379	337	44	UMa	24	0	55	26	14	4
380	338	45	UMa	25	1	36	24	54	4
381	339	46	UMa	26	3	10	54	18	2
382	340	47	UMa	27	9	56.5	56	22	2
383	341	48	UMa	28	21	12	54	25	2
384	342	49	UMa	29	17	43.5	40	6	2
385	0	50	UMa	30	28	10	41	30	4
386	0	51	UMa	31	21	2	33	1	5
387	0	52	UMa	32	6	17	17	55	3
388	0	53	UMa	33	8	10	20	42	4
389	0	54	UMa	34	5	0	20	5	4
390	0	55	UMa	35	1	57	20	51	4
391	0	56	UMa	36	29	42	23	41	4
392	0	57	UMa	37	14	12	21	53	4
393	0	58	UMa	38	18	55	25	4	4
394	0	59	UMa	39	19	57	24	50	3
395	0	60	UMa	40	23	22	21	28	3
396	0	61	UMa	41	26	9	20	44	3
397	0	62	UMa	42	25	19	24	58	4
398	0	63	UMa	43	12	16	40	30	5
399	0	64	UMa	44	21	29	58	8	6
400	0	65	UMa	45	23	55	47	14	6
401	0	66	UMa	46	19	49	47	30	6
402	0	67	UMa	47	23	17	46	50	6
403	0	68	UMa	48	3	58	47	55	6

(continued)

Table 4.1 (continued)

1598 list	1602 list	1627 list	Const	Star seq #	Ecliptic longitude Degree	Ecliptic longitude Arcmin	Ecliptic latitude Degree	Ecliptic latitude Arcmin	Brahe's V Mag
404	0	69	UMa	49	6	0	48	40	6
405	0	70	UMa	50	6	30	49	42	6
406	0	71	UMa	51	6	19	49	42	6
407	0	72	UMa	52	19	5	49	0	6
408	0	73	UMa	53	18	1	49	27	6
409	0	74	UMa	54	25	42	48	11	6
410	0	75	UMa	55	16	2	52	25	6
411	0	76	UMa	56	1	41	35	40	6
412	343	77	Dra	1	18	56.5	76	17	4
413	344	78	Dra	2	4	14.5	78	15.5	4
414	345	79	Dra	3	6	19.5	75	21	3
415	346	80	Dra	4	19	3	80	21.5	4
416	347	81	Dra	5	22	24	75	3.5	3
417	348	82	Dra	6	17	4	81	53	5
418	349	83	Dra	7	24	31	77	57	5
419	350	84	Dra	8	20	33.5	79	51.5	5
420	351	85	Dra	9	9	29	80	53.5	4
421	352	86	Dra	10	28	33	81	51	4
422	353	87	Dra	11	12	26.5	82	49	3
423	354	88	Dra	12	15	21	78	9.5	4
424	355	89	Dra	13	27	47	79	25	3
425	356	90	Dra	14	15	18	83	5	4
426	357	91	Dra	15	19	40.5	80	38	4
427	358	92	Dra	16	26	44	80	54	4
428	359	93	Dra	17	6	34.5	83	4.5	4
429	360	94	Dra	18	1	28	83	28.5	4
430	361	95	Dra	19	5	31	84	48	4
431	362	96	Dra	20	29	44.5	81	4.5	3
432	363	97	Dra	21	6	26	86	53	4
433	364	98	Dra	22	28	21	83	18	5
434	365	99	Dra	23	28	22	81	41	5
435	366	100	Dra	24	26	51.5	84	46	3
436	367	101	Dra	25	7	55	78	32	3
437	368	102	Dra	26	12	28.5	74	11.5	3
438	369	103	Dra	27	29	22	71	4	3
439	370	104	Dra	28	29	17	65	18	5
440	371	105	Dra	29	2	10.5	66	36	2
441	372	106	Dra	30	10	26	61	33	3
442	373	107	Dra	31	4	37.5	57	7	3
443	374	108	Dra	32	1	4	77	31.5	5
444	375	109	Cep	1	0	13	71	7	3
445	376	110	Cep	2	7	13	68	54	3
446	377	111	Cep	3	27	53.5	62	35	4
447	378	112	Cep	4	8	29	61	3	4

(continued)

Table 4.1 (continued)

1598 list	1602 list	1627 list	Const	Star seq #	Ecliptic longitude Degree	Ecliptic longitude Arcmin	Ecliptic latitude Degree	Ecliptic latitude Arcmin	Brahe's V Mag
448	0	113	Cep	5	7	53.5	59	59	4
449	0	114	Cep	6	13	39	58	46	4
450	0	115	Cep	7	29	21	71	49	4
451	0	116	Cep	8	29	54	74	0.5	4
452	0	117	Cep	9	18	46	65	42	5
453	0	118	Cep	10	27	33	75	27	4
454	0	119	Cep	11	24	23	64	28	3
455	379	120	Boo	1	24	9.5	58	53	4
456	380	121	Boo	2	25	33	58	51	4
457	381	122	Boo	3	26	59.5	60	5	4
458	382	123	Boo	4	1	18	54	40	4
459	383	124	Boo	5	13	5.5	49	33.5	3
460	384	125	Boo	6	18	43.5	54	15.5	3
461	385	126	Boo	7	27	29.5	49	1	3
462	386	127	Boo	8	22	29.5	40	40	3
463	387	128	Boo	9	18	16	42	11	4
464	388	129	Boo	10	17	17.5	42	35.5	4
465	389	130	Boo	11	27	26.5	27	57	3
466	390	131	Boo	12	13	42	28	9	3
467	391	132	Boo	13	12	25	26	33	4
468	392	133	Boo	14	13	37	25	14	4
469	393	134	Boo	15	18	39	31	2.5	1
470	394	135	Boo	16	26	13.5	30	27.5	4
471	395	136	Boo	17	27	11	31	22	4
472	396	137	Boo	18	27	52	33	52	4
473	0	138	Boo	19	28	11	40	14.5	5
474	0	139	Boo	20	29	40	40	31.5	5
475	0	140	Boo	21	27	53	42	16	5
476	0	141	Boo	22	29	16	41	55	6
477	0	142	Boo	23	29	34.5	45	6	5
478	0	143	Boo	24	1	26.5	46	52	5
479	0	144	Boo	25	27	32	53	27.5	4
480	0	145	Boo	26	2	35	54	0	4
481	0	146	Boo	27	11	49	60	40	6
482	0	147	Boo	28	12	33	60	57	6
497	397	148	CrB	1	6	38.5	44	23	2
498	398	149	CrB	2	3	37	46	8	4
499	399	150	CrB	3	3	10.5	48	25	5
500	400	151	CrB	4	8	2	50	21	6
501	401	152	CrB	5	9	14.5	44	33	4
502	402	153	CrB	6	11	25	44	52	4
503	403	154	CrB	7	13	32	46	9.5	4
504	404	155	CrB	8	13	2	48	24	6
505	405	156	Her	1	10	31	37	23	3

(continued)

Table 4.1 (continued)

1598 list	1602 list	1627 list	Const	Star seq #	Ecliptic longitude Degree	Arcmin	Ecliptic latitude Degree	Arcmin	Brahe's V Mag
506	406	157	Her	2	25	27.5	42	48	3
507	407	158	Her	3	23	36	40	5.5	3
508	408	159	Her	4	20	6.5	37	19	4
509	409	160	Her	5	9	10	47	47	3
510	410	161	Her	6	14	22	49	23	4
511	411	162	Her	7	19	36	51	16.5	4
512	412	163	Her	8	27	19	52	19	4
513	413	164	Her	9	23	57	53	46	4
514	414	165	Her	10	23	38	52	47	4
515	415	166	Her	11	26	2	53	10.5	3
516	416	167	Her	12	2	45.5	53	21	3
517	417	168	Her	13	6	21.5	59	38	4
518	418	169	Her	14	7	19	60	11.5	4
519	419	170	Her	15	9	47.5	60	13.5	4
520	420	171	Her	16	22	56	60	47	3
521	421	172	Her	17	14	17	69	22	3
522	422	173	Her	18	7	5.5	71	20	6
523	423	174	Her	19	11	7	71	13.5	6
524	424	175	Her	20	18	0	71	5	9
525	425	176	Her	21	23	8.5	60	22.5	3
526	426	177	Her	22	17	39.5	63	14	4
527	427	178	Her	23	8	43.5	65	55	4
528	428	179	Her	24	5	57	63	51	4
529	429	180	Her	25	2	43	64	23	4
530	430	181	Her	26	16	32	62	29	5
531	431	182	Her	27	2	28.5	60	15.5	4
532	432	183	Her	28	27	6	57	15.5	4
533	433	184	Lyr	1	9	43	61	47.5	1
534	434	185	Lyr	2	13	14	62	27	5
535	435	186	Lyr	3	12	26	60	26	5
536	436	187	Lyr	4	16	10.5	59	26	4
537	437	188	Lyr	5	24	32.5	60	46	5
538	438	189	Lyr	6	25	2	59	41	5
539	439	190	Lyr	7	13	16.5	56	5	3
540	440	191	Lyr	8	13	3.5	55	16	6
541	441	192	Lyr	9	16	11	55	6	3
542	442	193	Lyr	10	16	20	54	31.5	6
543	443	194	Lyr	11	20	52	58	6	5
544	444	195	Cyg	1	25	44	49	2	3
545	445	196	Cyg	2	29	20	50	42	5
546	446	197	Cyg	3	7	33	54	19	4
547	447	198	Cyg	4	19	25	57	9.5	3
548	448	199	Cyg	5	29	53.5	59	56.5	2
549	449	200	Cyg	6	10	53	64	28	3

(continued)

Table 4.1 (continued)

1598 list	1602 list	1627 list	Const	Star seq #	Ecliptic longitude		Ecliptic latitude		Brahe's V Mag
					Degree	Arcmin	Degree	Arcmin	
550	450	201	Cyg	7	13	21	69	42	4
551	451	202	Cyg	8	12	39.5	71	31	4
552	452	203	Cyg	9	9	36.5	73	50.5	4
553	453	204	Cyg	10	22	9.5	49	26	3
554	454	205	Cyg	11	24	18	51	41.5	4
555	455	206	Cyg	12	27	43	43	44	3
556	456	207	Cyg	13	0	32	54	59	4
557	457	208	Cyg	14	5	21.5	56	36	4
558	458	209	Cyg	15	22	50	63	37	4
559	459	210	Cyg	16	24	34.5	64	17.5	4
560	460	211	Cyg	17	3	3.5	50	33	4
561	461	212	Cyg	18	4	53.5	51	31	4
562	0	213	Cyg	19	4	33	38	39	3
563	0	214	Cyg	20	19	57	66	15	4
564	0	215	Cyg	21	24	49.5	68	52	4
565	0	216	Cyg	22	13	31	69	35	4
566	0	217	Cyg	23	28	44	25	11	6
567	0	218	Cyg	24	28	22	35	35	6
568	0	219	Cyg	25	18	15	53	12	6
569	0	220	Cyg	26	13	18	69	42	6
0	0	221	Cyg	27	16	15	55	30	0
570	462	222	Cas	1	29	35	44	40.5	4
571	463	223	Cas	2	2	17.5	46	35.5	3
572	464	224	Cas	3	4	38	47	5	4
573	465	225	Cas	4	8	27.5	48	46	3
574	466	226	Cas	5	12	21	46	22	3
575	467	227	Cas	6	19	13.5	47	29	3
576	468	228	Cas	7	26	39	48	54	4
577	469	229	Cas	8	6	14.5	43	6.5	4
578	470	230	Cas	9	5	16	43	28	5
579	471	231	Cas	10	24	39	49	24.5	6
580	472	232	Cas	11	7	6	52	14	4
581	473	233	Cas	12	29	35.5	51	14.5	3
582	474	234	Cas	13	25	34	51	8	6
583	475	235	Cas	14	25	32	52	39	6
584	476	236	Cas	15	19	28	52	48	6
585	477	237	Cas	16	22	21	56	13	6
586	478	238	Cas	17	22	23	54	27	6
587	479	239	Cas	18	21	58	52	8.5	6
588	480	240	Cas	19	12	57.5	44	57.5	6
589	481	241	Cas	20	10	0	45	4.5	6
590	482	242	Cas	21	6	52	47	31.5	6
591	483	243	Cas	22	29	10	45	38	6
592	484	244	Cas	23	29	32	41	15	6

(continued)

Table 4.1 (continued)

1598 list	1602 list	1627 list	Const	Star seq #	Ecliptic longitude Degree	Arcmin	Ecliptic latitude Degree	Arcmin	Brahe's V Mag
593	485	245	Cas	24	27	57	41	25.5	6
594	486	246	Cas	25	26	56	39	15.5	6
595	487	247	Cas	26	25	54.5	38	9	6
596	0	248	Cas	27	1	46	53	16	6
597	0	249	Cas	28	6	12	53	32	6
598	0	250	Cas	29	0	11	52	4	6
599	0	251	Cas	30	6	45	59	8	6
600	0	252	Cas	31	17	17	35	50	6
601	0	253	Cas	32	27	19	35	48	6
602	0	254	Cas	33	2	33	34	49	6
603	0	255	Cas	34	3	0	30	22	6
604	0	256	Cas	35	0	45	44	10	6
605	0	257	Cas	36	0	57	45	32	6
606	0	258	Cas	37	26	15	45	32	6
607	0	259	Cas	38	0	10	54	43	6
608	0	260	Cas	39	27	45	56	15	6
609	0	261	Cas	40	4	13	56	55	6
610	0	262	Cas	41	29	58	59	18	6
611	0	263	Cas	42	7	54	60	7	6
612	0	264	Cas	43	10	14	62	4	6
613	0	265	Cas	44	9	37	62	46	6
614	0	266	Cas	45	20	58	63	17	6
0	0	267	Cas	46	6	54	53	45	0
615	488	268	Per	1	18	31	39	0.5	6
616	489	269	Per	2	23	9.5	37	28.5	4
617	490	270	Per	3	24	26.5	34	30	3
618	491	271	Per	4	19	4.5	31	34.5	4
619	492	272	Per	5	21	50	34	26	5
620	493	273	Per	6	23	33	30	36.5	4
621	494	274	Per	7	26	17	30	5	2
622	495	275	Per	8	27	4.5	27	59	5
623	496	276	Per	9	28	13.5	27	55	5
624	497	277	Per	10	29	15	27	14	3
625	498	278	Per	11	22	6	26	4	4
626	499	279	Per	12	20	37	22	22	3
627	500	280	Per	13	20	31	20	54	5
628	501	281	Per	14	19	18	20	33	4
629	502	282	Per	15	18	20	21	35	4
630	503	283	Per	16	6	13.5	28	22.5	5
631	504	284	Per	17	4	11.5	28	50	4
632	505	285	Per	18	3	55	26	11	5
633	506	286	Per	19	5	14	26	39	4
634	507	287	Per	20	6	0	24	35	6
635	508	288	Per	21	8	1	18	56	5

(continued)

Table 4.1 (continued)

1598 list	1602 list	1627 list	Const	Star seq #	Ecliptic longitude Degree	Arcmin	Ecliptic latitude Degree	Arcmin	Brahe's V Mag
636	509	289	Per	22	28	11	22	6	4
637	510	290	Per	23	0	8	19	4	3
638	511	291	Per	24	29	23.5	14	53.5	5
639	512	292	Per	25	25	33	12	8	4
640	513	293	Per	26	27	36	11	17.5	3
641	514	294	Per	27	26	45	42	26	5
642	515	295	Per	28	2	32	29	31	5
643	516	296	Per	29	16	16	20	53	4
644	0	297	Per	30	2	18	45	10	6
645	0	298	Per	31	4	12	48	7	6
646	0	299	Per	32	4	41	49	27	6
647	0	300	Per	33	6	25	53	37	6
648	517	301	Aur	1	23	38	32	15	6
649	518	302	Aur	2	24	14	30	50	4
650	519	303	Aur	3	16	16	22	50.5	1
651	520	304	Aur	4	25	52	21	27.5	2
652	521	305	Aur	5	24	28	13	44	4
653	522	306	Aur	6	13	9	20	52	4
654	523	307	Aur	7	13	5.5	18	8.5	4
655	524	308	Aur	8	13	49.5	18	11.5	4
656	525	309	Aur	9	11	4.5	10	22	4
657	0	310	Aur	10	24	25.5	27	27	5
658	0	311	Aur	11	16	52.5	18	34.5	6
659	0	312	Aur	12	16	6	16	59	5
660	0	313	Aur	13	14	58	15	21.5	5
661	0	314	Aur	14	17	9	14	4	6
662	0	315	Aur	15	12	0	15	3	5
663	0	316	Aur	16	22	12.5	15	42.5	5
664	0	317	Aur	17	22	24	15	43	5
665	0	318	Aur	18	22	35	13	49	6
666	0	319	Aur	19	16	39.5	11	15	5
667	0	320	Aur	20	18	34	8	51	5
668	0	321	Aur	21	10	4.5	14	51	5
669	0	322	Aur	22	10	31	14	2	5
670	0	323	Aur	23	27	47	6	4	4
671	0	324	Aur	24	22	58	4	6	4
672	0	325	Aur	25	23	58	2	26	4
673	0	326	Aur	26	19	52.5	2	28	4
674	0	327	Aur	27	21	55	1	6	4
688	540	328	Oph	1	16	50	35	57	3
689	541	329	Oph	2	19	45	28	1	3
690	542	330	Oph	3	21	5	26	11	3
691	543	331	Oph	4	4	59.5	32	35.5	3
692	544	332	Oph	5	6	16	31	56	4

(continued)

Table 4.1 (continued)

1598 list	1602 list	1627 list	Const	Star seq #	Ecliptic longitude Degree	Arcmin	Ecliptic latitude Degree	Arcmin	Brahe's V Mag
693	545	333	Oph	6	0	3	23	39.5	4
694	546	334	Oph	7	26	44.5	17	19	3
695	547	335	Oph	8	27	57	16	30.5	3
696	548	336	Oph	9	19	33	15	19	4
697	549	337	Oph	10	24	13.5	13	47	4
698	550	338	Oph	11	25	14.5	15	20	5
699	551	339	Oph	12	12	24	7	18	3
700	552	340	Oph	13	3	39	11	30	3
701	553	341	Oph	14	14	23	2	12	3
702	554	342	Oph	15	26	31	33	2.5	4
703	0	343	Oph	16	16	48	26	36.5	4
704	0	344	Oph	17	14	49	10	21	4
705	0	345	Oph	18	18	57	8	4	3
706	0	346	Oph	19	19	48	10	35	4
707	0	347	Oph	20	18	45	15	18	4
708	0	348	Oph	21	0	57	13	19	5
709	0	349	Oph	22	24	30	27	55	4
710	0	350	Oph	23	24	38	26	23	4
711	0	351	Oph	24	24	53	24	50	4
712	0	352	Oph	25	25	58	26	10	4
676	0	353	Oph	26	14	1	2	16	3
677	0	354	Oph	27	15	42	1	32	4
678	0	355	Oph	28	16	23	0	20	4
679	0	356	Oph	29	17	12	0	29	5
680	0	357	Oph	30	17	36	0	58	5
681	0	358	Oph	31	16	50	7	10	5
682	0	359	Oph	32	21	45	4	20	6
683	0	360	Oph	33	0	7	23	34	5
684	0	361	Oph	34	15	0	10	18	5
685	0	362	Oph	35	19	2	8	5	4
686	0	363	Oph	36	20	4	10	40	5
687	0	364	Oph	37	19	5	15	6	5
713	555	365	Ser	1	11	35	38	12	5
714	556	366	Ser	2	14	24.5	39	6.5	3
715	557	367	Ser	3	17	6.5	35	25	3
716	558	368	Ser	4	14	21.5	34	27.5	3
717	559	369	Ser	5	15	10	37	28.5	4
718	560	370	Ser	6	16	32	42	37	4
719	561	371	Ser	7	12	46.5	28	58	3
720	562	372	Ser	8	16	30	25	35	2
721	563	373	Ser	9	18	46.5	24	5.5	3
722	564	374	Ser	10	20	26.5	16	26.5	4
723	565	375	Ser	11	24	34.5	19	57	3
724	566	376	Ser	12	0	12.5	20	37.5	3

(continued)

Table 4.1 (continued)

1598 list	1602 list	1627 list	Const	Star seq #	Ecliptic longitude Degree	Arcmin	Ecliptic latitude Degree	Arcmin	Brahe's V Mag
725	567	377	Ser	13	10	10	26	59	3
726	568	378	Sge	1	1	32	39	13	4
727	569	379	Sge	2	27	55	38	58.5	5
728	570	380	Sge	3	28	31	39	31	6
729	571	381	Sge	4	25	30.5	38	53	4
730	572	382	Sge	5	25	39	38	18	4
731	0	383	Sge	6	0	13	42	43	4
732	0	384	Sge	7	1	36	44	2	4
733	0	385	Sge	8	23	57	46	3	4
734	573	386	Aql	1	29	28.5	27	8.5	6
735	574	387	Aql	2	26	53	26	49.5	3
736	575	388	Aql	3	26	9	29	21.5	2
737	576	389	Aql	4	25	33	30	54.5	6
738	577	390	Aql	5	25	26	31	18	3
739	578	391	Aql	6	26	8.5	31	59	5
740	579	392	Aql	7	21	16.5	28	46.5	4
741	580	393	Aql	8	22	14	26	35	5
742	581	394	Aql	9	14	15.5	36	16.5	3
743	582	395	Aql	10	12	44	37	40	3
744	583	396	Aql	11	9	12	43	32.5	4
745	584	397	Aql	12	9	17.5	41	5	4
746	585	398	Atn	1	29	21.5	18	48	3
747	586	399	Atn	2	20	17.5	20	14.5	3
748	587	400	Atn	3	19	17	14	28	3
749	0	401	Atn	4	18	1	24	56	3
750	0	402	Atn	5	24	50	21	38	3
751	0	403	Atn	6	11	46	17	41	3
752	0	404	Atn	7	10	29	16	57	4
753	588	405	Del	1	8	32	29	8	3
754	589	406	Del	2	9	48	28	52.5	6
755	590	407	Del	3	9	42	27	34	6
756	591	408	Del	4	10	56	31	57.5	3
757	592	409	Del	5	11	50.5	33	5	3
758	593	410	Del	6	13	36.5	32	0	3
759	594	411	Del	7	13	52	32	47	3
760	595	412	Del	8	10	17	32	8.5	5
761	596	413	Del	9	9	18	30	41.5	6
762	597	414	Del	10	10	42	30	41	6
763	598	415	Equ	1	17	32.5	20	12.5	4
764	599	416	Equ	2	19	54.5	21	6	4
765	600	417	Equ	3	17	54	25	16	4
766	601	418	Equ	4	18	54.5	24	52	4
767	602	419	Peg	1	26	22	22	7.5	3
768	603	420	Peg	2	1	15.5	16	25	4

(continued)

Table 4.1 (continued)

1598 list	1602 list	1627 list	Const	Star seq #	Ecliptic longitude Degree	Arcmin	Ecliptic latitude Degree	Arcmin	Brahe's V Mag
769	604	421	Peg	3	29	45.5	15	43	5
770	605	422	Peg	4	13	0	14	30.5	6
771	606	423	Peg	5	12	44	15	43.5	6
772	607	424	Peg	6	10	39.5	17	41	3
773	608	425	Peg	7	12	25	18	29	5
774	609	426	Peg	8	3	23	36	42.5	4
775	610	427	Peg	9	8	50	34	19	4
776	611	428	Peg	10	14	3	41	0.5	4
777	612	429	Peg	11	17	29.5	28	49	4
778	613	430	Peg	12	18	53.5	29	24.5	4
779	614	431	Peg	13	20	10.5	35	7.5	3
780	615	432	Peg	14	19	25	34	24.5	5
781	616	433	Peg	15	25	33	25	35	6
782	617	434	Peg	16	27	6	24	50.5	6
783	618	435	Peg	17	17	56.5	19	26	2
784	619	436	Peg	18	23	49	31	7.5	2
785	620	437	Peg	19	3	38	12	35	2
786	0	438	Peg	20	6	28	20	51	4
787	0	439	Peg	21	24	51	33	21	4
788	0	440	Peg	22	28	47	36	11	4
789	0	441	Peg	23	15	15	23	16	4
790	621	442	And	1	8	47	25	42	2
791	622	443	And	2	17	6.5	27	6.5	5
792	623	444	And	3	15	25	23	3.5	4
793	624	445	And	4	14	58	31	33	5
794	625	446	And	5	15	45.5	33	20.5	4
795	626	447	And	6	16	7	32	14.5	5
796	627	448	And	7	10	28	40	56.5	4
797	628	449	And	8	11	46	41	44	4
798	629	450	And	9	14	23	42	8	5
799	630	451	And	10	12	47	43	49.5	4
800	631	452	And	11	15	9	17	48	4
801	632	453	And	12	16	53.5	15	58	5
802	633	454	And	13	24	49	25	59	2
803	634	455	And	14	24	6.5	30	33.5	4
804	635	456	And	15	23	36	32	30.5	4
805	636	457	And	16	8	39	27	46.5	2
806	637	458	And	17	9	6.5	36	49.5	5
807	638	459	And	18	6	52	35	21.5	4
808	639	460	And	19	5	6	28	59	5
809	640	461	And	20	3	23	27	54.5	5
810	641	462	And	21	0	56	36	20	5
811	642	463	And	22	24	0	57	19	4
812	643	464	And	23	16	19.5	24	20	3

(continued)

Table 4.1 (continued)

1598 list	1602 list	1627 list	Const	Star seq #	Ecliptic longitude Degree	Arcmin	Ecliptic latitude Degree	Arcmin	Brahe's V Mag
813	644	465	Tri	1	1	19	16	49.5	4
814	645	466	Tri	2	6	49.5	20	33	4
815	646	467	Tri	3	7	59	19	29	5
816	647	468	Tri	4	7	58	18	57	4
483	527	469	Com	1	18	17	28	25	3
484	540	470	Com	2	28	15	28	32	5
485	528	471	Com	3	18	42	27	23.5	4
486	529	472	Com	4	18	46	27	20	4
487	530	473	Com	5	19	19	27	7	4
488	531	474	Com	6	18	25	25	51	4
489	532	475	Com	7	18	48.5	26	7	4
490	533	476	Com	8	18	0	23	30	4
491	534	477	Com	9	21	10	25	16	4
492	535	478	Com	10	20	51	24	56	4
493	536	479	Com	11	22	52	24	0.5	4
494	537	480	Com	12	28	58.5	32	46	4
495	538	481	Com	13	27	49.5	31	42	4
496	539	482	Com	14	24	17	30	16	4
0	0	483	Com	15	28	15	28	32	5
1	1	484	Ari	1	27	37	7	8.5	4
2	2	485	Ari	2	28	23	8	29	4
3	3	486	Ari	3	2	6	9	57	3
4	4	487	Ari	4	2	34	7	23	6
5	5	488	Ari	5	3	20	5	42.5	6
6	6	489	Ari	6	27	57	5	24	5
7	7	490	Ari	7	8	36	6	7	6
8	8	491	Ari	8	12	57	4	8.5	5
9	9	492	Ari	9	15	15	1	46.5	4
10	10	493	Ari	10	16	24	2	50	5
11	11	494	Ari	11	17	50.5	2	36	6
12	12	495	Ari	12	11	22	1	12	6
13	13	496	Ari	13	9	35	1	7	6
14	14	497	Ari	14	9	23	1	30	6
15	15	498	Ari	15	7	52	0	39	6
16	16	499	Ari	16	8	46	4	1	6
17	17	500	Ari	17	1	41	9	13	6
18	18	501	Ari	18	10	35	10	50.5	5
19	19	502	Ari	19	11	23	11	16	4
20	20	503	Ari	20	12	40	10	24	3
21	21	504	Ari	21	12	51	12	25.5	4
22	22	505	Tau	1	18	0	5	57	5
23	23	506	Tau	2	17	30	7	29	6
24	24	507	Tau	3	16	18	8	49.5	4
25	25	508	Tau	4	15	35.5	9	22.5	4

(continued)

Table 4.1 (continued)

1598 list	1602 list	1627 list	Const	Star seq #	Ecliptic longitude Degree	Arcmin	Ecliptic latitude Degree	Arcmin	Brahe's V Mag
26	26	509	Tau	5	21	46	8	41	5
27	27	510	Tau	6	25	1	8	3	4
28	28	511	Tau	7	27	59	12	13.5	4
29	29	512	Tau	8	24	19	14	30.5	4
30	30	513	Tau	9	4	9	9	32	5
31	31	514	Tau	10	3	11	11	48	5
32	32	515	Tau	11	0	12	5	46.5	3
33	33	516	Tau	12	1	16.5	4	2	3
34	34	517	Tau	13	2	22	5	53	4
35	35	518	Tau	14	4	12.5	5	31	1
36	36	519	Tau	15	2	53	2	36.5	3
37	37	520	Tau	16	8	12	3	40	6
38	38	521	Tau	17	12	13.5	2	30.5	6
39	39	522	Tau	18	11	4	1	49.5	4
40	40	523	Tau	19	19	12	2	14	3
41	41	524	Tau	20	6	35	0	40	5
42	42	525	Tau	21	16	59.5	5	20	2
43	43	526	Tau	22	2	54	1	4	5
44	44	527	Tau	23	2	38	0	35	4
45	45	528	Tau	24	27	51	1	12	5
46	46	529	Tau	25	0	28.5	0	46.5	6
47	47	530	Tau	26	0	4	5	16	6
48	48	531	Tau	27	29	45.5	7	55	5
49	49	532	Tau	28	2	34	3	57	5
50	50	533	Tau	29	2	25.5	5	45.5	5
51	51	534	Tau	30	23	13.5	4	11	5
52	52	535	Tau	31	24	3	4	2	6
53	53	536	Tau	32	24	24	4	0	3
54	54	537	Tau	33	24	47	3	55	5
55	55	538	Tau	34	19	57	13	30	6
56	56	539	Tau	35	0	10	12	2	6
57	57	540	Tau	36	1	58.5	8	41	5
58	58	541	Tau	37	1	42	6	56.5	5
59	59	542	Tau	38	3	28	7	4.5	5
60	60	543	Tau	39	4	55	6	17.5	5
61	61	544	Tau	40	15	2	1	4	6
62	62	545	Tau	41	16	55.5	1	20	6
63	63	546	Tau	42	17	33	9	34.5	6
64	64	547	Tau	43	29	22.5	6	33	5
65	65	548	Gem	1	14	41	10	2	2
66	66	549	Gem	2	17	43	6	38	2
67	67	550	Gem	3	5	32	10	58	5
68	68	551	Gem	4	9	54	7	43	4
69	69	552	Gem	5	13	24	5	42.5	4

(continued)

Table 4.1 (continued)

1598 list	1602 list	1627 list	Const	Star seq #	Ecliptic longitude Degree	Arcmin	Ecliptic latitude Degree	Arcmin	Brahe's V Mag
70	70	553	Gem	6	15	47	5	10	5
71	71	554	Gem	7	18	6	3	3	4
72	72	555	Gem	8	13	18	2	56	6
73	73	556	Gem	9	14	10	6	0.5	6
74	74	557	Gem	10	4	22	2	11	3
75	75	558	Gem	11	9	26	2	6.5	3
76	76	559	Gem	12	12	56	0	13.5	3
77	77	560	Gem	13	13	13	5	41	4
78	78	561	Gem	14	27	53	0	58	4
79	79	562	Gem	15	29	44	0	53	3
80	80	563	Gem	16	1	14	3	8	4
81	81	564	Gem	17	3	31	6	48.5	2
82	82	565	Gem	18	5	29.5	10	9	4
83	83	566	Gem	19	7	56	9	41	6
84	84	567	Gem	20	6	23	1	12	6
85	85	568	Gem	21	8	37.5	1	31	6
86	86	569	Gem	22	19	42	5	44	6
87	87	570	Gem	23	17	4.5	7	24	5
88	88	571	Gem	24	13	29	9	42	5
89	89	572	Gem	25	25	22	0	13	4
90	0	573	Gem	26	17	2.5	5	52	6
91	0	574	Gem	27	18	6	3	48.5	6
92	0	575	Gem	28	19	30.5	2	42	6
93	0	576	Gem	29	21	28	0	57.5	6
95	90	577	Cnc	1	1	46.5	1	14	9
96	91	578	Cnc	2	29	49	1	31.5	5
97	92	579	Cnc	3	0	9.5	0	47.5	5
98	93	580	Cnc	4	1	57	3	8	4
99	94	581	Cnc	5	3	8	0	4	4
100	95	582	Cnc	6	8	3.5	5	8	3
101	96	583	Cnc	7	0	44	10	23	5
102	97	584	Cnc	8	23	56	1	15.5	5
103	98	585	Cnc	9	25	4	7	5	5
104	99	586	Cnc	10	25	45.5	2	18.5	4
105	100	587	Cnc	11	28	12.5	1	4	6
106	101	588	Cnc	12	6	47.5	1	54	6
107	102	589	Cnc	13	10	36	5	36	5
108	103	590	Cnc	14	5	27	7	14	6
109	104	591	Cnc	15	7	36.5	5	20	6
110	105	592	Leo	1	9	41.5	10	23	4
111	106	593	Leo	2	12	16.5	7	52	4
112	107	594	Leo	3	15	51	12	21	4
113	108	595	Leo	4	15	5	9	40	3
114	109	596	Leo	5	21	57.5	11	50	3

(continued)

Table 4.1 (continued)

1598 list	1602 list	1627 list	Const	Star seq #	Ecliptic longitude		Ecliptic latitude		Brahe's V Mag
					Degree	Arcmin	Degree	Arcmin	
115	110	597	Leo	6	23	59	8	47	2
116	111	598	Leo	7	22	20	4	52	3
117	112	599	Leo	8	24	17	0	26.5	1
118	113	600	Leo	9	24	50.5	1	25.5	5
119	114	601	Leo	10	21	43.5	0	0.5	4
120	115	602	Leo	11	17	54.5	0	16	5
121	116	603	Leo	12	16	7	3	10	4
122	117	604	Leo	13	18	40	3	47	4
123	118	605	Leo	14	23	46	3	55	4
124	119	606	Leo	15	0	48	0	8	4
125	120	607	Leo	16	22	24	2	10	6
126	121	608	Leo	17	2	6	5	56	6
127	122	609	Leo	18	4	5	2	49.5	6
128	123	610	Leo	19	3	14	12	53	5
129	124	611	Leo	20	5	41	14	20	2
130	125	612	Leo	21	7	50	9	41.5	3
131	126	613	Leo	22	9	8	7	50.5	6
132	127	614	Leo	23	11	58.5	6	7	3
133	128	615	Leo	24	13	8.5	1	40	4
134	129	616	Leo	25	15	57	0	33	4
135	130	617	Leo	26	19	27	3	2.5	4
136	131	618	Leo	27	16	3	12	18	1
137	132	619	Leo	28	16	32	4	48	6
138	133	620	Leo	29	16	1.5	5	43	5
139	134	621	Leo	30	0	14	10	17	6
140	0	622	Leo	31	16	13	10	47.5	6
141	0	623	Leo	32	15	53	7	39	4
142	0	624	Leo	33	18	50	5	41	5
143	0	625	Leo	34	26	22.5	17	40	5
144	0	626	Leo	35	29	57	16	30	5
145	0	627	Leo	36	4	54.5	16	47	5
146	0	628	Leo	37	13	22	17	19	4
147	0	629	Leo	38	8	58	1	20.5	4
148	0	630	Leo	39	8	30	0	9.5	5
149	0	631	Leo	40	9	20	2	29	5
150	135	632	Vir	1	17	44	6	6.5	5
151	136	633	Vir	2	18	33	4	37	5
152	137	634	Vir	3	22	7	8	33.5	5
153	138	635	Vir	4	21	58	6	10	5
154	139	636	Vir	5	21	32	0	43	3
155	140	637	Vir	6	29	16	1	25	4
156	141	638	Vir	7	4	35.5	2	50	3
157	142	639	Vir	8	9	28.5	2	23.5	6
158	143	640	Vir	9	12	37	1	45	4

(continued)

Table 4.1 (continued)

1598 list	1602 list	1627 list	Const	Star seq #	Ecliptic longitude Degree	Arcmin	Ecliptic latitude Degree	Arcmin	Brahe's V Mag
159	144	641	Vir	10	5	55	8	41	3
160	145	642	Vir	11	29	53	13	36.5	5
161	146	643	Vir	12	1	52	11	37	6
162	147	644	Vir	13	4	23.5	16	15.5	3
163	148	645	Vir	14	18	16	1	59	1
164	149	646	Vir	15	15	22.5	8	10	3
165	150	647	Vir	16	17	58.5	3	11	6
166	151	648	Vir	17	21	9.5	1	45.5	6
167	152	649	Vir	18	19	44	0	19.5	6
168	153	650	Vir	19	24	44	2	24.5	6
169	154	651	Vir	20	27	49	11	2.5	5
170	155	652	Vir	21	28	9	7	18.5	4
171	156	653	Vir	22	28	51	2	57.5	4
172	157	654	Vir	23	29	51.5	11	48	4
173	158	655	Vir	24	1	22	0	31.5	4
174	159	656	Vir	25	4	30	9	49	4
175	160	657	Vir	26	1	21	10	26	6
176	161	658	Vir	27	21	37.5	9	40.5	6
177	162	659	Vir	28	27	45.5	4	59.5	6
178	163	660	Vir	29	8	25	16	14	6
179	164	661	Vir	30	10	11	12	40.5	5
180	165	662	Vir	31	14	46	12	34.5	6
181	166	663	Vir	32	22	11	13	7.5	5
182	167	664	Vir	33	22	56.5	3	22.5	6
183	0	665	Vir	34	6	38	3	25	5
184	0	666	Vir	35	10	39	3	23	5
185	0	667	Vir	36	14	8.5	3	13.5	5
186	0	668	Vir	37	17	13	7	51	5
187	0	669	Vir	38	19	35	9	16	5
188	0	670	Vir	39	20	35.5	6	16	5
189	168	671	Lib	1	9	31	0	26	2
190	169	672	Lib	2	8	42	1	55	5
191	170	673	Lib	3	13	48	8	35	2
192	171	674	Lib	4	9	40.5	8	18.5	4
193	172	675	Lib	5	12	26.5	1	14	5
194	173	676	Lib	6	16	19	2	58.5	6
195	174	677	Lib	7	19	33	4	28	3
196	175	678	Lib	8	21	48.5	4	4	4
197	176	679	Lib	9	19	27	2	21	4
198	177	680	Lib	10	15	46	8	7	4
199	0	681	Lib	11	22	11	0	2.5	4
200	0	682	Lib	12	25	3.5	0	7	4
201	0	683	Lib	13	24	16	3	33	4
202	0	684	Lib	14	24	48	6	10	4

(continued)

Table 4.1 (continued)

1598 list	1602 list	1627 list	Const	Star seq #	Ecliptic longitude		Ecliptic latitude		Brahe's V Mag
					Degree	Arcmin	Degree	Arcmin	
203	0	685	Lib	15	25	41.5	9	19	4
204	0	686	Lib	16	27	19	10	57	5
205	0	687	Lib	17	15	27	7	37	3
206	0	688	Lib	18	15	17	1	48	3
207	178	689	Sco	1	27	36	1	5	2
208	179	690	Sco	2	26	59	1	54.5	3
209	180	691	Sco	3	27	25	5	22.5	3
210	181	692	Sco	4	27	43.5	8	27.5	4
211	182	693	Sco	5	29	3.5	1	42	4
212	183	694	Sco	6	28	7	0	14	5
213	184	695	Sco	7	2	11	3	55	4
214	185	696	Sco	8	4	13	4	27	1
215	186	697	Sco	9	5	53	5	50	4
216	187	698	Sco	10	0	46.5	6	37.5	5
217	188	699	Sgr	1	0	47.5	2	0	4
218	189	700	Sgr	2	27	41.5	2	27.5	4
219	190	701	Sgr	3	6	51	3	31	4
220	191	702	Sgr	4	4	40	3	50	5
221	192	703	Sgr	5	7	56.5	1	44.5	4
222	193	704	Sgr	6	9	28	0	59	4
223	194	705	Sgr	7	10	43	1	31	4
224	195	706	Sgr	8	12	44	3	6.5	6
225	196	707	Sgr	9	13	54.5	4	17	4
226	197	708	Sgr	10	14	11	6	9.5	5
227	198	709	Sgr	11	19	8.5	5	8	6
228	199	710	Sgr	12	22	52.5	5	12	6
229	200	711	Sgr	13	19	24	1	25	6
230	201	712	Sgr	14	16	26	3	8	6
231	202	713	Cap	1	28	18	7	2	3
232	203	714	Cap	2	28	51	6	53	6
233	204	715	Cap	3	28	31	4	41	3
234	205	716	Cap	4	27	8	7	16	6
235	206	717	Cap	5	28	57	0	48.5	9
236	207	718	Cap	6	29	41	0	28	9
237	208	719	Cap	7	29	37	1	20	6
238	209	720	Cap	8	27	13	0	24	9
239	210	721	Cap	9	2	49	3	25	6
240	211	722	Cap	10	2	6	0	15	6
241	212	723	Cap	11	1	47	6	58	6
242	213	724	Cap	12	2	28	9	2	6
243	214	725	Cap	13	6	13	8	8	6
244	215	726	Cap	14	11	24.5	6	56	5
245	216	727	Cap	15	12	0	6	29	6
246	217	728	Cap	16	9	23	4	25	6

(continued)

Table 4.1 (continued)

1598 list	1602 list	1627 list	Const	Star seq #	Ecliptic longitude Degree	Arcmin	Ecliptic latitude Degree	Arcmin	Brahe's V Mag
247	218	729	Cap	17	7	31	4	27	6
248	219	730	Cap	18	7	18	3	1	5
249	220	731	Cap	19	8	21	0	29	5
250	221	732	Cap	20	12	7	1	16.5	5
251	222	733	Cap	21	14	25	4	48	4
252	223	734	Cap	22	16	6	4	49	5
253	224	735	Cap	23	16	14	2	26	3
254	225	736	Cap	24	18	0	2	29	3
255	226	737	Cap	25	18	14	2	22	5
256	227	738	Cap	26	20	27	0	14.5	5
257	228	739	Cap	27	20	16	0	10	6
258	229	740	Cap	28	19	54	4	17	6
259	230	741	Aqr	1	22	26.5	15	23	6
260	231	742	Aqr	2	27	49.5	10	42	3
261	232	743	Aqr	3	26	36	9	11.5	5
262	233	744	Aqr	4	17	51	8	42	3
263	234	745	Aqr	5	18	38	6	0.5	5
264	235	746	Aqr	6	10	51	4	50	5
265	236	747	Aqr	7	7	28.5	8	19	5
266	237	748	Aqr	8	6	12	8	10	4
267	238	749	Aqr	9	1	10	8	17.5	3
268	239	750	Aqr	10	3	4.5	10	31	5
269	240	751	Aqr	11	3	23	8	52.5	4
270	241	752	Aqr	12	4	53	8	10	4
271	242	753	Aqr	13	27	45	2	46	4
272	243	754	Aqr	14	28	31	2	29.5	6
273	244	755	Aqr	15	29	53	1	10	5
274	245	756	Aqr	16	23	13	2	0	4
275	246	757	Aqr	17	3	22	8	10	3
276	247	758	Aqr	18	3	5	5	37	5
277	248	759	Aqr	19	29	40	5	40	6
278	249	760	Aqr	20	26	55.5	10	48.5	5
279	250	761	Aqr	21	29	50	9	57.5	6
280	251	762	Aqr	22	3	52	4	8.5	4
281	252	763	Aqr	23	6	4	0	19.5	4
282	253	764	Aqr	24	9	0	1	24	6
283	254	765	Aqr	25	11	38	1	0	5
284	255	766	Aqr	26	11	33	2	49	5
285	256	767	Aqr	27	10	43	3	58.5	5
286	257	768	Aqr	28	11	11	4	10.5	5
287	258	769	Aqr	29	11	14.5	4	44	5
288	259	770	Aqr	30	14	7	10	59	5
289	260	771	Aqr	31	14	38	11	33	5
290	261	772	Aqr	32	13	3	14	29	5

(continued)

Table 4.1 (continued)

1598 list	1602 list	1627 list	Const	Star seq #	Ecliptic longitude Degree	Arcmin	Ecliptic latitude Degree	Arcmin	Brahe's V Mag
291	262	773	Aqr	33	13	46	15	16.5	6
292	263	774	Aqr	34	14	44	16	23	6
293	264	775	Aqr	35	7	54.5	14	45	5
294	265	776	Aqr	36	8	21	15	30	5
295	266	777	Aqr	37	9	50	16	31	5
296	267	778	Aqr	38	4	25	14	25.5	5
297	268	779	Aqr	39	4	2	15	40	5
298	269	780	Aqr	40	3	17	15	53	5
299	270	781	Aqr	41	28	11.5	21	0	1
300	271	782	Psc	1	13	2	9	4	5
301	272	783	Psc	2	15	50.5	7	17.5	4
302	273	784	Psc	3	17	30.5	8	54.5	6
303	274	785	Psc	4	19	42	9	3	5
304	275	786	Psc	5	21	56.5	7	13.5	5
305	276	787	Psc	6	17	21	4	27	5
306	277	788	Psc	7	21	5	3	25	5
307	278	789	Psc	8	27	2	6	23.5	5
308	279	790	Psc	9	28	27	7	27	6
309	280	791	Psc	10	2	29	5	28	6
310	281	792	Psc	11	8	36	2	11	4
311	282	793	Psc	12	11	58	1	5.5	4
312	283	794	Psc	13	14	19	0	57.5	4
313	284	795	Psc	14	12	25	1	31	6
314	285	796	Psc	15	13	46	4	19.5	6
315	286	797	Psc	16	17	33	3	3	5
316	287	798	Psc	17	19	56	4	40.5	5
317	288	799	Psc	18	21	57.5	7	56	5
318	289	800	Psc	19	23	47.5	9	4.5	3
319	290	801	Psc	20	22	12	1	38.5	5
320	291	802	Psc	21	21	16	1	51.5	5
321	292	803	Psc	22	21	16	5	21	4
322	293	804	Psc	23	21	36.5	9	24	5
323	294	805	Psc	24	23	15	22	0	6
324	295	806	Psc	25	22	49.5	20	43	5
325	296	807	Psc	26	19	22.5	20	55	6
326	297	808	Psc	27	18	6.5	19	24	6
327	298	809	Psc	28	17	3.5	20	24	6
328	299	810	Psc	29	17	56.5	13	21	5
329	300	811	Psc	30	18	2.5	12	21.5	6
330	301	812	Psc	31	18	9	11	21	6
331	302	813	Psc	32	23	18	17	26	5
332	303	814	Psc	33	20	58.5	15	30	5
333	304	815	Psc	34	19	0	12	27.5	5
334	305	816	Psc	35	24	11	18	31	6

(continued)

Table 4.1 (continued)

1598 list	1602 list	1627 list	Const	Star seq #	Ecliptic longitude		Ecliptic latitude		Brahe's V Mag
					Degree	Arcmin	Degree	Arcmin	
335	306	817	Psc	36	21	41	23	3	6
817	648	818	Cet	1	9	31	7	50	4
818	649	819	Cet	2	8	47	12	37	2
819	650	820	Cet	3	3	53.5	12	2.5	3
820	651	821	Cet	4	2	2	14	32	3
821	652	822	Cet	5	1	54	5	52	4
822	653	823	Cet	6	6	7	5	36	4
823	654	824	Cet	7	28	29.5	4	19	4
824	655	825	Cet	8	24	9	25	17	4
825	656	826	Cet	9	24	32.5	28	31	4
826	657	827	Cet	10	28	11.5	28	16.5	4
827	658	828	Cet	11	27	47.5	25	58	3
828	659	829	Cet	12	12	25	25	1	4
829	660	830	Cet	13	13	50	31	4	4
830	661	831	Cet	14	16	25	20	19	3
831	662	832	Cet	15	10	42.5	15	46.5	3
832	663	833	Cet	16	6	11.5	16	55	3
833	664	834	Cet	17	25	23	10	1	3
834	665	835	Cet	18	26	56	20	47	2
835	666	836	Cet	19	12	45	14	30	5
836	667	837	Cet	20	15	4.5	21	55	5
837	668	838	Cet	21	2	49.5	9	12.5	4
838	669	839	Ori	1	18	11.5	13	26	4
839	670	840	Ori	2	18	6.5	13	54	5
840	671	841	Ori	3	18	33.5	14	4.5	5
841	672	842	Ori	4	23	12	16	6	2
842	673	843	Ori	5	15	23	16	53	2
843	674	844	Ori	6	16	47	17	22	5
844	675	845	Ori	7	25	4.5	14	51	4
845	676	846	Ori	8	28	30.5	11	30	6
846	677	847	Ori	9	27	23.5	9	15	4
847	678	848	Ori	10	26	21	8	44	4
848	679	849	Ori	11	27	22	7	20.5	6
849	680	850	Ori	12	28	8.5	7	19	6
850	681	851	Ori	13	23	9	3	12.5	5
851	682	852	Ori	14	25	21.5	3	21	5
852	683	853	Ori	15	18	56.5	19	17.5	5
853	684	854	Ori	16	17	40	19	36.5	6
854	685	855	Ori	17	16	46	19	52.5	6
855	686	856	Ori	18	15	34	20	8.5	5
856	687	857	Ori	19	7	53	8	17	4
857	688	858	Ori	20	8	48	9	7	4
858	689	859	Ori	21	8	10	11	6	6
859	690	860	Ori	22	8	0.5	12	25.5	4

(continued)

Table 4.1 (continued)

1598 list	1602 list	1627 list	Const	Star seq #	Ecliptic longitude Degree	Arcmin	Ecliptic latitude Degree	Arcmin	Brahe's V Mag
860	691	861	Ori	23	6	49	13	3.5	4
861	692	862	Ori	24	6	23	15	27	4
862	693	863	Ori	25	6	33	16	50	4
863	694	864	Ori	26	6	58	20	2	4
864	695	865	Ori	27	7	57	20	55.5	4
865	696	866	Ori	28	16	50.5	23	38	2
866	697	867	Ori	29	17	54	24	33.5	2
867	698	868	Ori	30	19	6.5	25	21.5	2
868	699	869	Ori	31	14	37.5	25	36.5	3
869	700	870	Ori	32	17	28	28	9.5	5
870	701	871	Ori	33	17	24.5	28	45	3
871	702	872	Ori	34	17	27.5	29	17	3
872	703	873	Ori	35	16	20	30	37.5	4
873	704	874	Ori	36	18	23	30	38	5
874	705	875	Ori	37	11	17	31	11.5	1
875	706	876	Ori	38	12	15.5	29	53	4
876	707	877	Ori	39	14	2	31	0	5
877	708	878	Ori	40	20	49.5	33	8	3
878	709	879	Ori	41	18	39	26	0.5	4
879	710	880	Ori	42	14	34	19	40	6
880	0	881	Ori	43	14	45	24	6	6
881	0	882	Ori	44	13	59	23	32	5
882	0	883	Ori	45	14	57	21	23	5
883	0	884	Ori	46	11	58	20	8	4
884	0	885	Ori	47	19	45	21	58	5
885	0	886	Ori	48	22	25.5	21	39	5
886	0	887	Ori	49	24	10	22	57	5
887	0	888	Ori	50	13	36.5	11	45	6
888	0	889	Ori	51	11	33.5	13	8	6
889	0	890	Ori	52	11	0	14	24	6
890	0	891	Ori	53	28	44	29	31	4
891	0	892	Ori	54	2	43	29	49	4
892	0	893	Ori	55	2	22	28	4	5
893	0	894	Ori	56	1	8	18	47	4
894	0	895	Ori	57	2	58	15	56.5	4
895	0	896	Ori	58	4	50	13	15	4
896	0	897	Ori	59	2	58	18	24	5
897	0	898	Ori	60	6	36	14	59	5
898	0	899	Ori	61	7	14.5	20	33	4
899	0	900	Ori	62	14	0	22	47	4
900	711	901	Eri	1	9	40	31	35.5	4
901	712	902	Eri	2	9	42	27	54.5	3
902	713	903	Eri	3	7	39	29	52	5
903	714	904	Eri	4	5	29.5	27	51.5	5

(continued)

Table 4.1 (continued)

1598 list	1602 list	1627 list	Const	Star seq #	Ecliptic longitude Degree	Ecliptic longitude Arcmin	Ecliptic latitude Degree	Ecliptic latitude Arcmin	Brahe's V Mag
904	715	905	Eri	5	3	45.5	25	34	4
905	716	906	Eri	6	1	14.5	25	11.5	4
906	717	907	Eri	7	18	18	33	13.5	3
907	718	908	Eri	8	15	22.5	31	9	4
908	719	909	Eri	9	15	7	28	46.5	3
909	720	910	Eri	10	12	45	27	47	3
910	0	911	Eri	11	3	10	24	34	3
911	0	912	Eri	12	5	36	23	58.5	4
912	0	913	Eri	13	8	16	25	59	3
914	0	914	Eri	14	23	49	30	25	5
915	0	915	Eri	15	23	53	27	32	4
916	0	916	Eri	16	24	58	28	9.5	4
917	0	917	Eri	17	27	46	25	3	5
918	0	918	Eri	18	16	25.5	18	26	4
919	0	919	Eri	19	20	7	22	45	4
920	721	920	Lep	1	10	14.5	34	34	5
921	722	921	Lep	2	10	20.5	35	54	5
922	723	922	Lep	3	12	27	35	18	6
923	724	923	Lep	4	12	14	36	14	5
924	725	924	Lep	5	9	49	39	4	5
925	726	925	Lep	6	6	25.5	45	0	4
926	727	926	Lep	7	15	49.5	41	5.5	3
927	728	927	Lep	8	14	6.5	43	57.5	3
928	729	928	Lep	9	19	21.5	45	49.5	3
929	730	929	Lep	10	21	36	44	18	3
930	731	930	Lep	11	20	26.5	38	16	4
931	732	931	Lep	12	23	27.5	37	40.5	4
932	733	932	Lep	13	26	22	38	26	4
933	734	933	CMa	1	8	35.5	39	30	1
934	735	934	CMa	2	11	1.5	34	50	4
935	736	935	CMa	3	11	27	36	43	5
936	737	936	CMa	4	14	6	38	2.5	3
937	738	937	CMa	5	12	3	39	30	4
938	739	938	CMa	6	6	32.5	42	12.5	5
939	740	939	CMa	7	1	42.5	41	18.5	2
940	741	940	CMa	8	15	30.5	46	9.5	5
941	742	941	CMa	9	12	36.5	46	39.5	5
942	743	942	CMa	10	17	55	48	30	3
943	744	943	CMa	11	15	21.5	51	24.5	3
944	745	944	CMa	12	1	7	51	46.5	3
945	746	945	CMa	13	24	11.5	51	24.5	3
946	747	946	CMi	1	16	39.5	13	33.5	3
947	748	947	CMi	2	20	18.5	15	57	2
948	0	948	CMi	3	16	49	12	51	6

(continued)

Table 4.1 (continued)

1598 list	1602 list	1627 list	Const	Star seq #	Ecliptic longitude Degree	Ecliptic longitude Arcmin	Ecliptic latitude Degree	Ecliptic latitude Arcmin	Brahe's V Mag
949	0	949	CMi	4	16	42.5	9	46	6
950	0	950	CMi	5	20	57.5	10	19.5	5
951	749	951	Arg	1	5	53.5	43	18.5	3
952	750	952	Arg	2	0	35.5	44	58.5	3
953	751	953	Arg	3	28	0	47	28	3
954	0	954	Arg	4	4	6.5	32	7	4
955	0	955	Arg	5	4	27	38	31	4
956	0	956	Arg	6	12	26.5	32	56	6
957	0	957	Arg	7	12	51.5	30	18	4
958	0	958	Arg	8	10	1.5	24	29.5	4
959	0	959	Arg	9	29	26	21	39.5	4
960	0	960	Arg	10	4	20.5	22	29.5	3
961	0	961	Arg	11	23	44	30	30	3
962	752	962	Hya	1	5	39.5	14	37	5
963	753	963	Hya	2	6	46	14	16.5	4
964	754	964	Hya	3	6	48	11	8	4
965	755	965	Hya	4	7	22.5	11	36	5
966	756	966	Hya	5	9	0.5	11	1	4
967	757	967	Hya	6	11	51.5	11	5.5	6
968	758	968	Hya	7	14	41.5	13	5	4
969	759	969	Hya	8	20	11.5	15	0	5
970	760	970	Hya	9	22	4	14	17.5	4
971	761	971	Hya	10	19	53.5	16	46	5
972	762	972	Hya	11	21	45.5	22	24	1
973	763	973	Hya	12	27	12	26	33.5	4
974	764	974	Hya	13	0	9	26	12	5
975	765	975	Hya	14	2	48	23	13	5
976	766	976	Hya	15	3	53	21	51	4
977	767	977	Hya	16	9	31.5	24	38	4
978	768	978	Hya	17	12	41.5	23	31	5
979	769	979	Hya	18	14	51	21	48.5	4
980	770	980	Hya	19	4	45.5	12	27	4
981	0	981	Hya	20	23	1.5	25	36	4
982	0	982	Hyd	21	23	49	30	17	5
983	0	983	Hya	22	21	24	13	43	3
984	0	984	Hya	23	19	24	14	37	6
985	0	985	Hya	24	28	44	10	19	3
986	771	986	Crt	1	18	13	22	41	4
987	772	987	Crt	2	23	43	19	39	4
988	773	988	Crt	3	21	10.5	17	25	4
989	0	989	Crt	4	20	27	13	10	4
990	0	990	Crt	5	23	2	11	17	4
991	0	991	Crt	6	28	30	18	10	4
992	0	992	Crt	7	0	33	16	2	4

(continued)

Table 4.1 (continued)

1598 list	1602 list	1627 list	Const	Star seq #	Ecliptic longitude Degree	Ecliptic longitude Arcmin	Ecliptic latitude Degree	Ecliptic latitude Arcmin	Brahe's V Mag
993	0	993	Crt	8	24	55	14	9	5
994	774	994	Crv	1	6	8	19	39	4
995	775	995	Crv	2	5	13	14	25	3
996	776	996	Crv	3	7	55	12	7	3
997	777	997	Crv	4	11	49	17	59	3
978	768	978	Hya	17	12	41.5	23	31	5
979	769	979	Hya	18	14	51	21	48.5	4
980	770	980	Hya	19	4	45.5	12	27	4
981	0	981	Hya	20	23	1.5	25	36	4
982	0	982	Hyd	21	23	49	30	17	5
983	0	983	Hya	22	21	24	13	43	3
984	0	984	Hya	23	19	24	14	37	6
985	0	985	Hya	24	28	44	10	19	3
986	771	986	Crt	1	18	13	22	41	4
987	772	987	Crt	2	23	43	19	39	4
988	773	988	Crt	3	21	10.5	17	25	4
989	0	989	Crt	4	20	27	13	10	4
990	0	990	Crt	5	23	2	11	17	4
991	0	991	Crt	6	28	30	18	10	4
992	0	992	Crt	7	0	33	16	2	4
993	0	993	Crt	8	24	55	14	9	5
994	774	994	Crv	1	6	8	19	39	4
995	775	995	Crv	2	5	13	14	25	3
996	776	996	Crv	3	7	55	12	7	3
997	777	997	Crv	4	11	49	17	59	3
998	0	998	Crv	5	6	38	21	46	4
999	0	999	Crv	6	8	14	18	14	5
1000	0	1000	Crv	7	8	21.5	11	28	5
1001	0	1001	Cen	1	1	27	21	49	5
1002	0	1002	Cen	2	0	59	19	8	5
1003	0	1003	Cen	3	0	12	20	51	5
1004	0	1004	Cen	4	1	3	20	12	5
94	0	0	Gem	30	23	54	1	18.5	6
675	0	0	Oph	38	12	24	7	18	3
913	0	0	Eri	20	15	23	31	9	4

The first catalog contains 1,044 stars from Tycho Brahe's original manuscript of 1598 and published by J. L. Dreyer in the *Tychonis Brahe Dani Opera Volume III* (Kopenhagen, 1916), the second catalog contains Tycho Brahe's edited edition containing 777 stars published in the *Astronomiae Instauratae Progymnasmata* 1602, and the third star catalog is the full listing of 1,044 stars (with variants) published by Johannes Kepler in the *Tabulae Rudolphinae* in 1627. This historical catalog data was provided by the Centre de Données astronomiques de Strasbourg

Otte Brahe became enraged with his brother threatening him to get his first son back. After a few years had passed, and the birth of several more children, Otte and Beate Brahe relinquished and allowed the childless Jørgen Brahe and his wife to adopt Tycho.

When Tycho was seven his uncle began reading him for a career in the court. At the age of 13 Tycho enrolled in the University of Copenhagen to study philosophy and law. During his time studying at the University a predicted partial solar eclipse occurred on schedule, this event made a deep impression on his life. Brahe became obsessed with Astronomy. He purchased a copy of Ptolemy's *Almagest* written in Latin. He also purchased the Alfonsine tables[1] and a set of tables based on Copernicus' theory so he could also predict the occurrence of astronomical events.

At the age of 16 Tycho spent much of the time from 1562 to 1576 travelling in Germany with his tutor Anders Vedel,[2] studying law at the Universities of Leipzig, Wittenberg, and Rostock. During this time Tycho would hide his books and his instruments from his tutor and stayed up late into the night observing the stars. When he was 17 there was another predicted astronomical event, the planets Saturn and Jupiter were supposed to pass very near each other. To Tycho's surprise the event occurred on August 17th, 1563, the Alfonsine tables were several months off and the Copernican tables were several days off. He found these errors to be totally unacceptable and decided to take it upon himself to create more accurate tables. Tycho undertook the study astronomy with Bartholomew Schultz at Leipzig. Schultz would teach Tycho how to obtain very accurate observations. Tycho learned that accurate observations required high quality instruments. Meanwhile, Tycho's adopted father died of pneumonia. His real father and mother would inherit his uncle's estate since he was still underage. When he returned home his family realized he had given up his career in law and were very bitter towards him. He then returned to Germany and his studies in Astronomy. During his return an unfortunate incident occurred. During his visit to the University of Rostock he got into an argument with another young Danish student by the name of Manderupius. The argument was reported to be about who was a better mathematician. After a very heated argument they begin the trade insults. A duel was arranged with swords. Manderupius was swinging wildly when he swung his sword and cut off Tycho's nose. He made two fake noses to hide his disfigurement, one with brass and putty and one with silver and gold. He must have been a very colorful figure with his red hair, long red mustache and beard, blue eyes and gold nose.

When Tycho returned to Germany he convinced a group of amateur astronomers that more accurate observations were necessary. With their help he constructed a large quadrant with a 19 ft radius capable of measuring a position within a 60th of

---

[1] The Alfonsine tables containing data for computing the position of the Sun, planets and the moon. They were prepared by groups of astronomers and scholars around 1,252–1,270. They are named after Alfonso X of Castille who ordered their creation.

[2] Anders Vedel became Demark's first great historian.

a degree. This was the start of Tyco's fame in compiling the world's most accurate measurements of his time.

In 1572 a "new star" (supernova[3]) appeared in the night sky brighter than Venus in the constellation of Cassiopeia. The star was so bright it could be seen during the day. It would appear in the sky for 18 months. Tycho was not the first to see the star but he did provide the most accurate measurement. He had just finished the construction of a new sextant with arms that were 5½ft long. It was built with a huge brass hinge and a scale calibrated to a 60th of a degree. Tycho further refined the measurement by keeping a table of errors found during testing of the device to be used to adjust the final measurements. His equipment was so far advanced above others he was able to settle an argument to whether the new star moved against the background of other stars. The star prompted Tycho's first publication to describe his observations. In 1576, he was ready to move to Basle, to continue his astronomical studies, when he received a generous offer from King Frederick II. He received the island of Hven for his own private fiefdom. With royal backing, Tycho would construct a home and observatory he called Uraniborg. With the aid of several skilled craftsmen a variety of enormous instruments were constructed with remarkable precision. These instruments were used by Tycho to make his observations of the comets, stars, and planets. Uraniborg would even include its own press for printing the scientific works produced by Tycho. The collection of instruments and visiting astronomers and scholars would make Uraniborg the foremost scientific institute in the world.

In 1588, Tycho used his private press to publish the book, *De Mundi Aetherei Recentioribus Phaenomenis*, on the comet that had appeared. The eighth chapter of this book also contained Tycho's view of the system of the world. In Tycho's view he had the earth as the unmoving center of the universe, but had all of the other planets orbiting of the Sun. Some of the other works Tycho created are the *Astronomiae instauratae mechanica* (1598), an illustrated description of how he constructed his instruments and observatories, and the *Astronomiae instauratae progymnasmata* (1602), containing his theory of lunar and solar motions, a portion of his star catalog of stars, and a his investigation of the supernova of 1572.

Tycho's royal support changed when King Ferdinand died in 1588. His replacement, King Christian was very young and did not see the importance of supporting an astronomer on a distant island. Over the years Tycho's funding and royal diminished as King Christian IV's gained his political majority forced Tycho to pack as many of his instruments he could and he and his family left Denmark in 1597. By1599 he had settled near Prague, after receiving an appointment as an Imperial Mathematician by Emperor Rudolph II. At this same time a mathematician was going through similar difficulties. The Archduke Ferdinand of Hapsburg was closing down Lutheran institutions to cleanse the Austrian provinces of heresy.

---

[3] The supernova SN 1572 is known today as Tycho's Supernova or Tycho's Nova. It can be found at 0 h 25.3 min right ascension and +64° 09′ declination in the constellation of Cassiopeia.

Tycho Brahe would invite one of the students by the name of Johannes Kepler to work with him in Prague.

When Kepler arrived in Prague he found it was not easy working with the great astronomer. Tycho was very secretive about his data and had his own strong ideas about the model of the solar system. But Kepler did not have to wait long. In the next year Tycho attended a banquet and became very ill. A few days later Tycho would die in agony. His death is still a controversy today. According to Kepler's he died because he refused to insult his host by excusing himself to use the bathroom and he ruptured his bladder. Modern day forensics performed on his hair indicates he died of a massive dose of mercury.

Chapter 5

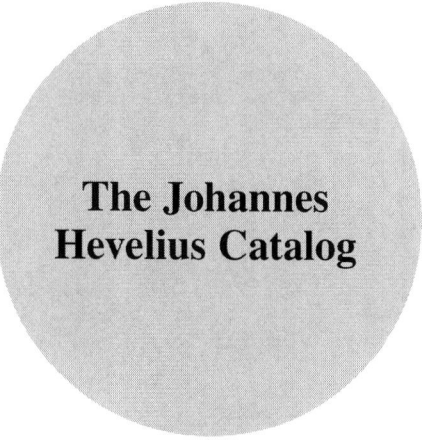

The Johannes Hevelius Catalog

There are many forms of Johannes Hevelius's name, including Hevel, Hövel, Hewel, Höfelcke, Hewelcke, Höwelcke, Höfelius, Hövelius, Höwelius, and Heweliusz. Today he is known as Johannes Hevelius (1611–1687). Johannes Hevelius's father was Abraham Höwelcke (1576–1649) and his mother was Cordelia Hecker (1576–1655). His family were the owners of a profitable brewery and several houses. He was a member of a large family, with three brothers and six sisters. He was the only son to survive childhood. The family lived in the city of Danzig under control of Poland.

Johannes entered the Gymnasium (European secondary school) in Danzig when he was 7 years old. While enrolled in the Gymnasium he showed an exceptional talent for mathematics. His mathematics teacher, Peter Krüger, also had an interest in astronomy. Hevelius soon began to take private lesson from Krüger learning theoretical astronomy as well as the construction of astronomical instruments.

When Hevelius reached the age of 19 he left for University of Leyden in Holland to study law. During his voyage to Holland there was an eclipse of the Sun. Later in his life he would publish the observations he made of the eclipse. While he studied law at Leyden he took advantage of the opportunity to study optics and mechanics. After studying a year at Leyden he began to travel through Europe, visiting many well-known astronomers. In 1634 Hevelius's plans were canceled when his parents sent a message to him to return home. Being the only remaining son his parents wanted him to take over the family brewery.

Upon returning to his hometown of Danzig Hevelius undertook the studies he needed to be admitted to the Brewer's Guild and took over running the family business. In 1635 he married Katharina Rebeschke, the daughter of a wealthy merchant. As he settled down to family life and running the brewery his old teacher Krüger was nearing death. He visited his teacher, who encouraged him not to give up astronomy. Shortly after his visit in June 1639 there was another eclipse of the Sun.

J.D. Cavin, *The Amateur Astronomer's Guide to the Deep-Sky Catalogs*,
Patrick Moore's Practical Astronomy Series, DOI 10.1007/978-1-4614-0656-3_5,
© Springer Science+Business Media, LLC 2012

Fig. 5.1 An engraving of Johannes Hevelius from book edited by Charles Leeson Prince, *The Illustrated Account Given by Hevelius in his "Machina Celestis" of the Method and Mounting his Telescopes and Erecting an Observatory* (1882)

The visit to his teacher and the solar eclipse fully revived his interest. For the rest of his life he remained dedicated to astronomy.[1]

The day-to-day duties of running a brewery took up much of his time. To devote more time to astronomy his wife Katharina stepped in to run the brewery. Hevelius began to construct the equipment that turned the observatory in his home into a world-class astronomical facility.

One of the first objects he observed with his instruments was the Moon. He used his newly constructed 12-foot telescope to create drawings of the Moon's surface. He also discovered and documented the Moon's libration in longitude. The libration in longitude is caused by the eccentricity of the Moon's orbit around Earth, so that the rotation of the Moon sometimes leads and sometimes lags in its orbital position. In 1647 he published the *Selenographia* containing 60 drawings of the Moon. The drawings of the Moon's features were so detailed astronomers used Hevelius's book as a reference for over 100 years.

On September 26, 1679, Hevelius's home and observatory were destroyed in a fire. This would be severe blow to anyone who had spent a lifetime of constructing his own instruments and collecting data. By August of 1681 Hevelius quickly rebuilt his observatory so he could continue his work. He would also dedicate the new constellation of Sextans to the memory of his destroyed instruments. A description of the fire can be found in the preface of his book *Annus climactericus,* published in 1685. The book contained a description of the observations he made with his wife after the observatory was rebuilt. He then continued working on the second volume of *Machina Coelestis*, published in 1679, as his copies were destroyed in the fire. Luckily he had sent a few copies to his friends.

In March 1662 Hevelius's first wife Katharina died. The next year he married the 16-year-old daughter of a fellow Danzig merchant, Catherina Elisabetha Koopman. Unlike his first wife, Elisabetha had a real passion for astronomy. She was able to make use of his instruments to take measurements and perform calculations with the observed data.

Hevelius gained a reputation for his observations across Europe through his correspondence with the leading astronomers in Europe, including Flamsteed and Halley in England and Gassendi and Boulliau in France. In 1663 he was awarded a pension by the French government and was elected into the Royal Society of London in 1664.

At one point Hevelius became involved in a dispute with Robert Hooke, the English natural philosopher. The argument concerned Hevelius's observations without the use of a telescope sight and whether they had any scientific value. Hooke learned of Hevelius's methods of using naked-eye observations in the *Machina Coelestis* (1673). The feud between Hevelius and Hooke became so intense that on May 26, 1679, the Royal Society sent a young Edmond Halley to try and resolve the dispute. Upon arrival Hevelius gave Halley a warm welcome and together they set about planning how to demonstrate the accuracy of Hevelius's naked-eye observations. To perform his measurements Halley used a 2-foot quadrant with telescopic sights. The results of the test were inconclusive. The dispute would continue between Hooke and Hevelius. Although Hooke was right scientifically, and we now recognize that telescopic sights provide far more accurate coordinates than naked-eye observations, the manner in which Hooke expressed himself in the argument did not win him many friends. At the end two of the secretaries of the Royal Society resigned because of the Hooke and Hevelius feud that appeared publically in the *Philosophical Transactions*.

Hevelius became ill in November 1686. He died 12 weeks later, on his birthday, and was buried in St Catherine's Church in Danzig. Catherina published three books posthumously containing Hevelius's notes and observations. The first of the books, *Prodromus astronomiae* (1690), contained previously unpublished observations. The second book, *Catalogus Stellarum Fixarum*, contained his catalog of 1,564 stars. The third book, *Firmamentum Sobiescianum sive Uranographia* (1687), was an atlas of constellations listed in his catalog (Table 5.1).

Table 5.1 The Hevelius catalog as published in *Prodromus astronomiae* (1690) by his wife
Elisabeth Koopman

Hevelius number (1690)	Constellation	Sequence number	V Mag (Hevelius)	V Mag (modern)	RA (2000) "hour:min"	Dec (2000) "degree:min"
1	And	1	2	2.1	00:08.3	+29:09
2	And	2	2	2.1	01:09.8	+35:39
3	And	3	2	2.1	02:03.9	+42:21
4	And	4	3	3.3	00:39.4	+30:54
5	And	5	4	4.3	00:36.8	+33:45
6	And	6	4	4.3	00:38.7	+29:21
7	And	7	5	4.5	00:18.4	+36:48
8	And	8	5	5.2	00:21.2	+38:02
9	And	9	5	4.6	00:17.3	+38:36
10	And	10	4	4.3	23:38.2	+43:17
11	And	11	4	4.2	23:40.6	+44:24
12	And	12	4	3.8	23:37.5	+46:30
13	And	13	6	5.0	23:46.4	+46:26
14	And	14	4	3.6	23:02.2	+42:23
15	And	15	4	4.1	00:46.5	+24:17
16	And	16	4	4.4	00:57.2	+23:25
17	And	17	4	4.5	00:49.7	+41:06
18	And	18	3	3.9	00:56.7	+38:32
19	And	19	5	5.0	01:40.4	+40:36
20	And	20	5	5.0	01:41.5	+42:38
21	And	21	5	4.3	01:09.6	+47:16
22	And	22	4	3.6	01:38.0	+48:38
23	And	23	4	4.0	01:43.8	+50:41
24	And	24	6	4.8	00:14.8	+20:13
25	And	25	6	4.6	23:57.8	+25:06
26	And	26	6	4.9	23:44.2	+29:24
27	And	27	5	4.9	22:51.7	+43:20
28	And	28	6	4.6	23:04.2	+50:05
29	And	29	6	4.5	23:12.3	+49:27
30	And	30	6	5.4	23:19.9	+48:44
31	And	31	6	6.1	00:43.1	+24:12
32	And	32	9	3.4	00:42.7	+41:20
33	And	33	6	5.3	00:41.0	+39:30
34	And	34	5	5.0	01:08.0	+43:55
35	And	35	6	5.7	01:10.7	+42:07
36	And	36	6	5.9	01:03.0	+41:22
37	And	37	5	4.9	01:22.5	+45:32
38	And	38	5	4.8	01:27.4	+45:26
39	And	39	6	5.3	01:31.1	+46:42
40	And	40	6	4.1	01:36.9	+41:29
41	And	41	6	5.4	01:53.3	+40:45
42	And	42	6	6.1	02:10.8	+39:03

(continued)

Table 5.1 (continued)

Hevelius number (1690)	Constellation	Sequence number	V Mag (Hevelius)	V Mag (modern)	RA (2000) "hour:min"	Dec (2000) "degree:min"
43	And	43	6	6.0	01:50.8	+51:39
44	And	44	6	5.8	02:23.1	+41:20
45	And	45	6	5.2	00:29.2	+44:25
46	And	46	6	5.1	00:36.7	+44:31
47	And	47	6	5.2	23:31.1	+39:19
48	Atn	1	3	3.2	20:11.2	−00:46
49	Atn	2	3	3.4	19:25.4	+03:10
50	Atn	3	3	3.4	19:06.2	−04:50
51	Atn	4	4	3.9	19:52.4	+01:04
52	Atn	5	4	4.4	19:36.7	−01:13
53	Atn	6	4	4.9	19:36.8	−07:08
54	Atn	7	4	4.0	19:01.7	−05:41
55	Atn	8	5	4.6	19:26.4	+00:21
56	Atn	9	6	5.0	19:20.5	−05:29
57	Atn	10	5	5.7	19:54.5	−08:10
58	Atn	11	6	5.1	19:35.6	−10:23
59	Atn	12	5	5.4	19:51.0	−10:41
60	Atn	13	6	5.0	19:30.6	−02:48
61	Atn	14	5	4.9	20:29.6	−02:49
62	Atn	15	5	4.9	20:37.8	−02:11
63	Atn	16	4	4.3	20:38.3	−01:03
64	Atn	17	6	5.2	20:39.4	+00:33
65	Atn	18	4	4.4	20:47.7	−04:58
66	Atn	19	5	5.6	20:51.8	−05:28
67	Aqr	1	3	3.0	22:05.7	−00:16
68	Aqr	2	3	2.9	21:31.5	−05:31
69	Aqr	3	4	3.8	20:47.8	−09:26
70	Aqr	4	5	4.7	20:52.7	−08:55
71	Aqr	5	5	4.5	21:09.4	−11:18
72	Aqr	6	5	4.7	21:37.7	−07:49
73	Aqr	7	6	5.1	21:39.5	+02:19
74	Aqr	8	5	4.7	22:03.2	−02:06
75	Aqr	9	3	3.9	22:21.6	−01:20
76	Aqr	10	5	4.8	22:25.6	+01:36
77	Aqr	11	4	3.7	22:28.8	+00:02
78	Aqr	12	4	4.0	22:35.4	−00:03
79	Aqr	13	4	4.2	22:16.7	−07:44
80	Aqr	14	5	5.3	22:20.1	−07:46
81	Aqr	15	4	4.3	22:06.4	−13:48
82	Aqr	16	5	4.8	22:30.6	−10:36
83	Aqr	17	0			
84	Aqr	18	5	5.2	22:34.6	−20:37
85	Aqr	19	6	4.7	22:45.4	−18:43

(continued)

Table 5.1 (continued)

Hevelius number (1690)	Constellation	Sequence number	V Mag (Hevelius)	V Mag (modern)	RA (2000) "hour:min"	Dec (2000) "degree:min"
86	Aqr	20	5	4.1	22:49.5	−13:31
87	Aqr	21	3	3.3	22:54.5	−15:45
88	Aqr	22	5	5.0	22:37.8	−04:10
89	Aqr	23	4	3.7	22:52.6	−07:32
90	Aqr	24	6	5.4	23:05.1	−07:39
91	Aqr	25	5	4.2	23:14.3	−05:59
92	Aqr	26	6	4.9	23:16.9	−07:40
93	Aqr	27	5	4.2	23:15.7	−09:03
94	Aqr	28	5	4.4	23:17.8	−09:07
95	Aqr	29	5	5.0	23:19.1	−09:34
96	Aqr	30	5	5.0	23:39.8	−14:09
97	Aqr	31	5	4.5	23:42.7	−14:28
98	Aqr	32	5	4.8	23:44.2	−17:45
99	Aqr	33	6	5.3	23:48.0	−18:37
100	Aqr	34	6	5.2	23:53.7	−18:50
101	Aqr	35	5	4.0	23:25.3	−20:00
102	Aqr	36	5	4.4	23:26.6	−20:34
103	Aqr	37	5	4.7	23:35.7	−20:49
104	Aqr	38	5	3.7	23:11.7	−21:06
105	Aqr	39	5	4.7	23:12.4	−22:22
106	Aqr	40	5	4.5	23:09.0	−23:40
107	Aqr	41	1	1.2	22:57.2	−29:32
108	Aqr	42	4	4.2	22:39.0	−26:39
109	Aqr	43	6	5.2	23:19.0	−13:22
110	Aqr	44	6	5.2	23:22.5	−14:57
111	Aqr	45	6	5.6	21:47.6	+02:45
112	Aqr	46	6	5.7	21:42.6	+01:45
113	Aqr	47	6	5.4	22:10.6	−11:30
114	Aqr	48	7	5.3	22:16.7	−12:46
115	Aql	1	1	0.8	19:50.5	+08:52
116	Aql	2	3	2.7	19:46.2	+10:41
117	Aql	3	4	3.7	19:55.2	+06:30
118	Aql	4	6	5.5	20:04.2	+07:21
119	Aql	5	6	5.1	19:50.8	+10:28
120	Aql	6	6	5.8	19:48.8	+11:56
121	Aql	7	4	4.4	19:33.9	+07:26
122	Aql	8	5	5.2	19:39.1	+05:28
123	Aql	9	4	4.0	18:59.5	+15:08
124	Aql	10	4	4.3	18:46.8	+18:14
125	Aql	11	4	4.2	18:45.5	+20:37
126	Aql	12	3	3.0	19:05.3	+13:55
127	Aql	13	5	4.7	19:54.1	+08:32
128	Aql	14	6	5.2	19:24.7	+11:57

(continued)

Table 5.1 (continued)

Hevelius number (1690)	Constellation	Sequence number	V Mag (Hevelius)	V Mag (modern)	RA (2000) "hour:min"	Dec (2000) "degree:min"
129	Aql	15	6	5.3	19:17.9	+11:38
130	Aql	16	6	5.5	19:19.2	+13:25
131	Aql	17	6	5.1	19:07.0	+11:04
132	Aql	18	6	5.3	18:59.2	+13:47
133	Aql	19	6	5.3	20:23.5	+05:25
134	Aql	20	6	5.3	19:42.4	+11:53
135	Aql	21	6	5.3	19:56.4	+11:38
136	Aql	22	5	4.6	18:54.8	+22:41
137	Aql	23	5	4.9	20:14.1	+15:13
138	Ari	1	2	2.0	02:07.1	+23:29
139	Ari	2	3	2.6	01:54.6	+20:50
140	Ari	3	4	3.9	01:53.5	+19:19
141	Ari	4	5	5.1	01:57.2	+17:50
142	Ari	5	5	5.0	02:06.4	+22:39
143	Ari	6	6	5.2	02:12.6	+21:13
144	Ari	7	5	5.6	02:18.0	+19:55
145	Ari	8	6	5.4	02:38.7	+21:56
146	Ari	9	6	5.7	02:42.3	+20:00
147	Ari	10	6	5.3	02:49.4	+17:25
148	Ari	11	6	5.6	02:56.2	+18:04
149	Ari	12	6	5.8	02:44.6	+15:21
150	Ari	13	6	5.5	02:51.6	+15:07
151	Ari	14	5	4.6	02:59.1	+21:22
152	Ari	15	4	4.3	03:11.4	+19:44
153	Ari	16	5	4.9	03:14.8	+21:04
154	Ari	17	6	5.3	03:21.2	+21:09
155	Ari	18	5	5.3	02:41.0	+27:04
156	Ari	19	4	4.7	02:43.4	+27:43
157	Ari	20	4	4.5	02:47.8	+29:16
158	Ari	21	3	3.6	02:49.9	+27:16
159	Ari	22	6	4.8	01:57.8	+23:37
160	Ari	23	6	5.0	02:09.5	+25:58
161	Ari	24	6	5.1	03:22.7	+20:46
162	Ari	25	6	5.2	01:42.8	+20:20
163	Ari	26	6	5.9	01:48.2	+17:01
164	Ari	27	6	5.2	02:44.0	+13:24
165	Aur	1	1	0.1	05:16.6	+46:03
166	Aur	2	2	1.9	05:59.4	+44:57
167	Aur	3	3	2.7	05:59.5	+37:14
168	Aur	4	3	2.7	04:57.0	+33:10
169	Aur	5	4	3.7	05:02.6	+41:04
170	Aur	6	4	3.2	05:06.4	+41:14
171	Aur	7	4	3.0	05:01.9	+43:49

(continued)

Table 5.1 (continued)

Hevelius number (1690)	Constellation	Sequence number	V Mag (Hevelius)	V Mag (modern)	RA (2000) "hour:min"	Dec (2000) "degree:min"
172	Aur	8	6	5.0	05:55.0	+55:42
173	Aur	9	4	3.7	05:59.2	+54:16
174	Aur	10	5	4.3	05:59.9	+45:55
175	Aur	11	6	5.2	05:22.4	+41:48
176	Aur	12	5	4.7	05:18.9	+40:10
177	Aur	13	5	4.8	05:13.6	+38:30
178	Aur	14	5	5.0	05:24.8	+37:20
179	Aur	15	5	4.9	04:59.2	+37:55
180	Aur	16	5	4.9	04:50.6	+37:30
181	Aur	17	5	4.8	04:52.6	+36:44
182	Aur	18	5	4.5	05:49.8	+39:12
183	Aur	19	5	4.0	05:51.5	+39:11
184	Aur	20	6	4.7	05:51.8	+37:18
185	Aur	21	5	5.1	05:27.6	+34:31
186	Aur	22	5	4.7	05:32.8	+32:14
187	Aur	23	4	4.3	06:15.4	+29:33
188	Aur	24	5	5.2	05:40.0	+25:52
189	Aur	25	5	4.9	05:49.1	+24:33
190	Aur	26	5	4.6	05:53.4	+27:36
191	Aur	27	5	4.8	05:58.1	+25:56
192	Aur	28	5	5.0	05:07.0	+51:37
193	Aur	29	6	5.5	05:28.0	+25:58
194	Aur	30	6	6.1	06:02.5	+42:34
195	Aur	31	6	6.0	05:16.6	+34:25
196	Aur	32	6	5.4	05:19.7	+33:48
197	Aur	33	6	4.5	05:18.7	+33:32
198	Aur	34	6	5.0	05:16.4	+31:20
199	Aur	35	6	5.4	05:38.9	+30:13
200	Aur	36	6	5.7	05:21.9	+29:50
201	Aur	37	4	4.0	05:03.0	+60:25
202	Aur	38	5	5.2	05:06.0	+58:59
203	Aur	39	6	5.2	05:23.2	+57:33
204	Aur	40	6	4.8	06:39.3	+42:29
205	Boo	1	1	−0.1	14:16.1	+19:25
206	Boo	2	3	2.3	14:45.0	+27:07
207	Boo	3	3	2.7	13:54.7	+18:28
208	Boo	4	3	3.5	15:02.0	+40:24
209	Boo	5	3	3.5	15:15.4	+33:22
210	Boo	6	3	3.0	14:32.1	+38:20
211	Boo	7	3	3.8	14:41.1	+13:46
212	Boo	8	4	4.5	14:13.1	+51:48
213	Boo	9	4	4.8	14:16.4	+51:18
214	Boo	10	4	4.0	14:25.1	+51:54

(continued)

Table 5.1 (continued)

Hevelius number (1690)	Constellation	Sequence number	V Mag (Hevelius)	V Mag (modern)	RA (2000) "hour:min"	Dec (2000) "degree:min"
215	Boo	11	4	4.2	14:16.5	+46:09
216	Boo	12	4	3.6	14:31.7	+30:26
217	Boo	13	4	4.5	14:34.5	+29:47
218	Boo	14	4	4.1	13:49.5	+15:49
219	Boo	15	4	4.5	13:47.5	+17:28
220	Boo	16	4	4.5	14:51.3	+19:09
221	Boo	17	4	4.6	14:45.2	+17:00
222	Boo	18	3	4.5	14:40.6	+16:27
223	Boo	19	5	4.5	15:04.5	+27:00
224	Boo	20	6	5.7	15:08.2	+26:13
225	Boo	21	5	4.8	15:02.0	+25:04
226	Boo	22	5	4.9	15:07.2	+24:56
227	Boo	23	4	4.3	15:24.4	+37:23
228	Boo	24	4	4.6	15:39.2	+36:41
229	Boo	25	5	5.3	15:14.5	+29:11
230	Boo	26	5	5.0	15:22.9	+30:20
231	Boo	27	6	5.5	15:19.8	+29:36
232	Boo	28	5	5.4	15:21.7	+32:58
233	Boo	29	5	5.1	14:08.8	+43:54
234	Boo	30	5	4.9	14:23.4	+08:28
235	Boo	31	5	5.1	14:24.3	+05:52
236	Boo	32	6	5.6	14:59.6	+39:16
237	Boo	33	6	5.5	14:49.7	+37:17
238	Boo	34	6	5.5	15:02.9	+35:13
239	Boo	35	5	5.8	14:49.8	+28:39
240	Boo	36	6	4.8	14:43.2	+26:34
241	Boo	37	6	5.7	13:40.8	+19:59
242	Boo	38	5	4.9	13:49.7	+21:17
243	Boo	39	6	5.8	13:58.8	+21:43
244	Boo	40	4	3.7	14:46.3	+01:55
245	Boo	41	4	4.4	15:02.8	+02:07
246	Boo	42	5	5.0	15:19.0	+01:51
247	Boo	43	6	5.2	15:28.5	+01:55
248	Boo	44	5	5.4	14:26.3	+19:16
249	Boo	45	5	4.8	14:19.6	+16:20
250	Boo	46	6	5.4	14:19.1	+13:03
251	Boo	47	6	5.3	14:14.9	+10:10
252	Boo	48	5	4.8	14:10.2	+25:09
253	Boo	49	7	4.9	14:41.5	+08:12
254	Boo	50	6	5.6	13:39.7	+10:49
255	Boo	51	5	5.7	15:16.2	+20:34
256	Boo	52	5	4.8	14:17.9	+35:38
257	Cnc	1	3	4.3	08:58.4	+11:51

(continued)

Table 5.1 (continued)

Hevelius number (1690)	Constellation	Sequence number	V Mag (Hevelius)	V Mag (modern)	RA (2000) "hour:min"	Dec (2000) "degree:min"
258	Cnc	2	3	3.5	08:16.5	+09:11
259	Cnc	3	9		08:40.1	+19:42
260	Cnc	4	4	4.7	08:43.2	+21:27
261	Cnc	5	4	3.9	08:44.5	+18:04
262	Cnc	6	6	5.4	09:02.6	+24:27
263	Cnc	7	6	5.2	09:09.3	+22:02
264	Cnc	8	5	4.0	08:46.6	+28:48
265	Cnc	9	5	5.3	08:07.5	+21:35
266	Cnc	10	4	4.7	08:12.1	+17:39
267	Cnc	11	5	5.3	08:32.6	+20:26
268	Cnc	12	5	5.3	08:31.6	+18:05
269	Cnc	13	6	5.9	08:23.6	+18:18
270	Cnc	14	6	5.2	08:57.1	+15:20
271	Cnc	15	5	5.2	09:07.7	+10:40
272	Cnc	16	5	5.1	08:04.9	+13:08
273	Cnc	17	5	5.2	07:46.2	+10:46
274	Cnc	18	6		08:41.3	+28:01
275	Cnc	19	6	5.6	08:44.1	+10:06
276	Cnc	20	6	5.9	08:37.5	+09:37
277	Cnc	21	6	5.1	08:25.8	+07:36
278	Cnc	22	6	4.6	07:29.7	+12:02
279	Cnc	23	6	5.4	07:25.1	+11:44
280	Cnc	24	7	5.7	08:10.3	+25:38
281	Cnc	25	6	5.1	08:20.0	+27:15
282	Cnc	26	6	5.6	08:26.7	+26:57
283	Cnc	27	6	5.7	08:36.9	+24:00
284	Cnc	28	6	5.4	09:17.5	+14:56
285	Cnc	29	6	5.2	08:59.3	+32:17
286	CMa	1	1	−1.4	06:45.5	−16:35
287	CMa	2	2	2.0	06:22.9	−17:56
288	CMa	3	3	4.1	07:04.0	−15:36
289	CMa	4	2	1.8	07:08.7	−26:22
290	CMa	5	2	1.5	06:58.6	−28:56
291	CMa	6	2	2.5	07:24.1	−29:16
292	CMa	7	2	3.0	06:20.6	−30:02
293	CMa	8	5	4.1	06:54.5	−12:01
294	CMa	9	4	5.0	06:56.2	−14:00
295	CMa	10	4	4.4	06:56.4	−17:02
296	CMa	11	5	4.7	06:55.7	−20:10
297	CMa	12	5	4.0	06:36.8	−19:13
298	CMa	13	5	4.3	06:32.1	−23:24
299	CMa	14	5	4.5	06:35.3	−22:56
300	CMa	15	4	3.0	07:03.2	−23:48

(continued)

Table 5.1 (continued)

Hevelius number (1690)	Constellation	Sequence number	V Mag (Hevelius)	V Mag (modern)	RA (2000) "hour:min"	Dec (2000) "degree:min"
301	CMa	16	5	3.9	06:54.5	−24:08
302	CMa	17	0			
303	CMa	18	5	4.4	06:38.1	−18:12
304	CMa	19	4	3.5	07:01.8	−27:54
305	CMa	20	5	5.3	06:46.7	−14:27
306	CMa	21	5	4.4	07:18.9	−24:54
307	CMa	22	5	4.0	07:14.8	−26:38
308	CMi	1	1	0.4	07:39.6	+05:20
309	CMi	2	3	2.9	07:27.3	+08:19
310	CMi	3	6	4.3	07:28.0	+08:56
311	CMi	4	6	5.0	07:25.6	+09:15
312	CMi	5	6	5.2	07:32.0	+01:56
313	CMi	6	6	5.8	07:34.2	+03:16
314	CMi	7	6		07:55.7	+05:32
315	CMi	8	5		08:03.6	+00:04
316	CMi	9	6	5.2	07:28.0	+06:57
317	CMi	10	5	5.1	07:51.6	+01:46
318	CMi	11	6	5.3	07:58.3	+02:14
319	CMi	12	5	4.4	08:02.1	+02:19
320	CMi	13	5	5.9	07:56.2	+08:47
321	CVn	1	2	2.9	12:56.0	+38:19
322	CVn	2	5	4.2	12:33.8	+41:20
323	CVn	3	6	5.7	12:15.9	+40:33
324	CVn	4	5	5.0	12:25.7	+39:01
325	CVn	5	5	5.0	12:16.2	+33:03
326	CVn	6	6	5.4	12:35.9	+33:17
327	CVn	7	6	5.9	12:50.0	+37:28
328	CVn	8	5	5.2	13:05.0	+35:58
329	CVn	9	6	6.0	13:19.5	+35:09
330	CVn	10	6	5.9	13:10.5	+38:32
331	CVn	11	5	4.9	13:14.6	+40:12
332	CVn	12	5	6.0	13:46.5	+25:52
333	CVn	13	6	4.7	13:17.4	+40:36
334	CVn	14	6	5.6	13:19.6	+40:12
335	CVn	15	5	4.9	13:34.7	+37:12
336	CVn	16	6	4.8	13:37.6	+36:17
337	CVn	17	6	4.8	13:51.9	+34:29
338	CVn	18	6	5.5	13:46.7	+38:33
339	CVn	19	6	5.9	13:45.4	+41:13
340	CVn	20	4	5.1	13:18.1	+49:42
341	CVn	21	4	4.7	13:34.2	+49:01
342	CVn	22	5	5.0	13:56.5	+27:32
343	CVn	23	4	4.9	13:00.4	+30:49

(continued)

Table 5.1 (continued)

Hevelius number (1690)	Constellation	Sequence number	V Mag (Hevelius)	V Mag (modern)	RA (2000) "hour:min"	Dec (2000) "degree:min"
344	Cam	1	5	5.4	12:49.0	+83:24
345	Cam	2	6	5.2	10:30.4	+82:34
346	Cam	3	6	5.5	10:26.2	+84:27
347	Cam	4	6	5.4	07:53.4	+79:42
348	Cam	5	6	4.9	07:28.6	+82:26
349	Cam	6	5	5.1	12:12.0	+77:37
350	Cam	7	6	5.4	06:45.7	+79:38
351	Cam	8	5	4.6	06:59.7	+76:59
352	Cam	9	5	5.1	05:22.0	+79:13
353	Cam	10	6	5.4	05:11.7	+73:50
354	Cam	11	4	4.4	04:57.3	+53:45
355	Cam	12	5	5.3	04:47.8	+56:46
356	Cam	13	5	5.1	04:39.7	+53:08
357	Cam	14	6	5.2	04:16.5	+53:20
358	Cam	15	5	5.1	03:30.1	+55:27
359	Cam	16	4	4.6	03:29.7	+58:52
360	Cam	17	4	4.2	03:30.0	+60:03
361	Cam	18	5	4.7	03:19.8	+65:37
362	Cam	19	5	4.4	03:49.5	+65:31
363	Cam	20	6	5.3	04:22.2	+65:09
364	Cam	21	5	4.8	03:45.8	+63:18
365	Cam	22	5	4.9	03:57.0	+63:02
366	Cam	23	6	5.0	03:57.6	+60:59
367	Cam	24	6	5.0	04:07.6	+59:25
368	Cam	25	6	5.4	04:22.1	+60:47
369	Cam	26	6	5.7	04:16.7	+61:40
370	Cam	27	4	4.3	04:53.5	+66:18
371	Cam	28	5	4.6	03:50.0	+71:19
372	Cam	29	6	5.4	06:14.7	+65:16
373	Cam	30	5	4.8	06:19.7	+69:25
374	Cam	31	5	5.1	06:56.2	+69:08
375	Cam	32	5	5.1	06:51.9	+67:37
376	Cap	1	3	3.6	20:17.9	−12:28
377	Cap	2	3	3.0	20:20.9	−14:43
378	Cap	3	3	3.7	21:39.9	−16:34
379	Cap	4	3	2.8	21:46.8	−16:01
380	Cap	5	9	5.3	20:19.8	−19:00
381	Cap	6	9	5.1	20:27.4	−18:06
382	Cap	7	9	5.9	20:29.8	−18:30
383	Cap	8	6	4.8	20:29.1	−17:43
384	Cap	9	6	5.2	20:39.3	−14:53
385	Cap	10	6	5.2	20:40.6	−18:01
386	Cap	11	5	4.1	20:46.0	−25:11

(continued)

Table 5.1 (continued)

Hevelius number (1690)	Constellation	Sequence number	V Mag (Hevelius)	V Mag (modern)	RA (2000) "hour:min"	Dec (2000) "degree:min"
387	Cap	12	6	4.1	20:51.9	−26:51
388	Cap	13	6	4.5	21:07.2	−24:53
389	Cap	14	5	4.1	21:05.8	−17:09
390	Cap	15	5	4.3	21:22.1	−16:45
391	Cap	16	5	4.8	21:04.3	−19:46
392	Cap	17	6	5.3	21:08.6	−21:07
393	Cap	18	6	5.2	21:17.3	−20:34
394	Cap	19	5	3.8	21:26.5	−22:21
395	Cap	20	6	4.5	21:26.0	−21:46
396	Cap	21	4	4.5	21:36.9	−19:24
397	Cap	22	5	4.7	21:42.3	−18:48
398	Cap	23	6	5.1	21:44.9	−09:01
399	Cap	24	5	5.6	21:46.6	−11:17
400	Cap	25	0			
401	Cap	26	5	5.1	21:53.1	−13:29
402	Cap	27	6	4.8	20:20.6	−12:38
403	Cap	28	6	5.8	20:12.4	−12:03
404	Cap	29	4	4.3	20:17.5	−12:25
405	Cap	30	6	5.2	21:41.7	−13:56
406	Cas	1	3	2.3	00:08.8	+59:10
407	Cas	2	3	2.2	00:40.3	+56:33
408	Cas	3	3	2.2	00:56.6	+60:43
409	Cas	4	3	2.7	01:25.5	+60:15
410	Cas	5	3	3.3	01:54.0	+63:39
411	Cas	6	4	3.7	00:37.2	+53:54
412	Cas	7	4	3.5	00:48.5	+57:52
413	Cas	8	4	4.5	02:29.2	+67:23
414	Cas	9	4	4.2	00:33.2	+62:56
415	Cas	10	0			
416	Cas	11	6	4.9	00:43.4	+47:02
417	Cas	12	6	4.5	00:44.8	+48:18
418	Cas	13	6	4.9	00:49.0	+50:59
419	Cas	14	6	4.8	00:42.2	+50:31
420	Cas	15	6	4.7	00:32.2	+54:35
421	Cas	16	6	4.9	23:59.0	+55:46
422	Cas	17	6	4.9	23:47.5	+58:37
423	Cas	18	6	4.5	23:54.6	+57:31
424	Cas	19	5	5.2	01:07.7	+55:05
425	Cas	20	4	4.3	01:11.1	+55:10
426	Cas	21	6	4.6	00:56.7	+59:13
427	Cas	22	6	4.9	01:20.2	+58:14
428	Cas	23	6	4.7	01:34.1	+59:13
429	Cas	24	4	4.0	02:03.8	+72:26

(continued)

Table 5.1 (continued)

Hevelius number (1690)	Constellation	Sequence number	V Mag (Hevelius)	V Mag (modern)	RA (2000) "hour:min"	Dec (2000) "degree:min"
430	Cas	25	4	4.5	02:02.1	+70:52
431	Cas	26	5	5.2	01:42.9	+70:37
432	Cas	27	6	5.3	01:38.3	+73:01
433	Cas	28	5	5.2	02:37.6	+72:48
434	Cas	29	6	4.8	03:11.8	+74:24
435	Cas	30	6	4.9	23:30.1	+58:32
436	Cas	31	7		00:35.6	+62:01
437	Cas	32	6		00:36.8	+64:26
438	Cas	33	7		01:21.5	+63:31
439	Cas	34	6	4.8	00:52.7	+61:12
440	Cas	35	5	5.3	01:10.8	+68:48
441	Cas	36	5	4.7	01:25.9	+68:04
442	Cas	37	6	5.6	01:33.8	+68:20
443	Cas	38	5	5.0	01:55.8	+68:41
444	Cep	1	3	2.5	21:18.4	+62:36
445	Cep	2	3	3.2	21:28.8	+70:35
446	Cep	3	3	3.2	23:39.8	+77:38
447	Cep	4	4	4.2	22:15.0	+57:05
448	Cep	5	4	3.4	22:10.8	+58:14
449	Cep	6	4	4.1	22:29.3	+58:28
450	Cep	7	4	4.2	20:29.6	+63:04
451	Cep	8	4	3.4	20:45.4	+61:49
452	Cep	9	4	3.5	22:50.0	+66:13
453	Cep	10	5	4.3	22:03.8	+64:38
454	Cep	11	4	4.4	20:08.9	+77:43
455	Cep	12	5	4.4	22:21.9	+52:00
456	Cep	13	4	3.8	22:31.4	+50:20
457	Cep	14	5	4.6	22:37.5	+51:36
458	Cep	15	5	4.6	21:41.9	+71:17
459	Cep	16	6	5.2	21:43.5	+72:07
460	Cep	17	5	4.8	22:11.6	+72:24
461	Cep	18	6	5.0	21:59.7	+73:12
462	Cep	19	6	5.4	00:48.0	+74:53
463	Cep	20	6	5.7	21:54.7	+56:04
464	Cep	21	6	5.7	21:39.6	+57:33
465	Cep	22	6	5.1	22:07.5	+58:44
466	Cep	23	6	4.8	23:06.9	+59:24
467	Cep	24	6	5.1	19:54.1	+57:49
468	Cep	25	6	5.0	19:56.3	+58:40
469	Cep	26	6	5.1	04:51.0	+81:05
470	Cep	27	6	5.1	07:33.5	+87:04
471	Cep	28	6	5.3	02:07.3	+77:22
472	Cep	29	6	5.6	01:12.6	+79:39

(continued)

Table 5.1 (continued)

Hevelius number (1690)	Constellation	Sequence number	V Mag (Hevelius)	V Mag (modern)	RA (2000) "hour:min"	Dec (2000) "degree:min"
473	Cep	30	6	5.8	02:34.5	+81:23
474	Cep	31	4	4.3	20:13.4	+56:38
475	Cep	32	5	4.5	20:45.6	+57:40
476	Cep	33	6	5.4	21:28.8	+66:53
477	Cep	34	6	5.2	21:20.3	+64:58
478	Cep	35	6	4.8	21:37.8	+61:55
479	Cep	36	5	4.2	21:45.7	+61:12
480	Cep	37	5	5.1	22:05.4	+62:19
481	Cep	38	5	5.2	22:39.0	+63:39
482	Cep	39	5	4.8	23:19.3	+68:08
483	Cep	40	5	5.1	23:47.7	+67:49
484	Cep	41	6	5.3	22:11.5	+86:07
485	Cep	42	6	5.6	23:23.5	+87:19
486	Cep	43	5	4.2	01:09.2	+86:17
487	Cep	44	6	4.7	22:53.3	+84:22
488	Cep	45	6	4.8	22:46.8	+83:11
489	Cep	46	7	5.5	03:03.5	+79:03
490	Cep	47	6	5.4	03:17.6	+77:30
491	Cep	48	6	5.1	04:09.7	+80:40
492	Cep	49	5		18:22.9	+89:02
493	Cep	50	6	4.4	23:07.5	+75:24
494	Cep	51	6	5.1	22:36.4	+73:40
495	Cer	1	5	4.3	18:01.4	+21:39
496	Cer	2	5	4.7	18:00.0	+16:51
497	Cer	3	4	3.8	18:23.5	+21:50
498	Cer	4	5	4.4	18:08.7	+20:52
499	Cet	1	2	2.5	03:02.3	+04:05
500	Cet	2	3	3.5	02:43.3	+03:16
501	Cet	3	3	4.1	02:39.3	+00:22
502	Cet	4	3	3.6	01:24.0	−08:09
503	Cet	5	3	3.5	01:08.4	−10:08
504	Cet	6	3	3.6	00:19.4	−08:47
505	Cet	7	2	2.0	00:43.4	−17:57
506	Cet	8	4	4.7	02:59.7	+08:56
507	Cet	9	4	4.3	02:46.0	+09:09
508	Cet	10	4	4.3	02:28.1	+08:28
509	Cet	11	4	4.9	02:35.8	+05:36
510	Cet	12	4	4.4	02:12.9	+08:52
511	Cet	13	4	4.9	02:26.1	−12:19
512	Cet	14	4	4.7	02:32.2	−15:11
513	Cet	15	3	4.8	02:39.7	−11:49
514	Cet	16	3	4.2	02:44.2	−13:50
515	Cet	17	3	3.7	01:51.4	−10:18

(continued)

Table 5.1 (continued)

Hevelius number (1690)	Constellation	Sequence number	V Mag (Hevelius)	V Mag (modern)	RA (2000) "hour:min"	Dec (2000) "degree:min"
516	Cet	18	3	3.5	01:44.7	−15:57
517	Cet	19	0			
518	Cet	20	4	4.0	02:00.0	−21:01
519	Cet	21	5	4.8	03:19.4	+03:22
520	Cet	22	4	4.3	03:37.2	+00:27
521	Cet	23	6	4.5	03:54.4	−02:57
522	Cet	24	6	5.3	04:01.6	−01:33
523	Cet	25	2	6.5	02:19.3	−02:55
524	Cet	26	6	5.3	02:31.6	+02:14
525	Cet	27	6	5.4	02:32.2	−01:07
526	Cet	28	6	5.4	02:22.2	−00:51
527	Cet	29	6	5.3	02:21.5	+00:28
528	Cet	30	6	5.4	02:03.2	+00:06
529	Cet	31	6	5.7	02:12.7	−02:16
530	Cet	32	5	4.6	00:03.6	−17:15
531	Cet	33	5	4.9	00:11.6	−15:24
532	Cet	34	5	4.4	00:15.9	−18:51
533	Cet	35	5	4.4	00:01.9	−05:58
534	Cet	36	4	4.6	00:05.3	−05:38
535	Cet	37	5	4.9	23:58.9	−03:31
536	Cet	38	5	5.1	00:01.7	−02:59
537	Cet	39	6	5.5	02:25.0	+10:37
538	Cet	40	6	5.5	03:40.5	−05:05
539	Cet	41	5	4.7	01:49.5	−10:37
540	Cet	42	5	4.9	01:25.6	−14:33
541	Cet	43	5	4.8	00:44.2	−10:33
542	Cet	44	5	5.2	00:50.3	−10:35
543	Cet	45	6	5.3	00:56.4	−11:13
544	Cet	46	5	5.1	01:14.3	−07:54
545	Com	1	4	4.3	12:26.9	+28:19
546	Com	2	5	4.9	12:26.9	+27:16
547	Com	3	5	5.0	12:26.9	+26:51
548	Com	4	5	5.3	12:28.9	+25:57
549	Com	5	5	5.2	12:24.5	+25:56
550	Com	6	5	4.8	12:22.6	+25:51
551	Com	7	5	4.9	12:16.4	+23:57
552	Com	8	5	5.5	12:31.0	+24:38
553	Com	9	5	5.5	12:30.1	+24:08
554	Com	10	4	4.8	12:35.0	+22:39
555	Com	11	5	4.9	12:51.6	+27:36
556	Com	12	5	4.8	13:07.4	+27:41
557	Com	13	5		13:03.1	+24:59
558	Com	14	4	4.2	13:12.2	+27:50

(continued)

Table 5.1 (continued)

Hevelius number (1690)	Constellation	Sequence number	V Mag (Hevelius)	V Mag (modern)	RA (2000) "hour:min"	Dec (2000) "degree:min"
559	Com	15	6	5.6	13:41.4	+22:41
560	Com	16	6	5.7	13:38.2	+24:42
561	Com	17	6	5.8	13:25.7	+23:53
562	Com	18	4	4.9	12:53.4	+21:16
563	Com	19	6	5.0	12:35.4	+18:25
564	Com	20	7		12:41.5	+16:34
565	Com	21	6	5.0	13:28.4	+13:53
566	CrB	1	2	2.2	15:34.6	+26:46
567	CrB	2	4	3.7	15:27.8	+29:07
568	CrB	3	4	4.1	15:32.8	+31:24
569	CrB	4	5	5.6	15:43.6	+32:32
570	CrB	5	5	5.0	16:01.4	+29:55
571	CrB	6	4	3.8	15:42.7	+26:20
572	CrB	7	5	4.6	15:49.5	+26:07
573	CrB	8	4	4.1	15:57.5	+26:56
574	Crv	1	3	2.6	12:16.0	−17:29
575	Crv	2	3	2.9	12:30.0	−16:27
576	Crv	3	3	2.7	12:34.3	−23:20
577	Crv	4	4	3.0	12:10.1	−22:35
578	Crv	5	4	4.0	12:08.6	−24:40
579	Crv	6	5	5.2	12:20.7	−22:08
580	Crv	7	5	4.3	12:32.2	−16:06
581	Crv	8	6	5.2	12:41.5	−13:01
582	Crt	1	4	4.1	10:59.9	−18:16
583	Crt	2	4	3.6	11:19.3	−14:45
584	Crt	3	4	4.1	11:24.9	−17:36
585	Crt	4	4	4.8	11:24.5	−10:50
586	Crt	5	4	4.7	11:36.6	−09:47
587	Crt	6	4	4.7	11:44.7	−18:17
588	Crt	7	4	5.2	11:56.1	−17:07
589	Crt	8	5	5.5	11:38.6	−13:12
590	Crt	9	6	5.1	11:23.6	−18:43
591	Crt	10	6	5.3	12:00.9	−19:37
592	Cyg	1	3	3.0	19:30.7	+28:02
593	Cyg	2	3	4.8	20:17.6	+38:06
594	Cyg	3	3	2.2	20:22.1	+40:19
595	Cyg	4	2	1.2	20:41.4	+45:19
596	Cyg	5	3	2.9	19:44.8	+45:07
597	Cyg	6	3	2.5	20:46.0	+34:00
598	Cyg	7	3	3.2	21:12.8	+30:18
599	Cyg	8	5	4.7	19:39.3	+30:13
600	Cyg	9	4	3.9	19:56.3	+35:09
601	Cyg	10	4	3.8	19:17.0	+53:22

(continued)

Table 5.1 (continued)

Hevelius number (1690)	Constellation	Sequence number	V Mag (Hevelius)	V Mag (modern)	RA (2000) "hour:min"	Dec (2000) "degree:min"
602	Cyg	11	4	3.8	19:29.5	+51:43
603	Cyg	12	4	4.5	19:36.3	+50:13
604	Cyg	13	4	3.8	20:13.6	+46:49
605	Cyg	14	4	4.0	20:15.5	+47:46
606	Cyg	15	4	3.9	20:57.1	+41:14
607	Cyg	16	4	3.7	21:04.9	+43:58
608	Cyg	17	4	4.5	20:47.3	+36:34
609	Cyg	18	4	4.2	21:17.5	+39:27
610	Cyg	19	4	3.7	21:14.7	+38:04
611	Cyg	20	4	4.4	21:17.9	+34:57
612	Cyg	21	3	4.5	21:44.0	+28:50
613	Cyg	22	4	4.4	20:23.8	+32:15
614	Cyg	23	4	4.0	20:29.3	+30:26
615	Cyg	24	4	4.2	20:45.7	+30:47
616	Cyg	25	4	4.6	20:34.2	+35:19
617	Cyg	26	5	5.2	21:05.3	+38:32
618	Cyg	27	6	4.9	20:30.1	+49:02
619	Cyg	28	6		20:58.6	+54:24
620	Cyg	29	5	4.8	20:48.9	+46:10
621	Cyg	30	5	4.8	20:53.2	+44:26
622	Cyg	31	5	5.1	20:49.1	+44:04
623	Cyg	32	5	4.7	20:59.9	+47:35
624	Cyg	33	5	4.6	21:06.7	+47:41
625	Cyg	34	6	5.0	21:18.5	+44:00
626	Cyg	35	5	4.0	21:34.0	+45:39
627	Cyg	36	5	4.7	21:42.0	+51:14
628	Cyg	37	5	4.2	21:46.8	+49:20
629	Cyg	38	5	5.0	19:24.0	+29:41
630	Cyg	39	5	5.0	19:46.3	+33:51
631	Cyg	40	5	4.9	19:56.1	+38:35
632	Cyg	41	5	5.2	19:59.9	+37:05
633	Cyg	42	5	5.4	20:06.3	+36:03
634	Cyg	43	5	4.9	20:09.5	+36:54
635	Cyg	44	5	4.9	20:14.5	+36:52
636	Cyg	45	5	4.9	19:44.2	+38:08
637	Cyg	46	6	5.1	21:49.6	+30:15
638	Cyg	47	6	5.5	21:52.8	+28:53
639	Del	1	3	4.3	20:46.6	+16:12
640	Del	2	3	3.6	20:37.4	+14:40
641	Del	3	3	3.8	20:39.4	+15:58
642	Del	4	4	4.4	20:43.3	+15:08
643	Del	5	3	4.0	20:33.2	+11:21
644	Del	6	5	4.6	20:35.2	+14:45

(continued)

Table 5.1 (continued)

Hevelius number (1690)	Constellation	Sequence number	V Mag (Hevelius)	V Mag (modern)	RA (2000) "hour:min"	Dec (2000) "degree:min"
645	Del	7	6	5.4	20:33.9	+13:05
646	Del	8	6	5.7	20:39.0	+13:21
647	Del	9	6	5.4	20:37.4	+11:26
648	Del	10	6	5.1	20:39.3	+10:09
649	Del	11	6	5.2	20:47.4	+13:07
650	Del	12	6	5.5	20:58.5	+10:53
651	Del	13	6	5.5	20:55.7	+12:39
652	Del	14	5	5.6	20:47.8	+06:04
653	Dra	1	3	2.8	17:30.3	+52:19
654	Dra	2	3	2.2	17:56.5	+51:31
655	Dra	3	3	3.1	19:13.1	+67:45
656	Dra	4	3	3.2	17:09.3	+65:48
657	Dra	5	3	2.7	16:23.9	+61:35
658	Dra	6	3	4.0	16:02.1	+58:35
659	Dra	7	3	3.3	15:24.9	+59:00
660	Dra	8	2	3.7	14:04.3	+64:25
661	Dra	9	3	3.8	12:33.4	+69:49
662	Dra	10	3	3.8	11:30.9	+69:20
663	Dra	11	4	4.9	17:05.1	+54:30
664	Dra	12	4	4.9	17:32.3	+55:14
665	Dra	13	4	3.7	17:53.6	+56:54
666	Dra	14	5	5.0	18:24.0	+58:53
667	Dra	15	5	5.0	18:42.7	+55:37
668	Dra	16	5	4.8	18:32.8	+57:09
669	Dra	17	4	4.6	18:50.9	+59:26
670	Dra	18	4	4.6	19:21.1	+65:46
671	Dra	19	5	5.2	20:01.4	+64:50
672	Dra	20	4	4.5	20:02.8	+67:52
673	Dra	21	4	4.7	19:31.2	+69:52
674	Dra	22	3	3.8	19:48.1	+70:18
675	Dra	23	4	4.8	18:54.0	+71:20
676	Dra	24	4	4.4	19:14.0	+72:52
677	Dra	25	4	4.2	18:20.5	+71:21
678	Dra	26	4	4.6	17:42.7	+72:13
679	Dra	27	4	3.5	18:20.5	+72:47
680	Dra	28	4	4.8	17:36.1	+68:35
681	Dra	29	3	4.9	16:28.2	+68:49
682	Dra	30	5	4.9	16:55.8	+65:10
683	Dra	31	5	4.8	16:40.9	+64:36
684	Dra	32	5	4.6	13:51.4	+64:44
685	Dra	33	5	4.3	09:35.2	+81:23
686	Dra	34	5	5.5	16:35.8	+52:55
687	Dra	35	6	5.2	15:46.6	+62:41

(continued)

Table 5.1 (continued)

Hevelius number (1690)	Constellation	Sequence number	V Mag (Hevelius)	V Mag (modern)	RA (2000) "hour:min"	Dec (2000) "degree:min"
688	Dra	36	5	4.9	10:34.6	+75:43
689	Dra	37	5	5.1	12:11.9	+77:38
690	Dra	38	6	5.5	12:19.1	+75:12
691	Dra	39	6	5.4	16:18.3	+66:27
692	Dra	40	6	4.6	14:57.8	+65:59
693	Equ	1	3	3.9	21:15.8	+05:18
694	Equ	2	4	5.2	21:22.6	+06:52
695	Equ	3	4	4.7	21:10.3	+10:13
696	Equ	4	4	4.5	21:14.4	+10:05
697	Equ	5	6	5.5	21:25.1	−03:23
698	Equ	6	5	5.2	20:59.2	+04:21
699	Eri	1	3	2.8	05:07.9	−05:04
700	Eri	2	4	4.2	05:09.3	−08:44
701	Eri	3	5	4.8	05:01.9	−07:10
702	Eri	4	5	4.4	04:52.9	−05:28
703	Eri	5	4	4.0	04:45.6	−03:14
704	Eri	6	4	3.9	04:36.5	−03:20
705	Eri	7	5	5.2	04:23.8	−03:44
706	Eri	8	4	4.4	04:16.3	−07:21
707	Eri	9	4	4.0	04:12.0	−06:50
708	Eri	10	5	4.9	04:14.5	−10:14
709	Eri	11	3	3.0	03:58.0	−13:28
710	Eri	12	4	4.4	03:46.3	−12:07
711	Eri	13	3	3.5	03:43.4	−09:50
712	Eri	14	3	3.7	03:33.4	−09:28
713	Eri	15	3	4.8	03:16.1	−08:50
714	Eri	16	4	5.3	03:03.1	−07:34
715	Eri	17	3	3.9	02:56.5	−08:51
716	Eri	18	4	4.5	02:45.0	−18:33
717	Eri	19	4	4.8	02:51.0	−20:57
718	Eri	20	4	4.1	03:02.9	−23:36
719	Eri	21	4	3.7	03:19.3	−21:43
720	Eri	22	4	4.3	03:33.8	−21:36
721	Eri	23	4	4.2	03:47.2	−23:10
722	Eri	24	5	5.2	03:48.8	−23:57
723	Eri	25	4	4.6	03:54.3	−23:36
724	Eri	26	4	4.6	04:00.2	−24:00
725	Eri	27	0			
726	Eri	28	0			
727	Eri	29	3	3.8	03:12.0	−28:58
728	Gem	1	2	1.6	07:34.6	+31:54
729	Gem	2	2	1.2	07:45.6	+28:03
730	Gem	3	2	1.9	06:37.6	+16:25

(continued)

Table 5.1 (continued)

Hevelius number (1690)	Constellation	Sequence number	V Mag (Hevelius)	V Mag (modern)	RA (2000) "hour:min"	Dec (2000) "degree:min"
731	Gem	4	3	2.9	06:23.0	+22:30
732	Gem	5	5	4.2	07:29.1	+31:48
733	Gem	6	5	4.2	07:43.2	+28:55
734	Gem	7	5	3.6	06:52.8	+33:59
735	Gem	8	4	4.4	07:11.1	+30:15
736	Gem	9	4	3.8	07:25.7	+27:48
737	Gem	10	6	5.1	07:29.8	+28:07
738	Gem	11	5	4.1	07:35.9	+26:55
739	Gem	12	4	3.6	07:44.4	+24:23
740	Gem	13	6	5.0	07:23.5	+25:03
741	Gem	14	5	5.2	07:03.0	+25:12
742	Gem	15	5	5.3	06:51.6	+21:45
743	Gem	16	4	3.5	07:20.1	+21:58
744	Gem	17	3	3.1	06:43.7	+25:06
745	Gem	18	3	4.0	07:04.2	+20:32
746	Gem	19	4	3.6	07:18.1	+16:32
747	Gem	20	6	5.3	07:33.6	+15:49
748	Gem	21	4	4.2	06:04.2	+23:14
749	Gem	22	4	3.3	06:14.9	+22:29
750	Gem	23	4	4.1	06:29.0	+20:12
751	Gem	24	4	3.3	06:45.2	+12:54
752	Gem	25	6	4.7	06:54.6	+13:11
753	Gem	26	6	5.0	07:39.5	+17:38
754	Gem	27	6	4.9	07:46.1	+18:30
755	Gem	28	6	5.4	07:55.6	+19:52
756	Gem	29	5	5.0	07:52.8	+26:47
757	Gem	30	6	5.2	07:30.4	+21:26
758	Gem	31	6	4.9	07:11.5	+39:20
759	Gem	32	6	5.2	07:24.5	+40:46
760	Gem	33	6	5.1	07:22.1	+36:47
761	Gem	34	6	4.9	07:39.1	+34:38
762	Gem	35	6	5.2	07:46.8	+36:36
763	Gem	36	6	5.1	07:47.5	+33:26
764	Gem	37	6	4.9	08:04.3	+27:48
765	Gem	38	6	5.6	08:13.3	+29:29
766	Her	1	3	2.8	17:14.6	+14:26
767	Her	2	3	2.8	16:30.1	+21:30
768	Her	3	3	3.1	17:14.9	+24:52
769	Her	4	4	2.8	16:41.6	+31:37
770	Her	5	3	3.7	16:22.0	+19:10
771	Her	6	3	3.9	17:00.3	+30:58
772	Her	7	4	3.2	17:15.0	+36:53
773	Her	8	3	3.5	16:43.0	+38:58

(continued)

Table 5.1 (continued)

Hevelius number (1690)	Constellation	Sequence number	V Mag (Hevelius)	V Mag (modern)	RA (2000) "hour:min"	Dec (2000) "degree:min"
774	Her	9	3	3.9	17:56.2	+37:16
775	Her	10	3	3.8	17:39.3	+46:03
776	Her	11	4	5.0	16:08.2	+17:07
777	Her	12	5	4.6	16:25.3	+14:04
778	Her	13	4	4.4	17:30.8	+26:12
779	Her	14	4	3.4	17:46.4	+27:49
780	Her	15	4	3.7	17:57.6	+29:19
781	Her	16	4	4.4	17:58.5	+30:16
782	Her	17	4	3.8	18:07.6	+28:50
783	Her	18	4	4.6	17:17.6	+37:21
784	Her	19	4	4.2	17:23.7	+37:13
785	Her	20	4	4.2	16:34.0	+42:28
786	Her	21	5	4.8	16:28.5	+41:56
787	Her	22	4	3.9	16:19.7	+46:20
788	Her	23	4	4.7	16:02.6	+46:04
789	Her	24	4	4.2	16:08.7	+44:57
790	Her	25	4	5.0	15:31.5	+40:53
791	Her	26	4	4.6	15:52.5	+42:22
792	Her	27	6	5.3	17:36.3	+48:33
793	Her	28	6	5.8	17:26.2	+48:13
794	Her	29	9		17:48.5	+48:03
795	Her	30	5	4.9	16:38.5	+48:56
796	Her	31	5	4.8	16:48.9	+46:01
797	Her	32	6	5.5	17:20.1	+46:29
798	Her	33	6	5.2	17:53.1	+40:01
799	Her	34	5	4.8	16:32.6	+11:33
800	Her	35	5	4.8	17:18.3	+33:44
801	Her	36	5	5.1	18:21.0	+28:55
802	Her	37	5	5.1	17:20.8	+24:34
803	Her	38	5	5.0	16:51.7	+24:42
804	Her	39	6	4.9	17:05.3	+12:49
805	Her	40	6	5.3	16:55.2	+18:21
806	Her	41	5	4.9	16:23.0	+30:15
807	Her	42	5	5.4	16:00.9	+33:26
808	Her	43	5	5.2	16:14.7	+33:54
809	Her	44	5	5.2	16:22.3	+33:48
810	Her	45	5	5.5	16:31.2	+37:30
811	Hya	1	1	2.0	09:27.6	−08:39
812	Hya	2	4	4.1	08:37.7	+05:42
813	Hya	3	4	3.4	08:46.9	+06:25
814	Hya	4	4	3.1	08:55.4	+05:58
815	Hya	5	5	4.4	08:39.1	+03:22
816	Hya	6	4	4.3	08:43.1	+03:25

(continued)

Table 5.1 (continued)

Hevelius number (1690)	Constellation	Sequence number	V Mag (Hevelius)	V Mag (modern)	RA (2000) "hour:min"	Dec (2000) "degree:min"
817	Hya	7	5	4.3	08:48.3	+05:52
818	Hya	8	6	5.0	09:05.9	+05:07
819	Hya	9	4	3.9	09:14.3	+02:21
820	Hya	10	5	4.6	09:29.0	−02:45
821	Hya	11	5	4.5	09:32.0	−01:10
822	Hya	12	4	3.9	09:39.8	−01:06
823	Hya	13	4	5.1	09:40.5	−14:17
824	Hya	14	5	4.1	09:51.5	−14:49
825	Hya	15	5	4.6	10:05.2	−13:03
826	Hya	16	4	3.6	10:10.8	−12:20
827	Hya	17	4	3.8	10:26.2	−16:48
828	Hya	18	5	4.9	10:38.7	−16:51
829	Hya	19	4	3.1	10:49.7	−16:11
830	Hya	20	4	4.5	11:11.7	−22:46
831	Hya	21	5	5.7	11:05.8	−27:14
832	Hya	22	6	4.9	13:09.1	−23:03
833	Hya	23	3	3.0	13:19.0	−23:07
834	Hya	24	4	3.2	14:06.4	−26:37
835	Hya	25	6	4.8	09:20.7	−09:33
836	Hya	26	6	4.8	09:19.8	−11:58
837	Hya	27	6	5.5	09:09.3	−08:44
838	Hya	28	6	6.4	13:29.8	−23:09
839	Hya	29	5	5.3	08:49.8	−03:26
840	Hya	30	6	5.5	09:16.7	−08:43
841	Hya	31	6	4.9	10:38.3	−13:32
842	Lac	1	5	4.1	22:16.0	+37:50
843	Lac	2	5	4.5	22:13.7	+39:46
844	Lac	3	5	4.5	22:30.6	+43:12
845	Lac	4	5	4.5	22:40.3	+44:19
846	Lac	5	5	4.6	22:20.9	+46:35
847	Lac	6	5	4.3	22:29.2	+47:43
848	Lac	7	6	4.6	22:24.3	+49:30
849	Lac	8	6	5.7	22:35.9	+39:41
850	Lac	9	6	4.9	22:39.3	+39:06
851	Lac	10	6	5.2	22:41.4	+40:16
852	Leo	1	1	1.4	10:08.4	+12:00
853	Leo	2	2	2.0	10:19.8	+19:52
854	Leo	3	3	2.6	11:14.0	+20:33
855	Leo	4	2	2.1	11:49.2	+14:37
856	Leo	5	3	3.0	09:45.8	+23:46
857	Leo	6	3	3.4	10:16.7	+23:26
858	Leo	7	3	3.5	10:07.3	+16:45
859	Leo	8	3	3.3	11:14.3	+15:26

(continued)

Table 5.1 (continued)

Hevelius number (1690)	Constellation	Sequence number	V Mag (Hevelius)	V Mag (modern)	RA (2000) "hour:min"	Dec (2000) "degree:min"
860	Leo	9	4	4.5	09:24.7	+26:12
861	Leo	10	4	4.3	09:31.7	+22:59
862	Leo	11	4	3.9	09:52.9	+26:01
863	Leo	12	6	5.3	09:51.8	+24:24
864	Leo	13	4	5.0	09:32.1	+11:20
865	Leo	14	5	5.4	09:43.6	+14:03
866	Leo	15	0			
867	Leo	16	4	5.3	09:58.1	+12:21
868	Leo	17	5	5.4	09:29.2	+09:05
869	Leo	18	6	5.1	09:32.1	+09:44
870	Leo	19	4	3.5	09:41.1	+09:56
871	Leo	20	4	4.7	10:00.2	+08:04
872	Leo	21	5	4.4	10:07.9	+10:01
873	Leo	22	4	3.8	10:32.9	+09:20
874	Leo	23	6	5.5	10:46.3	+18:54
875	Leo	24	6	5.5	10:46.4	+14:13
876	Leo	25	6	5.3	10:49.3	+10:34
877	Leo	26	5	4.4	11:02.1	+20:10
878	Leo	27	4	4.6	11:05.1	+07:22
879	Leo	28	5	5.0	11:00.6	+06:07
880	Leo	29	5	4.8	11:00.5	+03:39
881	Leo	30	6	5.3	11:16.2	+13:19
882	Leo	31	4	4.0	11:23.8	+10:33
883	Leo	32	4	4.1	11:21.2	+06:04
884	Leo	33	4	4.9	11:27.9	+02:52
885	Leo	34	4	4.4	11:16.6	−03:38
886	Leo	35	5	4.8	11:30.3	−02:59
887	Leo	36	5	5.1	10:43.4	+23:11
888	Leo	37	4	4.3	11:36.9	−00:49
889	Leo	38	5	4.3	10:55.6	+24:47
890	Leo	39	5	4.6	11:15.1	+23:07
891	Leo	40	5	4.5	11:48.0	+20:13
892	Leo	41	5	5.7	10:02.9	+21:56
893	Leo	42	6	5.4	10:53.7	−02:07
894	Leo	43	6	4.7	11:01.7	−01:28
895	Leo	44	6	5.4	10:17.2	+13:40
896	Leo	45	6	5.6	09:36.2	+30:58
897	Leo	46	6	5.6	09:42.8	+30:00
898	Leo	47	6	5.5	11:55.6	+15:39
899	Leo	48	5	5.3	11:40.8	+21:22
900	Leo	49	6	5.8	11:36.3	+27:47
901	Leo	50	6	5.6	10:26.6	+08:51
902	LMi	1	6	5.1	09:57.6	+41:03

(continued)

Table 5.1 (continued)

Hevelius number (1690)	Constellation	Sequence number	V Mag (Hevelius)	V Mag (modern)	RA (2000) "hour:min"	Dec (2000) "degree:min"
903	LMi	2	6	4.8	09:35.0	+39:39
904	LMi	3	6	4.5	09:34.2	+36:25
905	LMi	4	6	5.9	09:29.7	+34:22
906	LMi	5	4	4.5	10:07.1	+35:15
907	LMi	6	6	5.4	10:02.3	+31:57
908	LMi	7	6	5.5	10:16.9	+29:19
909	LMi	8	4	4.2	10:27.8	+36:42
910	LMi	9	3	4.7	10:25.9	+34:50
911	LMi	10	3	4.7	10:38.7	+31:59
912	LMi	11	6	5.6	10:33.4	+34:19
913	LMi	12	4	3.8	10:52.9	+33:31
914	LMi	13	3	5.4	10:45.9	+30:42
915	LMi	14	6	5.5	10:43.1	+26:20
916	LMi	15	5	4.3	10:55.6	+24:47
917	LMi	16	6	5.7	11:08.8	+24:39
918	LMi	17	5	4.6	11:15.1	+23:07
919	LMi	18	6	5.1	10:43.4	+23:11
920	Lep	1	3	2.6	05:32.8	−17:49
921	Lep	2	3	2.8	05:28.3	−20:43
922	Lep	3	4	3.6	05:45.6	−22:40
923	Lep	4	4	3.8	05:51.5	−20:48
924	Lep	5	5	4.4	05:12.4	−11:51
925	Lep	6	5	4.4	05:14.0	−12:56
926	Lep	7	5	5.3	05:20.2	−12:19
927	Lep	8	4	4.3	05:19.8	−13:09
928	Lep	9	4	3.3	05:13.0	−16:12
929	Lep	10	4	3.2	05:05.6	−22:21
930	Lep	11	4	3.5	05:47.1	−14:48
931	Lep	12	4	3.7	05:56.5	−14:10
932	Lep	13	4	4.7	06:06.4	−14:56
933	Lep	14	5	4.3	04:40.5	−19:38
934	Lep	15	4	3.9	04:38.4	−14:16
935	Lep	16	6	5.0	04:38.9	−12:04
936	Lib	1	2	2.8	14:50.9	−15:58
937	Lib	2	2	2.6	15:17.0	−09:20
938	Lib	3	5	5.3	14:49.2	−14:06
939	Lib	4	5	5.2	15:06.4	−16:11
940	Lib	5	5	4.9	15:00.8	−08:29
941	Lib	6	4	4.9	15:24.4	−10:15
942	Lib	7	6	5.5	15:33.0	−16:47
943	Lib	8	6	6.1	15:20.7	−15:22
944	Lib	9	6	3.9	15:35.4	−14:43
945	Lib	10	6	5.4	15:43.1	−15:37

(continued)

Table 5.1 (continued)

Hevelius number (1690)	Constellation	Sequence number	V Mag (Hevelius)	V Mag (modern)	RA (2000) "hour:min"	Dec (2000) "degree:min"
946	Lib	11	3	4.5	15:12.1	−19:44
947	Lib	12	3			
948	Lib	13	4	4.1	15:53.6	−16:40
949	Lib	14	4	4.9	15:58.0	−14:13
950	Lib	15	4	4.2	16:04.3	−11:21
951	Lib	16	5	4.6	15:33.9	−10:00
952	Lib	17	5	5.2	15:34.4	−09:07
953	Lib	18	7	5.5	14:56.0	−10:40
954	Lib	19	6		15:33.2	−19:55
955	Lib	20	6	5.5	15:32.8	−19:37
956	Lib	21	8	5.2	14:50.7	−15:54
957	Lyn	1	6	5.0	06:19.9	+61:41
958	Lyn	2	5	4.4	06:19.6	+59:03
959	Lyn	3	6	5.2	06:27.1	+58:30
960	Lyn	4	5	4.3	06:57.3	+58:27
961	Lyn	5	6	6.0	06:56.0	+57:46
962	Lyn	6	6	5.2	07:18.8	+59:46
963	Lyn	7	5	5.0	07:18.9	+49:33
964	Lyn	8	6	4.9	06:57.4	+45:06
965	Lyn	9	5	4.2	08:22.8	+43:12
966	Lyn	10	6	5.3	08:40.9	+45:50
967	Lyn	11	6	5.2	08:51.7	+43:38
968	Lyn	12	6	5.8	08:34.6	+36:25
969	Lyn	13	5	4.9	07:42.6	+58:43
970	Lyn	14	6	5.8	07:22.9	+55:19
971	Lyn	15	5	4.8	08:08.1	+51:30
972	Lyn	16	5	4.6	07:25.9	+49:02
973	Lyn	17	6	5.3	09:13.7	+43:14
974	Lyn	18	5	3.8	09:18.8	+36:50
975	Lyn	19	3	3.1	09:21.1	+34:25
976	Lyr	1	1		18:36.8	+38:48
977	Lyr	2	4	3.5	18:49.9	+33:25
978	Lyr	3	3	3.2	18:58.8	+32:45
979	Lyr	4	5	4.7	18:44.0	+39:39
980	Lyr	5	5	4.3	18:44.5	+37:37
981	Lyr	6	5	4.2	18:54.2	+36:55
982	Lyr	7	6	5.2	19:06.9	+36:08
983	Lyr	8	5	4.4	19:13.5	+39:11
984	Lyr	9	5	4.3	19:16.0	+38:10
985	Lyr	10	6	5.2	18:49.7	+32:38
986	Lyr	11	6	4.9	18:59.2	+32:36
987	Lyr	12	5	4.1	18:55.0	+43:57
988	Lyr	13	6	5.1	18:24.0	+39:33

(continued)

Table 5.1 (continued)

Hevelius number (1690)	Constellation	Sequence number	V Mag (Hevelius)	V Mag (modern)	RA (2000) "hour:min"	Dec (2000) "degree:min"
989	Lyr	14	6	5.2	19:26.0	+36:24
990	Lyr	15	5	4.7	19:31.5	+34:30
991	Lyr	16	5	4.3	18:19.6	+36:07
992	Lyr	17	6	5.0	19:01.4	+47:00
993	Mon	1	4	4.7	06:41.0	+09:53
994	Mon	2	5	4.8	06:47.5	+08:02
995	Mon	3	4	4.5	06:33.0	+07:21
996	Mon	4	4	4.4	06:24.0	+04:35
997	Mon	5	5	5.9	06:32.3	+04:53
998	Mon	6	4	4.5	06:48.0	+02:24
999	Mon	7	4	4.2	07:11.8	−00:32
1000	Mon	8	4	4.0	06:15.0	−06:14
1001	Mon	9	5	5.1	06:28.3	−04:46
1002	Mon	10	4	3.8	06:29.0	−07:01
1003	Mon	11	4	3.9	08:25.9	−03:53
1004	Mon	12	4	4.6	08:43.8	−07:13
1005	Mon	13	4	3.9	07:41.2	−09:32
1006	Mon	14	6	5.3	06:19.9	−07:48
1007	Mon	15	5	5.0	07:03.1	−04:14
1008	Mon	16	6	4.9	07:10.7	−04:14
1009	Mon	17	5	5.1	07:37.4	−04:10
1010	Mon	18	5	4.9	07:59.8	−03:37
1011	Mon	19	5	4.7	08:01.4	−01:24
1012	Arg	1	3	3.3	07:49.3	−24:51
1013	Arg	2	3	2.8	08:07.5	−24:17
1014	Arg	3	3	4.7	08:10.2	−12:55
1015	Arg	4	3	3.8	07:38.9	−26:48
1016	Arg	5	0			
1017	Ori	1	1	0.4	05:55.2	+07:25
1018	Ori	2	2	1.6	05:25.2	+06:20
1019	Ori	3	1	0.2	05:14.6	−08:11
1020	Ori	4	3	2.1	05:47.8	−09:40
1021	Ori	5	2	2.2	05:32.0	−00:17
1022	Ori	6	2	1.7	05:36.2	−01:12
1023	Ori	7	2	1.7	05:40.8	−01:56
1024	Ori	8	3	3.3	05:24.5	−02:23
1025	Ori	9	3	2.8	05:35.5	−05:56
1026	Ori	10	3	5.0	05:35.3	−05:23
1027	Ori	11	5	5.2	05:35.8	−04:50
1028	Ori	12	5	4.7	04:52.7	+14:16
1029	Ori	13	5	4.1	04:56.5	+13:33
1030	Ori	14	6	5.2	04:55.3	+11:24
1031	Ori	15	5	4.6	04:54.9	+10:10

(continued)

Table 5.1 (continued)

Hevelius number (1690)	Constellation	Sequence number	V Mag (Hevelius)	V Mag (modern)	RA (2000) "hour:min"	Dec (2000) "degree:min"
1032	Ori	16	4	4.3	04:50.7	+08:55
1033	Ori	17	4	3.2	04:49.7	+06:58
1034	Ori	18	4	3.7	04:51.3	+05:36
1035	Ori	19	4	3.7	04:54.3	+02:27
1036	Ori	20	5	4.5	04:58.6	+01:44
1037	Ori	21	6	5.4	05:07.7	+09:51
1038	Ori	22	6	5.3	05:08.0	+08:29
1039	Ori	23	6	5.5	05:16.1	+11:21
1040	Ori	24	5	4.5	05:13.5	+02:50
1041	Ori	25	4	3.4	05:35.2	+09:56
1042	Ori	26	5	4.4	05:34.8	+09:30
1043	Ori	27	5	4.1	05:36.8	+09:19
1044	Ori	28	5	4.2	05:30.8	+05:57
1045	Ori	29	4	4.1	06:02.5	+09:40
1046	Ori	30	6	5.0	05:22.7	+03:33
1047	Ori	31	5	4.6	05:26.8	+03:05
1048	Ori	32	6	5.5	05:31.3	+03:14
1049	Ori	33	6	5.3	05:34.3	+03:46
1050	Ori	34	5	4.5	05:39.2	+04:06
1051	Ori	35	5	4.9	05:24.8	+01:52
1052	Ori	36	5	4.7	05:21.8	−00:24
1053	Ori	37	6	5.1	05:24.4	−00:49
1054	Ori	38	4	3.8	05:38.7	−02:35
1055	Ori	39	5	4.9	05:42.5	+01:30
1056	Ori	40	5	4.8	05:52.4	+01:52
1057	Ori	41	5	5.2	05:59.0	+00:34
1058	Ori	42	6	5.0	06:16.1	+12:16
1059	Ori	43	5	4.4	06:07.6	+14:46
1060	Ori	44	5	4.4	06:12.1	+14:13
1061	Ori	45	6	4.9	06:12.1	+16:08
1062	Ori	46	6	5.3	06:15.4	+16:07
1063	Ori	47	5	4.4	05:54.5	+20:17
1064	Ori	48	7		06:00.1	+20:16
1065	Ori	49	5	4.6	06:04.0	+20:07
1066	Ori	50	4	3.6	05:17.8	−06:55
1067	Ori	51	5	4.1	05:24.0	−07:47
1068	Ori	52	4	4.6	05:32.0	−07:19
1069	Ori	53	5	4.8	05:38.9	−07:15
1070	Ori	54	6	5.4	06:17.3	+09:53
1071	Ori	55	5	4.7	05:04.6	+15:25
1072	Ori	56	5	4.8	05:09.8	+15:36
1073	Ori	57	6	6.0	05:55.6	−09:30
1074	Ori	58	6		06:14.1	−10:41

(continued)

Table 5.1 (continued)

Hevelius number (1690)	Constellation	Sequence number	V Mag (Hevelius)	V Mag (modern)	RA (2000) "hour:min"	Dec (2000) "degree:min"
1075	Ori	59	6	5.0	05:59.1	−09:30
1076	Ori	60	6	4.9	06:02.0	−10:36
1077	Ori	61	6	5.6	05:33.4	+14:22
1078	Ori	62	6	4.8	05:40.0	+16:33
1079	Peg	1	3	2.4	21:44.1	+09:56
1080	Peg	2	3	3.4	22:41.4	+10:53
1081	Peg	3	2	2.5	23:04.7	+15:15
1082	Peg	4	2	2.4	23:03.7	+28:07
1083	Peg	5	3	2.9	22:43.0	+30:16
1084	Peg	6	2	2.8	00:13.3	+15:13
1085	Peg	7	5	4.9	22:05.6	+05:06
1086	Peg	8	4	3.5	22:10.0	+06:15
1087	Peg	9	5	4.2	22:46.6	+12:17
1088	Peg	10	6	5.2	22:52.6	+09:54
1089	Peg	11	6	4.9	22:55.1	+08:53
1090	Peg	12	5	4.8	22:21.6	+12:15
1091	Peg	13	5	4.5	21:30.0	+23:41
1092	Peg	14	4	4.1	21:44.7	+25:43
1093	Peg	15	4	3.8	22:06.9	+25:24
1094	Peg	16	5	4.3	22:09.9	+33:14
1095	Peg	17	5	4.8	22:41.7	+29:23
1096	Peg	18	4	4.0	22:46.5	+23:37
1097	Peg	19	4	3.5	22:50.0	+24:38
1098	Peg	20	4	4.1	21:22.1	+19:51
1099	Peg	21	4	4.3	21:44.5	+17:25
1100	Peg	22	6	4.6	23:20.6	+23:45
1101	Peg	23	6	4.4	23:25.2	+23:24
1102	Peg	24	6	4.8	22:21.2	+28:23
1103	Peg	25	6	5.3	23:22.2	+31:53
1104	Peg	26	6	5.6	23:25.0	+32:27
1105	Peg	27	5	4.5	23:29.0	+12:48
1106	Peg	28	5	4.5	23:06.9	+09:27
1107	Peg	29	6	5.4	23:10.2	+09:52
1108	Peg	30	6	5.2	23:11.9	+08:45
1109	Peg	31	6	5.1	23:09.7	+08:43
1110	Peg	32	6	5.0	23:34.0	+31:23
1111	Peg	33	6	5.1	21:53.0	+25:58
1112	Peg	34	6	5.1	23:22.9	+12:22
1113	Peg	35	6	4.8	22:27.7	+04:46
1114	Peg	36	6	5.4	22:20.4	+05:51
1115	Peg	37	6	5.9	00:37.0	+15:15
1116	Peg	38	6	5.3	21:46.1	+23:01
1117	Per	1	2	1.8	03:24.2	+49:51

(continued)

Table 5.1 (continued)

Hevelius number (1690)	Constellation	Sequence number	V Mag (Hevelius)	V Mag (modern)	RA (2000) "hour:min"	Dec (2000) "degree:min"
1118	Per	2	3	2.9	03:04.7	+53:29
1119	Per	3	3	3.0	03:42.8	+47:46
1120	Per	4	3	2.9	03:57.9	+40:00
1121	Per	5	3	2.8	03:54.2	+31:54
1122	Per	6	2	2.1	03:08.2	+40:57
1123	Per	7	4	4.7	02:58.6	+39:42
1124	Per	8	4	3.3	03:05.0	+38:53
1125	Per	9	5	4.6	03:11.4	+39:39
1126	Per	10	4	4.2	02:50.5	+38:21
1127	Per	11	6	6.0	02:19.1	+56:57
1128	Per	12	4	3.8	02:50.8	+55:54
1129	Per	13	5	3.9	02:54.3	+52:45
1130	Per	14	4	4.1	02:44.1	+49:13
1131	Per	15	4	4.1	03:08.5	+49:37
1132	Per	16	4	3.8	03:09.4	+44:53
1133	Per	17	5	4.4	03:30.7	+47:59
1134	Per	18	5	4.3	03:36.5	+48:11
1135	Per	19	5	5.3	03:56.7	+50:36
1136	Per	20	4	4.2	04:06.6	+50:20
1137	Per	21	5	4.6	04:18.2	+50:16
1138	Per	22	5	4.0	04:08.6	+47:41
1139	Per	23	4	4.1	04:14.9	+48:24
1140	Per	24	6	4.8	04:21.5	+46:28
1141	Per	25	5	4.2	04:36.5	+41:15
1142	Per	26	4	3.8	03:45.2	+42:34
1143	Per	27	5	4.0	03:59.0	+35:47
1144	Per	28	4	3.8	03:44.2	+32:18
1145	Per	29	5	5.0	03:42.0	+34:08
1146	Per	30	6	5.5	01:51.7	+55:00
1147	Per	31	6	5.0	02:09.1	+54:19
1148	Per	32	6	4.7	02:24.3	+50:12
1149	Per	33	6	4.8	02:14.4	+44:53
1150	Per	34	6	5.1	03:49.5	+33:06
1151	Per	35	6	5.0	03:42.4	+33:53
1152	Per	36	5	4.7	04:15.0	+40:29
1153	Per	37	6	4.9	02:42.4	+40:12
1154	Per	38	5	4.8	03:05.7	+56:45
1155	Per	39	6	4.8	03:18.6	+34:13
1156	Per	40	6	4.6	02:51.4	+35:06
1157	Per	41	6	5.3	02:19.5	+47:16
1158	Per	42	5	5.5	03:14.9	+43:23
1159	Per	43	6	5.7	04:24.4	+34:05
1160	Per	44	6	5.3	04:26.0	+31:25

(continued)

Table 5.1 (continued)

Hevelius number (1690)	Constellation	Sequence number	V Mag (Hevelius)	V Mag (modern)	RA (2000) "hour:min"	Dec (2000) "degree:min"
1161	Per	45	6	4.9	03:02.1	+35:11
1162	Per	46	6	4.9	04:20.4	+34:34
1163	Per	47	5	4.8	03:05.7	+56:45
1164	Per	48	6	5.1	03:49.5	+33:06
1165	Psc	1	3	3.8	02:02.0	+02:47
1166	Psc	2	5	4.5	23:03.9	+03:53
1167	Psc	3	4	3.7	23:16.8	+03:19
1168	Psc	4	5	5.1	23:20.1	+05:26
1169	Psc	5	5	4.3	23:28.0	+06:25
1170	Psc	6	5	4.1	23:39.8	+05:41
1171	Psc	7	5	4.9	23:26.8	+01:19
1172	Psc	8	5	4.5	23:42.3	+01:47
1173	Psc	9	5	4.0	23:59.2	+06:54
1174	Psc	10	5	5.7	00:02.6	+08:30
1175	Psc	11	6	5.4	00:20.6	+08:11
1176	Psc	12	6	5.7	00:33.6	+07:00
1177	Psc	13	4	4.4	00:48.7	+07:36
1178	Psc	14	4	4.3	01:03.1	+07:55
1179	Psc	15	4	5.2	01:13.7	+07:36
1180	Psc	16	5	5.5	01:08.2	+05:40
1181	Psc	17	6	5.1	01:17.8	+03:39
1182	Psc	18	5	4.8	01:30.1	+06:11
1183	Psc	19	5	4.4	01:41.5	+05:29
1184	Psc	20	5	4.6	01:53.6	+03:14
1185	Psc	21	5	5.5	01:26.5	+19:13
1186	Psc	22	4	3.6	01:31.5	+15:22
1187	Psc	23	5	5.5	01:37.5	+12:09
1188	Psc	24	5	4.3	01:40.7	+12:13
1189	Psc	25	6	5.6	00:50.0	+27:46
1190	Psc	26	6	5.4	00:58.0	+28:57
1191	Psc	27	6	6.1	00:56.4	+27:17
1192	Psc	28	6	5.5	01:02.9	+31:50
1193	Psc	29	6	5.2	01:12.2	+31:28
1194	Psc	30	5	4.5	01:11.8	+30:07
1195	Psc	31	5	4.7	01:13.8	+24:36
1196	Psc	32	5	4.7	01:19.6	+27:17
1197	Psc	33	6	5.2	01:21.2	+28:48
1198	Psc	34	5	5.3	01:05.7	+21:30
1199	Psc	35	6	5.6	01:07.4	+20:49
1200	Psc	36	6	5.6	01:09.9	+19:41
1201	Psc	37	5	4.7	01:11.6	+21:05
1202	Psc	38	4	4.4	00:57.1	+23:25
1203	Psc	39	6	5.4	23:08.5	+02:10

(continued)

Table 5.1 (continued)

Hevelius number (1690)	Constellation	Sequence number	V Mag (Hevelius)	V Mag (modern)	RA (2000) "hour:min"	Dec (2000) "degree:min"
1204	Sge	1	4	3.5	19:58.6	+19:32
1205	Sge	2	4	3.7	19:47.3	+18:36
1206	Sge	3	6	5.0	19:48.8	+19:11
1207	Sge	4	4	4.4	19:40.9	+17:33
1208	Sge	5	4	4.4	19:39.9	+18:04
1209	Sgr	1	4	3.5	18:57.7	−21:01
1210	Sgr	2	4	3.8	19:04.7	−21:40
1211	Sgr	3	4	2.9	19:09.8	−20:56
1212	Sgr	4	4	2.0	18:55.2	−26:12
1213	Sgr	5	3	2.7	18:20.7	−29:41
1214	Sgr	6	4	3.8	18:14.1	−21:18
1215	Sgr	7	4	2.8	18:28.1	−25:19
1216	Sgr	8	0			
1217	Sgr	9	5	3.2	18:45.9	−26:55
1218	Sgr	10	0			
1219	Sgr	11	3	2.6	19:02.3	−29:37
1220	Sgr	12	4	3.3	19:06.9	−27:34
1221	Sgr	13	5	4.9	19:20.3	−25:28
1222	Sgr	14	6	4.6	19:37.1	−24:45
1223	Sgr	15	5	4.5	19:21.8	−15:52
1224	Sgr	16	4	3.9	19:21.7	−17:46
1225	Sgr	17	6	4.9	19:17.4	−18:48
1226	Sgr	18	6	5.5	19:41.2	−15:22
1227	Sgr	19	6	5.0	19:57.8	−15:23
1228	Sgr	20	6	4.9	19:46.4	−19:39
1229	Sgr	21	5	4.5	20:00.7	−27:11
1230	Sgr	22	0			
1231	Sgr	23	0			
1232	Sgr	24	5	4.4	20:03.6	−28:01
1233	Sgr	25	6	5.0	19:26.0	−24:21
1234	Sgr	26	6	5.0	18:58.7	−21:54
1235	Sco	1	1	1.1	16:29.2	−26:21
1236	Sco	2	2	2.6	16:05.3	−19:44
1237	Sco	3	3	2.3	16:00.2	−22:33
1238	Sco	4	3	2.9	15:58.8	−26:02
1239	Sco	5	4	2.9	16:20.9	−25:30
1240	Sco	6	4	2.8	16:35.7	−28:08
1241	Sco	7	3	3.2	15:04.0	−25:13
1242	Sco	8	5	4.9	16:11.7	−10:02
1243	Sco	9	4	5.0	15:53.2	−20:06
1244	Sco	10	4	4.0	16:11.8	−19:23
1245	Sco	11	5	3.9	16:07.0	−20:34
1246	Sco	12	4	3.9	15:56.6	−29:08

(continued)

Table 5.1 (continued)

Hevelius number (1690)	Constellation	Sequence number	V Mag (Hevelius)	V Mag (modern)	RA (2000) "hour:min"	Dec (2000) "degree:min"
1247	Sco	13	5	4.6	16:11.0	−27:38
1248	Sco	14	6	4.6	16:25.4	−23:22
1249	Sco	15	5	4.8	16:30.2	−25:02
1250	Sco	16	8		16:06.6	−20:39
1251	Sco	17	5	4.3	15:17.6	−30:05
1252	Sco	18	4	3.6	15:36.9	−28:06
1253	Sco	19	4	3.7	15:38.5	−29:42
1254	Sco	20	6	4.9	16:41.2	−17:40
1255	Sct	1	5	4.8	18:57.2	−05:47
1256	Sct	2	4	3.8	18:35.3	−08:11
1257	Sct	3	5	4.9	18:42.9	−07:58
1258	Sct	4	5	4.7	18:42.4	−09:03
1259	Sct	5	6	4.7	18:29.4	−14:13
1260	Sct	6	5	4.7	18:20.0	−08:24
1261	Sct	7	4	4.2	18:47.2	−04:41
1262	Oph	1	2	2.1	17:34.8	+12:38
1263	Oph	2	3	2.8	17:43.4	+04:36
1264	Oph	3	4	3.8	17:47.8	+02:44
1265	Oph	4	3	3.2	16:57.6	+09:24
1266	Oph	5	3	2.7	16:14.3	−03:37
1267	Oph	6	4	3.2	16:18.3	−04:39
1268	Oph	7	3	2.5	16:37.2	−10:32
1269	Oph	8	3	2.4	17:10.2	−15:40
1270	Oph	9	4	4.4	16:53.9	+10:12
1271	Oph	10	4	3.8	16:30.9	+02:02
1272	Oph	11	5	4.6	16:27.7	−08:18
1273	Oph	12	4	3.3	17:58.8	−09:42
1274	Oph	13	5	4.8	18:02.9	−08:05
1275	Oph	14	4	4.3	16:30.9	−16:32
1276	Oph	15	4	4.2	16:27.0	−18:22
1277	Oph	16	4	4.5	16:23.9	−19:58
1278	Oph	17	4	4.4	16:32.0	−21:24
1279	Oph	18	0			
1280	Oph	19	4	3.3	17:21.9	−24:56
1281	Oph	20	4	4.2	17:26.1	−24:05
1282	Oph	21	0			
1283	Oph	22	0			
1284	Oph	23	0			
1285	Oph	24	5	4.8	17:59.9	+04:24
1286	Oph	25	4	3.9	18:00.5	+02:58
1287	Oph	26	4	4.4	18:01.6	+01:23
1288	Oph	27	4	4.0	18:05.2	+02:39
1289	Oph	28	4	3.7	18:07.3	+09:37

(continued)

Table 5.1 (continued)

Hevelius number (1690)	Constellation	Sequence number	V Mag (Hevelius)	V Mag (modern)	RA (2000) "hour:min"	Dec (2000) "degree:min"
1290	Oph	29	5	4.6	18:07.4	+09:06
1291	Oph	30	5	4.5	17:26.4	−05:02
1292	Oph	31	5	4.6	16:49.7	−10:44
1293	Oph	32	5	5.8	16:40.5	+04:15
1294	Oph	33	6	5.2	16:47.8	+05:19
1295	Oph	34	6	5.3	17:12.3	+10:41
1296	Oph	35	6	5.0	17:18.5	+10:45
1297	Oph	36	6	5.8	17:34.5	+09:36
1298	Oph	37	5	4.3	17:26.3	+04:12
1299	Oph	38	6	4.8	16:22.1	+01:05
1300	Oph	39	5	5.5	16:53.2	+01:22
1301	Oph	40	5	5.5	16:50.1	+07:18
1302	Oph	41	6	5.2	16:45.6	+08:38
1303	Oph	42	5	4.7	17:16.3	−00:24
1304	Oph	43	5	4.8	17:00.8	−04:09
1305	Oph	44	6	5.2	16:54.9	−06:18
1306	Ser	1	2	2.6	15:44.3	+06:28
1307	Ser	2	4	4.8	16:02.4	+22:49
1308	Ser	3	5	4.5	15:41.5	+19:43
1309	Ser	4	4	4.7	15:51.2	+21:00
1310	Ser	5	4	4.1	15:48.6	+18:10
1311	Ser	6	3	3.8	15:56.2	+15:48
1312	Ser	7	3	3.7	15:46.1	+15:27
1313	Ser	8	3	3.8	15:34.8	+10:34
1314	Ser	9	4	4.4	15:46.6	+07:25
1315	Ser	10	3	3.7	15:50.5	+04:31
1316	Ser	11	4	3.5	15:49.6	−03:22
1317	Ser	12	4	4.3	17:20.7	−12:47
1318	Ser	13	4	3.5	17:37.4	−15:18
1319	Ser	14	5	4.2	17:41.2	−12:46
1320	Ser	15	4	4.6	17:37.8	−08:02
1321	Ser	16	3	4.6	18:00.4	−03:38
1322	Ser	17	3	3.2	18:21.4	−02:47
1323	Ser	18	3	4.6	18:56.2	+04:15
1324	Ser	19	6	5.2	15:50.2	+02:15
1325	Ser	20	6	5.9	15:45.1	+02:20
1326	Ser	21	5	4.8	16:22.1	+01:05
1327	Ser	22	6	5.2	15:27.3	+15:19
1328	Sex	1	6	5.1	09:52.5	−08:06
1329	Sex	2	6	5.6	10:10.9	−08:23
1330	Sex	3	6	5.2	10:16.3	−08:04
1331	Sex	4	7	5.8	10:50.3	−08:53
1332	Sex	5	5	4.7	09:38.8	+04:40

(continued)

Table 5.1 (continued)

Hevelius number (1690)	Constellation	Sequence number	V Mag (Hevelius)	V Mag (modern)	RA (2000) "hour:min"	Dec (2000) "degree:min"
1333	Sex	6	5	5.0	09:37.2	+06:52
1334	Sex	7	4	4.5	10:07.9	−00:21
1335	Sex	8	5	5.2	10:29.4	−02:42
1336	Sex	9	5	5.1	10:30.3	−00:34
1337	Sex	10	6	5.8	10:34.6	+04:44
1338	Sex	11	6	5.1	10:34.8	+06:58
1339	Sex	12	5	5.8	10:13.7	+04:36
1340	Tau	1	1	0.9	04:35.9	+16:32
1341	Tau	2	2	1.6	05:26.2	+28:37
1342	Tau	3	3	3.0	05:37.6	+21:07
1343	Tau	4	3	3.5	04:28.6	+19:11
1344	Tau	5	3	3.7	04:19.8	+15:39
1345	Tau	6	3	3.8	04:22.9	+17:33
1346	Tau	7	3	3.4	04:28.5	+15:54
1347	Tau	8	5	4.1	03:30.9	+12:57
1348	Tau	9	6	5.1	03:30.4	+11:21
1349	Tau	10	4	3.7	03:27.1	+09:45
1350	Tau	11	4	3.6	03:24.8	+09:02
1351	Tau	12	6	5.8	03:32.7	+09:23
1352	Tau	13	6	5.3	03:45.8	+06:04
1353	Tau	14	4	3.9	04:03.2	+06:00
1354	Tau	15	5	5.1	03:48.3	+11:09
1355	Tau	16	4	3.4	04:00.8	+12:30
1356	Tau	17	4	4.3	04:15.5	+08:55
1357	Tau	18	5	5.1	04:23.8	+09:29
1358	Tau	19	5	4.2	04:35.7	+10:10
1359	Tau	20	5	5.0	04:28.7	+13:03
1360	Tau	21	5	4.3	04:38.0	+12:31
1361	Tau	22	5	5.2	04:07.0	+29:00
1362	Tau	23	5	5.2	04:06.7	+27:38
1363	Tau	24	6	5.4	04:11.1	+26:29
1364	Tau	25	5	5.0	04:20.3	+27:20
1365	Tau	26	5	5.4	04:22.5	+25:39
1366	Tau	27	5	4.4	04:04.7	+22:05
1367	Tau	28	6	4.9	04:18.8	+20:34
1368	Tau	29	5	4.3	04:26.3	+22:47
1369	Tau	30	4	4.2	04:25.3	+22:17
1370	Tau	31	5	4.7	04:26.6	+14:43
1371	Tau	32	5	4.7	04:33.8	+14:50
1372	Tau	33	6	4.7	04:39.2	+15:55
1373	Tau	34	5	4.3	04:42.2	+22:58
1374	Tau	35	6	5.1	04:51.3	+18:51
1375	Tau	36	4	4.6	05:03.1	+21:35

(continued)

Table 5.1 (continued)

Hevelius number (1690)	Constellation	Sequence number	V Mag (Hevelius)	V Mag (modern)	RA (2000) "hour:min"	Dec (2000) "degree:min"
1376	Tau	37	6	5.3	05:08.0	+20:24
1377	Tau	38	6	5.0	05:19.2	+22:07
1378	Tau	39	6	4.9	05:27.5	+21:56
1379	Tau	40	3	2.8	03:47.5	+24:07
1380	Tau	41	5	4.3	03:44.9	+24:33
1381	Tau	42	6	3.6	03:49.2	+24:03
1382	Tau	43	6	3.9	03:46.1	+24:24
1383	Tau	44	5	3.7	03:45.0	+24:05
1384	Tau	45	5	4.1	03:46.0	+23:57
1385	Tau	46	6	4.3	04:25.4	+17:57
1386	Tau	47	6	4.9	05:07.2	+18:38
1387	Tau	48	5	4.3	05:32.4	+18:36
1388	Tau	49	6	5.0	05:24.4	+17:31
1389	Tau	50	6	5.2	03:50.3	+25:36
1390	Tau	51	6	5.3	04:46.2	+11:30
1391	Tri	1	4	3.4	01:53.1	+29:36
1392	Tri	2	4	3.0	02:09.5	+35:00
1393	Tri	3	4	4.0	02:17.3	+33:51
1394	Tri	4	5	4.8	02:16.3	+34:16
1395	Tri	5	6	5.2	02:15.0	+33:37
1396	Tri	6	6	5.6	01:41.9	+35:18
1397	Tri	7	6	5.9	01:55.8	+37:28
1398	Tri	8	5	4.8	02:08.5	+37:53
1399	Tri	9	6	5.5	02:03.1	+33:19
1400	TrM	1	6	5.3	02:18.7	+28:42
1401	TrM	2	6	4.9	02:12.3	+30:21
1402	TrM	3	6	5.3	02:28.3	+29:43
1403	Vir	1	1	1.0	13:28.0	−11:24
1404	Vir	2	3	2.8	13:02.3	+10:59
1405	Vir	3	3	3.6	11:50.3	+01:47
1406	Vir	4	3	3.9	12:20.0	−00:39
1407	Vir	5	3	2.7	12:41.8	−01:24
1408	Vir	6	3	3.4	12:55.7	+03:25
1409	Vir	7	5	4.8	11:45.1	+08:15
1410	Vir	8	5	4.0	11:45.8	+06:34
1411	Vir	9	5	4.1	12:05.1	+08:45
1412	Vir	10	5	4.7	12:00.9	+06:39
1413	Vir	11	5	5.4	11:59.8	+03:40
1414	Vir	12	5	4.9	12:41.8	+10:17
1415	Vir	13	6	5.2	12:45.7	+07:43
1416	Vir	14	6	5.6	12:41.7	+06:49
1417	Vir	15	6	5.2	13:17.1	+09:25
1418	Vir	16	5	4.8	13:17.7	+05:30

(continued)

Table 5.1 (continued)

Hevelius number (1690)	Constellation	Sequence number	V Mag (Hevelius)	V Mag (modern)	RA (2000) "hour:min"	Dec (2000) "degree:min"
1419	Vir	17	6	4.9	13:34.2	+03:41
1420	Vir	18	5	4.2	14:01.7	+01:34
1421	Vir	19	6	5.8	12:59.8	−03:53
1422	Vir	20	4	4.4	13:10.0	−05:30
1423	Vir	21	6	4.7	13:32.0	−06:13
1424	Vir	22	3	3.4	13:34.8	−00:33
1425	Vir	23	6	5.2	13:33.1	−10:09
1426	Vir	24	6	5.0	13:42.1	−08:40
1427	Vir	25	6	5.5	13:58.8	−09:16
1428	Vir	26	4	4.5	14:19.1	−13:18
1429	Vir	27	6	5.2	13:54.6	−01:27
1430	Vir	28	5	5.1	14:19.7	−02:16
1431	Vir	29	4	4.8	14:28.1	−02:11
1432	Vir	30	4	3.9	14:42.8	−05:35
1433	Vir	31	4	4.1	14:15.9	−05:56
1434	Vir	32	4	4.2	14:12.8	−10:15
1435	Vir	33	6	5.0	12:19.8	+03:11
1436	Vir	34	5	4.7	12:39.3	−07:58
1437	Vir	35	5	4.8	12:54.3	−09:32
1438	Vir	36	5	5.2	13:07.9	−10:42
1439	Vir	37	5	5.0	13:12.1	−16:08
1440	Vir	38	5	4.8	13:27.9	−15:56
1441	Vir	39	5	4.7	13:18.9	−18:10
1442	Vir	40	6	5.6	12:04.7	+20:29
1443	Vir	41	6	5.3	11:47.9	+08:15
1444	Vir	42	6	5.8	12:13.2	+10:15
1445	Vir	43	5	5.1	12:15.7	+14:56
1446	Vir	44	6	5.7	12:46.2	+09:33
1447	Vir	45	5	4.9	14:50.9	−02:15
1448	Vir	46	5	4.5	14:57.4	−04:17
1449	Vir	47	5	5.0	14:12.5	+02:17
1450	Vir	48	5	4.7	12:13.8	+17:48
1451	Vir	49	6	4.9	14:10.8	−16:15
1452	Vir	50	6	5.5	14:16.4	−18:10
1453	UMa	1	2	1.8	11:03.7	+61:47
1454	UMa	2	2	2.3	11:01.7	+56:25
1455	UMa	3	3	3.3	12:15.2	+57:04
1456	UMa	4	2	2.4	11:53.6	+53:44
1457	UMa	5	2	1.8	12:53.9	+55:59
1458	UMa	6	2	2.2	13:23.8	+54:58
1459	UMa	7	2	1.9	13:47.5	+49:21
1460	UMa	8	2	5.6	12:56.0	+38:19
1461	UMa	9	3	3.1	08:59.2	+48:03

(continued)

Table 5.1 (continued)

Hevelius number (1690)	Constellation	Sequence number	V Mag (Hevelius)	V Mag (modern)	RA (2000) "hour:min"	Dec (2000) "degree:min"
1462	UMa	10	3	3.6	09:03.4	+47:08
1463	UMa	11	3	3.2	09:33.3	+51:42
1464	UMa	12	4	4.0	09:00.7	+41:50
1465	UMa	13	4	4.6	09:06.4	+38:27
1466	UMa	14	4	3.3	08:30.3	+60:43
1467	UMa	15	5	5.7	08:53.1	+61:57
1468	UMa	16	4	4.7	09:10.5	+63:29
1469	UMa	17	5	5.5	08:36.4	+65:04
1470	UMa	18	4	4.6	08:39.9	+64:19
1471	UMa	19	4	4.7	09:03.6	+67:32
1472	UMa	20	5	4.8	09:10.5	+67:08
1473	UMa	21	5	5.2	09:14.1	+61:26
1474	UMa	22	4	3.7	09:31.1	+63:02
1475	UMa	23	5	4.5	09:34.7	+69:50
1476	UMa	24	4	3.8	09:51.2	+59:00
1477	UMa	25	4	4.6	09:51.8	+54:05
1478	UMa	26	6	4.5	09:35.0	+51:58
1479	UMa	27	5	4.8	09:15.9	+54:01
1480	UMa	28	5	4.5	09:08.7	+51:37
1481	UMa	29	4	3.5	10:16.8	+42:55
1482	UMa	30	4	3.1	10:22.0	+41:29
1483	UMa	31	5	4.7	10:53.9	+43:12
1484	UMa	32	4	3.0	11:09.3	+44:30
1485	UMa	33	4	3.7	11:47.1	+47:46
1486	UMa	34	4	3.5	11:17.9	+33:08
1487	UMa	35	4	3.8	11:18.4	+31:36
1488	UMa	36	5	4.2	12:33.8	+41:20
1489	UMa	37	6	4.6	13:40.6	+54:42
1490	UMa	38	6	5.5	13:39.4	+52:58
1491	UMa	39	6	5.7	13:53.9	+53:33
1492	UMa	40	4	4.8	12:25.9	+51:38
1493	UMa	41	5	5.3	12:19.7	+48:58
1494	UMa	42	5	4.9	13:00.5	+56:24
1495	UMa	43	5	5.4	13:27.5	+59:51
1496	UMa	44	6	5.4	12:30.2	+58:28
1497	UMa	45	5	5.6	07:31.3	+68:28
1498	UMa	46	6	5.3	08:12.0	+68:27
1499	UMa	47	5		08:09.7	+64:48
1500	UMa	48	5	6.0	08:04.4	+60:55
1501	UMa	49	6	5.4	09:28.5	+45:39
1502	UMa	50	6	5.1	09:48.2	+46:03
1503	UMa	51	5	5.0	11:00.6	+40:20
1504	UMa	52	5	4.8	11:18.9	+38:12

(continued)

Table 5.1 (continued)

Hevelius number (1690)	Constellation	Sequence number	V Mag (Hevelius)	V Mag (modern)	RA (2000) "hour:min"	Dec (2000) "degree:min"
1505	UMa	53	7	6.0	11:07.0	+38:52
1506	UMa	54	6	5.3	11:28.9	+39:16
1507	UMa	55	6	5.0	11:22.6	+43:30
1508	UMa	56	6	5.5	10:58.3	+45:44
1509	UMa	57	6	5.1	10:54.4	+55:09
1510	UMa	58	6	5.2	10:35.5	+57:04
1511	UMa	59	6	4.8	10:30.4	+55:57
1512	UMa	60	6	5.2	09:42.4	+72:19
1513	UMa	61	5	5.2	12:02.0	+43:03
1514	UMa	62	6	5.6	12:01.6	+36:02
1515	UMa	63	6	5.7	11:50.7	+35:07
1516	UMa	64	6	5.3	11:41.0	+34:15
1517	UMa	65	5	4.7	10:32.9	+40:25
1518	UMa	66	6	5.2	10:43.5	+46:08
1519	UMa	67	6	5.1	11:05.4	+40:19
1520	UMa	68	5	5.0	10:42.6	+69:00
1521	UMa	69	6	5.8	11:34.2	+61:46
1522	UMa	70	6	5.7	11:39.8	+31:14
1523	UMa	71	5	4.0	13:24.9	+54:58
1524	UMa	72	5	4.9	10:24.7	+65:29
1525	UMa	73	6	5.6	08:18.5	+59:32
1526	UMi	1	2	2.0	02:35.6	+89:16
1527	UMi	2	2	2.1	14:50.5	+74:12
1528	UMi	3	3	3.0	15:21.1	+71:52
1529	UMi	4	4	4.3	15:44.0	+77:49
1530	UMi	5	5	4.9	16:16.7	+75:41
1531	UMi	6	4	4.2	16:45.6	+82:04
1532	UMi	7	4	4.3	17:31.4	+86:35
1533	UMi	8	6	4.2	14:27.2	+75:52
1534	UMi	9	6	4.8	14:09.3	+77:31
1535	UMi	10	5	5.2	14:11.8	+69:24
1536	UMi	11	5	4.6	14:58.4	+65:50
1537	UMi	12	6	5.2	15:14.8	+67:16
1538	Vul	1	4	4.4	19:28.8	+24:45
1539	Vul	2	6	5.5	19:43.6	+25:50
1540	Vul	3	2	3.0	19:47.5	+27:20
1541	Vul	4	5	4.7	20:01.0	+27:48
1542	Vul	5	5	5.2	20:01.8	+24:55
1543	Vul	6	6	4.6	19:53.3	+24:08
1544	Vul	7	6	4.9	19:51.0	+22:39
1545	Vul	8	6	5.2	20:15.3	+23:34
1546	Vul	9	6	5.3	20:16.7	+24:43
1547	Vul	10	6	5.5	20:22.2	+24:31

(continued)

Table 5.1 (continued)

Hevelius number (1690)	Constellation	Sequence number	V Mag (Hevelius)	V Mag (modern)	RA (2000) "hour:min"	Dec (2000) "degree:min"
1548	Vul	11	5	4.8	20:15.2	+25:39
1549	Vul	12	6	5.6	20:36.9	+26:30
1550	Vul	13	5	4.5	20:15.8	+27:53
1551	Vul	14	6	4.9	20:44.4	+25:18
1552	Vul	15	6	5.1	20:38.3	+24:09
1553	Vul	16	5	4.8	20:38.3	+21:16
1554	Vul	17	6	5.3	20:58.2	+22:24
1555	Vul	18	6	4.6	20:52.0	+27:09
1556	Vul	19	5	5.0	20:55.1	+28:06
1557	Vul	20	5	4.5	21:30.0	+23:41
1558	Vul	21	5	5.1	20:06.8	+23:40
1559	Vul	22	6	5.2	19:22.8	+26:18
1560	Vul	23	5	4.8	19:16.2	+21:26
1561	Vul	24	6	5.6	19:25.7	+20:05
1562	Vul	25	5	5.0	19:34.6	+19:49
1563	Vul	26	6	5.1	20:04.9	+20:01
1564	Vul	27	6	5.5	19:17.7	+23:03

The material is from the SIMBAD object database of the Centre de Données Astronomiques de Strasbourg. The modern visual magnitude is provided to compare it to the naked-eye observations of Hevelius. The Epoch 2000 position for Right Ascension and Declination are also provided

Chapter 6

The Charles Messier Catalog

Charles Messier was born on June 26, 1730, in the principality of Salm in the town of Badonviller. He was the tenth of 12 children of Nicolas and Françoise Messier. Nicolas Messier held a high position in the dukedom's administration, and Charles enjoyed the benefits of being a member of an affluent household. Mortality during this time was high, though. Charles would lose some of his siblings in their early childhood, but this was a common during this time.

When Charles was 11 his father died, and as a result Charles had little opportunity for a formal education. Charles's older brother, Hyacinthe, stepped in to help the Messier family. Hyacinthe put Charles to work in his office, giving him paperwork that helped him develop good writing and drawing skills, which would be crucial for his future in astronomy.

The event that may have sparked his interest in astronomy was the appearance of the Great Comet of 1744. The comet was visible for several months over his hometown during 1743 and 1744. It was one of the brightest comets in history. The most remarkable characteristics of the comet were its six spectacular tails when it reached perihelion.

The year 1751 brought many changes to the life of the Messiers. The dukedom of Salm lost its independence, became part of Lorraine, and was annexed by France. The village Senones, outside of Badonviller, retained its independence and became the new home of the Messier family. That year also saw Charles Messier set out on his own at the age of 21. A family friend, who had contacts in Paris, helped him attain an assistantship at a new Naval Observatory. It was not Charles's interest in astronomy but his skills as an office assistant developed at his brother's office that helped him get the job. Messier learned to use the instruments at the observatory and became a very skilled observer. He was soon promoted to draftsman at the Marine Observatory at the Hotel de Cluny under Joseph-Nicholas de l'Isle, an astronomer for the French navy.

J.D. Cavin, *The Amateur Astronomer's Guide to the Deep-Sky Catalogs*,
Patrick Moore's Practical Astronomy Series, DOI 10.1007/978-1-4614-0656-3_6,
© Springer Science+Business Media, LLC 2012

Fig. 6.1 A portrait of Charles Messier (1730–1827) at the age of 41 painted by Nicolas Ansiaume in 1771

When Edmund Halley predicted that the comet of 1682 would return in late 1758 or early 1759, Messier began searching for it using charts that de l'Isle had incorrectly prepared. He located the comet on January 21, 1759, but de l'Isle initially refused to let Messier announce his discovery. (The comet was first sighted on Christmas night of 1758 by a German amateur astronomer named Palitzch.) From that time onward Messier devoted himself to searching for comets.

Charles Messier used many telescopes during his career. Owen Gingerich suggests that Messier's favorite telescope was a 32-ft Gregorian reflector with an aperture of 7.5 in. and a magnification of 104×. All of Messier's instruments were the best for his time, but they cannot compete with telescopes used by the average amateur astronomers today (Table 6.1).

During Messier's search for comets he found many faint patchy objects throughout the night sky. The nebulous objects continually hampered his comet search. He decided to create a list of these objects to avoid falsely identifying them as comets. On August 28, 1758, Messier recorded the first of these objects, which was in the constellation of Taurus. This object was the supernova remnant known today as the Crab Nebula (M1).

During 1764 Messier added 38 objects to his list, including M13, the great globular cluster in Hercules; M17, the Swan Nebula; and M31, the Andromeda Galaxy. The following year he added M41, the open cluster southwest of Sirius. Messier

Table 6.1 The telescopes used by Charles Messier between 1765 and 1769 (as published in Connaissance des Temps for 1807)

	Type of telescope	Focal length (ft)	Mag.	Owner
1	Ordinary refractor	25	138×	
2	Achromatic refractor	10.5	120×	M. de Courtanvaux
3	Achromatic refractor (Dollond)	3.25	120×	Duc de Chaulnes
4	Ordinary refractor	23	102×	
5	Ordinary refractor	30	117×	M. Baudouin
6	Campani refractor		64×	M. Maraldi
7	Gregorian reflector	6	110×	M. Lemonnier
8	Gregorian reflector	30	104×	
9	Newtonian reflector	4.5	60×	
10	Refractor (3" aperture)	1	44×	M. de Saron
11	Refractor	19	76×	Paris Observatory

also added the observations of other astronomers to his list. Only 17 of the 45 objects in the first publication of *Messier's Catalogue* (1774) were discovered by Messier. By 1780 there were 80 objects in the *Messier Catalogue*. By 1781 Messier added number 100 to the list. Three more observations by French astronomer Pierre Mechain were added in the final revision of the catalog, which was published in 1781. Forty of the 103 objects had been discovered by Messier. In November of 1781 Messier suffered a severe fall, and work on his catalog was discontinued.

The French Revolution encompassed a 10-year-period of major social and political upheaval in France. Messier and his associates would experience great losses during this time. In 1794 Messier lost his academic pension. The navy stopped paying his salary and the rent on his observatory. That year also saw his friend and mathematician President de Saron die on the guillotine only a few days before he had computed the orbit of a comet that Messier had observed. President de Saron was the first to verify that William Herschel's Uranus was a planet. It wouldn't be until the rise of Napoleon Bonaparte that the situation for Messier and Mechain would improve. Messier was admitted to the newly formed Academy of Sciences and the Bureau of Longitudes.

In 1798 Charles Messier made his last major discovery. He was forced to stop observing when he suffered a massive stroke. Two years later, on April 12, 1817, Charles Messier died at the age of 86.

There were seven additional objects known to have been logged by Messier; these were added to the Messier Catalog, and in 1967 the final entry, M110, was added.

Charles Messier may not have been a scientist, but he was an accomplished observational astronomer. He had discovered numerous comets and nebulous objects. The nebulous objects in the *Messier Catalogue* include many of the best deep-sky objects visible from the northern hemisphere (Table 6.2).

Table 6.2 Charles Messier's catalog of nebulous objects

Messier no.	NGC no.	Common name	Const	V Mag	RA (2000)	Dec (2000)
M1	NGC 1952	Crab Nebula	Taurus	6.6	05 h 34 m 31.9 s	+22° 00′ 52″
M2	NGC 7089		Aquarius	6.3	21 h 33 m 29.3 s	−00° 49′ 23″
M3	NGC 5272		Venatici	5.4	13 h 42 m 11.2 s	+28° 22′ 32″
M4	NGC 6121		Scorpius	5.7	16 h 23 m 35.4 s	−26° 31′ 31″
M5	NGC 5904		Serpens Caput	4.2	15 h 18 m 33.7 s	+02° 04′ 58″
M6	NGC 6405	Butterfly Cluster	Scorpius	4.2	17 h 40 m 20.7 s	−32° 15′ 15″
M7	NGC 6455	Ptolemy's Cluster	Scorpius		17 h 53 m 51.1 s	−34° 47′ 34″
M8	NGC 6433	Lagoon or Dragon Nebula	Sagittarius	7.8	18 h 04 m 04.0 s	−24° 23′ 49″
M9	NGC 6333		Ophiuchus	6.6	17 h 19 m 11.7 s	−18° 30′ 59″
M10	NGC 6254		Ophiuchus	5.8	16 h 57 m 09.0 s	−04° 04′ 58″
M11	NGC 6705	Wild Duck Cluster	Scutum	6.1	18 h 51 m 05.9 s	−06° 16′ 12″
M12	NGC 6218		Ophiuchus	5.8	16 h 47 m 14.4 s	−01° 56′ 52″
M13	NGC 6205	The Great Cluster in Hercules	Hercules	7.6	16 h 41 m 41.4 s	+36° 27′ 36″
M14	NGC 6402		Ophiuchus	6.3	17 h 37 m 36.1 s	−03° 14′ 46″
M15	NGC 7078		Pegasus	6.0	21 h 29 m 58.4 s	+12° 10′ 00″
M16	NGC 6611	Eagle Nebula	Serpens Cauda	6.3	18 h 18 m 48.1 s	−13° 48′ 26″
M17	NGC 6618	Checkmark Nebula	Sagittarius	6.9	18 h 20 m 47.1 s	−16° 10′ 18″
M18	NGC 6613		Sagittarius	6.8	18 h 19 m 58.4 s	−17° 06′ 07″
M19	NGC 6273		Ophiuchus		17 h 02 m 37.7 s	−26° 16′ 05″
M20	NGC 6514	Trifid Nebula	Sagittarius	5.9	17 h 02 m 37.7 s	−26° 16′ 05″
M21	NGC 6531		Sagittarius	5.2	18 h 04 m 13.4 s	−22° 29′ 24″
M22	NGC 6656		Sagittarius	5.5	18 h 36 m 24.1 s	−23° 54′ 12″
M23	NGC 6494		Sagittarius	11.1	17 h 57 m 04.7 s	−18° 59′ 07″
M24	NGC 6603	Delle Caustiche	Sagittarius	11.1	18 h 18 m 26.9 s	−18° 24′ 22″
M25	IC 4725		Sagittarius	8.0	18 h 31 m 36.0 s	−19° 15′ 00″
M26	NGC 6694		Scutum	7.4	18 h 45 m 18.6 s	−09° 23′ 01″
M27	NGC 6853	Dumbbell or Diablo Nebula	Vulpecula		19 h 59 m 36.33 s	+22° 43′ 16″

M28	NGC 6626		Sagittarius	6.9	18 h 24 m 32.8 s	−24° 52' 12"
M29	NGC 6913		Cygnus	6.6	20 h 23 m 57.7 s	+38° 30' 28"
M30	NGC 7099		Capricornus	6.9	21 h 40 m 21.9 s	−23° 10' 45"
M31	NGC 224	Andromeda Galaxy	Andromeda	3.6	00 h 42 m 44.3 s	+41° 16' 06"
M32	NGC 221		Andromeda	8.3	00 h 42 m 41.8 s	+40° 51' 52"
M33	NGC 598	Pinwheel or Triangulum Galaxy	Triangulum	5.8	01 h 33 m 50.8 s	+30° 39' 37"
M34	NGC 1039		Perseus	5.2	02 h 42 m 07.4 s	+42° 44' 46"
M35	NGC 2168		Gemini	5.1	06 h 08 m 55.9 s	+24° 21' 28"
M36	NGC 1960		Auriga	6.0	05 h 36 m 17.7 s	+34° 08' 27"
M37	NGC 2099		Auriga	5.6	05 h 52 m 18.3 s	+32° 33' 11"
M38	NGC 1912		Auriga	6.4	05 h 28 m 42.5 s	+35° 51' 18"
M39	NGC 7092		Cygnus	4.6	21 h 31 m 48.3 s	+48° 26' 55"
M40	Winnecke 4		Ursa Major		12 h 22 m 16.1 s	+58° 05' 04"
M41	NGC 2287		Canis Major	4.5	06 h 46 m 00.0 s	−20° 45' 15"
M42	NGC 1976	Orion Nebula	Orion		05 h 35 m 17.2 s	−05° 23' 27"
M43	NGC 1982	de Mairan's Nebula	Orion		05 h 35 m 31.3 s	−05° 16' 03"
M44	NGC 2632	Manger ("Praesepe") or Beehive Cluster	Cancer		08 h 40 m 22.2 s	+19° 40' 19"
M45	Melotte 22	Pleiades Cluster	Taurus		03 h 47 m 28 s	+24° 06' 18"
M46	NGC 2437		Puppis	6.1	07 h 41 m 46.8 s	+14° 48' 36"
M47	NGC 2422		Puppis	4.4	07 h 36 m 35.0 s	−14° 28' 57"
M48	NGC 2548		Hydra	5.8	08 h 13 m 43.1 s	−05° 45' 02"
M49	NGC 4472		Virgo	8.4	12 h 29 m 46.7 s	+08° 00' 00"
M50	NGC 2323		Monoceros	5.9	07 h 02 m 42.3 s	−08° 23' 26"
M51	NGC 5194	Whirlpool Galaxy	Canes Venatici	8.5	13 h 29 m 52.1 s	+47° 11' 43"
M52	NGC 7654		Cassiopeia	6.9	23 h 24 m 50.4 s	+61° 36' 24"
M53	NGC 5024		Coma Berenices	7.7	13 h 12 m 55.2 s	+18° 10' 08"
M54	NGC 6715		Sagittarius	7.7	18 h 55 m 03.3 s	−30° 28' 42"
M55	NGC 6809		Sagittarius	6.3	19 h 39 m 59.3 s	−30° 57' 44"
M56	NGC 6779		Lyra	8.4	19 h 16 m 35.4 s	+30° 11' 05"

(continued)

Table 6.2 (continued)

Messier no.	NGC no.	Common name	Const	V Mag	RA (2000)	Dec (2000)
M57	NGC 6720	Ring Nebula	Lyra	8.8	19 h 53 m 35.09 s	+33° 01' 44.5"
M58	NGC 4579		Virgo	10.1	12 h 37 m 43.5 s	+11° 49' 05"
M59	NGC 4621		Virgo	9.8	12 h 42 m 02.4 s	+11° 38' 48"
M60	NGC 4649		Virgo	8.8	12 h 43 m 39.7 s	+11° 33' 07"
M61	NGC 4303		Virgo	9.6	12 h 21 m 54.9 s	+04° 28' 24"
M62	NGC 6266		Ophiuchus	6.4	17 h 01 m 12.5 s	−30° 06' 44"
M63	NGC 5055	Sunflower Galaxy	Canes Venatici	9.0	13 h 15 m 49.1 s	+42° 01' 50"
M64	NGC 4826	Black-eye or Sleeping Beauty Galaxy	Coma Berenices	8.4	12 h 56 m 44.2 s	+21° 40' 58"
M65	NGC 3623		Leo	9.3	11 h 18 m 55.9 s	+13° 05' 37"
M66	NGC 3627		Leo	9.0	11 h 20 m 15.1 s	+12° 59' 28"
M67	NGC 2682		Cancer	6.9	08 h 51 m 20.1 s	+11° 48' 43"
M68	NGC 4590		Hydra	7.3	12 h 39 m 28.0 s	−26° 44' 34"
M69	NGC 6637		Sagittarius	7.6	18 h 31 m 23.2 s	−32° 20' 53"
M70	NGC 6681		Sagittarius	7.8	18 h 43 m 12.5 s	−32° 17' 31"
M71	NGC 6838		Sagittarius	8.4	19 h 53 m 46.1 s	+18° 46' 42"
M72	NGC 6981		Aquarius	9.2	20 h 53 m 27.9 s	−12° 32' 13"
M73	NGC 6994		Aquarius	8.9	20 h 58 m 55.9 s	−12° 38' 08"
M74	NGC 628		Pisces	9.5	01 h 36 m 41.6 s	+15° 47' 03"
M75	NGC 6264		Pisces	8.6	20 h 06 m 04.7 s	−21° 55' 17"
M76	NGC 650	Part of Little Dumbbell or Cork Nebula	Perseus	10.1	01 h 42 m 18.1 s	+51° 34' 16"
M77	NGC 1068	Seyfert Galaxy (brightest)	Cetus	9.1	02 h 42 m 40.7 s	−00° 00' 47"
M78	NGC 2068		Orion		05 h 46 m 45.8 s	−00° 04' 45"
M79	NGC 1904		Lepus	7.7	05 h 24 m 10.6 s	−24° 31' 27"
M80	NGC 6093		Scorpius	7.3	16 h 17 m 03.1 s	−22° 58' 30"
M81	NGC 3031	Bode's Nebula	Ursa Major	7.3	09 h 55 m 32.9 s	+69° 03' 55"
M82	NGC 3034	Cigar Nebula	Ursa Major	8.9	09 h 55 m 50.7 s	+69° 40' 43"
M83	NGC 5236	Southern Pinwheel Galaxy	Hydra	7.8	13 h 37 m 00.2 s	−29° 52' 04"

M84	NGC 4374	Markarian's Chain	Virgo	9.4	12 h 25 m 04.7 s	+12° 53' 13"
M85	NGC 4382		Coma Berenices	9.2	12 h 25 m 24.2 s	+18° 11' 27"
M86	NGC 4406	Member of Markarian's Chain	Virgo	9.0	12 h 26 m 11.9 s	+12° 56' 47"
M87	NGC 4486	Smoking Gun	Virgo	8.8	12 h 30 m 49.3 s	+12° 23' 26"
M88	NGC 4501		Coma Berenices	9.7	12 h 31 m 59.1 s	+14° 25' 15"
M89	NGC 4552		Virgo	9.8	12 h 35 m 39.3 s	+12° 33' 23"
M90	NGC 4569		Virgo	9.6	12 h 36 m 49.4 s	+13° 09' 45"
M91	NGC 4548		Coma Berenices	10.4	12 h 35 m 27.2 s	+14° 29' 48"
M92	NGC 6341		Hercules	6.5	17 h 17 m 07.5 s	+43°08' 11"
M93	NGC 2447		Puppis	6.2	07 h 44 m 29.2 s	−23° 51' 11'
M94	NGC 4736		Canes Venatici	8.0	12 h 50 m 53.0 s	+41° 07' 12"
M95	NGC 3351		Leo	10.0	10 h 43 m 57.7 s	+11° 42' 13"
M96	NGC 3368		Leo	9.3	10 h 46 m 45.9 s	+11° 49' 25"
M97	NGC 3587	Owl Nebula	Ursa Major	9.9	11 h 14 m 47.71 s	+55° 01' 07.7"
M98	NGC 4192		Coma Berenices	10.0	12 h 13 m 48.2 s	+14° 54' 00"
M99	NGC 4254		Coma Berenices	9.7	12 h 18 m 49.6 s	+14° 25' 01"
M100	NGC 4321		Coma Berenices	9.7	12 h 22 m 54.9 s	+15° 49' 21"
M101	NGC 5457	Pinwheel Galaxy	Ursa Major	7.5	14 h 03 m 12.4 s	+54° 20' 55"
M102		Possible duplication of M101				
M103	NGC 581		Cassiopeia	7.4	01 h 33 m 21.8 s	+60° 39' 29"
M104	NGC 4594	Sombrero Galaxy or Dark Lane Galaxy	Virgo	7.9	12 h 39 m 59.4 s	−11° 37' 23"
M105	NGC 3379		Leo	9.4	10 h 47 m 49.7 s	+12° 34' 53"
M106	NGC 4258		Canes Venatici	8.7	12 h 18 m 57.5 s	+47° 18' 15"
M107	NGC 6171		Ophiuchus	7.8	16 h 32 m 31.9 s	−13° 03' 10"
M108	NGC 3556		Ursa Major	10.2	11 h 11 m 31.2 s	+55° 40' 24"
M109	NGC 3992		Ursa Major	9.9	11 h 57 m 36.0 s	+53° 23' 28"
M110	NGC 205	Part of M31	Andromeda	8.2	00 h 40 m 21.9 s	+41° 41' 26"

Chapter 7

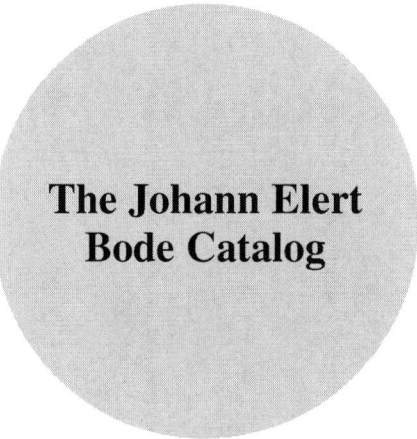

The Johann Elert Bode Catalog

Johann Elert Bode (1747–1826) was born in Hamburg, Germany, the oldest of nine brothers and sisters. As a child he was afflicted with an eye disease that significantly damaged his right eye. He continued to have eye problems the rest of his life. Although his father was a merchant he never attended a formal school. His father taught him the skills of a merchant, hoping he would enter the family business. During his father's illness in 1767 young Bode impressed the doctor, Heinrich Reimarus, with his mathematical skills. Dr. Reimarus introduced Johann Bode to Johann Georg Büsch, a professor of mathematics at the Academic Gymnasium in Hamburg. With the assistance of Professor Büsch and his library, Johann Bode would educate himself in mathematics, geography, and astronomy.

A great interest of Bode's was the observations of the planets of the Solar System. He collected observations made by many well-known astronomers throughout Europe. In 1772 he published the Titius-Bode law.[1] This law was based on a hypothesis that the orbits of the planets around the Sun follow an exponential relationship. The theory appeared to be correct until the outer planet Neptune was discovered. The orbit of Neptune showed the Titus-Bode law to be flawed.

When William Herschel discovered a new planet in March 1781 it was Bode who proposed the name "Uranus," which was adopted as the official name. With his collection of astronomical observations he discovered that the planet had been observed a number of times before Herschel. It had been observed by Tobias Mayer from 1756 and the earliest observation was by John Flamsteed, in December 1690. Flamsteed had cataloged the planet as the "star" 34 Tauri.

---

[1] The original publication made no mention of the contributions of Johann Titus. Only in later publications was he properly added to the credits.

J.D. Cavin, *The Amateur Astronomer's Guide to the Deep-Sky Catalogs*, Patrick Moore's Practical Astronomy Series, DOI 10.1007/978-1-4614-0656-3_7, © Springer Science+Business Media, LLC 2012

Fig. 7.1 German astronomer Johann Elert Bode

Bode combined his discoveries with other catalogs he had collected. He created a star catalog from the objects in Hevelius's catalog, the northern objects from Lacaille's catalog, the objects in the first edition of Messier's catalog, and other catalogs. This became Bode's "Complete Catalog of hitherto observed Nebulous Stars and Star Clusters," containing 75 entries, which he published in 1777 in the *Astronomisches Jahrbuch.* He continued to compile and publish several more catalogs and atlases. In 1782 he published the "Vorstellung der Gestirne," containing several new discoveries and meant for amateur astronomers. In 1801 Bode published his most famous and popular star atlas, the *Uranographia.* The *Uranographia* is one of the last great atlases. It provided accurate locations of the stars, the location of other nebula objects, and colorful, artistic representations of the constellations. In the *Uranographia*, Bode introduced several new constellations, but these are no longer in use today. After 40 years as the director of the Berlin Observatory, Bode retired from his post. In November 1826, the next year, Johann Elert Bode died in Berlin, Germany, at the age of 79.

The catalog following was published by Bode in his 1777 catalog of "Nebulous Stars and Star Clusters." Table 7.1 includes a column for Bode's original comments on the object, a column for positions in elliptical coordinates, a column for Bode's original positions converted to Right Ascension/Declination coordinates and precessed forward to Epoch 2000, and a column for the modern position of the object.

Table 7.1 Johann Elert Bode's catalog of "Nebulous Stars and Star Clusters" (1777)

Bode #	ID	Comments/Bode's description	Bode's original ecliptic coordinates		Bode's coordinates corrected for Epoch 2000		Epoch 2000 coordinates	
			Latitude	Longitude	RA	Dec	RA	Dec
1	IC 1434?	A star cluster	3:23	+57:30	22:09.1	+52:55	22:10.5	+52:50
2	M32	A small nebula	23:25	+32:24	00:39.9	+39:54	00:42.7	+40:52
3	M31	Nebula visible to the naked eye, 15' diameter	24:48	+33:22	00:42.6	+41:18	00:42.7	+41:16
4		A star cluster	29:37	+45:55	00:22.9	+53:57		
5	M33	A small dim nebula	29:53	+18:30	01:34.2	+29:46	01:33.9	+30:39
6	Fla 217	A nebulous star	38:25	+27:05	01:53.4	+40:44		
7	M34	A star cluster	48:56	+25:36	02:43.4	+42:48	02:42.0	+42:47
8	M45	A well-known cluster of small stars	56:55	+04:01	03:47.5	+24:06	03:47.0	+24:07
9	M38	A star cluster	79:05	+12:55	05:22.1	+36:08	05:28.4	+35:50
10	M42	Is the most remarkable nebula in sky, 6' large	79:49	−28:39	05:34.5	−05:20	05:35.4	−05:27
11	M1	A small nebula without stars	81:01	−01:23	05:34.5	+21:57	05:34.5	+22:01
12	M36	A small cluster of stars	81:05	+10:20	05:32.3	+33:40	05:36.1	+34:08
13	M37	A nebula	83:27	+08:56	05:43.7	+32:21	05:52.4	+32:33
14	M35	A nebula between small stars	89:17	+00:40	06:10.3	+24:07	06:08.9	+24:20
15	M41	A star cluster	98:49	−43:00	06:36.9	−19:55	06:47.0	−20:44
16	M50	A star cluster on a nebula	104:27	−30:36	07:00.7	−08:03	07:03.2	−08:20
17	M81	Two small nebulae 0.75° separated	116:19	+52:15	09:59.4	+69:40	09:55.6	+69:04
18	M82	Two small nebulae 0.75° separated	116:48	+51:21	09:56.6	+68:45	09:55.8	+69:41
19	Cr 140?	A star cluster	118:29	−55:13	07:24.5	−34:03	07:23.9	−32:12
20	M44	The well known star cluster Praesepe	124:21	+01:06	08:40.5	+19:30	08:40.1	+19:59
21	N 2477	A nebula visible to the naked eye	130:12	−57:59	07:50.8	−38:37	07:52.3	−38:33

(continued)

Table 7.1 (continued)

Bode #	ID	Comments/Bode's description	Bode's original ecliptic coordinates		Bode's coordinates corrected for Epoch 2000		Epoch 2000 coordinates	
			Latitude	Longitude	RA	Dec	RA	Dec
22	N 2546	A collection of small stars	136:04	−54:20	08:12.9	−36:24	08:12.4	−37:38
23	M40	Two small closely neighbored nebulous stars	147:59	+53:03	12:22.3	+58:05	12:22.4	+58:05
24	Hev 1496	A nebulous star	148:52	+53:59	12:30.0	+58:26		
25	M51	A small nebula	171:36	+51:06	13:29.2	+47:29	13:29.9	+47:12
26	M53	A rather conspicuous nebula	186:43	+23:36	13:14.0	+17:46	13:12.9	+18:10
27	M83	A small misshapen nebula	210:32	−18:20	13:37.1	−29:51	13:37.0	−29:52
28	Hev 953	A nebulous star	221:39	+05:52	14:56.0	−10:40		
29	M5	A nebula without stars	223:30	+19:39	15:18.5	+02:04	15:18.6	+02:05
30	M13	A rather vivid nebula	236:24	+57:55	16:41.5	+36:31	16:41.7	+36:28
31	M4	A nebula like the nucleus of a comet	245:23	−04:50	16:23.4	−26:31	16:23.6	−26:32
32	M12	Two nebulae without stars close together	247:56	+20:06	16:48.7	−02:12	16:47.2	−01:57
33	M10	Two nebulae without stars close together	250:20	+18:35	16:57.0	−03:59	16:57.1	−04:06
34	Hev 804	A nebulous star	250:30	+35:29	17:05.4	+12:47		
35	M19	A nebula	254:05	−03:28	17:02.7	−26:18	17:02.6	−26:16
36	M9	A nebula	257:13	+04:38	17:19.3	−18:29	17:19.2	−18:31
37	M14	A nebula	260:58	+20:05	17:37.6	−03:16	17:37.6	−03:15
38	Hev 794	A nebulous star	261:45	+71:50	17:50.3	+48:25		
39	M6	A cluster of small stars	262:40	−08:49	17:40.1	−32:13	17:40.1	−32:13
40	Hev 795	A nebulous star	264:00	+63:28	17:53.2	+40:01		
41	M7	A star cluster	265:38	−11:25	17:53.8	−34:53	17:53.9	−34:49
42	M23	A star cluster	266:09	+04:38	17:56.7	−18:50	17:56.8	−19:01

43	—	Small stars close together in a nebula	266:53	−00:20	17:59.8	−23:48	18:03.8	−24:23
44	M8 ???	Small stars close together in a nebula	266:59	−00:48	18:00.2	−24:16	18:03.8	−24:23
45	M8 ???	A nebula	267:11	−01:30	18:01.1	−24:58	18:31.4	−32:21
46	M69	Like a small comet's nucleus	267:27	−10:04	18:02.5	−33:32	18:02.6	−23:02
47	M20	A star cluster	267:31	+00:27	18:02.6	−23:01	18:04.8	−24:20
48	N6530	A star with a nebula	267:35	−00:48	18:02.9	−24:16	18:03.8	−24:23
49	M8	Three small stars with a nebula	267:48	−00:53	18:03.8	−24:21	18:04.6	−22:30
50	M21	A star cluster	267:56	+00:56	18:04.4	−22:32	18:16.9	−18:29
51	M24	A nebulous star cluster	270:39	+05:02	18:15.6	−18:23	18:18.8	−13:47
52	M16	Star cluster with nebula	271:26	+09:35	18:18.3	−13:49	18:19.9	−17:08
53	M18	Star cluster with nebula	271:45	+06:15	18:20.1	−17:08	18:20.8	−16:11
54	M17	A nebula	271:46	+07:06	18:20.0	−16:17	18:31.0	−13:10
55	Hev 1259 NGC6639?	A star cluster	274:09	+09:05	18:29.4	−14:12	18:31.6	−19:15
56	M25	A nebulous star cluster	274:25	+04:19	18:31.6	−18:57	18:36.4	−23:54
57	M22	A rather vivid nebula	275:15	−00:42	18:36.4	−23:54	18:24.5	−24:52
58	M28	A Nebula	277:54	−01:43	18:48.4	−24:43	18:45.2	−09:24
59	M26	Star cluster with nebula	278:25	+13:36	18:45.6	−09:25		
60	—	Two small nebulous stars	279:24	+00:09	18:54.2	−22:44		
61	—	Two small nebulous stars	279:37	+00:12	18:55.1	−22:40		
62	M11	Two small nebulous stars	280:12	+16:30	18:51.2	−06:23	18:51.1	−06:16
63	M55	A nebulous star cluster near a nebula	288:30	−09:20	19:40.1	−30:57	19:40.0	−30:58
64	Hev 380	Three small stars which Bayer and Hevel saw nebulous	299:39	+00:31	20:19.5	−19:04		
65	Hev 383	Three small stars which Bayer and Hevel saw nebulous	301:40	+00:59	20:27.4	−18:09		
66	Hev 382	Three small stars which Bayer and Hevel saw nebulous	302:09	+00:29	20:29.8	−18:31		
67	M27	A nebula	305:21	+42:15	19:59.6	+22:42	19:59.6	+22:43
68	M30	A nebula	316:38	−08:48	21:40.5	−23:16	21:40.4	−23:11

(continued)

Table 7.1 (continued)

Bode #	ID	Comments/Bode's description	Bode's original ecliptic coordinates		Bode's coordinates corrected for Epoch 2000		Epoch 2000 coordinates	
			Latitude	Longitude	RA	Dec	RA	Dec
69	M29	A star cluster	321:04	+55:28	20:23.9	+38:34	20:23.9	+38:32
70	M2	Like a comet without a tail	322:20	+13:09	21:33.1	−00:39	21:33.5	−00:49
71	M15	A small nebula	326:13	+25:30	21:30.2	+12:12	21:30.0	+12:10
72	Hev 618?	Nebulous stars	333:06	+61:06	20:39.8	+46:30		
73	Hev 1113	Nebulous stars	337:24	+13:23	22:27.7	+04:45		
74	Hev 619?	Nebulous stars	349:10	+67:37	20:49.0	+55:52		
75	M39	A star cluster	349:13	+57:30	21:31.8	+48:09	21:32.2	+48:26
76	M92	A nebula. More or less round w/pale glow	251:00	+66:00	17:24.9	+43:01	17:17.1	+43:08
77	M64	A small nebulous star	181:00	+26:00	12:58.3	+22:07	12:56.7	+21:41

#4 – An asterism near Zeta and Lambda Cas
#23 – Messier's position of M40
#24 – Hevelius' position of M40
? represent the modern name of objects near the position indicated that Johann Bode may have been viewing
??? represent objects that appear to have been repeated

Chapter 8

Christian Mayer Catalog

Christian Mayer (1719–1783) was a Czech teacher and astronomer. There is not much known about Mayer's early life. He was 20 when he decided to become a Jesuit and entered the Society of Jesus in Mannheim. After receiving his education he began teaching the humanities. His was appointed as a professor of Mathematics and Physics at Heidelberg. With his interest in Astronomy he was appointed the Court Astronomer at Mannheim. In 1771 he began the construction of the Mannheim Observatory that would be actively used for over 100 years. The observatory would be where Christian Mayer would make the observation that led to the first double star catalog (Table 8.1).

Mayer's observatory consisted of a wall-mounted brass quadrant 8 ft in radius built by John Bird of England. The quadrant was fitted with an achromatic telescope capable of 85×. The telescope was constructed by Peter Dolland, also of England.

Mayer's main focus of study was the proper motion of stars, but in his study he found that many stars appeared to have a close companion. Mayer's interest would make him the first astronomer to devote his resources to the observation of double stars.

In 1777, during a lecture about his observations of double stars to the Academy of Mannheim, he received harsh criticism from a fellow Jesuit astronomer by the name of Maximilian Hell, the director of the Imperial Observatory in Vienna. Hell believed the faint companions were not part of the same system, but rather background stars. Hell demanded that Mayer provide actual proof to back up his claims. Driven by Hell's demand Mayer would publish the first version of his double star catalog in 1779. Two years later Johann Bode would publish Mayer's double star list and add another 8 to bring the total to 80 double stars. Mayer described his theory of double stars in his book, *De novis in coelo sidereo phaenomenis in miris*

J.D. Cavin, *The Amateur Astronomer's Guide to the Deep-Sky Catalogs*,
Patrick Moore's Practical Astronomy Series, DOI 10.1007/978-1-4614-0656-3_8,
© Springer Science+Business Media, LLC 2012

Table 8.1 Christian Mayer's 1779 Double Star Catalog (Used with permission from J.S. Schlimmer)

No.	Mayer's description	Current description	WDS identifier	Sep 1778	Sep 2007	Pa 1778	Pa 2007	Mag WDS
1	Andromeda	HIP3617	00464+3057	45.9	47.1	238	46	7.25, 7.43
2	Androm.	74 Ψ Pisces	01057+2128	32.5	30.3	154	159	5.27, 5.45
3	zeta fische	ζ Pisces	01337+0735	24.3	23.3	67	63	5.22, 6.15
4	bei my Fishe	BSC 419	01269+0332	4.2	5.9	180	328	6.65, 9.51
5	Gamma Widder	γ Arietis	01535+1918	12.5	7.6	193	0	4.52, 4.58
6	Lambda Widder	λ Arietus	01579+2336	38.7	38.2	46	47	4.80, 6.65
7	Gamma Andromeda	γ Andromedae	02039+4220	12.2	9.6	62	63	2.31, 5.02
8	Alpha Widder	α Arietis	–	–	–	–	–	2.0
9	30 Widder	30 Arietis	02370+2439	41.4	39.0	276	275	6.50, 7.02
10	Wahlfisch	BCS 587	02004+0831	22.2	62.6	98	200	5.5, 9.8
11	Stier	BSC 1065	03313+2734	26.7	11.6	90	270	6.58, 6.93
12	tau Stier	τ 94 Taurus	04422+2257	60.3	63.0	210	214	4.24, 7.02
13	Stie	BSC 1600	04590+1433	43.6	40.3	303	305	6.06, 7.43
14	Orion	σ Orionis	05387−0236	25.8	13.4	36	84	3.76, 6.56
		σ Orionis	05387−0236	36.6	41.5	55	62	3.76, 6.34
15	Delta Orion	δ Orionis	05320−0018	50.0	53.3	0	1	2.41, 6.83
16	bei Zeta Orion	TYC 4771-01005-1	05441−0229	15.0	94.8	90	316	9.43, 9.77
17	Zwillinge	–	–	72.9	–	310	–	
18	Or. (11 Einh.)	Or. (11 Einh.)	06288−0702	9.6	7.1	322	133	4.62, 5.00
18	30 Zwillinge	HIP 31158	06323+1747	22.2	19.7	211	211	6.31, 6.88
20	Zwillinge	HIP 31323	06341+2207	47.0	53	62	244	7.17, 7.41
21	Castor	α Gemini	07346+3153	9.6	3.6	293	310	1.93, 2.97
22	Zeta Krebs	ζ Cancri	08122+1739	7.7	6.1	180	181	5.05, 6.20
23	2. Phi Krebs	2 Phi Canceri	08268+2656	8.8	5.0	131	215	6.16, 6.21
24	Krebs, dunkel	24 Canceri	08267+2432		5.4		50	6.92, 7.53
25	iota Krebs	ι Canceri	08467+2846	32.6	30.7	307	308	4.13, 5.99
26	bei pi Krebs	TYC0825-01529-1	09123+1500	6.0	20	0	216	6.56, 10.40

27	bei pi Krebs	—	—	—	—	—	—	—
28	54 Löwe	54 Leonis	10556+2445	7.2	6.5	110	112	4.48, 6.30
29	bei tau Löwe Nr.83	83 Leonis	11268+0301	28.9	28.6	51	150	6.55, 7.50
30	tau Löwe	τ Leonis	11279+0251	39.9	88.9	158	181	5.05, 7.47
31	a Wasserschlange	BSC4443	11323−2916	13.6	9.4	137	210	5.64, 5.73
32	Haar der Berenike	BSC4698	12207+2703	—	8.9	—	245	7.04, 7.13
33	gamma Jungfrau	γ Virginis	12417−0127	9.8	5.5	130	314	3.48, 3.53
34	12 Jagdhunde	α Canis Venatici	12560+3819	22.4	19.3	226	229	2.85, 5.52
35	54 Jungfrau	54 Virginis	13134−1850	21.5	5.4	266	34	6.78, 7.19
36	Jungfrau	HIP64638	13149−1122	30.3	107.6	77	—	7.11, 8.18
37	Zeta gr. Bären*	ζ Ursa Majoris	13239+5456	0.0	13.9	—	153	2.23, 3.88
38	Pi Bootes	π 29 Bootes	14407+1625	7.4	5.5	106	111	4.88, 5.79
39	Beta Scorpion	β Scorpii	16054−1948	15.4	13.1	40	24	2.59, 4.52
40	Ny Scorpion	ν Scorpii	16120−1928	42.0	41.2	336	338	4.35, 5.31
41	12 im Herkules	36 Hercules	16406+0413	70.3	69.1	229	229	5.76, 6.92
42	alpha Ophiuch.	α Ophiuchi	—	13.2	—	—	180	—
43	alpha Herkules	α Hercules	17146+1423	8.7	4.7	105	118	3.48, 5.40
44	39 Ophiuch.	39 Ophiuchi	17180−2417	14.0	10.3	352	180	5.23, 6.64
45	71 Herkules	70 (!) Hercules	17209+2430	2.4	224.7	56	90	5.12, 9.33
46	roh Herkules	ρ Hercules	17237+3709	7.6	4.1	291	319	4.50, 5.40
47	61 Ophiuch.	61 Ophiuchi	17446+0235	19.9	21.3	102	93	6.13, 6.47
48	b Schützen	β Sagittarius	19226−4428	—	28.6	—	76	3.98, 7.21
49	beim Oph.	—	—	1.4	—	90	—	—
50	Herkul.	95 Her	18015+2136	8.5	6.5	260	257	4.85, 5.20
51	rho Ophiuchi	ρ Ophiuchi	16256−2327	7.6	3.1	90	340	5.07, 5.74
52	Herkules	100 Her	18078+2606	17.5	14.3	180	183	5.81, 5.84
53	Schlange	HIP89489	18157−0321	0.0	3.3	90	6	7.48, 10.22
54	Schütze	BSC6848	18187−1837	9.9	17.5	0	51	6.86, 7.63
55	Ophiuchus	61 Serpentis	—	2.0	—	—	—	—

(continued)

Table 8.1 (continued)

No.	Mayer's description	Current description	WDS identifier	Sep 1778	Sep 2007	Pa 1778	Pa 2007	Mag WDS
56	zeta Leyer	ζ Lyrae	18448+3736	45.3	43.6	148	150	4.34, 5.62
57	epsilon Leyer	4 ε Lyrae	18443+3940	3.8	3.4	38	31	5.15, 6.10
58	5. Leyer	5 ε Lyrae	18443+3940	2.5	2.9	180	155	5.25, 5.38
59	beta Leyer	β Lyrae	18501+3322	48.4	47.4	151	149	3.63, 6.69
60	theta Schlange	θ Serpentis	18562+0412	23.4	23.0	106	104	4.59, 4.93
61	eta Leyer	η Lyrae	19138+3909	23.5	28.6	90	81	4.38, 8.58
62	beta Schwan	β Lyrae	19307+2758	33.9	35.3	54	54	3.19, 4.68
63	bei gamma Schwan	HIP104064	21050+3526	7.0	29.9	180	336	7.38, 10.74
64	omega Steinbock	ω Capricornus	20299−1835	25.6	22.6	242	239	5.91, 6.68
65	Delphin	BSC7840	20312+1116	15.2	17.8	103	255	7.12, 7.39
66	über beta Delphin	HIP101698	20368+1444	29.5	25.2	249	290	8.35, 8.42
67	gamma Delphin	γ Delphinus	20467+1607	17.5	9.1	278	266	4.36, 5.03
68	beim Füllen	ε Equuleus	20591+0418	13.8	10.7	78	67	5.30, 7.05
69	Schwan	HIP104064	21050+3526	15.3	29.9	126	336	7.38, 10.74
70	Schwan	61 Cygni	21069+3845	15.3	13.8	51	48	5.35, 6.10
71	Schwan	HIP104417	21091+3844	6.0	111.6	180	11	7.61, 9.57
72	my Schwan	μ Cygni	21441+2845	11.2	6.9	109	109	4.75, 6.18
73	Wassermann	HD 208718	21580+0556	15.5	10.9	75	57	7.21, 7.73
74	Zeta Wassermann	74 ψ Pisces	01057+2128	4.6	30.3	221	156	5.27, 5.45
75	Wassermann	HIP114702	23141−0855	26.0	24.4	180	177	7.60, 8.17
76	Wassermann	HD 220436	23238−0828	–	6.3	–	151	7.21, 7.67
77	Fische	HIP116035	23307+0515	–	10.4	–	184	7.77, 8.37
78	Andromeda	TYC2772-00004-1	23543+3154	0.0	36.5	–	334	8.25, 10.37
79	omega Fische	ω Piscis	–	–	–	–	–	–
80	Andromeda	BSC9075	23595+3343	4.3	3.7	180	179	6.46, 6.72

stellarum fixarum comitibus, and also introduced the possibility of smaller suns revolving around larger suns.

After hearing of Mayer's observations of double stars, William Herschel began his own study. After 25 years of observations of Castor (1778–1803), William Herschel finally provided the proof that double stars were part of the same system. He contended that the double stars were not an optical illusion and that their closeness was due to "a physical relationship" between the observed stars. Herschel would record over 600 double stars in three of his catalogs.

Chapter 9

The Herschel Catalogs

William Herschel

Friederich Wilhelm Herschel (William Herschel) was born to Isaac and Anna Herschel on November 15, 1738, in Hanover, Germany. The Herschel family were all musicians. At the age of 14 William and his brother Jacob followed their father into the Military Band of Hanover. The regiment the band was part of was eventually stationed in Kent, England. In England William would meet several families that would provide him with helpful contacts in the years later (Fig. 9.1).

When the Seven Year's War broke out the regiment was called home to Germany. William and Jacob experienced direct fire in the Battle of Hastenbeck. When the army showed little regard for the safety of the young musicians, William's father suggested he and Jacob leave the battlefield immediately. Although the band members were not considered members of the army their father secured a discharge signed by the colonel of the regiment. Although William had been too young to enter the army in 1782 George III signed a pardon for his "desertion."

Penniless at the age of 19 William immigrated to England with his brother Jacob. They earned a living copying music and playing at private concerts. His brother Jacob eventually returned to Germany, but William stayed in England. William became a professional musician, even traveling to Italy to sharpen his skills. In 1767 his father, Isaac Herschel, died. Being the eldest son still living at home Jacob took over his father's household. When William visited home he found that his sister Caroline had become the family servant. Jacob's view was that a woman's place in the world was doing housework; he was adamant that even teaching her music was beneath his dignity. William rescued his sister by paying his mother and brother a large amount of money to make up for the loss of their "household servant" and brought her to live with him in England.

J.D. Cavin, *The Amateur Astronomer's Guide to the Deep-Sky Catalogs*,
Patrick Moore's Practical Astronomy Series, DOI 10.1007/978-1-4614-0656-3_9,
© Springer Science+Business Media, LLC 2012

Fig. 9.1 Sir William Herschel, as depicted in the book *Sir William Herschel, His Life and Works* by Edward Singleton Holden (W. H. Allen & Co., 13 Waterloo Place, SW, London (1881), Frontispiece)

William began reading books on mathematics to help him with his harmony to improve his music. His interest quickly spread from mathematics to astronomy. Astronomy sparked a great passion in William. Telescopes at this time were very expensive. Since he was only earning a modest income he built his own. This period began the most productive time during his life. He would give 30–40 private music lessons a week, conduct concerts, and build many telescopes. He would enlist the help of his brother Alexander, an accomplished mechanic, and convince him to move from Germany to his home in England. During his career he would construct over 400 telescopes. With each telescope the Herschels constructed they would improve their techniques.

To the dismay of their sister Caroline William and Alexander turned every room of their home into a workshop. Caroline gave the details of one attempt to cast a 36-in. mirror that nearly proved fatal. The first attempt at pouring the cast resulted in a crack and it had to be re-melted. During the second attempt, the furnace broke, causing the metal to run across the flagstone floor of their basement. The heat from the 2,000° molten metal caused the flagstones to explode all over the room. They had run from the room to avoid the exploding flagstones.

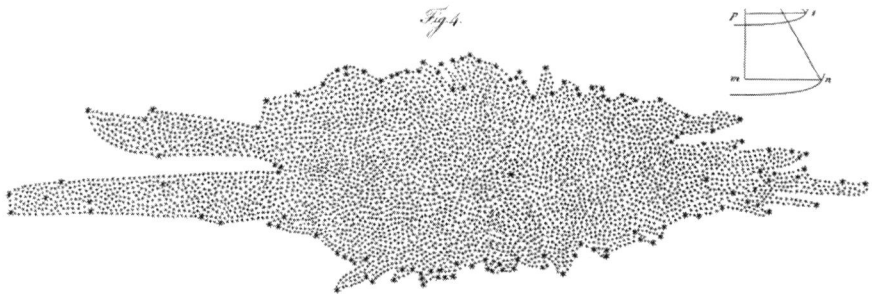

Fig. 9.2 William Herschel's map of the Milky Way Galaxy from the article "On the Construction of the Heavens," published in *Philosophical Transactions of the Royal Society of London,* Vol. 75 (1785), pp. 213–266

With the help of his brother and other assistance they would soon mass produce mirrors for telescopes that would be sold all over Europe. His sister Caroline would assist William by feeding him by hand while William would cast, grind, mounting the speculum metal mirrors.

On a spring night in 1781 William Hershel was surveying the night sky with his 7-ft telescope with a 6-in. mirror. He spotted an object that appeared to have an odd shape, slightly different from the shape of a star. He continued to record observations of the object thinking it might be a comet. With the help of Anders Lexell, they determined the object was a planet orbiting the Sun past the orbit of Saturn. Hershel would name this planet "Georgian star" (Georgium sidus) after King George III. However the name did not stick – a name more pleasing to the French was adopted. The name of the planet would become Uranus after a suggestion by another German astronomer, Johann Bode.

In 1785 Hershel published a paper describing a new statistical technique called 'star gauging'. Using this technique he counted the number of stars in 683 different regions of the sky. When they saw more stars they assumed the stars extended further into space. This is the first recorded attempt to map the star distribution of the Milky Way Galaxy. He found that the galaxy was disk-shaped from these observations and provided a drawing documenting the shape.

In 1788, William upset his sister Caroline when, at age of 50, he married a widow from Upton by the name of Mary Pitt. Caroline moved out of William's house and burned her diaries that she had kept for 10 years. Mary's kindness would eventually win over Caroline and they became friends. William and Mary would have one son, John, who later would also become an accomplished astronomer.

With a long list of remarkable accomplishments it is clear that the amateur astronomer William Herschel led the way into observational and statistical study of our universe. Those who knew him would write that he was a friendly and modest man, willing to talk to anyone about his discoveries. An inscription correctly attributed to William Herschel can be found on a monument erected at his observatory in Slough that reads "I have looked further into space than ever human being did before me," W. H. Herschel, 1813.

Fig. 9.3 Caroline and William Herschel, Image from *Child-life and Girlhood of Remarkable Women* by William Henry Davenport Adams (W. Swan Sonnenschein & Co., Paternoster Row, London, 1883, p. 273)

Caroline Herschel

Caroline was 22 when William moved her to his home in Bath, England. When Caroline arrived, William was working as an accomplished organist, a chorus director, and. a private music teacher. Caroline became devoted to supporting William's household in every way. She not only became William's housekeeper, but when William gave her voice lessons she became a prominent soprano singer.

When William received a pension from King George III he quit his job as a conductor to spend more time on his new passion, astronomy. Caroline helped

William began his new business of creating and selling telescopes. She would work alongside her brother spending long hours at grinding and polishing the mirrors for the telescopes. She became so familiar with the construction of telescopes that she took over running his business when William was absent traveling. She was given a pension of 50 lb by King George III. Caroline became the first woman to be recognized in a scientific position.

Caroline began observing the night sky when William gave her a small telescope. With the new telescope she began to look for comets and nebula, a common goal for many astronomers. When William traveled to Germany after his discovery of Uranus, Caroline made an important discovery herself – her first comet. As she collected more observations she would begin to catalog William and her observations for publication.

When William married he started spending less time at the telescope. This devastated Caroline; she felt she had lost her closest friend. She carried on her work as an observational astronomer without William. By the time of William's death, she had found seven more comets.

After her brother's death, she returned to Hanover and lived with her younger brother, Dietrich. During this period she continued to catalog the observations that they had made. When she had completed the catalog she gave it to the scientific community. As a reward for her effort she was proclaimed an honorary member of the Royal Astronomical Society. She was also honored by the King of Prussia when he gave her the Gold Medal of Science for her accomplishments.

During her retirement she remained in contact with many members of the scientific community. Among them was William's son John and his wife. Caroline was almost 98 when she died peacefully in her sleep on January 9, 1848. She was buried in the churchyard of her childhood near her parents. She was known and honored by the entire scientific community as a distinguished self-educated woman and the devoted sister of William Herschel.

John Herschel

John Fredrick William Herschel, the only son of Sir William Herschel, was born at Slough, Buckinghamshire, on March 7, 1792. He started his early education at Eton, but because of his mother's fears was quickly moved to the house of a private tutor. At the age of 17, he was sent to St. John's College in Cambridge. During his undergraduate years John Herschel won the highest academic distinction, graduating as "senior wrangler" in 1813. He would form a pact with two of his fellow students, Dean Peacock and Charles Babbage, to "do their best to leave the world wiser than they found it." He would continue working with these two individuals and publish mathematical papers that revolutionize branches of mathematical science in England.

After graduation John Herschel had intentions of being called to the bar; he had placed himself under the supervision of a well-known special pleader. An association with William Hyde Wollaston[1] soon changed the direction of his

---

[1] William Hyde Wollaston was an English chemist and natural philosopher, born at East Dereham, Norfolk. His grandfather, William Wollaston, was also an avid amateur astronomer.

profession. John Herschel soon started to experiment in physical optics and turned to the study of astronomy in 1816. By 1820, with the assistance of his father, he constructed a reflecting telescope containing a mirror with an 18-in. diameter and a focal length of 20 ft. This telescope became the instrument that enabled him to make his most important astronomical observations.

During 1821–1823 he became associated with Sir James South.[2] Together they re-examined his father's double stars and published a new double star catalog containing 380 stars. They made their observations for the new catalog with two refractors, one with a 7-ft focal length and the other with a 5-ft focal length.

His work in astronomy would win him many awards and honors. He was presented with the Astronomical Society's gold medal and in 1825 with the Lalande medal of the French Institute. His work in mathematics won him the Copley medal. In 1831 he was knighted by King William IV.

At this point in his life it is clear that John Herschel had become an astronomer from a sense of duty to his father. He began to complete the work begun at Slough by William Herschel. There his father had searched the northern heavens. John Herschel formulated a plan to re-explore the northern skies and to explore the southern. He and his family sailed for the Cape of Good Hope on November 13, 1833. They arrived at their destination in Table Bay on January 15, 1834. He would spend the next 4 years at Feldhausen with his family. There he would create several catalogs of southern nebula and clusters of stars; he cataloged the relative positions and magnitudes of double stars. He described his observations and the varying and relative brightness of the stars. Herschel returned to England in 1838 and was welcomed in a grand manner. He was created a baronet by the queen.

Sir John Herschel was the author of many books, including *Results of Astronomical Observations made at the Cape of Good Hope* (1847), *On the Study of Natural Philosophy* (1830), and *Outlines of Astronomy* (1849). Besides his many articles he contributed to the 8th edition of the *Encyclopaedia Britannica*. He was well known in his attempts to instill a love of knowledge among those not engaged in its pursuit. He never discouraged anyone's attempt at scientific discoveries and would willingly help in any way he could.

Sir John Herschel died at his residence near Hawkhurst in Kent, on May 11, 1871. He remains interred in Westminster Abbey near the grave of Sir Isaac Newton.

Alexander Herschel

Alexander Stewart Herschel was born during his father's visit to Feldhausen near Cape Town in South Africa. It appears as if Alexander's passion for astronomy was inherited from his father at an early age. Growing up in a house full of world-class

---

[2] Sir James South (October 1785–October 1867) was a British astronomer that helped to found the Astronomical Society of London. As president of the society from 1831 to 1832 he successfully petitioned to obtain a royal charter for what would become the Royal Astronomical Society.

astronomical equipment, family members recognized as noted astronomers and visiting scientists from all over the world have been a very strong influence. He was not an only child, like his father; Alexander had two brothers and nine sisters.

The three Herschel boys received an early education in science. Alexander would study at Trinity College and the Royal School of Mines in London. During his time in London he began observing the spectrum of meteors to determine their chemical composition.

In 1866 Alexander Herschel observed the spectra of meteors of the Perseid meteor shower. Later that year he published his observation of 17 meteors in *The Intellectual Observer.* He continued his observations of the spectra of meteors after being appointed as a Professor of Mechanical and Experimental Physics at the Andersonian University of Glasgow. His work in meteor spectroscopy and his deduction of the peak time of the Leonid meteor shower allowed Schiaparelli to identify the Leonids source as the Comet Tempel. He was appointed Professor of Physics and Experimental Philosophy at the University of Durham College of Physical Sciences at Newcastle. During his time at Durham he worked with R. S. Newell, who owned the largest telescope in the world during that time. At the young age of 50 Alexander Herschel retired from his position. He became so involved in his observations near the end of his life he became a recluse. His late nights finally ruined his health, and he died June 18, 1907.

The Herschel 400 was the first catalog created from a subset of William Herschel's original deep-sky catalog. The objects were selected by amateur astronomers Brenda F. Guzman (Branchett), Lydel Guzman, Paul Jones, James Morrison, Peggy Taylor, and Sara Saey of the Ancient City Astronomy Club in St. Augustine, Florida. They generated the catalog as a more difficult challenge after completing the Messier Catalogue.

The Herschel 400 is a list of 400 galaxies, nebulae, and star clusters, picked from the 2,500 deep-sky objects discovered and cataloged by Sir William Herschel and his sister Caroline. The Herschel 400 contains:

231 galaxies
107 open clusters
33 globular clusters
20 planetary nebulae
2 halves of a single planetary nebula
7 bright nebulae

The Herschel 400 is considered an advanced project for amateurs with a substantial amount of deep-sky experience, but it can still be accomplished with a modest 6-in. telescope.

The popularity of the Herschel 400 list has grown tremendously. The Herschel 400 Club is presently a part of the Astronomical League. The Astronomical League consists of more than 240 amateur astronomical societies across the United States. When one of their members completes observing all of the objects on the list the Astronomical League awards them with a Herschel 400 certificate. When an amateur astronomer completes the Herschel 400 list, they are considered to be an advanced observer (Table 9.1).

Table 9.1 The Herschel 400 list created by the members of the Ancient City Astronomers Club (Printed with permission from Brenda Branchett of the Astronomical League)

Object name	Type	RA (2000)	Dec (2000)	Const	Mag	Size
NGC 40	PN	00 h 13 m 01s	+72°31'20"	Cep	12.3	38"×35"
NGC 129	OC	00 h 29 m 54 s	+60°12'35"	Cas	6.5	21'
NGC 136	OC	00 h 31 m 30 s	+61°31'00"	Cas	11.3	1'.2
NGC 157	Glx	00 h 34 m 48 s	−08°24'00"	Cet	10.4	4'.3×2'.9
NGC 185	Glx	00 h 38 m 58 s	+48°20'27"	Cas	9.2	11'×9'.8
M 110	Glx	00 h 40 m 24 s	+41°41'00"	And	8.1	21'.9×10'.9
NGC 225	OC	00 h 43 m 32 s	+61°47'25"	Cas	7.0	12'
NGC 246	PN	00 h 47 m 03 s	−11°52'20"	Cet	10.9	4'.6×4'.1
NGC 247	Glx	00 h 47 m 08 s	−20°45'36"	Cet	9.2	21'.4×6'.9
NGC 253	Glx	00 h 47 m 33 s	−25°17'17"	Scl	7.8	17'.5×6'.8
NGC 278	Glx	00 h 52 m 05 s	+47°33'02"	Cas	10.9	2'.2×2'.1
NGC 288	GC	00 h 52 m 45 s	+26°34'43"	Scl	8.1	14'
NGC 381	OC	01 h 08 m 15 s	+61°35'00"	Cas	9.3	6'
NGC 404	Glx	01 h 09 m 30 s	+35°43'04"	And	10.0	3'.4×3'.4
NGC 436	OC	01 h 15 m 58 s	+58°49'02"	Cas	8.8	5'
NGC 457	OC	01 h 19 m 33 s	+58°17'27"	Cas	6.4	13'
NGC 488	Glx	01 h 21 m 47 s	+05°15'25"	Psc	10.3	5'.2×4'.1
NGC 524	Glx	01 h 24 m 48 s	+09°32'00"	Psc	10.3	2'.8
NGC 559	OC	01 h 29 m 29 s	+63°18'30"	Cas	9.5	4'.4
NGC 584	Glx	01 h 31 m 18 s	−06°52'00"	Cet	10.5	4'.2×2'.3
NGC 596	Glx	01 h 32 m 54 s	−07°02'00"	Cet	10.9	3'.2×2'.1
M 33	Glx	01 h 33 m 51 s	+30°39'37"	Tri	5.7	70'.8×41'.7
NGC 613	Glx	01 h 34 m 18 s	−29°25'00"	Scl	10.1	5'.5×4'.2
NGC 615	Glx	01 h 35 m 06 s	−07°20'00"	Cet	11.6	3'.6×1'.5
NGC 637	OC	01 h 43 m 03 s	+64°02'12"	Cas	8.2	3'.5
NGC 651	PN	01 h 42 m 24 s	+51°35'00"	Per	10.0	2'.7×1'.75
NGC 654	OC	01 h 43 m 59 s	+61°52'58"	Cas	6.5	5'
NGC 659	OC	01 h 44 m 23 s	+60°40'09"	Cas	7.9	5'
NGC 663	OC	01 h 46 m 16 s	+61°13'06"	Cas	7.1	16'
NGC 720	Glx	01 h 53 m 00 s	−13°44'00"	Cet	10.2	4'.7×2'.4
NGC 752	OC	01 h 57 m 48 s	+37°51'00"	And	5.7	50'
NGC 772	Glx	01 h 59 m 20 s	+19°00'22"	Ari	10.4	7'.2×4'.3
NGC 779	Glx	01 h 59 m 42 s	−05°58'00"	Cet	11.0	4'.1×1'.4
NGC 869	OC	02 h 19 m 04 s	+57°08'06"	Per	5.3	29'
NGC 884	OC	02 h 22 m 32 s	+57°08'39"	Per	6.1	29'
NGC 891	Glx	02 h 22 m 34 s	+42°21'03"	And	10.0	13'.5×2'.5
NGC 908	Glx	01 h 23 m 06 s	−21°14'00"	Cet	10.2	5'.5×2'.8
NGC 936	Glx	02 h 27 m 37 s	−01°09'23"	Cet	10.0	4'.7×4'.1
NGC 1022	Glx	02 h 38 m 30 s	−06°41'00"	Cet	11.4	2'.5×2'.1
NGC 1023	Glx	02 h 40 m 24 s	+39°03'48"	Per	9.5	8'.7×3'.0
NGC 1027	OC	02 h 42 m 35 s	+61°35'40"	Cas	6.7	20'
NGC 1052	Glx	02 h 41 m 06 s	−08°15'00"	Cet	10.5	2'.9×2'.0
NGC 1055	Glx	02 h 41 m 45 s	+00°26'32"	Cet	10.9	7'.6×2'.7
NGC 1084	Glx	02 h 46 m 00 s	−07°34'37"	Eri	11.1	3'.2×1'.8

(continued)

Table 9.1 (continued)

Object name	Type	RA (2000)	Dec (2000)	Const	Mag	Size
NGC 1245	OC	03 h 14 m 41 s	+47°14'20"	Per	8.4	10'
NGC 1342	OC	03 h 31 m 40 s	+37°22'28"	Per	6.7	14'
NGC 1407	Glx	03 h 40 m 12 s	−18°35'00"	Eri	9.8	2'.5
NGC 1444	OC	03 h 49 m 23 s	+52°39'44"	Per	6.6	4'
NGC 1501	PN	04 h 06 m 59 s	+60°55'15"	Cam	11.5	56"×48"
NGC 1502	OC	04 h 07 m 49 s	+62°19'54"	Cam	6.9	7'
NGC 1513	OC	04 h 09 m 55 s	+49°31'02"	Per	8.4	9'
NGC 1528	OC	04 h 15 m 18 s	+51°13'00"	Per	6.4	23'
NGC 1535	PN	04 h 14 m 16 s	−12°44'21"	Eri	9.6	48
NGC 1545	OC	04 h 20 m 56 s	+50°15'19"	Per	6.2	18'
NGC 1647	OC	04 h 45 m 54 s	+19°07'00"	Tau	6.4	45'
NGC 1664	OC	04 h 51 m 08 s	+43°40'28"	Aur	7.6	18'
NGC 1750	OC	05 h 04 m 00 s	+23°38'45"	Tau	9.8	2'.5
NGC 1788	NbDF	05 h 06 m 54 s	−03°20'00"	Ori	9.0	2'.0×2'.0
NGC 1817	OC	05 h 12 m 26 s	+16°41'03"	Tau	7.7	16'
NGC 1857	OC	05 h 20 m 05 s	+39°20'37"	Aur	7.0	5'
NGC 1907	OC	05 h 28 m 05 s	+35°19'32"	Aur	8.2	6'
NGC 1931	NbOC	05 h 31 m 26 s	+34°14'42"	Aur	10.1	1'
NGC 1961	Glx	05 h 42 m 05 s	+69°22'43"	Cam	11.2	4'.3×3'.0
NGC 1964	Glx	05 h 33 m 22 s	−21°56'48"	Lep	10.7	6'.2×2'.5
NGC 1980	NbOC	05 h 35 m 24 s	−05°55'00"	Ori	2.5	13'
NGC 1999	NbDF	05 h 36 m 24 s	−06°43'00"	Ori	9.0	2'.0×2'.0
NGC 2022	PN	05 h 42 m 06 s	+09°05'00"	Ori	11.9	28"×27"
NGC 2024	NbDF	05 h 41 m 42 s	−01°51'00"	Ori	–	30'×30'
NGC 2126	OC	06 h 02 m 33 s	+49°51'57"	Aur	10.2	6'
NGC 2129	OC	06 h 01 m 07 s	+23°19'20"	Gem	6.7	6'
NGC 2158	OC	06 h 07 m 24 s	+24°06'00"	Gem	8.6	5'
NGC 2169	OC	06 h 08 m 24 s	+13°57'53"	Ori	5.9	6'
NGC 2185	NbRF	06 h 11 m 01 s	−06°13'37"	Mon	12.0	4'.0×2'.0
NGC 2186	OC	06 h 12 m 07 s	+05°27'31"	Ori	8.7	4'
NGC 2194	OC	06 h 13 m 46 s	+12°48'24"	Ori	8.5	10'
NGC 2204	OC	06 h 15 m 30 s	−18°40'00"	CMa	8.6	12'
NGC 2215	OC	06 h 20 m 48 s	−07°17'00"	Mon	8.4	11'
NGC 2232	OC	06 h 28 m 00 s	−04°51'00"	Mon	3.9	29'
NGC 2244	OC	06 h 31 m 56 s	+04°56'35"	Mon	4.8	24'
NGC 2251	OC	06 h 34 m 38 s	+08°21'59"	Mon	7.3	10'
NGC 2264	NbOC	06 h 40 m 58 s	+09°53'44"	Mon	3.9	20'
NGC 2266	OC	06 h 43 m 19 s	+26°58'10"	Gem	9.5	6'
NGC 2281	OC	06 h 48 m 18 s	+41°04'44"	Aur	5.4	14'
NGC 2286	OC	06 h 47 m 42 s	−03°09'00"	Mon	7.5	14'
NGC 2301	OC	06 h 51 m 45 s	+00°27'33"	Mon	6.0	12'
NGC 2304	OC	06 h 55 m 12 s	+17°59'19"	Gem	10.0	5'
NGC 2311	OC	06 h 57 m 48 s	−04°37'00"	Mon	9.6	6'
NGC 2324	OC	07 h 04 m 08 s	+01°02'41"	Mon	8.4	7'
NGC 2335	OC	07 h 06 m 48 s	−10°02'00"	Mon	7.2	12'

(continued)

Table 9.1 (continued)

Object name	Type	RA (2000)	Dec (2000)	Const	Mag	Size
NGC 2343	OC	07 h 08 m 06 s	−10°37'00"	Mon	6.7	6'
NGC 2353	OC	07 h 14 m 30 s	−10°16'00"	Mon	7.1	20'
NGC 2354	OC	07 h 14 m 18 s	−25°42'00"	CMa	6.5	20'
NGC 2355	OC	07 h 16 m 59 s	+13°44'59"	Gem	9.7	9'
NGC 2360	OC	07 h 17 m 43 s	−15°38'29"	CMa	7.2	12'
NGC 2362	OC	07 h 18 m 42 s	−24°57'15"	CMa	3.8	8'
NGC 2371	PN	07 h 25 m 36 s	+29°30'00"	Gem	11.3	74"×54"
NGC 2372	PN	07 h 25 m 36 s	+29°30'00"	Gem	11.3	2'.2×0'.9
NGC 2392	PN	07 h 29 m 11 s	+20°54'42"	Gem	9.1	47"×43"
NGC 2395	OC	07 h 27 m 13 s	+13°36'29"	Gem	8.0	12'
NGC 2403	Glx	07 h 36 m 52 s	+65°36'13"	Cam	8.4	18'×11'
NGC 2419	GC	07 h 38 m 08 s	+38°52'53"	Lyn	10.3	4'.1
NGC 2420	OC	07 h 38 m 24 s	+21°34'27"	Gem	8.3	10'
NGC 2421	OC	07 h 17 m 00 s	+13°45'00"	Pup	9.7	9'
M 47	OC	07 h 36 m 36 s	−14°30'00"	Pup	4.5	30'
NGC 2423	OC	07 h 37 m 06 s	−13°52'00"	Pup	6.7	19
NGC 2438	PN	07 h 41 m 50 s	−14°44'09"	Pup	10.8	73"×68"
NGC 2440	PN	07 h 41 m 54 s	−18°13'00"	Pup	9.4	74"×42"
NGC 2479	OC	07 h 55 m 06 s	−17°42'00"	Pup	9.6	7'
NGC 2482	OC	07 h 55 m 10 s	−24°15'17"	Pup	7.3	12'
NGC 2489	OC	07 h 56 m 18 s	−30°04'00"	Pup	7.9	8'
NGC 2506	OC	08 h 00 m 02 s	−10°46'11"	Mon	7.6	8'
NGC 2509	OC	08 h 00 m 48 s	−19°03'00"	Pup	9.3	8'
NGC 2527	OC	08 h 05 m 00 s	−28°09'00"	Pup	6.5	22'
NGC 2539	OC	08 h 10 m 36 s	−12°49'00"	Pup	6.5	21'
M 48	OC	08 h 13 m 48 s	−05°36'00"	Hya	5.5	54'
NGC 2567	OC	08 h 18 m 30 s	−30°39'00"	Pup	7.4	10'
NGC 2571	OC	08 h 18 m 54 s	−29°45'00"	Pup	7.0	13'
NGC 2613	Glx	08 h 33 m 24 s	−22°58'00"	Pyx	10.4	7'.2×2'.1
NGC 2627	OC	08 h 37 m 12 s	−29°57'00"	Pyx	8.4	11'
NGC 2655	Glx	08 h 55 m 38 s	+78°13'24"	Cam	10.1	5'.1×4'.4
NGC 2681	Glx	08 h 53 m 33 s	+51°18'49"	UMa	10.3	3'.8×3'.5
NGC 2683	Glx	08 h 52 m 42 s	+33°25'20"	Lyn	9.7	9'.3×2'.5
NGC 2742	Glx	09 h 07 m 33 s	+60°28'45"	UMa	11.7	3'.1×1'.7
NGC 2768	Glx	09 h 11 m 37 s	+60°02'14"	UMa	10.0	6'.3×2'.8
NGC 2775	Glx	09 h 10 m 20 s	+07°02'16"	Cnc	10.4	4'.3×3'.3
NGC 2782	Glx	09 h 14 m 05 s	+40°06'48"	Lyn	11.5	3'.8×2'.9
NGC 2787	Glx	09 h 19 m 18 s	+69°12'11"	UMa	10.8	3'.4×2'.3
NGC 2811	Glx	09 h 16 m 11 s	−16°18'48"	Hya	11.3	2'.7×1'.0
NGC 2841	Glx	09 h 22 m 02 s	+50°58'44"	UMa	9.3	8'.1×3'.8
NGC 2859	Glx	09 h 24 m 18 s	+34°30'47"	LMi	10.7	4'.8×4'.2
NGC 2903	Glx	09 h 32 m 12 s	+21°30'00"	Leo	8.9	13'.0×6'.6
NGC 2950	Glx	09 h 42 m 35 s	+58°51'04"	UMa	11.0	3'.2×2'.1
NGC 2964	144Glx	09 h 42 m 54 s	+31°51'00"	Leo	11.3	3'.0×1'.7
NGC 2974	Glx	09 h 42 m 33 s	−03°41'58"	Sex	10.8	3'.4×2'.1

(continued)

Table 9.1 (continued)

Object name	Type	RA (2000)	Dec (2000)	Const	Mag	Size
NGC 2976	Glx	09 h 47 m 15 s	+67°54'59"	UMa	10.2	4'.9×2'.5
NGC 2985	Glx	09 h 50 m 22 s	+72°16'47"	UMa	10.5	4'.3×3'.4
M 82	Glx	09 h 55 m 48 s	+69°41'00"	UMa	8.4	11'×4'.6
NGC 3077	Glx	10 h 03 m 20 s	+68°44'02"	UMa	9.9	4'.6×3'.6
NGC 3079	Glx	10 h 01 m 57 s	+55°40'54"	UMa	10.8	7'.9×1'.4
NGC 3115	Glx	10 h 05 m 14 s	−07°43'09"	Sex	8.6	7'.2×2'.5
NGC 3147	Glx	10 h 16 m 54 s	+73°24'02"	Dra	10.7	4'.0×3'.5
NGC 3166	Glx	10 h 13 m 46 s	+03°25'31"	Sex	10.6	5'.2×2'.7
NGC 3169	Glx	10 h 14 m 15 s	+03°28'00"	Sex	10.5	4'.8×3'.2
NGC 3184	Glx	10 h 18 m 17 s	+41°25'27"	UMa	9.8	6'.9×6'.8
NGC 3190	Glx	10 h 18 m 06 s	+21°50'00"	Leo	11.9	4'.5×1'.7
NGC 3193	Glx	10 h 18 m 24 s	+21°54'00"	Leo	10.9	2'.8×2'.6
NGC 3198	Glx	10 h 19 m 55 s	+45°33'00"	UMa	10.4	8'.3×3'.7
NGC 3226	Glx	10 h 23 m 24 s	+19°54'00"	Leo	11.4	2'.8×2'.5
NGC 3227	Glx	10 h 23 m 30 s	+19°52'00"	Leo	10.8	5'.6×4'.0
M 101	Glx	14 h 03 m 18 s	+54°22'00"	UMa	8.5	22'
NGC 3245	Glx	10 h 27 m 18 s	+28°30'00"	LMi	10.8	3'.2×1'.9
NGC 3277	Glx	10 h 32 m 54 s	+28°31'00"	LMi	11.7	2'.0×1'.9
NGC 3294	Glx	10 h 36 m 12 s	+37°19'00"	LMi	11.7	3'.3×1'.8
NGC 3310	Glx	10 h 38 m 46 s	+53°30'10"	UMa	11.1	3'.6×3'.0
NGC 3344	Glx	10 h 43 m 30 s	+24°55'00"	LMi	10.0	6'.9×6'.5
NGC 3377	Glx	10 h 47 m 42 s	+13°59'00"	Leo	10.2	4'.4×2'.7
M 105	Glx	10 h 47 m 48 s	+12°35'00"	Leo	9.3	4'.5×4
NGC 3384	Glx	10 h 48 m 12 s	+12°38'00"	Leo	10.0	6'×3'
NGC 3395	Glx	10 h 49 m 48 s	+32°59'00"	LMi	12.1	1'.9×1'.2
NGC 3412	Glx	10 h 50 m 54 s	+13°25'00"	Leo	10.6	3'.6×2'.0
NGC 3414	Glx	10 h 51 m 18 s	+27°59'00"	LMi	10.8	3'.6×2'.7
NGC 3432	Glx	10 h 52 m 30 s	+37°10'00"	LMi	11.3	6'.2×1'.5
NGC 3486	Glx	11 h 00 m 24 s	+28°58'29"	LMi	10.3	6'.9×5'.4
NGC 3489	Glx	11 h 00 m 18 s	+13°54'00"	Leo	10.3	3'.7×2'.1
NGC 3504	Glx	11 h 03 m 11 s	+27°58'20"	LMi	11.1	2'.7×2'.2
NGC 3521	Glx	11 h 05 m 48 s	−00°02'00"	Leo	8.9	9'.5×5'.0
M 108	Glx	11 h 11 m 30 s	+55°40'00"	UMa	9.9	8'.3×2'.5
NGC 3593	Glx	11 h 14 m 36 s	+12°49'00"	Leo	11.0	5'.8×2'.5
NGC 3607	Glx	11 h 16 m 55 s	+18°03'05"	Leo	10.0	3'.7×3'.2
NGC 3608	Glx	11 h 16 m 59 s	+18°08'54"	Leo	11.0	3'.0×2'.5
NGC 3610	Glx	11 h 18 m 25 s	+58°47'10"	UMa	10.5	2'.7×2'.3
NGC 3613	Glx	11 h 18 m 36 s	+57°59'59"	UMa	11.6	3'.6×2'.0
NGC 3619	Glx	11 h 19 m 22 s	+57°45'29"	UMa	12.6	3'.1×2'.6
NGC 3621	Glx	11 h 18 m 17 s	−32°48'50"	Hya	8.9	10'.0×6'.5
NGC 3626	Glx	11 h 20 m 04 s	+18°21'23"	Leo	10.6	2'.7×1'.9
NGC 3628	Glx	11 h 20 m 17 s	+13°35'20"	Leo	9.4	15'×3.6'
NGC 3631	Glx	11 h 21 m 03 s	+53°10'09"	UMa	10.6	5'.0×4'.8
NGC 3640	Glx	11 h 21 m 07 s	+03°14'05"	Leo	10.3	4'.1×3'.4
NGC 3655	Glx	11 h 22 m 55 s	+16°35'24"	Leo	11.6	1'.6×1'.1

(continued)

Table 9.1 (continued)

Object name	Type	RA (2000)	Dec (2000)	Const	Mag	Size
NGC 3665	Glx	11 h 24 m 44 s	+38°45'47"	UMa	10.8	2'.5×2'.0
NGC 3675	Glx	11 h 26 m 08 s	+43°35'10"	UMa	10.2	5'.9×3'.1
NGC 3686	Glx	11 h 27 m 44 s	+17°13'25"	Leo	11.4	3'.3×2'.6
NGC 3726	Glx	11 h 33 m 21 s	+47°01'44"	UMa	10.5	6'.2×4'.3
NGC 3729	Glx	11 h 33 m 49 s	+53°07'32"	UMa	11.4	3'.1×2'.1
NGC 3810	Glx	11 h 40 m 59 s	+11°28'17"	Leo	10.7	4'.3×3'.1
NGC 3813	Glx	11 h 41 m 19 s	+36°32'47"	UMa	11.7	2'.3×1'.2
NGC 3877	Glx	11 h 46 m 08 s	+47°29'41"	UMa	11.8	5'.4×1'.5
NGC 3893	Glx	11 h 48 m 38 s	+48°42'38"	UMa	10.4	4'.5×2'.8
NGC 3898	Glx	11 h 49 m 15 s	+56°05'04"	UMa	10.8	4'.4×2'.6
NGC 3900	Glx	11 h 49 m 09 s	+27°01'18"	Leo	11.4	3'.2×1'.7
NGC 3912	Glx	11 h 50 m 05 s	+26°28'46"	Leo	12.6	1'.5×0'.9
NGC 3938	Glx	11 h 52 m 49 s	+44°07'13"	UMa	10.4	5'.4×4'.9
NGC 3941	Glx	11 h 52 m 55 s	+36°59'10"	UMa	11.3	3'.8×2'.5
NGC 3945	Glx	11 h 53 m 13 s	+60°40'31"	UMa	10.6	5'.5×3'.6
NGC 3949	Glx	11 h 53 m 42 s	+47°51'30"	UMa	11.0	3'.0×1'.8
NGC 3953	Glx	11 h 53 m 49 s	+52°19'35"	UMa	10.1	6'.6×3'.6
NGC 3962	Glx	11 h 54 m 40 s	−13°58'29"	Crt	10.6	2'.9×2'.6
NGC 3982	Glx	11 h 56 m 28 s	+55°07'29"	UMa	11.6	2'.5×2'.2
M 109	Glx	11 h 57 m 36 s	+53°23'00"	UMa	11.0	7'×4'
NGC 3998	Glx	11 h 57 m 56 s	+55°27'12"	UMa	10.6	3'.1×2'.5
NGC 4026	Glx	11 h 59 m 25 s	+50°57'42"	UMa	11.5	5'.1×1'.4
NGC 4027	Glx	11 h 59 m 30 s	−19°15'44"	Crv	11.1	3'.0×2'.3
NGC 4030	Glx	12 h 00 m 24 s	−01°06'00"	Vir	12.4	4'.3×3'.2
NGC 4036	Glx	12 h 01 m 27 s	+61°53'44"	UMa	10.7	4'.3×1'.7
NGC 4038	Glx	12 h 01 m 53 s	−18°52'07"	Crv	10.5	5'.2×3'.1
NGC 4041	Glx	12 h 02 m 12 s	+62°08'14"	UMa	11.3	2'.7×2'.5
NGC 4051	Glx	12 h 03 m 10 s	+44°31'53"	UMa	10.2	5'.2×3'.9
NGC 4085	Glx	12 h 05 m 23 s	+50°21'11"	UMa	12.4	2'.8×0'.8
NGC 4088	Glx	12 h 05 m 34 s	+50°32'23"	UMa	10.6	5'.8×2'.2
NGC 4102	Glx	12 h 06 m 23 s	+52°42'40"	UMa	11.8	3'.2×1'.9
NGC 4111	Glx	12 h 07 m 03 s	+43°03'57"	CVn	10.7	4'.6×1'.0
NGC 4143	Glx	12 h 09 m 36 s	+42°32'02"	CVn	10.7	2'.3×1'.4
NGC 4147	GC	12 h 10 m 06 s	+18°32'32"	Com	10.2	4'
NGC 4150	Glx	12 h 10 m 34 s	+30°24'06"	Com	11.7	2'.3×1'.6
NGC 4151	Glx	12 h 10 m 32 s	+39°24'21"	CVn	10.8	6'.3×4'.5
NGC 4179	Glx	12 h 12 m 52 s	+01°17'57"	Vir	11.0	4'.0×1'.1
NGC 4203	Glx	12 h 15 m 05 s	+33°11'52"	Com	10.9	3'.4×3'.2
NGC 4214	Glx	12 h 15 m 40 s	+36°19'35"	CVn	9.7	7'.9×6'.3
NGC 4216	Glx	12 h 15 m 54 s	+13°08'59"	Vir	10.0	8'.1×1'.8
NGC 4245	Glx	12 h 17 m 37 s	+29°36'28"	Com	11.4	2'.9×2'.2
NGC 4251	Glx	12 h 18 m 08 s	+28°10'30"	Com	11.5	4'.2×1'.9
M 106	Glx	12 h 18 m 54 s	+14°19'00"	CVn	9.5	19'×8'
NGC 4261	Glx	12 h 19 m 23 s	+05°49'30"	Vir	10.4	4'.1×3'.6
NGC 4273	Glx	12 h 19 m 56 s	+05°20'36"	Vir	11.9	2'.3×1'.5

(continued)

Table 9.1 (continued)

Object name	Type	RA (2000)	Dec (2000)	Const	Mag	Size
NGC 4274	Glx	12 h 19 m 50 s	+29°36'50"	Com	10.4	6'.8×2'.5
NGC 4278	Glx	12 h 20 m 07 s	+29°16'50"	Com	10.2	4'.1×3'.8
NGC 4281	Glx	12 h 20 m 22 s	+05°23'10"	Vir	11.3	3'.0×1'.5
NGC 4293	Glx	12 h 21 m 13 s	+18°22'56"	Com	10.4	5'.6×2'.6
M 61	Glx	12 h 36 m 54 s	+04°29'00"	Vir	10.5	6'×5'.5
NGC 4314	Glx	12 h 22 m 32 s	+29°53'43"	Com	10.6	4'.2×3'.7
NGC 4346	Glx	12 h 23 m 28 s	+46°59'37"	CVn	12.1	3'.3×1'.3
NGC 4350	Glx	12 h 23 m 58 s	+16°41'37"	Com	11.0	3'.0×1'.4
NGC 4361	PN	12 h 24 m 31 s	−18°47'05"	Crv	10.9	1'.9×1'.9
NGC 4365	Glx	12 h 24 m 28 s	+07°19'03"	Vir	9.6	6'.9×5'.0
NGC 4371	Glx	12 h 24 m 55 s	+11°42'14"	Vir	10.8	4'.0×2'.2
NGC 4394	Glx	12 h 25 m 54 s	+18°13'00"	Com	10.9	3'
NGC 4414	Glx	12 h 26 m 27 s	+31°13'24"	Com	10.3	3'.6×2'.2
NGC 4419	Glx	12 h 26 m 56 s	+15°02'50"	Com	11.0	3'.4×1'.3
NGC 4429	Glx	12 h 27 m 27 s	+11°06'26"	Vir	10.2	5'.5×2'.6
NGC 4435	Glx	12 h 27 m 41 s	+13°04'44"	Vir	10.8	3'.0×1'.9
NGC 4438	Glx	12 h 27 m 46 s	+13°00'31"	Vir	10.0	9'.3×3'.9
NGC 4442	Glx	12 h 28 m 04 s	+09°48'14"	Vir	10.4	4'.6×1'.9
NGC 4448	Glx	12 h 28 m 15 s	+28°37'14"	Com	11.1	4'.0×1'.6
NGC 4449	Glx	12 h 28 m 11 s	+44°05'36"	CVn	9.8	6'.2×4'.4
NGC 4450	Glx	12 h 28 m 30 s	+17°05'06"	Com	10.1	4'.8×3'.5
NGC 4459	Glx	12 h 29 m 00 s	+13°58'41"	Com	10.4	3'.8×2'.8
NGC 4473	Glx	12 h 29 m 49 s	+13°25'49"	Com	10.2	4'.5×2'.6
NGC 4477	Glx	12 h 30 m 02 s	+13°38'12"	Com	10.4	4'.0×3'.5
NGC 4478	Glx	12 h 30 m 17 s	+12°19'43"	Vir	11.2	2'.0×1'.8
NGC 4485	Glx	12 h 30 m 31 s	+41°42'03"	CVn	11.7	2'.4×1'.8
NGC 4490	Glx	12 h 30 m 36 s	+41°38'34"	CVn	9.5	6'.4×3'.2
NGC 4494	Glx	12 h 31 m 24 s	+25°46'31"	Com	9.9	4'.8×3'.8
NGC 4526	Glx	12 h 34 m 03 s	+07°41'58"	Vir	9.6	7'.2×2'.3
NGC 4527	Glx	12 h 34 m 08 s	+02°39'09"	Vir	10.4	6'.3×2'.3
NGC 4535	Glx	12 h 34 m 20 s	+08°11'52"	Vir	10.0	6'.8×5'.0
NGC 4536	Glx	12 h 34 m 27 s	+02°11'16"	Vir	10.4	7'.4×3'.5
NGC 4546	Glx	12 h 35 m 29 s	−03°47'38"	Vir	10.3	3'.5×1'.7
M 91	Glx	12 h 35 m 30 s	+14°30'00"	Com	11.5	5'×4'
NGC 4550	Glx	12 h 35 m 30 s	+12°13'14"	Vir	11.5	3'.5×1'.1
NGC 4559	Glx	12 h 35 m 58 s	+27°57'35"	Com	9.8	10'.7×4'.4
NGC 4565	Glx	12 h 36 m 21 s	+25°59'19"	Com	9.5	15'.8×2'.1
NGC 4570	Glx	12 h 36 m 54 s	+07°14'47"	Vir	10.8	4'.1×1'.3
M 104	Glx	12 h 39 m 59 s	−11°37'23"	Vir	7.9	8'.7×3'.5
NGC 4596	Glx	12 h 39 m 56 s	+10°10'33"	Vir	10.5	3'.9×2'.8
NGC 4618	Glx	12 h 41 m 32 s	+41°09'02"	CVn	10.8	4'.4×3'.8
NGC 4631	Glx	12 h 42 m 08 s	+32°32'26"	CVn	9.1	15'.5×2'.7
NGC 4636	Glx	12 h 42 m 50 s	+02°41'16"	Vir	9.6	6'.2×5'.0
NGC 4643	Glx	12 h 43 m 20 s	+01°58'42"	Vir	10.6	3'.4×2'.7
NGC 4654	Glx	12 h 43 m 56 s	+13°07'32"	Vir	10.5	4'.7×3'.0

(continued)

Table 9.1 (continued)

Object name	Type	RA (2000)	Dec (2000)	Const	Mag	Size
NGC 4656	Glx	12 h 43 m 58 s	+32°10'13"	CVn	10.3	15'.3×2'.4
NGC 4660	Glx	12 h 44 m 32 s	+11°11'26"	Vir	10.9	2'.8×1'.9
NGC 4665	Glx	12 h 45 m 06 s	+03°03'27"	Vir	12.4	3'.5×3'.5
NGC 4666	Glx	12 h 45 m 08 s	−00°27'44"	Vir	10.8	4'.5×1'.5
NGC 4689	Glx	12 h 47 m 46 s	+13°45'44"	Com	10.9	4'.0×3'.5
NGC 4697	Glx	12 h 48 m 36 s	−05°48'03"	Vir	9.2	7'.2×4'.7
NGC 4698	Glx	12 h 48 m 23 s	+08°29'16"	Vir	10.5	4'.3×2'.5
NGC 4699	Glx	12 h 49 m 02 s	−08°39'52"	Vir	9.6	3'.5×2'.7
NGC 4725	Glx	12 h 50 m 26 s	+25°30'01"	Com	9.2	11'.0×7'.9
NGC 4753	Glx	12 h 52 m 22 s	−01°12'16"	Vir	9.9	5'.4×2'.9
NGC 4754	Glx	12 h 52 m 18 s	+11°18'49"	Vir	10.5	4'.7×2'.6
NGC 4762	Glx	12 h 52 m 56 s	+11°13'52"	Vir	10.2	8'.7×1'.6
NGC 4781	Glx	12 h 54 m 23 s	−10°32'09"	Vir	12.8	3'.5×1'.8
NGC 4800	Glx	12 h 54 m 38 s	+46°31'52"	CVn	12.0	1'.8×1'.4
NGC 4845	Glx	12 h 58 m 01 s	+01°34'32"	Vir	12.9	5'.0×1'.6
NGC 4856	Glx	12 h 59 m 21 s	−15°02'32"	Vir	10.4	4'.6×1'.6
NGC 4866	Glx	12 h 59 m 27 s	+14°10'17"	Vir	11.2	6'.5×1'.5
NGC 4900	Glx	13 h 00 m 39 s	+02°29'59"	Vir	11.5	2'.3×2'.2
NGC 4958	Glx	13 h 05 m 49 s	−08°01'15"	Vir	10.5	4'.1×1'.4
NGC 4995	Glx	13 h 09 m 41 s	−07°49'59"	Vir	11.0	2'.5×1'.7
NGC 5005	Glx	13 h 10 m 57 s	+37°03'33"	CVn	10.3	5'.8×2'.8
NGC 5033	Glx	13 h 13 m 28 s	+36°35'36"	CVn	10.7	10'.0×5'.6
NGC 5054	Glx	13 h 16 m 58 s	−16°38'07"	Vir	12.0	5'.0×3'.1
NGC 5195	Glx	13 h 30 m 00 s	+14°16'00"	CVn	9.6	6'×5'
NGC 5248	Glx	13 h 37 m 32 s	+08°53'08"	Boo	10.0	6'.2×4'.5
NGC 5273	Glx	13 h 42 m 08 s	+35°39'15"	CVn	11.6	2'.8×2'.5
NGC 5322	Glx	13 h 49 m 15 s	+60°11'25"	UMa	10.1	6'.0×4'.1
NGC 5363	Glx	13 h 56 m 07 s	+05°15'19"	Vir	10.2	4'.2×2'.7
NGC 5364	Glx	13 h 56 m 12 s	+05°00'52"	Vir	11.2	6'.8×4'.4
NGC 5466	GC	14 h 05 m 27 s	+28°32'04"	Boo	9.0	11'
NGC 5473	Glx	14 h 04 m 43 s	+54°53'32"	UMa	11.4	2'.6×1'.8
NGC 5474	Glx	14 h 05 m 01 s	+53°39'44"	UMa	10.9	4'.5×4'.2
NGC 5557	Glx	14 h 18 m 26 s	+36°29'36"	Boo	11.1	2'.4×2'.2
NGC 5566	Glx	14 h 20 m 20 s	+03°56'02"	Vir	10.5	6'.5×2'.4
NGC 5576	Glx	14 h 21 m 04 s	+03°16'15"	Vir	10.9	3'.2×2'.2
NGC 5361	Glx	13 h 54 m 35 s	+38°26'58"	CVn	13.9	0'.8×0'.4
NGC 5634	GC	14 h 29 m 37 s	−05°58'36"	Vir	9.4	4'.9
NGC 5676	Glx	14 h 32 m 47 s	+49°27'26"	Boo	10.9	3'.9×2'.0
NGC 5689	Glx	14 h 35 m 30 s	+48°44'29"	Boo	11.9	3'.7×1'.2
NGC 5694	GC	14 h 39 m 36 s	−26°32'18"	Hya	10.2	3'.6
NGC 5746	Glx	14 h 44 m 56 s	+01°57'21"	Vir	10.5	7'.9×1'.7
NGC 5846	Glx	15 h 06 m 29 s	+01°36'22"	Vir	10.2	3'.4×3'.2
M 102	Glx	15 h 06 m 30 s	+55°45'47"	Dra	10.5	6.5'×3.0'
NGC 5897	GC	15 h 17 m 24 s	−21°00'37"	Lib	8.6	13'
NGC 5907	Glx	15 h 15 m 54 s	+56°19'45"	Dra	11.4	12'.8×1'.8

(continued)

Table 9.1 (continued)

Object name	Type	RA (2000)	Dec (2000)	Const	Mag	Size
NGC 5982	Glx	15 h 38 m 40 s	+59°21'21"	Dra	11.1	2'.9×2'.2
NGC 6118	Glx	16 h 21 m 48 s	−02°17'01"	Ser	11.6	4'.7×2'.0
NGC 6144	GC	16 h 27 m 14 s	−26°01'29"	Sco	9.0	9'.3
M 107	GC	16 h 32 m 32 s	−13°03'13"	Oph	8.9	10'
NGC 6207	Glx	16 h 43 m 04 s	+36°49'56"	Her	11.6	3'.0×1'.4
NGC 6217	Glx	16 h 32 m 40 s	+78°11'51"	UMi	11.2	3'.1×2'.7
NGC 6229	GC	16 h 46 m 59 s	+47°31'40"	Her	9.4	4'.5
NGC 6235	GC	16 h 53 m 25 s	−22°10'38"	Oph	10.0	5'.0
NGC 6284	GC	17 h 04 m 29 s	−24°45'53"	Oph	8.8	5'.6
NGC 6287	GC	17 h 05 m 09 s	−22°42'29"	Oph	9.3	5'.1
NGC 6293	GC	17 h 10 m 10 s	−26°34'57"	Oph	8.2	7'.9
NGC 6304	GC	17 h 14 m 32 s	−29°27'44"	Oph	8.4	6'.8
NGC 6316	GC	17 h 16 m 37 s	−28°08'24"	Oph	8.8	4'.9
NGC 6342	GC	17 h 21 m 10 s	−19°35'14"	Oph	9.8	4'.4
NGC 6355	GC	17 h 23 m 59 s	−26°21'13"	Oph	9.7	5'.0
NGC 6356	GC	17 h 23 m 36 s	−17°49'00"	Oph	8.9	7'.2
NGC 6369	PN	17 h 29 m 20 s	−23°45'35"	Oph	11.4	58"×34"
NGC 6401	GC	17 h 38 m 36 s	−23°55'00"	Oph	9.5	5'.6
NGC 6426	GC	17 h 44 m 54 s	+03°00'00"	Oph	11.1	3'.2
NGC 6440	GC	17 h 48 m 53 s	−20°21'34"	Sgr	9.3	5.4'
NGC 6445	PN	17 h 49 m 15 s	−20°00'34"	Sgr	11.2	2'.8×0'.9
NGC 6451	OC	17 h 50 m 42 s	−30°13'00"	Sco	8.2	7'
M 20	NbOC	18 h 02 m 23 s	−23°01'48"	Sgr	6.3	30'×20'
NGC 6517	GC	18 h 01 m 48 s	−08°58'00"	Oph	10.3	4'.3
NGC 6520	OC	18 h 03 m 25 s	−27°53'28"	Sgr	7.6	6'
NGC 6522	GC	18 h 03 m 36 s	−30°02'00"	Sgr	8.6	5'.6
NGC 6528	GC	18 h 04 m 48 s	−30°03'00"	Sgr	9.5	3'.7
NGC 6540	GC	18 h 06 m 08 s	−27°45'55"	Sgr	14.6	1'.5
NGC 6543	PN	17 h 58 m 33 s	+66°37'59"	Dra	8.1	23"×17"
NGC 6544	GC	18 h 07 m 18 s	−25°00'00"	Sgr	8.3	8'.9
NGC 6553	GC	18 h 09 m 18 s	−25°54'28"	Sgr	8.1	9'.2
NGC 6568	OC	18 h 12 m 48 s	−21°36'00"	Sgr	8.6	12'
NGC 6569	GC	18 h 13 m 36 s	−31°50'00"	Sgr	8.7	5'.8
NGC 6583	OC	18 h 15 m 48 s	−22°08'00"	Sgr	10.0	2'.8
NGC 6624	GC	18 h 23 m 42 s	−30°22'00"	Sgr	9.1	5'.9
NGC 6629	PN	18 h 25 m 42 s	−23°12'00"	Sgr	11.6	16"×14"
NGC 6633	OC	18 h 27 m 15 s	+06°30'30"	Oph	4.6	20'
NGC 6638	GC	18 h 30 m 54 s	−25°30'00"	Sgr	9.2	5'
NGC 6642	GC	18 h 31 m 54 s	−23°29'00"	Sgr	8.8	4'.5
NGC 6645	OC	18 h 32 m 36 s	−16°45'00"	Sgr	8.5	10'
NGC 6664	OC	18 h 36 m 42 s	−08°13'00"	Sct	7.8	16'
NGC 6712	GC	18 h 53 m 06 s	−08°42'00"	Sct	8.2	7'.2
NGC 6755	OC	19 h 07 m 48 s	+04°14'00"	Aql	7.5	14'
NGC 6756	OC	19 h 08 m 42 s	+04°41'00"	Aql	10.6	4'
NGC 6781	PN	19 h 18 m 30 s	+06°32'00"	Aql	11.4	1'.9×1'.8

(continued)

Table 9.1 (continued)

Object name	Type	RA (2000)	Dec (2000)	Const	Mag	Size
NGC 6802	OC	19 h 30 m 35 s	+20°15'39"	Vul	8.8	3'.2
NGC 6818	PN	19 h 43 m 58 s	−14°09'12"	Sgr	9.3	22"×15"
NGC 6823	OC	19 h 43 m 06 s	+23°18'00"	Vul	7.1	12'
NGC 6826	PN	19 h 44 m 48 s	+50°31'30"	Cyg	8.8	27"×24"
NGC 6830	OC	19 h 51 m 00 s	+23°04'00"	Vul	7.9	12'
NGC 6834	OC	19 h 52 m 12 s	+29°25'00"	Cyg	7.8	5'
NGC 6866	OC	20 h 03 m 42 s	+44°00'00"	Cyg	7.6	6'
Collinder	OC	20 h 11 m 35 s	+26°32'04"	Vul	5.6	8'
NGC 6885	OC	20 h 12 m 01 s	+26°28'42"	Vul	8.1	20'
NGC 6905	PN	20 h 22 m 23 s	+20°06'17"	Del	11.1	42"×35"
NGC 6910	OC	20 h 23 m 06 s	+40°47'00"	Cyg	7.4	7'
NGC 6934	GC	20 h 34 m 12 s	+07°24'16"	Del	8.9	7'.0
NGC 6939	OC	20 h 31 m 24 s	+60°38'00"	Cep	7.8	7'
NGC 6940	OC	20 h 34 m 27 s	+28°16'58"	Vul	6.3	31'
NGC 6946	Glx	20 h 34 m 53 s	+60°09'11"	Cep	9.1	11'.5×9'.8
NGC 7000	NbEM	21 h 01 m 48 s	+44°12'00"	Cyg	4.5	100'×60'
NGC 7006	GC	21 h 01 m 29 s	+16°11'16"	Del	10.6	2'.8
NGC 7008	PN	21 h 00 m 33 s	+54°32'35"	Cyg	10.7	98"×75"
NGC 7009	PN	21 h 04 m 11 s	−11°21'49"	Aqr	8.0	44"×23"
NGC 7044	OC	21 h 13 m 09 s	+42°29'46"	Cyg	12.0	3'.5
NGC 7062	OC	21 h 23 m 27 s	+46°22'43"	Cyg	8.3	6'
NGC 7086	OC	21 h 30 m 28 s	+51°36'08"	Cyg	8.4	9'
NGC 7128	OC	21 h 43 m 58 s	+53°42'55"	Cyg	9.7	3'.1
NGC 7142	OC	21 h 45 m 09 s	+65°46'28"	Cep	9.3	4'.3
NGC 7160	OC	21 h 53 m 40 s	+62°36'12"	Cep	6.1	7'
NGC 7209	OC	22 h 05 m 12 s	+46°30'00"	Lac	7.7	25'
NGC 7217	Glx	22 h 07 m 54 s	+31°22'00"	Peg	10.2	3'.7×3'.2
NGC 7243	OC	22 h 15 m 08 s	+49°53'51"	Lac	6.4	21'
NGC 7296	OC	22 h 28 m 12 s	+52°17'00"	Lac	9.7	4'
NGC 7331	Glx	22 h 37 m 05 s	+34°25'01"	Peg	9.7	10'.5×3'.7
NGC 7380	NbOC	22 h 47 m 00 s	+58°06'00"	Cep	7.2	12'
NGC 7448	Glx	23 h 00 m 06 s	+15°59'00"	Peg	11.7	2'.7×1'.3
NGC 7479	Glx	23 h 04 m 56 s	+12°19'00"	Peg	11.0	4'.1×3'.2
NGC 7510	OC	23 h 11 m 04 s	+60°34'15"	Cep	7.9	7'
NGC 7606	Glx	23 h 19 m 06 s	−08°29'00"	Aqr	10.8	5'.8×2'.6
NGC 7662	PN	23 h 25 m 54 s	+42°32'06"	And	8.3	32"×28"
NGC 7686	OC	23 h 30 m 12 s	+49°08'00"	And	5.6	14'
NGC 7723	Glx	23 h 38 m 54 s	−12°58'00"	Aqr	11.1	3'.6×2'.6
NGC 7727	Glx	23 h 39 m 54 s	−12°18'00"	Aqr	10.7	4'.2×3'.4
NGC 7789	OC	23 h 57 m 00 s	+56°44'00"	Cas	6.7	16'
NGC 7790	OC	23 h 58 m 24 s	+61°13'00"	Cas	8.5	17'

The Herschel II list was created by the Rose City Astronomers of Portland, Oregon, in the Pacific Northwest. Many of their members had completed the Messier Observing List as well as the Herschel 400 list and were looking for a new challenge. A collaboration among these members led to the Herschel II list as a new challenge (Table 9.2).

Table 9.2 The objects of the Herschel II list (With permission from Carol Cole Huston and Candace Pratt)

Object name	Object type	Const	V Mag	RA (2000)	Dec (2000)
NGC 23	Galaxy	Pegasus	12.0	00 h 09 m 53.0 s	+25° 55' 13"
NGC 24	Galaxy	Sculptor	11.6	00 h 09 m 55.7 s	−24° 57' 39"
NGC 125	Galaxy	Pisces	12.1	00 h 28 m 49.7 s	+02° 50' 18"
NGC 151	Galaxy	Cetus	11.6	00 h 34 m 02.0 s	−09° 42' 22"
NGC 175	Galaxy	Cetus	12.2	00 h 37 m 21.1 s	−19° 56' 11"
NGC 198	Galaxy	Pisces	13.1	00 h 39 m 22.3 s	+02° 47' 46"
NGC 206	Bright nebula	Andromeda		00 h 40 m 35.3 s	+40° 43' 47"
NGC 214	Galaxy	Andromeda	12.3	00 h 41 m 27.5 s	+25° 29' 49"
NGC 217	Galaxy	Cetus		00 h 41 m 33.1 s	−10° 01' 22"
NGC 315	Galaxy	Pisces	11.2	00 h 57 m 48.3 s	+30° 20' 59"
NGC 337	Galaxy	Cetus	11.6	00 h 59 m 49.1 s	−07° 34' 44"
NGC 357	Galaxy	Cetus	12.0	01 h 03 m 21.0 s	−06° 20' 26"
NGC 410	Galaxy	Pisces	11.5	01 h 10 m 57.5 s	+33° 08' 55"
NGC 428	Galaxy	Cetus	11.5	01 h 12 m 54.3 s	+00° 58' 54"
NGC 499	Galaxy	Pisces	12.1	01 h 23 m 10.6 s	+33° 27' 25"
NGC 513	Galaxy	Andromeda	12.9	01 h 24 m 26.0 s	+33° 47' 45"
NGC 514	Galaxy	Pisces	11.6	01 h 24 m 03.1 s	+12° 54' 52"
NGC 604	Bright nebula	Triangulum		01 h 34 m 29.1 s	+30° 47' 49"
NGC 636	Galaxy	Cetus	11.5	01 h 39 m 05.6 s	−07° 30' 50"
NGC 660	Galaxy	Pisces	11.2	01 h 43 m 00.5 s	+13° 38' 29"
NGC 665	Galaxy	Pisces	12.1	01 h 44 m 55.1 s	+10° 25' 14"
NGC 672	Galaxy	Triangulum	10.9	01 h 47 m 52.5 s	+27° 25' 51"
NGC 706	Galaxy	Pisces	12.5	01 h 51 m 49.3 s	+06° 17' 37"
NGC 718	Galaxy	Pisces	11.7	01 h 53 m 11.5 s	+04° 11' 41"
NGC 741	Galaxy	Pisces	11.1	01 h 56 m 19.9 s	+05° 37' 38"
NGC 821	Galaxy	Aries	10.7	02 h 08 m 20.0 s	+10° 59' 36"
NGC 890	Galaxy	Triangulum	11.2	02 h 21 m 59.9 s	+33° 15' 47"
NGC 896	Bright nebula	Cassiopeia		02 h 26 m 34.1 s	+62° 06' 49"
NGC 925	Galaxy	Triangulum	10.1	02 h 27 m 15.7 s	+33° 34' 32"
NGC 991	Galaxy	Cetus	11.7	02 h 35 m 31.5 s	−07° 09' 26"
NGC 1003	Galaxy	Perseus	11.4	02 h 39 m 15.2 s	+40° 52' 13"
NGC 1012	Galaxy	Aries	12.0	02 h 39 m 13.9 s	+30° 09' 01"
NGC 1032	Galaxy	Cetus	11.6	02 h 39 m 22.5 s	+01° 05' 29"
NGC 1035	Galaxy	Cetus	12.2	02 h 39 m 28.1 s	−08° 08' 07"
NGC 1045	Galaxy	Cetus	13.0	02 h 40 m 28.0 s	−11° 16' 47"
NGC 1058	Galaxy	Perseus	11.2	02 h 43 m 28.7 s	+37° 20' 21"
NGC 1060	Galaxy	Triangulum	11.8	02 h 43 m 13.9 s	+32° 25' 20"

(continued)

Table 9.2 (continued)

Object name	Object type	Const	V Mag	RA (2000)	Dec (2000)
NGC 1070	Galaxy	Cetus	11.9	02 h 43 m 21.2 s	+04° 57' 56"
NGC 1073	Galaxy	Cetus	11.0	02 h 43 m 39.2 s	+01° 22' 29"
NGC 1087	Galaxy	Cetus	10.9	02 h 46 m 23.8 s	−00° 29' 53"
NGC 1090	Galaxy	Cetus	11.8	02 h 46 m 32.3 s	−00° 14' 54"
NGC 1114	Galaxy	Eridanus	13.0	02 h 49 m 05.8 s	−16° 59' 41"
NGC 1156	Galaxy	Aries	11.7	02 h 59 m 41.4 s	+25° 14' 06"
NGC 1161	Galaxy	Perseus	11.0	03 h 01 m 13.0 s	+44° 53' 42"
NGC 1162	Galaxy	Eridanus	13.0	02 h 58 m 54.7 s	−12° 24' 02"
NGC 1169	Galaxy	Perseus	11.3	03 h 03 m 33.6 s	+46° 22' 55"
NGC 1172	Galaxy	Eridanus	11.9	03 h 01 m 34.8 s	−14° 50' 18"
NGC 1175	Galaxy	Perseus	12.9	03 h 04 m 31.3 s	+42° 20' 14"
NGC 1184	Galaxy	Cepheus	12.4	03 h 16 m 37.8 s	+80° 47' 30"
NGC 1187	Galaxy	Eridanus	10.8	03 h 02 m 36.2 s	−22° 52' 09"
NGC 1193	Open cluster	Perseus	12.6	03 h 05 m 46.6 s	+44° 22' 51"
NGC 1199	Galaxy	Eridanus	11.4	03 h 03 m 37.2 s	−15° 36' 55"
NGC 1207	Galaxy	Perseus	12.6	03 h 08 m 13.8 s	+38° 22' 45"
NGC 1209	Galaxy	Eridanus	11.4	03 h 06 m 01.8 s	−15° 36' 48"
NGC 1325	Galaxy	Eridanus	11.5	03 h 24 m 23.8 s	−21° 32' 39"
NGC 1332	Galaxy	Eridanus	10.3	03 h 26 m 16.0 s	−21° 20' 15"
NGC 1348	Open cluster	Perseus		03 h 33 m 46.3 s	+51° 25' 53"
NGC 1353	Galaxy	Eridanus	11.4	03 h 32 m 01.8 s	−20° 49' 11"
NGC 1400	Galaxy	Eridanus	11.0	03 h 39 m 29.9 s	−18° 41' 26"
NGC 1421	Galaxy	Eridanus	11.4	03 h 42 m 28.2 s	−13° 29' 25"
NGC 1491	Bright nebula	Perseus		04 h 03 m 16.3 s	+51° 17' 54"
NGC 1507	Galaxy	Eridanus	12.3	04 h 04 m 26.1 s	−02° 11' 25"
NGC 1514	Planetary nebula	a Taurus	10.8	04 h 09 m 13.4 s	+30° 46' 45"
NGC 1579	Bright nebula	Perseus		04 h 30 m 04.7 s	+35° 15' 54"
NGC 1582	Open cluster	Perseus	7.0	04 h 31 m 58.5 s	+43° 50' 55"
NGC 1587	Galaxy	Taurus	11.7	04 h 30 m 38.4 s	+00° 39' 35"
NGC 1600	Galaxy	Eridanus	10.9	04 h 31 m 38.7 s	−05° 05' 25"
NGC 1605	Open cluster	Perseus	10.7	04 h 34 m 58.5 s	+45° 14' 55"
NGC 1618	Galaxy	Eridanus	12.7	04 h 36 m 05.5 s	−03° 09' 06"
NGC 1624	Bright nebula	Perseus	11.8	04 h 40 m 22.3 s	+50° 26' 56"
NGC 1637	Galaxy	Eridanus	10.8	04 h 41 m 27.0 s	−02° 51' 39"
NGC 1662	Open cluster	Orion	6.4	04 h 48 m 28.9 s	+10° 55' 53"
NGC 1663	Open cluster	Orion		04 h 48 m 34.9 s	+13° 09' 53"
NGC 1700	Galaxy	Eridanus	11.2	04 h 56 m 55.1 s	−04° 52' 12"
NGC 1750	Open cluster	Taurus		05 h 03 m 52.8 s	+23° 38' 54"
NGC 1762	Galaxy	Orion	12.6	05 h 03 m 35.5 s	+01° 34' 16"
NGC 1778	Open cluster	Auriga	7.7	05 h 08 m 04.7 s	+37° 02' 56"
NGC 1779	Galaxy	Eridanus	12.1	05 h 05 m 16.9 s	−09° 09' 00"
NGC 1832	Galaxy	Lepus	11.3	05 h 12 m 02.2 s	−15° 41' 26"
NGC 1883	Open cluster	Auriga	12.0	05 h 25 m 52.5 s	+46° 32' 58"
NGC 1977	Open cluster	Orion		05 h 35 m 22.9 s	−04° 50' 09"
NGC 1990	Bright nebula	Orion		05 h 36 m 16.9 s	−01° 12' 09"

(continued)

Table 9.2 (continued)

Object name	Object type	Const	V Mag	RA (2000)	Dec (2000)
NGC 2023	Bright nebula	Orion		05 h 41 m 35.0 s	−02° 13' 09"
NGC 2071	Bright nebula	Orion		05 h 47 m 11.0 s	+00° 17' 51'
NGC 2112	Open cluster	Orion	9.1	05 h 53 m 53.0 s	+00° 23' 51"
NGC 2139	Galaxy	Lepus	11.6	06 h 01 m 06.9 s	−23° 40' 46"
NGC 2170	Bright nebula	Monoceros		06 h 07 m 35.0 s	−06° 23' 10"
NGC 2182	Bright nebula	Monoceros		06 h 09 m 29.0 s	−06° 19' 10"
NGC 2192	Open cluster	Auriga	10.9	06 h 15 m 10.8 s	+39° 50' 59"
NGC 2196	Galaxy	Lepus	11.0	06 h 12 m 09.0 s	−21° 48' 37"
NGC 2236	Open cluster	Monoceros	8.5	06 h 29 m 41.1 s	+06° 49' 53"
NGC 2245	Bright nebula	Monoceros		06 h 32 m 41.1 s	+10° 09' 53"
NGC 2252	Open cluster	Monoceros	7.7	06 h 34 m 59.1 s	+05° 22' 52"
NGC 2253	Open cluster	Camelopardalis		06 h 42 m 21.7 s	+66° 20' 06"
NGC 2254	Open cluster	Monoceros	9.1	06 h 35 m 59.1 s	+07° 39' 53"
NGC 2259	Open cluster	Monoceros	10.8	06 h 38 m 35.1 s	+10° 52' 54"
NGC 2261	Bright nebula	Monoceros		06 h 39 m 11.1 s	+08° 43' 53"
NGC 2269	Open cluster	Monoceros	10.0	06 h 43 m 53.2 s	+04° 33' 52"
NGC 2274	Galaxy	Gemi	12.1	06 h 47 m 16.2 s	+33° 34' 00"
NGC 2283	Galaxy	Canis Major	12.2	06 h 45 m 52.4 s	−18° 12' 52"
NGC 2302	Open cluster	Monoceros	8.9	06 h 51 m 53.2 s	−07° 04' 11"
NGC 2309	Open cluster	Monoceros	10.5	06 h 56 m 11.2 s	−07° 12' 11"
NGC 2316	Bright nebula	Monoceros		06 h 59 m 41.2 s	−07° 44' 11"
NGC 2331	Open cluster	Gemi	8.5	07 h 07 m 11.1 s	+27° 20' 58"
NGC 2339	Galaxy	Gemi	11.8	07 h 08 m 19.9 s	+18° 46' 50"
NGC 2346	Planetary nebula	a Monoceros	12.5	07 h 09 m 21.9 s	−00° 48' 33"
NGC 2347	Galaxy	Camelopardalis	12.4	07 h 16 m 02.4 s	+64° 42' 44"
NGC 2359	Bright nebula	Canis Major		07 h 17 m 47.3 s	−13° 13' 12"
NGC 2366	Galaxy	Camelopardalis	11.1	07 h 28 m 52.1 s	+69° 13' 01"
NGC 2367	Open cluster	Canis Major	7.9	07 h 20 m 05.3 s	−21° 56' 14"
NGC 2374	Open cluster	Canis Major	8.0	07 h 23 m 59.3 s	−13° 16' 12"
NGC 2396	Open cluster	Puppis	7.4	07 h 28 m 05.3 s	−11° 44' 12"
NGC 2414	Open cluster	Puppis	7.9	07 h 33 m 17.4 s	−15° 27' 13"
NGC 2415	Galaxy	Lynx	12.4	07 h 36 m 55.9 s	+35° 14' 35"
NGC 2432	Open cluster	Puppis	10.2	07 h 40 m 53.4 s	−19° 05' 14"
NGC 2467	Bright nebula	Puppis	7.1	07 h 52 m 29.4 s	−26° 23' 16"
NGC 2493	Galaxy	Lynx	12.0	08 h 00 m 23.0 s	+39° 49' 52"
NGC 2500	Galaxy	Lynx	11.6	08 h 01 m 52.6 s	+50° 44' 24"
NGC 2525	Galaxy	Puppis	11.6	08 h 05 m 37.6 s	−11° 25' 53"
NGC 2541	Galaxy	Lynx	11.8	08 h 14 m 39.4 s	+49° 03' 51"
NGC 2610	Planetary nebul	a Hydra	13.0	08 h 33 m 23.0 s	−16° 09' 13"
NGC 2639	Galaxy	Ursa Major	11.6	08 h 43 m 37.3 s	+50° 12' 31"
NGC 2756	Galaxy	Ursa Major	12.4	09 h 09 m 01.0 s	+53° 51' 05"
NGC 2765	Galaxy	Hydra	12.1	09 h 07 m 36.3 s	+03° 23' 32"
NGC 2781	Galaxy	Hydra	11.6	09 h 11 m 27.5 s	−14° 49' 14"
NGC 2784	Galaxy	Hydra	10.2	09 h 12 m 18.6 s	−24° 10' 38"
NGC 2805	Galaxy	Ursa Major	11.0	09 h 20 m 19.7 s	+64° 06' 24"

(continued)

Table 9.2 (continued)

Object name	Object type	Const	V Mag	RA (2000)	Dec (2000)
NGC 2855	Galaxy	Hydra	11.6	09 h 21 m 27.4 s	−11° 54' 47"
NGC 2880	Galaxy	Ursa Major	11.5	09 h 29 m 34.5 s	+62° 29' 37"
NGC 2889	Galaxy	Hydra	11.7	09 h 27 m 12.6 s	−11° 38' 49"
NGC 2986	Galaxy	Hydra	10.8	09 h 44 m 16.4 s	−21° 16' 57"
NGC 3065	Galaxy	Ursa Major	12.5	10 h 01 m 52.7 s	+72° 10' 27"
NGC 3067	Galaxy	Leo	12.1	09 h 58 m 21.9 s	+32° 22' 17"
NGC 3073	Galaxy	Ursa Major	13.4	10 h 00 m 51.7 s	+55° 37' 24"
NGC 3078	Galaxy	Hydra	11.1	09 h 58 m 24.8 s	−26° 55' 50"
NGC 3107	Galaxy	Leo	13.4	10 h 04 m 22.0 s	+13° 37' 14"
NGC 3145	Galaxy	Hydra	11.7	10 h 10 m 10.2 s	−12° 26' 13"
NGC 3156	Galaxy	Sextans	12.3	10 h 12 m 41.3 s	+03° 07' 44"
NGC 3158	Galaxy	Leo Mor	11.9	10 h 13 m 50.8 s	+38° 46' 01"
NGC 3162	Galaxy	Leo	11.6	10 h 13 m 32.1 s	+22° 44' 24"
NGC 3177	Galaxy	Leo	12.4	10 h 16 m 34.6 s	+21° 07' 29"
NGC 3225	Galaxy	Ursa Major	12.6	10 h 25 m 10.4 s	+58° 09' 13"
NGC 3254	Galaxy	Leo Mor	11.7	10 h 29 m 20.2 s	+29° 29' 31"
NGC 3274	Galaxy	Leo	12.8	10 h 32 m 16.9 s	+27° 40' 16"
NGC 3301	Galaxy	Leo	11.4	10 h 36 m 55.9 s	+21° 52' 59"
NGC 3319	Galaxy	Ursa Major	11.1	10 h 39 m 10.2 s	+41° 41' 25"
NGC 3338	Galaxy	Leo	11.1	10 h 42 m 07.8 s	+13° 44' 50"
NGC 3359	Galaxy	Ursa Major	10.6	10 h 46 m 37.9 s	+63° 13' 35"
NGC 3424	Galaxy	Leo Mor	12.4	10 h 51 m 46.9 s	+32° 54' 06"
NGC 3430	Galaxy	Leo Mor	11.6	10 h 52 m 11.2 s	+32° 57' 14"
NGC 3507	Galaxy	Leo	10.9	11 h 03 m 26.0 s	+18° 08' 14"
NGC 3511	Galaxy	Crater	11.0	11 h 03 m 24.4 s	−23° 05' 24"
NGC 3513	Galaxy	Crater	11.5	11 h 03 m 46.4 s	−23° 14' 54"
NGC 3516	Galaxy	Ursa Major	11.7	11 h 06 m 47.9 s	+72° 34' 27"
NGC 3524	Galaxy	Leo	12.8	11 h 06 m 32.7 s	+11° 23' 02"
NGC 3547	Galaxy	Leo	12.8	11 h 09 m 55.9 s	+10° 43' 20"
NGC 3583	Galaxy	Ursa Major	11.1	11 h 14 m 12.0 s	+48° 19' 22"
NGC 3585	Galaxy	Hydra	9.9	11 h 13 m 17.6 s	−26° 45' 34"
NGC 3596	Galaxy	Leo	11.3	11 h 15 m 05.7 s	+14° 47' 15"
NGC 3599	Galaxy	Leo	11.9	11 h 15 m 27.8 s	+18° 06' 46"
NGC 3605	Galaxy	Leo	12.3	11 h 16 m 47.2 s	+18° 01' 04"
NGC 3611	Galaxy	Leo	12.1	11 h 17 m 30.1 s	+04° 33' 13"
NGC 3622	Galaxy	Ursa Major	13.2	11 h 20 m 13.4 s	+67° 14' 41"
NGC 3636	Galaxy	Crater	12.4	11 h 20 m 25.9 s	−10° 17' 04"
NGC 3637	Galaxy	Crater	12.7	11 h 20 m 40.3 s	−10° 15' 36"
NGC 3642	Galaxy	Ursa Major	11.2	11 h 22 m 19.0 s	+59° 04' 46"
NGC 3646	Galaxy	Leo	11.1	11 h 21 m 43.8 s	+20° 10' 17"
NGC 3652	Galaxy	Ursa Major	12.2	11 h 22 m 40.0 s	+37° 46' 06"
NGC 3659	Galaxy	Leo	12.3	11 h 23 m 45.7 s	+17° 48' 59"
NGC 3666	Galaxy	Leo	12.0	11 h 24 m 26.2 s	+11° 20' 32"
NGC 3668	Galaxy	Ursa Major	12.3	11 h 25 m 32.0 s	+63° 26' 54"
NGC 3669	Galaxy	Ursa Major	12.4	11 h 25 m 27.9 s	+57° 43' 33"

(continued)

Table 9.2 (continued)

Object name	Object type	Const	V Mag	RA (2000)	Dec (2000)
NGC 3672	Galaxy	Crater	11.4	11 h 25 m 03.1 s	−09° 47' 52"
NGC 3681	Galaxy	Leo	11.2	11 h 26 m 30.0 s	+16° 51' 51"
NGC 3682	Galaxy	Draco	12.5	11 h 27 m 43.4 s	+66° 35' 39"
NGC 3683	Galaxy	Ursa Major	12.4	11 h 27 m 32.6 s	+56° 52' 54"
NGC 3689	Galaxy	Leo	12.3	11 h 28 m 10.8 s	+25° 39' 46"
NGC 3693	Galaxy	Crater	13.0	11 h 28 m 12.4 s	−13° 11' 53"
NGC 3705	Galaxy	Leo	11.1	11 h 30 m 07.4 s	+09° 16' 33"
NGC 3732	Galaxy	Crater	12.5	11 h 34 m 14.7 s	−09° 50' 54"
NGC 3756	Galaxy	Ursa Major	11.5	11 h 36 m 49.2 s	+54° 17' 57"
NGC 3887	Galaxy	Crater	10.6	11 h 47 m 05.5 s	−16° 51' 27"
NGC 3892	Galaxy	Crater	11.5	11 h 48 m 01.9 s	−10° 57' 52"
NGC 4013	Galaxy	Ursa Major	11.2	11 h 58 m 32.4 s	+43° 56' 56"
NGC 4024	Galaxy	Corvus	11.7	11 h 58 m 32.5 s	−18° 20' 53"
NGC 4039	Galaxy	Corvus	10.6	12 h 01 m 54.8 s	−18° 53' 17"
NGC 4045	Galaxy	Virgo	12.0	12 h 02 m 42.8 s	+01° 58' 39"
NGC 4047	Galaxy	Ursa Major	12.2	12 h 02 m 52.0 s	+48° 38' 22"
NGC 4062	Galaxy	Ursa Major	11.1	12 h 04 m 04.2 s	+31° 53' 45"
NGC 4073	Galaxy	Virgo	11.4	12 h 04 m 27.3 s	+01° 53' 43"
NGC 4096	Galaxy	Ursa Major	10.9	12 h 06 m 01.8 s	+47° 28' 40"
NGC 4100	Galaxy	Ursa Major	11.2	12 h 06 m 09.4 s	+49° 35' 03"
NGC 4105	Galaxy	Hydra	10.6	12 h 06 m 41.8 s	−29° 45' 55"
NGC 4124	Galaxy	Virgo	11.4	12 h 08 m 10.0 s	+10° 22' 43"
NGC 4133	Galaxy	Draco	12.3	12 h 08 m 52.9 s	+74° 54' 32"
NGC 4136	Galaxy	Coma Berenices	11.0	12 h 09 m 18.8 s	+29° 55' 40"
NGC 4138	Galaxy	Canes Venatici	11.3	12 h 09 m 31.6 s	+43° 41' 23"
NGC 4144	Galaxy	Ursa Major	11.6	12 h 10 m 00.2 s	+46° 27' 34"
NGC 4152	Galaxy	Coma Berenices	12.2	12 h 10 m 38.4 s	+16° 02' 01"
NGC 4157	Galaxy	Ursa Major	11.4	12 h 11 m 05.8 s	+50° 29' 19"
NGC 4168	Galaxy	Virgo	11.2	12 h 12 m 17.3 s	+13° 12' 23"
NGC 4169	Galaxy	Coma Berenices	12.2	12 h 12 m 19.8 s	+29° 10' 53"
NGC 4185	Galaxy	Coma Berenices	12.1	12 h 13 m 22.8 s	+28° 30' 45"
NGC 4189	Galaxy	Coma Berenices	11.7	12 h 13 m 47.6 s	+13° 25' 35"
NGC 4212	Galaxy	Coma Berenices	11.2	12 h 15 m 39.9 s	+13° 54' 04"
NGC 4217	Galaxy	Canes Venatici	11.2	12 h 15 m 51.8 s	+47° 05' 40"
NGC 4220	Galaxy	Canes Venatici	11.4	12 h 16 m 12.7 s	+47° 53' 14"
NGC 4224	Galaxy	Virgo	11.8	12 h 16 m 34.4 s	+07° 27' 37"
NGC 4233	Galaxy	Virgo	11.9	12 h 17 m 07.4 s	+07° 37' 20"
NGC 4235	Galaxy	Virgo	11.6	12 h 17 m 09.8 s	+07° 11' 28"
NGC 4236	Galaxy	Draco	9.6	12 h 16 m 44.9 s	+69° 28' 09"
NGC 4237	Galaxy	Coma Berenices	11.6	12 h 17 m 11.5 s	+15° 19' 28"
NGC 4241	Galaxy	Virgo	11.9	12 h 17 m 26.2 s	+06° 41' 23"
NGC 4244	Galaxy	Canes Venatici	10.4	12 h 17 m 30.8 s	+37° 48' 32"
NGC 4248	Galaxy	Canes Venatici	12.5	12 h 17 m 51.1 s	+47° 24' 45"
NGC 4250	Galaxy	Draco	11.8	12 h 17 m 28.0 s	+70° 48' 20"
NGC 4256	Galaxy	Draco	11.9	12 h 18 m 45.5 s	+65° 54' 11"

(continued)

Table 9.2 (continued)

Object name	Object type	Const	V Mag	RA (2000)	Dec (2000)
NGC 4260	Galaxy	Virgo	11.8	12 h 19 m 22.9 s	+06° 05' 58"
NGC 4264	Galaxy	Virgo	12.8	12 h 19 m 36.5 s	+05° 50' 48"
NGC 4267	Galaxy	Virgo	10.9	12 h 19 m 46.5 s	+12° 47' 56"
NGC 4270	Galaxy	Virgo	12.2	12 h 19 m 49.5 s	+05° 27' 49"
NGC 4271	Galaxy	Ursa Major	12.6	12 h 19 m 34.6 s	+56° 44' 25"
NGC 4290	Galaxy	Ursa Major	11.8	12 h 20 m 49.5 s	+58° 05' 42"
NGC 4291	Galaxy	Draco	11.5	12 h 20 m 20.2 s	+75° 22' 34"
NGC 4294	Galaxy	Virgo	12.1	12 h 21 m 18.2 s	+11° 30' 38"
NGC 4298	Galaxy	Coma Berenices	11.3	12 h 21 m 33.5 s	+14° 36' 24"
NGC 4299	Galaxy	Virgo	12.5	12 h 21 m 41.5 s	+11° 30' 13"
NGC 4302	Galaxy	Coma Berenices	11.6	12 h 21 m 43.3 s	+14° 36' 04"
NGC 4310	Galaxy	Coma Berenices	12.2	12 h 22 m 27.3 s	+29° 12' 36"
NGC 4312	Galaxy	Coma Berenices	11.7	12 h 22 m 32.3 s	+15° 32' 20"
NGC 4313	Galaxy	Virgo	11.6	12 h 22 m 38.9 s	+11° 48' 12'
NGC 4319	Galaxy	Draco	11.9	12 h 21 m 46.0 s	+75° 19' 42"

When Tom Hoffelder, an amateur astronomer from Pennsylvania, completed his certificate for the Herschel 400 he looked forward to the next challenge. He found the challenge in a new list being compiled by the Rose City Astronomers. Hoffelder, with his knowledge of Herschel objects, suggested a number of objects for the new Herschel II list. After observing both of the Herschel lists Tom looked for another challenge. He noticed that the remaining objects of Herschel's original catalog were comprised mostly of galaxies. It was obvious that what was needed was a new Hershel catalog comprised entirely of galaxies for galaxy hunters. This catalog became the Hershel 3 catalog found below.

Using the copy of the NGC objects containing all of the Hershel objects he marked out those contained in the first two lists. This left about 1,700 objects. Hoffelder picked about 350 galaxies based on the NGC descriptions. He narrowed the list down to 300, based on the image of the object and to match the same number of objects in the other Herschel lists (Table 9.3).

Table 9.3 The Herschel 3 catalog compiled exclusively for galaxy hunters (With permission from Tom Hoffelder)

Object name	Const	V Mag	Size	RA (2000)	Dec (2000)
NGC16	And	13	2.1	00 h 09 m 04.6 s	+27° 43' 50"
NGC 95	Peg	13.5	1.9	00 h 22 m 14.1 s	+10° 29' 28"
NGC 128	Psc	13	3.4	00 h 29 m 15.1 s	+02° 51' 51"
NGC 210	Cet	12	5.4	00 h 40 m 34.8 s	−13° 52' 21"
NGC 255	Cet	12.5	3.1	00 h 47 m 47.3 s	−11° 28' 07"
NGC 266	And	12.5	3.2	00 h 49 m 48.1 s	+32° 16' 43"
NGC 274	Cet	13	1.7	00 h 51 m 01.8 s	−07° 03' 26"
NGC 268	–	13.5	1.7	00 h 50 m 09.4 s	−05° 11' 38"
NGC 245	–	13	1.4	00 h 46 m 05.4 s	−01° 43' 28"
NGC 450	Cet	12.5	3.2	01 h 15 m 30.4 s	−00° 51' 40"
NGC 520	Psc	12.5	4.8	01 h 24 m 34.3 s	+03° 47' 42"
NGC 474	–	12	7.9	01 h 20 m 06.6 s	+03° 24' 55"
NGC 470	–	12.5	3	01 h 19 m 44.9 s	+03° 24' 35"
NGC 533	–	13	3.7	01 h 25 m 31.4 s	+01° 45' 32"
NGC 536	And	13	3.7	01 h 26 m 21.7 s	+34° 42' 13"
NGC 600	Cet	12.5	3.5	01 h 33 m 05.3 s	−07° 18' 46"
NGC 670	Tri	13	2.5	01 h 47 m 24.9 s	+27° 53' 08"
NGC 673	Psc	13.5	2.4	01 h 48 m 22.7 s	+11° 31' 23"
NGC 676	Psc	10.5	4.3	01 h 48 m 57.4 s	+05° 54' 27"
NGC 680	Ari	13	2.9	01 h 49 m 47.3 s	+21° 58' 14"
NGC 678	–	13.5	5	01 h 49 m 25.2 s	+21° 59' 51"
NGC 697	–	12.5	4.7	01 h 51 m 17.5 s	+22° 21' 28"
NGC 681	Cet	13	2.8	01 h 49 m 10.8 s	−10° 25' 40"
NGC 705	Tri	14.5	~ 1	01 h 52 m 41.5 s	+36° 08' 39"
NGC 701	Cet	13	2.5	01 h 51 m 03.8 s	−09° 42' 14"
NGC 788	–	13.5	1.8	02 h 01 m 06.5 s	−06° 48' 56"
NGC 748	–	12	2.9	01 h 56 m 21.7 s	−04° 28' 04"
NGC 777	Tri	13	3	02 h 00 m 15.0 s	+31° 25' 47"
NGC 750	–	13	1.6	01 h 57 m 32.5 s	+33° 12' 36"
NGC 864	Cet	12	4.6	02 h 15 m 27.3 s	+06° 00' 05"
NGC 877	Cet	12.5	2.3	02 h 17 m 58.7 s	+14° 32' 50"
NGC 922	Eri	12.5	1.9	02 h 25 m 04.7 s	−24° 47' 25"
NGC 945	Cet	12	2.5	02 h 28 m 37.2 s	−10° 32' 22"
NGC 949	Per	13	2.8	02 h 30 m 48.8 s	+37° 08' 09"
NGC 955	Cet	13.5	3	02 h 30 m 33.4 s	−01° 06' 32"
NGC 941	–	13	2.8	02 h 28 m 27.9 s	−01° 09' 05"
NGC 958	–	13	2.8	02 h 30 m 42.7 s	−02° 56' 21"
NGC 895	–	12	3.6	02 h 21 m 36.5 s	−05° 31' 18"
NGC 972	Ari	12.5	3.6	02 h 34 m 13.3 s	+29° 18' 42"
NGC 1097	For	10.5	9.3	02 h 46 m 19.0 s	−30° 16' 28"
NGC 1140	Eri	13.5	1.4	02 h 54 m 33.5 s	−10° 01' 44"
NGC 1186	Per	12.5	3.3	03 h 05 m 30.9 s	+42° 50' 07"
NGC 1201	Eri	12	4.4	03 h 04 m 07.9 s	−26° 04' 10"
NGC 1232	Eri	10.5	7.8	03 h 09 m 45.1 s	−20° 34' 45"

(continued)

Table 9.3 (continued)

Object name	Const	V Mag	Size	RA (2000)	Dec (2000)
NGC 1241	Eri	13	3	03 h 11 m 14.7 s	−08° 55' 20"
NGC 1247	–	13	3.6	03 h 12 m 14.1 s	−10° 28' 50"
NGC 1309	Eri	12.5	2.3	03 h 22 m 06.5 s	−15° 24' 02"
NGC 1344	For	12	3.9	03 h 28 m 19.7 s	−30° 04' 07"
NGC 1357	Eri	12.5	2.4	03 h 33 m 17.2 s	−13° 39' 53"
NGC 1358	Eri	12.5	2.8	03 h 33 m 39.7 s	−05° 05' 22"
NGC 1376	–	13	2	03 h 37 m 05.9 s	−05° 02' 35"
NGC 1417	–	13	2.8	03 h 41 m 57.0 s	−04° 42' 21"
NGC 1453	–	13	2.1	03 h 46 m 27.2 s	−03° 58' 09"
NGC 1415	Eri	12.5	3.6	03 h 40 m 56.9 s	−22° 33' 53"
NGC 1395	–	11.5	3.2	03 h 38 m 29.7 s	−23° 01' 40"
NGC 1385	–	12	3	03 h 37 m 28.8 s	−24° 30' 07"
NGC 1371	–	12	5.4	03 h 35 m 01.2 s	−24° 56' 02"
NGC 1425	For	12	5.4	03 h 42 m 11.4 s	−29° 53' 36"
NGC 1426	Eri	13	2.1	03 h 42 m 49.3 s	−22° 06' 29"
NGC 1439	–	13	2.3	03 h 44 m 49.7 s	−21° 55' 15"
NGC 1452	–	13	1.7	03 h 45 m 22.3 s	−18° 38' 00"
NGC 1440	–	13	2.3	03 h 45 m 02.9 s	−18° 15' 58"
NGC 1461	–	13	3.3	03 h 48 m 27.1 s	−16° 23' 34"
NGC 1569	Cam	12.5	2.9	04 h 30 m 49.2 s	+64° 50' 51"
NGC 1638	Eri	13	2.5	04 h 41 m 36.3 s	−01° 48' 33"
NGC 1620	–	13.5	3	04 h 36 m 37.3 s	−00° 08' 35"
NGC 1589	–	14	3.1	04 h 30 m 45.7 s	+00° 51' 48"
NGC 1659	Eri	13	1.7	04 h 46 m 29.8 s	−04° 47' 18"
NGC 1888	Lup	13	3	05 h 22 m 34.7 s	−11° 29' 58"
NGC 2507	Cnc	14	2.5	08 h 01 m 37.1 s	+15° 42' 34"
NGC 2537	Lyn	12.5	1.7	08 h 13 m 14.5 s	+45° 59' 31"
NGC 2543	Lyn	12.5	2.5	08 h 12 m 57.9 s	+36° 15' 14"
NGC 2545	Cnc	13.5	2.2	08 h 14 m 14.2 s	+21° 21' 19"
NGC 2566	Pup	13	3.1	08 h 18 m 45.5 s	−25° 30' 04"
NGC 2608	Cnc	13	2.5	08 h 35 m 17.2 s	+25° 28' 24"
NGC 2672	Cnc	13	2.6	08 h 49 m 21.9 s	+19° 04' 29"
NGC 2693	UMa	13	2.2	08 h 56 m 59.3 s	+51° 20' 49"
NGC 2701	UMa	12.5	2.1	08 h 59 m 05.4 s	+53° 46' 14"
NGC 2718	Hya	13.5	2.5	08 h 58 m 50.4 s	+06° 17' 34"
NGC 2763	Hya	12.5	2.3	09 h 06 m 49.1 s	−15° 30' 04"
NGC 2848	–	12.5	2.7	09 h 20 m 09.6 s	−16° 31' 34"
NGC 2907	–	13.5	1.9	09 h 31 m 36.5 s	−16° 44' 07"
NGC 2770	Lyn	12	3.7	09 h 09 m 33.7 s	+33° 07' 25"
NGC 2798	Lyn	13	2.8	09 h 17 m 22.8 s	+41° 59' 59"
NGC 2815	Pyx	13	3.5	09 h 16 m 19.5 s	−23° 38' 00"
NGC 2820	UMa	13	4.3	09 h 21 m 49.9 s	+64° 15' 41"
NGC 2872	Leo	13	2.1	09 h 25 m 42.6 s	+11° 25' 54"
NGC 2874	–	13.5	2.5	09 h 25 m 47.3 s	+11° 25' 28"
NGC 2935	Hya	12	3.5	09 h 36 m 44.9 s	−21° 07' 46"

(continued)

Table 9.3 (continued)

Object name	Const	V Mag	Size	RA (2000)	Dec (2000)
NGC 2983	–	13	2.6	09 h 43 m 41.1 s	−20° 28' 40"
NGC 3052	–	13	2.1	09 h 54 m 28.0 s	−18° 38' 24"
NGC 2955	Lyn	13	1.8	09 h 41 m 16.6 s	+35° 52' 56"
NGC 2968	–	13	2.2	09 h 43 m 13.0 s	+31° 55' 44"
NGC 3003	–	12	5.9	09 h 48 m 36.5 s	+33° 25' 22"
NGC 2967	Hya	12	3	09 h 42 m 03.5 s	+00° 20' 09"
NGC 3044	–	12.5	4.8	09 h 53 m 41.1 s	+01° 34' 41"
NGC 2997	Ant	10.5	8.1	09 h 45 m 38.5 s	−31° 11' 25"
NGC 2998	UMa	12.5	3	09 h 48 m 43.6 s	+44° 04' 51"
NGC 3020	Leo	13	3.2	09 h 50 m 06.6 s	+12° 48' 50"
NGC 3024	–	13.5	2.2	09 h 50 m 27.5 s	+12° 45' 54"
NGC 3027	UMa	12.5	4.7	09 h 55 m 40.4 s	+72° 12' 13"
NGC 3041	Leo	12.5	3.7	09 h 53 m 07.1 s	+16° 40' 38"
NGC 3081	Hya	12.5	2.2	09 h 59 m 29.5 s	−22° 49' 34"
NGC 3091	–	13	2.2	10 h 00 m 14.1 s	−19° 38' 11"
NGC 3370	Leo	12.5	3.1	10 h 47 m 04.4 s	+17° 16' 24"
NGC 3346	–	12.5	2.8	10 h 43 m 38.9 s	+14° 52' 18"
NGC 3367	–	12.5	2.3	10 h 46 m 34.8 s	+13° 45' 02"
NGC 3389	–	12.5	2.7	10 h 48 m 28.0 s	+12° 31' 58"
NGC 3433	–	12.5	3.5	10 h 52 m 03.9 s	+10° 08' 53"
NGC 3423	–	12	3.9	10 h 51 m 14.2 s	+05° 50' 23"
NGC 3478	UMa	13	2.8	10 h 59 m 27.3 s	+46° 07' 20"
NGC 3320	–	13	2.2	10 h 39 m 36.5 s	+47° 23' 52"
NGC 3348	Dra	12.5	2.2	10 h 47 m 10.1 s	+72° 50' 23"
NGC 3403	–	13	3.1	10 h 53 m 55.1 s	+73° 41' 23"
NGC 3381	LMi	13	2.4	10 h 48 m 24.5 s	+34° 42' 42"
NGC 3396	LMi	12.5	2.8	10 h 49 m 55.3 s	+32° 59' 26"
NGC 3437	Leo	13	2.6	10 h 52 m 35.7 s	+22° 56' 04"
NGC 3455	–	13	2.8	10 h 54 m 31.0 s	+17° 17' 03"
NGC 3448	UMa	12.5	5.4	10 h 54 m 39.0 s	+54° 18' 18"
NGC 3549	–	13	3.2	11 h 10 m 56.7 s	+53° 23' 16"
NGC 3485	Leo	13	2.5	11 h 00 m 02.3 s	+14° 50' 28"
NGC 3495	Leo	12.5	4.6	11 h 01 m 16.3 s	+03° 37' 41"
NGC 3510	UMa	13	3.8	11 h 03 m 43.5 s	+28° 53' 05"
NGC 3512	–	13	1.7	11 h 04 m 02.9 s	+28° 02' 11"
NGC 3571	Crt	13	3.3	11 h 11 m 30.3 s	−18° 17' 23"
NGC 3614	UMa	12.5	4.6	11 h 18 m 21.3 s	+45° 44' 52"
NGC 3660	Crt	12.5	2.8	11 h 23 m 32.2 s	−08° 39' 31"
NGC 3690	UMa	12.5	2.4	11 h 28 m 31.9 s	+58° 33' 45"
NGC 3718	UMa	11.5	8.7	11 h 32 m 35.0 s	+53° 04' 06"
NGC 3738	–	12.5	2.6	11 h 35 m 48.5 s	+54° 31' 27"
NGC 3780	–	12.5	3.1	11 h 39 m 22.3 s	+56° 16' 14"
NGC 3735	Dra	12.5	4.2	11 h 36 m 57.3 s	+70° 32' 08"
NGC 3769	Uma	12.5	3.2	11 h 37 m 44.1 s	+47° 53' 34"
NGC 3801	Leo	13.5	3.2	11 h 40 m 16.7 s	+17° 43' 41"

(continued)

Table 9.3 (continued)

Object name	Const	V Mag	Size	RA (2000)	Dec (2000)
NGC 3802	–	14.5	1.4	11 h 40 m 18.7 s	+17° 45' 56"
NGC 3894	UMa	13	2.4	11 h 48 m 50.3 s	+59° 24' 55"
NGC 3895	–	14	1.4	11 h 49 m 04.0 s	+59° 25' 56"
NGC 3904	Hya	12.5	2.2	11 h 49 m 13.2 s	+29° 16' 35"
NGC 3923	–	11.5	2.9	11 h 51 m 01.8 s	−28° 48' 21"
NGC 3885	–	13	1.7	11 h 46 m 46.4 s	−27° 55' 17"
NGC 3917	UMa	12.5	4.9	11 h 50 m 45.3 s	+51° 49' 28"
NGC 3955	Crv	13	3.2	11 h 53 m 57.1 s	−23° 09' 52"
NGC 3956	–	12.5	3.5	11 h 54 m 00.7 s	−20° 34' 03"
NGC 3957	–	13	3.5	11 h 54 m 01.7 s	−19° 34' 14"
NGC 3981	–	12.5	3.9	11 h 56 m 07.1 s	−19° 53' 49"
NGC 4033	–	13	2.5	12 h 00 m 34.8 s	−17° 50' 34"
NGC 4050	–	12.5	3.1	12 h 02 m 54.0 s	−16° 22' 27"
NGC 3968	Leo	13.5	3	11 h 55 m 28.7 s	+11° 58' 05"
NGC 3972	UMa	13	4	11 h 55 m 45.2 s	+55° 19' 12"
NGC 3963	–	12.5	2.8	11 h 54 m 58.6 s	+58° 29' 36"
NGC 3976	Vir	12.5	3.9	11 h 55 m 57.3 s	+06° 44' 56"
NGC 4017	Com	13.5	1.8	11 h 58 m 45.8 s	+27° 27' 10"
NGC 4123	Vir	12	4.5	12 h 08 m 11.1 s	+02° 52' 41"
NGC 4128	Dra	13	2.8	12 h 08 m 32.4 s	+68° 46' 03"
NGC 4145	CVn	11.5	5.8	12 h 10 m 01.6 s	+39° 52' 58"
NGC 4197	Vir	14	3.5	12 h 14 m 38.5 s	+05° 48' 21"
NGC 4215	–	13	1.9	12 h 15 m 54.5 s	+06° 24' 04"
NGC 4206	Vir	14	5.2	12 h 15 m 16.7 s	+13° 01' 26"
NGC 4388	–	12	5.1	12 h 25 m 46.8 s	+12° 39' 43"
NGC 4425	–	13	3.4	12 h 27 m 13.4 s	+12° 44' 04"
NGC 4461	–	12.5	3.7	12 h 29 m 03.0 s	+13° 11' 01"
NGC 4452	–	13	2.4	12 h 28 m 43.3 s	+11° 45' 18"
NGC 4283	Com	13	1.4	12 h 20 m 20.8 s	+29° 18' 38"
NGC 4348	Vir	13	3.5	12 h 23 m 53.9 s	−03° 26' 34"
NGC 4454	–	13	2.2	12 h 28 m 50.8 s	−01° 56' 21"
NGC 4389	UMa	13	2.7	12 h 25 m 35.3 s	+45° 41' 03"
NGC 4460	–	12.5	4.4	12 h 28 m 45.6 s	+44° 51' 51"
NGC 4417	Vir	12.5	3.6	12 h 26 m 50.6 s	+09° 35' 03"
NGC 4469	–	12.5	3.9	12 h 29 m 28.1 s	+08° 45' 01"
NGC 4421	Com	13	2.7	12 h 27 m 02.5 s	+15° 27' 41"
NGC 4474	–	13	2.3	12 h 29 m 53.6 s	+14° 04' 07"
NGC 4479	–	14	1.8	12 h 30 m 18.4 s	+13° 34' 40"
NGC 4455	Com	13	2.8	12 h 28 m 44.0 s	+22° 49' 20"
NGC 4457	Vir	12	3	12 h 28 m 59.1 s	+03° 34' 13"
NGC 4412	–	13	1.5	12 h 26 m 36.1 s	+03° 57' 53"
NGC 4378	–	12.5	3.3	12 h 25 m 18.1 s	+04° 55' 29"
NGC 4496	–	12	3.9	12 h 31 m 39.6 s	+03° 56' 27"
NGC 4462	Crv	13	3.7	12 h 29 m 21.1 s	−23° 10' 00"
NGC 4504	Vir	12	4	12 h 32 m 17.3 s	−07° 33' 50"

(continued)

Table 9.3 (continued)

Object name	Const	V Mag	Size	RA (2000)	Dec (2000)
NGC 4521	Dra	13	1	12 h 32 m 47.7 s	+63° 56' 20"
NGC 4532	Vir	12.5	2.9	12 h 34 m 19.3 s	+06° 28' 07"
NGC 4580	–	13	2.4	12 h 37 m 48.4 s	+05° 22' 05"
NGC 4561	Com	13.5	1.5	12 h 36 m 08.2 s	+19° 19' 19"
NGC 4568	Vir	12	4.6	12 h 36 m 34.3 s	+11° 14' 17"
NGC 4567	–	12.5	3	12 h 36 m 32.8 s	+11° 15' 28"
NGC 4564	–	12.5	3.1	12 h 36 m 26.9 s	+11° 26' 21"
NGC 4503	–	12.5	3.5	12 h 32 m 06.2 s	+11° 10' 35"
NGC 4592	Vir	12.5	4.6	12 h 39 m 18.9 s	–00° 31' 54"
NGC 4517	–	11.5	10.2	12 h 32 m 45.5 s	+00° 06' 48"
NGC 4597	Vir	12.5	3.6	12 h 40 m 12.8 s	–05° 48' 00"
NGC 4593	–	12	4	12 h 39 m 39.4 s	–05° 20' 39"
NGC 4602	–	12.5	3.6	12 h 40 m 36.7 s	–05° 07' 55"
NGC 4623	Vir	13	2.6	12 h 42 m 10.7 s	+07° 40' 37"
NGC 4632	Vir	12.5	3.2	12 h 42 m 32.8 s	–00° 04' 47"
NGC 4653	–	13	2.6	12 h 43 m 50.8 s	–00° 33' 40'
NGC 4634	Vir	13.5	2.4	12 h 42 m 41.0 s	+14° 17' 45"
NGC 4658	Vir	13	2.2	12 h 44 m 37.7 s	–10° 05' 02"
NGC 4684	Vir	12.5	2.9	12 h 47 m 17.4 s	–12° 43' 37"
NGC 4694	Vir	12.5	3.6	12 h 48 m 15.1 s	+10° 59' 01"
NGC 4710	–	12	5.1	12 h 49 m 38.7 s	+15° 09' 54"
NGC 4750	Dra	12.5	2.3	12 h 50 m 07.0 s	+72° 52' 27"
NGC 4772	Vir	12.5	3.3	12 h 53 m 29.1 s	+02° 10' 06"
NGC 4771	–	13	4	12 h 53 m 21.1 s	+01° 16' 10"
NGC 4779	Vir	13.5	2.3	12 h 53 m 50.9 s	+09° 42' 35"
NGC 4795	–	13.5	1.7	12 h 55 m 02.9 s	+08° 03' 56"
NGC 4790	Vir	13	1.8	12 h 54 m 51.9 s	–10° 14' 52"
NGC 4700	–	12.5	3	12 h 49 m 07.7 s	–11° 24' 39"
NGC 4782	–	13	1.5	12 h 54 m 35.9 s	–12° 34' 08"
NGC 4783	–	13	1.7	12 h 54 m 36.3 s	–12° 33' 30"
NGC 4825	–	13	2	12 h 57 m 12.2 s	–13° 39' 55"
NGC 4899	–	12.5	2.7	13 h 00 m 56.5 s	–13° 56' 41"
NGC 4877	–	13	2.7	13 h 00 m 26.3 s	–15° 17' 00"
NGC 4808	Vir	12.5	2.7	13 h 55 m 49.1 s	+04° 18' 15"
NGC 4814	UMa	12.5	3.2	12 h 55 m 21.9 s	+58° 20' 38"
NGC 4818	Vir	12	4.5	12 h 56 m 48.9 s	–08° 31' 32"
NGC 4868	CVn	13	1.7	12 h 59 m 08.9 s	+37° 18' 37"
NGC 4861	–	12.5	4.1	12 h 59 m 02.5 s	+34° 51' 50"
NGC 4889	Com	12.5	3	13 h 00 m 08.1 s	+27° 58' 37"
NGC 4793	–	12.5	2.9	12 h 54 m 40.6 s	+28° 56' 17"
NGC 4747	–	12.5	3.6	12 h 51 m 45.5 s	+25° 46' 30"
NGC 4933	Vir	13	2.5	13 h 03 m 56.9 s	–11° 29' 52"
NGC 4951	Vir	12.5	3.3	13 h 05 m 07.7 s	–06° 29' 41"
NGC 4775	–	12	2.2	12 h 53 m 45.9 s	–06° 37' 21"
NGC 4731	–	11.5	6.5	12 h 51 m 01.2 s	–06° 23' 34"

(continued)

Table 9.3 (continued)

Object name	Const	V Mag	Size	RA (2000)	Dec (2000)
NGC 5015	Vir	13	1.9	13 h 12 m 22.6 s	–04° 20' 17"
NGC 5016	Com	13	1.9	13 h 12 m 06.7 s	+24° 05' 42"
NGC 5073	Vir	13	3.5	13 h 19 m 20.9 s	–14° 50' 41"
NGC 5085	Hya	12	3.4	13 h 20 m 17.4 s	–24° 26' 26"
NGC 5101	–	12	5.5	13 h 21 m 46.1 s	–27° 25' 47"
NGC 5112	CVn	12	3.9	13 h 21 m 56.4 s	+38° 44' 02"
NGC 5147	Vir	12.5	1.8	13 h 26 m 19.8 s	+02° 06' 02"
NGC 5170	Vir	12	8.1	13 h 29 m 48.7 s	–17° 57' 57"
NGC 5247	–	11.5	5.4	13 h 38 m 03.0 s	–17° 53' 03"
NGC 5230	Boo	13	2.2	13 h 35 m 31.8 s	+13° 40' 33"
NGC 5253	Cen	11	4	13 h 39 m 55.9 s	–31° 38' 31"
NGC 5300	Vir	12	3.9	13 h 48 m 15.9 s	+03° 57' 02"
NGC 5301	UMa	12.5	4.4	13 h 46 m 24.5 s	+46° 06' 25"
NGC 5297	–	12.5	5.6	13 h 46 m 23.7 s	+43° 52' 18"
NGC 5290	–	13	3.7	13 h 45 m 18.9 s	+41° 42' 47"
NGC 5326	–	13.5	2.5	13 h 50 m 50.7 s	+39° 34' 29"
NGC 5324	Vir	12.5	2.4	13 h 52 m 05.9 s	–06° 03' 30"
NGC 5328	Hya	13	1.7	13 h 52 m 53.0 s	–28° 29' 23"
NGC 5334	Vir	12	4.4	13 h 52 m 54.4 s	–01° 06' 52"
NGC 5347	CVn	13	1.9	13 h 53 m 17.8 s	+33° 29' 27"
NGC 5351	CVn	13	3.1	13 h 53 m 27.7 s	+37° 54' 51"
NGC 5395	–	13	3.1	13 h 58 m 37.5 s	+37° 25' 32"
NGC 5406	–	13	2.1	14 h 00 m 20.1 s	+38° 54' 55"
NGC 5353	–	12.5	2.8	13 h 53 m 26.7 s	+40° 16' 58"
NGC 5350	–	12.5	3.2	13 h 53 m 21.5 s	+40° 21' 49"
NGC 5377	UMa	12	4.6	13 h 56 m 16.6 s	+47° 14' 07"
NGC 5422	UMa	13	3.9	14 h 00 m 42.0 s	+55° 09' 52"
NGC 5376	–	13	2.1	13 h 55 m 15.9 s	+59° 30' 23"
NGC 5379	–	14	2.2	13 h 55 m 34.4 s	+59° 44' 34"
NGC 5389	–	13	4.1	13 h 56 m 06.5 s	+59° 44' 30"
NGC 5468	Vir	12.5	2.5	14 h 06 m 34.9 s	–05° 27' 11"
NGC 5427	–	12	2.5	14 h 03 m 26.1 s	–06° 01' 51"
NGC 5574	Vir	13.5	1.6	14 h 20 m 56.1 s	+03° 14' 17"
NGC 5669	Boo	12.5	4.1	14 h 32 m 43.9 s	+09° 53' 31"
NGC 5665	–	13	2.1	14 h 32 m 25.6 s	+08° 04' 45"
NGC 5645	–	13	2.6	14 h 30 m 39.5 s	+07° 16' 28"
NGC 5678	Dra	12.5	3.2	14 h 32 m 05.8 s	+57° 55' 16"
NGC 5690	Vir	13	3.5	14 h 37 m 40.9 s	+02° 17' 27"
NGC 5701	–	12	4.7	14 h 39 m 11.1 s	+05° 21' 48"
NGC 5729	Vir	13	3	14 h 42 m 06.7 s	–09° 00' 31"
NGC 5740	Vir	12.5	3.1	14 h 44 m 24.3 s	+01° 40' 47"
NGC 5713	–	12	2.8	14 h 40 m 11.3 s	–00° 17' 27"
NGC 5719	–	14	3.4	14 h 40 m 56.4 s	–00° 19' 05"
NGC 5757	Lib	13	2.1	14 h 47 m 46.2 s	–19° 04' 39"
NGC 5792	Vir	12	7.2	14 h 58 m 22.7 s	–01° 05' 27"

(continued)

Table 9.3 (continued)

Object name	Const	V Mag	Size	RA (2000)	Dec (2000)
NGC 5885	Lib	12	3.5	15 h 15 m 04.0 s	−10° 05' 11"
NGC 5898	Lib	13	1.7	15 h 18 m 13.7 s	−24° 05' 54"
NGC 5903	–	13	2	15 h 18 m 36.9 s	−24° 03' 59"
NGC 5905	Dra	12.5	4.2	15 h 15 m 23.3 s	+55° 31' 01"
NGC 5908	–	13	3.2	15 h 16 m 43.3 s	+55° 24' 35"
NGC 5875	–	13.5	2.6	15 h 09 m 13.1 s	+52° 31' 42"
NGC 5921	Ser	12	4.9	15 h 21 m 56.5 s	+05° 04' 13"
NGC 5949	UMi	13	2.4	15 h 28 m 00.6 s	+64° 45' 47"
NGC 5962	Ser	12.5	2.8	15 h 36 m 31.7 s	+16° 36' 29"
NGC 5965	Dra	13.5	5.4	15 h 34 m 01.9 s	+56° 41' 09"
NGC 5984	Ser	13	3	15 h 42 m 53.3 s	+14° 13' 51"
NGC 6412	Dra	12.5	2.3	17 h 29 m 37.5 s	+75° 42' 15"
NGC 6926	Aql	12.5	2.1	20 h 33 m 05.9 s	−02° 01' 41"
NGC 7013	Cyg	13	4.9	21 h 03 m 33.5 s	+29° 53' 40"
NGC 7137	Peg	13.5	1.5	21 h 48 m 12.9 s	+22° 09' 34"
NGC 7252	PsA	13	2.2	15 h 21 m 56.5 s	+05° 04' 13"
NGC 7302	Aqr	13.5	1.9	22 h 20 m 44.9 s	−24° 40' 43"
NGC 7309	Aqr	13	2.1	15 h 21 m 56.5 s	+05° 04' 13"
NGC 7371	Aqr	13	2.1	22 h 34 m 20.3 s	−10° 21' 29"
NGC 7385	Peg	14.5	2	22 h 49 m 54.6 s	+11° 36' 30"
NGC 7497	Peg	13.5	5	23 h 09 m 03.3 s	+18° 10' 35"
NGC 7625	–	13	1.8	15 h 21 m 56.5 s	+05° 04' 13"
NGC 7585	Aqr	13	2.3	23 h 20 m 30.1 s	+17° 13' 32"
NGC 7678	Peg	13	2.3	23 h 28 m 27.9 s	+22° 25' 16"
NGC 7721	Aqr	12.5	3.4	23 h 38 m 48.7 s	−06° 31' 06"
NGC 7741	Peg	12	4	23 h 43 m 54.3 s	+26° 04' 31"
NGC 7753	And	13	3.4	23 h 47 m 04.9 s	+29° 29' 00"
NGC 7769	Peg	13	1.8	23 h 51 m 03.9 s	+20° 08' 59"
NGC 7771	–	13	2.7	23 h 51 m 24.6 s	+20° 06' 41"
NGC 7782	Psc	13	2.4	23 h 53 m 53.9 s	+07° 58' 14"
NGC 7743	–	12.5	3.1	23 h 44 m 21.1 s	+09° 56' 03"

Chapter 10

J. L. E. Dreyer and the NGC Catalog

Fig. 10.1 John Louis Emil Dreyer (1852–1926)

J.D. Cavin, *The Amateur Astronomer's Guide to the Deep-Sky Catalogs*,
Patrick Moore's Practical Astronomy Series, DOI 10.1007/978-1-4614-0656-3_10,
© Springer Science+Business Media, LLC 2012

John Louis Emil Dreyer (1852–1926) was born in Copenhagen into a family with a long line of military tradition. His father, a former Lieutenant General in the Danish army, was the Danish Minister for War and the Navy. His grandfather was a staff officer in Napoleon's Army. His great grandfather was a Quarter Master General in the Danish army.

At the age of fourteen John was given a book on Tycho Brahe's observatories, and from that time on he knew he had to become an astronomer "and nothing else." He attended the University of Copenhagen, where he studied astronomy under the guidance of Heinrich Louis d'Arrest and Hans Schjellerup. At the University of Copenhagen Dreyer excelled in history, mathematics, and physics. At the age of 22 he was appointed as the assistant to Earl of Rosse, in Parsonstown, Ireland. For the next four years he used the largest telescope of his time, the 72-in. Leviathan of Parsonstown built by Lord Rosse. The observations he made were primarily of nebula and star clusters. The original drawings based on his observations can be seen at the Armagh Observatory.

Dreyer moved in 1878 to the Trinity College Observatory of Dublin University to work for Robert Stawell Ball. At the observatory he used his knowledge of the meridian circle (or meridian telescope) to observe when an object passed the meridian and at the same time measure the star's angular distance from the zenith. Using this data along with the latitude and longitude of the observer the object's right ascension and declination can then be computed. The concept of the meridian circle had been known since ancient times, but it was Tycho Brahe that constructed and used the first meridian quadrant.

In 1882 Dreyer became the director of Armagh Observatory in Northern Ireland. He would work at Armagh until he retired. The equipment at Armagh Observatory consisted of a 10-in. refractor and a micrometer. With these instruments Dreyer took measurements of stars, comets, nebulas, planets, and planetary occultations. He compared the measurements he took with the measurements in existing catalogs to correct the errors. One of the catalogs he used to compare his measurements with was the *General Catalogue of Nebula* (1864) by John and William Herschel.

Dreyer had been correcting measurements and recording new objects since working with the Earl of Rosse. Some of these had previously been published by Lord Rosse in 1879 as a supplement to the "General Catalogue of Nebula" in the *Scientific Transactions of the Royal Irish Academy*. Dreyer had compiled a second supplement of his observations and those of others and prepared it for publication. The Council of Royal Astronomers asked Dreyer to combine all of the catalogues into one. The *General Catalogue of Nebula* of John and William Herschel and the first and second supplement compiled by Dreyer and others became the source of the *New General Catalog of Nebula and Star Clusters* published in 1888 as a memoir for the Royal Astronomical Society.

Of the 7,840 NGC objects about 78% of the *New General Catalog* contained objects from Herschel's *General Catalogue of Nebula.* Over 400 of the objects attributed to Lord Rosse were actually observed by his assistant (Dreyer). Over 4,000 of the observations were made by a variety of other astronomers.

Over the next several years Dreyer continued to take measurements of new objects and correcting the measurements of objects already cataloged. In 1895 he published his *Index Catalogue of Nebulae found in the years 1888 to 1894*. The *Index Catalogue* (IC) contained 1,529 objects, with many corrections to the NGC catalog. In 1908 Dreyer published the *Second Index Catalogue* containing 3,857 nebula and many corrections to the NGC and IC catalogs.

The book on Tycho Brahe Dreyer had read as a young man continued to influence him throughout his life. In 1890 Dreyer published a book on the life and work of Brahe that remains even today as the standard biography of Tycho Brahe. In 1906 Dreyer published another classic in the history of astronomy, *The History of the Planetary System from Thales to Kepler*.

Dreyer retired as the director of Armagh Observatory in 1916, and he and his wife Kate moved to Oxford. During his retirement Dreyer worked on his last great work, a 15-volume edition of the works of Tycho Brahe. The last volume would be published after his death.

The following tables of the NGC catalog are provided by the NGC/IC Project LLC and its major contributors, Dr. Harold G. Corwin, Jr., Steve Gottlieb, Malcolm Thomson, Robert Erdmann, and others (Tables 10.1–10.4).

Table 10.1 These NGC objects are found in the constellations of Andromeda through Coma Berenices (Copyright © The NGC/IC Project, LLC – All rights reserved. Used with permission)

NGC #	Const	Type	Size	V Mag	RA (2000)	Dec (2000)
5	And	Gxy	1.2'×0.7'	13.4	00 h 07 m 48.8 s	+35° 21' 44"
6	And	Gxy	1.7'×1.5'	13.1	00 h 09 m 32.8 s	+33° 18' 34"
11	And	Gxy	1.5'×0.3'	13.9	00 h 08 m 42.4 s	+37° 26' 52"
13	And	Gxy	2.5'×0.6'	13.4	00 h 08 m 48.0 s	+33° 26' 03"
19	And	Gxy	1.1'×0.6'	13.3	00 h 10 m 40.9 s	+32° 58' 58"
20	And	Gxy	1.5'×1.5'	13.1	00 h 09 m 32.8 s	+33° 18' 34"
21	And	Gxy	1.6'×0.7'	12.9	00 h 10 m 46.9 s	+33° 21' 10"
27	And	Gxy	1.2'×0.5'	13.7	00 h 10 m 32.9 s	+28° 59' 50"
29	And	Gxy	1.7'×0.8'	12.9	00 h 10 m 46.9 s	+33° 21' 10"
39	And	Gxy	1.1'×1.0'	13.7	00 h 12 m 18.9 s	+31° 03' 40"
43	And	Gxy	1.6'×1.5'	12.7	00 h 13 m 00.9 s	+30° 54' 57"
44	And	**	…	…	00 h 13 m 13.3 s	+31° 17' 11"
48	And	Gxy	1.4'×0.9'	13.8	00 h 14 m 02.0 s	+48° 14' 06"
49	And	Gxy	1.1'×1.0'	13.8	00 h 14 m 22.3 s	+48° 14' 48"
51	And	Gxy	1.3'×1.0'	13.2	00 h 14 m 34.7 s	+48° 15' 20"
67	And	Gxy	0.4'×0.3'	14.8	00 h 18 m 12.5 s	+30° 03' 19"
68	And	Gxy	2.0'×1.8'	13	00 h 18 m 18.5 s	+30° 04' 17"
69	And	Gxy	0.9'×0.8'	14.4	00 h 18 m 20.5 s	+30° 02' 23"
70	And	Gxy	1.4'×1.2'	13.7	00 h 18 m 22.5 s	+30° 04' 46"
71	And	Gxy	1.2'×1.1'	13.5	00 h 18 m 23.5 s	+30° 03' 47"
72	And	Gxy	1.1'×0.9'	14	00 h 18 m 28.3 s	+30° 02' 26"
74	And	Gxy	0.8'×0.3'	15.3	00 h 18 m 49.3 s	+30° 03' 41"
76	And	Gxy	1.0'×0.9'	13	00 h 19 m 37.7 s	+29° 56' 01"
79	And	Gxy	0.6'×0.6'	14	00 h 21 m 02.9 s	+22° 34' 00"
80	And	Gxy	1.8'×1.7'	12.5	00 h 21 m 10.9 s	+22° 21' 26"
81	And	Gxy	0.2'×0.1'	15.7	00 h 21 m 13.3 s	+22° 22' 58"
82	And	*	…	…	00 h 21 m 17.4 s	+22° 27' 41"
83	And	Gxy	1.5'×1.5'	13	00 h 21 m 22.5 s	+22° 26' 01"
84	And	*	…	…	00 h 21 m 21.3 s	+22° 37' 03"
85	And	Gxy	0.6'×0.5'	14.8	00 h 21 m 25.6 s	+22° 30' 43"
86	And	Gxy	0.7'×0.3'	14.8	00 h 21 m 28.7 s	+22° 33' 23"
90	And	Gxy	2.2'×0.9'	13.8	00 h 21 m 51.4 s	+22° 24' 00"
91	And	*	1.9'×0.8'	…	00 h 21 m 51.7 s	+22° 22' 06"
93	And	Gxy	1.4'×0.6'	13.6	00 h 22 m 03.3 s	+22° 24' 29"
94	And	Gxy	0.4'×0.2'	14.6	00 h 22 m 13.5 s	+22° 28' 59"
96	And	Gxy	0.6'×0.6'	14.6	00 h 22 m 17.7 s	+22° 32' 47"
97	And	Gxy	1.5'×1.3'	12.4	00 h 22 m 30.0 s	+29° 44' 43"
108	And	Gxy	2.0'×1.6'	12.2	00 h 25 m 59.7 s	+29° 12' 43"
109	And	Gxy	1.3'×1.1'	14	00 h 26 m 14.5 s	+21° 48' 26"
112	And	Gxy	1.1'×0.5'	13.8	00 h 26 m 48.5 s	+31° 42' 14"
140	And	Gxy	1.5'×1.3'	13.4	00 h 31 m 20.3 s	+30° 47' 32"
149	And	Gxy	1.3'×0.8'	13.8	00 h 33 m 50.3 s	+30° 43' 25"
160	And	Gxy	3.0'×1.7'	12.4	00 h 36 m 04.1 s	+23° 57' 29"

(continued)

Table 10.1 (continued)

NGC #	Const	Type	Size	V Mag	RA (2000)	Dec (2000)
162	And	*	...	...	00 h 36 m 09.2 s	+23° 57' 45"
169	And	Gxy	2.6'×0.7'	12.4	00 h 36 m 51.9 s	+23° 59' 23"
181	And	Gxy	0.5'×0.2'	14.9	00 h 38 m 23.6 s	+29° 28' 15"
183	And	Gxy	2.1'×1.6'	12.5	00 h 38 m 29.7 s	+29° 30' 35"
184	And	Gxy	0.7'×0.2'	14.6	00 h 38 m 36.2 s	+29° 26' 47"
205	And	Gxy	21.9'×11.0'	8.2	00 h 40 m 21.9 s	+41° 41' 26"
206	And	GxyCld	...	...	00 h 40 m 31.3 s	+40° 44' 22"
214	And	Gxy	1.9'×1.4'	12.4	00 h 41 m 28.2 s	+25° 30' 00"
218	And	Gxy	1.1'×1.1'	14.3	00 h 41 m 44.9 s	+36° 21' 26"
221	And	Gxy	8.7'×6.5'	8.3	00 h 42 m 41.8 s	+40° 51' 52"
224	And	Gxy	190.5'×61.7'	3.6	00 h 42 m 44.3 s	+41° 16' 06"
226	And	Gxy	0.9'×0.9'	13.4	00 h 42 m 54.1 s	+32° 34' 51"
228	And	Gxy	1.2'×1.1'	13.9	00 h 42 m 54.5 s	+23° 30' 10"
229	And	Gxy	0.9'×0.2'	14.1	00 h 43 m 04.7 s	+23° 30' 32"
233	And	Gxy	1.7'×1.5'	12.5	00 h 43 m 36.5 s	+30° 35' 15"
243	And	Gxy	0.7'×0.3'	13.7	00 h 46 m 01.3 s	+29° 57' 31"
252	And	Gxy	1.5'×1.1'	12.5	00 h 48 m 01.5 s	+27° 37' 26"
258	And	Gxy	0.5'×0.4'	14.2	00 h 48 m 12.8 s	+27° 39' 24"
260	And	Gxy	0.9'×0.9'	13.7	00 h 48 m 34.7 s	+27° 41' 33"
262	And	Gxy	1.1'×1.1'	13.9	00 h 48 m 47.2 s	+31° 57' 26"
272	And	OC	...	8.5	00 h 51 m 25.1 s	+35° 49' 18"
280	And	Gxy	1.7'×1.1'	13.5	00 h 52 m 29.9 s	+24° 21' 01"
304	And	Gxy	1.1'×0.7'	13.1	00 h 56 m 06.5 s	+24° 07' 30"
317	And	Gxy	1.0'×0.5'	13.5	00 h 57 m 40.4 s	+43° 47' 31"
389	And	Gxy	1.3'×0.4'	13.9	01 h 08 m 29.7 s	+39° 41' 40"
393	And	Gxy	1.7'×1.4'	12.2	01 h 08 m 36.7 s	+39° 38' 35"
404	And	Gxy	3.5'×3.5'	10.2	01 h 09 m 26.9 s	+35° 43' 04"
425	And	Gxy	1.0'×1.0'	12.7	01 h 13 m 02.5 s	+38° 46' 07"
431	And	Gxy	1.4'×0.9'	13	01 h 14 m 04.7 s	+33° 42' 13"
464	And	**	...	...	01 h 19 m 26.7 s	+34° 57' 20"
477	And	Gxy	2.2'×1.2'	13.2	01 h 21 m 20.3 s	+40° 29' 17"
512	And	Gxy	1.6'×0.4'	13.4	01 h 23 m 59.8 s	+33° 54' 29"
513	And	Gxy	0.7'×0.3'	12.9	01 h 24 m 26.8 s	+33° 47' 58"
523	And	Gxy	2.5'×0.8'	13	01 h 25 m 20.8 s	+34° 01' 31"
528	And	Gxy	1.7'×1.1'	12.6	01 h 25 m 33.6 s	+33° 40' 17"
529	And	Gxy	2.4'×2.1'	11.9	01 h 25 m 40.3 s	+34° 42' 47"
531	And	Gxy	1.9'×0.5'	13.9	01 h 26 m 18.8 s	+34° 45' 15"
536	And	Gxy	3.0'×1.1'	12.4	01 h 26 m 21.7 s	+34° 42' 13"
537	And	Gxy	2.5'×0.8'	13	01 h 25 m 20.8 s	+34° 01' 31"
542	And	Gxy	1.0'×0.2'	14.8	01 h 26 m 30.8 s	+34° 40' 32"
551	And	Gxy	1.8'×0.8'	12.9	01 h 27 m 40.5 s	+37° 10' 58"
561	And	Gxy	1.6'×1.5'	13.1	01 h 28 m 18.7 s	+34° 18' 31"
562	And	Gxy	1.3'×1.0'	13.5	01 h 28 m 29.2 s	+48° 23' 15"
573	And	Gxy	0.4'×0.4'	12.8	01 h 30 m 49.3 s	+41° 15' 25"
590	And	Gxy	2.6'×1.3'	13.1	01 h 33 m 40.9 s	+44° 55' 46"

(continued)

Table 10.1 (continued)

NGC #	Const	Type	Size	V Mag	RA (2000)	Dec (2000)
591	And	Gxy	1.3'×1.0'	13.1	01 h 33 m 31.3 s	+35° 40' 05"
605	And	Gxy	2.2'×1.1'	13	01 h 35 m 02.3 s	+41° 14' 54"
620	And	Gxy	1.1'×1.0'	12.9	01 h 36 m 59.7 s	+42° 19' 20"
653	And	Gxy	1.5'×0.2'	13.6	01 h 42 m 25.8 s	+35° 38' 19"
662	And	Gxy	0.8'×0.5'	13.7	01 h 44 m 35.3 s	+37° 41' 49"
668	And	Gxy	1.8'×1.2'	12.8	01 h 46 m 22.6 s	+36° 27' 39"
679	And	Gxy	2.1'×2.1'	11.9	01 h 49 m 43.7 s	+35° 47' 04"
687	And	Gxy	1.4'×1.4'	12.3	01 h 50 m 33.2 s	+36° 22' 15"
700	And	Gxy	0.9'×0.7'	14.6	01 h 52 m 16.8 s	+36° 02' 11"
703	And	Gxy	1.2'×0.9'	13.3	01 h 52 m 39.6 s	+36° 10' 18"
704	And	Gxy	0.6'×0.6'	13	01 h 52 m 37.7 s	+36° 07' 33"
705	And	Gxy	1.2'×0.3'	13.8	01 h 52 m 41.5 s	+36° 08' 39"
708	And	Gxy	3.0'×2.5'	12.8	01 h 52 m 46.3 s	+36° 09' 08"
709	And	Gxy	...'×...'	14.2	01 h 52 m 50.5 s	+36° 13' 24"
710	And	Gxy	1.3'×1.2'	13.6	01 h 52 m 53.9 s	+36° 03' 12"
712	And	Gxy	1.3'×1.0'	12.9	01 h 53 m 08.5 s	+36° 49' 12"
714	And	Gxy	1.5'×0.4'	13.2	01 h 53 m 29.6 s	+36° 13' 17"
717	And	Gxy	1.3'×0.2'	14	01 h 53 m 54.9 s	+36° 13' 45"
721	And	Gxy	1.7'×1.0'	13.1	01 h 54 m 45.5 s	+39° 22' 55"
732	And	Gxy	1.4'×1.0'	13.6	01 h 56 m 27.7 s	+36° 48' 05"
746	And	Gxy	1.9'×1.3'	13.1	01 h 57 m 51.3 s	+44° 55' 05"
752	And	OC	50'	5.7	01 h 57 m 47.9 s	+37° 51' 00"
753	And	Gxy	2.5'×1.9'	11.9	01 h 57 m 42.4 s	+35° 54' 58"
759	And	Gxy	1.6'×1.4'	12.5	01 h 57 m 50.3 s	+36° 20' 33"
797	And	Gxy	1.6'×1.3'	12.7	02 h 03 m 27.9 s	+38° 07' 01"
801	And	Gxy	3.2'×0.7'	12.9	02 h 03 m 44.9 s	+38° 15' 32"
812	And	Gxy	3.2'×1.5'	11.6	02 h 06 m 52.2 s	+44° 34' 12"
818	And	Gxy	3.0'×1.3'	12.1	02 h 08 m 44.4 s	+38° 46' 36"
828	And	Gxy	2.9'×2.2'	12	02 h 10 m 09.2 s	+39° 11' 29"
834	And	Gxy	1.1'×0.5'	13	02 h 11 m 01.2 s	+37° 39' 59"
841	And	Gxy	1.8'×1.0'	12.3	02 h 11 m 17.1 s	+37° 29' 50"
845	And	Gxy	1.7'×0.4'	13.8	02 h 12 m 19.5 s	+37° 28' 36"
846	And	Gxy	1.9'×1.7'	12.3	02 h 12 m 12.1 s	+44° 34' 06"
847	And	Gxy	2.3'×2.1'	12.3	02 h 12 m 12.1 s	+44° 34' 06"
891	And	Gxy	13.5'×2.5'	10	02 h 22 m 33.5 s	+42° 21' 03"
898	And	Gxy	1.9'×0.5'	13.1	02 h 23 m 20.2 s	+41° 57' 05"
906	And	Gxy	1.8'×1.6'	13.3	02 h 25 m 16.2 s	+42° 05' 24"
909	And	Gxy	0.9'×0.9'	13.4	02 h 25 m 22.7 s	+42° 02' 08"
910	And	Gxy	2.0'×2.0'	12.2	02 h 25 m 26.9 s	+41° 49' 27"
911	And	Gxy	1.7'×0.9'	12.8	02 h 25 m 42.3 s	+41° 57' 23"
912	And	Gxy	0.9'×0.8'	14.1	02 h 25 m 42.7 s	+41° 46' 38"
913	And	Gxy	0.5'×0.2'	15	02 h 25 m 44.7 s	+41° 47' 54"
914	And	Gxy	1.8'×1.3'	13.2	02 h 26 m 05.2 s	+42° 08' 39"
920	And	Gxy	1.5'×1.1'	14.1	02 h 27 m 51.9 s	+45° 56' 49"
923	And	Gxy	0.8'×0.5'	13.9	02 h 27 m 34.6 s	+41° 58' 42"

(continued)

Table 10.1 (continued)

NGC #	Const	Type	Size	V Mag	RA (2000)	Dec (2000)
933	And	Gxy	1.3'×0.9'	14.3	02 h 29 m 17.5 s	+45° 54' 40"
937	And	Gxy	1.1'×0.5'	14.4	02 h 29 m 28.1 s	+42° 15' 00"
946	And	Gxy	1.4'×1.0'	13.3	02 h 30 m 38.5 s	+42° 13' 59"
956	And	OC	7.0'	...	02 h 32 m 30.9 s	+44° 35' 37"
980	And	Gxy	1.7'×0.9'	13.1	02 h 35 m 18.6 s	+40° 55' 35"
982	And	Gxy	1.5'×0.6'	12.7	02 h 35 m 24.9 s	+40° 52' 11"
995	And	Gxy	1.7'×1.2'	13.5	02 h 38 m 31.9 s	+41° 31' 46"
996	And	Gxy	1.4'×1.4'	13.1	02 h 38 m 39.9 s	+41° 38' 51"
999	And	Gxy	0.9'×0.8'	13.7	02 h 38 m 47.5 s	+41° 40' 14"
1000	And	Gxy	0.6'×0.6'	14.6	02 h 38 m 49.8 s	+41° 27' 35"
7440	And	Gxy	1.4'×1.1'	13.6	22 h 58 m 32.6 s	+35° 48' 09"
7445	And	Gxy	0.7'×0.2'	14.6	22 h 59 m 22.4 s	+39° 06' 27"
7446	And	Gxy	0.9'×0.9'	14.2	22 h 59 m 28.8 s	+39° 04' 58"
7449	And	Gxy	1.0'×0.7'	14.1	22 h 59 m 37.5 s	+39° 08' 44"
7618	And	Gxy	1.2'×1.0'	13.1	23 h 19 m 47.3 s	+42° 51' 09"
7640	And	Gxy	10.5'×2.0'	10.9	23 h 22 m 06.5 s	+40° 50' 44"
7662	And	PN	32"×28"	8.3	23 h 25 m 53.9 s	+42° 32' 06"
7686	And	OC	14'	5.6	23 h 30 m 07.3 s	+49° 08' 03"
7707	And	Gxy	1.3'×1.1'	13.5	23 h 34 m 51.4 s	+44° 18' 14"
7831	And	Gxy	1.7'×0.4'	13	00 h 07 m 19.5 s	+32° 36' 34"
7836	And	Gxy	0.9'×0.5'	13.8	00 h 08 m 01.6 s	+33° 04' 15"
2904	Ant	Gxy	1.5'×1.0'	12.8	09 h 30 m 16.9 s	−30° 23' 02"
2973	Ant	***	...	...	09 h 41 m 34.7 s	−30° 02' 54"
2997	Ant	Gxy	8.9'×6.8'	9.6	09 h 45 m 38.5 s	−31° 11' 25"
3001	Ant	Gxy	2.9'×1.9'	11.8	09 h 46 m 18.7 s	−30° 26' 18"
3038	Ant	Gxy	2.5'×1.3'	12	09 h 51 m 15.3 s	−32° 45' 15"
3046	Ant	Gxy	2.1'×1.9'	11.8	09 h 53 m 58.9 s	−27° 17' 12"
3051	Ant	Gxy	2.1'×1.9'	11.9	09 h 53 m 58.9 s	−27° 17' 12"
3056	Ant	Gxy	1.8'×1.1'	11.7	09 h 54 m 32.7 s	−28° 17' 55"
3082	Ant	Gxy	1.8'×0.7'	12.6	09 h 58 m 52.9 s	−30° 21' 28"
3084	Ant	Gxy	1.7'×1.2'	12.5	09 h 59 m 06.1 s	−27° 07' 43"
3087	Ant	Gxy	2.5'×2.0'	11.9	09 h 59 m 08.6 s	−34° 13' 34"
3089	Ant	Gxy	1.8'×1.0'	12.6	09 h 59 m 36.6 s	−28° 19' 54"
3095	Ant	Gxy	3.5'×2.0'	11.9	10 h 00 m 05.7 s	−31° 33' 14"
3100	Ant	Gxy	3.2'×1.6'	11.3	10 h 00 m 40.9 s	−31° 39' 53"
3103	Ant	Gxy	2.8'×1.7'	11.1	10 h 00 m 40.9 s	−31° 39' 53"
3108	Ant	Gxy	2.5'×1.8'	11.6	10 h 02 m 29.0 s	−31° 40' 34"
3113	Ant	Gxy	3.3'×1.2'	12.8	10 h 04 m 26.0 s	−28° 26' 42"
3120	Ant	Gxy	1.8'×1.3'	12.9	10 h 05 m 22.7 s	−34° 13' 16"
3125	Ant	Gxy	1.5'×0.8'	12.4	10 h 06 m 33.2 s	−29° 56' 10"
3137	Ant	Gxy	6.3'×2.2'	11.7	10 h 09 m 07.3 s	−29° 03' 52"
3157	Ant	Gxy	2.5'×0.5'	13.4	10 h 11 m 42.3 s	−31° 38' 35"
3173	Ant	Gxy	2.1'×1.7'	13	10 h 14 m 34.9 s	−27° 41' 38"
3175	Ant	Gxy	5.0'×1.3'	11.3	10 h 14 m 42.3 s	−28° 52' 17"
3223	Ant	Gxy	4.1'×2.5'	11.2	10 h 21 m 35.7 s	−34° 16' 03"

(continued)

Table 10.1 (continued)

NGC #	Const	Type	Size	V Mag	RA (2000)	Dec (2000)
3224	Ant	Gxy	1.9' × 1.5'	12.2	10 h 21 m 41.1 s	−34° 41' 47"
3241	Ant	Gxy	2.2' × 1.5'	12.4	10 h 24 m 16.9 s	−32° 29' 01"
3244	Ant	Gxy	2.0' × 1.5'	12.6	10 h 25 m 28.7 s	−39° 49' 43"
3249	Ant	Gxy	1.6' × 1.3'	13.3	10 h 26 m 22.0 s	−34° 57' 52"
3250	Ant	Gxy	2.8' × 2.0'	11.2	10 h 26 m 32.4 s	−39° 56' 42"
3257	Ant	Gxy	1.0' × 0.9'	13.1	10 h 28 m 47.0 s	−35° 39' 30"
3258	Ant	Gxy	2.9' × 2.5'	11.7	10 h 28 m 53.3 s	−35° 36' 19"
3260	Ant	Gxy	1.2' × 1.0'	12.8	10 h 29 m 06.1 s	−35° 35' 47"
3267	Ant	Gxy	1.9' × 1.1'	12.6	10 h 29 m 48.4 s	−35° 19' 17"
3268	Ant	Gxy	3.2' × 2.6'	11.6	10 h 30 m 00.5 s	−35° 19' 32"
3269	Ant	Gxy	2.5' × 1.1'	12.3	10 h 29 m 56.9 s	−35° 13' 28"
3271	Ant	Gxy	3.1' × 1.8'	11.9	10 h 30 m 26.4 s	−35° 21' 34"
3273	Ant	Gxy	1.7' × 0.8'	12.6	10 h 30 m 29.1 s	−35° 36' 38"
3275	Ant	Gxy	2.8' × 2.1'	11.8	10 h 30 m 51.7 s	−36° 44' 15"
3276	Ant	Gxy	1.0' × 0.6'	13.5	10 h 31 m 09.1 s	−39° 56' 43"
3278	Ant	Gxy	1.3' × 0.9'	12.5	10 h 31 m 35.5 s	−39° 57' 23"
3281	Ant	Gxy	3.3' × 1.7'	12	10 h 31 m 52.1 s	−34° 51' 17"
3289	Ant	Gxy	2.2' × 0.6'	12.8	10 h 34 m 07.3 s	−35° 19' 24"
3302	Ant	Gxy	1.7' × 1.2'	12.6	10 h 35 m 47.4 s	−32° 21' 32"
3333	Ant	Gxy	2.0' × 0.4'	13.3	10 h 39 m 49.7 s	−36° 02' 12"
3347	Ant	Gxy	3.6' × 2.1'	11.6	10 h 42 m 46.6 s	−36° 21' 11"
3354	Ant	Gxy	0.8' × 0.7'	12.8	10 h 43 m 02.8 s	−36° 21' 46"
3358	Ant	Gxy	3.3' × 1.9'	11.6	10 h 43 m 32.9 s	−36° 24' 37"
3378	Ant	Gxy	1.5' × 1.4'	13	10 h 46 m 43.3 s	−40° 00' 57"
3449	Ant	Gxy	3.3' × 1.0'	12.3	10 h 52 m 53.1 s	−32° 55' 34"
5612	Aps	Gxy	1.9' × 1.0'	12.3	14 h 34 m 02.0 s	−78° 23' 18"
5799	Aps	Gxy	1.3' × 1.0'	13.4	15 h 05 m 36.1 s	−72° 25' 57"
5833	Aps	Gxy	3.1' × 2.3'	12	15 h 11 m 54.1 s	−72° 51' 31"
5967	Aps	Gxy	2.9' × 1.7'	12.2	15 h 48 m 15.8 s	−75° 40' 24"
6101	Aps	GC	10.7'	9.2	16 h 25 m 48.5 s	−72° 12' 06"
6151	Aps	Ast	0.5'	…	16 h 38 m 24.1 s	−73° 15' 09"
6209	Aps	Gxy	2.0' × 1.6'	11.8	16 h 54 m 58.9 s	−72° 35' 11"
6392	Aps	Gxy	1.3' × 1.3'	11.7	17 h 43 m 30.4 s	−69° 47' 06"
6709	Aql	OC	13'	6.7	18 h 51 m 18.9 s	+10° 19' 07"
6724	Aql	OC	5' × 3'	…	18 h 56 m 46.8 s	+10° 25' 43"
6735	Aql	OC	10'	…	19 h 00 m 37.3 s	−00° 28' 31"
6738	Aql	OC	15'	8.3	19 h 01 m 21.5 s	+11° 36' 56"
6741	Aql	PN	9" × 7"	11.5	19 h 02 m 37.1 s	−00° 26' 58"
6749	Aql	GC	6.3'	12.4	19 h 05 m 15.6 s	+01° 54' 02"
6751	Aql	PN	21" × 21"	11.9	19 h 05 m 55.4 s	−05° 59' 32"
6755	Aql	OC	14'	7.5	19 h 07 m 49.0 s	+04° 15' 59"
6756	Aql	OC	4.0'	10.6	19 h 08 m 42.5 s	+04° 42' 21"
6760	Aql	GC	6.6'	9	19 h 11 m 12.0 s	+01° 01' 50"
6772	Aql	PN	70" × 56"	12.7	19 h 14 m 36.3 s	−02° 42' 25"
6773	Aql	OC	10'	…	19 h 15 m 08.4 s	+04° 51' 24"

(continued)

Table 10.1 (continued)

NGC #	Const	Type	Size	V Mag	RA (2000)	Dec (2000)
6775	Aql	OC	7'	...	19 h 16 m 42.8 s	−00° 56' 00"
6778	Aql	PN	25"×19"	12.3	19 h 18 m 24.8 s	−01° 35' 47"
6781	Aql	PN	1.9'×1.8'	11.4	19 h 18 m 28.0 s	+06° 32' 19"
6785	Aql	PN	25"×19"	12.3	19 h 18 m 24.8 s	−01° 35' 47"
6790	Aql	PN	9.6"×5.4"	10.5	19 h 22 m 56.9 s	+01° 30' 47"
6795	Aql	Ast	...	...	19 h 26 m 21.9 s	+03° 30' 52"
6803	Aql	PN	5"×5"	11.4	19 h 31 m 16.5 s	+10° 03' 22"
6804	Aql	PN	62"×49"	12	19 h 31 m 35.3 s	+09° 13' 31"
6807	Aql	PN	2"	12	19 h 34 m 33.5 s	+05° 41' 03"
6814	Aql	Gxy	3.0'×2.8'	11.3	19 h 42 m 40.6 s	−10° 19' 24"
6821	Aql	Gxy	1.2'×1.0'	13.1	19 h 44 m 24.1 s	−06° 50' 02"
6828	Aql	OC	12'×10'	...	19 h 50 m 17.5 s	+07° 54' 09"
6837	Aql	OC	6'×3'	12	19 h 53 m 08.6 s	+11° 41' 56"
6840	Aql	OC	8'×10'	...	19 h 55 m 16.2 s	+12° 06' 53"
6843	Aql	OC	6'×8'	...	19 h 56 m 06.2 s	+12° 09' 50"
6852	Aql	PN	28"	12.6	20 h 00 m 39.1 s	+01° 43' 41"
6858	Aql	OC	5'	...	20 h 02 m 59.3 s	+11° 15' 34"
6859	Aql	***	...	...	20 h 03 m 49.5 s	+00° 26' 41"
6863	Aql	Ast	...	...	20 h 05 m 07.2 s	−03° 33' 19"
6865	Aql	Gxy	0.8'×0.6'	15	20 h 05 m 56.5 s	−09° 02' 29"
6900	Aql	Gxy	0.7'×0.6'	13.5	20 h 21 m 35.0 s	−02° 34' 12"
6901	Aql	Gxy	1.4'×0.6'	13.9	20 h 22 m 21.6 s	+06° 25' 48"
6906	Aql	Gxy	1.7'×0.8'	13	20 h 23 m 34.0 s	+06° 26' 36"
6915	Aql	Gxy	1.4'×1.0'	12.3	20 h 27 m 46.0 s	−03° 04' 39"
6922	Aql	Gxy	1.3'×1.0'	13.6	20 h 29 m 52.9 s	−02° 11' 29"
6926	Aql	Gxy	1.9'×1.3'	12.7	20 h 33 m 05.9 s	−02° 01' 41"
6929	Aql	Gxy	0.8'×0.7'	13.8	20 h 33 m 21.7 s	−02° 02' 15"
6941	Aql	Gxy	2.0'×1.4'	12.8	20 h 36 m 23.3 s	−04° 37' 06"
6945	Aqr	Gxy	1.0'×0.5'	13.4	20 h 39 m 00.5 s	−04° 58' 22"
6959	Aqr	Gxy	0.6'×0.3'	13.8	20 h 47 m 07.1 s	+00° 25' 48"
6961	Aqr	Gxy	0.6'×0.5'	13.8	20 h 47 m 10.5 s	+00° 21' 48"
6962	Aqr	Gxy	2.9'×2.3'	12.3	20 h 47 m 19.0 s	+00° 19' 13"
6963	Aqr	**	...	...	20 h 47 m 19.3 s	+00° 30' 32"
6964	Aqr	Gxy	1.7'×1.3'	12.8	20 h 47 m 24.3 s	+00° 18' 03"
6965	Aqr	Gxy	0.7'×0.7'	14.1	20 h 47 m 20.4 s	+00° 29' 01"
6966	Aqr	**	...	...	20 h 47 m 26.7 s	+00° 22' 04"
6967	Aqr	Gxy	1.0'×0.7'	13.3	20 h 47 m 34.1 s	+00° 24' 40"
6968	Aqr	Gxy	1.4'×1.1'	13.3	20 h 48 m 32.3 s	−08° 21' 35"
6973	Aqr	*	...	...	20 h 52 m 05.9 s	−05° 53' 41"
6975	Aqr	Gxy	1.0'×0.2'	14	20 h 52 m 25.9 s	−05° 46' 19"
6976	Aqr	Gxy	1.3'×1.1'	14	20 h 52 m 25.9 s	−05° 46' 19"
6977	Aqr	Gxy	1.2'×0.9'	13.3	20 h 52 m 29.7 s	−05° 44' 46"
6978	Aqr	Gxy	1.5'×0.7'	13.3	20 h 52 m 35.3 s	−05° 42' 39"
6980	Aqr	*	...	...	20 h 52 m 48.9 s	−05° 50' 16"
6981	Aqr	GC	5.9'	9.2	20 h 53 m 27.9 s	−12° 32' 13"

(continued)

Table 10.1 (continued)

NGC #	Const	Type	Size	V Mag	RA (2000)	Dec (2000)
6985	Aqr	Gxy	1.5'×0.8'	13.8	20 h 45 m 02.9 s	−11° 06' 15"
6994	Aqr	Ast	2.8'	8.9	20 h 58 m 55.9 s	−12° 38' 08"
7001	Aqr	Gxy	1.4'×1.1'	13.2	21 h 01 m 07.8 s	−00° 11' 45"
7005	Aqr	Ast	...	...	21 h 01 m 57.0 s	−12° 52' 50"
7009	Aqr	PN	44"×23"	8	21 h 04 m 10.7 s	−11° 21' 49"
7010	Aqr	Gxy	1.9'×1.0'	13	21 h 04 m 39.5 s	−12° 20' 18"
7047	Aqr	Gxy	1.2'×0.7'	13.6	21 h 16 m 27.5 s	−00° 49' 34"
7051	Aqr	Gxy	1.3'×1.1'	13	21 h 19 m 51.3 s	−08° 46' 59"
7065	Aqr	Gxy	1.0'×0.9'	13.7	21 h 26 m 42.4 s	−06° 59' 42"
7069	Aqr	Gxy	1.3'×0.9'	13.5	21 h 28 m 05.9 s	−01° 38' 48"
7077	Aqr	Gxy	0.8'×0.7'	13.2	21 h 29 m 59.5 s	+02° 24' 50"
7081	Aqr	Gxy	1.1'×1.1'	12.9	21 h 31 m 24.1 s	+02° 29' 28"
7088	Aqr	NF	...	...	21 h 33 m 22.1 s	−00° 22' 57"
7089	Aqr	GC	13'	6.6	21 h 33 m 29.3 s	−00° 49' 23"
7108	Aqr	Gxy	0.5'×0.3'	14.1	21 h 41 m 53.7 s	−06° 42' 33"
7111	Aqr	Gxy	0.5'×0.3'	14.1	21 h 41 m 53.7 s	−06° 42' 33"
7120	Aqr	Gxy	0.8'×0.4'	14.4	21 h 44 m 33.1 s	−06° 31' 26"
7121	Aqr	Gxy	1.1'×0.5'	13.8	21 h 44 m 52.6 s	−03° 37' 12"
7164	Aqr	Gxy	0.9'×0.6'	14.3	21 h 56 m 23.6 s	+01° 21' 50"
7165	Aqr	Gxy	0.9'×0.8'	13.7	21 h 59 m 26.1 s	−16° 30' 44"
7167	Aqr	Gxy	1.9'×1.4'	12.7	22 h 00 m 30.7 s	−24° 38' 01"
7170	Aqr	Gxy	1.2'×0.8'	13.5	22 h 01 m 26.3 s	−05° 25' 58"
7171	Aqr	Gxy	2.6'×1.5'	12	22 h 01 m 01.9 s	−13° 16' 11"
7180	Aqr	Gxy	1.6'×0.7'	12.7	22 h 02 m 18.5 s	−20° 32' 54"
7181	Aqr	Gxy	0.9'×0.8'	14.1	22 h 01 m 43.5 s	−01° 57' 37"
7182	Aqr	Gxy	0.9'×0.4'	14.5	22 h 01 m 51.6 s	−02° 11' 48"
7183	Aqr	Gxy	3.8'×1.1'	12.1	22 h 02 m 21.7 s	−18° 55' 03"
7184	Aqr	Gxy	6.0'×1.5'	11.4	22 h 02 m 39.8 s	−20° 48' 46"
7185	Aqr	Gxy	2.3'×1.5'	12.2	22 h 02 m 56.7 s	−20° 28' 18"
7188	Aqr	Gxy	1.6'×0.7'	13.4	22 h 03 m 29.0 s	−20° 19' 02"
7189	Aqr	Gxy	1.0'×0.7'	13.7	22 h 03 m 16.0 s	+00° 34' 16"
7198	Aqr	Gxy	1.5'×1.0'	13.4	22 h 05 m 14.1 s	−00° 38' 54"
7211	Aqr	Gxy	1.0'×0.7'	14.3	22 h 06 m 21.8 s	−08° 05' 24"
7215	Aqr	Gxy	1.0'×0.4	14	22 h 08 m 34.4 s	+00° 30' 41"
7218	Aqr	Gxy	2.5'×1.1'	12.2	22 h 10 m 11.7 s	−16° 39' 40"
7220	Aqr	Gxy	0.9'×0.7'	13.6	22 h 11 m 30.9 s	−22° 57' 13"
7222	Aqr	Gxy	1.2'×1.2'	14	22 h 10 m 51.7 s	+02° 06' 19"
7230	Aqr	Gxy	0.9'×0.9'	14.1	22 h 14 m 13.1 s	−17° 04' 29"
7239	Aqr	Gxy	1.0'×0.6'	14.8	22 h 15 m 01.3 s	−05° 03' 12"
7246	Aqr	Gxy	1.4'×0.7'	14.5	22 h 17 m 42.7 s	−15° 34' 15"
7247	Aqr	Gxy	1.4'×0.9'	12.8	22 h 17 m 41.1 s	−23° 43' 52"
7251	Aqr	Gxy	1.9'×1.7'	12.7	22 h 20 m 27.1 s	−15° 46' 27"
7252	Aqr	Gxy	1.9'×1.6'	11.7	22 h 20 m 44.9 s	−24° 40' 43"
7254	Aqr	Gxy	1.4'×0.7'	13.2	22 h 22 m 35.9 s	−21° 44' 13"
7255	Aqr	Gxy	1.3'×0.4'	14.1	22 h 23 m 08.1 s	−15° 32' 29"

(continued)

Table 10.1 (continued)

NGC #	Const	Type	Size	V Mag	RA (2000)	Dec (2000)
7256	Aqr	Gxy	1.4'×0.6'	13.3	22 h 22 m 35.9 s	−21° 44' 13"
7257	Aqr	Gxy	1.9'×1.4'	12.9	22 h 22 m 36.3 s	−04° 07' 15"
7260	Aqr	Gxy	1.9'×1.4'	12.9	22 h 22 m 36.3 s	−04° 07' 15"
7266	Aqr	Gxy	0.9'×0.7'	13.9	22 h 23 m 58.9 s	−04° 04' 25"
7269	Aqr	Gxy	1.0'×0.7'	13.7	22 h 25 m 46.5 s	−13° 10' 01"
7284	Aqr	Gxy	2.1'×1.5'	12.1	22 h 28 m 35.8 s	−24° 50' 42"
7285	Aqr	Gxy	2.3'×1.4'	12.1	22 h 28 m 37.8 s	−24° 50' 28"
7287	Aqr	***	...	...	22 h 27 m 16.9 s	−22° 07' 00"
7288	Aqr	Gxy	2.3'×1.5'	13	22 h 28 m 14.9 s	−02° 53' 04"
7293	Aqr	PN	17.57'	7.3	22 h 29 m 38.5 s	−20° 50' 14"
7298	Aqr	Gxy	1.3'×1.1'	13.7	22 h 30 m 50.7 s	−14° 11' 19"
7300	Aqr	Gxy	2.0'×1.0'	12.6	22 h 31 m 00.0 s	−14° 00' 13"
7301	Aqr	Gxy	1.0'×0.5'	13.4	22 h 30 m 34.7 s	−17° 34' 27"
7302	Aqr	Gxy	1.8'×1.1'	12	22 h 32 m 23.9 s	−14° 07' 16"
7308	Aqr	Gxy	1.0'×0.8'	13.7	22 h 34 m 32.2 s	−12° 56' 02"
7309	Aqr	Gxy	1.9'×1.8'	12.3	22 h 34 m 20.3 s	−10° 21' 29"
7310	Aqr	Gxy	0.9'×0.7'	14.1	22 h 34 m 36.7 s	−22° 29' 08"
7341	Aqr	Gxy	2.4'×1.0'	12.6	22 h 39 m 05.6 s	−22° 40' 01"
7344	Aqr	Gxy	1.3'×0.7'	13.7	22 h 39 m 36.1 s	−04° 09' 33"
7349	Aqr	Gxy	1.0'×0.5'	14.4	22 h 41 m 14.7 s	−21° 47' 49"
7351	Aqr	Gxy	1.8'×1.0	12.9	22 h 41 m 26.8 s	−04° 26' 42"
7359	Aqr	Gxy	2.3'×0.6'	12.5	22 h 44 m 47.8 s	−23° 41' 18"
7364	Aqr	Gxy	1.7'×1.0'	12.8	22 h 44 m 24.3 s	−00° 09' 45"
7365	Aqr	Gxy	1.5'×1.0'	12.6	22 h 45 m 10.1 s	−19° 57' 07"
7371	Aqr	Gxy	2.1'×2.0'	11.9	22 h 46 m 03.7 s	−11° 00' 04"
7377	Aqr	Gxy	3.0'×2.5'	11.1	22 h 47 m 47.4 s	−22° 18' 43"
7378	Aqr	Gxy	1.4'×0.9'	12.7	22 h 47 m 47.7 s	−11° 49' 01"
7381	Aqr	Gxy	0.7'×0.4'	14.4	22 h 50 m 08.0 s	−19° 43' 27"
7391	Aqr	Gxy	1.7'×1.5'	12.4	22 h 50 m 36.1 s	−01° 32' 41"
7392	Aqr	Gxy	2.1'×1.3'	12.1	22 h 51 m 48.5 s	−20° 36' 21"
7393	Aqr	Gxy	1.8'×0.7'	12.9	22 h 51 m 38.3 s	−05° 33' 26"
7399	Aqr	Gxy	0.8'×0.3'	13.7	22 h 52 m 39.2 s	−09° 16' 05"
7406	Aqr	Gxy	1.1'×0.6'	14.1	22 h 53 m 56.3 s	−06° 34' 48"
7416	Aqr	Gxy	3.2'×0.9'	12.4	22 h 55 m 41.5 s	−05° 29' 42"
7425	Aqr	Gxy	0.7'×0.4'	14.7	22 h 57 m 15.5 s	−10° 57' 01"
7441	Aqr	Gxy	1.3'×0.8'	13.8	22 h 56 m 41.4 s	−07° 22' 45"
7443	Aqr	Gxy	1.4'×0.5'	12.9	23 h 00 m 08.9 s	−12° 48' 29"
7444	Aqr	Gxy	2.0'×0.9'	12.9	23 h 00 m 09.0 s	−12° 50' 04"
7447	Aqr	NF	...	...	23 h 00 m 26.0 s	−10° 31' 41"
7450	Aqr	Gxy	1.8'×1.8'	13.2	23 h 00 m 47.8 s	−12° 55' 06"
7453	Aqr	***	...	...	23 h 01 m 25.4 s	−06° 21' 24"
7471	Aqr	NF	...	...	23 h 03 m 53.8 s	−22° 54' 25"
7481	Aqr	NF	...	...	23 h 05 m 51.6 s	−19° 56' 23"
7491	Aqr	Gxy	1.0'×0.7'	13.8	23 h 08 m 06.0 s	−05° 58' 02"
7492	Aqr	GC	6.2'	11.2	23 h 08 m 26.6 s	−15° 36' 41"

(continued)

Table 10.1 (continued)

NGC #	Const	Type	Size	V Mag	RA (2000)	Dec (2000)
7494	Aqr	Gxy	0.8' × 0.7'	14.5	23 h 08 m 58.3 s	−24° 22' 11"
7498	Aqr	Gxy	1.0' × 0.5'	14.4	23 h 09 m 55.4 s	−24° 25' 31"
7502	Aqr	***	…	…	23 h 10 m 19.8 s	−21° 44' 15"
7520	Aqr	Gxy	1.3' × 0.8'	…	23 h 12 m 53.1 s	−23° 28' 10"
7522	Aqr	*	…	…	23 h 15 m 36.3 s	−22° 53' 42"
7526	Aqr	Ast	…	…	23 h 14 m 02.3 s	−09° 13' 18"
7573	Aqr	Gxy	1.3' × 1.1'	13.2	23 h 16 m 26.3 s	−22° 09' 16"
7576	Aqr	Gxy	1.2' × 1.0'	12.9	23 h 17 m 22.6 s	−04° 43' 42"
7585	Aqr	Gxy	2.3' × 2.0'	11.7	23 h 18 m 01.3 s	−04° 39' 04"
7592	Aqr	Gxy	1.2' × 0.9'	13.5	23 h 18 m 22.1 s	−04° 25' 01"
7596	Aqr	Gxy	1.0' × 0.5'	14.2	23 h 17 m 11.9 s	−06° 54' 41"
7600	Aqr	Gxy	2.5' × 1.1'	12	23 h 18 m 53.8 s	−07° 34' 50"
7606	Aqr	Gxy	5.4' × 2.1'	10.9	23 h 19 m 04.8 s	−08° 29' 08"
7646	Aqr	Gxy	0.9' × 0.6'	14	23 h 24 m 06.8 s	−11° 51' 37"
7656	Aqr	Gxy	1.2' × 1.1'	13.4	23 h 24 m 31.5 s	−19° 03' 33"
7663	Aqr	Gxy	0.9' × 0.4'	14.5	23 h 26 m 45.1 s	−04° 57' 58"
7665	Aqr	Gxy	0.7' × 0.7'	13.3	23 h 27 m 14.8 s	−09° 23' 15"
7666	Aqr	NF	…	…	23 h 27 m 24.5 s	−04° 11' 11"
7692	Aqr	Gxy	0.5' × 0.4'	14.8	23 h 32 m 46.7 s	−05° 35' 48"
7709	Aqr	Gxy	2.3' × 0.6'	12.7	23 h 35 m 27.4 s	−16° 42' 19"
7717	Aqr	Gxy	1.4' × 1.1'	12.8	23 h 37 m 43.4 s	−15° 07' 09"
7719	Aqr	Gxy	0.9' × 0.7'	14.2	23 h 38 m 02.5 s	−22° 58' 31"
7721	Aqr	Gxy	3.5' × 1.4'	11.8	23 h 38 m 48.7 s	−06° 31' 06"
7723	Aqr	Gxy	3.5' × 2.3'	11.2	23 h 38 m 57.0 s	−12° 57' 41"
7724	Aqr	Gxy	1.4' × 1.0'	13.5	23 h 39 m 07.0 s	−12° 13' 29"
7725	Aqr	Gxy	0.8' × 0.7'	13.8	23 h 39 m 14.7 s	−04° 32' 22"
7727	Aqr	Gxy	4.7' × 3.5'	10.8	23 h 39 m 53.7 s	−12° 17' 36"
7730	Aqr	Gxy	0.7' × 0.3'	14.3	23 h 40 m 45.8 s	−20° 30' 34"
7736	Aqr	Gxy	1.7' × 1.5'	12.7	23 h 42 m 25.7 s	−19° 27' 11"
7754	Aqr	Gxy	0.6' × 0.3'	14.6	23 h 49 m 11.2 s	−16° 36' 04"
7758	Aqr	Gxy	0.7' × 0.5'	14.5	23 h 48 m 55.1 s	−22° 01' 27"
7759	Aqr	Gxy	1.3' × 1.0'	13.5	23 h 48 m 54.7 s	−16° 32' 28"
7761	Aqr	Gxy	1.2' × 1.2'	13.1	23 h 51 m 28.8 s	−13° 22' 54"
7763	Aqr	Gxy	0.7' × 0.4'	14.3	23 h 50 m 15.6 s	−16° 35' 24"
7776	Aqr	Gxy	0.8' × 0.2'	14.5	23 h 54 m 16.5 s	−13° 35' 12"
6188	Ara	Neb	20' × 12'	…	16 h 40 m 05.8 s	−48° 39' 44"
6193	Ara	OC	14'	5.2	16 h 41 m 20.2 s	−48° 45' 45"
6200	Ara	OC	12'	7.4	16 h 44 m 07.3 s	−47° 27' 45"
6204	Ara	OC	5'	8.2	16 h 46 m 09.5 s	−47° 01' 01"
6208	Ara	OC	15'	7.2	16 h 49 m 28.1 s	−53° 43' 42"
6215	Ara	Gxy	2.1' × 1.8'	11.2	16 h 51 m 06.8 s	−58° 59' 33"
6221	Ara	Gxy	3.5' × 2.5'	10.6	16 h 52 m 46.2 s	−59° 13' 00"
6250	Ara	OC	7'	5.9	16 h 57 m 56.0 s	−45° 56' 12"
6253	Ara	OC	5'	10.2	16 h 59 m 05.1 s	−52° 42' 32"
6300	Ara	Gxy	4.5' × 3.0'	10.3	17 h 17 m 00.1 s	−62° 49' 15"

(continued)

Table 10.1 (continued)

NGC #	Const	Type	Size	V Mag	RA (2000)	Dec (2000)
6305	Ara	Gxy	1.8'×1.2'	12.1	17 h 18 m 00.7 s	−59° 10' 17"
6326	Ara	PN	15"×10"	...	17 h 20 m 46.3 s	−51° 45' 16"
6328	Ara	Gxy	2.4'×1.4'	12.5	17 h 23 m 40.9 s	−65° 00' 37"
6352	Ara	GC	7.1'	7.8	17 h 25 m 29.1 s	−48° 25' 22"
6362	Ara	GC	10.7'	8.1	17 h 31 m 54.7 s	−67° 02' 53"
6397	Ara	GC	25.7'	5.3	17 h 40 m 41.2 s	−53° 40' 25"
671	Ari	Gxy	1.5'×0.5'	13.3	01 h 46 m 59.4 s	+13° 07' 35"
673	Ari	Gxy	2.1'×1.7'	12.5	01 h 48 m 22.7 s	+11° 31' 23"
674	Ari	Gxy	4.5'×1.5'	12	01 h 51 m 17.5 s	+22° 21' 28"
675	Ari	Gxy	1.2'×0.5'	14.4	01 h 49 m 08.7 s	+13° 03' 33"
677	Ari	Gxy	2.0'×2.0'	12.7	01 h 49 m 14.5 s	+13° 03' 18"
678	Ari	Gxy	4.5'×0.8'	12.5	01 h 49 m 25.2 s	+21° 59' 51"
680	Ari	Gxy	1.9'×1.6'	11.9	01 h 49 m 47.3 s	+21° 58' 14"
683	Ari	Gxy	1.0'×1.0'	13.7	01 h 49 m 46.6 s	+11° 42' 04"
691	Ari	Gxy	3.5'×2.6'	12.3	01 h 50 m 41.7 s	+21° 45' 35"
694	Ari	Gxy	3.8'×2.5'	12.3	01 h 50 m 58.6 s	+21° 59' 55"
695	Ari	Gxy	0.8'×0.7'	12.9	01 h 51 m 13.9 s	+22° 34' 59"
697	Ari	Gxy	4.5'×1.5'	12	01 h 51 m 17.5 s	+22° 21' 28"
711	Ari	Gxy	1.6'×0.8'	13.2	01 h 52 m 27.7 s	+17° 30' 47"
716	Ari	Gxy	1.8'×0.8'	13.1	01 h 52 m 59.7 s	+12° 42' 30"
719	Ari	Gxy	1.4'×1.1'	13.3	01 h 53 m 38.8 s	+19° 50' 24"
722	Ari	Gxy	1.7'×0.5'	13.6	01 h 54 m 46.9 s	+20° 41' 53"
765	Ari	Gxy	2.8'×2.8'	13	01 h 58 m 47.6 s	+24° 53' 30"
770	Ari	Gxy	1.2'×0.9'	13	01 h 59 m 13.1 s	+18° 57' 19"
772	Ari	Gxy	7.2'×4.3'	10.4	01 h 59 m 20.3 s	+19° 00' 22"
774	Ari	Gxy	1.5'×1.2'	13.1	01 h 59 m 34.9 s	+14° 00' 30"
776	Ari	Gxy	1.7'×1.7'	12.6	01 h 59 m 54.7 s	+23° 38' 37"
781	Ari	Gxy	1.5'×0.4'	13.3	02 h 00 m 09.2 s	+12° 39' 19"
786	Ari	Gxy	0.7'×0.6'	13.5	02 h 01 m 24.6 s	+15° 38' 48"
792	Ari	Gxy	1.7'×1.0'	13.2	02 h 02 m 15.2 s	+15° 42' 44"
794	Ari	Gxy	1.3'×1.1'	12.8	02 h 02 m 29.4 s	+18° 22' 23"
803	Ari	Gxy	3.0'×1.3'	12.6	02 h 03 m 45.1 s	+16° 01' 54"
810	Ari	Gxy	1.7'×1.3'	13.9	02 h 05 m 28.5 s	+13° 15' 05"
817	Ari	Gxy	0.7'×0.3'	13.2	02 h 07 m 33.9 s	+17° 12' 05"
820	Ari	Gxy	1.3'×0.8'	13	02 h 08 m 25.2 s	+14° 20' 57"
821	Ari	Gxy	2.6'×1.6'	11.2	02 h 08 m 21.0 s	+10° 59' 44"
870	Ari	Gxy	0.1'×0.1'	15.5	02 h 17 m 09.1 s	+14° 31' 20"
871	Ari	Gxy	1.2'×0.5'	13.2	02 h 17 m 10.6 s	+14° 32' 52"
876	Ari	Gxy	2.1'×0.4'	13.8	02 h 17 m 53.3 s	+14° 31' 16"
877	Ari	Gxy	2.4'×1.8'	11.8	02 h 17 m 58.7 s	+14° 32' 50"
882	Ari	Gxy	1.1'×0.6'	13.7	02 h 19 m 39.9 s	+15° 48' 51"
900	Ari	Gxy	1.1'×0.7'	13.8	02 h 23 m 32.2 s	+26° 30' 42"
901	Ari	Gxy	1.1'×0.7'	14.8	02 h 23 m 34.1 s	+26° 33' 25"
903	Ari	Gxy	0.5'×0.3'	15.7	02 h 24 m 00.9 s	+27° 21' 23"
904	Ari	Gxy	1.2'×0.9'	13.7	02 h 24 m 05.6 s	+27° 20' 32"

(continued)

Table 10.1 (continued)

NGC #	Const	Type	Size	V Mag	RA (2000)	Dec (2000)
915	Ari	Gxy	0.7'×0.7'	14.2	02 h 25 m 45.6 s	+27° 13' 15"
916	Ari	Gxy	0.9'×0.3'	15.3	02 h 25 m 47.7 s	+27° 14' 33"
918	Ari	Gxy	3.5'×2.0'	13.1	02 h 25 m 50.5 s	+18° 29' 50"
919	Ari	Gxy	1.3'×0.3'	14.7	02 h 26 m 16.7 s	+27° 12' 43"
924	Ari	Gxy	2.3'×1.3'	12.5	02 h 26 m 46.5 s	+20° 29' 55"
927	Ari	Gxy	1.2'×1.2'	13.6	02 h 26 m 37.3 s	+12° 09' 21"
928	Ari	Gxy	0.7'×0.3'	14	02 h 27 m 41.0 s	+27° 13' 15"
930	Ari	NF	...	...	02 h 27 m 51.4 s	+20° 20' 30"
932	Ari	Gxy	1.8'×1.7'	12.5	02 h 27 m 54.7 s	+20° 19' 56"
935	Ari	Gxy	1.7'×1.1'	13.1	02 h 28 m 11.2 s	+19° 35' 59"
938	Ari	Gxy	1.6'×1.2'	12.5	02 h 28 m 33.5 s	+20° 17' 02"
962	Ari	Gxy	...'×...'	13	02 h 32 m 40.2 s	+28° 04' 13"
972	Ari	Gxy	3.3'×1.7'	11.4	02 h 34 m 13.3 s	+29° 18' 42"
976	Ari	Gxy	1.5'×1.2'	12.3	02 h 33 m 59.9 s	+20° 58' 35"
984	Ari	Gxy	3.0'×2.0'	12.9	02 h 34 m 43.1 s	+23° 24' 47"
990	Ari	Gxy	1.8'×1.5'	12.5	02 h 36 m 18.3 s	+11° 38' 30"
992	Ari	Gxy	0.9'×0.7'	14.2	02 h 37 m 25.5 s	+21° 06' 02"
1012	Ari	Gxy	2.5'×1.1'	12.2	02 h 39 m 14.9 s	+30° 09' 05"
1024	Ari	Gxy	3.9'×1.4'	12.6	02 h 39 m 11.9 s	+10° 50' 49"
1028	Ari	Gxy	0.8'×0.3'	14.8	02 h 39 m 37.2 s	+10° 50' 38"
1029	Ari	Gxy	1.4'×0.4'	13.3	02 h 39 m 36.5 s	+10° 47' 37"
1030	Ari	Gxy	1.6'×0.7'	13.5	02 h 39 m 50.5 s	+18° 01' 23"
1036	Ari	Gxy	1.4'×1.0'	13	02 h 40 m 29.0 s	+19° 17' 49"
1054	Ari	Gxy	0.9'×0.5'	13.9	02 h 42 m 15.7 s	+18° 13' 03"
1056	Ari	Gxy	2.3'×1.1'	12.6	02 h 42 m 48.6 s	+28° 34' 26"
1059	Ari	**	...	...	02 h 42 m 35.6 s	+17° 59' 48"
1088	Ari	Gxy	1.2'×0.7'	13.6	02 h 47 m 03.9 s	+16° 11' 59"
1109	Ari	Gxy	1.3'×1.0'	14.2	02 h 47 m 43.5 s	+13° 15' 20"
1111	Ari	Gxy	0.5'×0.2'	15.2	02 h 48 m 39.2 s	+13° 15' 31"
1112	Ari	Gxy	1.2'×0.9'	13.9	02 h 49 m 00.3 s	+13° 13' 25"
1113	Ari	*	...	...	02 h 50 m 05.1 s	+13° 19' 39"
1115	Ari	Gxy	0.6'×0.2'	14.8	02 h 50 m 25.3 s	+13° 15' 59"
1116	Ari	Gxy	1.5'×0.3'	14.5	02 h 50 m 35.7 s	+13° 20' 05"
1117	Ari	Gxy	0.5'×0.3'	14.3	02 h 51 m 13.1 s	+13° 11' 07"
1127	Ari	Gxy	1.0'×1.0'	14.6	02 h 52 m 51.8 s	+13° 15' 24"
1134	Ari	Gxy	2.5'×0.9'	12.3	02 h 53 m 41.2 s	+13° 00' 53"
1156	Ari	Gxy	3.3'×2.5'	11.7	02 h 59 m 42.1 s	+25° 14' 11"
1166	Ari	Gxy	1.2'×1.1'	14.2	03 h 00 m 34.9 s	+11° 50' 34"
1168	Ari	Gxy	1.3'×0.8'	14.4	03 h 00 m 46.9 s	+11° 46' 17"
1170	Ari	NF	...	...	03 h 02 m 26.7 s	+27° 04' 19"
1236	Ari	Gxy	0.5'×0.3'	14.7	03 h 11 m 28.0 s	+10° 48' 30"
1240	Ari	**	...	...	03 h 13 m 26.7 s	+30° 30' 26"
1664	Aur	OC	18'	7.6	04 h 51 m 08.4 s	+43° 40' 28"
1724	Aur	OC	2.0'	...	05 h 03 m 33.3 s	+49° 29' 30"
1778	Aur	OC	6'	7.7	05 h 08 m 05.7 s	+37° 01' 22"

(continued)

Table 10.1 (continued)

NGC #	Const	Type	Size	V Mag	RA (2000)	Dec (2000)
1790	Aur	Ast	…	…	05 h 10 m 56.2 s	+52° 03' 35"
1798	Aur	OC	5'	10	05 h 11 m 40.3 s	+47° 41' 20"
1857	Aur	OC	5'	7	05 h 20 m 05.5 s	+39° 20' 37"
1883	Aur	OC	2.5'	12	05 h 25 m 54.2 s	+46° 29' 25"
1893	Aur	OC+Neb	11'	7.5	05 h 22 m 45.1 s	+33° 25' 13"
1907	Aur	OC	6'	8.2	05 h 28 m 04.5 s	+35° 19' 32"
1912	Aur	OC	21'	6.4	05 h 28 m 42.5 s	+35° 51' 18"
1931	Aur	OC+Neb	1.0'	10.1	05 h 31 m 25.8 s	+34° 14' 42"
1960	Aur	OC	12'	6	05 h 36 m 17.7 s	+34° 08' 27"
1985	Aur	Neb	70"	…	05 h 37 m 47.8 s	+31° 59' 20"
2013	Aur	OC	…	…	05 h 44 m 01.7 s	+55° 47' 37"
2099	Aur	OC	23'	5.6	05 h 52 m 18.3 s	+32° 33' 11"
2126	Aur	OC	6'	10.2	06 h 02 m 33.0 s	+49° 51' 57"
2165	Aur	OC	6'	…	06 h 11 m 04.2 s	+51° 40' 38"
2192	Aur	OC	5'	10.9	06 h 15 m 17.4 s	+39° 51' 19"
2208	Aur	Gxy	1.7'×1.0'	12.9	06 h 22 m 34.6 s	+51° 54' 33"
2240	Aur	OC	…	…	06 h 33 m 10.5 s	+35° 15' 01"
2242	Aur	PN	0.5'	15	06 h 34 m 07.2 s	+44° 46' 38"
2281	Aur	OC	14'	5.4	06 h 48 m 17.8 s	+41° 04' 44"
2303	Aur	Gxy	1.5'×1.5'	12.7	06 h 56 m 17.5 s	+45° 29' 34"
2387	Aur	Gxy	0.5'×0.4'	14.5	07 h 28 m 57.8 s	+36° 52' 46"
5239	Boo	Gxy	1.8'×1.2'	13	13 h 36 m 26.2 s	+07° 22' 10"
5248	Boo	Gxy	6.2'×4.5'	10	13 h 37 m 32.0 s	+08° 53' 08"
5249	Boo	Gxy	1.5'×1.1'	13	13 h 37 m 37.5 s	+15° 58' 19"
5251	Boo	Gxy	0.7'×0.7'	13.9	13 h 37 m 24.7 s	+27° 25' 09"
5293	Boo	Gxy	1.9'×1.5'	13.1	13 h 46 m 52.7 s	+16° 16' 22"
5332	Boo	Gxy	0.9'×0.9'	13.3	13 h 52 m 07.9 s	+16° 58' 10"
5405	Boo	Gxy	0.8'×0.8'	13.8	14 h 01 m 09.3 s	+07° 42' 07"
5409	Boo	Gxy	1.7'×1.1'	13.2	14 h 01 m 46.1 s	+09° 29' 24"
5411	Boo	Gxy	1.4'×0.8'	13.4	14 h 01 m 58.9 s	+08° 56' 16"
5414	Boo	Gxy	1.0'×0.8'	12.8	14 h 02 m 03.5 s	+09° 55' 46"
5416	Boo	Gxy	1.4'×0.8'	13.2	14 h 02 m 11.3 s	+09° 26' 23"
5417	Boo	Gxy	1.5'×0.6'	13.2	14 h 02 m 13.0 s	+08° 02' 12"
5418	Boo	Gxy	1.1'×0.5'	13.6	14 h 02 m 17.5 s	+07° 41' 02"
5423	Boo	Gxy	1.5'×0.9'	12.7	14 h 02 m 48.6 s	+09° 20' 28"
5424	Boo	Gxy	1.6'×1.3'	12.7	14 h 02 m 55.7 s	+09° 25' 15"
5431	Boo	Gxy	0.7'×0.5'	14.3	14 h 03 m 07.1 s	+09° 21' 47"
5434	Boo	Gxy	1.8'×1.8'	12.9	14 h 03 m 23.1 s	+09° 26' 57"
5436	Boo	Gxy	1.1'×0.5'	14	14 h 03 m 40.9 s	+09° 34' 24"
5437	Boo	Gxy	0.8'×0.4'	14.3	14 h 03 m 47.3 s	+09° 31' 26"
5438	Boo	Gxy	0.8'×0.8'	13.6	14 h 03 m 48.0 s	+09° 36' 38"
5446	Boo	Gxy	0.8'×0.8'	13.6	14 h 03 m 48.0 s	+09° 36' 38"
5454	Boo	Gxy	1.9'×1.3'	12.8	14 h 04 m 45.7 s	+14° 22' 55"
5456	Boo	Gxy	1.2'×1.0'	13	14 h 04 m 58.8 s	+11° 52' 17"
5459	Boo	Gxy	1.1'×1.0'	13.2	14 h 05 m 00.1 s	+13° 07' 54"

(continued)

Table 10.1 (continued)

NGC #	Const	Type	Size	V Mag	RA (2000)	Dec (2000)
5463	Boo	Gxy	1.2'×0.5'	12.9	14 h 06 m 10.5 s	+09° 21' 11"
5466	Boo	GC	11'	9.2	14 h 05 m 27.9 s	+28° 31' 49"
5469	Boo	Gxy	0.5'×0.4'	13.9	14 h 12 m 29.9 s	+08° 38' 55"
5482	Boo	Gxy	1.2'×0.9'	13	14 h 08 m 30.7 s	+08° 55' 55"
5487	Boo	Gxy	0.9'×0.4'	14	14 h 09 m 43.8 s	+08° 04' 08"
5490	Boo	Gxy	2.4'×2.0'	11.9	14 h 09 m 57.3 s	+17° 32' 44"
5492	Boo	Gxy	2.4'×0.7'	13	14 h 10 m 35.3 s	+19° 36' 43"
5497	Boo	Gxy	1.2'×0.7'	14.3	14 h 10 m 31.5 s	+38° 53' 36"
5498	Boo	Gxy	1.0'×0.8'	13.7	14 h 11 m 04.5 s	+25° 41' 53"
5499	Boo	Gxy	0.9'×0.6'	13.6	14 h 10 m 47.7 s	+35° 54' 47"
5500	Boo	Gxy	1.0'×0.9'	13.4	14 h 10 m 15.3 s	+48° 32' 45"
5504	Boo	Gxy	1.3'×1.1'	13.2	14 h 12 m 15.8 s	+15° 50' 30"
5505	Boo	Gxy	1.0'×0.8'	13.4	14 h 12 m 31.7 s	+13° 18' 17"
5508	Boo	Gxy	1.1'×0.8'	13.2	14 h 12 m 29.0 s	+24° 38' 07"
5509	Boo	Gxy	1.1'×0.8'	14.1	14 h 12 m 39.5 s	+20° 23' 12"
5511	Boo	Gxy	0.4'×0.2'	14.5	14 h 13 m 05.3 s	+08° 37' 55"
5512	Boo	Gxy	0.5'×0.3'	14.4	14 h 12 m 41.1 s	+30° 51' 18"
5513	Boo	Gxy	1.9'×1.1'	12.7	14 h 13 m 08.7 s	+20° 24' 58"
5514	Boo	Gxy	2.2'×1.1'	13.3	14 h 13 m 38.7 s	+07° 39' 36"
5515	Boo	Gxy	1.3'×0.7'	13.1	14 h 12 m 38.1 s	+39° 18' 35"
5517	Boo	Gxy	1.0'×0.8'	14	14 h 12 m 51.3 s	+35° 42' 38"
5518	Boo	Gxy	0.7'×0.7'	14	14 h 13 m 47.6 s	+20° 50' 54"
5519	Boo	Gxy	2.1'×1.2'	13.3	14 h 14 m 20.8 s	+07° 30' 56"
5520	Boo	Gxy	2.0'×1.1'	12.7	14 h 12 m 22.8 s	+50° 20' 53"
5522	Boo	Gxy	1.9'×0.4'	13.7	14 h 14 m 50.3 s	+15° 08' 48"
5523	Boo	Gxy	4.6'×1.3'	12.3	14 h 14 m 51.4 s	+25° 19' 03"
5524	Boo	**	...	...	14 h 13 m 48.9 s	+36° 22' 53"
5525	Boo	Gxy	1.4'×0.9'	13	14 h 15 m 39.2 s	+14° 16' 57"
5527	Boo	Gxy	0.8'×0.7'	14.4	14 h 14 m 27.3 s	+36° 24' 15"
5528	Boo	Gxy	0.8'×0.3'	14.1	14 h 16 m 19.9 s	+08° 17' 34"
5529	Boo	Gxy	6.2'×0.8'	12.2	14 h 15 m 34.0 s	+36° 13' 35"
5531	Boo	Gxy	0.3'×0.2'	13.7	14 h 16 m 43.3 s	+10° 53' 06"
5532	Boo	Gxy	1.6'×1.6'	11.9	14 h 16 m 52.9 s	+10° 48' 26"
5533	Boo	Gxy	3.1'×1.9'	11.9	14 h 16 m 07.7 s	+35° 20' 37"
5535	Boo	Gxy	0.3'×0.3'	15	14 h 17 m 31.2 s	+08° 12' 29"
5536	Boo	Gxy	0.9'×0.9'	13.7	14 h 16 m 23.7 s	+39° 30' 08"
5538	Boo	Gxy	0.8'×0.2'	14.7	14 h 17 m 42.4 s	+07° 28' 35"
5539	Boo	Gxy	0.4'×0.3'	14.6	14 h 17 m 37.7 s	+08° 10' 46"
5541	Boo	Gxy	0.8'×0.5'	12.7	14 h 16 m 31.8 s	+39° 35' 19"
5542	Boo	Gxy	0.4'×0.3'	14.3	14 h 17 m 53.2 s	+07° 33' 31"
5543	Boo	Gxy	0.6'×0.3'	14.7	14 h 18 m 04.1 s	+07° 39' 16"
5544	Boo	Gxy	1.0'×1.0'	13.3	14 h 17 m 02.5 s	+36° 34' 16"
5545	Boo	Gxy	1.3'×0.4'	14.1	14 h 17 m 05.3 s	+36° 34' 31"
5546	Boo	Gxy	1.3'×1.1'	12.8	14 h 18 m 09.2 s	+07° 33' 51"
5548	Boo	Gxy	1.9'×1.7'	12.1	14 h 17 m 59.6 s	+25° 08' 12"

(continued)

Table 10.1 (continued)

NGC #	Const	Type	Size	V Mag	RA (2000)	Dec (2000)
5550	Boo	Gxy	1.2'×0.8'	13.3	14 h 18 m 28.0 s	+12° 52' 58"
5553	Boo	Gxy	1.3'×0.3'	14.3	14 h 18 m 29.6 s	+26° 17' 14"
5557	Boo	Gxy	2.3'×1.9'	11.2	14 h 18 m 25.7 s	+36° 29' 36"
5559	Boo	Gxy	1.4'×0.4'	14.2	14 h 19 m 12.7 s	+24° 47' 55"
5562	Boo	Gxy	0.7'×0.7'	13.6	14 h 20 m 11.0 s	+10° 15' 46"
5567	Boo	Gxy	1.1'×0.9'	14	14 h 19 m 17.5 s	+35° 08' 17"
5568	Boo	Gxy	0.8'×0.6'	15	14 h 19 m 21.3 s	+35° 05' 32"
5570	Boo	Gxy	2.1'×1.2'	13.3	14 h 14 m 20.8 s	+07° 30' 56"
5571	Boo	Ast	...	...	14 h 19 m 32.0 s	+35° 09' 03"
5572	Boo	Gxy	1.2'×0.9'	14.4	14 h 19 m 35.3 s	+36° 08' 25"
5579	Boo	Gxy	1.9'×1.4'	13.8	14 h 20 m 26.5 s	+35° 11' 17"
5580	Boo	Gxy	1.8'×1.8'	12.3	14 h 21 m 38.3 s	+35° 12' 16"
5581	Boo	Gxy	0.8'×0.7'	14.1	14 h 21 m 16.3 s	+23° 28' 47"
5582	Boo	Gxy	2.8'×1.7'	11.8	14 h 20 m 43.1 s	+39° 41' 36"
5583	Boo	Gxy	0.8'×0.6'	13.5	14 h 21 m 40.5 s	+13° 13' 55"
5586	Boo	NF	...	...	14 h 22 m 07.6 s	+13° 11' 03"
5587	Boo	Gxy	2.6'×0.8'	12.7	14 h 22 m 10.8 s	+13° 55' 03"
5588	Boo	Gxy	1.3'×1.3'	13.3	14 h 21 m 25.2 s	+35° 16' 13"
5589	Boo	Gxy	1.1'×1.1'	13.5	14 h 21 m 25.2 s	+35° 16' 13"
5590	Boo	Gxy	1.8'×1.8'	12.4	14 h 21 m 38.3 s	+35° 12' 16"
5591	Boo	Gxy	1.5'×0.7'	13.4	14 h 22 m 33.4 s	+13° 43' 00"
5594	Boo	Gxy	1.0'×0.5'	14.2	14 h 23 m 10.2 s	+26° 15' 56"
5596	Boo	Gxy	1.0'×0.8'	13.5	14 h 22 m 28.7 s	+37° 07' 19"
5598	Boo	Gxy	1.5'×1.0'	13.1	14 h 22 m 28.3 s	+40° 19' 11"
5600	Boo	Gxy	1.4'×1.4'	12	14 h 23 m 49.5 s	+14° 38' 20"
5601	Boo	Gxy	0.9'×0.3'	14.7	14 h 22 m 53.3 s	+40° 18' 34"
5602	Boo	Gxy	1.4'×0.8'	12.9	14 h 22 m 18.9 s	+50° 30' 07"
5603	Boo	Gxy	1.1'×1.1'	13	14 h 23 m 01.6 s	+40° 22' 38"
5608	Boo	Gxy	2.6'×1.3'	13.5	14 h 23 m 17.7 s	+41° 46' 31"
5609	Boo	Gxy	0.3'×0.2'	15.7	14 h 23 m 48.2 s	+34° 50' 35"
5610	Boo	Gxy	2.0'×0.7'	13.4	14 h 24 m 22.9 s	+24° 36' 51"
5611	Boo	Gxy	1.3'×0.6'	12.7	14 h 24 m 04.7 s	+33° 02' 51"
5613	Boo	Gxy	1.0'×0.8'	14.3	14 h 24 m 05.9 s	+34° 53' 32"
5614	Boo	Gxy	2.5'×2.0'	11.8	14 h 24 m 07.6 s	+34° 51' 34"
5615	Boo	Gxy	0.2'×0.2'	14.7	14 h 24 m 06.3 s	+34° 51' 55"
5616	Boo	Gxy	2.1'×0.4'	14	14 h 24 m 20.6 s	+36° 27' 42"
5621	Boo	**	...	...	14 h 27 m 49.7 s	+08° 14' 25"
5622	Boo	Gxy	1.7'×1.0'	13.4	14 h 26 m 12.2 s	+48° 33' 50"
5623	Boo	Gxy	1.6'×1.1'	12.6	14 h 27 m 08.7 s	+33° 15' 09"
5624	Boo	Gxy	1.1'×0.7'	13.2	14 h 26 m 35.4 s	+51° 35' 09"
5625	Boo	Gxy	1.2'×0.8'	14	14 h 27 m 02.3 s	+39° 57' 26"
5627	Boo	Gxy	1.6'×0.9'	13	14 h 28 m 34.3 s	+11° 22' 42"
5628	Boo	Gxy	1.1'×0.7'	13.4	14 h 28 m 25.7 s	+17° 55' 28"
5629	Boo	Gxy	1.8'×1.8'	12.6	14 h 28 m 16.3 s	+25° 50' 55"
5630	Boo	Gxy	2.2'×0.7'	13.1	14 h 27 m 36.6 s	+41° 15' 27"

(continued)

Table 10.1 (continued)

NGC #	Const	Type	Size	V Mag	RA (2000)	Dec (2000)
5633	Boo	Gxy	2.0' × 1.2'	12.2	14 h 27 m 28.5 s	+46° 08' 48"
5635	Boo	Gxy	2.3' × 1.1'	12.9	14 h 28 m 31.7 s	+27° 24' 32"
5637	Boo	Gxy	0.9' × 0.5'	13.9	14 h 28 m 59.5 s	+23° 11' 29"
5639	Boo	Gxy	1.4' × 0.9'	13.7	14 h 28 m 46.6 s	+30° 24' 46"
5641	Boo	Gxy	2.5' × 1.3'	12.4	14 h 29 m 16.7 s	+28° 49' 17"
5642	Boo	Gxy	1.8' × 1.3'	12.7	14 h 29 m 13.3 s	+30° 01' 34"
5644	Boo	Gxy	1.4' × 1.4'	12.6	14 h 30 m 25.6 s	+11° 55' 41"
5646	Boo	Gxy	1.5' × 0.4'	14.4	14 h 29 m 34.0 s	+35° 27' 42"
5647	Boo	Gxy	1.1' × 0.3'	14.2	14 h 30 m 36.0 s	+11° 52' 35"
5648	Boo	Gxy	1.1' × 0.8'	13.3	14 h 30 m 32.7 s	+14° 01' 22"
5649	Boo	Gxy	1.0' × 0.7'	13.3	14 h 30 m 32.7 s	+14° 01' 22"
5653	Boo	Gxy	1.7' × 1.3'	12.3	14 h 30 m 10.3 s	+31° 12' 54"
5654	Boo	Gxy	1.6' × 1.0'	13.1	14 h 30 m 01.5 s	+36° 21' 35"
5655	Boo	Gxy	1.1' × 0.9'	13.3	14 h 30 m 50.9 s	+13° 58' 08"
5656	Boo	Gxy	1.9' × 1.5'	12	14 h 30 m 25.4 s	+35° 19' 15"
5657	Boo	Gxy	1.9' × 0.8'	13.5	14 h 30 m 43.5 s	+29° 10' 50"
5659	Boo	Gxy	1.6' × 0.4'	14.1	14 h 31 m 06.1 s	+25° 21' 19"
5660	Boo	Gxy	2.8' × 2.5'	11.6	14 h 29 m 49.8 s	+49° 37' 20"
5665	Boo	Gxy	1.9' × 1.3'	12.2	14 h 32 m 25.6 s	+08° 04' 45"
5666	Boo	Gxy	0.9' × 0.8'	12.8	14 h 33 m 09.1 s	+10° 30' 38"
5669	Boo	Gxy	4.0' × 2.8'	11.5	14 h 32 m 43.9 s	+09° 53' 31"
5672	Boo	Gxy	0.9' × 0.6'	14.1	14 h 32 m 38.5 s	+31° 40' 11"
5673	Boo	Gxy	2.5' × 0.6'	13.4	14 h 31 m 30.9 s	+49° 57' 30"
5675	Boo	Gxy	2.8' × 1.0'	13	14 h 32 m 39.9 s	+36° 18' 07"
5676	Boo	Gxy	4.0' × 1.9'	11.1	14 h 32 m 46.7 s	+49° 27' 26"
5677	Boo	Gxy	0.9' × 0.7'	13.9	14 h 34 m 12.7 s	+25° 28' 03"
5681	Boo	Gxy	0.9' × 0.6'	13.5	14 h 35 m 42.9 s	+08° 18' 01"
5682	Boo	Gxy	1.7' × 0.6'	14.2	14 h 34 m 45.0 s	+48° 40' 15"
5683	Boo	Gxy	1.3' × 0.6'	14.4	14 h 34 m 52.5 s	+48° 39' 41"
5684	Boo	Gxy	1.5' × 1.3'	13	14 h 35 m 50.1 s	+36° 32' 34"
5685	Boo	Gxy	1.1' × 1.0'	13.4	14 h 36 m 15.3 s	+29° 54' 30"
5686	Boo	Gxy	0.4' × 0.4'	14.4	14 h 36 m 02.6 s	+36° 30' 10"
5687	Boo	Gxy	2.6' × 1.9'	11.8	14 h 34 m 52.3 s	+54° 28' 32"
5689	Boo	Gxy	3.5' × 1.0'	11.7	14 h 35 m 29.7 s	+48° 44' 29"
5693	Boo	Gxy	1.8' × 1.5'	13.6	14 h 36 m 11.3 s	+48° 35' 06"
5695	Boo	Gxy	1.5' × 1.1'	12.7	14 h 37 m 22.1 s	+36° 34' 03"
5696	Boo	Gxy	2.0' × 1.5'	13.2	14 h 36 m 57.1 s	+41° 49' 40"
5697	Boo	Gxy	0.9' × 0.5'	13.9	14 h 36 m 31.9 s	+41° 41' 07"
5698	Boo	Gxy	1.9' × 0.9'	13.2	14 h 37 m 14.8 s	+38° 27' 17"
5699	Boo	Gxy	0.4' × 0.3'	14.8	14 h 38 m 42.4 s	+30° 27' 59"
5700	Boo	Gxy	1.0' × 0.4'	14.4	14 h 37 m 01.6 s	+48° 32' 41"
5702	Boo	Gxy	1.1' × 0.8'	13.4	14 h 38 m 55.0 s	+20° 30' 22"
5703	Boo	Gxy	1.6' × 0.4'	13.6	14 h 38 m 49.9 s	+30° 26' 32"
5704	Boo	Gxy	1.7' × 0.6'	13.3	14 h 38 m 16.3 s	+40° 27' 24"
5706	Boo	Gxy	0.4' × 0.3'	14.8	14 h 38 m 42.4 s	+30° 27' 59"

(continued)

Table 10.1 (continued)

NGC #	Const	Type	Size	V Mag	RA (2000)	Dec (2000)
5707	Boo	Gxy	2.6'×0.5'	12.7	14 h 37 m 30.6 s	+51° 33' 42"
5708	Boo	Gxy	1.6'×0.6'	13.4	14 h 38 m 16.3 s	+40° 27' 24"
5709	Boo	Gxy	1.6'×0.4'	13.8	14 h 38 m 49.9 s	+30° 26' 32"
5710	Boo	Gxy	1.2'×1.1'	13	14 h 39 m 16.1 s	+20° 02' 35"
5711	Boo	Gxy	1.0'×0.6'	14.1	14 h 39 m 22.5 s	+19° 59' 25"
5714	Boo	Gxy	3.2'×0.4'	13.6	14 h 38 m 11.6 s	+46° 38' 17"
5717	Boo	Gxy	0.7'×0.6'	14.8	14 h 38 m 37.7 s	+46° 39' 46"
5720	Boo	Gxy	2.2'×1.5'	13.6	14 h 38 m 33.3 s	+50° 48' 53"
5721	Boo	Gxy	0.3'×0.2'	14.9	14 h 38 m 53.0 s	+46° 40' 29"
5722	Boo	Gxy	0.3'×0.3'	14.7	14 h 38 m 54.3 s	+46° 39' 56"
5723	Boo	Gxy	0.6'×0.2'	15	14 h 38 m 57.9 s	+46° 41' 23"
5724	Boo	Gxy	0.1'×0.1'	15.9	14 h 39 m 02.1 s	+46° 41' 32"
5727	Boo	Gxy	2.2'×1.2'	13.7	14 h 40 m 26.0 s	+33° 59' 21"
5730	Boo	Gxy	1.8'×0.4'	14.2	14 h 39 m 52.1 s	+42° 44' 31"
5731	Boo	Gxy	1.6'×0.4'	13.3	14 h 40 m 09.2 s	+42° 46' 45"
5732	Boo	Gxy	1.3'×0.7'	13.7	14 h 40 m 38.9 s	+38° 38' 15"
5735	Boo	Gxy	2.4'×1.9'	12.5	14 h 42 m 33.3 s	+28° 43' 34"
5736	Boo	Gxy	1.0'×0.6'	14.3	14 h 43 m 30.7 s	+11° 12' 10"
5737	Boo	Gxy	1.3'×0.8'	13.7	14 h 43 m 11.9 s	+18° 52' 47"
5739	Boo	Gxy	2.3'×2.1'	12.3	14 h 42 m 28.8 s	+41° 50' 32"
5747	Boo	Gxy	1.0'×0.9'	13.6	14 h 44 m 20.8 s	+12° 07' 54"
5748	Boo	Gxy	0.8'×0.5'	14.5	14 h 45 m 05.1 s	+21° 54' 58"
5751	Boo	Gxy	1.5'×0.8'	13.4	14 h 43 m 49.1 s	+53° 24' 02"
5752	Boo	Gxy	0.6'×0.3'	14.1	14 h 45 m 14.1 s	+38° 43' 44"
5753	Boo	Gxy	0.5'×0.3'	15	14 h 45 m 18.8 s	+38° 48' 21"
5754	Boo	Gxy	2.0'×1.8'	13.1	14 h 45 m 19.6 s	+38° 43' 54"
5755	Boo	Gxy	1.7'×1.3'	13.5	14 h 45 m 24.5 s	+38° 46' 47"
5758	Boo	Gxy	1.0'×0.9'	13.6	14 h 47 m 02.1 s	+13° 40' 05"
5759	Boo	Gxy	0.5'×0.3'	14.3	14 h 47 m 14.7 s	+13° 27' 24"
5760	Boo	Gxy	1.5'×0.7'	13.5	14 h 47 m 42.3 s	+18° 30' 06"
5762	Boo	Gxy	1.5'×1.3'	13.1	14 h 48 m 42.6 s	+12° 27' 26"
5763	Boo	Gxy	0.5'×0.5'	14.2	14 h 48 m 58.7 s	+12° 29' 25"
5767	Boo	Gxy	0.9'×0.6'	14.3	14 h 49 m 34.3 s	+47° 22' 33"
5769	Boo	Gxy	0.3'×0.3'	14.4	14 h 52 m 41.5 s	+07° 55' 54"
5771	Boo	Gxy	0.4'×0.3'	13.6	14 h 52 m 14.3 s	+29° 50' 43"
5772	Boo	Gxy	2.1'×1.3'	13	14 h 51 m 38.9 s	+40° 35' 56"
5773	Boo	Gxy	0.9'×0.9'	13.5	14 h 52 m 30.4 s	+29° 48' 26"
5778	Boo	Gxy	1.1'×0.7'	13.9	14 h 54 m 31.5 s	+18° 38' 32"
5780	Boo	Gxy	0.8'×0.4'	14.1	14 h 54 m 22.6 s	+28° 56' 22"
5782	Boo	Gxy	0.7'×0.7'	13.7	14 h 55 m 55.2 s	+11° 51' 41"
5783	Boo	Gxy	2.9'×1.8'	13	14 h 53 m 28.1 s	+52° 04' 33"
5784	Boo	Gxy	1.9'×1.8'	12.5	14 h 54 m 16.3 s	+42° 33' 28"
5785	Boo	Gxy	2.8'×1.6'	12.9	14 h 53 m 28.1 s	+52° 04' 33"
5787	Boo	Gxy	1.1'×1.0'	13.2	14 h 55 m 15.5 s	+42° 30' 24"
5788	Boo	Gxy	0.6'×0.4'	14.9	14 h 53 m 16.7 s	+52° 02' 38"

(continued)

Table 10.1 (continued)

NGC #	Const	Type	Size	V Mag	RA (2000)	Dec (2000)
5789	Boo	Gxy	0.9' × 0.8'	13.7	14 h 56 m 35.6 s	+30° 14' 03"
5790	Boo	Gxy	1.1' × 0.8'	13.8	14 h 57 m 35.9 s	+08° 17' 07"
5794	Boo	Gxy	1.0' × 1.0'	13.6	14 h 55 m 53.7 s	+49° 43' 32"
5795	Boo	Gxy	1.6' × 0.3'	14	14 h 56 m 19.3 s	+49° 23' 55"
5797	Boo	Gxy	1.1' × 0.7'	13	14 h 56 m 24.0 s	+49° 41' 46"
5798	Boo	Gxy	1.4' × 1.0'	13.2	14 h 57 m 37.7 s	+29° 58' 07"
5804	Boo	Gxy	1.3' × 1.1'	13.3	14 h 57 m 06.9 s	+49° 40' 07"
5805	Boo	Gxy	0.4' × 0.3'	15	14 h 57 m 11.7 s	+49° 37' 44"
5818	Boo	Gxy	1.2' × 0.9'	13.8	14 h 58 m 58.3 s	+49° 49' 16"
5820	Boo	Gxy	1.7' × 1.1'	11.7	14 h 58 m 39.7 s	+53° 53' 09"
5821	Boo	Gxy	1.4' × 0.8'	13.7	14 h 58 m 59.7 s	+53° 55' 23"
5825	Boo	Gxy	1.1' × 0.7'	13.9	14 h 54 m 31.5 s	+18° 38' 32"
5827	Boo	Gxy	1.1' × 0.9'	13.2	15 h 01 m 53.6 s	+25° 57' 52"
5828	Boo	Gxy	0.6' × 0.5'	13.6	15 h 00 m 46.1 s	+49° 59' 35"
5829	Boo	Gxy	1.8' × 1.5'	13.1	15 h 02 m 42.0 s	+23° 20' 00"
5830	Boo	Gxy	1.0' × 0.7'	14.4	15 h 01 m 50.9 s	+47° 52' 30"
5835	Boo	Gxy	1.1' × 0.9'	14.6	15 h 02 m 25.5 s	+48° 52' 38"
5837	Boo	Gxy	1.0' × 0.6'	13.7	15 h 04 m 40.6 s	+12° 38' 00"
5840	Boo	NF	...	...	15 h 04 m 20.5 s	+29° 30' 22"
5842	Boo	Gxy	0.4' × 0.4'	14.3	15 h 04 m 52.1 s	+21° 04' 10"
5851	Boo	Gxy	1.1' × 0.3'	14.3	15 h 06 m 53.4 s	+12° 51' 31"
5852	Boo	Gxy	0.9' × 0.7'	13.7	15 h 06 m 56.4 s	+12° 50' 48"
5853	Boo	Gxy	1.5' × 1.0'	13.7	15 h 05 m 53.3 s	+39° 31' 20"
5856	Boo	**	...	...	15 h 07 m 20.2 s	+18° 26' 33"
5857	Boo	Gxy	1.2' × 0.6'	13.4	15 h 07 m 27.2 s	+19° 35' 51"
5859	Boo	Gxy	2.9' × 0.9'	12.5	15 h 07 m 34.7 s	+19° 34' 56"
5860	Boo	Gxy	0.7' × 0.7'	13.2	15 h 06 m 33.7 s	+42° 38' 30"
5874	Boo	Gxy	2.3' × 1.6'	13.2	15 h 07 m 51.8 s	+54° 45' 09"
5875	Boo	Gxy	2.5' × 1.2'	12.6	15 h 09 m 13.1 s	+52° 31' 42"
5876	Boo	Gxy	2.4' × 1.2'	12.9	15 h 09 m 31.5 s	+54° 30' 22"
5884	Boo	**	...	...	15 h 13 m 09.1 s	+31° 51' 42"
5886	Boo	Gxy	0.9' × 0.4'	14	15 h 12 m 45.5 s	+41° 14' 00"
5888	Boo	Gxy	1.3' × 0.8'	13.6	15 h 13 m 07.3 s	+41° 15' 52"
5889	Boo	Gxy	1.4' × 0.8'	15.4	15 h 13 m 15.7 s	+41° 19' 41"
5893	Boo	Gxy	1.3' × 1.1'	13.4	15 h 13 m 34.1 s	+41° 57' 32"
5895	Boo	Gxy	0.8' × 0.2'	14.3	15 h 13 m 49.9 s	+42° 00' 28"
5896	Boo	Gxy	0.3' × 0.2'	15	15 h 13 m 50.5 s	+42° 01' 27"
5899	Boo	Gxy	3.2' × 1.2'	12	15 h 15 m 03.1 s	+42° 02' 58"
5900	Boo	Gxy	1.7' × 0.5'	14.2	15 h 15 m 05.0 s	+42° 12' 33"
5901	Boo	*	...	...	15 h 15 m 02.7 s	+42° 13' 50"
5902	Boo	Gxy	1.0' × 0.9'	13.2	15 h 14 m 22.3 s	+50° 19' 48"
5914	Boo	Gxy	0.7' × 0.4'	14.4	15 h 18 m 43.7 s	+41° 51' 56"
5918	Boo	Gxy	1.9' × 0.8'	13.4	15 h 19 m 25.3 s	+45° 52' 48"
5922	Boo	**	...	...	15 h 21 m 08.9 s	+41° 40' 21"
5923	Boo	Gxy	1.8' × 1.8'	13.6	15 h 21 m 14.1 s	+41° 43' 33"

(continued)

Table 10.1 (continued)

NGC #	Const	Type	Size	V Mag	RA (2000)	Dec (2000)
5929	Boo	Gxy	1.0'×0.9'	13.3	15 h 26 m 06.1 s	+41° 40' 12"
5930	Boo	Gxy	1.7'×0.9'	12.2	15 h 26 m 07.9 s	+41° 40' 32"
5932	Boo	Gxy	0.7'×0.7'	14.5	15 h 26 m 48.1 s	+48° 36' 53"
5933	Boo	Gxy	0.3'×0.2'	14.8	15 h 27 m 01.4 s	+48° 36' 48"
5934	Boo	Gxy	0.6'×0.3'	13.8	15 h 28 m 12.7 s	+42° 55' 48"
5935	Boo	Gxy	0.7'×0.3'	14.8	15 h 28 m 16.5 s	+42° 56' 39"
5943	Boo	Gxy	1.3'×1.3'	13.3	15 h 29 m 44.0 s	+42° 46' 40"
5945	Boo	Gxy	2.1'×1.9'	13	15 h 29 m 45.1 s	+42° 55' 06"
5947	Boo	Gxy	1.2'×1.1'	13.9	15 h 30 m 36.6 s	+42° 43' 00"
5950	Boo	Gxy	1.5'×0.8'	13.9	15 h 31 m 30.7 s	+40° 25' 47"
5966	Boo	Gxy	1.8'×1.1'	12.7	15 h 35 m 52.0 s	+39° 46' 07"
5992	Boo	Gxy	0.9'×0.7'	13.5	15 h 44 m 21.5 s	+41° 05' 12"
5993	Boo	Gxy	1.2'×0.9'	13.3	15 h 44 m 27.5 s	+41° 07' 16"
1567	Cae	Gxy	1.3'×1.3'	12.4	04 h 21 m 08.7 s	−48° 15' 19"
1570	Cae	Gxy	3.0'×2.5'	12.3	04 h 22 m 08.9 s	−43° 37' 50"
1571	Cae	Gxy	2.0'×1.5'	12.3	04 h 22 m 08.9 s	−43° 37' 50"
1572	Cae	Gxy	2.5'×1.2'	12.9	04 h 22 m 42.7 s	−40° 36' 06"
1585	Cae	Gxy	1.1'×0.7'	13.7	04 h 27 m 33.0 s	−42° 09' 55"
1595	Cae	Gxy	1.3'×0.9'	12.8	04 h 28 m 21.6 s	−47° 48' 57"
1598	Cae	Gxy	1.4'×0.8'	13.4	04 h 28 m 33.4 s	−47° 46' 59"
1616	Cae	Gxy	1.8'×0.9'	13	04 h 32 m 41.9 s	−43° 42' 56"
1658	Cae	Gxy	1.4'×0.5'	13.7	04 h 44 m 01.1 s	−41° 27' 50"
1660	Cae	Gxy	1.0'×0.5'	14.1	04 h 44 m 11.1 s	−41° 29' 54"
1668	Cae	Gxy	1.6'×0.9'	12.8	04 h 46 m 05.9 s	−44° 43' 59"
1679	Cae	Gxy	2.7'×2.0'	11.6	04 h 49 m 55.4 s	−31° 58' 03"
1687	Cae	Gxy	1.3'×0.5'	14.1	04 h 51 m 21.0 s	−33° 56' 22"
1701	Cae	Gxy	1.2'×0.9'	13	04 h 55 m 51.1 s	−29° 53' 03"
1759	Cae	Gxy	1.4'×1.3'	13.2	05 h 00 m 49.0 s	−38° 40' 26"
1469	Cam	Gxy	1.9'×0.8'	12.8	04 h 00 m 27.8 s	+68° 34' 37"
1485	Cam	Gxy	2.1'×0.7'	12.8	04 h 05 m 03.7 s	+70° 59' 44"
1501	Cam	PN	56"×48"	11.5	04 h 06 m 59.3 s	+60° 55' 15"
1502	Cam	OC	7.0'	6.9	04 h 07 m 49.3 s	+62° 19' 54"
1530	Cam	Gxy	4.9'×2.9'	12.3	04 h 23 m 26.9 s	+75° 17' 45"
1560	Cam	Gxy	10.0'×1.8'	11.3	04 h 32 m 48.9 s	+71° 53' 00"
1569	Cam	Gxy	2.9'×1.5'	11.3	04 h 30 m 49.2 s	+64° 50' 51"
1573	Cam	Gxy	1.9'×1.3'	12.2	04 h 35 m 04.1 s	+73° 15' 45"
1708	Cam	OC	20'×15'	…	05 h 03 m 21.9 s	+52° 49' 56"
1961	Cam	Gxy	4.6'×3.0'	11.3	05 h 42 m 04.5 s	+69° 22' 43"
2128	Cam	Gxy	1.5'×1.1'	12.7	06 h 04 m 34.3 s	+57° 37' 39"
2146	Cam	Gxy	6.0'×3.8'	10.3	06 h 18 m 39.0 s	+78° 21' 28"
2253	Cam	NF	…	…	06 h 41 m 59.9 s	+65° 50' 35"
2256	Cam	Gxy	2.3'×2.0'	12.6	06 h 47 m 13.9 s	+74° 14' 12"
2258	Cam	Gxy	2.3'×1.5'	12.1	06 h 47 m 46.5 s	+74° 28' 52"
2268	Cam	Gxy	3.4'×2.2'	11.5	07 h 14 m 17.0 s	+84° 22' 59"
2314	Cam	Gxy	1.7'×1.4'	11.9	07 h 10 m 32.7 s	+75° 19' 36"

(continued)

Table 10.1 (continued)

NGC #	Const	Type	Size	V Mag	RA (2000)	Dec (2000)
2336	Cam	Gxy	7.1'×3.9'	10.7	07 h 27 m 03.7 s	+80° 10' 42"
2347	Cam	Gxy	1.8'×1.3'	12.4	07 h 16 m 04.0 s	+64° 42' 40"
2363	Cam	Gxy	1.7'×1.1'	14.9	07 h 28 m 29.9 s	+69° 11' 36"
2366	Cam	Gxy	8.1'×3.3'	11	07 h 28 m 43.0 s	+69° 11' 22"
2403	Cam	Gxy	18'×11'	8.6	07 h 36 m 51.8 s	+65° 36' 13"
2404	Cam	GxyCld	...	...	07 h 37 m 07.1 s	+65° 36' 40"
2408	Cam	OC	10'	...	07 h 40 m 31.9 s	+71° 40' 05"
2441	Cam	Gxy	2.0'×1.7'	12	07 h 51 m 54.9 s	+73° 00' 54"
2460	Cam	Gxy	2.5'×1.9'	11.7	07 h 56 m 52.4 s	+60° 20' 57"
2523	Cam	Gxy	3.0'×1.8'	11.8	08 h 15 m 00.1 s	+73° 34' 44"
2544	Cam	Gxy	1.1'×0.8'	12.9	08 h 21 m 40.4 s	+73° 59' 18"
2550	Cam	Gxy	1.0'×0.4'	13	08 h 24 m 34.5 s	+74° 00' 43"
2551	Cam	Gxy	1.7'×1.1'	12.3	08 h 24 m 50.3 s	+73° 24' 44"
2591	Cam	Gxy	3.0'×0.6'	12.5	08 h 37 m 25.9 s	+78° 01' 34"
2633	Cam	Gxy	2.6'×1.7'	11.9	08 h 48 m 04.3 s	+74° 05' 55"
2634	Cam	Gxy	1.7'×1.6'	11.7	08 h 48 m 25.2 s	+73° 58' 00"
2636	Cam	Gxy	0.6'×0.6'	13.6	08 h 48 m 24.5 s	+73° 40' 15"
2646	Cam	Gxy	1.3'×1.3'	12.2	08 h 50 m 22.1 s	+73° 27' 45"
2653	Cam	**	...	...	08 h 54 m 55.6 s	+78° 23' 37"
2655	Cam	Gxy	4.9'×4.1'	10.1	08 h 55 m 38.4 s	+78° 13' 24"
2715	Cam	Gxy	4.9'×1.7'	11.3	09 h 08 m 06.5 s	+78° 05' 06"
2732	Cam	Gxy	2.1'×0.8'	12	09 h 13 m 24.8 s	+79° 11' 14"
2748	Cam	Gxy	3.1'×1.3'	11.6	09 h 13 m 42.9 s	+76° 28' 32"
2760	Cam	Gxy	0.5'×0.4'	...	09 h 24 m 13.7 s	+76° 31' 52"
3901	Cam	Gxy	1.9'×0.8'	13.9	11 h 42 m 49.2 s	+77° 22' 22"
4127	Cam	Gxy	3.2'×1.6'	12.9	12 h 08 m 26.3 s	+76° 48' 14"
5295	Cam	Gxy	0.4'×0.4'	14.3	13 h 38 m 39.3 s	+79° 27' 31"
5640	Cam	Gxy	0.9'×0.4'	14.7	14 h 20 m 41.1 s	+80° 07' 21"
6897	Cap	Gxy	0.9'×0.5'	14	20 h 21 m 01.3 s	−12° 15' 17"
6898	Cap	Gxy	1.1'×0.6'	13.6	20 h 21 m 08.0 s	−12° 21' 32"
6903	Cap	Gxy	2.7'×2.5'	12	20 h 23 m 44.9 s	−19° 19' 32"
6907	Cap	Gxy	3.3'×2.7'	11.3	20 h 25 m 06.7 s	−24° 48' 30"
6908	Cap	Gxy	0.3'×0.1'	14.6	20 h 25 m 08.7 s	−24° 48' 02"
6912	Cap	Gxy	1.4'×1.1'	13.8	20 h 26 m 52.1 s	−18° 37' 05"
6924	Cap	Gxy	2.0'×1.7'	12.6	20 h 33 m 19.2 s	−25° 28' 28"
6931	Cap	Gxy	1.0'×0.4'	13.7	20 h 33 m 41.3 s	−11° 22' 07"
6936	Cap	Gxy	1.9'×1.0'	12.9	20 h 35 m 56.2 s	−25° 16' 50"
6986	Cap	Gxy	1.0'×0.6'	13.7	20 h 56 m 30.6 s	−18° 34' 00"
6993	Cap	Gxy	1.2'×1.0'	13.1	20 h 53 m 54.0 s	−25° 28' 22"
7016	Cap	Gxy	0.8'×0.8'	13.7	21 h 07 m 16.1 s	−25° 28' 09"
7017	Cap	Gxy	0.8'×0.6'	14.4	21 h 07 m 20.5 s	−25° 29' 16"
7018	Cap	Gxy	0.9'×0.6'	13.5	21 h 07 m 25.5 s	−25° 25' 43"
7019	Cap	Gxy	0.8'×0.4'	14.1	21 h 06 m 26.0 s	−24° 24' 45"
7030	Cap	Gxy	0.9'×0.7'	13.9	21 h 11 m 13.3 s	−20° 29' 11"
7035	Cap	Gxy	0.6'×0.5'	14.6	21 h 10 m 45.5 s	−23° 08' 08"

(continued)

Table 10.1 (continued)

NGC #	Const	Type	Size	V Mag	RA (2000)	Dec (2000)
7073	Cap	Gxy	0.9'×0.8'	13.7	21 h 29 m 25.9 s	−11° 29' 21"
7099	Cap	GC	11'	6.9	21 h 40 m 21.9 s	−23° 10' 45"
7103	Cap	Gxy	1.3'×1.2'	14.3	21 h 39 m 51.3 s	−22° 28' 26"
7104	Cap	Gxy	0.3'×0.2'	13.8	21 h 40 m 03.1 s	−22° 25' 29"
7105	Cap	Gxy	1.0'×0.5'	13.3	21 h 41 m 41.3 s	−10° 38' 08"
7122	Cap	**	...	...	21 h 45 m 47.8 s	−08° 49' 47"
7131	Cap	Gxy	1.7'×1.0'	13.9	21 h 47 m 36.0 s	−13° 10' 58"
7134	Cap	Ast	...	...	21 h 48 m 56.2 s	−12° 58' 23"
7136	Cap	**	...	...	21 h 49 m 43.1 s	−11° 47' 36"
7158	Cap	***	...	...	21 h 57 m 28.1 s	−11° 35' 34"
2191	Car	Gxy	1.7'×0.9'	12.3	06 h 08 m 23.7 s	−52° 30' 45"
2369	Car	Gxy	3.5'×1.1'	12.2	07 h 16 m 38.5 s	−62° 20' 40"
2381	Car	Gxy	1.6'×1.4'	12.7	07 h 19 m 57.5 s	−63° 04' 03"
2417	Car	Gxy	2.8'×1.9'	12.6	07 h 30 m 12.2 s	−62° 15' 10"
2502	Car	Gxy	2.0'×1.3'	12.2	07 h 55 m 51.5 s	−52° 18' 27"
2516	Car	OC	29'	3.8	07 h 58 m 07.1 s	−60° 45' 12"
2609	Car	OC	6'	...	08 h 29 m 29.7 s	−61° 06' 37"
2640	Car	Gxy	2.2'×1.9'	11.7	08 h 37 m 24.7 s	−55° 07' 27"
2714	Car	Gxy	1.4'×1.4'	13	08 h 53 m 29.7 s	−59° 13' 04"
2788	Car	Gxy	1.8'×0.4'	12.5	09 h 09 m 03.3 s	−67° 55' 57"
2808	Car	GC	13.8'	6.2	09 h 12 m 02.6 s	−64° 51' 47"
2822	Car	Gxy	3.3'×2.2'	10.8	09 h 13 m 49.8 s	−69° 38' 41"
2836	Car	Gxy	2.6'×1.9'	12	09 h 13 m 44.7 s	−69° 20' 05"
2842	Car	Gxy	1.5'×1.3'	12.5	09 h 15 m 36.4 s	−63° 04' 16"
2867	Car	PN	12.0"	9.7	09 h 21 m 25.3 s	−58° 18' 42"
2887	Car	Gxy	2.1'×1.5'	11.6	09 h 23 m 24.1 s	−63° 48' 44"
3036	Car	OC	4'	...	09 h 49 m 15.9 s	−62° 40' 32"
3059	Car	Gxy	3.6'×3.2'	11.1	09 h 50 m 08.1 s	−73° 55' 18"
3114	Car	OC	35'	4.2	10 h 02 m 42.7 s	−60° 06' 32"
3136	Car	Gxy	3.1'×2.1'	11	10 h 05 m 48.1 s	−67° 22' 40"
3199	Car	Neb	22'×22'	...	10 h 17 m 24.4 s	−57° 55' 20"
3211	Car	PN	14"	10.7	10 h 17 m 50.4 s	−62° 40' 14"
3247	Car	OC	6'	...	10 h 25 m 52.0 s	−57° 55' 35"
3255	Car	OC	2.0'	11	10 h 26 m 31.3 s	−60° 40' 42"
3293	Car	OC	5'	4.7	10 h 35 m 53.8 s	−58° 14' 10"
3324	Car	OC+Neb	5'	...	10 h 37 m 18.7 s	−58° 39' 36"
3372	Car	Neb	120'×120'	...	10 h 45 m 08.5 s	−59° 52' 04"
3496	Car	OC	9'	8.2	10 h 59 m 33.8 s	−60° 20' 12"
3503	Car	OC+Neb	3'×3'	...	11 h 01 m 17.3 s	−59° 50' 50"
3519	Car	OC	4.0'	7.7	11 h 04 m 02.8 s	−61° 22' 05"
3532	Car	OC	55'×50'	3	11 h 05 m 47.5 s	−58° 46' 13"
3572	Car	OC	6'	6.6	11 h 10 m 19.2 s	−60° 14' 54"
3576	Car	Neb	1.0'	...	11 h 11 m 32.8 s	−61° 21' 44"
3579	Car	Neb	20'×15'	...	11 h 11 m 59.6 s	−61° 14' 35"
3581	Car	Neb	...	...	11 h 12 m 01.9 s	−61° 18' 06"

(continued)

Table 10.1 (continued)

NGC #	Const	Type	Size	V Mag	RA (2000)	Dec (2000)
3582	Car	Neb	...	...	11 h 12 m 12.0 s	−61° 16′ 25″
3584	Car	Neb	...	...	11 h 12 m 19.8 s	−61° 13′ 43″
3586	Car	Neb	...	...	11 h 12 m 29.9 s	−61° 21′ 08″
3590	Car	OC	4.0′	8.2	11 h 12 m 59.0 s	−60° 47′ 20″
3603	Car	OC	2.5′	9.1	11 h 15 m 06.6 s	−61° 15′ 40″
103	Cas	OC	5′	9.8	00 h 25 m 17.4 s	+61° 19′ 19″
110	Cas	OC	...	...	00 h 27 m 25.0 s	+71° 23′ 30″
129	Cas	OC	21′	6.5	00 h 29 m 54.1 s	+60° 12′ 35″
133	Cas	OC	7′	9.4	00 h 31 m 16.9 s	+63° 21′ 10″
136	Cas	OC	1.2′	...	00 h 31 m 30.7 s	+61° 30′ 34″
146	Cas	OC	6′	9.1	00 h 33 m 03.9 s	+63° 18′ 33″
147	Cas	Gxy	13.2′×7.8′	9.6	00 h 33 m 12.0 s	+48° 30′ 31″
185	Cas	Gxy	11.7′×10.0′	9.3	00 h 38 m 58.1 s	+48° 20′ 27″
189	Cas	OC	3.7′	8.8	00 h 39 m 35.7 s	+61° 05′ 40″
225	Cas	OC	12′	7	00 h 43 m 32.3 s	+61° 47′ 25″
278	Cas	Gxy	2.1′×2.0′	10.9	00 h 52 m 04.5 s	+47° 33′ 02″
281	Cas	OC+Neb	25′×30′	...	00 h 52 m 59.3 s	+56° 37′ 19″
358	Cas	Ast	3.0′	...	01 h 05 m 10.9 s	+62° 01′ 14″
366	Cas	OC	3.0′	...	01 h 06 m 25.9 s	+62° 13′ 44″
381	Cas	OC	6′	9.3	01 h 08 m 14.9 s	+61° 35′ 00″
433	Cas	OC	2.5′	...	01 h 15 m 09.2 s	+60° 07′ 33″
436	Cas	OC	5.0′	8.8	01 h 15 m 57.7 s	+58° 49′ 02″
457	Cas	OC	13′	6.4	01 h 19 m 32.6 s	+58° 17′ 27″
559	Cas	OC	4.4′	9.5	01 h 29 m 29.1 s	+63° 18′ 30″
581	Cas	OC	6′	7.4	01 h 33 m 21.8 s	+60° 39′ 29″
609	Cas	OC	3.0′	11	01 h 36 m 23.7 s	+64° 32′ 12″
629	Cas	Ast	...	...	01 h 38 m 58.5 s	+72° 52′ 02″
637	Cas	OC	3.5′	8.2	01 h 43 m 03.1 s	+64° 02′ 12″
654	Cas	OC	5.0′	6.5	01 h 43 m 59.4 s	+61° 52′ 58″
657	Cas	OC	3.0′×4.0′	...	01 h 43 m 20.8 s	+55° 50′ 11″
659	Cas	OC	5′	7.9	01 h 44 m 23.0 s	+60° 40′ 09″
663	Cas	OC	16′	7.1	01 h 46 m 16.0 s	+61° 13′ 06″
743	Cas	OC	5′	...	01 h 58 m 31.3 s	+60° 09′ 59″
771	Cas	*	...	...	02 h 03 m 26.6 s	+72° 25′ 15″
886	Cas	OC	20.0′	...	02 h 23 m 11.7 s	+63° 46′ 44″
896	Cas	Neb	27′×13′	...	02 h 25 m 27.9 s	+62° 01′ 10″
1027	Cas	OC	20′	6.7	02 h 42 m 35.1 s	+61° 35′ 40″
1343	Cas	Gxy	2.6′×1.6′	12.9	03 h 37 m 50.2 s	+72° 34′ 15″
7438	Cas	OC	30′×10′	...	22 h 57 m 20.3 s	+54° 18′ 34″
7635	Cas	Neb	15.0′×8.0′	...	23 h 20 m 45.5 s	+61° 12′ 45″
7654	Cas	OC	16.0′	6.9	23 h 24 m 50.4 s	+61° 36′ 24″
7788	Cas	OC	4′	9.4	23 h 56 m 45.5 s	+61° 24′ 00″
7789	Cas	OC	25′	6.7	23 h 57 m 24.0 s	+56° 42′ 30″
7790	Cas	OC	5′	8.5	23 h 58 m 24.2 s	+61° 12′ 30″
7795	Cas	OC	10′	...	23 h 58 m 37.4 s	+60° 02′ 06″

(continued)

Table 10.1 (continued)

NGC #	Const	Type	Size	V Mag	RA (2000)	Dec (2000)
7801	Cas	Ast	...	...	00 h 00 m 20.0 s	+50° 44' 20"
3533	Cen	Gxy	2.8'×0.6'	13.1	11 h 07 m 07.4 s	−37° 10' 25"
3557	Cen	Gxy	4.1'×3.0'	10.6	11 h 09 m 57.5 s	−37° 32' 22"
3564	Cen	Gxy	1.8'×0.8'	12.4	11 h 10 m 36.2 s	−37° 32' 53"
3568	Cen	Gxy	2.5'×0.8'	12.5	11 h 10 m 48.1 s	−37° 26' 56"
3573	Cen	Gxy	3.6'×1.0'	12.4	11 h 11 m 18.9 s	−36° 52' 25"
3680	Cen	OC	12'	7.6	11 h 25 m 37.1 s	−43° 15' 00"
3699	Cen	PN	70"	11.3	11 h 27 m 58.1 s	−59° 57' 37"
3706	Cen	Gxy	3.0'×1.8'	11.3	11 h 29 m 44.4 s	−36° 23' 29"
3742	Cen	Gxy	2.4'×1.7'	12.2	11 h 35 m 32.1 s	−37° 57' 23"
3749	Cen	Gxy	3.2'×0.8'	12.5	11 h 35 m 53.0 s	−37° 59' 52"
3766	Cen	OC	12.0'	5.3	11 h 36 m 14.4 s	−61° 36' 36"
3778	Cen	Gxy	1.2'×0.9'	13.2	11 h 38 m 21.3 s	−50° 42' 57"
3783	Cen	Gxy	1.9'×1.7'	11.9	11 h 39 m 01.7 s	−37° 44' 19"
3882	Cen	Gxy	2.3'×1.3'	12.5	11 h 46 m 06.1 s	−56° 23' 28"
3903	Cen	Gxy	1.1'×1.0'	13	11 h 49 m 04.0 s	−37° 30' 59"
3909	Cen	OC	15' to 20'	...	11 h 50 m 08.1 s	−48° 14' 17"
3918	Cen	PN	13"	8.1	11 h 50 m 17.8 s	−57° 10' 56"
3960	Cen	OC	6'	8.3	11 h 50 m 33.2 s	−55° 40' 35"
4112	Cen	Gxy	1.6'×0.9'	12.2	12 h 07 m 09.1 s	−40° 12' 29"
4219	Cen	Gxy	4.3'×1.3'	11.9	12 h 16 m 27.9 s	−43° 19' 28"
4230	Cen	OC	7'×5'	9.4	12 h 17 m 09.3 s	−55° 17' 10"
4373	Cen	Gxy	3.4'×2.5'	10.9	12 h 25 m 18.5 s	−39° 45' 37"
4444	Cen	Gxy	2.5'×2.3'	12.5	12 h 28 m 36.1 s	−43° 15' 45"
4499	Cen	Gxy	1.8'×1.3'	13.1	12 h 32 m 05.2 s	−39° 58' 57"
4507	Cen	Gxy	1.7'×1.3'	12.2	12 h 35 m 36.5 s	−39° 54' 33"
4553	Cen	Gxy	2.1'×1.0'	12.4	12 h 36 m 08.0 s	−39° 26' 25"
4573	Cen	Gxy	2.6'×2.0'	12.7	12 h 37 m 43.7 s	−43° 37' 12"
4574	Cen	Gxy	1.7'×1.1'	13.2	12 h 37 m 43.5 s	−35° 31' 05"
4575	Cen	Gxy	2.0'×1.3'	12.5	12 h 37 m 51.6 s	−40° 32' 15"
4601	Cen	Gxy	1.8'×0.5'	13.6	12 h 40 m 46.8 s	−40° 53' 34"
4603	Cen	Gxy	3.4'×2.5'	11.8	12 h 40 m 55.3 s	−40° 58' 33"
4616	Cen	Gxy	0.9'×0.8'	13.2	12 h 42 m 16.5 s	−40° 38' 30"
4622	Cen	Gxy	1.7'×1.6'	12.6	12 h 42 m 37.4 s	−40° 44' 44"
4645	Cen	Gxy	2.2'×1.4'	11.9	12 h 44 m 09.9 s	−41° 44' 53"
4650	Cen	Gxy	2.6'×1.8'	12	12 h 44 m 19.5 s	−40° 43' 55"
4661	Cen	Gxy	1.1'×0.4'	13.6	12 h 45 m 14.9 s	−40° 49' 29"
4672	Cen	Gxy	2.0'×0.6'	13.4	12 h 46 m 15.5 s	−41° 42' 22"
4677	Cen	Gxy	1.7'×0.7'	12.8	12 h 46 m 57.0 s	−41° 34' 59"
4679	Cen	Gxy	2.4'×1.0'	12.7	12 h 47 m 30.3 s	−39° 34' 15"
4681	Cen	Gxy	1.3'×1.2'	12.8	12 h 47 m 28.7 s	−43° 20' 06"
4683	Cen	Gxy	1.4'×0.8'	13.1	12 h 47 m 42.1 s	−41° 31' 45"
4696	Cen	Gxy	4.5'×3.2'	10.5	12 h 48 m 49.1 s	−41° 18' 41"
4706	Cen	Gxy	1.4'×0.6'	13.1	12 h 49 m 54.0 s	−41° 16' 50"
4709	Cen	Gxy	2.4'×2.0'	11.6	12 h 50 m 03.8 s	−41° 22' 56"

(continued)

Table 10.1 (continued)

NGC #	Const	Type	Size	V Mag	RA (2000)	Dec (2000)
4729	Cen	Gxy	1.5' × 1.4'	12.6	12 h 51 m 46.4 s	−41° 07' 54"
4730	Cen	Gxy	1.2' × 0.9'	13	12 h 52 m 00.5 s	−41° 08' 48"
4743	Cen	Gxy	1.3' × 0.5'	13.1	12 h 52 m 15.6 s	−41° 23' 24"
4744	Cen	Gxy	2.1' × 1.0'	12.7	12 h 52 m 19.5 s	−41° 03' 36"
4751	Cen	Gxy	3.0' × 1.1'	12.1	12 h 52 m 51.4 s	−42° 39' 35"
4767	Cen	Gxy	2.6' × 1.2'	11.6	12 h 53 m 53.3 s	−39° 42' 52"
4785	Cen	Gxy	1.9' × 1.0'	12.6	12 h 53 m 26.8 s	−48° 44' 59"
4811	Cen	Gxy	1.3' × 0.9'	13.2	12 h 56 m 52.0 s	−41° 47' 49"
4812	Cen	Gxy	1.1' × 0.4'	13.2	12 h 56 m 53.0 s	−41° 48' 43"
4832	Cen	Gxy	1.9' × 1.2'	12.3	12 h 57 m 47.6 s	−39° 45' 42"
4835	Cen	Gxy	4.0' × 0.9'	11.9	12 h 58 m 07.5 s	−46° 15' 53"
4852	Cen	OC	11'	8.9	13 h 00 m 14.5 s	−59° 36' 58"
4903	Cen	Gxy	1.5' × 1.3'	13.2	13 h 01 m 23.1 s	−30° 56' 02"
4905	Cen	Gxy	1.4' × 1.3'	13.3	13 h 01 m 30.6 s	−30° 52' 04"
4909	Cen	Gxy	1.9' × 1.7'	12.9	13 h 02 m 01.8 s	−42° 46' 19"
4930	Cen	Gxy	4.5' × 3.7'	11.8	13 h 04 m 04.5 s	−41° 24' 41"
4936	Cen	Gxy	2.7' × 2.3'	10.9	13 h 04 m 16.9 s	−30° 31' 36"
4937	Cen	Ast	0.5'	…	13 h 04 m 51.5 s	−47° 13' 09"
4940	Cen	Gxy	1.2' × 1.1'	13.2	13 h 05 m 00.1 s	−47° 14' 11"
4945	Cen	Gxy	20.0' × 3.8'	8.9	13 h 05 m 27.1 s	−49° 28' 03"
4946	Cen	Gxy	1.5' × 1.2'	12.4	13 h 05 m 29.3 s	−43° 35' 28"
4947	Cen	Gxy	2.4' × 1.3'	12	13 h 05 m 20.7 s	−35° 20' 15"
4950	Cen	Gxy	0.9' × 0.7'	14	13 h 05 m 36.4 s	−43° 30' 04"
4953	Cen	Gxy	1.1' × 0.9'	13.3	13 h 06 m 10.3 s	−37° 35' 09"
4976	Cen	Gxy	5.6' × 3.0'	10.2	13 h 08 m 37.4 s	−49° 30' 21"
4988	Cen	Gxy	1.7' × 0.5'	12.9	13 h 09 m 54.3 s	−43° 06' 22"
5011	Cen	Gxy	2.4' × 2.0'	11.4	13 h 12 m 51.7 s	−43° 05' 46"
5026	Cen	Gxy	3.2' × 2.0'	11.7	13 h 14 m 13.3 s	−42° 57' 42"
5043	Cen	OC	10' × 7'	…	13 h 16 m 09.7 s	−60° 04' 26"
5045	Cen	OC	…	…	13 h 17 m 04.6 s	−63° 24' 48"
5062	Cen	Gxy	2.2' × 0.7'	12.3	13 h 18 m 23.5 s	−35° 27' 33"
5063	Cen	Gxy	2.3' × 1.8'	12.6	13 h 18 m 25.7 s	−35° 21' 12"
5064	Cen	Gxy	2.6' × 1.3'	11.9	13 h 19 m 00.0 s	−47° 54' 36"
5082	Cen	Gxy	1.7' × 1.0'	12.9	13 h 20 m 39.8 s	−43° 42' 02"
5086	Cen	**	…	…	13 h 20 m 59.3 s	−43° 44' 01"
5090	Cen	Gxy	2.9' × 2.4'	11.6	13 h 21 m 12.7 s	−43° 42' 19"
5091	Cen	Gxy	1.8' × 0.5'	13.3	13 h 21 m 17.9 s	−43° 43' 14"
5102	Cen	Gxy	8.7' × 2.8'	9.1	13 h 21 m 57.5 s	−36° 37' 49"
5108	Cen	Gxy	1.2' × 0.3'	14.4	13 h 23 m 18.7 s	−32° 20' 32"
5114	Cen	Gxy	1.7' × 1.0'	12.6	13 h 24 m 01.6 s	−32° 20' 40"
5120	Cen	OC	2.8'	…	13 h 25 m 40.4 s	−63° 27' 30"
5121	Cen	Gxy	2.3' × 2.0'	11.8	13 h 24 m 45.4 s	−37° 40' 59"
5124	Cen	Gxy	2.2' × 0.7'	12.3	13 h 24 m 50.1 s	−30° 18' 31"
5126	Cen	Gxy	1.4' × 0.4'	13.3	13 h 24 m 53.4 s	−30° 20' 00"
5128	Cen	Gxy	25.7' × 20.0'	6.8	13 h 25 m 29.0 s	−43° 01' 00"

(continued)

Table 10.1 (continued)

NGC #	Const	Type	Size	V Mag	RA (2000)	Dec (2000)
5138	Cen	OC	7'	7.6	13 h 27 m 15.2 s	−59° 02' 27"
5139	Cen	GC	36'	3.9	13 h 26 m 45.9 s	−47° 28' 37"
5140	Cen	Gxy	2.0'×1.7'	11.9	13 h 26 m 21.7 s	−33° 52' 08"
5155	Cen	OC	...	...	13 h 29 m 35.4 s	−63° 25' 29"
5156	Cen	Gxy	2.3'×2.0'	12	13 h 28 m 44.1 s	−48° 55' 02"
5161	Cen	Gxy	5.6'×2.2'	11.6	13 h 29 m 14.1 s	−33° 10' 29"
5168	Cen	OC	4.0'	9.1	13 h 31 m 07.3 s	−60° 56' 21"
5188	Cen	Gxy	3.0'×1.1'	12.1	13 h 31 m 27.8 s	−34° 47' 43"
5193	Cen	Gxy	1.9'×1.7'	11.7	13 h 31 m 53.5 s	−33° 14' 06"
5206	Cen	Gxy	3.7'×3.2'	10.7	13 h 33 m 43.8 s	−48° 09' 04"
5215	Cen	Gxy	0.8'×0.3'	13.2	13 h 35 m 07.9 s	−33° 29' 04"
5219	Cen	Gxy	2.3'×1.6'	12.8	13 h 38 m 41.9 s	−45° 51' 16"
5220	Cen	Gxy	2.3'×0.7'	12.4	13 h 35 m 56.7 s	−33° 27' 17"
5234	Cen	Gxy	1.3'×0.8'	13.2	13 h 37 m 29.7 s	−49° 50' 16"
5237	Cen	Gxy	1.9'×1.6'	12.8	13 h 37 m 38.9 s	−42° 50' 53"
5244	Cen	Gxy	1.7'×0.6'	12.7	13 h 38 m 41.9 s	−45° 51' 16"
5253	Cen	Gxy	5.0'×1.9'	10.7	13 h 39 m 55.9 s	−31° 38' 31"
5266	Cen	Gxy	3.2'×2.1'	11.2	13 h 43 m 01.8 s	−48° 10' 11"
5269	Cen	OC	3.0'	...	13 h 44 m 44.2 s	−62° 54' 57"
5281	Cen	OC	8.0'	5.9	13 h 46 m 35.1 s	−62° 54' 59"
5284	Cen	OC	30'×20'	...	13 h 47 m 23.3 s	−59° 08' 58"
5286	Cen	GC	9.1'	7.4	13 h 46 m 26.5 s	−51° 22' 24"
5291	Cen	Gxy	1.1'×0.7'	14.2	13 h 47 m 24.4 s	−30° 24' 28"
5292	Cen	Gxy	1.8'×1.5'	12.1	13 h 47 m 39.5 s	−30° 56' 27"
5298	Cen	Gxy	1.4'×0.7'	13.2	13 h 48 m 36.4 s	−30° 25' 43"
5299	Cen	OC	30.0'	...	13 h 50 m 26.3 s	−60° 26' 40"
5302	Cen	Gxy	1.8'×1.1'	12.3	13 h 48 m 49.7 s	−30° 30' 40"
5304	Cen	Gxy	1.5'×1.0'	12.7	13 h 50 m 01.5 s	−30° 34' 46"
5307	Cen	PN	0.5'	11.2	13 h 51 m 03.1 s	−51° 12' 20"
5316	Cen	OC	13'	6	13 h 53 m 57.2 s	−61° 52' 09"
5333	Cen	Gxy	1.9'×1.0'	11.9	13 h 54 m 24.1 s	−48° 30' 47"
5357	Cen	Gxy	1.5'×1.3'	12.2	13 h 55 m 59.5 s	−30° 20' 30"
5365	Cen	Gxy	3.0'×1.9'	11.4	13 h 57 m 50.5 s	−43° 55' 56"
5367	Cen	Neb	13'×8'	...	13 h 57 m 43.8 s	−39° 58' 42"
5381	Cen	OC	13'	...	14 h 00 m 41.9 s	−59° 35' 12"
5397	Cen	Gxy	1.4'×1.0'	12.9	14 h 01 m 10.6 s	−33° 56' 45"
5398	Cen	Gxy	2.8'×1.7'	12.4	14 h 01 m 21.9 s	−33° 03' 45"
5408	Cen	Gxy	2.8'×1.5'	12.4	14 h 03 m 20.9 s	−41° 22' 44"
5419	Cen	Gxy	4.2'×3.3'	10.8	14 h 03 m 38.7 s	−33° 58' 43"
5460	Cen	OC	35'	5.6	14 h 07 m 27.7 s	−48° 20' 33"
5483	Cen	Gxy	3.7'×3.4'	11.4	14 h 10 m 25.2 s	−43° 19' 32"
5488	Cen	Gxy	3.4'×1.0'	12.1	14 h 08 m 03.1 s	−33° 18' 53"
5489	Cen	Gxy	1.5'×1.0'	12.4	14 h 12 m 01.0 s	−46° 05' 22"
5494	Cen	Gxy	2.2'×1.9'	12.1	14 h 12 m 24.0 s	−30° 38' 39"
5516	Cen	Gxy	1.8'×1.2'	12.1	14 h 15 m 54.7 s	−48° 06' 54"

(continued)

Table 10.1 (continued)

NGC #	Const	Type	Size	V Mag	RA (2000)	Dec (2000)
5606	Cen	OC	3.0'	7.7	14 h 27 m 47.2 s	−59° 37' 56"
5617	Cen	OC	10'	6.3	14 h 29 m 44.0 s	−60° 42' 39"
5662	Cen	OC	30'	5.5	14 h 35 m 37.5 s	−56° 37' 05"
5786	Cen	Gxy	2.3'×1.1'	11.2	14 h 58 m 56.7 s	−42° 00' 45"
40	Cep	PN	38"×35"	12.3	00 h 13 m 00.9 s	+72° 31' 20"
188	Cep	OC	13'	8.1	00 h 47 m 29.7 s	+85° 14' 29"
1184	Cep	Gxy	2.8'×0.7'	12.6	03 h 16 m 45.4 s	+80° 47' 32"
1544	Cep	Gxy	1.7'×1.1'	13.7	05 h 02 m 36.0 s	+86° 13' 21"
2276	Cep	Gxy	2.6'×2.5'	11.8	07 h 27 m 10.2 s	+85° 45' 19"
2300	Cep	Gxy	2.8'×2.0'	11.1	07 h 32 m 19.7 s	+85° 42' 33"
6939	Cep	OC	7'	7.8	20 h 31 m 30.1 s	+60° 39' 44"
6949	Cep	Gxy	1.7'×1.6'	13.8	20 h 35 m 07.2 s	+64° 48' 10"
6951	Cep	Gxy	3.8'×3.3'	11.3	20 h 37 m 14.1 s	+66° 06' 20"
6952	Cep	Gxy	3.8'×3.3'	11.3	20 h 37 m 14.1 s	+66° 06' 20"
6953	Cep	OC	1.5'	…	20 h 37 m 46.1 s	+65° 45' 54"
7023	Cep	OC+Neb	5'	…	21 h 01 m 35.5 s	+68° 10' 11"
7055	Cep	OC	4.0'	…	21 h 19 m 25.9 s	+57° 35' 26"
7076	Cep	PN	2.0'	…	21 h 26 m 23.3 s	+62° 53' 35"
7129	Cep	Neb	2.7'	…	21 h 42 m 58.9 s	+66° 06' 47"
7133	Cep	NF	…	…	21 h 44 m 26.7 s	+66° 10' 07"
7139	Cep	PN	86"×70"	13.3	21 h 46 m 08.7 s	+63° 47' 30"
7142	Cep	OC	4.3'	9.3	21 h 45 m 09.4 s	+65° 46' 28"
7160	Cep	OC	7'	6.1	21 h 53 m 40.2 s	+62° 36' 12"
7226	Cep	OC	1.8'	9.6	22 h 10 m 26.9 s	+55° 23' 55"
7234	Cep	OC	4.0'	…	22 h 12 m 24.9 s	+57° 16' 17"
7235	Cep	OC	4.0'	…	22 h 12 m 24.9 s	+57° 16' 17"
7261	Cep	OC	5'	8.4	22 h 20 m 11.3 s	+58° 07' 19"
7281	Cep	OC	15'×9'	…	22 h 25 m 20.9 s	+57° 49' 16"
7352	Cep	OC	…	…	22 h 39 m 43.7 s	+57° 23' 40"
7354	Cep	PN	22"×18"	12.2	22 h 40 m 19.8 s	+61° 17' 08"
7380	Cep	OC+Neb	12'	7.2	22 h 47 m 20.9 s	+58° 07' 57"
7419	Cep	OC	6.0'	13	22 h 54 m 20.0 s	+60° 48' 56"
7423	Cep	OC	5.0'	…	22 h 55 m 08.5 s	+57° 05' 49"
7429	Cep	OC	14'	…	22 h 56 m 00.5 s	+59° 58' 26"
7510	Cep	OC	7.0'	7.9	23 h 11 m 03.7 s	+60° 34' 15"
7538	Cep	Neb	8'×7'	…	23 h 13 m 38.6 s	+61° 30' 45"
7708	Cep	OC	15'	…	23 h 35 m 01.4 s	+72° 50' 00"
7748	Cep	*	…	…	23 h 44 m 56.6 s	+69° 45' 18"
7762	Cep	OC	15'	10	23 h 50 m 01.7 s	+68° 02' 17"
7822	Cep	Neb	20'×4'	…	00 h 02 m 09.3 s	+67° 25' 12"
17	Cet	Gxy	0.4'×0.3'	14.4	00 h 11 m 06.6 s	−12° 06' 26"
34	Cet	Gxy	2.2'×0.8'	14.4	00 h 11 m 06.6 s	−12° 06' 26"
35	Cet	Gxy	0.7'×0.5'	14.2	00 h 11 m 10.2 s	−12° 01' 15"
45	Cet	Gxy	8.5'×5.9'	10.8	00 h 14 m 03.9 s	−23° 10' 52"
47	Cet	Gxy	2.2'×2.1'	13.1	00 h 14 m 30.9 s	−07° 10' 05"

(continued)

Table 10.1 (continued)

NGC #	Const	Type	Size	V Mag	RA (2000)	Dec (2000)
50	Cet	Gxy	2.3'×1.7'	12.3	00 h 14 m 44.6 s	−07° 20' 44"
54	Cet	Gxy	1.3'×0.4'	13.7	00 h 15 m 07.6 s	−07° 06' 27"
58	Cet	Gxy	2.3'×2.0'	13.1	00 h 14 m 30.9 s	−07° 10' 05"
59	Cet	Gxy	2.6'×1.3'	12.2	00 h 15 m 25.1 s	−21° 26' 38"
62	Cet	Gxy	1.0'×0.8'	13.5	00 h 17 m 05.5 s	−13° 29' 12"
64	Cet	Gxy	1.5'×1.1'	13.2	00 h 17 m 30.3 s	−06° 49' 32"
65	Cet	Gxy	0.8'×0.7'	14.5	00 h 18 m 59.5 s	−22° 52' 51"
66	Cet	Gxy	1.2'×0.7'	13.6	00 h 19 m 04.5 s	−22° 56' 15"
73	Cet	Gxy	1.6'×1.1'	13.7	00 h 18 m 39.4 s	−15° 19' 19"
77	Cet	Gxy	0.4'×0.4'	15.8	00 h 20 m 01.5 s	−22° 31' 57"
102	Cet	Gxy	1.0'×0.7'	13.5	00 h 24 m 36.3 s	−13° 57' 26"
107	Cet	Gxy	0.9'×0.6'	15.2	00 h 25 m 42.1 s	−08° 17' 03"
111	Cet	NF	…	…	00 h 26 m 38.4 s	−02° 37' 30"
113	Cet	Gxy	1.4'×1.0'	13.1	00 h 26 m 54.6 s	−02° 30' 03"
114	Cet	Gxy	0.9'×0.7'	13.8	00 h 26 m 58.7 s	−01° 47' 05"
116	Cet	Gxy	0.7'×0.4'	…	00 h 27 m 05.2 s	−07° 40' 09"
117	Cet	Gxy	0.7'×0.4'	14.4	00 h 27 m 10.8 s	+01° 20' 02"
118	Cet	Gxy	0.7'×0.5'	14.4	00 h 27 m 16.2 s	−01° 46' 44"
120	Cet	Gxy	1.5'×0.6'	13.5	00 h 27 m 30.1 s	−01° 30' 48"
122	Cet	*	…	…	00 h 27 m 38.3 s	−01° 38' 25"
123	Cet	*	…	…	00 h 27 m 40.0 s	−01° 37' 39"
124	Cet	Gxy	1.4'×0.9'	13.2	00 h 27 m 52.3 s	−01° 48' 38"
132	Cet	Gxy	1.9'×1.4'	12.9	00 h 30 m 10.6 s	+02° 05' 35"
135	Cet	Gxy	0.5'×0.5'	15.1	00 h 31 m 45.9 s	−13° 20' 16"
142	Cet	Gxy	1.1'×0.6'	14	00 h 31 m 09.1 s	−22° 37' 08"
143	Cet	Gxy	1.0'×0.2'	14.6	00 h 31 m 16.1 s	−22° 33' 38"
144	Cet	Gxy	0.8'×0.8'	13.9	00 h 31 m 21.1 s	−22° 38' 44"
145	Cet	Gxy	1.8'×1.3'	12.7	00 h 31 m 45.6 s	−05° 09' 09"
151	Cet	Gxy	3.7'×1.7'	11.6	00 h 34 m 02.7 s	−09° 42' 18"
153	Cet	Gxy	3.7'×1.9'	11.6	00 h 34 m 02.7 s	−09° 42' 18"
154	Cet	Gxy	1.1'×0.9'	14	00 h 34 m 19.3 s	−12° 39' 20"
155	Cet	Gxy	1.7'×1.3'	13.3	00 h 34 m 40.1 s	−10° 45' 59"
156	Cet	**	…	…	00 h 34 m 35.7 s	−08° 20' 24"
157	Cet	Gxy	4.2'×2.7'	10.4	00 h 34 m 46.7 s	−08° 23' 48"
158	Cet	**	…	…	00 h 35 m 05.5 s	−08° 20' 45"
161	Cet	Gxy	1.3'×0.8'	13.4	00 h 35 m 33.8 s	−02° 50' 55"
163	Cet	Gxy	1.5'×1.2'	12.7	00 h 35 m 59.7 s	−10° 07' 19"
165	Cet	Gxy	1.5'×1.3'	13.1	00 h 36 m 28.9 s	−10° 06' 24"
166	Cet	Gxy	0.9'×0.3'	14.5	00 h 35 m 48.5 s	−13° 36' 36"
167	Cet	Gxy	1.0'×0.7'	13.8	00 h 35 m 23.4 s	−23° 22' 35"
168	Cet	Gxy	1.2'×0.2'	13.9	00 h 36 m 39.4 s	−22° 35' 36"
170	Cet	Gxy	0.4'×0.3'	14.5	00 h 36 m 45.7 s	+01° 53' 11"
171	Cet	Gxy	2.6'×2.4'	12.2	00 h 37 m 21.6 s	−19° 56' 05"
172	Cet	Gxy	2.0'×0.3'	14	00 h 37 m 14.3 s	−22° 35' 06"
173	Cet	Gxy	3.2'×2.6'	13.2	00 h 37 m 12.5 s	+01° 56' 32"

(continued)

Table 10.1 (continued)

NGC #	Const	Type	Size	V Mag	RA (2000)	Dec (2000)
175	Cet	Gxy	2.1'×1.9'	12	00 h 37 m 21.6 s	−19° 56' 05"
177	Cet	Gxy	2.2'×0.5'	13.4	00 h 37 m 35.3 s	−22° 32' 54"
178	Cet	Gxy	2.0'×1.0'	12.4	00 h 39 m 08.4 s	−14° 10' 21"
179	Cet	Gxy	0.9'×0.8'	13.1	00 h 37 m 46.4 s	−17° 50' 57"
187	Cet	Gxy	1.3'×0.5'	13.2	00 h 39 m 30.3 s	−14° 39' 23"
191	Cet	Gxy	1.5'×1.2'	13.4	00 h 38 m 59.3 s	−09° 00' 10"
192	Cet	Gxy	1.9'×0.9'	13	00 h 39 m 13.6 s	+00° 51' 50"
195	Cet	Gxy	1.1'×0.8'	13.7	00 h 39 m 35.8 s	−09° 11' 42"
196	Cet	Gxy	1.3'×0.8'	13.6	00 h 39 m 17.9 s	+00° 54' 45"
197	Cet	Gxy	1.2'×0.8'	13.7	00 h 39 m 18.9 s	+00° 53' 29"
201	Cet	Gxy	1.8'×1.4'	13.6	00 h 39 m 35.1 s	+00° 51' 34"
207	Cet	Gxy	0.9'×0.4	13.7	00 h 39 m 40.7 s	−14° 14' 16"
209	Cet	Gxy	1.4'×1.1'	13	00 h 39 m 03.6 s	−18° 36' 28"
210	Cet	Gxy	5.0'×3.3'	11.1	00 h 40 m 34.8 s	−13° 52' 21"
216	Cet	Gxy	2.0'×0.7'	12.6	00 h 41 m 26.8 s	−21° 02' 43"
217	Cet	Gxy	2.6'×0.6'	12.6	00 h 41 m 33.8 s	−10° 01' 20"
219	Cet	Gxy	0.5'×0.5'	14.4	00 h 42 m 11.3 s	+00° 54' 14"
223	Cet	Gxy	1.3'×0.9'	13.4	00 h 42 m 15.8 s	+00° 50' 44"
227	Cet	Gxy	1.6'×1.3'	12.1	00 h 42 m 36.6 s	−01° 31' 43"
230	Cet	Gxy	1.1'×0.2'	14.9	00 h 42 m 27.5 s	−23° 37' 46"
232	Cet	Gxy	1.0'×0.8'	13.7	00 h 42 m 45.5 s	−23° 33' 40"
235	Cet	Gxy	1.3'×0.7'	13.3	00 h 42 m 52.4 s	−23° 32' 28"
237	Cet	Gxy	1.6'×0.9'	12.8	00 h 43 m 27.7 s	−00° 07' 32"
239	Cet	Gxy	1.0'×0.5'	14	00 h 44 m 37.5 s	−03° 45' 35"
244	Cet	Gxy	1.2'×1.0'	12.9	00 h 45 m 46.2 s	−15° 35' 50"
245	Cet	Gxy	1.4'×1.2'	12.4	00 h 46 m 05.4 s	−01° 43' 28"
246	Cet	PN	4.6'×4.1'	10.9	00 h 47 m 03.2 s	−11° 52' 20"
247	Cet	Gxy	21.4'×6.9'	9.2	00 h 47 m 08.4 s	−20° 45' 36"
255	Cet	Gxy	3.0'×2.5'	11.7	00 h 47 m 47.3 s	−11° 28' 07"
259	Cet	Gxy	2.8'×0.6'	12.8	00 h 48 m 03.3 s	−02° 46' 35"
263	Cet	Gxy	0.6'×0.3'	14.3	00 h 48 m 48.3 s	−13° 06' 28"
268	Cet	Gxy	1.6'×1.1'	12.6	00 h 50 m 09.4 s	−05° 11' 38"
270	Cet	Gxy	1.7'×1.5'	12.9	00 h 50 m 32.4 s	−08° 39' 08"
271	Cet	Gxy	2.1'×1.7'	12.2	00 h 50 m 41.8 s	−01° 54' 39"
273	Cet	Gxy	2.2'×0.7'	12.9	00 h 50 m 48.5 s	−06° 53' 10"
274	Cet	Gxy	1.5'×1.5'	11.9	00 h 51 m 01.8 s	−07° 03' 26"
275	Cet	Gxy	1.5'×1.1'	12.7	00 h 51 m 04.3 s	−07° 03' 56"
276	Cet	Gxy	1.0'×0.4'	15.1	00 h 52 m 06.5 s	−22° 40' 48"
277	Cet	Gxy	1.4'×1.2'	13.7	00 h 51 m 17.1 s	−08° 35' 49"
279	Cet	Gxy	1.6'×1.2'	12.8	00 h 52 m 08.9 s	−02° 13' 06"
283	Cet	Gxy	1.6'×1.0'	14.1	00 h 53 m 13.2 s	−13° 09' 52"
284	Cet	Gxy	0.7'×0.6'	14.4	00 h 53 m 24.2 s	−13° 09' 33"
285	Cet	Gxy	0.7'×0.4'	14.7	00 h 53 m 29.9 s	−13° 09' 42"
286	Cet	Gxy	1.3'×0.9'	14.1	00 h 53 m 30.4 s	−13° 06' 46"
291	Cet	Gxy	1.1'×0.5'	13.9	00 h 53 m 29.7 s	−08° 46' 09"

(continued)

Table 10.1 (continued)

NGC #	Const	Type	Size	V Mag	RA (2000)	Dec (2000)
293	Cet	Gxy	1.1'×0.9'	14.2	00 h 54 m 15.9 s	−07° 14' 08"
297	Cet	Gxy	0.3'×0.2'	15.5	00 h 54 m 58.9 s	−07° 20' 59"
298	Cet	Gxy	1.7'×0.4'	13.8	00 h 55 m 02.3 s	−07° 20' 00"
301	Cet	Gxy	0.7'×0.5'	14.8	00 h 56 m 18.4 s	−10° 40' 27"
302	Cet	*	...	...	00 h 56 m 25.3 s	−10° 39' 46"
303	Cet	Gxy	1.7'×1.4'	14.3	00 h 54 m 54.5 s	−16° 39' 08"
307	Cet	Gxy	1.6'×0.7'	13	00 h 56 m 32.6 s	−01° 46' 19"
308	Cet	*	...	...	00 h 56 m 34.4 s	−01° 47' 04"
309	Cet	Gxy	3.0'×2.5'	11.6	00 h 56 m 42.5 s	−09° 54' 50"
310	Cet	*	...	...	00 h 56 m 47.9 s	−01° 45' 55"
320	Cet	Gxy	0.9'×0.5'	13.7	00 h 58 m 46.3 s	−20° 50' 25"
321	Cet	Gxy	0.7'×0.5'	15	00 h 57 m 39.3 s	−05° 05' 11"
325	Cet	Gxy	1.4'×0.2'	14.8	00 h 57 m 47.8 s	−05° 06' 45"
327	Cet	Gxy	1.6'×0.7'	13.5	00 h 57 m 55.2 s	−05° 07' 50"
329	Cet	Gxy	1.6'×0.6'	13.5	00 h 58 m 01.4 s	−05° 04' 17"
331	Cet	Gxy	0.0'×0.6'	14.6	00 h 47 m 06.9 s	−02° 43' 52"
333	Cet	Gxy	1.6'×0.9'	13.9	00 h 58 m 51.3 s	−16° 28' 11"
335	Cet	Gxy	1.1'×0.3'	14.5	00 h 59 m 19.4 s	−18° 14' 01"
336	Cet	Gxy	0.7'×0.2'	14.7	00 h 58 m 03.2 s	−18° 23' 12"
337	Cet	Gxy	2.9'×1.8'	11.6	00 h 59 m 50.0 s	−07° 34' 40"
340	Cet	Gxy	0.9'×0.3'	13.7	01 h 00 m 34.8 s	−06° 52' 00"
341	Cet	Gxy	1.2'×1.1'	13	01 h 00 m 45.7 s	−09° 11' 08"
342	Cet	Gxy	0.6'×0.3'	14.4	01 h 00 m 49.8 s	−06° 46' 21"
343	Cet	Gxy	0.7'×0.3'	15.5	00 h 58 m 23.9 s	−23° 13' 30"
344	Cet	Gxy	0.4'×0.2'	16.1	00 h 58 m 25.4 s	−23° 13' 46"
345	Cet	Gxy	1.2'×0.8'	13.9	01 h 01 m 22.0 s	−06° 53' 04"
347	Cet	Gxy	0.6'×0.4'	14.8	01 h 01 m 35.0 s	−06° 44' 01"
349	Cet	Gxy	1.3'×0.9'	13.1	01 h 01 m 50.7 s	−06° 48' 00"
350	Cet	Gxy	0.3'×0.2'	14.9	01 h 01 m 56.6 s	−06° 47' 45"
351	Cet	Gxy	1.4'×0.8'	13.4	01 h 01 m 57.8 s	−01° 56' 13"
352	Cet	Gxy	2.4'×0.9'	12.7	01 h 02 m 09.2 s	−04° 14' 45"
353	Cet	Gxy	1.3'×0.4'	13.9	01 h 02 m 24.5 s	−01° 57' 28"
355	Cet	Gxy	1.0'×0.4'	15.3	01 h 03 m 06.9 s	−06° 19' 28"
356	Cet	Gxy	1.5'×0.8'	13.1	01 h 03 m 06.9 s	−06° 59' 17"
357	Cet	Gxy	2.4'×1.7'	11.9	01 h 03 m 21.7 s	−06° 20' 22"
359	Cet	Gxy	1.5'×1.1'	13.4	01 h 04 m 16.9 s	−00° 45' 53"
363	Cet	Gxy	0.6'×0.5'	14.7	01 h 06 m 15.9 s	−16° 32' 35"
364	Cet	Gxy	1.4'×1.3'	13.2	01 h 04 m 40.8 s	−00° 48' 10"
367	Cet	Gxy	0.9'×0.5'	14.7	01 h 05 m 48.9 s	−12° 07' 44"
369	Cet	Gxy	1.0'×0.8'	14	01 h 05 m 08.8 s	−17° 45' 34"
377	Cet	Gxy	0.8'×0.3'	14.9	01 h 06 m 35.1 s	−20° 19' 56"
391	Cet	Gxy	0.9'×0.7'	13.5	01 h 07 m 22.5 s	+00° 55' 32"
412	Cet	NF	...	...	01 h 10 m 20.3 s	−20° 00' 54"
413	Cet	Gxy	1.1'×0.7'	14.2	01 h 12 m 31.5 s	−02° 47' 38"
417	Cet	Gxy	0.6'×0.5'	14.2	01 h 11 m 05.5 s	−18° 08' 54"

(continued)

Table 10.1 (continued)

NGC #	Const	Type	Size	V Mag	RA (2000)	Dec (2000)
426	Cet	Gxy	1.4'×1.0'	13.1	01 h 12 m 48.6 s	−00° 17' 25"
428	Cet	Gxy	4.1'×3.1'	11.2	01 h 12 m 55.5 s	+00° 58' 55"
429	Cet	Gxy	1.4'×0.3'	13.5	01 h 12 m 57.3 s	−00° 20' 42"
430	Cet	Gxy	1.3'×1.1'	12.5	01 h 12 m 59.9 s	−00° 15' 10"
435	Cet	Gxy	1.1'×0.4'	14.3	01 h 13 m 59.8 s	+02° 04' 14"
442	Cet	Gxy	1.0'×0.5'	13.7	01 h 14 m 38.7 s	−01° 01' 16"
445	Cet	Gxy	0.8'×0.6'	14.2	01 h 14 m 52.6 s	+01° 55' 04"
448	Cet	Gxy	1.6'×0.8'	12.2	01 h 15 m 16.5 s	−01° 37' 36"
450	Cet	Gxy	3.1'×2.3'	11.7	01 h 15 m 30.4 s	−00° 51' 40"
478	Cet	Gxy	0.9'×0.7'	14.3	01 h 20 m 09.4 s	−22° 22' 41"
480	Cet	Gxy	0.5'×0.2'	15.2	01 h 20 m 34.3 s	−09° 52' 49"
481	Cet	Gxy	1.7'×1.3'	13.5	01 h 21 m 12.5 s	−09° 12' 40"
487	Cet	Gxy	1.1'×0.7'	13.4	01 h 21 m 55.1 s	−16° 22' 17"
493	Cet	Gxy	3.4'×1.0'	12.3	01 h 22 m 09.1 s	+00° 56' 44"
497	Cet	Gxy	2.1'×0.9'	13.2	01 h 22 m 23.9 s	−00° 52' 33"
519	Cet	Gxy	0.5'×0.3'	14.4	01 h 24 m 28.7 s	−01° 38' 30"
521	Cet	Gxy	3.2'×2.9'	11.8	01 h 24 m 33.8 s	+01° 43' 52"
530	Cet	Gxy	1.5'×0.4'	13.2	01 h 24 m 41.7 s	−01° 35' 16"
533	Cet	Gxy	3.8'×2.3'	11.4	01 h 25 m 31.4 s	+01° 45' 32"
535	Cet	Gxy	1.0'×0.3'	14	01 h 25 m 31.1 s	−01° 24' 33"
538	Cet	Gxy	1.0'×0.5'	13.9	01 h 25 m 26.1 s	−01° 33' 05"
539	Cet	Gxy	1.5'×1.3'	13.8	01 h 25 m 21.7 s	−18° 09' 51"
540	Cet	Gxy	0.8'×0.5'	14.9	01 h 27 m 08.8 s	−20° 02' 10"
541	Cet	Gxy	1.8'×1.7'	12.6	01 h 25 m 44.4 s	−01° 22' 49"
543	Cet	Gxy	0.6'×0.3'	14.2	01 h 25 m 50.0 s	−01° 17' 37"
545	Cet	Gxy	2.4'×1.6'	12.3	01 h 25 m 59.1 s	−01° 20' 24"
547	Cet	Gxy	3.0'×2.0'	12.3	01 h 26 m 00.6 s	−01° 20' 42"
548	Cet	Gxy	0.9'×0.8'	13.8	01 h 26 m 02.6 s	−01° 13' 36"
550	Cet	Gxy	1.5'×0.6'	12.9	01 h 26 m 42.7 s	+02° 01' 20"
554	Cet	Gxy	0.7'×0.4'	14.1	01 h 27 m 09.4 s	−22° 43' 27"
555	Cet	Gxy	0.7'×0.6'	14.3	01 h 27 m 11.4 s	−22° 45' 45"
556	Cet	Gxy	0.4'×0.3'	14.5	01 h 27 m 13.4 s	−22° 41' 52"
557	Cet	Gxy	1.4'×0.8'	13.7	01 h 26 m 25.2 s	−01° 38' 22"
558	Cet	Gxy	0.4'×0.2'	14.4	01 h 27 m 16.1 s	−01° 58' 18"
560	Cet	Gxy	1.9'×0.6'	12.9	01 h 27 m 25.4 s	−01° 54' 51"
563	Cet	Gxy	1.5'×1.3'	13.4	01 h 25 m 21.7 s	−18° 09' 51"
564	Cet	Gxy	1.4'×1.2'	12.6	01 h 27 m 48.2 s	−01° 52' 48"
565	Cet	Gxy	1.3'×0.4'	13.7	01 h 28 m 10.1 s	−01° 18' 26"
567	Cet	Gxy	0.8'×0.4'	14.2	01 h 27 m 02.5 s	−10° 15' 56"
570	Cet	Gxy	1.5'×1.3'	13.1	01 h 28 m 58.5 s	−00° 57' 00"
577	Cet	Gxy	1.8'×1.4'	13.1	01 h 30 m 40.7 s	−01° 59' 44"
578	Cet	Gxy	4.9'×3.1'	11.1	01 h 30 m 28.7 s	−22° 40' 00"
580	Cet	Gxy	2.2'×1.8'	13	01 h 30 m 40.7 s	−01° 59' 44"
583	Cet	Gxy	0.7'×0.6'	14.2	01 h 29 m 44.1 s	−18° 20' 22"
584	Cet	Gxy	4.2'×2.3'	10.3	01 h 31 m 20.8 s	−06° 52' 06"

(continued)

Table 10.1 (continued)

NGC #	Const	Type	Size	V Mag	RA (2000)	Dec (2000)
585	Cet	Gxy	2.1'×0.5'	13.3	01 h 31 m 42.1 s	−00° 56' 03"
586	Cet	Gxy	1.6'×0.8'	13.2	01 h 31 m 36.9 s	−06° 53' 41"
589	Cet	Gxy	1.1'×0.9'	14.1	01 h 32 m 39.9 s	−12° 02' 34"
593	Cet	Gxy	1.2'×0.2'	14.6	01 h 32 m 20.7 s	−12° 21' 26"
594	Cet	Gxy	1.3'×0.6'	13.5	01 h 32 m 56.8 s	−16° 32' 11"
596	Cet	Gxy	3.2'×2.1'	10.8	01 h 32 m 51.9 s	−07° 01' 57"
599	Cet	Gxy	1.4'×1.3'	13.5	01 h 32 m 53.8 s	−12° 11' 28"
600	Cet	Gxy	3.3'×2.8'	12.4	01 h 33 m 05.3 s	−07° 18' 46"
601	Cet	Gxy	0.2'×0.2'	15.1	01 h 33 m 06.6 s	−12° 12' 33"
607	Cet	**	...	...	01 h 34 m 16.3 s	−07° 24' 46"
610	Cet	NF	...	...	01 h 34 m 18.0 s	−20° 08' 39"
611	Cet	NF	...	...	01 h 34 m 18.0 s	−20° 07' 39"
615	Cet	Gxy	3.6'×1.4'	11.7	01 h 35 m 05.7 s	−07° 20' 28"
617	Cet	Gxy	0.5'×0.5'	14.5	01 h 34 m 02.5 s	−09° 46' 27"
622	Cet	Gxy	1.8'×1.3'	13.1	01 h 36 m 00.2 s	+00° 39' 47"
624	Cet	Gxy	1.5'×0.9'	13.3	01 h 35 m 51.1 s	−10° 00' 10"
635	Cet	Gxy	0.5'×0.5'	14.5	01 h 38 m 17.7 s	−22° 55' 44"
636	Cet	Gxy	2.8'×2.1'	11.4	01 h 39 m 06.5 s	−07° 30' 46"
640	Cet	Gxy	0.6'×0.4'	15	01 h 39 m 25.0 s	−09° 24' 06"
647	Cet	Gxy	1.5'×1.1'	13.5	01 h 39 m 56.3 s	−09° 14' 33"
648	Cet	Gxy	1.0'×0.5'	14.4	01 h 38 m 39.8 s	−17° 49' 53"
649	Cet	Gxy	0.9'×0.3'	14.3	01 h 40 m 07.5 s	−09° 16' 21"
655	Cet	Gxy	1.1'×0.8'	14	01 h 41 m 55.2 s	−13° 04' 55"
667	Cet	Gxy	0.6'×0.5'	14.3	01 h 44 m 57.3 s	−22° 55' 17"
681	Cet	Gxy	2.6'×1.6'	12	01 h 49 m 10.8 s	−10° 25' 40"
682	Cet	Gxy	1.4'×1.1'	13.4	01 h 49 m 04.5 s	−14° 58' 30"
690	Cet	Gxy	1.2'×0.8'	14.5	01 h 47 m 48.1 s	−16° 43' 17"
699	Cet	Gxy	1.5'×0.3'	14.1	01 h 50 m 44.0 s	−12° 02' 10"
701	Cet	Gxy	2.5'×1.2'	12.3	01 h 51 m 03.8 s	−09° 42' 14"
702	Cet	Gxy	1.5'×1.1'	13.1	01 h 51 m 18.3 s	−04° 03' 21"
707	Cet	Gxy	1.3'×0.8'	13.6	01 h 51 m 27.1 s	−08° 30' 20"
713	Cet	Gxy	1.0'×0.3'	14.4	01 h 55 m 21.5 s	−09° 05' 01"
715	Cet	Gxy	0.6'×0.4'	14	01 h 53 m 12.4 s	−12° 52' 26"
720	Cet	Gxy	4.7'×2.4'	10.3	01 h 53 m 00.4 s	−13° 44' 21"
723	Cet	Gxy	1.5'×1.3'	12.6	01 h 53 m 45.8 s	−23° 45' 28"
724	Cet	Gxy	1.7'×1.5'	12.6	01 h 53 m 45.8 s	−23° 45' 28"
725	Cet	Gxy	0.7'×0.6'	14.3	01 h 52 m 35.6 s	−16° 31' 05"
726	Cet	Gxy	1.2'×0.6'	14.3	01 h 55 m 31.9 s	−10° 48' 00"
731	Cet	Gxy	1.7'×1.7'	12.1	01 h 54 m 56.1 s	−09° 00' 39"
734	Cet	Gxy	0.7'×0.4'	15.9	01 h 54 m 57.3 s	−17° 04' 47"
747	Cet	Gxy	1.0'×0.4'	14	01 h 57 m 30.5 s	−09° 27' 45"
748	Cet	Gxy	2.3'×1.1'	12.6	01 h 56 m 21.7 s	−04° 28' 04"
755	Cet	Gxy	3.4'×1.1'	12.6	01 h 56 m 22.5 s	−09° 03' 43"
756	Cet	Gxy	0.4'×0.4'	14.5	01 h 54 m 29.1 s	−16° 42' 28"
757	Cet	Gxy	1.7'×1.7'	12.1	01 h 54 m 56.1 s	−09° 00' 39"

(continued)

Table 10.1 (continued)

NGC #	Const	Type	Size	V Mag	RA (2000)	Dec (2000)
758	Cet	Gxy	0.9' × 0.7'	14.4	01 h 55 m 42.1 s	−03° 04' 00"
762	Cet	Gxy	1.3' × 1.1'	13	01 h 56 m 57.7 s	−05° 24' 10"
763	Cet	Gxy	3.6' × 1.1'	12.6	01 h 56 m 22.5 s	−09° 03' 43"
764	Cet	**	...	...	01 h 57 m 03.3 s	−16° 03' 45"
767	Cet	Gxy	1.1' × 0.4'	14.5	01 h 58 m 50.9 s	−09° 35' 13"
768	Cet	Gxy	1.7' × 0.8'	13.4	01 h 58 m 40.9 s	+00° 31' 45"
773	Cet	Gxy	1.3' × 0.7'	13.1	01 h 58 m 52.2 s	−11° 30' 54"
779	Cet	Gxy	4.0' × 1.2'	11.3	01 h 59 m 42.5 s	−05° 57' 54"
787	Cet	Gxy	2.5' × 1.9'	12.9	02 h 00 m 48.6 s	−09° 00' 09"
788	Cet	Gxy	1.9' × 1.4'	12.3	02 h 01 m 06.5 s	−06° 48' 56"
790	Cet	Gxy	1.3' × 1.3'	13	02 h 01 m 21.6 s	−05° 22' 16"
799	Cet	Gxy	2.0' × 1.7'	13.4	02 h 02 m 12.3 s	−00° 06' 04"
800	Cet	Gxy	1.0' × 0.9'	13.6	02 h 02 m 11.8 s	−00° 07' 51"
806	Cet	Gxy	1.2' × 0.4'	14.1	02 h 03 m 31.5 s	−09° 55' 56"
808	Cet	Gxy	1.4' × 0.7'	13.6	02 h 03 m 57.1 s	−23° 18' 50"
809	Cet	Gxy	1.5' × 1.0'	13.8	02 h 04 m 18.9 s	−08° 44' 07"
811	Cet	Gxy	0.3' × 0.2'	...	02 h 04 m 00.0 s	−09° 06' 20"
814	Cet	Gxy	1.4' × 0.5'	13.8	02 h 10 m 37.7 s	−15° 46' 25"
815	Cet	Gxy	0.3' × 0.1'	15.5	02 h 10 m 39.3 s	−15° 48' 47"
825	Cet	Gxy	2.2' × 0.9'	13.4	02 h 08 m 32.3 s	+06° 19' 27"
827	Cet	Gxy	2.2' × 0.8'	13	02 h 08 m 56.3 s	+07° 58' 19"
829	Cet	Gxy	1.2' × 0.7'	13.7	02 h 08 m 42.2 s	−07° 47' 28"
830	Cet	Gxy	1.4' × 0.9'	13.3	02 h 08 m 58.7 s	−07° 46' 04"
831	Cet	Gxy	0.4' × 0.3'	14.4	02 h 09 m 34.5 s	+06° 05' 46"
833	Cet	Gxy	1.5' × 0.7'	12.8	02 h 09 m 20.6 s	−10° 07' 59"
835	Cet	Gxy	1.3' × 1.0'	12.2	02 h 09 m 24.7 s	−10° 08' 10"
836	Cet	Gxy	1.3' × 0.9'	13.4	02 h 10 m 25.2 s	−22° 03' 17"
837	Cet	Gxy	0.9' × 0.4'	14.2	02 h 10 m 15.9 s	−22° 25' 47"
838	Cet	Gxy	1.1' × 0.9'	13	02 h 09 m 38.6 s	−10° 08' 47"
839	Cet	Gxy	1.4' × 0.7'	13.1	02 h 09 m 42.9 s	−10° 11' 02"
840	Cet	Gxy	1.8' × 1.0'	13.6	02 h 10 m 16.1 s	+07° 50' 43"
842	Cet	Gxy	0.9' × 0.5'	12.7	02 h 09 m 50.7 s	−07° 45' 44"
844	Cet	Gxy	0.03' × 0.03'	14	02 h 10 m 14.3 s	+06° 03' 00"
848	Cet	Gxy	1.5' × 1.0'	13	02 h 10 m 17.3 s	−10° 19' 12"
849	Cet	Gxy	0.5' × 0.3'	14.4	02 h 10 m 11.0 s	−22° 19' 22"
850	Cet	Gxy	1.1' × 1.1'	13	02 h 11 m 13.7 s	−01° 29' 11"
851	Cet	Gxy	1.0' × 0.6'	13.6	02 h 11 m 12.1 s	+03° 46' 46"
853	Cet	Gxy	1.5' × 1.3'	12.9	02 h 11 m 41.2 s	−09° 18' 20"
856	Cet	Gxy	1.3' × 0.9'	13.4	02 h 13 m 38.4 s	−00° 43' 02"
858	Cet	Gxy	1.3' × 1.1'	13.7	02 h 12 m 29.7 s	−22° 28' 16"
859	Cet	Gxy	1.2' × 0.9'	13.3	02 h 13 m 38.4 s	−00° 43' 02"
863	Cet	Gxy	1.1' × 1.0'	13.2	02 h 14 m 33.5 s	−00° 46' 00"
864	Cet	Gxy	4.7' × 3.5'	11.1	02 h 15 m 27.3 s	+06° 00' 05"
866	Cet	Gxy	1.4' × 1.3'	13	02 h 14 m 33.5 s	−00° 46' 00"
867	Cet	Gxy	1.5' × 1.5'	13.1	02 h 17 m 04.7 s	+01° 14' 37"

(continued)

Table 10.1 (continued)

NGC #	Const	Type	Size	V Mag	RA (2000)	Dec (2000)
868	Cet	Gxy	1.3'×1.0'	14	02 h 15 m 58.5 s	−00° 42' 49"
872	Cet	Gxy	1.5'×0.8'	14	02 h 15 m 25.3 s	−17° 46' 54"
873	Cet	Gxy	1.6'×1.3'	12.6	02 h 16 m 32.3 s	−11° 20' 56"
874	Cet	Gxy	0.9'×0.5'	14.4	02 h 16 m 02.2 s	−23° 18' 22"
875	Cet	Gxy	1.1'×1.1'	13.1	02 h 17 m 04.7 s	+01° 14' 37"
878	Cet	Gxy	0.8'×0.5'	13.9	02 h 17 m 54.4 s	−23° 22' 59"
879	Cet	Gxy	0.8'×0.6'	14.7	02 h 16 m 51.3 s	−08° 57' 49"
880	Cet	Gxy	0.4'×0.3'	14.6	02 h 18 m 27.1 s	−04° 12' 21"
881	Cet	Gxy	2.2'×1.5'	12.5	02 h 18 m 45.3 s	−06° 38' 24"
883	Cet	Gxy	1.7'×1.3'	12.6	02 h 19 m 05.1 s	−06° 47' 29"
885	Cet	Gxy	1.4'×1.3'	13	02 h 14 m 33.5 s	−00° 46' 00"
887	Cet	Gxy	1.9'×1.5'	12.7	02 h 19 m 32.6 s	−16° 04' 13"
892	Cet	Gxy	0.6'×0.3'	15.1	02 h 20 m 52.3 s	−23° 06' 48"
894	Cet	GxyCld	...	...	02 h 21 m 34.5 s	−05° 30' 37"
895	Cet	Gxy	3.6'×2.6'	11.4	02 h 21 m 36.5 s	−05° 31' 18"
899	Cet	Gxy	1.6'×1.1'	12.7	02 h 21 m 54.0 s	−20° 49' 21"
902	Cet	Gxy	0.5'×0.4'	14	02 h 22 m 21.9 s	−16° 40' 44"
905	Cet	Gxy	0.5'×0.3'	15.1	02 h 22 m 43.5 s	−08° 43' 09"
907	Cet	Gxy	1.8'×0.6'	12.8	02 h 23 m 02.0 s	−20° 42' 46"
908	Cet	Gxy	6.0'×2.6'	10.5	02 h 23 m 04.7 s	−21° 14' 03"
921	Cet	Gxy	1.3'×0.6'	14.3	02 h 26 m 33.5 s	−15° 50' 51"
926	Cet	Gxy	1.8'×1.0'	13	02 h 26 m 07.0 s	−00° 19' 51"
929	Cet	Gxy	1.0'×0.6'	14.2	02 h 27 m 18.1 s	−12° 05' 13"
934	Cet	Gxy	1.3'×0.9'	13.1	02 h 27 m 32.9 s	−00° 14' 40"
936	Cet	Gxy	4.7'×4.1'	10	02 h 27 m 37.5 s	−01° 09' 23"
941	Cet	Gxy	2.6'×1.9'	12.4	02 h 28 m 27.9 s	−01° 09' 05"
942	Cet	Gxy	...'×...'	13	02 h 29 m 10.2 s	−10° 50' 09"
943	Cet	Gxy	...'×...'	12.9	02 h 29 m 09.5 s	−10° 49' 40"
944	Cet	Gxy	1.1'×0.3'	14	02 h 26 m 41.7 s	−14° 30' 57"
945	Cet	Gxy	2.4'×2.0'	12.1	02 h 28 m 37.2 s	−10° 32' 22"
947	Cet	Gxy	2.0'×1.1'	12.9	02 h 28 m 33.1 s	−19° 02' 32"
948	Cet	Gxy	1.5'×1.2'	13.7	02 h 28 m 45.4 s	−10° 30' 51"
950	Cet	Gxy	1.3'×0.8'	13.8	02 h 29 m 11.6 s	−11° 01' 33"
951	Cet	Gxy	1.0'×0.6'	14.8	02 h 28 m 56.9 s	−22° 20' 59"
955	Cet	Gxy	2.8'×0.7'	12.3	02 h 30 m 33.4 s	−01° 06' 32"
958	Cet	Gxy	2.9'×1.0'	12.5	02 h 30 m 42.7 s	−02° 56' 21"
960	Cet	Gxy	1.2'×0.3'	14.2	02 h 31 m 41.3 s	−09° 18' 01"
961	Cet	Gxy	2.1'×1.8'	12.8	02 h 41 m 02.4 s	−06° 56' 10"
963	Cet	Gxy	0.6'×0.6'	13.9	02 h 30 m 31.1 s	−04° 13' 01"
965	Cet	Gxy	1.0'×0.8'	14.4	02 h 32 m 25.1 s	−18° 38' 24"
966	Cet	Gxy	1.0'×0.9'	13.4	02 h 31 m 47.0 s	−19° 52' 58"
967	Cet	Gxy	1.6'×1.0'	12.6	02 h 32 m 12.7 s	−17° 13' 01"
975	Cet	Gxy	1.1'×0.8'	13.3	02 h 33 m 22.7 s	+09° 36' 05"
977	Cet	Gxy	1.9'×1.6'	13.5	02 h 33 m 03.4 s	−10° 45' 36"
981	Cet	Gxy	1.0'×0.6'	14	02 h 32 m 59.9 s	−10° 58' 23"

(continued)

Table 10.1 (continued)

NGC #	Const	Type	Size	V Mag	RA (2000)	Dec (2000)
985	Cet	Gxy	1.0'×0.9'	13.5	02 h 34 m 37.4 s	−08° 47' 10"
988	Cet	Gxy	4.6'×2.5'	11	02 h 35 m 29.7 s	−09° 21' 34"
989	Cet	Gxy	0.7'×0.6'	14.1	02 h 33 m 46.0 s	−16° 30' 40"
991	Cet	Gxy	2.7'×2.4'	11.9	02 h 35 m 32.6 s	−07° 09' 20"
993	Cet	Gxy	0.9'×0.8'	13.7	02 h 36 m 45.9 s	+02° 02' 58"
994	Cet	Gxy	1.1'×1.0'	13.7	02 h 36 m 45.9 s	+02° 02' 58"
997	Cet	Gxy	1.2'×1.2'	13.4	02 h 37 m 14.4 s	+07° 18' 23"
998	Cet	Gxy	0.6'×0.5'	14.8	02 h 37 m 16.5 s	+07° 20' 09"
1004	Cet	Gxy	1.4'×1.3'	13	02 h 37 m 41.7 s	+01° 58' 30"
1006	Cet	Gxy	0.9'×0.8'	14.1	02 h 37 m 34.8 s	−11° 01' 33"
1007	Cet	Gxy	0.7'×0.2'	15.8	02 h 37 m 52.2 s	+02° 09' 21"
1008	Cet	Gxy	0.8'×0.6'	13.7	02 h 37 m 55.3 s	+02° 04' 47"
1009	Cet	Gxy	1.4'×0.2'	14.6	02 h 38 m 18.9 s	+02° 18' 35"
1010	Cet	Gxy	0.9'×0.8'	14.1	02 h 37 m 34.8 s	−11° 01' 33"
1011	Cet	Gxy	0.6'×0.6'	14.3	02 h 37 m 38.9 s	−11° 00' 20"
1013	Cet	Gxy	0.8'×0.6'	14	02 h 37 m 50.5 s	−11° 30' 26"
1014	Cet	**	...	...	02 h 38 m 00.8 s	−09° 34' 24"
1015	Cet	Gxy	2.6'×2.6'	12.3	02 h 38 m 11.6 s	−01° 19' 10"
1016	Cet	Gxy	2.4'×2.4'	11.9	02 h 38 m 19.5 s	+02° 07' 09"
1017	Cet	Gxy	0.7'×0.6'	14.4	02 h 37 m 49.9 s	−11° 00' 40"
1018	Cet	Gxy	1.0'×0.6'	14.8	02 h 38 m 10.3 s	−09° 32' 38"
1019	Cet	Gxy	1.0'×0.9'	13.7	02 h 38 m 27.5 s	+01° 54' 25"
1020	Cet	Gxy	0.8'×0.2'	14.2	02 h 38 m 44.3 s	+02° 13' 50"
1021	Cet	Gxy	0.6'×0.5'	14.4	02 h 38 m 47.9 s	+02° 13' 01"
1022	Cet	Gxy	2.4'×2.0'	11.3	02 h 38 m 32.7 s	−06° 40' 41"
1026	Cet	Gxy	2.0'×1.8'	12.7	02 h 39 m 19.3 s	+06° 32' 38"
1032	Cet	Gxy	3.3'×1.1'	12.1	02 h 39 m 23.7 s	+01° 05' 35"
1033	Cet	Gxy	1.3'×1.1'	13.8	02 h 40 m 16.1 s	−08° 46' 38"
1034	Cet	Gxy	1.0'×0.7'	14	02 h 38 m 13.9 s	−15° 48' 34"
1035	Cet	Gxy	2.2'×0.7'	12.6	02 h 39 m 29.3 s	−08° 08' 01"
1037	Cet	NF	...	...	02 h 39 m 58.3 s	−01° 44' 03"
1038	Cet	Gxy	1.2'×0.4'	13.6	02 h 40 m 06.3 s	+01° 30' 29"
1041	Cet	Gxy	1.7'×1.2'	13.3	02 h 40 m 25.2 s	−05° 26' 26"
1042	Cet	Gxy	4.7'×3.6'	11.3	02 h 40 m 23.7 s	−08° 26' 02"
1043	Cet	Gxy	0.9'×0.2'	15	02 h 40 m 46.5 s	+01° 20' 35"
1044	Cet	Gxy	0.6'×0.6'	13.4	02 h 41 m 06.1 s	+08° 44' 16"
1045	Cet	Gxy	2.3'×1.2'	12.4	02 h 40 m 29.1 s	−11° 16' 42"
1046	Cet	Gxy	0.3'×0.3'	13.9	02 h 41 m 12.8 s	+08° 43' 09"
1047	Cet	Gxy	1.3'×0.6'	13.5	02 h 40 m 32.9 s	−08° 08' 52"
1048	Cet	Gxy	1.0'×0.2'	14.6	02 h 40 m 37.9 s	−08° 32' 01"
1051	Cet	Gxy	2.1'×1.5'	12.8	02 h 41 m 02.4 s	−06° 56' 10"
1052	Cet	Gxy	3.0'×2.1'	10.6	02 h 41 m 04.7 s	−08° 15' 21"
1055	Cet	Gxy	7.6'×2.7'	10.9	02 h 41 m 45.4 s	+00° 26' 32"
1063	Cet	Gxy	1.4'×0.5'	13.8	02 h 42 m 10.0 s	−05° 34' 08"
1064	Cet	Gxy	1.1'×1.0'	14.3	02 h 42 m 23.5 s	−09° 21' 45"

(continued)

Table 10.1 (continued)

NGC #	Const	Type	Size	V Mag	RA (2000)	Dec (2000)
1065	Cet	Gxy	0.5'×0.5'	14.1	02 h 42 m 06.2 s	−15° 05' 30"
1068	Cet	Gxy	7.1'×6.0'	9.1	02 h 42 m 40.7 s	−00° 00' 47"
1069	Cet	Gxy	1.4'×0.9'	13.8	02 h 42 m 59.9 s	−08° 17' 22"
1070	Cet	Gxy	2.3'×1.9'	12.1	02 h 43 m 22.0 s	+04° 58' 05"
1071	Cet	Gxy	1.1'×0.5'	14.4	02 h 43 m 07.7 s	−08° 46' 29"
1072	Cet	Gxy	1.5'×0.5'	13.6	02 h 43 m 31.3 s	+00° 18' 23"
1073	Cet	Gxy	4.9'×4.5'	11.1	02 h 43 m 40.6 s	+01° 22' 31"
1074	Cet	Gxy	1.9'×1.2'	14.3	02 h 43 m 36.1 s	−16° 17' 49"
1075	Cet	Gxy	0.8'×0.6'	14.6	02 h 43 m 33.5 s	−16° 12' 04"
1076	Cet	Gxy	1.9'×1.1'	12.7	02 h 43 m 29.1 s	−14° 45' 18"
1078	Cet	Gxy	0.3'×0.3'	14.3	02 h 44 m 08.0 s	−09° 27' 08"
1080	Cet	Gxy	1.0'×0.8'	13.4	02 h 45 m 09.9 s	−04° 42' 38"
1085	Cet	Gxy	3.0'×2.1'	12.5	02 h 46 m 25.3 s	+03° 36' 26"
1087	Cet	Gxy	3.7'×2.2'	10.8	02 h 46 m 25.2 s	−00° 29' 59"
1090	Cet	Gxy	4.0'×1.7'	11.9	02 h 46 m 34.1 s	−00° 14' 51"
1094	Cet	Gxy	1.3'×1.0'	12.7	02 h 47 m 27.8 s	−00° 17' 07"
1095	Cet	Gxy	1.3'×0.8'	13.5	02 h 47 m 37.7 s	+04° 38' 15"
1101	Cet	Gxy	1.4'×1.2'	13.4	02 h 48 m 14.8 s	+04° 34' 41"
1104	Cet	Gxy	1.2'×0.9'	13.8	02 h 48 m 38.7 s	−00° 16' 18"
1107	Cet	Gxy	1.8'×1.5'	12.7	02 h 49 m 19.5 s	+08° 05' 34"
1128	Cet	Gxy	0.7'×0.3'	14.5	02 h 57 m 41.5 s	+06° 01' 29"
1137	Cet	Gxy	2.1'×1.3'	12.6	02 h 54 m 02.7 s	+02° 57' 42"
1141	Cet	Gxy	1.1'×1.1'	13.1	02 h 55 m 09.7 s	−00° 10' 44"
1142	Cet	Gxy	0.9'×0.6'	12.8	02 h 55 m 12.0 s	−00° 11' 03"
1143	Cet	Gxy	0.9'×0.8'	13.1	02 h 55 m 09.7 s	−00° 10' 44"
1144	Cet	Gxy	1.1'×0.7'	12.8	02 h 55 m 12.0 s	−00° 11' 03"
1149	Cet	Gxy	0.6'×0.5'	14.1	02 h 57 m 23.8 s	−00° 18' 34"
1153	Cet	Gxy	1.3'×1.2'	12.5	02 h 58 m 10.3 s	+03° 21' 42"
1194	Cet	Gxy	1.8'×1.0'	13.4	03 h 03 m 49.3 s	−01° 06' 14"
1211	Cet	Gxy	2.1'×1.8'	12.3	03 h 06 m 52.5 s	−00° 47' 42"
1218	Cet	Gxy	1.3'×1.0'	13	03 h 08 m 26.1 s	+04° 06' 39"
1219	Cet	Gxy	1.2'×1.2'	12.8	03 h 08 m 28.1 s	+02° 06' 29"
1251	Cet	**	...	...	03 h 14 m 09.0 s	+01° 27' 24"
1254	Cet	Gxy	0.8'×0.7'	14.2	03 h 14 m 23.9 s	+02° 40' 41"
1280	Cet	Gxy	0.9'×0.8'	13.4	03 h 17 m 57.1 s	−00° 10' 11"
7807	Cet	Gxy	0.6'×0.5'	15	00 h 00 m 26.6 s	−18° 50' 32"
7808	Cet	Gxy	1.3'×1.3'	13.5	00 h 03 m 32.1 s	−10° 44' 41"
7813	Cet	Gxy	0.4'×0.2'	14.2	00 h 04 m 09.3 s	−11° 59' 04"
7821	Cet	Gxy	1.4'×0.5'	13.3	00 h 05 m 16.6 s	−16° 28' 39"
7826	Cet	Ast	12.0'	...	00 h 05 m 14.3 s	−20° 42' 54"
7828	Cet	Gxy	0.9'×0.5'	13.9	00 h 06 m 27.1 s	−13° 24' 58"
7829	Cet	Gxy	0.7'×0.7'	13.9	00 h 06 m 28.9 s	−13° 25' 15"
2915	Cha	Gxy	1.5'×1.0'	12.5	09 h 26 m 10.5 s	−76° 37' 35"
3149	Cha	Gxy	2.2'×2.0'	12.3	10 h 03 m 43.4 s	−80° 25' 21"
3195	Cha	PN	40"×30"	11.6	10 h 09 m 21.1 s	−80° 51' 30"

(continued)

Table 10.1 (continued)

NGC #	Const	Type	Size	V Mag	RA (2000)	Dec (2000)
3620	Cha	Gxy	3.0'×0.9'	13.5	11 h 16 m 06.0 s	−76° 12' 58"
5288	Cir	OC	4.0'	11.8	13 h 48 m 44.9 s	−64° 41' 07"
5315	Cir	PN	0.23'	9.8	13 h 53 m 57.0 s	−66° 30' 50"
5359	Cir	OC	8.0'	…	14 h 00 m 09.5 s	−70° 23' 32"
5715	Cir	OC	5.0'	9.8	14 h 43 m 29.7 s	−57° 34' 37"
5823	Cir	OC	12'	7.9	15 h 05 m 30.6 s	−55° 36' 13"
2204	CMa	OC	12'	8.6	06 h 15 m 32.2 s	−18° 39' 57"
2206	CMa	Gxy	2.4'×1.3'	12.4	06 h 15 m 59.7 s	−26° 45' 57"
2207	CMa	Gxy	4.3'×2.8'	11	06 h 16 m 21.8 s	−21° 22' 24"
2211	CMa	Gxy	1.4'×0.7'	12.8	06 h 18 m 30.3 s	−18° 32' 14"
2212	CMa	Gxy	1.5'×0.8'	13.6	06 h 18 m 35.7 s	−18° 31' 10"
2216	CMa	Gxy	1.4'×1.1'	13	06 h 21 m 30.7 s	−22° 05' 15"
2217	CMa	Gxy	4.5'×4.2'	10.4	06 h 21 m 39.7 s	−27° 14' 04"
2223	CMa	Gxy	3.2'×2.8'	11.8	06 h 24 m 35.7 s	−22° 50' 21"
2227	CMa	Gxy	2.1'×1.1'	12.9	06 h 25 m 57.9 s	−22° 00' 19"
2243	CMa	OC	5'	9.4	06 h 29 m 34.5 s	−31° 16' 53"
2263	CMa	Gxy	2.6'×1.5'	12.3	06 h 38 m 28.2 s	−24° 50' 49"
2267	CMa	Gxy	1.7'×1.3'	12.3	06 h 40 m 51.6 s	−32° 28' 55"
2271	CMa	Gxy	2.1'×1.4'	12.3	06 h 42 m 52.9 s	−23° 28' 35"
2272	CMa	Gxy	2.5'×1.6'	11.8	06 h 42 m 41.2 s	−27° 27' 36"
2280	CMa	Gxy	6.3'×3.1'	11	06 h 44 m 49.1 s	−27° 38' 21"
2283	CMa	Gxy	3.6'×2.8'	12.4	06 h 45 m 52.7 s	−18° 12' 38"
2287	CMa	OC	38'	4.5	06 h 46 m 00.0 s	−20° 45' 15"
2292	CMa	Gxy	4.1'×3.6'	10.9	06 h 47 m 39.8 s	−26° 44' 49"
2293	CMa	Gxy	4.2'×3.3'	10.8	06 h 47 m 43.0 s	−26° 45' 19"
2295	CMa	Gxy	2.1'×0.6'	12.9	06 h 47 m 23.1 s	−26° 44' 11"
2296	CMa	Neb	0.6'×0.4'	…	06 h 48 m 39.1 s	−16° 54' 05"
2318	CMa	OC	…	…	06 h 59 m 27.0 s	−13° 41' 54"
2325	CMa	Gxy	3.3'×1.9'	11.5	07 h 02 m 40.5 s	−28° 41' 51"
2327	CMa	Neb	20.0'	…	07 h 04 m 07.2 s	−11° 18' 51"
2345	CMa	OC	12'	7.7	07 h 08 m 18.8 s	−13° 11' 38"
2352	CMa	Ast	…	…	07 h 13 m 06.0 s	−24° 02' 00"
2354	CMa	OC	20'	6.5	07 h 14 m 15.3 s	−25° 41' 33"
2358	CMa	OC	20'×15'	…	07 h 16 m 56.3 s	−17° 06' 50"
2359	CMa	Neb	10'×5'	…	07 h 18 m 30.9 s	−13° 13' 38"
2360	CMa	OC	12'	7.2	07 h 17 m 43.1 s	−15° 38' 29"
2361	CMa	Neb	…	…	07 h 18 m 23.8 s	−13° 12' 34"
2362	CMa	OC	8'	3.8	07 h 18 m 41.5 s	−24° 57' 15"
2367	CMa	OC	3.5'	7.9	07 h 20 m 04.5 s	−21° 53' 03"
2374	CMa	OC	19'	8	07 h 23 m 56.1 s	−13° 15' 48"
2380	CMa	Gxy	2.0'×1.9'	11.6	07 h 23 m 55.2 s	−27° 31' 43"
2382	CMa	Gxy	1.3'×1.2'	11.3	07 h 23 m 55.2 s	−27° 31' 43"
2383	CMa	OC	5'	8.4	07 h 24 m 39.9 s	−20° 56' 51"
2384	CMa	OC	2.5'	7.4	07 h 25 m 11.8 s	−21° 01' 24"
2350	CMi	Gxy	1.3'×0.7'	12.5	07 h 13 m 12.2 s	+12° 15' 57"

(continued)

Table 10.1 (continued)

NGC #	Const	Type	Size	V Mag	RA (2000)	Dec (2000)
2394	CMi	OC	10'	...	07 h 28 m 36.5 s	+07° 05' 12"
2399	CMi	***	...	...	07 h 29 m 50.5 s	−00° 12' 45"
2400	CMi	***	...	...	07 h 29 m 55.5 s	−00° 12' 53"
2402	CMi	Gxy	1.2'×1.2'	14.1	07 h 30 m 46.5 s	+09° 38' 51"
2412	CMi	*	...	...	07 h 34 m 21.5 s	+08° 32' 52"
2416	CMi	Gxy	1.0'×0.7'	13.6	07 h 35 m 41.5 s	+11° 36' 42"
2433	CMi	***	...	...	07 h 42 m 43.6 s	+09° 15' 33"
2459	CMi	OC	<1'	...	07 h 52 m 01.7 s	+09° 33' 27"
2470	CMi	Gxy	1.9'×0.6'	13.2	07 h 54 m 20.7 s	+04° 27' 35"
2485	CMi	Gxy	1.6'×1.6'	12.4	07 h 56 m 48.7 s	+07° 28' 39"
2491	CMi	Gxy	0.3'×0.2'	14.8	07 h 58 m 27.5 s	+07° 58' 59"
2496	CMi	Gxy	1.1'×0.9'	13.1	07 h 58 m 37.4 s	+08° 01' 44"
2499	CMi	Gxy	1.0'×0.6'	14.1	07 h 58 m 51.7 s	+07° 29' 33"
2504	CMi	Gxy	0.5'×0.4'	13	07 h 59 m 52.3 s	+05° 36' 29"
2508	CMi	Gxy	1.4'×1.1'	12.8	08 h 01 m 57.1 s	+08° 33' 06"
2510	CMi	Gxy	1.0'×0.7'	13.5	08 h 02 m 10.6 s	+09° 29' 10"
2511	CMi	Gxy	0.9'×0.3'	14.3	08 h 02 m 14.9 s	+09° 23' 40"
2538	CMi	Gxy	1.4'×1.2'	12.8	08 h 11 m 23.0 s	+03° 37' 56"
2503	Cnc	Gxy	1.1'×1.1'	13.9	08 h 00 m 36.7 s	+22° 23' 59"
2507	Cnc	Gxy	2.5'×1.8'	12.4	08 h 01 m 37.1 s	+15° 42' 34"
2512	Cnc	Gxy	1.4'×0.9'	13.4	08 h 03 m 07.7 s	+23° 23' 30"
2513	Cnc	Gxy	2.5'×2.0'	11.9	08 h 02 m 24.6 s	+09° 24' 49"
2514	Cnc	Gxy	1.3'×1.2'	13.4	08 h 02 m 49.7 s	+15° 48' 30"
2515	Cnc	**	...	...	08 h 03 m 21.3 s	+20° 11' 17"
2522	Cnc	Gxy	1.0'×0.3'	14	08 h 06 m 13.3 s	+17° 42' 24"
2526	Cnc	Gxy	0.9'×0.5'	13.9	08 h 06 m 58.6 s	+08° 00' 15"
2529	Cnc	NF	...	...	08 h 07 m 49.1 s	+17° 49' 15"
2530	Cnc	Gxy	1.4'×1.0'	13.7	08 h 07 m 55.5 s	+17° 49' 06"
2531	Cnc	NF	...	...	08 h 08 m 01.1 s	+17° 49' 14"
2535	Cnc	Gxy	2.5'×1.2'	12.7	08 h 11 m 13.5 s	+25° 12' 23"
2536	Cnc	Gxy	0.9'×0.6'	14.3	08 h 11 m 16.0 s	+25° 10' 44"
2540	Cnc	Gxy	1.3'×0.9'	13.7	08 h 12 m 46.5 s	+26° 21' 41"
2545	Cnc	Gxy	2.0'×1.1'	12.3	08 h 14 m 14.2 s	+21° 21' 19"
2553	Cnc	Gxy	0.9'×0.7'	14.1	08 h 17 m 34.9 s	+20° 54' 10"
2554	Cnc	Gxy	3.2'×2.3'	11.9	08 h 17 m 53.6 s	+23° 28' 19"
2556	Cnc	Gxy	0.4'×0.2'	14.3	08 h 19 m 00.9 s	+20° 56' 13"
2557	Cnc	Gxy	1.4'×1.3'	13.3	08 h 19 m 10.8 s	+21° 26' 08"
2558	Cnc	Gxy	2.2'×1.7'	13.2	08 h 19 m 12.8 s	+20° 30' 37"
2560	Cnc	Gxy	1.7'×0.5'	14.1	08 h 19 m 51.9 s	+20° 59' 06"
2562	Cnc	Gxy	1.0'×0.7'	13.5	08 h 20 m 23.8 s	+21° 07' 52"
2563	Cnc	Gxy	2.1'×1.5'	12.3	08 h 20 m 35.7 s	+21° 04' 03"
2565	Cnc	Gxy	1.9'×0.9'	13.1	08 h 19 m 48.3 s	+22° 01' 50"
2569	Cnc	Gxy	0.4'×0.3'	14.3	08 h 21 m 21.1 s	+20° 52' 03"
2570	Cnc	Gxy	1.3'×0.7'	14.2	08 h 21 m 22.6 s	+20° 54' 37"
2572	Cnc	Gxy	1.3'×0.5'	14	08 h 21 m 24.7 s	+19° 08' 51"

(continued)

Table 10.1 (continued)

NGC #	Const	Type	Size	V Mag	RA (2000)	Dec (2000)
2575	Cnc	Gxy	2.3'×1.9'	12.9	08 h 22 m 45.0 s	+24° 17' 49"
2576	Cnc	Gxy	1.7'×0.3'	14.5	08 h 22 m 57.7 s	+25° 44' 20"
2577	Cnc	Gxy	1.8'×1.1'	12.6	08 h 22 m 43.4 s	+22° 33' 11"
2581	Cnc	Gxy	1.1'×0.8'	13.5	08 h 24 m 31.0 s	+18° 35' 48"
2582	Cnc	Gxy	1.2'×1.2'	13.3	08 h 25 m 12.1 s	+20° 20' 05"
2592	Cnc	Gxy	1.7'×1.4'	12.4	08 h 27 m 08.1 s	+25° 58' 12"
2593	Cnc	Gxy	0.9'×0.5'	14.2	08 h 26 m 47.9 s	+17° 22' 28"
2594	Cnc	Gxy	0.6'×0.4'	14.1	08 h 27 m 17.1 s	+25° 52' 44"
2595	Cnc	Gxy	3.2'×2.4'	12.4	08 h 27 m 42.0 s	+21° 28' 44"
2596	Cnc	Gxy	1.5'×0.6'	13.6	08 h 27 m 26.6 s	+17° 17' 02"
2597	Cnc	**	...	...	08 h 29 m 57.4 s	+21° 30' 07"
2598	Cnc	Gxy	1.1'×0.4'	14.3	08 h 30 m 02.6 s	+21° 29' 19"
2599	Cnc	Gxy	1.9'×1.7'	12.4	08 h 32 m 11.3 s	+22° 33' 37"
2604	Cnc	Gxy	2.1'×2.1'	12.5	08 h 33 m 22.9 s	+29° 32' 16"
2607	Cnc	Gxy	0.9'×0.8'	14	08 h 33 m 56.6 s	+26° 58' 21"
2608	Cnc	Gxy	2.3'×1.4'	12.3	08 h 35 m 17.2 s	+28° 28' 24"
2611	Cnc	Gxy	0.7'×0.2'	14.5	08 h 35 m 29.3 s	+25° 01' 38"
2619	Cnc	Gxy	2.3'×1.4'	12.6	08 h 37 m 32.7 s	+28° 42' 17"
2620	Cnc	Gxy	2.0'×0.5'	13.7	08 h 37 m 28.3 s	+24° 56' 48"
2621	Cnc	Gxy	0.7'×0.4'	14.9	08 h 37 m 37.0 s	+24° 59' 59"
2622	Cnc	Gxy	0.8'×0.4'	14.1	08 h 38 m 11.0 s	+24° 53' 43"
2623	Cnc	Gxy	2.4'×0.7'	13.2	08 h 38 m 24.1 s	+25° 45' 15"
2624	Cnc	Gxy	0.8'×0.7'	13.9	08 h 38 m 09.5 s	+19° 43' 31"
2625	Cnc	Gxy	0.3'×0.3'	14.3	08 h 38 m 23.1 s	+19° 42' 56"
2628	Cnc	Gxy	1.1'×1.1'	13.5	08 h 40 m 22.7 s	+23° 32' 21"
2632	Cnc	OC	95'	3.1	08 h 40 m 22.2 s	+19° 40' 19"
2637	Cnc	Gxy	0.4'×0.5'	...	08 h 41 m 13.3 s	+19° 41' 26"
2643	Cnc	Gxy	0.6'×0.4'	...	08 h 41 m 51.7 s	+19° 42' 08"
2647	Cnc	Gxy	0.8'×0.5'	14.1	08 h 42 m 43.1 s	+19° 39' 01"
2648	Cnc	Gxy	3.2'×1.1'	12	08 h 42 m 39.9 s	+14° 17' 09"
2651	Cnc	Gxy	0.6'×0.5'	14.5	08 h 43 m 55.2 s	+11° 46' 15"
2657	Cnc	Gxy	1.3'×1.3'	13.1	08 h 45 m 15.9 s	+09° 38' 44"
2661	Cnc	Gxy	1.4'×1.3'	13	08 h 45 m 59.5 s	+12° 37' 14"
2664	Cnc	OC	5'	...	08 h 47 m 07.0 s	+12° 36' 21"
2667	Cnc	Gxy	0.7'×0.2'	14	08 h 48 m 27.3 s	+19° 01' 20"
2672	Cnc	Gxy	3.0'×2.8'	11.2	08 h 49 m 21.9 s	+19° 04' 29"
2673	Cnc	Gxy	1.2'×1.2'	13	08 h 49 m 24.1 s	+19° 04' 27"
2677	Cnc	Gxy	0.3'×0.2'	14.6	08 h 50 m 01.3 s	+19° 00' 34"
2678	Cnc	OC	...	...	08 h 50 m 02.7 s	+11° 20' 17"
2679	Cnc	Gxy	1.8'×1.8'	13.1	08 h 51 m 33.1 s	+30° 51' 55"
2680	Cnc	Gxy	0.2'×0.15'	14.5	08 h 51 m 33.4 s	+30° 51' 56"
2682	Cnc	OC	29'	6.9	08 h 51 m 20.1 s	+11° 48' 43"
2711	Cnc	Gxy	0.9'×0.6'	13.8	08 h 57 m 23.6 s	+17° 17' 17"
2720	Cnc	Gxy	1.2'×1.1'	12.9	08 h 59 m 08.1 s	+11° 08' 57"
2725	Cnc	Gxy	0.7'×0.6'	13.3	09 h 01 m 03.2 s	+11° 05' 50"

(continued)

Table 10.1 (continued)

NGC #	Const	Type	Size	V Mag	RA (2000)	Dec (2000)
2728	Cnc	Gxy	1.1'×0.8'	13.8	09 h 01 m 40.9 s	+11° 04' 58"
2730	Cnc	Gxy	1.7'×1.3'	13	09 h 02 m 15.8 s	+16° 50' 19"
2731	Cnc	Gxy	0.8'×0.5'	13.7	09 h 02 m 08.4 s	+08° 18' 05"
2734	Cnc	Gxy	0.6'×0.5'	15.5	09 h 03 m 01.7 s	+16° 51' 49"
2735	Cnc	Gxy	1.2'×0.4'	13.3	09 h 02 m 38.7 s	+25° 56' 05"
2737	Cnc	Gxy	1.1'×0.4'	14.3	09 h 03 m 59.7 s	+21° 54' 23"
2738	Cnc	Gxy	1.5'×0.7'	13.2	09 h 04 m 00.5 s	+21° 58' 03"
2741	Cnc	Gxy	0.4'×0.2'	15	09 h 03 m 16.5 s	+18° 15' 41"
2743	Cnc	Gxy	1.2'×0.8'	13.7	09 h 04 m 54.3 s	+25° 00' 14"
2744	Cnc	Gxy	1.7'×1.1'	13.1	09 h 04 m 39.1 s	+18° 27' 49"
2745	Cnc	Gxy	0.4'×0.2'	14.3	09 h 04 m 39.3 s	+18° 15' 25"
2747	Cnc	Gxy	0.4'×0.2'	14.5	09 h 05 m 18.3 s	+18° 26' 30"
2749	Cnc	Gxy	1.7'×1.4'	12	09 h 05 m 21.4 s	+18° 18' 47"
2750	Cnc	Gxy	2.2'×1.9'	12.1	09 h 05 m 47.9 s	+25° 26' 12"
2751	Cnc	Gxy	1.1'×0.7'	14.3	09 h 05 m 32.4 s	+18° 15' 44"
2752	Cnc	Gxy	1.9'×0.4'	13.9	09 h 05 m 43.1 s	+18° 20' 23"
2753	Cnc	Gxy	0.8'×0.5'	14.5	09 h 07 m 08.3 s	+25° 20' 30"
2761	Cnc	Gxy	0.6'×0.4'	14.3	09 h 07 m 30.7 s	+18° 26' 04"
2764	Cnc	Gxy	1.5'×0.9'	12.7	09 h 08 m 17.5 s	+21° 26' 36"
2766	Cnc	Gxy	1.3'×0.5'	13.8	09 h 08 m 47.5 s	+29° 51' 52"
2773	Cnc	Gxy	0.7'×0.3'	13.6	09 h 09 m 44.2 s	+07° 10' 25"
2774	Cnc	Gxy	0.8'×0.8'	13.9	09 h 10 m 39.9 s	+18° 41' 47"
2775	Cnc	Gxy	4.3'×3.3'	10.4	09 h 10 m 20.3 s	+07° 02' 16"
2777	Cnc	Gxy	0.9'×0.6'	13.5	09 h 10 m 41.9 s	+07° 12' 23"
2783	Cnc	Gxy	2.1'×1.5'	12.4	09 h 13 m 39.5 s	+29° 59' 34"
2786	Cnc	Gxy	0.9'×0.6'	12.8	09 h 13 m 35.9 s	+12° 26' 25"
2789	Cnc	Gxy	1.9'×1.9'	12.4	09 h 14 m 59.7 s	+29° 43' 49"
2790	Cnc	Gxy	0.4'×0.3'	14.5	09 h 15 m 02.8 s	+19° 41' 49"
2791	Cnc	Gxy	0.8'×0.3'	14.7	09 h 15 m 02.0 s	+17° 35' 34"
2794	Cnc	Gxy	1.2'×1.2'	13.2	09 h 16 m 01.8 s	+17° 35' 23"
2795	Cnc	Gxy	1.4'×1.0'	12.9	09 h 16 m 03.9 s	+17° 37' 41"
2796	Cnc	Gxy	1.1'×0.7'	13.5	09 h 16 m 41.8 s	+30° 54' 55"
2797	Cnc	Gxy	0.6'×0.5'	13.5	09 h 16 m 21.7 s	+17° 43' 38"
2801	Cnc	Gxy	1.1'×1.0'	14.2	09 h 16 m 44.2 s	+19° 56' 09"
2802	Cnc	Gxy	1.1'×0.6'	14.1	09 h 16 m 41.5 s	+18° 57' 49"
2803	Cnc	Gxy	1.2'×1.2'	14.1	09 h 16 m 43.9 s	+18° 57' 16"
2804	Cnc	Gxy	2.2'×2.0'	12.3	09 h 16 m 50.0 s	+20° 11' 54"
2806	Cnc	*	…	…	09 h 16 m 56.8 s	+20° 04' 14"
2807	Cnc	Gxy	0.4'×0.2'	15	09 h 17 m 00.7 s	+20° 02' 12"
2809	Cnc	Gxy	1.3'×1.2'	12.8	09 h 17 m 06.9 s	+20° 04' 10"
2812	Cnc	Gxy	0.6'×0.1'	14.9	09 h 17 m 40.7 s	+19° 55' 08"
2813	Cnc	Gxy	1.0'×0.8'	13.6	09 h 17 m 45.4 s	+19° 54' 24"
2819	Cnc	Gxy	1.4'×1.3'	12.9	09 h 18 m 09.3 s	+16° 11' 52"
2824	Cnc	Gxy	0.9'×0.6'	13.5	09 h 19 m 02.3 s	+26° 16' 11"
2843	Cnc	Gxy	0.4'×0.2'	15.5	09 h 20 m 28.8 s	+18° 55' 35"

(continued)

Table 10.1 (continued)

NGC #	Const	Type	Size	V Mag	RA (2000)	Dec (2000)
1792	Col	Gxy	5.2'×2.6'	10.2	05 h 05 m 14.1 s	−37° 58' 48"
1800	Col	Gxy	2.0'×1.1'	12.6	05 h 06 m 25.4 s	−31° 57' 18"
1808	Col	Gxy	6.5'×3.9'	10	05 h 07 m 42.3 s	−37° 30' 47"
1811	Col	Gxy	1.7'×0.4'	13.7	05 h 08 m 42.5 s	−29° 16' 36"
1812	Col	Gxy	1.2'×0.9'	12.9	05 h 08 m 52.8 s	−29° 15' 08"
1827	Col	Gxy	3.0'×0.7'	12.7	05 h 10 m 04.4 s	−36° 57' 39"
1851	Col	GC	11'	7.1	05 h 14 m 06.3 s	−40° 02' 50"
1879	Col	Gxy	2.5'×1.7'	12.7	05 h 19 m 48.1 s	−32° 08' 35"
1891	Col	Ast	5.0'	…	05 h 21 m 23.9 s	−35° 42' 59"
1963	Col	OC	…	…	05 h 32 m 16.8 s	−36° 23' 55"
1989	Col	Gxy	1.5'×1.1'	13.2	05 h 34 m 21.0 s	−30° 48' 04"
1992	Col	Gxy	1.0'×0.7'	13.9	05 h 34 m 32.0 s	−30° 53' 47"
2049	Col	Gxy	2.0'×1.0'	13	05 h 43 m 15.1 s	−30° 04' 43"
2061	Col	Ast	8.0'×8.0'	…	05 h 42 m 42.0 s	−33° 59' 59"
2090	Col	Gxy	4.9'×2.4'	11.3	05 h 47 m 01.6 s	−34° 15' 04"
2188	Col	Gxy	4.4'×1.1'	11.8	06 h 10 m 09.7 s	−34° 06' 19"
2255	Col	Gxy	1.5'×0.7'	13.6	06 h 33 m 58.6 s	−34° 48' 43"
4014	Com	Gxy	2.2'×1.3'	12.5	11 h 58 m 35.8 s	+16° 10' 37"
4015	Com	Gxy	1.4'×1.4'	13.5	11 h 58 m 42.6 s	+25° 02' 11"
4017	Com	Gxy	1.8'×1.4'	12.7	11 h 58 m 45.8 s	+27° 27' 10"
4018	Com	Gxy	1.7'×0.3'	14	11 h 58 m 40.7 s	+25° 18' 59"
4019	Com	Gxy	2.4'×0.3'	13.2	12 h 01 m 10.5 s	+14° 06' 18"
4021	Com	Gxy	0.9'×0.7'	14.8	11 h 59 m 02.5 s	+25° 05' 00"
4022	Com	Gxy	1.2'×1.2'	13.1	11 h 59 m 00.9 s	+25° 13' 22"
4023	Com	Gxy	0.9'×0.7'	13.7	11 h 59 m 05.4 s	+24° 59' 20"
4028	Com	Gxy	2.2'×1.3'	12.4	11 h 58 m 35.8 s	+16° 10' 37"
4032	Com	Gxy	1.9'×1.8'	12.2	12 h 00 m 32.9 s	+20° 04' 33"
4037	Com	Gxy	2.5'×2.0'	12.1	12 h 01 m 23.7 s	+13° 24' 05"
4040	Com	Gxy	1.9'×1.3'	13.4	12 h 02 m 05.5 s	+17° 49' 24"
4042	Com	Gxy	0.1'×0.1'	16.4	12 h 02 m 46.7 s	+20° 09' 49"
4048	Com	Gxy	0.6'×0.5'	14	12 h 02 m 50.0 s	+18° 00' 57"
4049	Com	Gxy	0.9'×0.6'	13.9	12 h 02 m 54.7 s	+18° 45' 09"
4053	Com	Gxy	1.1'×0.4'	13.9	12 h 03 m 11.5 s	+19° 43' 43"
4055	Com	Gxy	1.2'×0.9'	13.2	12 h 04 m 01.4 s	+20° 13' 57"
4056	Com	Gxy	0.2'×0.2'	15.5	12 h 03 m 57.7 s	+20° 18' 41"
4057	Com	Gxy	1.1'×1.0'	12.7	12 h 04 m 06.1 s	+20° 14' 07"
4059	Com	Gxy	1.0'×1.0'	13.2	12 h 04 m 11.3 s	+20° 24' 36"
4060	Com	Gxy	0.4'×0.3'	14.6	12 h 04 m 01.0 s	+20° 20' 15"
4061	Com	Gxy	1.2'×0.9'	13.2	12 h 04 m 01.4 s	+20° 13' 57"
4064	Com	Gxy	4.4'×1.7'	11.5	12 h 04 m 11.3 s	+18° 26' 34"
4065	Com	Gxy	1.1'×1.0'	13	12 h 04 m 06.1 s	+20° 14' 07"
4066	Com	Gxy	1.2'×1.2'	13.1	12 h 04 m 09.4 s	+20° 20' 53"
4069	Com	Gxy	0.3'×0.2'	15.2	12 h 04 m 06.0 s	+20° 19' 26"
4070	Com	Gxy	1.0'×1.0'	13.2	12 h 04 m 11.3 s	+20° 24' 36"
4072	Com	Gxy	0.5'×0.2'	14.8	12 h 04 m 13.8 s	+20° 12' 35"

(continued)

Table 10.1 (continued)

NGC #	Const	Type	Size	V Mag	RA (2000)	Dec (2000)
4074	Com	Gxy	0.4'×0.2'	14.5	12 h 04 m 29.7 s	+20° 18' 59"
4076	Com	Gxy	0.9'×0.9'	13.5	12 h 04 m 32.6 s	+20° 12' 17"
4080	Com	Gxy	1.2'×0.5'	13.9	12 h 04 m 51.9 s	+26° 59' 33"
4084	Com	Gxy	0.4'×0.4'	14.5	12 h 05 m 15.0 s	+21° 12' 55"
4086	Com	Gxy	1.4'×1.0'	13.7	12 h 05 m 29.3 s	+20° 14' 49"
4089	Com	Gxy	0.8'×0.8'	13.7	12 h 05 m 37.5 s	+20° 33' 20"
4090	Com	Gxy	1.2'×0.5'	14.2	12 h 05 m 27.9 s	+20° 18' 32"
4091	Com	Gxy	1.1'×0.3'	14.4	12 h 05 m 40.3 s	+20° 33' 20"
4092	Com	Gxy	1.0'×1.0'	13.5	12 h 05 m 49.9 s	+20° 28' 37"
4093	Com	Gxy	0.8'×0.7'	14.4	12 h 05 m 51.4 s	+20° 31' 18"
4095	Com	Gxy	...'×...'	13.6	12 h 05 m 54.3 s	+20° 34' 22"
4098	Com	Gxy	0.8'×0.4'	13.6	12 h 06 m 03.7 s	+20° 36' 24"
4099	Com	Gxy	1.1'×1.1'	13.6	12 h 06 m 03.7 s	+20° 36' 24"
4101	Com	Gxy	1.1'×0.9'	13.7	12 h 06 m 10.5 s	+25° 33' 24"
4104	Com	Gxy	2.6'×1.5'	12.2	12 h 06 m 38.9 s	+28° 10' 25"
4110	Com	Gxy	1.3'×0.7'	13.9	12 h 07 m 03.4 s	+18° 31' 53"
4115	Com	*	...	...	12 h 07 m 09.5 s	+14° 24' 23"
4126	Com	Gxy	1.0'×0.7'	13.5	12 h 08 m 37.4 s	+16° 08' 33"
4131	Com	Gxy	1.3'×0.7'	13.2	12 h 08 m 47.3 s	+29° 18' 16"
4132	Com	Gxy	1.1'×0.3'	14	12 h 09 m 01.3 s	+29° 15' 00"
4134	Com	Gxy	2.2'×0.9'	13	12 h 09 m 09.9 s	+29° 10' 36"
4136	Com	Gxy	4.0'×3.7'	11.2	12 h 09 m 17.7 s	+29° 55' 39"
4146	Com	Gxy	1.4'×1.3'	12.9	12 h 10 m 18.3 s	+26° 25' 51"
4147	Com	GC	4.0'	10.4	12 h 10 m 06.3 s	+18° 32' 32"
4150	Com	Gxy	2.3'×1.6'	11.6	12 h 10 m 33.6 s	+30° 24' 06"
4152	Com	Gxy	2.2'×1.7'	12	12 h 10 m 37.5 s	+16° 01' 59"
4153	Com	GC	4.0'	10.4	12 h 10 m 06.3 s	+18° 32' 32"
4155	Com	Gxy	1.1'×1.0'	13.4	12 h 10 m 45.6 s	+19° 02' 27"
4158	Com	Gxy	1.9'×1.7'	12.3	12 h 11 m 10.1 s	+20° 10' 32"
4162	Com	Gxy	2.3'×1.4'	12	12 h 11 m 52.5 s	+24° 07' 26"
4166	Com	Gxy	1.2'×1.0'	13.2	12 h 12 m 09.7 s	+17° 45' 24"
4169	Com	Gxy	1.8'×0.9'	12.3	12 h 12 m 18.7 s	+29° 10' 46"
4170	Com	*	...	...	12 h 12 m 12.9 s	+29° 10' 01"
4171	Com	*	...	...	12 h 12 m 38.4 s	+29° 13' 28"
4173	Com	Gxy	5.0'×0.7'	12.7	12 h 12 m 20.2 s	+29° 12' 42"
4174	Com	Gxy	0.8'×0.3'	13.5	12 h 12 m 26.8 s	+29° 08' 57"
4175	Com	Gxy	1.8'×0.4'	13.4	12 h 12 m 31.1 s	+29° 10' 06"
4185	Com	Gxy	2.6'×1.9'	12.3	12 h 13 m 22.2 s	+28° 30' 38"
4186	Com	Gxy	1.1'×0.9'	14	12 h 14 m 06.5 s	+14° 43' 33"
4189	Com	Gxy	2.4'×1.7'	11.9	12 h 13 m 47.5 s	+13° 25' 32"
4192	Com	Gxy	9.8'×2.8'	10	12 h 13 m 48.2 s	+14° 54' 00"
4196	Com	Gxy	1.2'×0.9'	12.9	12 h 14 m 29.7 s	+28° 25' 24"
4203	Com	Gxy	3.4'×3.2'	10.7	12 h 15 m 05.0 s	+33° 11' 52"
4204	Com	Gxy	3.6'×2.9'	12.4	12 h 15 m 14.3 s	+20° 39' 31"
4208	Com	Gxy	3.2'×2.0'	11.1	12 h 15 m 39.3 s	+13° 54' 06"

(continued)

Table 10.1 (continued)

NGC #	Const	Type	Size	V Mag	RA (2000)	Dec (2000)
4209	Com	Gxy	2.9'×2.2'	...	12 h 13 m 22.2 s	+28° 30' 38"
4211	Com	Gxy	1.0'×1.0'	13.5	12 h 15 m 35.9 s	+28° 10' 39"
4212	Com	Gxy	3.2'×1.9'	11.3	12 h 15 m 39.3 s	+13° 54' 06"
4213	Com	Gxy	1.7'×1.7'	12.6	12 h 15 m 37.5 s	+23° 58' 55"
4222	Com	Gxy	3.3'×0.5'	13.5	12 h 16 m 22.7 s	+13° 18' 25"
4237	Com	Gxy	2.1'×1.3'	11.9	12 h 17 m 11.4 s	+15° 19' 26"
4239	Com	Gxy	1.8'×1.2'	12.4	12 h 17 m 14.9 s	+16° 31' 52"
4245	Com	Gxy	2.9'×2.2'	11.3	12 h 17 m 36.8 s	+29° 36' 28"
4251	Com	Gxy	3.6'×1.5'	10.6	12 h 18 m 08.3 s	+28° 10' 30"
4253	Com	Gxy	1.0'×0.8'	13.3	12 h 18 m 26.5 s	+29° 48' 47"
4254	Com	Gxy	5.4'×4.7'	9.7	12 h 18 m 49.6 s	+14° 25' 01"
4262	Com	Gxy	1.9'×1.7'	11.5	12 h 19 m 30.5 s	+14° 52' 39"
4272	Com	Gxy	1.0'×0.9'	13.2	12 h 19 m 47.5 s	+30° 20' 19"
4274	Com	Gxy	6.8'×2.5'	10.4	12 h 19 m 50.5 s	+29° 36' 50"
4275	Com	Gxy	0.8'×0.7'	12.8	12 h 19 m 52.6 s	+27° 37' 16"
4278	Com	Gxy	4.1'×3.8'	10.2	12 h 20 m 06.7 s	+29° 16' 50"
4283	Com	Gxy	1.5'×1.5'	12.1	12 h 20 m 20.8 s	+29° 18' 38"
4286	Com	Gxy	1.6'×1.0'	13.3	12 h 20 m 42.1 s	+29° 20' 44"
4293	Com	Gxy	5.6'×2.6'	10.5	12 h 21 m 12.6 s	+18° 22' 56"
4295	Com	Gxy	0.5'×0.3'	13.9	12 h 21 m 09.8 s	+28° 09' 55"
4298	Com	Gxy	3.2'×1.8'	11.6	12 h 21 m 32.8 s	+14° 36' 24"
4302	Com	Gxy	5.5'×1.0'	12.2	12 h 21 m 42.3 s	+14° 35' 59"
4308	Com	Gxy	0.8'×0.7'	13.5	12 h 21 m 56.7 s	+30° 04' 26"
4310	Com	Gxy	2.2'×1.2'	12.3	12 h 22 m 26.3 s	+29° 12' 30"
4311	Com	NF	...	...	12 h 22 m 26.3 s	+29° 12' 16"
4312	Com	Gxy	4.6'×1.1'	12.1	12 h 22 m 31.4 s	+15° 32' 15"
4314	Com	Gxy	4.2'×3.7'	10.5	12 h 22 m 31.9 s	+29° 53' 43"
4317	Com	NF	...	...	12 h 22 m 41.9 s	+31° 02' 16"
4321	Com	Gxy	7.4'×6.3'	9.7	12 h 22 m 54.9 s	+15° 49' 21"
4322	Com	*	...	...	12 h 22 m 42.1 s	+15° 54' 12"
4327	Com	*	...	...	12 h 23 m 07.5 s	+15° 44' 11"
4328	Com	Gxy	1.3'×1.2'	13.4	12 h 23 m 20.1 s	+15° 49' 13"
4336	Com	Gxy	2.0'×0.9'	12.7	12 h 23 m 29.7 s	+19° 25' 36"
4338	Com	Gxy	2.3'×1.2'	12.3	12 h 22 m 26.3 s	+29° 12' 30"
4340	Com	Gxy	3.5'×2.8'	11	12 h 23 m 35.2 s	+16° 43' 20"
4344	Com	Gxy	1.7'×1.6'	12.4	12 h 23 m 37.6 s	+17° 32' 29"
4350	Com	Gxy	3.0'×1.4'	11.1	12 h 23 m 57.9 s	+16° 41' 37"
4359	Com	Gxy	3.5'×0.8'	12.9	12 h 24 m 11.4 s	+31° 31' 18"
4375	Com	Gxy	1.6'×1.4'	13	12 h 25 m 00.5 s	+28° 33' 30"
4377	Com	Gxy	1.7'×1.4'	11.9	12 h 25 m 12.4 s	+14° 45' 43"
4379	Com	Gxy	1.9'×1.6'	11.6	12 h 25 m 14.7 s	+15° 36' 26"
4382	Com	Gxy	7.1'×5.5'	9.2	12 h 25 m 24.2 s	+18° 11' 27"
4383	Com	Gxy	1.9'×1.0'	12	12 h 25 m 25.4 s	+16° 28' 11"
4393	Com	Gxy	3.2'×3.0'	12.2	12 h 25 m 51.3 s	+27° 33' 44"
4394	Com	Gxy	3.6'×3.2'	10.9	12 h 25 m 55.6 s	+18° 12' 50"

(continued)

Table 10.1 (continued)

NGC #	Const	Type	Size	V Mag	RA (2000)	Dec (2000)
4396	Com	Gxy	3.3'×1.0'	12.7	12 h 25 m 59.5 s	+15° 40' 12"
4397	Com	***	...	...	12 h 25 m 58.1 s	+18° 18' 04"
4405	Com	Gxy	1.8'×1.1'	12.2	12 h 26 m 07.2 s	+16° 10' 52"
4408	Com	Gxy	...'×...'	14.1	12 h 26 m 17.1 s	+27° 52' 17"
4414	Com	Gxy	3.6'×2.0'	10.7	12 h 26 m 27.1 s	+31° 13' 24"
4419	Com	Gxy	3.3'×1.1'	11.4	12 h 26 m 56.3 s	+15° 02' 50"
4421	Com	Gxy	2.7'×2.0'	11.8	12 h 27 m 02.5 s	+15° 27' 41"
4426	Com	**	...	...	12 h 27 m 10.5 s	+27° 50' 17"
4427	Com	**	...	...	12 h 27 m 10.5 s	+27° 50' 17"
4446	Com	Gxy	1.1'×0.9'	14.1	12 h 28 m 06.8 s	+13° 54' 43"
4447	Com	Gxy	0.9'×0.7'	14	12 h 28 m 12.5 s	+13° 53' 57"
4448	Com	Gxy	3.9'×1.4'	11.2	12 h 28 m 15.4 s	+28° 37' 14"
4450	Com	Gxy	5.2'×3.9'	10.3	12 h 28 m 29.5 s	+17° 05' 06"
4455	Com	Gxy	2.8'×0.8'	12.4	12 h 28 m 44.0 s	+22° 49' 20"
4459	Com	Gxy	3.8'×2.8'	10.4	12 h 29 m 00.1 s	+13° 58' 41"
4468	Com	Gxy	1.4'×1.1'	13	12 h 29 m 30.9 s	+14° 02' 56"
4473	Com	Gxy	4.5'×2.5'	10.1	12 h 29 m 48.9 s	+13° 25' 49"
4474	Com	Gxy	2.4'×1.4'	11.8	12 h 29 m 53.6 s	+14° 04' 07"
4475	Com	Gxy	2.0'×1.0'	13.5	12 h 29 m 47.5 s	+27° 14' 35"
4477	Com	Gxy	3.8'×3.5'	10.4	12 h 30 m 02.3 s	+13° 38' 12"
4479	Com	Gxy	1.5'×1.3'	12.6	12 h 30 m 18.4 s	+13° 34' 40"
4489	Com	Gxy	1.7'×1.5'	12.1	12 h 30 m 52.3 s	+16° 45' 31"
4494	Com	Gxy	4.8'×3.5'	9.7	12 h 31 m 24.3 s	+25° 46' 31"
4495	Com	Gxy	1.4'×0.8'	13.4	12 h 31 m 22.9 s	+29° 08' 10"
4498	Com	Gxy	3.0'×1.6'	12.2	12 h 31 m 39.6 s	+16° 51' 11"
4501	Com	Gxy	6.9'×3.7'	9.7	12 h 31 m 59.1 s	+14° 25' 15"
4502	Com	Gxy	1.1'×0.6'	14.1	12 h 32 m 03.3 s	+16° 41' 15"
4506	Com	Gxy	1.6'×1.1'	12.9	12 h 32 m 10.5 s	+13° 25' 10"
4514	Com	Gxy	1.2'×1.0'	13.4	12 h 32 m 43.0 s	+29° 42' 44"
4515	Com	Gxy	1.3'×1.1'	12.5	12 h 33 m 05.0 s	+16° 15' 55"
4516	Com	Gxy	1.7'×1.0'	13	12 h 33 m 07.5 s	+14° 34' 29"
4523	Com	Gxy	2.0'×1.9'	13.5	12 h 33 m 47.9 s	+15° 10' 05"
4525	Com	Gxy	2.6'×1.3'	12.4	12 h 33 m 51.0 s	+30° 16' 39"
4529	Com	Gxy	2.0'×0.4'	14.3	12 h 32 m 51.6 s	+20° 10' 59"
4539	Com	Gxy	3.3'×1.3'	12.3	12 h 34 m 34.9 s	+18° 12' 09"
4540	Com	Gxy	1.9'×1.5'	12	12 h 34 m 50.8 s	+15° 33' 04"
4548	Com	Gxy	5.4'×4.3'	10.4	12 h 35 m 27.2 s	+14° 29' 48"
4555	Com	Gxy	1.9'×1.6'	12.2	12 h 35 m 41.3 s	+26° 31' 22"
4556	Com	Gxy	1.2'×1.0'	13.2	12 h 35 m 45.9 s	+26° 54' 31"
4557	Com	***	...	...	12 h 35 m 50.3 s	+27° 03' 11"
4558	Com	Gxy	0.8'×0.5'	14.7	12 h 35 m 52.7 s	+26° 59' 31"
4559	Com	Gxy	10.7'×4.4'	9.8	12 h 35 m 57.7 s	+27° 57' 35"
4561	Com	Gxy	1.5'×1.3'	12.6	12 h 36 m 08.2 s	+19° 19' 19"
4562	Com	Gxy	2.5'×0.8'	13.5	12 h 35 m 34.8 s	+25° 50' 59"
4563	Com	Gxy	0.5'×0.4'	14.7	12 h 36 m 12.6 s	+26° 56' 28"

(continued)

Table 10.1 (continued)

NGC #	Const	Type	Size	V Mag	RA (2000)	Dec (2000)
4565	Com	Gxy	15.8'×2.1'	9.5	12 h 36 m 20.7 s	+25° 59' 19"
4571	Com	Gxy	3.6'×3.2'	11.5	12 h 36 m 56.5 s	+14° 13' 02"
4585	Com	Gxy	0.7'×0.4'	14.1	12 h 38 m 13.1 s	+28° 56' 13"
4595	Com	Gxy	1.7'×1.1'	12.3	12 h 39 m 51.8 s	+15° 17' 51"
4611	Com	Gxy	1.2'×0.2'	14.3	12 h 41 m 25.4 s	+13° 43' 44"
4613	Com	Gxy	0.5'×0.4'	15.2	12 h 41 m 29.0 s	+26° 05' 18"
4614	Com	Gxy	1.0'×0.8'	13.5	12 h 41 m 31.5 s	+26° 02' 33"
4615	Com	Gxy	1.6'×0.7'	13.3	12 h 41 m 37.3 s	+26° 04' 21"
4633	Com	Gxy	2.1'×0.9'	13.4	12 h 42 m 37.0 s	+14° 21' 20"
4634	Com	Gxy	2.6'×0.7'	13	12 h 42 m 41.0 s	+14° 17' 45"
4635	Com	Gxy	2.0'×1.4'	12.7	12 h 42 m 39.0 s	+19° 56' 39"
4651	Com	Gxy	4.0'×2.6'	10.8	12 h 43 m 42.7 s	+16° 23' 35"
4659	Com	Gxy	1.7'×1.3'	12.3	12 h 44 m 29.5 s	+13° 29' 54"
4670	Com	Gxy	1.4'×1.1'	12.1	12 h 45 m 16.9 s	+27° 07' 30"
4673	Com	Gxy	1.0'×0.9'	13	12 h 45 m 34.5 s	+27° 03' 38"
4676	Com	Gxy	2.2'×0.4'	13.5	12 h 46 m 10.6 s	+30° 43' 35"
4685	Com	Gxy	1.6'×1.0'	12.7	12 h 47 m 11.5 s	+19° 27' 51"
4689	Com	Gxy	4.3'×3.5'	11	12 h 47 m 45.7 s	+13° 45' 44"
4692	Com	Gxy	1.3'×1.3'	12.8	12 h 47 m 55.3 s	+27° 13' 20"
4702	Com	Gxy	1.1'×1.0'	15.7	12 h 47 m 55.3 s	+27° 13' 20"
4710	Com	Gxy	4.9'×1.2'	10.9	12 h 49 m 38.7 s	+15° 09' 54"
4712	Com	Gxy	2.5'×1.1'	12.5	12 h 49 m 34.1 s	+25° 28' 11"
4715	Com	Gxy	1.6'×1.3'	13.2	12 h 49 m 57.9 s	+27° 49' 20"
4721	Com	Gxy	0.6'×0.2'	14.5	12 h 50 m 19.8 s	+27° 19' 26"
4725	Com	Gxy	10.7'×7.6'	9.2	12 h 50 m 26.5 s	+25° 30' 01"
4728	Com	Gxy	1.2'×1.2'	13.6	12 h 50 m 28.0 s	+27° 26' 06"
4735	Com	Gxy	0.6'×0.3'	14.6	12 h 51 m 01.7 s	+28° 55' 40"
4738	Com	Gxy	2.2'×0.3'	14.2	12 h 51 m 08.8 s	+28° 47' 16"
4745	Com	Gxy	0.6'×0.6'	15.1	12 h 51 m 26.1 s	+27° 25' 16"
4747	Com	Gxy	3.5'×1.2'	12.3	12 h 51 m 45.5 s	+25° 46' 30"
4752	Com	Gxy	0.7'×0.3'	14.5	12 h 51 m 29.1 s	+13° 46' 55"
4758	Com	Gxy	3.1'×0.7'	13.4	12 h 52 m 44.1 s	+15° 50' 53"
4787	Com	Gxy	1.1'×0.3'	14.6	12 h 54 m 05.5 s	+27° 04' 06"
4788	Com	Gxy	0.8'×0.3'	14.5	12 h 54 m 15.9 s	+27° 18' 12"
4789	Com	Gxy	1.9'×1.5'	12.2	12 h 54 m 18.9 s	+27° 04' 04"
4793	Com	Gxy	2.8'×1.5'	11.8	12 h 54 m 40.6 s	+28° 56' 17"
4797	Com	Gxy	1.0'×0.7'	13.2	12 h 54 m 55.1 s	+27° 24' 44"
4798	Com	Gxy	1.2'×0.9'	13.2	12 h 54 m 55.1 s	+27° 24' 44"
4805	Com	*	...	...	12 h 55 m 24.1 s	+27° 58' 53"
4807	Com	Gxy	1.0'×0.8'	13.4	12 h 55 m 29.1 s	+27° 31' 16"
4816	Com	Gxy	1.3'×1.1'	13.3	12 h 56 m 12.1 s	+27° 44' 42"
4817	Com	Gxy	0.5'×0.5'	14.7	12 h 56 m 29.8 s	+27° 56' 23"
4819	Com	Gxy	1.2'×0.9'	13.3	12 h 56 m 27.9 s	+26° 59' 14"
4821	Com	Gxy	0.5'×0.3'	14.6	12 h 56 m 29.1 s	+26° 57' 25"
4824	Com	*	...	...	12 h 56 m 36.4 s	+27° 25' 57"

(continued)

Table 10.1 (continued)

NGC #	Const	Type	Size	V Mag	RA (2000)	Dec (2000)
4826	Com	Gxy	10.0'×5.4'	8.4	12 h 56 m 44.2 s	+21° 40' 58"
4827	Com	Gxy	1.4'×1.3'	12.8	12 h 56 m 43.6 s	+27° 10' 42"
4828	Com	Gxy	0.5'×0.5'	14.6	12 h 56 m 42.9 s	+28° 01' 13"
4839	Com	Gxy	4.0'×1.9'	11.8	12 h 57 m 24.2 s	+27° 29' 54"
4840	Com	Gxy	0.7'×0.7'	13.9	12 h 57 m 32.9 s	+27° 36' 36"
4841	Com	Gxy	1.6'×1.0'	13.1	12 h 57 m 31.9 s	+28° 28' 36"
4842	Com	Gxy	0.6'×0.4'	14	12 h 57 m 35.9 s	+27° 29' 35"
4848	Com	Gxy	1.6'×0.5'	13.6	12 h 58 m 05.7 s	+28° 14' 31"
4849	Com	Gxy	1.7'×1.3'	13	12 h 58 m 12.6 s	+26° 23' 49"
4850	Com	Gxy	0.7'×0.7'	14.2	12 h 58 m 21.7 s	+27° 58' 05"
4851	Com	Gxy	0.4'×0.2'	14.2	12 h 58 m 21.7 s	+28° 08' 57"
4853	Com	Gxy	0.8'×0.8'	13.5	12 h 58 m 35.1 s	+27° 35' 46"
4854	Com	Gxy	1.2'×0.8'	14.2	12 h 58 m 47.3 s	+27° 40' 28"
4858	Com	Gxy	0.5'×0.5'	14.9	12 h 59 m 01.8 s	+28° 06' 56"
4859	Com	Gxy	1.6'×0.7'	13.6	12 h 59 m 01.9 s	+26° 48' 56"
4860	Com	Gxy	1.4'×1.2'	13.1	12 h 59 m 03.7 s	+28° 07' 27"
4864	Com	Gxy	0.6'×0.3'	14.3	12 h 59 m 13.0 s	+27° 58' 39"
4865	Com	Gxy	0.9'×0.5'	13.8	12 h 59 m 19.7 s	+28° 05' 05"
4867	Com	Gxy	0.7'×0.7'	14.3	12 h 59 m 15.2 s	+27° 58' 14"
4869	Com	Gxy	0.7'×0.7'	13.8	12 h 59 m 23.2 s	+27° 54' 42"
4871	Com	Gxy	0.7'×0.5'	14.3	12 h 59 m 29.9 s	+27° 57' 23"
4872	Com	Gxy	1.5'×1.5'	13.3	12 h 59 m 34.1 s	+27° 56' 48"
4873	Com	Gxy	0.8'×0.6'	14.1	12 h 59 m 32.7 s	+27° 59' 01"
4874	Com	Gxy	1.9'×1.9'	12.2	12 h 59 m 35.6 s	+27° 57' 33"
4875	Com	Gxy	…'×…'	14.6	12 h 59 m 37.9 s	+27° 54' 26"
4876	Com	Gxy	0.5'×0.5'	14.4	12 h 59 m 44.3 s	+27° 54' 44"
4881	Com	Gxy	1.0'×1.0'	13.4	12 h 59 m 57.7 s	+28° 14' 45"
4882	Com	Gxy	0.8'×0.8'	13.9	13 h 00 m 04.5 s	+27° 59' 15"
4883	Com	Gxy	…'×…'	14.4	12 h 59 m 55.9 s	+28° 02' 04"
4884	Com	Gxy	2.8'×2.0'	11.5	13 h 00 m 08.1 s	+27° 58' 37"
4886	Com	Gxy	1.1'×1.1'	13.5	13 h 00 m 04.5 s	+27° 59' 15"
4889	Com	Gxy	2.9'×1.9'	11.5	13 h 00 m 08.1 s	+27° 58' 37"
4892	Com	Gxy	1.3'×0.3'	13.8	13 h 00 m 03.5 s	+26° 53' 53"
4894	Com	Gxy	…'×…'	14.7	13 h 00 m 16.4 s	+27° 58' 03"
4895	Com	Gxy	1.8'×0.6'	13.2	13 h 00 m 17.9 s	+28° 12' 08"
4896	Com	Gxy	1.0'×0.6'	13.9	13 h 00 m 30.8 s	+28° 20' 46"
4898	Com	Gxy	…'×…'	13.6	13 h 00 m 17.7 s	+27° 57' 20"
4906	Com	Gxy	…'×…'	14.1	13 h 00 m 39.7 s	+27° 55' 25"
4907	Com	Gxy	1.1'×1.0'	13.7	13 h 00 m 48.7 s	+28° 09' 29"
4908	Com	Gxy	1.1'×1.0'	13.5	13 h 00 m 54.4 s	+28° 00' 27"
4911	Com	Gxy	1.4'×1.3'	13	13 h 00 m 55.9 s	+27° 47' 28"
4919	Com	Gxy	1.1'×0.7'	13.7	13 h 01 m 17.6 s	+27° 48' 32"
4921	Com	Gxy	2.5'×2.2'	12.3	13 h 01 m 26.1 s	+27° 53' 08"
4922	Com	Gxy	1.8'×1.3'	13.2	13 h 01 m 24.8 s	+29° 18' 39"
4923	Com	Gxy	0.8'×0.8'	13.7	13 h 01 m 31.7 s	+27° 50' 51"

(continued)

Table 10.1   (continued)

NGC #	Const	Type	Size	V Mag	RA (2000)	Dec (2000)
4926	Com	Gxy	1.2'×1.1'	13	13 h 01 m 53.7 s	+27° 37' 28"
4927	Com	Gxy	...'×...'	13.8	13 h 01 m 57.5 s	+28° 00' 21"
4929	Com	Gxy	1.0'×1.0'	13.5	13 h 02 m 44.3 s	+28° 02' 43"
4931	Com	Gxy	1.7'×0.7'	13.2	13 h 03 m 00.9 s	+28° 01' 56"
4934	Com	Gxy	1.0'×0.2'	14.2	13 h 03 m 16.2 s	+28° 01' 48"
4935	Com	Gxy	1.2'×1.0'	13.2	13 h 03 m 21.1 s	+14° 22' 39"
4943	Com	Gxy	0.5'×0.3'	14.6	13 h 03 m 44.9 s	+28° 05' 03"
4944	Com	Gxy	1.7'×0.6'	12.9	13 h 03 m 50.0 s	+28° 11' 09"
4949	Com	Gxy	0.6'×0.3'	14.9	13 h 04 m 17.9 s	+29° 01' 45"
4952	Com	Gxy	1.8'×1.1'	12.5	13 h 04 m 58.4 s	+29° 07' 19"
4957	Com	Gxy	1.2'×1.0'	13.1	13 h 05 m 12.3 s	+27° 34' 11"
4960	Com	Gxy	1.6'×1.1'	13.1	13 h 05 m 47.5 s	+27° 44' 01"
4961	Com	Gxy	1.6'×1.1'	12.8	13 h 05 m 47.5 s	+27° 44' 01"
4962	Com	Gxy	1.8'×1.1'	12.6	13 h 04 m 58.4 s	+29° 07' 19"
4966	Com	Gxy	1.0'×0.5'	13.2	13 h 06 m 17.3 s	+29° 03' 46"
4971	Com	Gxy	1.0'×1.0'	14.2	13 h 06 m 55.0 s	+28° 32' 53"
4978	Com	Gxy	1.5'×0.8'	13.3	13 h 07 m 50.5 s	+18° 24' 56"
4979	Com	Gxy	1.0'×0.7'	14.3	13 h 07 m 42.8 s	+24° 48' 38"
4983	Com	Gxy	1.1'×0.7'	14	13 h 08 m 27.3 s	+28° 19' 14"
5000	Com	Gxy	1.7'×1.4'	13	13 h 09 m 47.6 s	+28° 54' 22"
5004	Com	Gxy	1.4'×1.1'	13	13 h 11 m 01.5 s	+29° 38' 11"
5008	Com	Gxy	2.4'×1.4'	13.4	14 h 10 m 57.3 s	+25° 29' 48"
5012	Com	Gxy	2.9'×1.7'	12.4	13 h 11 m 36.8 s	+22° 54' 56"
5016	Com	Gxy	1.7'×1.3'	13	13 h 12 m 06.7 s	+24° 05' 42"
5024	Com	GC	13'	7.7	13 h 12 m 55.2 s	+18° 10' 08"
5032	Com	Gxy	2.1'×1.1'	12.8	13 h 13 m 27.1 s	+27° 48' 09"
5041	Com	Gxy	1.7'×1.5'	13.1	13 h 14 m 32.4 s	+30° 42' 20"
5052	Com	Gxy	1.4'×0.9'	13.4	13 h 15 m 34.9 s	+29° 40' 32"
5053	Com	GC	11'	9	13 h 16 m 27.0 s	+17° 41' 52"
5056	Com	Gxy	1.7'×1.0'	13	13 h 16 m 12.3 s	+30° 57' 00"
5057	Com	Gxy	1.2'×1.1'	13.1	13 h 16 m 27.7 s	+31° 01' 52"
5065	Com	Gxy	1.3'×0.8'	13.7	13 h 17 m 30.6 s	+31° 05' 33"
5081	Com	Gxy	2.2'×0.8'	13.3	13 h 19 m 08.1 s	+28° 30' 24"
5089	Com	Gxy	1.7'×0.9'	13.5	13 h 19 m 39.3 s	+30° 15' 23"
5092	Com	Gxy	1.0'×1.0'	13.4	13 h 19 m 51.7 s	+22° 59' 58"
5116	Com	Gxy	2.0'×0.7'	12.9	13 h 22 m 55.6 s	+26° 58' 51"
5151	Com	Gxy	0.7'×0.7'	14	13 h 26 m 40.9 s	+16° 52' 26"
5158	Com	Gxy	1.3'×1.2'	13	13 h 27 m 46.9 s	+17° 46' 43"
5172	Com	Gxy	3.3'×1.7'	11.8	13 h 29 m 19.1 s	+17° 03' 07"
5180	Com	Gxy	1.4'×1.0'	13.1	13 h 29 m 26.9 s	+16° 49' 32"
5190	Com	Gxy	1.0'×0.8'	13.4	13 h 30 m 38.5 s	+18° 08' 04"
5217	Com	Gxy	1.5'×1.4'	12.7	13 h 34 m 05.9 s	+17° 51' 23"

Table 10.2 These NGC objects are found in the constellations of Corona Austrina through Hydra (Copyright © The NGC/IC Project, LLC – All rights reserved. Used with permission)

NGC #	Const	Type	Size	V Mag	RA (2000)	Dec (2000)
6541	CrA	GC	13'	6.3	18 h 08 m 02.3 s	−43° 42' 57"
6726	CrA	Neb	2.0'×2.0'	…	19 h 01 m 38.7 s	−36° 53' 27"
6727	CrA	Neb	2.0'×2.0'	…	19 h 01 m 40.6 s	−36° 52' 36"
6729	CrA	Neb	VAR'	…	19 h 01 m 54.1 s	−36° 57' 12"
6768	CrA	Gxy	1.2'×1.1'	12.2	19 h 16 m 32.7 s	−40° 12' 33"
5924	CrB	Gxy	0.6'×0.2'	14.6	15 h 22 m 01.9 s	+31° 13' 58"
5958	CrB	Gxy	1.0'×1.0'	12.7	15 h 34 m 49.1 s	+28° 39' 17"
5961	CrB	Gxy	0.8'×0.3'	13.6	15 h 35 m 16.1 s	+30° 51' 51"
5974	CrB	Gxy	0.6'×0.3'	14	15 h 39 m 02.3 s	+31° 45' 35"
6001	CrB	Gxy	1.0'×1.0'	13.7	15 h 47 m 45.9 s	+28° 38' 32"
6002	CrB	*	…	…	15 h 47 m 44.4 s	+28° 36' 35"
6016	CrB	Gxy	1.0'×0.4'	14.5	15 h 55 m 54.8 s	+26° 58' 00"
6038	CrB	Gxy	1.1'×1.1'	13.7	16 h 02 m 40.5 s	+37° 21' 32"
6069	CrB	Gxy	0.7'×0.5'	14.3	16 h 07 m 41.7 s	+38° 55' 50"
6076	CrB	Gxy	1.2'×0.4'	14.4	16 h 11 m 13.3 s	+26° 52' 21"
6077	CrB	Gxy	1.4'×1.1'	13.4	16 h 11 m 14.1 s	+26° 55' 24"
6085	CrB	Gxy	1.5'×1.2'	13.2	16 h 12 m 35.3 s	+29° 21' 54"
6086	CrB	Gxy	1.8'×1.5'	13.1	16 h 12 m 35.5 s	+29° 29' 05"
6089	CrB	Gxy	0.7'×0.6'	14.1	16 h 12 m 40.2 s	+33° 02' 06"
6092	CrB	**	…	…	16 h 14 m 04.5 s	+28° 07' 32"
6096	CrB	Gxy	0.8'×0.4'	14.4	16 h 14 m 46.8 s	+26° 33' 32"
6097	CrB	Gxy	1.0'×0.5'	13.9	16 h 14 m 26.1 s	+35° 06' 32"
6102	CrB	Gxy	1.2'×0.8'	14.4	16 h 15 m 36.8 s	+28° 09' 30"
6103	CrB	Gxy	0.7'×0.5'	13.7	16 h 15 m 44.5 s	+31° 57' 50"
6104	CrB	Gxy	0.8'×0.7'	13.3	16 h 16 m 30.7 s	+35° 42' 28"
6105	CrB	Gxy	0.6'×0.5'	14.8	16 h 17 m 09.2 s	+34° 52' 44"
6107	CrB	Gxy	0.9'×0.7'	13.7	16 h 17 m 19.9 s	+34° 54' 04"
6108	CrB	Gxy	0.7'×0.5'	14.5	16 h 17 m 25.5 s	+35° 08' 09"
6109	CrB	Gxy	1.0'×1.0'	13	16 h 17 m 40.5 s	+35° 00' 16"
6110	CrB	Gxy	0.6'×0.3'	14.8	16 h 17 m 43.9 s	+35° 05' 13"
6112	CrB	Gxy	0.6'×0.5'	14.4	16 h 18 m 00.5 s	+35° 06' 36"
6114	CrB	Gxy	0.8'×0.5'	14.4	16 h 18 m 23.6 s	+35° 10' 26"
6116	CrB	Gxy	1.1'×0.6'	14.5	16 h 18 m 54.5 s	+35° 09' 14"
6117	CrB	Gxy	1.2'×1.2'	13.8	16 h 19 m 18.2 s	+37° 05' 41"
6119	CrB	Gxy	0.8'×0.5'	15.3	16 h 19 m 41.9 s	+37° 48' 23"
6120	CrB	Gxy	0.6'×0.4'	14.2	16 h 19 m 48.0 s	+37° 46' 27"
6122	CrB	Gxy	0.9'×0.3'	14.6	16 h 20 m 09.5 s	+37° 47' 54"
6126	CrB	Gxy	0.8'×0.8'	13.6	16 h 21 m 27.9 s	+36° 22' 36"
6129	CrB	Gxy	0.6'×0.6'	14	16 h 21 m 43.3 s	+37° 59' 46"
6131	CrB	Gxy	1.0'×1.0'	13.6	16 h 21 m 52.2 s	+38° 55' 58"
6137	CrB	Gxy	1.9'×1.2'	12.8	16 h 23 m 03.1 s	+37° 55' 20"
6142	CrB	Gxy	1.9'×0.5'	14	16 h 23 m 21.1 s	+37° 15' 29"
3431	Crt	Gxy	1.3'×0.3'	13.6	10 h 51 m 15.0 s	−17° 00' 30"
3452	Crt	Gxy	1.0'×0.3'	14.1	10 h 54 m 14.1 s	−11° 24' 20"

(continued)

Table 10.2 (continued)

NGC #	Const	Type	Size	V Mag	RA (2000)	Dec (2000)
3456	Crt	Gxy	1.9' × 1.3'	12.6	10 h 54 m 03.3 s	−16° 01' 41"
3459	Crt	Gxy	1.6' × 0.5'	13.4	10 h 54 m 44.5 s	−17° 02' 32"
3469	Crt	Gxy	1.7' × 1.2'	13.1	10 h 56 m 57.7 s	−14° 18' 03"
3472	Crt	Gxy	0.8' × 0.6'	14.6	11 h 57 m 28.1 s	−19° 37' 26"
3479	Crt	Gxy	1.7' × 1.2'	13	10 h 58 m 55.4 s	−14° 57' 41"
3481	Crt	Gxy	0.7' × 0.6'	13.8	10 h 59 m 26.6 s	−07° 32' 40"
3497	Crt	Gxy	3.5' × 1.8'	11.6	11 h 07 m 18.3 s	−19° 28' 21"
3502	Crt	Gxy	1.8' × 1.3'	13	10 h 58 m 55.4 s	−14° 57' 41"
3505	Crt	Gxy	1.1' × 0.9'	13.2	11 h 02 m 59.7 s	−16° 17' 21"
3508	Crt	Gxy	1.1' × 0.9'	13.2	11 h 02 m 59.7 s	−16° 17' 21"
3511	Crt	Gxy	5.8' × 2.0'	11.1	11 h 03 m 23.7 s	−23° 05' 11"
3513	Crt	Gxy	2.8' × 2.2'	11.6	11 h 03 m 45.9 s	−23° 14' 39"
3514	Crt	Gxy	1.3' × 1.1'	12.8	11 h 03 m 59.9 s	−18° 46' 51"
3520	Crt	Gxy	1.3' × 1.0'	13.3	11 h 07 m 08.7 s	−18° 01' 28"
3525	Crt	Gxy	3.5' × 1.8'	11.6	11 h 07 m 18.3 s	−19° 28' 21"
3528	Crt	Gxy	2.6' × 1.4'	11.6	11 h 07 m 18.3 s	−19° 28' 21"
3529	Crt	Gxy	1.0' × 0.8'	13.1	11 h 07 m 19.3 s	−19° 33' 22"
3537	Crt	Gxy	1.2' × 1.2'	12.8	11 h 08 m 26.6 s	−10° 15' 25"
3541	Crt	Gxy	1.4' × 1.3'	14.5	11 h 08 m 32.1 s	−10° 29' 31"
3544	Crt	Gxy	3.3' × 1.3'	12.2	11 h 11 m 30.3 s	−18° 17' 23"
3546	Crt	Gxy	1.3' × 0.7'	13.4	11 h 09 m 46.7 s	−13° 22' 50"
3565	Crt	Gxy	1.0' × 0.6'	14.6	11 h 07 m 47.9 s	−20° 01' 20"
3566	Crt	Gxy	0.8' × 0.5'	14.6	11 h 07 m 47.9 s	−20° 01' 20"
3571	Crt	Gxy	3.0' × 1.0'	12.3	11 h 11 m 30.3 s	−18° 17' 23"
3578	Crt	**	…	…	11 h 12 m 49.9 s	−15° 56' 43"
3591	Crt	Gxy	1.2' × 0.9'	13.3	11 h 14 m 03.3 s	−14° 05' 14"
3597	Crt	Gxy	1.9' × 1.5'	12.7	11 h 14 m 41.9 s	−23° 43' 46"
3634	Crt	Gxy	0.8' × 0.5'	14.5	11 h 20 m 30.3 s	−09° 00' 50"
3635	Crt	Gxy	1.3' × 0.9'	14.5	11 h 20 m 31.3 s	−09° 00' 49"
3636	Crt	Gxy	1.3' × 1.3'	12.4	11 h 20 m 25.2 s	−10° 16' 55"
3637	Crt	Gxy	1.6' × 1.5'	11.9	11 h 20 m 39.6 s	−10° 15' 27"
3638	Crt	Gxy	2.2' × 0.7'	14.2	11 h 20 m 10.0 s	−08° 06' 20"
3660	Crt	Gxy	2.7' × 2.2'	12.2	11 h 23 m 32.2 s	−08° 39' 31"
3661	Crt	Gxy	1.6' × 0.6'	13.1	11 h 23 m 38.4 s	−13° 49' 53"
3663	Crt	Gxy	1.9' × 1.3'	12.5	11 h 24 m 00.0 s	−12° 17' 45"
3667	Crt	Gxy	1.5' × 1.0'	13	11 h 24 m 16.9 s	−13° 51' 26"
3672	Crt	Gxy	4.2' × 1.9'	11.2	11 h 25 m 02.3 s	−09° 47' 43"
3676	Crt	Gxy	0.8' × 0.7'	13.7	11 h 25 m 37.4 s	−11° 08' 23"
3688	Crt	Gxy	1.3' × 0.9'	14.3	11 h 27 m 44.4 s	−09° 09' 57"
3693	Crt	Gxy	3.2' × 0.7'	12.7	11 h 28 m 11.5 s	−13° 11' 42"
3696	Crt	Gxy	1.2' × 1.0'	14.1	11 h 28 m 44.0 s	−11° 16' 58"
3702	Crt	Gxy	1.3' × 0.7'	13.1	11 h 30 m 13.4 s	−08° 51' 46"
3703	Crt	Gxy	0.8' × 0.4'	14.7	11 h 29 m 09.3 s	−08° 26' 46"
3704	Crt	Gxy	1.6' × 1.4'	12.9	11 h 30 m 04.6 s	−11° 32' 47"
3707	Crt	Gxy	0.6' × 0.4'	14.7	11 h 30 m 11.5 s	−11° 32' 37"

(continued)

Table 10.2 (continued)

NGC #	Const	Type	Size	V Mag	RA (2000)	Dec (2000)
3711	Crt	Gxy	0.8'×0.4'	14.7	11 h 29 m 26.0 s	−11° 04' 27"
3715	Crt	Gxy	1.3'×0.9'	12.5	11 h 31 m 32.2 s	−14° 13' 57"
3721	Crt	Gxy	0.7'×0.7'	…	11 h 31 m 53.5 s	−09° 31' 57"
3722	Crt	Gxy	0.4'×0.4'	…	11 h 31 m 45.9 s	−09° 40' 24"
3723	Crt	Gxy	1.0'×0.9'	13.3	11 h 32 m 30.5 s	−09° 58' 10"
3724	Crt	Gxy	1.7'×1.1'	…	11 h 31 m 54.7 s	−09° 43' 34"
3727	Crt	Gxy	0.6'×0.9'	14.1	11 h 33 m 40.9 s	−13° 52' 44"
3730	Crt	Gxy	1.0'×0.9'	13	11 h 34 m 16.8 s	−09° 34' 34"
3732	Crt	Gxy	1.2'×1.2'	12.4	11 h 34 m 13.9 s	−09° 50' 44"
3734	Crt	Gxy	1.3'×1.0'	13.9	11 h 34 m 40.6 s	−14° 04' 55"
3763	Crt	Gxy	1.2'×1.2'	12.8	11 h 36 m 30.2 s	−09° 50' 47"
3771	Crt	Gxy	1.3'×1.3'	12.6	11 h 39 m 06.1 s	−09° 20' 54"
3774	Crt	Gxy	1.0'×0.5'	13.8	11 h 38 m 30.3 s	−08° 58' 34"
3775	Crt	Gxy	1.1'×0.4'	13.8	11 h 38 m 26.7 s	−10° 38' 19"
3777	Crt	Gxy	1.1'×0.6'	13.5	11 h 36 m 06.9 s	−12° 34' 09"
3779	Crt	Gxy	1.9'×1.0'	13.8	11 h 38 m 50.8 s	−10° 35' 02"
3789	Crt	Gxy	1.3'×0.6'	13.4	11 h 38 m 09.0 s	−09° 36' 26"
3791	Crt	Gxy	1.3'×1.0'	13.4	11 h 39 m 41.6 s	−09° 22' 02"
3823	Crt	Gxy	1.5'×1.2'	12.7	11 h 42 m 15.2 s	−13° 52' 00"
3831	Crt	Gxy	2.7'×0.6'	12.7	11 h 43 m 18.5 s	−12° 52' 41"
3836	Crt	Gxy	1.4'×1.3'	12.9	11 h 43 m 29.9 s	−16° 47' 46"
3854	Crt	Gxy	2.3'×1.7'	12.2	11 h 44 m 52.1 s	−09° 13' 58"
3858	Crt	Gxy	1.9'×1.3'	13.2	11 h 45 m 11.7 s	−09° 18' 51"
3865	Crt	Gxy	2.0'×1.5'	12.2	11 h 44 m 52.1 s	−09° 13' 58"
3866	Crt	Gxy	1.4'×0.8'	13.2	11 h 45 m 11.7 s	−09° 18' 51"
3887	Crt	Gxy	3.3'×2.5'	10.9	11 h 47 m 04.7 s	−16° 51' 17"
3892	Crt	Gxy	3.0'×2.2'	11.6	11 h 48 m 01.1 s	−10° 57' 44"
3905	Crt	Gxy	1.9'×1.4'	12.8	11 h 49 m 04.9 s	−09° 43' 46"
3942	Crt	Gxy	1.4'×0.8'	13.8	11 h 51 m 29.9 s	−11° 25' 27"
3955	Crt	Gxy	2.9'×0.9'	11.7	11 h 53 m 57.1 s	−23° 09' 52"
3956	Crt	Gxy	3.4'×1.0'	12.3	11 h 54 m 00.7 s	−20° 34' 03"
3957	Crt	Gxy	3.1'×0.7'	11.9	11 h 54 m 01.7 s	−19° 34' 14"
3959	Crt	Gxy	1.1'×0.9'	13.7	11 h 54 m 37.9 s	−07° 45' 25"
3962	Crt	Gxy	2.6'×2.2'	10.9	11 h 54 m 39.9 s	−13° 58' 29"
3965	Crt	Gxy	0.5'×0.3'	…	11 h 54 m 23.1 s	−10° 51' 59"
3967	Crt	Gxy	1.5'×1.1'	13.4	11 h 55 m 10.4 s	−07° 50' 36"
3969	Crt	Gxy	1.4'×0.9'	13.2	11 h 55 m 09.2 s	−18° 55' 40"
3970	Crt	Gxy	1.1'×0.5'	13.5	11 h 55 m 28.1 s	−12° 03' 40"
3974	Crt	Gxy	1.1'×1.1'	13.4	11 h 55 m 40.1 s	−12° 01' 39"
3981	Crt	Gxy	5.2'×2.3'	11.5	11 h 56 m 07.1 s	−19° 53' 49"
4052	Cru	OC	7' to 10'	8.8	12 h 02 m 05.2 s	−63° 13' 24"
4103	Cru	OC	6'	7.4	12 h 06 m 39.5 s	−61° 15' 00"
4184	Cru	OC	5'	…	12 h 13 m 33.0 s	−62° 42' 29"
4337	Cru	OC	3.5'	8.9	12 h 24 m 03.3 s	−58° 07' 25"
4349	Cru	OC	15'	7.4	12 h 24 m 06.0 s	−61° 52' 13"

(continued)

Table 10.2 (continued)

NGC #	Const	Type	Size	V Mag	RA (2000)	Dec (2000)
4439	Cru	OC	4.0'	8.4	12 h 28 m 26.3 s	−60° 06' 11"
4609	Cru	OC	5'	6.9	12 h 42 m 19.9 s	−62° 59' 38"
4755	Cru	OC	10.0'	4.2	12 h 53 m 37.1 s	−60° 21' 22"
4024	Crv	Gxy	1.9'×1.5'	11.9	11 h 58 m 31.3 s	−18° 20' 49"
4027	Crv	Gxy	3.2'×2.4'	11.2	11 h 59 m 30.5 s	−19° 15' 44"
4033	Crv	Gxy	2.6'×1.1'	11.7	12 h 00 m 34.8 s	−17° 50' 34"
4035	Crv	Gxy	1.2'×1.1'	13.4	12 h 00 m 29.4 s	−15° 56' 53"
4038	Crv	Gxy	5.2'×3.1'	10.5	12 h 01 m 53.0 s	−18° 52' 07"
4039	Crv	Gxy	3.1'×1.6'	10.7	12 h 01 m 53.6 s	−18° 53' 10"
4050	Crv	Gxy	3.1'×2.1'	11.5	12 h 02 m 54.0 s	−16° 22' 27"
4094	Crv	Gxy	4.2'×1.5'	12	12 h 05 m 53.9 s	−14° 31' 35"
4114	Crv	Gxy	1.9'×0.9'	13.1	12 h 07 m 12.2 s	−14° 11' 08"
4177	Crv	Gxy	1.7'×1.1'	12.5	12 h 12 m 41.2 s	−14° 00' 52"
4188	Crv	Gxy	0.7'×0.6'	13.7	12 h 14 m 07.5 s	−12° 35' 12"
4225	Crv	Gxy	0.7'×0.5'	14	12 h 16 m 38.3 s	−12° 19' 40"
4263	Crv	Gxy	1.2'×0.5'	12.6	12 h 19 m 42.2 s	−12° 13' 29"
4265	Crv	Gxy	1.2'×0.5'	12.6	12 h 19 m 42.2 s	−12° 13' 29"
4329	Crv	Gxy	2.4'×1.4'	13.7	12 h 23 m 20.7 s	−12° 33' 32"
4361	Crv	PN	1.9'×1.9'	10.9	12 h 24 m 30.7 s	−18° 47' 05"
4462	Crv	Gxy	3.2'×1.2'	12	12 h 29 m 21.1 s	−23° 10' 00"
4524	Crv	Gxy	1.2'×0.8'	13.7	12 h 33 m 54.3 s	−12° 01' 39"
4714	Crv	Gxy	1.6'×1.2'	12.7	12 h 50 m 19.3 s	−13° 19' 33"
4722	Crv	Gxy	1.8'×0.7'	12.8	12 h 51 m 32.3 s	−13° 19' 49"
4723	Crv	Gxy	1.1'×0.6'	14.5	12 h 51 m 02.9 s	−13° 14' 16"
4724	Crv	Gxy	1.1'×0.8'	12.7	12 h 50 m 53.5 s	−14° 19' 56"
4726	Crv	Gxy	0.7'×0.2'	14.8	12 h 50 m 45.9 s	−14° 16' 05"
4727	Crv	Gxy	1.4'×1.1'	11.9	12 h 50 m 57.2 s	−14° 19' 59"
4740	Crv	Gxy	1.6'×1.4'	11.9	12 h 50 m 57.2 s	−14° 19' 59"
4748	Crv	Gxy	0.8'×0.7'	13.7	12 h 52 m 12.7 s	−13° 24' 49"
4756	Crv	Gxy	1.6'×1.3'	12.2	12 h 52 m 52.6 s	−15° 24' 48"
4763	Crv	Gxy	1.5'×1.1'	12.6	12 h 53 m 27.2 s	−17° 00' 20"
4782	Crv	Gxy	1.8'×1.7'	11.8	12 h 54 m 35.9 s	−12° 34' 08"
4783	Crv	Gxy	1.8'×1.7'	11.6	12 h 54 m 36.3 s	−12° 33' 30"
4792	Crv	Gxy	0.8'×0.4'	14.2	12 h 55 m 03.6 s	−12° 29' 52"
4794	Crv	Gxy	1.9'×0.8'	13.6	12 h 55 m 10.8 s	−12° 36' 41"
4802	Crv	Gxy	2.4'×1.6'	11.6	12 h 55 m 49.7 s	−12° 03' 19"
4804	Crv	Gxy	0.9'×0.6'	11.6	12 h 55 m 49.7 s	−12° 03' 19"
4109	CVn	Gxy	1.0'×0.9'	14.1	12 h 06 m 51.1 s	+42° 59' 43"
4111	CVn	Gxy	4.6'×1.0'	10.7	12 h 07 m 02.7 s	+43° 03' 57"
4113	CVn	Gxy	0.7'×0.4'	14.7	12 h 07 m 08.5 s	+32° 59' 45"
4117	CVn	Gxy	1.8'×0.9'	13.1	12 h 07 m 46.1 s	+43° 07' 34"
4118	CVn	Gxy	0.7'×0.4'	14.7	12 h 07 m 52.8 s	+43° 06' 41"
4122	CVn	Gxy	0.7'×0.4'	14.7	12 h 07 m 08.5 s	+32° 59' 45"
4135	CVn	Gxy	0.9'×0.6'	14.2	12 h 09 m 08.9 s	+44° 00' 11"
4137	CVn	Gxy	1.1'×0.7'	14.3	12 h 09 m 17.5 s	+44° 05' 23"

(continued)

Table 10.2 (continued)

NGC #	Const	Type	Size	V Mag	RA (2000)	Dec (2000)
4138	CVn	Gxy	2.6'×1.7'	11.2	12 h 09 m 30.0 s	+43° 41' 04"
4143	CVn	Gxy	2.3'×1.4'	11.1	12 h 09 m 36.1 s	+42° 32' 02"
4145	CVn	Gxy	5.9'×4.3'	11.2	12 h 10 m 01.6 s	+39° 52' 58"
4148	CVn	Gxy	1.5'×1.0'	13.4	12 h 10 m 07.9 s	+35° 52' 38"
4151	CVn	Gxy	6.3'×4.5'	10.2	12 h 10 m 32.5 s	+39° 24' 21"
4156	CVn	Gxy	1.4'×1.1'	13.5	12 h 10 m 49.4 s	+39° 28' 23"
4160	CVn	NF	…	…	12 h 12 m 11.7 s	+43° 44' 18"
4163	CVn	Gxy	1.8'×1.6'	13.1	12 h 12 m 09.1 s	+36° 10' 09"
4167	CVn	Gxy	1.8'×1.6'	13.4	12 h 12 m 09.1 s	+36° 10' 09"
4181	CVn	Gxy	1.0'×0.7'	14	12 h 12 m 49.0 s	+52° 54' 11"
4183	CVn	Gxy	5.2'×0.8'	12.4	12 h 13 m 16.9 s	+43° 41' 52"
4187	CVn	Gxy	1.3'×1.0'	13.3	12 h 13 m 29.3 s	+50° 44' 29"
4190	CVn	Gxy	1.7'×1.5'	12.9	12 h 13 m 44.5 s	+36° 38' 05"
4214	CVn	Gxy	8.5'×6.6'	9.8	12 h 15 m 39.5 s	+36° 19' 35"
4217	CVn	Gxy	5.2'×1.5'	11.6	12 h 15 m 51.3 s	+47° 05' 28"
4218	CVn	Gxy	1.2'×0.7'	12.7	12 h 15 m 46.3 s	+48° 07' 53"
4220	CVn	Gxy	3.9'×1.4'	11.3	12 h 16 m 11.7 s	+47° 52' 59"
4226	CVn	Gxy	1.3'×0.6'	13.7	12 h 16 m 26.2 s	+47° 01' 31"
4227	CVn	Gxy	1.5'×0.9'	12.8	12 h 16 m 33.7 s	+33° 31' 19"
4228	CVn	Gxy	7.9'×6.3'	9.6	12 h 15 m 39.5 s	+36° 19' 35"
4229	CVn	Gxy	1.3'×0.9'	13.3	12 h 16 m 38.7 s	+33° 33' 38"
4231	CVn	Gxy	1.2'×1.1'	13.4	12 h 16 m 48.7 s	+47° 27' 24"
4232	CVn	Gxy	1.4'×0.7'	13.8	12 h 16 m 48.7 s	+47° 26' 18"
4242	CVn	Gxy	5.0'×3.8'	11.1	12 h 17 m 30.1 s	+45° 37' 08"
4244	CVn	Gxy	16.6'×1.9'	10	12 h 17 m 29.7 s	+37° 48' 24"
4248	CVn	Gxy	3.0'×1.1'	13.2	12 h 17 m 50.3 s	+47° 24' 32"
4258	CVn	Gxy	18.6'×7.2'	8.7	12 h 18 m 57.5 s	+47° 18' 15"
4288	CVn	Gxy	2.1'×1.6'	12.9	12 h 20 m 38.0 s	+46° 17' 32"
4346	CVn	Gxy	3.3'×1.3'	11.3	12 h 23 m 27.9 s	+46° 59' 37"
4357	CVn	Gxy	3.6'×1.3'	12.6	12 h 23 m 58.9 s	+48° 46' 47"
4369	CVn	Gxy	2.1'×2.0'	11.5	12 h 24 m 36.3 s	+39° 22' 58"
4381	CVn	Gxy	3.9'×1.5'	12.4	12 h 23 m 58.9 s	+48° 46' 47"
4389	CVn	Gxy	2.6'×1.3'	12.1	12 h 25 m 35.3 s	+45° 41' 03"
4392	CVn	Gxy	0.6'×0.5'	13.7	12 h 25 m 18.7 s	+45° 50' 50"
4395	CVn	Gxy	13.2'×11.0'	10.5	12 h 25 m 48.9 s	+33° 32' 50"
4399	CVn	HIIRgn	…	…	12 h 25 m 43.0 s	+33° 30' 57"
4400	CVn	HIIRgn	…	…	12 h 25 m 55.9 s	+33° 30' 55"
4401	CVn	HIIRgn	…	…	12 h 25 m 57.5 s	+33° 31' 42"
4449	CVn	Gxy	6.2'×4.4'	9.8	12 h 28 m 11.1 s	+44° 05' 36"
4460	CVn	Gxy	4.0'×1.2'	11.4	12 h 28 m 45.6 s	+44° 51' 51"
4485	CVn	Gxy	2.3'×1.6'	12.1	12 h 30 m 31.4 s	+41° 42' 00"
4490	CVn	Gxy	6.3'×3.1'	9.8	12 h 30 m 36.1 s	+41° 38' 33"
4509	CVn	Gxy	0.9'×0.6'	13.7	12 h 33 m 06.8 s	+32° 05' 31"
4530	CVn	*	…	…	12 h 33 m 47.6 s	+41° 21' 12"
4534	CVn	Gxy	2.6'×2.1'	12.4	12 h 34 m 05.5 s	+35° 31' 05"

(continued)

Table 10.2 (continued)

NGC #	Const	Type	Size	V Mag	RA (2000)	Dec (2000)
4537	CVn	Gxy	1.0'×0.5'	14.4	12 h 34 m 48.9 s	+50° 48' 19"
4542	CVn	Gxy	1.0'×0.5'	14.4	12 h 34 m 48.9 s	+50° 48' 19"
4583	CVn	Gxy	1.0'×0.9'	13.6	12 h 38 m 04.5 s	+33° 27' 32"
4617	CVn	Gxy	3.0'×0.5'	13.4	12 h 41 m 05.7 s	+50° 23' 35"
4618	CVn	Gxy	4.2'×3.4'	10.9	12 h 41 m 32.5 s	+41° 09' 02"
4619	CVn	Gxy	1.3'×1.3'	12.9	12 h 41 m 44.3 s	+35° 03' 45"
4625	CVn	Gxy	2.2'×1.9'	12.5	12 h 41 m 52.7 s	+41° 16' 26"
4627	CVn	Gxy	2.6'×1.8'	12	12 h 41 m 59.6 s	+32° 34' 26"
4631	CVn	Gxy	15.5'×2.7'	9.1	12 h 42 m 08.1 s	+32° 32' 26"
4655	CVn	Gxy	0.9'×0.9'	14.4	12 h 43 m 36.5 s	+41° 01' 06"
4656	CVn	Gxy	15.1'×3.0'	10.2	12 h 43 m 58.3 s	+32° 10' 13"
4657	CVn	Gxy	1.3'×0.6'	12.4	12 h 44 m 08.3 s	+32° 12' 32"
4662	CVn	Gxy	1.9'×1.6'	13.2	12 h 44 m 26.3 s	+37° 07' 16"
4687	CVn	Gxy	0.9'×0.8'	13.3	12 h 47 m 23.8 s	+35° 21' 06"
4704	CVn	Gxy	1.0'×0.9'	14	12 h 48 m 46.5 s	+41° 55' 15"
4707	CVn	Gxy	2.2'×2.1'	14.1	12 h 48 m 23.1 s	+51° 09' 55"
4711	CVn	Gxy	1.5'×0.9'	13.6	12 h 48 m 45.9 s	+35° 19' 57"
4719	CVn	Gxy	1.4'×1.1'	13.4	12 h 50 m 08.6 s	+33° 09' 34"
4736	CVn	Gxy	11.2'×9.1'	8	12 h 50 m 53.0 s	+41° 07' 12"
4737	CVn	Gxy	0.5'×0.4'	14.3	12 h 50 m 52.8 s	+34° 09' 24"
4741	CVn	Gxy	1.3'×0.7'	13.9	12 h 50 m 59.4 s	+47° 40' 17"
4774	CVn	Gxy	0.6'×0.4'	14.3	12 h 53 m 06.7 s	+36° 49' 06"
4800	CVn	Gxy	1.6'×1.2'	11.7	12 h 54 m 37.7 s	+46° 31' 52"
4834	CVn	Gxy	0.8'×0.3'	14.8	12 h 56 m 25.1 s	+52° 17' 45"
4837	CVn	Gxy	1.1'×0.5'	13.4	12 h 56 m 48.6 s	+48° 17' 49"
4846	CVn	Gxy	1.3'×0.6'	13.7	12 h 57 m 47.6 s	+36° 22' 14"
4861	CVn	Gxy	4.0'×1.5'	12.1	12 h 59 m 02.5 s	+34° 51' 50"
4868	CVn	Gxy	1.6'×1.5'	12.4	12 h 59 m 08.9 s	+37° 18' 37"
4870	CVn	Gxy	0.9'×0.3'	14.6	12 h 59 m 17.7 s	+37° 02' 54"
4893	CVn	Gxy	1.0'×0.6'	14.5	12 h 59 m 59.6 s	+37° 11' 24"
4901	CVn	Gxy	1.2'×1.2'	14.6	12 h 59 m 56.3 s	+47° 12' 21"
4912	CVn	NF	…	…	13 h 00 m 46.5 s	+37° 22' 39"
4913	CVn	NF	…	…	13 h 00 m 46.5 s	+37° 20' 39"
4914	CVn	Gxy	3.5'×1.9'	11.5	13 h 00 m 42.9 s	+37° 18' 54"
4916	CVn	NF	…	…	13 h 00 m 54.5 s	+37° 21' 40"
4917	CVn	Gxy	1.4'×1.0'	14	13 h 00 m 55.6 s	+47° 13' 20"
4932	CVn	Gxy	1.8'×1.4'	13.8	13 h 02 m 37.6 s	+50° 26' 19"
4938	CVn	Gxy	0.7'×0.6'	14.4	13 h 02 m 57.5 s	+51° 19' 07"
4956	CVn	Gxy	1.5'×1.5'	12.4	13 h 05 m 00.9 s	+35° 10' 40"
4959	CVn	Gxy	0.7'×0.7'	14.6	13 h 05 m 40.9 s	+33° 10' 43"
4963	CVn	Gxy	0.8'×0.8'	13.4	13 h 05 m 52.0 s	+41° 43' 18"
4985	CVn	Gxy	1.3'×1.1'	13.8	13 h 08 m 12.1 s	+41° 40' 34"
4986	CVn	Gxy	1.7'×0.9'	13.4	13 h 08 m 24.3 s	+35° 12' 24"
4987	CVn	Gxy	1.2'×0.7'	13.5	13 h 07 m 59.1 s	+51° 55' 45"
4998	CVn	Gxy	0.9'×0.8'	14.5	13 h 08 m 10.3 s	+50° 39' 50"

(continued)

Table 10.2 (continued)

NGC #	Const	Type	Size	V Mag	RA (2000)	Dec (2000)
5002	CVn	Gxy	1.7'×1.0'	14	13 h 10 m 38.2 s	+36° 38' 02"
5003	CVn	Gxy	1.0'×0.8'	14.3	13 h 08 m 37.9 s	+43° 44' 15"
5005	CVn	Gxy	5.8'×2.8'	10.3	13 h 10 m 56.6 s	+37° 03' 33"
5009	CVn	Gxy	1.1'×0.7'	14.7	13 h 10 m 47.1 s	+50° 05' 33"
5014	CVn	Gxy	1.7'×0.6'	12.9	13 h 11 m 31.4 s	+36° 16' 54"
5021	CVn	Gxy	1.5'×0.7'	13.6	13 h 12 m 06.0 s	+46° 11' 43"
5023	CVn	Gxy	6.0'×0.8'	12.5	13 h 12 m 11.9 s	+44° 02' 17"
5025	CVn	Gxy	2.2'×0.6'	13.6	13 h 12 m 44.7 s	+31° 48' 33"
5029	CVn	Gxy	1.7'×1.1'	13.2	13 h 12 m 37.5 s	+47° 03' 47"
5033	CVn	Gxy	10.7'×5.0'	10.1	13 h 13 m 27.7 s	+36° 35' 40"
5040	CVn	Gxy	0.9'×0.4'	14.3	13 h 13 m 32.7 s	+51° 15' 30"
5055	CVn	Gxy	12.6'×7.2'	9	13 h 15 m 49.1 s	+42° 01' 50"
5074	CVn	Gxy	0.9'×0.8'	13.6	13 h 18 m 25.7 s	+31° 28' 07"
5083	CVn	Gxy	1.4'×1.2'	14.4	13 h 19 m 03.1 s	+39° 35' 20"
5093	CVn	Gxy	1.4'×0.7'	13.9	13 h 19 m 37.8 s	+40° 23' 09"
5096	CVn	Gxy	1.4'×0.7'	14	13 h 20 m 08.5 s	+33° 05' 20"
5098	CVn	Gxy	0.7'×0.7'	14.1	13 h 20 m 14.6 s	+33° 08' 36"
5103	CVn	Gxy	1.4'×1.0'	12.7	13 h 20 m 30.1 s	+43° 05' 01"
5107	CVn	Gxy	1.7'×0.5'	13.3	13 h 21 m 24.5 s	+38° 32' 19"
5112	CVn	Gxy	4.0'×2.8'	11.7	13 h 21 m 56.4 s	+38° 44' 02"
5117	CVn	Gxy	2.2'×1.0'	13.4	13 h 22 m 57.1 s	+28° 18' 59"
5123	CVn	Gxy	1.3'×1.1'	13	13 h 23 m 10.5 s	+43° 05' 09"
5127	CVn	Gxy	2.8'×2.2'	12	13 h 23 m 45.1 s	+31° 33' 55"
5131	CVn	Gxy	2.1'×0.3'	13.7	13 h 23 m 56.9 s	+30° 59' 16"
5141	CVn	Gxy	1.3'×1.0'	12.9	13 h 24 m 51.3 s	+36° 22' 42"
5142	CVn	Gxy	1.0'×0.7'	13.2	13 h 25 m 01.1 s	+36° 23' 57"
5143	CVn	Gxy	0.4'×0.2'	15.8	13 h 25 m 01.3 s	+36° 26' 14"
5145	CVn	Gxy	2.0'×1.8'	12.6	13 h 25 m 14.0 s	+43° 16' 02"
5149	CVn	Gxy	1.5'×0.9'	13.2	13 h 26 m 09.1 s	+35° 56' 02"
5154	CVn	Gxy	1.3'×1.3'	14	13 h 26 m 28.6 s	+36° 00' 35"
5157	CVn	Gxy	1.3'×0.9'	13.5	13 h 27 m 16.9 s	+32° 01' 50"
5166	CVn	Gxy	2.3'×0.4'	13.7	13 h 28 m 15.1 s	+32° 01' 55"
5169	CVn	Gxy	2.3'×0.9'	13.7	13 h 28 m 10.1 s	+46° 40' 19"
5173	CVn	Gxy	1.8'×1.7'	12.3	13 h 28 m 25.2 s	+46° 35' 30"
5187	CVn	Gxy	1.3'×0.8'	13.7	13 h 29 m 48.1 s	+31° 07' 48"
5194	CVn	Gxy	11'×7.8'	8.5	13 h 29 m 52.1 s	+47° 11' 43"
5195	CVn	Gxy	5.8'×4.6'	10.2	13 h 29 m 58.8 s	+47° 15' 59"
5198	CVn	Gxy	2.1'×1.8'	11.9	13 h 30 m 11.3 s	+46° 40' 15"
5199	CVn	Gxy	1.0'×1.0'	13.7	13 h 30 m 42.7 s	+34° 49' 49"
5214	CVn	Gxy	1.2'×0.9'	13.8	13 h 32 m 48.7 s	+41° 52' 19"
5223	CVn	Gxy	1.5'×1.3'	13.1	13 h 34 m 25.1 s	+34° 41' 25"
5225	CVn	Gxy	0.6'×0.6'	13.7	13 h 33 m 20.1 s	+51° 29' 25"
5228	CVn	Gxy	1.0'×0.9'	13.4	13 h 34 m 35.1 s	+34° 46' 39"
5229	CVn	Gxy	3.3'×0.6'	13.8	13 h 34 m 02.6 s	+47° 54' 56"
5233	CVn	Gxy	1.1'×0.5'	14.1	13 h 35 m 13.4 s	+34° 40' 37"

(continued)

Table 10.2 (continued)

NGC #	Const	Type	Size	V Mag	RA (2000)	Dec (2000)
5238	CVn	Gxy	1.7'×1.4'	13.4	13 h 34 m 42.6 s	+51° 36' 48"
5240	CVn	Gxy	1.9'×1.4'	13.3	13 h 35 m 55.1 s	+35° 35' 16"
5243	CVn	Gxy	1.5'×0.4'	13.3	13 h 36 m 15.1 s	+38° 20' 37"
5259	CVn	Gxy	1.1'×0.7'	14.2	13 h 39 m 24.6 s	+30° 59' 26"
5263	CVn	Gxy	1.6'×0.4'	13.4	13 h 39 m 55.5 s	+28° 24' 00"
5265	CVn	Gxy	0.7'×0.6'	14.2	13 h 40 m 08.9 s	+36° 51' 41"
5267	CVn	Gxy	1.4'×0.5'	13.7	13 h 40 m 39.9 s	+38° 47' 38"
5271	CVn	Gxy	0.8'×0.7'	14.3	13 h 41 m 42.4 s	+30° 07' 31"
5272	CVn	GC	18'	6.3	13 h 42 m 11.2 s	+28° 22' 32"
5273	CVn	Gxy	2.8'×2.5'	11.5	13 h 42 m 08.4 s	+35° 39' 15"
5274	CVn	Gxy	0.4'×0.4'	14.6	13 h 42 m 23.3 s	+29° 50' 50"
5275	CVn	Gxy	0.6'×0.6'	14.2	13 h 42 m 23.5 s	+29° 49' 29"
5276	CVn	Gxy	1.0'×0.5'	14	13 h 42 m 22.0 s	+35° 37' 25"
5277	CVn	Gxy	0.6'×0.4'	14.6	13 h 42 m 38.3 s	+29° 57' 15"
5280	CVn	Gxy	0.8'×0.8'	13.9	13 h 42 m 55.4 s	+29° 52' 06"
5282	CVn	Gxy	1.1'×0.8'	13.8	13 h 43 m 24.9 s	+30° 04' 10"
5287	CVn	Gxy	0.5'×0.2'	15.3	13 h 44 m 52.7 s	+29° 46' 17"
5289	CVn	Gxy	1.9'×0.6'	12.9	13 h 45 m 08.7 s	+41° 30' 12"
5290	CVn	Gxy	3.5'×0.9'	12.3	13 h 45 m 18.9 s	+41° 42' 47"
5296	CVn	Gxy	1.0'×0.5'	14	13 h 46 m 18.5 s	+43° 51' 04"
5297	CVn	Gxy	5.6'×1.3'	11.8	13 h 46 m 23.7 s	+43° 52' 18"
5301	CVn	Gxy	4.2'×0.9'	12.2	13 h 46 m 24.5 s	+46° 06' 25"
5303	CVn	Gxy	0.9'×0.4'	13	13 h 47 m 44.9 s	+38° 18' 16"
5305	CVn	Gxy	1.5'×1.1'	13.8	13 h 47 m 55.7 s	+37° 49' 34"
5311	CVn	Gxy	2.6'×2.2'	12.5	13 h 48 m 56.1 s	+39° 59' 08"
5312	CVn	Gxy	0.7'×0.4'	13.9	13 h 49 m 50.4 s	+33° 37' 18"
5313	CVn	Gxy	1.9'×1.1'	12.2	13 h 49 m 44.2 s	+39° 59' 04"
5318	CVn	Gxy	1.5'×0.9'	12.4	13 h 50 m 35.9 s	+33° 42' 17"
5319	CVn	Gxy	1.5'×0.9'	15.5	13 h 50 m 40.5 s	+33° 45' 43"
5320	CVn	Gxy	3.4'×1.7'	12.3	13 h 50 m 20.3 s	+41° 21' 59"
5321	CVn	Gxy	0.6'×0.5'	14.4	13 h 50 m 43.6 s	+33° 37' 57"
5325	CVn	Gxy	0.8'×0.7'	15.3	13 h 50 m 54.1 s	+38° 16' 26"
5326	CVn	Gxy	2.2'×1.1'	12.2	13 h 50 m 50.7 s	+39° 34' 29"
5336	CVn	Gxy	1.6'×1.3'	13	13 h 52 m 09.7 s	+43° 14' 33"
5337	CVn	Gxy	1.7'×0.8'	12.5	13 h 52 m 22.9 s	+39° 41' 14"
5341	CVn	Gxy	1.3'×0.5'	13.6	13 h 52 m 32.0 s	+37° 49' 02"
5346	CVn	Gxy	2.0'×0.8'	14	13 h 53 m 01.9 s	+39° 34' 50"
5347	CVn	Gxy	1.7'×1.3'	12.5	13 h 53 m 17.8 s	+33° 29' 27"
5349	CVn	Gxy	1.7'×0.5'	14.3	13 h 53 m 13.1 s	+37° 52' 58"
5350	CVn	Gxy	3.2'×2.3'	11.7	13 h 53 m 21.5 s	+40° 21' 49"
5351	CVn	Gxy	3.0'×1.5'	12.2	13 h 53 m 27.7 s	+37° 54' 51"
5352	CVn	Gxy	1.2'×1.0'	13.1	13 h 53 m 38.3 s	+36° 08' 02"
5353	CVn	Gxy	2.2'×1.1'	11.2	13 h 53 m 26.7 s	+40° 16' 58"
5354	CVn	Gxy	1.4'×1.3'	11.6	13 h 53 m 26.7 s	+40° 18' 09"
5355	CVn	Gxy	1.2'×0.7'	13.1	13 h 53 m 45.5 s	+40° 20' 19"

(continued)

Table 10.2 (continued)

NGC #	Const	Type	Size	V Mag	RA (2000)	Dec (2000)
5358	CVn	Gxy	1.4'×0.4'	13.9	13 h 54 m 00.3 s	+40° 16' 38"
5361	CVn	Gxy	0.8'×0.4'	13.9	13 h 54 m 35.1 s	+38° 26' 58"
5362	CVn	Gxy	2.3'×1.0'	12.5	13 h 54 m 53.3 s	+41° 18' 46"
5371	CVn	Gxy	4.4'×3.5'	10.7	13 h 55 m 39.9 s	+40° 27' 41"
5375	CVn	Gxy	3.2'×2.8'	11.7	13 h 56 m 56.0 s	+29° 09' 51"
5377	CVn	Gxy	3.7'×2.1'	11.5	13 h 56 m 16.6 s	+47° 14' 07"
5378	CVn	Gxy	2.6'×2.1'	12.7	13 h 56 m 50.9 s	+37° 47' 49"
5380	CVn	Gxy	1.7'×1.7'	12.1	13 h 56 m 56.7 s	+37° 36' 36"
5383	CVn	Gxy	3.2'×2.7'	11.6	13 h 57 m 05.0 s	+41° 50' 47"
5390	CVn	Gxy	4.4'×3.6'	10.5	13 h 55 m 39.9 s	+40° 27' 41"
5391	CVn	NF	…	…	13 h 57 m 37.5 s	+46° 19' 31"
5394	CVn	Gxy	1.7'×1.0'	13	13 h 58 m 33.7 s	+37° 27' 13"
5395	CVn	Gxy	2.9'×1.5'	12	13 h 58 m 37.5 s	+37° 25' 32"
5396	CVn	Gxy	3.5'×2.7'	12	13 h 56 m 56.0 s	+29° 09' 51"
5399	CVn	Gxy	1.2'×0.3'	14	13 h 59 m 31.4 s	+34° 46' 25"
5401	CVn	Gxy	1.5'×0.3'	13.9	13 h 59 m 43.3 s	+36° 14' 16"
5403	CVn	Gxy	3.1'×0.9'	13.8	13 h 59 m 50.8 s	+38° 10' 58"
5406	CVn	Gxy	1.9'×1.4'	12.4	14 h 00 m 20.1 s	+38° 54' 55"
5407	CVn	Gxy	1.4'×1.0'	13.1	14 h 00 m 50.1 s	+39° 09' 22"
5410	CVn	Gxy	1.5'×0.8'	13.2	14 h 00 m 54.6 s	+40° 59' 18"
5421	CVn	Gxy	1.2'×0.9'	13.2	14 h 01 m 41.3 s	+33° 49' 35"
5433	CVn	Gxy	1.6'×0.4'	13.6	14 h 02 m 36.1 s	+32° 30' 36"
5439	CVn	Gxy	1.1'×0.3'	13.9	14 h 01 m 57.5 s	+46° 18' 42"
5440	CVn	Gxy	3.3'×1.4'	12.5	14 h 03 m 01.0 s	+34° 45' 27"
5441	CVn	Gxy	0.5'×0.5'	15.6	14 h 03 m 01.0 s	+34° 45' 27"
5444	CVn	Gxy	2.4'×2.1'	11.6	14 h 03 m 24.1 s	+35° 07' 54"
5445	CVn	Gxy	1.5'×0.7'	13.1	14 h 03 m 31.5 s	+35° 01' 30"
6764	Cyg	Gxy	2.3'×1.3'	12.5	19 h 08 m 16.3 s	+50° 56' 00"
6766	Cyg	PN	5.6"×5.0"	10.9	20 h 10 m 23.5 s	+46° 27' 39"
6783	Cyg	Gxy	0.3'×0.3'	14.4	19 h 16 m 47.6 s	+46° 01' 02"
6798	Cyg	Gxy	1.6'×0.9'	13.3	19 h 24 m 03.1 s	+53° 37' 28"
6801	Cyg	Gxy	1.3'×0.7'	14.1	19 h 27 m 35.8 s	+54° 22' 22"
6811	Cyg	OC	12'	6.8	19 h 37 m 17.9 s	+46° 23' 20"
6819	Cyg	OC	5'	7.3	19 h 41 m 18.0 s	+40° 11' 12"
6824	Cyg	Gxy	1.7'×1.2'	12.4	19 h 43 m 40.9 s	+56° 06' 33"
6826	Cyg	PN	27"×24"	8.8	19 h 44 m 48.0 s	+50° 31' 31"
6833	Cyg	PN	2"	12.1	19 h 49 m 46.5 s	+48° 57' 40"
6834	Cyg	OC	5'	7.8	19 h 52 m 12.5 s	+29° 24' 29"
6846	Cyg	OC	0.5'	14.2	19 h 56 m 28.0 s	+32° 20' 59"
6847	Cyg	OC+Neb	…	…	19 h 56 m 37.7 s	+30° 12' 46"
6856	Cyg	OC	3.0'	…	19 h 59 m 17.1 s	+56° 07' 51"
6857	Cyg	Neb	3'×3'	…	20 h 01 m 48.1 s	+33° 31' 33"
6866	Cyg	OC	6'	…	20 h 03 m 55.1 s	+44° 09' 33"
6871	Cyg	OC	30'	5.2	20 h 05 m 59.3 s	+35° 46' 38"
6874	Cyg	OC	7.0'	…	20 h 07 m 33.0 s	+38° 14' 46"

(continued)

Table 10.2 (continued)

NGC #	Const	Type	Size	V Mag	RA (2000)	Dec (2000)
6881	Cyg	PN	5"×5"	13.9	20 h 10 m 52.3 s	+37° 24' 42"
6883	Cyg	OC	35'	8	20 h 11 m 19.7 s	+35° 49' 56"
6884	Cyg	PN	5.6"×5.0"	10.9	20 h 10 m 23.5 s	+46° 27' 39"
6888	Cyg	Neb	20'×10'	…	20 h 12 m 06.4 s	+38° 21' 18"
6894	Cyg	PN	44"×39"	12.3	20 h 16 m 23.9 s	+30° 33' 55"
6895	Cyg	OC	20'×18'	…	20 h 16 m 32.3 s	+50° 14' 26"
6896	Cyg	**	…	…	20 h 18 m 04.1 s	+30° 38' 20"
6910	Cyg	OC	7'	7.4	20 h 23 m 12.0 s	+40° 46' 43"
6913	Cyg	OC	10'	6.6	20 h 23 m 57.7 s	+38° 30' 28"
6914	Cyg	Neb	3'	…	20 h 24 m 43.3 s	+42° 28' 58"
6916	Cyg	Gxy	2.0'×1.4'	13.7	20 h 23 m 33.1 s	+58° 20' 39"
6946	Cyg	Gxy	11.5'×9.8'	9.1	20 h 34 m 52.7 s	+60° 09' 11"
6960	Cyg	SNR	70.0'×6.0'	…	20 h 45 m 58.1 s	+30° 35' 43"
6974	Cyg	SNR	…	…	20 h 51 m 04.3 s	+31° 49' 41"
6979	Cyg	SNR	…	…	20 h 50 m 27.9 s	+32° 01' 33"
6989	Cyg	OC	10.0'	…	20 h 54 m 06.9 s	+45° 14' 21"
6991	Cyg	OC	6'×4' & 12'×8'	…	20 h 55 m 17.9 s	+47° 22' 00"
6992	Cyg	SNR	60'×8.0'	…	20 h 56 m 19.0 s	+31° 44' 34"
6995	Cyg	SNR	12.0'	…	20 h 57 m 10.7 s	+31° 14' 07"
6996	Cyg	OC	5.0'	10	20 h 56 m 29.9 s	+45° 28' 23"
6997	Cyg	OC	6'	10	20 h 56 m 39.4 s	+44° 37' 54"
7000	Cyg	Neb	100'×60'	…	21 h 01 m 48.0 s	+44° 12' 00"
7008	Cyg	PN	98"×75"	10.7	21 h 00 m 32.8 s	+54° 32' 35"
7011	Cyg	Ast	…	…	21 h 01 m 49.3 s	+47° 21' 16"
7013	Cyg	Gxy	4.0'×1.4'	11.3	21 h 03 m 33.5 s	+29° 53' 50"
7024	Cyg	OC	10'	…	21 h 06 m 06.9 s	+41° 29' 40"
7026	Cyg	PN	27"×11"	10.9	21 h 06 m 18.4 s	+47° 51' 08"
7027	Cyg	PN	18"×10"	8.5	21 h 07 m 01.5 s	+42° 14' 10"
7031	Cyg	OC	5'	9.1	21 h 07 m 12.5 s	+50° 52' 32"
7037	Cyg	OC	7'	…	21 h 10 m 53.9 s	+33° 45' 50"
7039	Cyg	OC	25'	7.6	21 h 10 m 47.7 s	+45° 37' 19"
7044	Cyg	OC	3.5'	12	21 h 13 m 09.3 s	+42° 29' 46"
7048	Cyg	PN	1.02'	12.1	21 h 14 m 14.1 s	+46° 17' 28"
7050	Cyg	OC	5'×2'	…	21 h 15 m 15.9 s	+36° 10' 19"
7054	Cyg	NF	20'	…	21 h 20 m 43.4 s	+39° 10' 20"
7058	Cyg	OC	10'	…	21 h 21 m 53.5 s	+50° 49' 09"
7062	Cyg	OC	6'	8.3	21 h 23 m 27.4 s	+46° 22' 43"
7063	Cyg	OC	7'	7	21 h 24 m 21.7 s	+36° 29' 15"
7067	Cyg	OC	3.0'	9.7	21 h 24 m 23.1 s	+48° 00' 34"
7071	Cyg	OC	4.0'	…	21 h 26 m 39.3 s	+47° 55' 11"
7082	Cyg	OC	25'	7.2	21 h 29 m 17.7 s	+47° 07' 35"
7086	Cyg	OC	9'	8.4	21 h 30 m 27.5 s	+51° 36' 08"
7092	Cyg	OC	31'	4.6	21 h 31 m 48.3 s	+48° 26' 55"
7093	Cyg	OC	…	…	21 h 34 m 21.7 s	+45° 57' 54"

(continued)

Table 10.2 (continued)

NGC #	Const	Type	Size	V Mag	RA (2000)	Dec (2000)
7114	Cyg	*	...	...	21 h 41 m 43.9 s	+42° 50' 28"
7116	Cyg	Gxy	1.1'×0.4'	13.5	21 h 42 m 40.3 s	+28° 56' 48"
7127	Cyg	OC	2.8'	...	21 h 43 m 41.9 s	+54° 37' 48"
7128	Cyg	OC	3.1'	9.7	21 h 43 m 57.7 s	+53° 42' 55"
7143	Cyg	Ast	...	...	21 h 48 m 53.6 s	+29° 57' 18"
7150	Cyg	Ast	...	...	21 h 50 m 23.9 s	+49° 45' 22"
7175	Cyg	Ast	29'×21'	...	21 h 58 m 53.9 s	+54° 50' 00"
6891	Del	PN	74"×62"	10.5	20 h 15 m 08.7 s	+12° 42' 15"
6905	Del	PN	42"×35"	11.1	20 h 22 m 22.9 s	+20° 06' 16"
6917	Del	Gxy	1.6'×1.1'	13.2	20 h 27 m 28.4 s	+08° 05' 52"
6927	Del	Gxy	0.9'×0.4'	14.4	20 h 32 m 38.1 s	+09° 55' 00"
6928	Del	Gxy	2.0'×0.6'	12.9	20 h 32 m 50.2 s	+09° 55' 38"
6930	Del	Gxy	1.3'×0.5'	13.5	20 h 32 m 58.7 s	+09° 52' 27"
6933	Del	**	...	...	20 h 33 m 38.1 s	+07° 23' 14"
6934	Del	GC	7.0'	8.9	20 h 34 m 11.5 s	+07° 24' 16"
6944	Del	Gxy	1.5'×0.6'	13.4	20 h 38 m 23.9 s	+06° 59' 47"
6950	Del	OC	5'	...	20 h 41 m 05.5 s	+16° 37' 20"
6954	Del	Gxy	1.0'×0.6'	13.3	20 h 44 m 03.1 s	+03° 12' 34"
6955	Del	Gxy	1.4'×1.3'	13.8	20 h 44 m 17.9 s	+02° 35' 41"
6956	Del	Gxy	1.9'×1.9'	12.5	20 h 43 m 53.7 s	+12° 30' 42"
6957	Del	Gxy	0.7'×0.7'	14.6	20 h 44 m 47.5 s	+02° 34' 51"
6969	Del	Gxy	1.1'×0.3'	14.2	20 h 48 m 27.6 s	+07° 44' 22"
6971	Del	Gxy	1.1'×0.9'	13.9	20 h 49 m 23.8 s	+05° 59' 43"
6972	Del	Gxy	1.1'×0.5'	13.5	20 h 49 m 59.0 s	+09° 53' 56"
6988	Del	Gxy	0.5'×0.5'	13.9	20 h 55 m 48.9 s	+10° 30' 27"
7003	Del	Gxy	1.1'×0.8'	13.2	21 h 00 m 42.2 s	+17° 48' 14"
7006	Del	GC	2.8'	10.6	21 h 01 m 29.3 s	+16° 11' 16"
7025	Del	Gxy	1.9'×1.3'	13	21 h 07 m 47.3 s	+16° 20' 08"
7028	Del	Gxy	0.8'×0.4'	...	21 h 05 m 50.1 s	+18° 28' 06"
1500	Dor	Gxy	1.1'×0.9'	13.5	03 h 58 m 14.0 s	−52° 19' 44"
1506	Dor	Gxy	1.2'×0.9'	13.6	04 h 00 m 21.6 s	−52° 34' 26"
1515	Dor	Gxy	5.2'×1.1'	11.4	04 h 04 m 02.1 s	−54° 06' 02"
1522	Dor	Gxy	1.2'×0.8'	13.2	04 h 06 m 07.7 s	−52° 40' 12"
1523	Dor	Ast	0.5'×0.4'	...	04 h 06 m 11.1 s	−54° 05' 18"
1533	Dor	Gxy	2.8'×2.3'	10.9	04 h 09 m 51.9 s	−56° 07' 06"
1546	Dor	Gxy	3.0'×1.7'	11.4	04 h 14 m 36.9 s	−56° 03' 38"
1549	Dor	Gxy	4.9'×4.1'	9.7	04 h 15 m 45.0 s	−55° 35' 30"
1553	Dor	Gxy	4.5'×2.8'	9.4	04 h 16 m 10.6 s	−55° 46' 48"
1556	Dor	Gxy	1.7'×0.5'	12.8	04 h 17 m 44.7 s	−50° 09' 53"
1566	Dor	Gxy	7.0'×4.7'	9.8	04 h 20 m 00.7 s	−54° 56' 16"
1578	Dor	Gxy	1.2'×1.1'	13.1	04 h 23 m 46.7 s	−51° 36' 00"
1581	Dor	Gxy	1.8'×0.7'	12.4	04 h 24 m 44.9 s	−54° 56' 34"
1596	Dor	Gxy	3.7'×1.0'	11.2	04 h 27 m 38.0 s	−55° 01' 37"
1602	Dor	Gxy	1.9'×1.1'	13.4	04 h 27 m 54.4 s	−55° 03' 26"
1617	Dor	Gxy	4.3'×2.1'	10.7	04 h 31 m 39.3 s	−54° 36' 06"

(continued)

Table 10.2 (continued)

NGC #	Const	Type	Size	V Mag	RA (2000)	Dec (2000)
1641	Dor	OC	9.0'	...	04 h 35 m 42.9 s	−65° 46' 23"
1644	Dor	OC	...	...	04 h 37 m 39.6 s	−66° 11' 49"
1649	Dor	OC	1.5'	12.9	04 h 38 m 22.9 s	−68° 40' 23"
1652	Dor	OC	15'	13.1	04 h 38 m 22.9 s	−68° 40' 23"
1669	Dor	Gxy	0.7'×0.4'	14.1	04 h 43 m 00.0 s	−65° 48' 53"
1672	Dor	Gxy	6.6'×5.5'	10.2	04 h 45 m 42.9 s	−59° 14' 52"
1676	Dor	OC	1.0'	...	04 h 43 m 54.2 s	−68° 49' 40"
1688	Dor	Gxy	2.4'×1.9'	12.1	04 h 48 m 23.5 s	−59° 48' 01"
1693	Dor	OC	0.7'	12.9	04 h 47 m 38.8 s	−69° 20' 37"
1695	Dor	OC	1.5'	12.2	04 h 47 m 44.5 s	−69° 22' 26"
1696	Dor	OC	0.9'	14	04 h 48 m 30.0 s	−68° 14' 35"
1697	Dor	GC	2.5'	12.6	04 h 48 m 36.4 s	−68° 33' 29"
1698	Dor	GC	1.5'	12.1	04 h 49 m 04.5 s	−69° 06' 49"
1703	Dor	Gxy	3.0'×2.6'	11.7	04 h 52 m 51.9 s	−59° 44' 36"
1704	Dor	OC	1.7'	11.5	04 h 49 m 55.5 s	−69° 45' 23"
1706	Dor	Gxy	1.4'×1.0'	13.3	04 h 52 m 31.1 s	−62° 59' 12"
1712	Dor	OC+Neb	2.6'	...	04 h 50 m 58.5 s	−69° 24' 27"
1714	Dor	OC	1.1'	11.6	04 h 52 m 08.4 s	−66° 55' 31"
1715	Dor	Neb	...	...	04 h 52 m 10.5 s	−66° 54' 32"
1718	Dor	OC	2.0'	12.3	04 h 52 m 25.5 s	−67° 03' 03"
1722	Dor	OC+Neb	1.0'	13.2	04 h 51 m 43.4 s	−69° 23' 54"
1727	Dor	OC+Neb	2.5'×1.1'	11.1	04 h 52 m 12.8 s	−69° 20' 20"
1731	Dor	OC+Neb	8.0'	9.9	04 h 53 m 32.1 s	−66° 55' 31"
1732	Dor	OC	0.9'	12.3	04 h 53 m 10.7 s	−68° 39' 00"
1733	Dor	OC	1.1'	13.3	04 h 54 m 04.8 s	−66° 40' 58"
1734	Dor	OC	1.4'	13.1	04 h 53 m 33.5 s	−68° 46' 08"
1735	Dor	OC	1.3'	10.8	04 h 54 m 19.7 s	−67° 05' 59"
1736	Dor	Neb	...	...	04 h 53 m 01.6 s	−68° 03' 11"
1737	Dor	Neb	...	...	04 h 53 m 57.9 s	−69° 10' 16"
1743	Dor	Neb	2.0'×2.0'	...	04 h 54 m 03.2 s	−69° 11' 57"
1745	Dor	OC+Neb	2.0'×2.0'	...	04 h 54 m 20.7 s	−69° 09' 32"
1747	Dor	OC+Neb	10'.0×12.0'	9.4	04 h 55 m 11.0 s	−67° 10' 08"
1748	Dor	OC+Neb	0.5'×0.5'	...	04 h 54 m 26.0 s	−69° 11' 03"
1749	Dor	OC	1.1'	13.6	04 h 54 m 56.1 s	−68° 11' 20"
1755	Dor	OC	2.0'	9.9	04 h 55 m 14.9 s	−68° 12' 15"
1756	Dor	OC	1.0'×1.0'	12.2	04 h 54 m 49.9 s	−69° 14' 15"
1760	Dor	Neb	...	...	04 h 56 m 44.4 s	−66° 31' 39"
1761	Dor	OC+Neb	1.2'	9.9	04 h 56 m 37.8 s	−66° 28' 44"
1763	Dor	Neb	...	...	04 h 56 m 49.3 s	−66° 24' 33"
1764	Dor	OC	0.5'×0.5'	12.6	04 h 56 m 27.8 s	−67° 41' 38"
1765	Dor	Gxy	1.2'×1.0'	13	04 h 58 m 24.5 s	−62° 01' 42"
1767	Dor	OC+Neb	1.6'	10.6	04 h 56 m 27.3 s	−69° 24' 02"
1768	Dor	OC	0.9'	12.8	04 h 57 m 00.2 s	−68° 14' 58"
1769	Dor	Neb	...	...	04 h 57 m 44.7 s	−66° 27' 49"
1770	Dor	OC+Neb	1.6'	...	04 h 57 m 15.8 s	−68° 25' 05"

(continued)

Table 10.2 (continued)

NGC #	Const	Type	Size	V Mag	RA (2000)	Dec (2000)
1771	Dor	Gxy	1.9'×0.5'	13.8	04 h 58 m 55.6 s	−63° 17' 56"
1772	Dor	OC+Neb	1.5'	11	04 h 56 m 52.9 s	−69° 33' 22"
1773	Dor	Neb	...	...	04 h 58 m 11.3 s	−66° 21' 33"
1774	Dor	OC	1.8'	10.8	04 h 58 m 06.9 s	−67° 14' 33"
1776	Dor	OC	1.1'	13	04 h 58 m 39.8 s	−66° 25' 47"
1782	Dor	OC+Neb	1.2'	10.5	04 h 57 m 51.1 s	−69° 23' 32"
1783	Dor	OC	2.9'	10.9	04 h 59 m 08.8 s	−65° 59' 07"
1785	Dor	Ast	3.0'	...	04 h 58 m 35.4 s	−68° 50' 40"
1786	Dor	OC	1.5'	10.9	04 h 59 m 07.9 s	−67° 44' 43"
1787	Dor	OC	23'	10.9	05 h 01 m 42.5 s	−65° 49' 23"
1793	Dor	OC	0.5'×0.7'	12.4	04 h 59 m 38.3 s	−69° 33' 28"
1795	Dor	OC	1.5'×1.0'	12.4	04 h 59 m 46.9 s	−69° 48' 04"
1796	Dor	Gxy	1.9'×1.0'	12.5	05 h 02 m 42.8 s	−61° 08' 25"
1801	Dor	OC	2.1'	12.2	05 h 00 m 34.6 s	−69° 36' 50"
1804	Dor	OC	1.0'	11.9	05 h 01 m 03.3 s	−69° 04' 58"
1805	Dor	OC	2.2'	10.6	05 h 02 m 21.4 s	−66° 06' 44"
1806	Dor	OC	2.2'	...	05 h 02 m 11.4 s	−67° 59' 02"
1809	Dor	Gxy	3.2'×0.8'	14.8	05 h 02 m 05.3 s	−69° 34' 08"
1810	Dor	OC	1.2'	11.9	05 h 03 m 23.3 s	−66° 22' 55"
1814	Dor	OC+Neb	1.0'	9	05 h 03 m 46.5 s	−67° 18' 03"
1816	Dor	OC+Neb	1.0'	9	05 h 03 m 50.8 s	−67° 15' 39"
1818	Dor	OC	2.5'	9.7	05 h 04 m 14.8 s	−66° 26' 04"
1820	Dor	OC	10.0'	9	05 h 04 m 01.7 s	−67° 15' 58"
1822	Dor	OC	0.8'	13.2	05 h 05 m 09.3 s	−66° 12' 38"
1824	Dor	Gxy	3.2'×0.9'	12.8	05 h 06 m 56.0 s	−59° 43' 32"
1825	Dor	GC	0.5'	12	05 h 04 m 19.1 s	−68° 55' 35"
1826	Dor	OC	1.0'	13.3	05 h 05 m 34.1 s	−66° 13' 52"
1828	Dor	GC	0.7'	12.5	05 h 04 m 20.9 s	−69° 23' 18"
1829	Dor	OC+Neb	2.1'	...	05 h 04 m 57.3 s	−68° 03' 20"
1830	Dor	OC	0.8'	12.6	05 h 04 m 38.3 s	−69° 20' 25"
1831	Dor	OC	4.0'	11.2	05 h 06 m 16.7 s	−64° 55' 07"
1834	Dor	GC	...	...	05 h 05 m 11.5 s	−69° 12' 27"
1835	Dor	GC	1.2'	10.6	05 h 05 m 05.7 s	−69° 24' 15"
1836	Dor	OC	0.8'	12.2	05 h 05 m 34.5 s	−68° 37' 41"
1838	Dor	OC	0.7'	12.9	05 h 06 m 07.9 s	−68° 26' 43"
1839	Dor	OC	0.8'	11.8	05 h 06 m 02.4 s	−68° 37' 37"
1842	Dor	OC	0.9'	14	05 h 07 m 18.1 s	−67° 16' 24"
1844	Dor	OC	1.4'	12.1	05 h 07 m 30.7 s	−67° 19' 25"
1846	Dor	OC	2.7'	11.3	05 h 07 m 34.7 s	−67° 27' 31"
1847	Dor	GC	1.0'	11.1	05 h 07 m 08.2 s	−68° 58' 17"
1849	Dor	OC	1.3'	12.8	05 h 09 m 34.9 s	−66° 18' 57"
1850	Dor	GC	3.4'	9	05 h 08 m 44.8 s	−68° 45' 42"
1852	Dor	OC	1.9'	12	05 h 09 m 23.9 s	−67° 46' 39"
1853	Dor	Gxy	2.0'×0.7'	13.1	05 h 12 m 16.5 s	−57° 23' 55"
1854	Dor	GC	1.0'	10.4	05 h 09 m 20.3 s	−68° 50' 56"

(continued)

Table 10.2 (continued)

NGC #	Const	Type	Size	V Mag	RA (2000)	Dec (2000)
1855	Dor	OC	5.0'	...	05 h 09 m 17.0 s	−68° 50' 45"
1856	Dor	GC	1.8'	10.1	05 h 09 m 29.5 s	−69° 07' 40"
1858	Dor	OC+Neb	5.0'	9.9	05 h 09 m 56.4 s	−68° 53' 59"
1859	Dor	OC	2.1'	12.3	05 h 11 m 31.9 s	−65° 14' 59"
1860	Dor	OC	0.8'	11	05 h 10 m 39.6 s	−68° 45' 08"
1862	Dor	OC	0.4'	13.3	05 h 12 m 34.6 s	−66° 09' 16"
1863	Dor	GC	0.5'	11	05 h 11 m 39.7 s	−68° 43' 37"
1864	Dor	OC	0.9'	12.9	05 h 12 m 40.7 s	−67° 37' 17"
1865	Dor	OC	0.8'	12.9	05 h 12 m 25.1 s	−68° 46' 16"
1866	Dor	OC	5.1'	9.7	05 h 13 m 39.1 s	−65° 27' 56"
1867	Dor	OC	1.3'	13.4	05 h 13 m 42.5 s	−66° 17' 39"
1868	Dor	OC	4.0'	11.6	05 h 14 m 36.5 s	−63° 57' 18"
1869	Dor	OC+Neb	14'	...	05 h 13 m 56.4 s	−67° 22' 41"
1870	Dor	GC	0.5'	11.3	05 h 13 m 09.9 s	−69° 07' 01"
1871	Dor	OC+Neb	2.1'	10.1	05 h 13 m 51.8 s	−67° 27' 10"
1872	Dor	GC	1.0'	11	05 h 13 m 11.5 s	−69° 18' 46"
1873	Dor	OC+Neb	3.5'	10.4	05 h 13 m 55.7 s	−67° 20' 04"
1874	Dor	OC+Neb	1.0'	...	05 h 13 m 12.9 s	−69° 22' 31"
1876	Dor	OC+Neb	...	11.7	05 h 13 m 19.3 s	−69° 21' 41"
1877	Dor	OC+Neb	3.0'	...	05 h 13 m 22.2 s	−69° 22' 38"
1880	Dor	Neb	...	...	05 h 13 m 38.8 s	−69° 23' 02"
1881	Dor	OC+Neb	...	...	05 h 13 m 37.2 s	−69° 17' 57"
1882	Dor	OC	1.2'	12.3	05 h 15 m 33.4 s	−66° 07' 47"
1884	Dor	NF	...	...	05 h 15 m 48.3 s	−66° 16' 12"
1885	Dor	GC	0.8'	12	05 h 15 m 05.9 s	−68° 58' 39"
1887	Dor	OC	1.2'	12.7	05 h 16 m 06.0 s	−66° 19' 07"
1892	Dor	Gxy	3.3'×1.0'	12.2	05 h 17 m 09.5 s	−64° 57' 38"
1894	Dor	OC	0.8'	12.2	05 h 15 m 51.3 s	−69° 28' 07"
1895	Dor	Neb	...	...	05 h 16 m 52.4 s	−67° 19' 47"
1897	Dor	OC	1.0'	13.5	05 h 17 m 32.4 s	−67° 26' 56"
1898	Dor	OC	...	11.9	05 h 16 m 42.5 s	−69° 39' 23"
1899	Dor	Neb	0.7'×0.7'	...	05 h 17 m 48.7 s	−67° 54' 03"
1900	Dor	OC	1.7'	13.6	05 h 19 m 09.4 s	−63° 01' 25"
1901	Dor	OC	15'×15'	...	05 h 18 m 15.2 s	−68° 26' 11"
1902	Dor	OC	1.6'	11.8	05 h 18 m 19.2 s	−66° 37' 39"
1903	Dor	GC	1.0'	11.9	05 h 17 m 22.3 s	−69° 20' 07"
1905	Dor	OC	1.0'	13.2	05 h 18 m 23.6 s	−67° 16' 41"
1910	Dor	OC+Neb	9.0'	11.2	05 h 18 m 43.1 s	−69° 13' 55"
1911	Dor	OC	...	...	05 h 20 m 33.1 s	−66° 46' 43"
1913	Dor	OC	0.6'	11.1	05 h 18 m 19.3 s	−69° 32' 12"
1915	Dor	OC	3'×2'	...	05 h 19 m 44.7 s	−66° 48' 33"
1916	Dor	GC	1.1'	10.4	05 h 18 m 36.5 s	−69° 24' 25"
1917	Dor	GC	1.5'	12.3	05 h 19 m 01.2 s	−68° 59' 56"
1918	Dor	SNR	7' to 8'	...	05 h 19 m 07.1 s	−69° 39' 45"
1919	Dor	OC+Neb	1.7'	...	05 h 20 m 17.1 s	−66° 52' 53"

(continued)

Table 10.2 (continued)

NGC #	Const	Type	Size	V Mag	RA (2000)	Dec (2000)
1920	Dor	OC+Neb	...	...	05 h 20 m 33.1 s	−66° 46' 43"
1921	Dor	GC	...	...	05 h 19 m 22.8 s	−69° 47' 16"
1922	Dor	OC	0.5'	11.5	05 h 19 m 49.0 s	−69° 26' 54"
1923	Dor	OC+Neb	1.0'	11.2	05 h 21 m 34.1 s	−65° 29' 12"
1925	Dor	OC+Neb	10.0'	...	05 h 21 m 44.0 s	−65° 47' 37"
1926	Dor	GC	1.0'	11.8	05 h 20 m 34.2 s	−69° 31' 27"
1928	Dor	OC	0.5'	11.9	05 h 20 m 56.5 s	−69° 28' 41"
1929	Dor	Neb	20'×15'	...	05 h 21 m 38.2 s	−67° 54' 43"
1932	Dor	*	...	...	05 h 22 m 17.3 s	−66° 09' 16"
1933	Dor	OC	...	...	05 h 22 m 27.3 s	−66° 09' 07"
1934	Dor	OC	...	...	05 h 21 m 47.9 s	−67° 56' 14"
1935	Dor	Neb	...	...	05 h 21 m 58.8 s	−67° 57' 27"
1936	Dor	Neb	...	...	05 h 22 m 14.6 s	−67° 58' 34"
1937	Dor	OC	...	...	05 h 22 m 29.2 s	−67° 53' 41"
1940	Dor	OC	1.0'	11.9	05 h 22 m 43.8 s	−67° 11' 12"
1941	Dor	OC+Neb	...	...	05 h 23 m 07.7 s	−66° 22' 43"
1942	Dor	OC	1.2'	13.5	05 h 24 m 44.6 s	−63° 56' 32"
1945	Dor	Neb	...	...	05 h 24 m 55.0 s	−66° 27' 27"
1946	Dor	OC	1.0'	12.6	05 h 25 m 16.4 s	−66° 23' 41"
1947	Dor	Gxy	3.0'×2.6'	10.9	05 h 26 m 47.5 s	−63° 45' 38"
1948	Dor	OC+Neb	6.0'	11.6	05 h 25 m 46.3 s	−66° 16' 01"
1949	Dor	OC+Neb	...	...	05 h 25 m 05.1 s	−68° 28' 16"
1951	Dor	OC	2.0'	10.6	05 h 26 m 06.9 s	−66° 35' 50"
1953	Dor	GC	0.8'	11.7	05 h 25 m 27.9 s	−68° 50' 18"
1955	Dor	OC+Neb	1.9'	9	05 h 26 m 10.0 s	−67° 29' 51"
1958	Dor	GC	0.8'	13	05 h 25 m 30.6 s	−69° 50' 13"
1962	Dor	OC	0.5'	11.5	05 h 26 m 17.8 s	−68° 50' 16"
1965	Dor	OC	...	...	05 h 26 m 29.1 s	−68° 48' 23"
1966	Dor	OC	13'×12'	...	05 h 26 m 45.9 s	−68° 49' 12"
1967	Dor	OC	0.4'	10.8	05 h 26 m 43.4 s	−69° 06' 06"
1968	Dor	OC+Neb	1.1'	9	05 h 27 m 22.2 s	−67° 27' 50"
1969	Dor	OC	0.8'	12.5	05 h 26 m 32.4 s	−69° 50' 29"
1970	Dor	OC	0.5'	...	05 h 38 m 38.4 s	−69° 05' 39"
1971	Dor	OC	0.8'	11.9	05 h 26 m 45.0 s	−69° 51' 06"
1972	Dor	OC	0.9'	12.6	05 h 38 m 24.5 s	−70° 14' 03"
1974	Dor	OC+Neb	1.7'	9	05 h 27 m 54.4 s	−67° 25' 34"
1978	Dor	OC	4.2'×1.9'	10.7	05 h 28 m 45.2 s	−66° 14' 07"
1983	Dor	OC	0.6'	9.9	05 h 27 m 44.3 s	−68° 59' 10"
1984	Dor	OC	1.0'	10	05 h 27 m 41.0 s	−69° 08' 04"
1991	Dor	OC+Neb	1.7'	9	05 h 28 m 00.4 s	−67° 25' 23"
1994	Dor	OC	0.6'	9.8	05 h 28 m 21.9 s	−69° 08' 31"
1997	Dor	OC	1.3'	13.4	05 h 30 m 34.4 s	−63° 12' 15"
2001	Dor	OC	1.7'	...	05 h 29 m 02.1 s	−68° 46' 10"
2002	Dor	OC	2.1'	10.1	05 h 30 m 20.4 s	−66° 53' 03"
2003	Dor	OC	2.2'	11.3	05 h 30 m 54.4 s	−66° 27' 59"

(continued)

Table 10.2 (continued)

NGC #	Const	Type	Size	V Mag	RA (2000)	Dec (2000)
2004	Dor	OC	2.7'	9.6	05 h 30 m 40.3 s	−67° 17' 10"
2005	Dor	GC	1.8'	11.6	05 h 30 m 10.9 s	−69° 45' 09"
2006	Dor	OC	15'×7'	11.5	05 h 31 m 19.6 s	−66° 58' 19"
2009	Dor	OC	0.8'	11	05 h 30 m 59.2 s	−69° 10' 54"
2011	Dor	OC	1.0'	10.6	05 h 32 m 20.3 s	−67° 31' 24"
2014	Dor	OC+Neb	1.8'	9	05 h 32 m 19.9 s	−67° 41' 24"
2015	Dor	OC	5.5'	...	05 h 32 m 06.5 s	−69° 14' 35"
2020	Dor	Neb	2.0'	...	05 h 33 m 12.7 s	−67° 42' 57"
2021	Dor	OC	0.9'	12.1	05 h 33 m 30.7 s	−67° 27' 11"
2027	Dor	OC	0.7'	11.9	05 h 34 m 59.7 s	−66° 54' 59"
2029	Dor	Neb	...	...	05 h 34 m 59.8 s	−67° 33' 23"
2030	Dor	OC+Neb	...	...	05 h 35 m 40.1 s	−66° 02' 08"
2032	Dor	Neb	...	...	05 h 35 m 20.7 s	−67° 34' 07"
2033	Dor	OC+Neb	0.6'	11.6	05 h 34 m 40.5 s	−69° 44' 52"
2034	Dor	OC	...	...	05 h 35 m 32.8 s	−66° 54' 13"
2035	Dor	Neb	3'×3'	...	05 h 35 m 32.5 s	−67° 35' 08"
2037	Dor	OC	0.5'	10.3	05 h 35 m 00.7 s	−69° 43' 54"
2040	Dor	Neb	3'×3'	...	05 h 36 m 07.6 s	−67° 34' 04"
2041	Dor	OC	0.7'	10.4	05 h 36 m 28.1 s	−66° 59' 23"
2042	Dor	OC	8.5'	9.5	05 h 35 m 53.1 s	−68° 55' 33"
2044	Dor	OC	2.1'	10.6	05 h 35 m 56.7 s	−69° 11' 40"
2048	Dor	OC+Neb	0.5'	...	05 h 35 m 55.4 s	−69° 38' 58"
2050	Dor	OC	1.0'	9.3	05 h 36 m 38.9 s	−69° 23' 01"
2052	Dor	Neb	18'×12'	...	05 h 37 m 11.1 s	−69° 46' 22"
2053	Dor	OC	1.2'	12.2	05 h 37 m 39.7 s	−67° 24' 47"
2055	Dor	OC	0.6'	8.4	05 h 36 m 44.8 s	−69° 29' 55"
2060	Dor	OC+Neb	3.5'	9.6	05 h 37 m 48.7 s	−69° 10' 24"
2062	Dor	OC	0.9'	12.7	05 h 40 m 02.7 s	−66° 52' 33"
2069	Dor	Neb	40'	...	05 h 38 m 46.5 s	−68° 58' 28"
2070	Dor	Neb	40'×25'	...	05 h 38 m 38.4 s	−69° 05' 39"
2074	Dor	Neb	16'×10'	...	05 h 39 m 03.7 s	−69° 29' 53"
2077	Dor	Neb	15'×15'	...	05 h 39 m 36.1 s	−69° 39' 26"
2078	Dor	Neb	15'×11'	...	05 h 39 m 39.3 s	−69° 44' 38"
2079	Dor	Neb	15'×11'	...	05 h 39 m 39.9 s	−69° 46' 26"
2080	Dor	Neb	15'×11'	...	05 h 39 m 44.2 s	−69° 38' 44"
2081	Dor	OC+Neb	16'×10'	...	05 h 39 m 59.5 s	−69° 24' 21"
2082	Dor	Gxy	1.8'×1.7'	12.2	05 h 41 m 51.3 s	−64° 18' 07"
2083	Dor	Neb	...	...	05 h 39 m 59.3 s	−69° 44' 16"
2084	Dor	Neb	...	...	05 h 40 m 07.1 s	−69° 45' 34"
2085	Dor	Neb	...	...	05 h 40 m 09.9 s	−69° 40' 22"
2086	Dor	OC+Neb	...	...	05 h 40 m 12.9 s	−69° 40' 05"
2088	Dor	OC	1.7'	12.5	05 h 40 m 59.9 s	−68° 27' 56"
2091	Dor	OC	1.7'	12.1	05 h 40 m 58.1 s	−69° 26' 14"
2092	Dor	OC	1.2'	...	05 h 41 m 22.1 s	−69° 13' 27"
2093	Dor	OC+Neb	1.7'	11.6	05 h 41 m 49.8 s	−68° 55' 17"

(continued)

Table 10.2 (continued)

NGC #	Const	Type	Size	V Mag	RA (2000)	Dec (2000)
2094	Dor	OC	...	...	05 h 42 m 12.8 s	−68° 55' 07"
2095	Dor	OC	0.8'	13.1	05 h 42 m 36.2 s	−67° 19' 08"
2096	Dor	OC	1.2'	11.3	05 h 42 m 17.9 s	−68° 27' 31"
2097	Dor	OC	1.7'	13.7	05 h 44 m 16.1 s	−62° 47' 08"
2098	Dor	OC	1.6'	10.7	05 h 42 m 30.4 s	−68° 16' 32"
2100	Dor	OC	2.3'	9.6	05 h 42 m 09.1 s	−69° 12' 43"
2102	Dor	OC	0.5'×0.5'	11.4	05 h 42 m 20.5 s	−69° 29' 14"
2105	Dor	OC	1.0'×0.9'	12.2	05 h 44 m 19.2 s	−66° 55' 04"
2108	Dor	OC	1.0'×1.0'	12.8	05 h 43 m 55.3 s	−69° 10' 51"
2109	Dor	OC	1.2'×1.0'	12.2	05 h 44 m 23.0 s	−68° 32' 52"
2113	Dor	OC+Neb	1.5'×1.2'	12.3	05 h 45 m 24.7 s	−69° 46' 27"
2114	Dor	OC	1.0'×0.5'	12.5	05 h 46 m 12.2 s	−68° 02' 54"
2116	Dor	OC	1.2'×0.5'	12.9	05 h 47 m 15.2 s	−68° 30' 29"
2117	Dor	OC	0.8'×0.7'	12.7	05 h 47 m 46.0 s	−67° 27' 01"
2118	Dor	OC	0.6'×0.6'	12	05 h 47 m 39.6 s	−69° 07' 55"
2120	Dor	OC	1.0' × 1.0'	12.7	05 h 50 m 34.7 s	−63° 40' 30"
2123	Dor	OC	0.4'×0.4'	12.6	05 h 51 m 43.4 s	−65° 19' 18"
2125	Dor	OC	0.4'×0.3'	...	05 h 50 m 54.3 s	−69° 28' 45"
2127	Dor	OC	0.6'×0.5'	11.6	05 h 51 m 22.4 s	−69° 21' 41"
2130	Dor	OC	0.4'×0.4'	12	05 h 52 m 23.8 s	−67° 20' 03"
2135	Dor	OC	0.3'×0.3'	12.1	05 h 53 m 35.0 s	−67° 25' 38"
2136	Dor	OC	1.0'×1.0'	10.5	05 h 52 m 59.1 s	−69° 29' 36"
2137	Dor	OC	0.5'×0.5'	12.7	05 h 53 m 13.2 s	−69° 28' 55"
2138	Dor	OC	0.3'×0.3'	13.8	05 h 54 m 48.8 s	−65° 50' 07"
2140	Dor	OC	0.6'×0.3'	12.4	05 h 54 m 16.3 s	−68° 36' 00"
2147	Dor	OC+Neb	0.2'×0.2'	12.9	05 h 55 m 45.7 s	−68° 12' 06"
2150	Dor	Gxy	1.1'×0.9'	13.4	05 h 55 m 46.4 s	−69° 33' 40"
2151	Dor	OC	0.3'×0.2'	...	05 h 56 m 20.5 s	−69° 01' 03"
2153	Dor	OC	0.4'×0.3'	13.1	05 h 57 m 51.8 s	−66° 24' 03"
2154	Dor	OC	1.0' × 1.0'	11.8	05 h 57 m 38.3 s	−67° 15' 44"
2155	Dor	OC	1.8'×1.7'	12.6	05 h 58 m 33.3 s	−65° 28' 35"
2156	Dor	OC	0.8'×0.5'	11.4	05 h 57 m 50.5 s	−68° 27' 39"
2157	Dor	OC	1.0'×1.0'	10.2	05 h 57 m 34.9 s	−69° 11' 50"
2159	Dor	OC	1.0'×1.0'	11.4	05 h 58 m 03.0 s	−68° 37' 28"
2160	Dor	OC	0.8'×0.4'	12.2	05 h 58 m 12.9 s	−68° 17' 23"
2162	Dor	OC	2.0'×1.8'	...	06 h 00 m 30.4 s	−63° 43' 20"
2164	Dor	OC	1.2'×1.1'	10.3	05 h 58 m 54.9 s	−68° 30' 56"
2166	Dor	OC	0.5'×0.5'	12.9	05 h 59 m 33.8 s	−67° 56' 28"
2172	Dor	OC	0.7'×0.6'	11.8	06 h 00 m 05.9 s	−68° 38' 13"
2176	Dor	OC	0.6'×0.6'	...	06 h 01 m 19.5 s	−66° 51' 12"
2177	Dor	OC	0.7'×0.5'	12.8	06 h 01 m 16.5 s	−67° 44' 00"
2181	Dor	OC	0.7'×0.7'	13.6	06 h 02 m 43.7 s	−65° 15' 54"
2187	Dor	Gxy	2.5'×2.1'	12.6	06 h 03 m 48.3 s	−69° 35' 00"
2193	Dor	OC	0.5'×0.5'	13.4	06 h 06 m 17.9 s	−65° 05' 57"
2197	Dor	OC	0.8'×0.8'	13.5	06 h 06 m 08.7 s	−67° 05' 51"

(continued)

Table 10.2 (continued)

NGC #	Const	Type	Size	V Mag	RA (2000)	Dec (2000)
2210	Dor	OC	1.7'	10.2	06 h 11 m 31.9 s	−69° 07' 15"
2214	Dor	OC	3.5'	10.9	06 h 12 m 57.5 s	−68° 15' 33"
2228	Dor	Gxy	0.8'×0.7'	13.7	06 h 21 m 15.9 s	−64° 27' 34"
2229	Dor	Gxy	1.4'×0.4'	13.6	06 h 21 m 23.8 s	−64° 57' 26"
2230	Dor	Gxy	1.1'×0.9'	13.2	06 h 21 m 27.9 s	−64° 59' 35"
2231	Dor	OC	2.1'	13.2	06 h 20 m 43.0 s	−67° 31' 07"
2233	Dor	Gxy	0.8'×0.2'	13.9	06 h 21 m 40.1 s	−65° 02' 01"
2235	Dor	Gxy	1.3'×1.0'	13.1	06 h 22 m 22.2 s	−64° 56' 06"
2241	Dor	OC	1.2'	13.3	06 h 22 m 53.3 s	−68° 55' 29"
2249	Dor	OC	1.0'×1.0'	12.2	06 h 25 m 49.7 s	−68° 55' 11"
2257	Dor	OC	3.4'	13.5	06 h 30 m 13.1 s	−64° 19' 29"
2908	Dra	Gxy	1.0'×1.0'	13.3	09 h 43 m 31.5 s	+79° 42' 04"
2938	Dra	Gxy	1.7'×1.0'	13.7	09 h 38 m 25.0 s	+76° 19' 09"
2957	Dra	Gxy	1.2'×0.6'	14.8	09 h 47 m 15.9 s	+72° 59' 10"
2963	Dra	Gxy	1.2'×0.6'	13.7	09 h 47 m 50.5 s	+72° 57' 51"
2977	Dra	Gxy	1.8'×0.8'	12.4	09 h 43 m 46.7 s	+74° 51' 35"
3057	Dra	Gxy	2.2'×1.3'	13.4	10 h 05 m 39.5 s	+80° 17' 09"
3061	Dra	Gxy	1.7'×1.5'	13.2	09 h 56 m 12.0 s	+75° 51' 57"
3144	Dra	Gxy	1.4'×0.8'	13.6	10 h 15 m 32.1 s	+74° 13' 13"
3147	Dra	Gxy	3.9'×3.5'	10.8	10 h 16 m 53.6 s	+73° 24' 02"
3155	Dra	Gxy	1.6'×1.1'	12.9	10 h 17 m 39.8 s	+74° 20' 51"
3174	Dra	Gxy	1.4'×0.8'	13.6	10 h 15 m 32.1 s	+74° 13' 13"
3183	Dra	Gxy	2.4'×1.3'	12	10 h 21 m 48.7 s	+74° 10' 34"
3194	Dra	Gxy	1.6'×1.1'	12.9	10 h 17 m 39.8 s	+74° 20' 51"
3197	Dra	Gxy	1.3'×1.0'	13.7	10 h 14 m 27.7 s	+77° 49' 12"
3210	Dra	**	...	...	10 h 27 m 59.3 s	+79° 49' 57"
3212	Dra	Gxy	1.7'×0.9'	13.7	10 h 28 m 16.5 s	+79° 49' 24"
3215	Dra	Gxy	1.1'×1.0'	13.2	10 h 28 m 40.6 s	+79° 48' 47"
3218	Dra	Gxy	2.4'×1.3'	12	10 h 21 m 48.7 s	+74° 10' 34"
3252	Dra	Gxy	2.0'×0.6'	13.6	10 h 34 m 22.7 s	+73° 45' 50"
3329	Dra	Gxy	1.8'×1.0'	12.4	10 h 44 m 39.4 s	+76° 48' 34"
3343	Dra	Gxy	1.3'×0.9'	13.5	10 h 46 m 10.4 s	+73° 21' 10"
3397	Dra	Gxy	2.1'×1.3'	12.2	10 h 44 m 39.4 s	+76° 48' 34"
3403	Dra	Gxy	3.0'×1.2'	12.4	10 h 53 m 55.1 s	+73° 41' 23"
3465	Dra	Gxy	1.5'×1.2'	13.7	10 h 59 m 31.2 s	+75° 11' 29"
3484	Dra	NF	...	...	11 h 03 m 05.5 s	+75° 49' 08"
3500	Dra	Gxy	1.5'×1.2'	14.2	11 h 01 m 52.0 s	+75° 12' 06"
3523	Dra	Gxy	1.6'×1.5'	13.1	11 h 03 m 06.3 s	+75° 06' 55"
3538	Dra	**	...	...	11 h 11 m 34.4 s	+75° 34' 11"
3562	Dra	Gxy	...'×...'	12.3	11 h 12 m 58.8 s	+72° 52' 44"
3682	Dra	Gxy	2.3'×1.6'	12.6	11 h 27 m 41.3 s	+66° 35' 24"
3735	Dra	Gxy	4.2'×1.0'	12	11 h 35 m 57.3 s	+70° 32' 08"
3736	Dra	Gxy	1.4'×0.8'	14.9	11 h 35 m 41.7 s	+73° 27' 06"
3747	Dra	Gxy	0.7'×0.3'	15	11 h 32 m 31.0 s	+74° 22' 43"
3752	Dra	Gxy	1.8'×0.8'	13.1	11 h 32 m 32.3 s	+74° 37' 38"

(continued)

Table 10.2 (continued)

NGC #	Const	Type	Size	V Mag	RA (2000)	Dec (2000)
3879	Dra	Gxy	2.6'×0.5'	13.1	11 h 46 m 49.3 s	+69° 23' 01"
3890	Dra	Gxy	1.0'×1.0'	13.4	11 h 49 m 19.7 s	+74° 18' 08"
3939	Dra	Gxy	1.0'×1.0'	13.4	11 h 49 m 19.7 s	+74° 18' 08"
3961	Dra	Gxy	1.4'×1.4'	13.7	11 h 54 m 57.5 s	+69° 19' 48"
4034	Dra	Gxy	1.7'×1.1'	13.7	12 h 01 m 29.3 s	+69° 19' 25"
4108	Dra	Gxy	1.7'×1.4'	12.5	12 h 06 m 44.5 s	+67° 09' 46"
4120	Dra	Gxy	1.8'×0.4'	13.7	12 h 08 m 31.0 s	+69° 32' 34"
4121	Dra	Gxy	1.0'×0.8'	13.5	12 h 07 m 56.6 s	+65° 06' 50"
4125	Dra	Gxy	5.8'×3.2'	9.9	12 h 08 m 05.6 s	+65° 10' 28"
4128	Dra	Gxy	2.6'×0.9'	11.9	12 h 08 m 32.4 s	+68° 46' 03"
4133	Dra	Gxy	2.0'×1.6'	12.5	12 h 08 m 49.9 s	+74° 54' 14"
4159	Dra	Gxy	1.3'×0.5'	13.5	12 h 10 m 53.5 s	+76° 07' 35"
4205	Dra	Gxy	1.7'×0.6'	13	12 h 14 m 55.3 s	+63° 46' 56"
4210	Dra	Gxy	2.0'×1.5'	12.7	12 h 15 m 15.9 s	+65° 59' 07"
4221	Dra	Gxy	1.9'×1.3'	12.4	12 h 15 m 59.9 s	+66° 13' 50"
4236	Dra	Gxy	21.9'×7.2'	9.9	12 h 16 m 42.1 s	+69° 27' 45"
4238	Dra	Gxy	1.8'×0.5'	13.7	12 h 16 m 55.8 s	+63° 24' 36"
4250	Dra	Gxy	2.7'×2.1'	11.9	12 h 17 m 26.2 s	+70° 48' 08"
4256	Dra	Gxy	4.6'×1.0'	12.1	12 h 18 m 42.9 s	+65° 53' 54"
4291	Dra	Gxy	1.9'×1.6'	11.4	12 h 20 m 17.7 s	+75° 22' 14"
4319	Dra	Gxy	3.0'×2.3'	12.1	12 h 21 m 43.8 s	+75° 19' 19"
4331	Dra	Gxy	2.2'×0.6'	14.2	12 h 22 m 35.5 s	+76° 10' 22"
4332	Dra	Gxy	2.1'×1.5'	12.4	12 h 22 m 46.7 s	+65° 50' 37"
4345	Dra	Gxy	3.1'×2.5'	...	12 h 21 m 43.8 s	+75° 19' 19"
4363	Dra	Gxy	1.5'×1.5'	13.7	12 h 23 m 28.4 s	+74° 57' 08"
4386	Dra	Gxy	2.5'×1.3'	11.6	12 h 24 m 28.3 s	+75° 31' 43"
4391	Dra	Gxy	1.1'×1.1'	12.8	12 h 25 m 18.8 s	+64° 55' 59"
4441	Dra	Gxy	3.2'×2.5'	12.2	12 h 27 m 20.3 s	+64° 48' 05"
4481	Dra	Gxy	0.7'×0.3'	14.2	12 h 29 m 48.7 s	+64° 01' 59"
4510	Dra	Gxy	1.5'×0.9'	13.1	12 h 31 m 47.1 s	+64° 14' 01"
4512	Dra	Gxy	2.7'×0.6'	12.4	12 h 32 m 47.7 s	+63° 56' 20"
4513	Dra	Gxy	1.4'×0.9'	13.1	12 h 32 m 01.5 s	+66° 19' 56"
4521	Dra	Gxy	2.5'×0.5'	12.4	12 h 32 m 47.7 s	+63° 56' 20"
4545	Dra	Gxy	2.5'×1.5'	12.5	12 h 34 m 34.2 s	+63° 31' 30"
4572	Dra	Gxy	1.8'×1.0'	14	12 h 35 m 45.8 s	+74° 14' 44"
4589	Dra	Gxy	3.2'×2.6'	10.9	12 h 37 m 25.0 s	+74° 11' 30"
4648	Dra	Gxy	2.2'×1.7'	11.6	12 h 41 m 44.3 s	+74° 25' 16"
4693	Dra	Gxy	2.5'×0.5'	13.5	12 h 47 m 09.1 s	+71° 10' 34"
4749	Dra	Gxy	1.7'×0.4'	13.7	12 h 51 m 12.1 s	+71° 38' 08"
4750	Dra	Gxy	2.0'×1.9'	11.4	12 h 50 m 07.0 s	+72° 52' 27"
4857	Dra	Gxy	1.3'×0.6'	14	12 h 57 m 18.3 s	+70° 12' 12"
4954	Dra	Gxy	0.8'×0.6'	13.2	13 h 02 m 19.9 s	+75° 24' 15"
4972	Dra	Gxy	0.8'×0.6'	13.2	13 h 02 m 19.9 s	+75° 24' 15"
5283	Dra	Gxy	1.1'×1.0'	13.3	13 h 41 m 05.7 s	+67° 40' 20"
5413	Dra	Gxy	1.2'×1.0'	13.3	13 h 57 m 53.4 s	+64° 54' 39"

(continued)

Table 10.2 (continued)

NGC #	Const	Type	Size	V Mag	RA (2000)	Dec (2000)
5667	Dra	Gxy	1.7'×1.1'	12.7	14 h 30 m 22.9 s	+59° 28' 11"
5678	Dra	Gxy	3.3'×1.6'	11.5	14 h 32 m 05.8 s	+57° 55' 16"
5777	Dra	Gxy	3.1'×0.4'	13.5	14 h 51 m 17.9 s	+58° 58' 40"
5779	Dra	Gxy	0.4'×0.4'	15.2	14 h 52 m 09.4 s	+55° 53' 56"
5807	Dra	Gxy	0.5'×0.5'	14.2	14 h 55 m 48.7 s	+63° 54' 12"
5826	Dra	Gxy	1.2'×0.9'	14	15 h 06 m 33.9 s	+55° 28' 44"
5862	Dra	Gxy	0.5'×0.5'	14.9	15 h 06 m 03.2 s	+55° 34' 25"
5866	Dra	Gxy	4.7'×1.9'	10.2	15 h 06 m 29.3 s	+55° 45' 47"
5867	Dra	Gxy	0.1'×0.1'	15.5	15 h 06 m 24.2 s	+55° 43' 53"
5870	Dra	Gxy	1.2'×0.9'	14	15 h 06 m 33.9 s	+55° 28' 44"
5879	Dra	Gxy	4.2'×1.3'	11.3	15 h 09 m 46.9 s	+57° 00' 02"
5881	Dra	Gxy	1.0'×0.7'	13.3	15 h 06 m 20.6 s	+62° 58' 52"
5894	Dra	Gxy	3.0'×0.4'	12.9	15 h 11 m 41.1 s	+59° 48' 32"
5905	Dra	Gxy	4.0'×2.6'	11.9	15 h 15 m 23.3 s	+55° 31' 01"
5906	Dra	GxyCld	...	...	15 h 15 m 52.1 s	+56° 19' 48"
5907	Dra	Gxy	12.6'×1.4'	10.7	15 h 15 m 54.0 s	+56° 19' 45"
5908	Dra	Gxy	3.2'×1.2'	12.4	15 h 16 m 43.3 s	+55° 24' 35"
5949	Dra	Gxy	2.4'×1.2'	12.2	15 h 28 m 00.6 s	+64° 45' 47"
5963	Dra	Gxy	3.3'×2.6'	12.3	15 h 33 m 27.9 s	+56° 33' 33"
5965	Dra	Gxy	5.2'×0.7'	12.6	15 h 34 m 01.9 s	+56° 41' 09"
5969	Dra	Gxy	0.5'×0.3'	14.6	15 h 34 m 51.0 s	+56° 27' 03"
5971	Dra	Gxy	1.6'×0.6'	14	15 h 35 m 36.9 s	+56° 27' 40"
5976	Dra	Gxy	0.9'×0.4'	14.6	15 h 36 m 47.9 s	+59° 23' 52"
5981	Dra	Gxy	2.8'×0.5'	13.6	15 h 37 m 53.4 s	+59° 23' 29"
5982	Dra	Gxy	2.6'×1.9'	11.3	15 h 38 m 39.9 s	+59° 21' 21"
5985	Dra	Gxy	5.5'×3.0'	11.1	15 h 39 m 37.0 s	+59° 19' 54"
5987	Dra	Gxy	4.2'×1.3'	12.4	15 h 39 m 57.1 s	+58° 04' 47"
5989	Dra	Gxy	0.9'×0.9'	13.3	15 h 41 m 32.7 s	+59° 45' 17"
6015	Dra	Gxy	5.4'×2.1'	11.2	15 h 51 m 25.3 s	+62° 18' 35"
6019	Dra	Gxy	0.3'×0.3'	14.7	15 h 52 m 09.3 s	+64° 50' 26"
6024	Dra	Gxy	0.7'×0.5'	14.5	15 h 53 m 07.8 s	+64° 55' 04"
6079	Dra	Gxy	1.4'×1.0'	12.8	16 h 04 m 29.1 s	+69° 40' 05"
6088	Dra	Gxy	0.4'×0.3'	15.2	16 h 10 m 42.5 s	+57° 27' 59"
6090	Dra	Gxy	1.7'×0.7'	13.7	16 h 11 m 40.7 s	+52° 27' 24"
6095	Dra	Gxy	1.8'×1.6'	13.1	16 h 11 m 10.9 s	+61° 16' 04"
6111	Dra	Gxy	1.6'×0.4'	14.2	16 h 14 m 22.4 s	+63° 15' 39"
6123	Dra	Gxy	0.8'×0.3'	14	16 h 17 m 19.6 s	+61° 56' 21"
6125	Dra	Gxy	1.4'×1.4'	12	16 h 19 m 11.5 s	+57° 59' 02"
6127	Dra	Gxy	1.4'×1.4'	12.1	16 h 19 m 11.5 s	+57° 59' 02"
6128	Dra	Gxy	1.4'×1.4'	12	16 h 19 m 11.5 s	+57° 59' 02"
6130	Dra	Gxy	1.0'×0.7'	13.7	16 h 19 m 33.3 s	+57° 36' 55"
6133	Dra	***	...	...	16 h 20 m 17.1 s	+56° 39' 09"
6135	Dra	Gxy	0.9'×0.3'	14.2	16 h 14 m 25.0 s	+64° 58' 59"
6136	Dra	Gxy	0.8'×0.3'	14.8	16 h 20 m 59.3 s	+55° 58' 13"
6140	Dra	Gxy	6.3'×4.6'	11.6	16 h 20 m 57.3 s	+65° 23' 24"

(continued)

Table 10.2 (continued)

NGC #	Const	Type	Size	V Mag	RA (2000)	Dec (2000)
6143	Dra	Gxy	1.0'×0.9'	13.4	16 h 21 m 42.3 s	+55° 05' 08"
6157	Dra	Gxy	0.3'×0.3'	14.5	16 h 25 m 48.4 s	+55° 21' 37"
6170	Dra	Gxy	0.7'×0.6'	13.8	16 h 27 m 36.5 s	+59° 33' 44"
6176	Dra	Gxy	0.7'×0.6'	13.8	16 h 27 m 36.5 s	+59° 33' 44"
6182	Dra	Gxy	1.9'×0.6'	13.7	16 h 29 m 33.9 s	+55° 31' 04"
6187	Dra	Gxy	0.4'×0.3'	14.6	16 h 31 m 36.7 s	+57° 42' 23"
6189	Dra	Gxy	1.9'×0.9'	12.9	16 h 31 m 40.9 s	+59° 37' 34"
6190	Dra	Gxy	1.4'×1.3'	12.8	16 h 32 m 06.6 s	+58° 26' 19"
6191	Dra	Gxy	1.8'×0.8'	12.6	16 h 31 m 40.9 s	+59° 37' 34"
6198	Dra	Gxy	1.0'×0.7'	13.7	16 h 35 m 30.7 s	+57° 29' 12"
6202	Dra	Gxy	0.8'×0.4'	13.1	16 h 43 m 23.1 s	+61° 59' 01"
6206	Dra	Gxy	0.7'×0.7'	13.7	16 h 40 m 07.8 s	+58° 37' 02"
6211	Dra	Gxy	1.7'×1.3'	12.7	16 h 41 m 27.7 s	+57° 47' 00"
6213	Dra	Gxy	0.6'×0.3'	14.8	16 h 41 m 37.3 s	+57° 48' 54"
6214	Dra	Gxy	1.1'×0.9'	13.5	16 h 39 m 32.0 s	+66° 02' 22"
6223	Dra	Gxy	3.5'×2.6'	12.3	16 h 43 m 04.4 s	+61° 34' 44"
6226	Dra	Gxy	0.8'×0.4'	13.1	16 h 43 m 23.1 s	+61° 59' 01"
6232	Dra	Gxy	1.6'×1.6'	12.7	16 h 43 m 20.1 s	+70° 37' 57"
6236	Dra	Gxy	3.0'×1.8'	12.1	16 h 44 m 34.3 s	+70° 46' 51"
6237	Dra	NF	...	...	16 h 44 m 07.4 s	+70° 38' 05"
6238	Dra	Gxy	0.5'×0.3'	13.9	16 h 47 m 16.6 s	+62° 08' 49"
6244	Dra	Gxy	1.5'×0.3'	13.7	16 h 48 m 03.7 s	+62° 12' 00"
6245	Dra	NF	...	...	16 h 45 m 22.4 s	+70° 48' 16"
6246	Dra	Gxy	1.5'×0.6'	13.6	16 h 49 m 53.1 s	+55° 32' 34"
6247	Dra	Gxy	1.1'×0.3'	12.7	16 h 48 m 20.1 s	+62° 58' 36"
6248	Dra	Gxy	3.2'×1.2'	13.2	16 h 46 m 22.7 s	+70° 21' 18"
6258	Dra	Gxy	0.9'×0.7'	13.5	16 h 52 m 29.8 s	+60° 30' 52"
6260	Dra	Gxy	0.8'×0.8'	14.1	16 h 51 m 50.1 s	+63° 42' 53"
6262	Dra	Gxy	0.6'×0.4'	13.7	16 h 58 m 42.8 s	+57° 05' 54"
6275	Dra	Gxy	...'×...'	14.3	16 h 55 m 33.2 s	+63° 14' 32"
6285	Dra	Gxy	0.9'×0.5'	13.6	16 h 58 m 24.1 s	+58° 57' 22"
6286	Dra	Gxy	1.3'×1.2'	13.5	16 h 58 m 31.9 s	+58° 56' 15"
6288	Dra	Gxy	0.8'×0.4'	14.5	16 h 57 m 24.8 s	+68° 27' 24"
6289	Dra	Gxy	0.8'×0.6'	14.5	16 h 57 m 44.9 s	+68° 30' 50"
6290	Dra	Gxy	1.2'×1.0'	13.5	17 h 00 m 56.4 s	+58° 58' 14"
6291	Dra	Gxy	0.5'×0.4'	14.1	17 h 00 m 55.7 s	+58° 56' 16"
6292	Dra	Gxy	1.8'×0.8'	13.7	17 h 03 m 03.3 s	+61° 02' 37"
6295	Dra	Gxy	0.9'×0.4'	14.9	17 h 03 m 15.3 s	+60° 20' 17"
6297	Dra	Gxy	0.7'×0.5'	13.7	17 h 03 m 36.3 s	+62° 01' 33"
6298	Dra	Gxy	0.7'×0.5'	15.7	17 h 03 m 36.3 s	+62° 01' 33"
6299	Dra	Gxy	0.6'×0.6'	14.1	17 h 05 m 04.4 s	+62° 27' 28"
6303	Dra	Gxy	1.3'×0.8'	13.8	17 h 05 m 02.7 s	+68° 49' 38"
6306	Dra	Gxy	1.0'×0.3'	13.9	17 h 07 m 36.8 s	+60° 43' 43"
6307	Dra	Gxy	1.3'×1.0'	13.2	17 h 07 m 40.4 s	+60° 45' 02"
6310	Dra	Gxy	2.0'×0.4'	13.3	17 h 07 m 57.3 s	+60° 59' 24"

(continued)

Table 10.2 (continued)

NGC #	Const	Type	Size	V Mag	RA (2000)	Dec (2000)
6317	Dra	Gxy	1.2'×0.4'	15.1	17 h 08 m 59.5 s	+62° 53' 54"
6319	Dra	Gxy	0.8'×0.8'	13.5	17 h 09 m 43.9 s	+62° 58' 22"
6338	Dra	Gxy	1.5'×1.0'	13.1	17 h 15 m 22.6 s	+57° 24' 40"
6340	Dra	Gxy	3.2'×3.0'	11.2	17 h 10 m 25.6 s	+72° 18' 22"
6345	Dra	Gxy	0.8'×0.2'	14.4	17 h 15 m 24.3 s	+57° 21' 00"
6346	Dra	Gxy	0.5'×0.4'	14.6	17 h 15 m 24.5 s	+57° 19' 20"
6358	Dra	Gxy	1.0'×0.5'	14.3	17 h 18 m 53.1 s	+52° 36' 55"
6359	Dra	Gxy	1.2'×0.9'	12.7	17 h 17 m 52.8 s	+61° 46' 50"
6361	Dra	Gxy	2.2'×0.6'	13.3	17 h 18 m 40.9 s	+60° 36' 29"
6365	Dra	Gxy	1.1'×0.2'	14.2	17 h 22 m 43.4 s	+62° 10' 23"
6370	Dra	Gxy	1.4'×1.4'	13	17 h 23 m 25.2 s	+56° 58' 30"
6373	Dra	Gxy	1.3'×1.0'	13.8	17 h 24 m 08.1 s	+58° 59' 40"
6376	Dra	Gxy	0.5'×0.3'	14.8	17 h 25 m 19.1 s	+58° 49' 00"
6377	Dra	Gxy	0.7'×0.3'	14.7	17 h 25 m 23.1 s	+58° 49' 21"
6381	Dra	Gxy	1.3'×1.0'	13.2	17 h 27 m 16.8 s	+60° 00' 49"
6382	Dra	Gxy	0.9'×0.7'	14.6	17 h 27 m 55.1 s	+56° 52' 06"
6385	Dra	Gxy	1.3'×1.3'	13.3	17 h 28 m 01.4 s	+57° 31' 19"
6386	Dra	Gxy	0.9'×0.9'	14.2	17 h 28 m 51.7 s	+52° 43' 24"
6387	Dra	Gxy	0.3'×0.2'	15.7	17 h 28 m 24.4 s	+57° 32' 38"
6390	Dra	Gxy	1.6'×0.3'	14	17 h 28 m 28.0 s	+60° 05' 38"
6391	Dra	Gxy	0.4'×0.3'	14.5	17 h 28 m 48.9 s	+58° 51' 02"
6393	Dra	Gxy	0.3'×0.3'	15.7	17 h 30 m 08.3 s	+59° 31' 54"
6394	Dra	Gxy	1.5'×0.4'	14.5	17 h 30 m 21.3 s	+59° 38' 23"
6395	Dra	Gxy	2.4'×0.7'	12.5	17 h 26 m 31.2 s	+71° 05' 46"
6399	Dra	Gxy	1.2'×0.6'	13.9	17 h 31 m 50.3 s	+59° 36' 56"
6409	Dra	Gxy	0.8'×0.6'	13.8	17 h 36 m 35.5 s	+50° 45' 57"
6410	Dra	**	...	...	17 h 35 m 20.5 s	+60° 47' 35"
6411	Dra	Gxy	2.3'×1.8'	12	17 h 35 m 32.7 s	+60° 48' 47"
6412	Dra	Gxy	2.5'×2.2'	12	17 h 29 m 37.5 s	+75° 42' 15"
6414	Dra	Gxy	1.2'×0.7'	15.3	17 h 30 m 36.7 s	+74° 22' 34"
6418	Dra	Gxy	0.5'×0.4'	14.4	17 h 38 m 09.2 s	+58° 42' 53"
6419	Dra	Gxy	1.1'×0.3'	14.8	17 h 36 m 05.8 s	+68° 09' 20"
6420	Dra	Gxy	0.6'×0.3'	14.5	17 h 36 m 16.1 s	+68° 03' 09"
6422	Dra	Gxy	0.6'×0.6'	14.8	17 h 36 m 29.8 s	+68° 03' 32"
6423	Dra	Gxy	0.8'×0.7'	14.8	17 h 36 m 53.0 s	+68° 10' 18"
6424	Dra	Gxy	0.9'×0.6'	14.1	17 h 36 m 12.0 s	+69° 59' 20"
6434	Dra	Gxy	2.3'×1.0'	12.6	17 h 36 m 48.8 s	+72° 05' 20"
6435	Dra	Gxy	1.1'×0.6'	13.7	17 h 40 m 11.0 s	+62° 38' 31"
6436	Dra	Gxy	1.5'×0.9'	14.2	17 h 41 m 13.1 s	+60° 27' 00"
6448	Dra	NF	...	...	17 h 44 m 20.5 s	+53° 32' 25"
6449	Dra	Gxy	1.1'×0.8'	14	17 h 43 m 46.3 s	+56° 48' 13"
6454	Dra	Gxy	0.9'×0.7'	13.7	17 h 44 m 56.5 s	+55° 42' 18"
6456	Dra	Gxy	0.8'×0.6'	14.7	17 h 42 m 31.7 s	+67° 35' 33"
6457	Dra	Gxy	1.2'×0.9'	14.3	17 h 42 m 52.7 s	+66° 28' 33"
6459	Dra	Gxy	0.6'×0.3'	14.4	17 h 45 m 46.9 s	+55° 46' 37"

(continued)

Table 10.2 (continued)

NGC #	Const	Type	Size	V Mag	RA (2000)	Dec (2000)
6461	Dra	Gxy	0.8'×0.4'	15.1	17 h 39 m 56.5 s	+74° 02' 04"
6462	Dra	Gxy	0.4'×0.4'	14	17 h 44 m 48.7 s	+61° 54' 38"
6463	Dra	Gxy	0.3'×0.3'	14.3	17 h 43 m 34.0 s	+67° 36' 11"
6464	Dra	Gxy	0.6'×0.6'	14.8	17 h 45 m 48.0 s	+60° 53' 47"
6466	Dra	Gxy	0.5'×0.3'	14.2	17 h 48 m 08.1 s	+51° 23' 58"
6470	Dra	Gxy	1.4'×0.2'	14.4	17 h 44 m 14.9 s	+67° 37' 09"
6471	Dra	Gxy	1.4'×0.3'	14.7	17 h 44 m 15.0 s	+67° 35' 33"
6472	Dra	Gxy	0.3'×0.2'	15.2	17 h 44 m 03.3 s	+67° 37' 48"
6473	Dra	***	...	...	17 h 46 m 58.1 s	+57° 14' 21"
6474	Dra	Gxy	0.9'×0.5'	14.2	17 h 47 m 05.3 s	+57° 18' 05"
6477	Dra	Gxy	0.4'×0.3'	15.3	17 h 44 m 30.2 s	+67° 36' 37"
6478	Dra	Gxy	1.9'×0.7'	13.5	17 h 48 m 38.3 s	+51° 09' 26"
6479	Dra	Gxy	1.0'×0.9'	13.9	17 h 48 m 21.6 s	+54° 08' 57"
6488	Dra	Gxy	0.7'×0.4'	14.5	17 h 49 m 20.8 s	+62° 13' 22"
6489	Dra	Gxy	0.4'×0.3'	15.3	17 h 50 m 01.3 s	+60° 05' 33"
6491	Dra	Gxy	1.2'×0.5'	13.8	17 h 50 m 00.6 s	+61° 31' 53"
6493	Dra	Gxy	1.2'×1.2'	14.6	17 h 50 m 22.5 s	+61° 33' 33"
6497	Dra	Gxy	1.4'×0.7'	13.7	17 h 51 m 17.9 s	+59° 28' 15"
6498	Dra	Gxy	1.5'×0.7'	13.4	17 h 51 m 17.9 s	+59° 28' 15"
6503	Dra	Gxy	6.2'×2.3'	10.6	17 h 49 m 26.2 s	+70° 08' 42"
6505	Dra	Gxy	1.1'×1.0'	14.6	17 h 51 m 07.5 s	+65° 31' 50"
6508	Dra	Gxy	1.3'×1.3'	12.9	17 h 49 m 46.3 s	+72° 01' 15"
6510	Dra	Gxy	1.0'×0.6'	13.7	17 h 54 m 39.3 s	+60° 49' 04"
6511	Dra	Gxy	1.0'×0.6'	13.7	17 h 54 m 39.3 s	+60° 49' 04"
6512	Dra	Gxy	0.5'×0.3'	13.9	17 h 54 m 50.2 s	+62° 38' 43"
6515	Dra	Gxy	1.6'×1.0'	13.1	17 h 57 m 25.1 s	+50° 43' 40"
6516	Dra	Gxy	0.5'×0.2'	15	17 h 55 m 16.4 s	+62° 40' 09"
6521	Dra	Gxy	1.6'×1.3'	13.1	17 h 55 m 48.1 s	+62° 36' 43"
6532	Dra	Gxy	2.1'×0.9'	14.1	17 h 59 m 13.8 s	+56° 13' 54"
6534	Dra	NF	...	...	17 h 56 m 08.7 s	+64° 16' 54"
6536	Dra	Gxy	1.2'×1.1'	13.6	17 h 57 m 16.3 s	+64° 56' 17"
6538	Dra	Gxy	1.1'×0.6'	13.4	17 h 54 m 17.1 s	+73° 25' 27"
6542	Dra	Gxy	1.3'×0.4'	13.5	17 h 59 m 38.6 s	+61° 21' 32"
6543	Dra	PN	23"×17"	8.1	17 h 58 m 33.3 s	+66° 37' 59"
6552	Dra	Gxy	1.0'×0.7'	13.8	18 h 00 m 07.2 s	+66° 36' 55"
6562	Dra	Gxy	0.6'×0.6'	13.9	18 h 05 m 00.8 s	+56° 15' 45"
6566	Dra	Gxy	0.7'×0.6'	14.7	18 h 07 m 00.5 s	+52° 15' 37"
6592	Dra	Gxy	0.5'×0.4'	14.6	18 h 09 m 50.7 s	+61° 25' 19"
6594	Dra	Gxy	1.0'×0.7'	14.5	18 h 10 m 05.5 s	+61° 08' 00"
6597	Dra	Gxy	0.8'×0.4'	14.8	18 h 11 m 13.4 s	+61° 10' 50"
6598	Dra	Gxy	2.0'×1.4'	13.1	18 h 08 m 55.6 s	+69° 04' 05"
6601	Dra	Gxy	0.5'×0.3'	14.7	18 h 11 m 44.2 s	+61° 27' 11"
6607	Dra	Gxy	0.6'×0.4'	15.3	18 h 12 m 14.8 s	+61° 19' 59"
6608	Dra	Gxy	1.0'×0.1'	14.9	18 h 12 m 28.9 s	+61° 17' 53"
6609	Dra	Gxy	0.7'×0.7'	14.5	18 h 12 m 33.5 s	+61° 19' 54"

(continued)

Table 10.2 (continued)

NGC #	Const	Type	Size	V Mag	RA (2000)	Dec (2000)
6617	Dra	Gxy	1.4'×1.1'	14.7	18 h 14 m 02.6 s	+61° 19' 11"
6621	Dra	Gxy	0.9'×0.4'	13.2	18 h 12 m 59.5 s	+68° 21' 15"
6622	Dra	Gxy	2.0'×0.9'	15	18 h 12 m 55.2 s	+68° 21' 48"
6636	Dra	Gxy	2.2'×0.4'	13.5	18 h 22 m 02.6 s	+66° 36' 58"
6643	Dra	Gxy	3.8'×1.9'	11.5	18 h 19 m 46.6 s	+74° 34' 08"
6648	Dra	**	…	…	18 h 25 m 37.7 s	+64° 58' 34"
6650	Dra	Gxy	0.4'×0.4'	13.9	18 h 25 m 28.1 s	+68° 00' 22"
6651	Dra	Gxy	1.6'×0.7'	13.3	18 h 24 m 19.6 s	+71° 36' 08"
6654	Dra	Gxy	2.8'×2.3'	11.8	18 h 24 m 07.5 s	+73° 10' 59"
6667	Dra	Gxy	2.3'×1.1'	12.9	18 h 30 m 40.0 s	+67° 59' 13"
6668	Dra	Gxy	3.2'×2.3'	13	18 h 30 m 40.0 s	+67° 59' 13"
6670	Dra	Gxy	1.0'×0.4'	14.3	18 h 33 m 37.4 s	+59° 53' 22"
6676	Dra	Gxy	1.6'×0.3'	14.6	18 h 33 m 09.9 s	+66° 57' 33"
6677	Dra	Gxy	0.9'×0.4'	13.6	18 h 33 m 35.9 s	+67° 06' 37"
6678	Dra	Gxy	3.2'×2.3'	13	18 h 30 m 40.0 s	+67° 59' 13"
6679	Dra	Gxy	0.3'×0.2'	15.1	18 h 33 m 30.3 s	+67° 08' 13"
6687	Dra	Gxy	1.8'×1.6'	14.1	18 h 37 m 22.0 s	+59° 38' 33"
6689	Dra	Gxy	4.2'×1.5'	12.3	18 h 34 m 49.9 s	+70° 31' 27"
6690	Dra	Gxy	3.8'×1.3'	12.1	18 h 34 m 49.9 s	+70° 31' 27"
6691	Dra	Gxy	1.6'×1.5'	13.2	18 h 39 m 12.3 s	+55° 38' 30"
6696	Dra	Gxy	0.8'×0.2'	15.2	18 h 40 m 06.1 s	+59° 20' 03"
6701	Dra	Gxy	1.5'×1.3'	12.3	18 h 43 m 12.3 s	+60° 39' 12"
6711	Dra	Gxy	1.3'×1.3'	13.3	18 h 49 m 00.9 s	+47° 39' 27"
6714	Dra	NF	…	…	18 h 45 m 49.7 s	+66° 43' 31"
6732	Dra	Gxy	1.2'×0.7'	13.5	18 h 56 m 24.0 s	+52° 22' 39"
6742	Dra	PN	31"×30"	13.4	18 h 59 m 19.9 s	+48° 27' 56"
6747	Dra	Gxy	0.5'×0.4'	14.1	18 h 55 m 21.7 s	+72° 46' 17"
6750	Dra	Gxy	1.0'×0.6'	13.1	19 h 00 m 36.0 s	+59° 10' 00"
6757	Dra	Gxy	1.4'×1.0'	13.1	19 h 05 m 06.1 s	+55° 43' 03"
6759	Dra	Gxy	0.9'×0.6'	14.3	19 h 06 m 57.2 s	+50° 20' 51"
6762	Dra	Gxy	1.4'×0.4'	13.6	19 h 05 m 37.1 s	+63° 56' 01"
6763	Dra	Gxy	1.6'×0.4'	13.4	19 h 05 m 37.1 s	+63° 56' 01"
6786	Dra	Gxy	1.1'×0.9'	13	19 h 10 m 53.6 s	+73° 24' 35"
6787	Dra	Gxy	1.2'×1.1'	14.1	19 h 16 m 10.5 s	+60° 25' 03"
6789	Dra	Gxy	1.3'×1.0'	13.4	19 h 16 m 41.9 s	+63° 58' 17"
6796	Dra	Gxy	1.9'×0.4'	13.1	19 h 21 m 30.8 s	+61° 08' 43"
6817	Dra	Gxy	0.6'×0.5'	15	19 h 37 m 23.4 s	+62° 23' 01"
6825	Dra	Gxy	0.5'×0.3'	14.3	19 h 41 m 54.7 s	+64° 04' 22"
6829	Dra	Gxy	1.6'×0.4'	14.3	19 h 47 m 07.5 s	+59° 54' 24"
6831	Dra	Gxy	1.6'×1.6'	13.4	19 h 47 m 57.2 s	+59° 53' 32"
6832	Dra	OC	15'	…	19 h 48 m 15.2 s	+59° 25' 16"
6869	Dra	Gxy	1.6'×1.3'	12	20 h 00 m 42.4 s	+66° 13' 39"
6911	Dra	Gxy	2.2'×1.2'	14.5	20 h 19 m 38.6 s	+66° 43' 43"
7015	Equ	Gxy	1.9'×1.6'	12.4	21 h 05 m 37.3 s	+11° 24' 50"
7040	Equ	Gxy	1.1'×0.8'	14	21 h 13 m 16.6 s	+08° 51' 54"

(continued)

Table 10.2 (continued)

NGC #	Const	Type	Size	V Mag	RA (2000)	Dec (2000)
7045	Equ	**	...	...	21 h 14 m 50.2 s	+04° 30' 25"
7046	Equ	Gxy	1.9'×1.3'	13.3	21 h 14 m 56.0 s	+02° 50' 05"
685	Eri	Gxy	3.7'×3.3'	11.5	01 h 47 m 43.1 s	−52° 45' 41"
745	Eri	Gxy	1.3'×0.8'	13.1	01 h 54 m 09.3 s	−56° 41' 24"
754	Eri	Gxy	0.5'×0.5'	14.3	01 h 54 m 21.1 s	−56° 45' 42"
782	Eri	Gxy	2.3'×2.0'	12	01 h 57 m 37.7 s	−57° 47' 25"
795	Eri	Gxy	1.2'×0.7'	13.3	01 h 59 m 49.7 s	−55° 49' 30"
852	Eri	Gxy	1.3'×1.0'	13.7	02 h 08 m 55.5 s	−56° 44' 14"
939	Eri	Gxy	1.2'×1.0'	13.1	02 h 26 m 21.5 s	−44° 26' 49"
954	Eri	Gxy	1.6'×0.8'	13.3	02 h 28 m 51.5 s	−41° 24' 11"
979	Eri	Gxy	1.2'×1.0'	12.9	02 h 31 m 38.7 s	−44° 31' 29"
1081	Eri	Gxy	1.5'×0.6'	13.3	02 h 45 m 05.4 s	−15° 35' 16"
1082	Eri	Gxy	0.9'×0.7'	14	02 h 45 m 41.2 s	−08° 10' 50"
1083	Eri	Gxy	1.6'×0.3'	14.4	02 h 45 m 40.7 s	−15° 21' 28"
1084	Eri	Gxy	3.2'×1.8'	11.1	02 h 46 m 00.0 s	−07° 34' 37"
1089	Eri	Gxy	0.9'×0.8'	13.6	02 h 46 m 10.1 s	−15° 04' 23"
1091	Eri	Gxy	0.9'×0.6'	14.1	02 h 45 m 22.7 s	−17° 32' 01"
1092	Eri	Gxy	0.9'×0.8'	13.5	02 h 45 m 29.6 s	−17° 32' 34"
1098	Eri	Gxy	1.8'×1.3'	12.7	02 h 44 m 53.6 s	−17° 39' 33"
1099	Eri	Gxy	1.8'×0.6'	13.4	02 h 45 m 18.0 s	−17° 42' 32"
1100	Eri	Gxy	1.7'×0.7'	13.3	02 h 45 m 36.2 s	−17° 41' 21"
1102	Eri	Gxy	0.8'×0.5'	14.9	02 h 47 m 12.7 s	−22° 12' 34"
1103	Eri	Gxy	2.1'×0.5'	12.9	02 h 48 m 06.0 s	−13° 57' 36"
1105	Eri	Gxy	0.7'×0.7'	14.3	02 h 43 m 41.9 s	−15° 42' 23"
1108	Eri	Gxy	0.8'×0.4'	13.9	02 h 48 m 38.5 s	−07° 57' 03"
1110	Eri	Gxy	2.9'×0.5'	13.9	02 h 49 m 09.5 s	−07° 50' 19"
1114	Eri	Gxy	1.7'×0.8'	12.9	02 h 49 m 07.1 s	−16° 59' 37"
1118	Eri	Gxy	2.1'×0.7'	13.2	02 h 49 m 58.7 s	−12° 09' 52"
1119	Eri	Gxy	0.5'×0.4'	13.9	02 h 48 m 17.1 s	−17° 59' 15"
1120	Eri	Gxy	1.3'×0.8'	13.6	02 h 49 m 04.1 s	−14° 28' 15"
1121	Eri	Gxy	0.9'×0.4'	13	02 h 50 m 39.1 s	−01° 44' 03"
1125	Eri	Gxy	1.8'×0.9'	12.6	02 h 51 m 40.0 s	−16° 39' 07"
1126	Eri	Gxy	0.7'×0.2'	14.8	02 h 52 m 18.6 s	−01° 17' 46"
1132	Eri	Gxy	2.5'×1.3'	12.5	02 h 52 m 51.8 s	−01° 16' 30"
1133	Eri	Gxy	0.9'×0.6'	14	02 h 52 m 42.2 s	−08° 48' 16"
1139	Eri	Gxy	1.2'×0.8'	15	02 h 52 m 46.7 s	−14° 31' 45"
1140	Eri	Gxy	1.7'×0.9'	12.6	02 h 54 m 33.5 s	−10° 01' 44"
1145	Eri	Gxy	3.2'×0.5'	13	02 h 54 m 33.5 s	−18° 38' 08"
1147	Eri	NF	...	...	02 h 55 m 09.3 s	−09° 07' 11"
1148	Eri	Gxy	1.4'×0.7'	14.9	02 h 57 m 04.4 s	−07° 41' 09"
1150	Eri	Gxy	0.7'×0.4'	14.1	02 h 57 m 01.4 s	−15° 02' 55"
1151	Eri	Gxy	0.4'×0.3'	15	02 h 57 m 04.6 s	−15° 00' 47"
1152	Eri	Gxy	0.7'×0.4'	14.5	02 h 57 m 33.7 s	−07° 45' 31"
1154	Eri	Gxy	1.0'×0.8'	13.6	02 h 58 m 07.7 s	−10° 21' 49"
1155	Eri	Gxy	0.1'×0.8'	13.4	02 h 58 m 13.1 s	−10° 21' 03"

(continued)

Table 10.2 (continued)

NGC #	Const	Type	Size	V Mag	RA (2000)	Dec (2000)
1157	Eri	Gxy	0.5' × 0.1'	14.8	02 h 58 m 06.5 s	−15° 07' 08"
1158	Eri	Gxy	0.9' × 0.6'	14.8	02 h 57 m 11.5 s	−14° 23' 44"
1162	Eri	Gxy	1.4' × 1.4'	12.5	02 h 58 m 55.9 s	−12° 23' 56"
1163	Eri	Gxy	2.2' × 0.3'	13.9	03 h 00 m 21.9 s	−17° 09' 11"
1172	Eri	Gxy	2.3' × 1.8'	11.7	03 h 01 m 35.9 s	−14° 50' 13"
1179	Eri	Gxy	4.9' × 3.8'	12	03 h 02 m 38.5 s	−18° 53' 55"
1180	Eri	Gxy	0.6' × 0.5'	14.9	03 h 01 m 51.1 s	−15° 01' 47"
1181	Eri	Gxy	0.8' × 0.2'	15.4	03 h 01 m 42.8 s	−15° 03' 10"
1182	Eri	Gxy	0.9' × 0.4'	14.7	03 h 03 m 28.5 s	−09° 40' 12"
1185	Eri	Gxy	1.2' × 0.4'	14.8	03 h 02 m 59.2 s	−09° 08' 01"
1187	Eri	Gxy	5.5' × 4.1'	10.7	03 h 02 m 37.7 s	−22° 52' 03"
1188	Eri	Gxy	1.0' × 0.5'	13.8	03 h 03 m 43.4 s	−15° 29' 05"
1189	Eri	Gxy	1.7' × 1.5'	13.8	03 h 03 m 24.3 s	−15° 37' 24"
1190	Eri	Gxy	0.9' × 0.3'	14.2	03 h 03 m 26.2 s	−15° 39' 43"
1191	Eri	Gxy	0.6' × 0.5'	14.3	03 h 03 m 30.9 s	−15° 41' 07"
1192	Eri	Gxy	0.7' × 0.3'	14.8	03 h 03 m 34.7 s	−15° 40' 44"
1195	Eri	Gxy	0.4' × 0.3'	14.5	03 h 03 m 32.8 s	−12° 02' 23"
1196	Eri	Gxy	1.4' × 1.4'	13.2	03 h 03 m 35.1 s	−12° 04' 34"
1199	Eri	Gxy	2.4' × 1.9'	11.5	03 h 03 m 38.5 s	−15° 36' 49"
1200	Eri	Gxy	3.0' × 1.5'	12.5	03 h 03 m 54.6 s	−11° 59' 30"
1202	Eri	Gxy	0.6' × 0.5'	14.7	03 h 05 m 02.4 s	−06° 29' 32"
1203	Eri	Gxy	0.4' × 0.4'	13.8	03 h 05 m 14.1 s	−14° 22' 50"
1204	Eri	Gxy	1.1' × 0.3'	13.3	03 h 04 m 40.0 s	−12° 20' 29"
1205	Eri	Gxy	0.9' × 0.4'	14.7	03 h 03 m 28.5 s	−09° 40' 12"
1206	Eri	Gxy	0.5' × 0.5'	14.5	03 h 06 m 09.7 s	−08° 50' 00"
1208	Eri	Gxy	1.8' × 0.9'	12.5	03 h 06 m 11.9 s	−09° 32' 28"
1209	Eri	Gxy	2.4' × 1.1'	11.3	03 h 06 m 03.1 s	−15° 36' 42"
1214	Eri	Gxy	1.3' × 0.3'	14.1	03 h 06 m 56.1 s	−09° 32' 37"
1215	Eri	Gxy	1.5' × 1.1'	14.1	03 h 07 m 09.5 s	−09° 35' 36"
1216	Eri	Gxy	0.8' × 0.2'	15.1	03 h 07 m 18.4 s	−09° 36' 48"
1221	Eri	Gxy	1.2' × 0.4'	14.1	03 h 08 m 15.5 s	−04° 15' 35"
1222	Eri	Gxy	1.1' × 0.9'	12.5	03 h 08 m 56.9 s	−02° 57' 17"
1223	Eri	Gxy	0.6' × 0.3'	14	03 h 08 m 19.9 s	−04° 08' 19"
1225	Eri	Gxy	1.0' × 0.6'	14.6	03 h 08 m 47.3 s	−04° 06' 06"
1228	Eri	Gxy	1.5' × 0.9'	13.3	03 h 08 m 11.8 s	−22° 55' 23"
1229	Eri	Gxy	1.4' × 0.9'	14.2	03 h 08 m 11.0 s	−22° 57' 37"
1230	Eri	Gxy	0.6' × 0.4'	14.5	03 h 08 m 16.4 s	−22° 59' 03"
1231	Eri	Gxy	0.8' × 0.8'	14.2	03 h 06 m 29.3 s	−15° 34' 11"
1232	Eri	Gxy	7.4' × 6.5'	10.1	03 h 09 m 45.1 s	−20° 34' 45"
1234	Eri	Gxy	1.4' × 1.0'	14.2	03 h 09 m 39.2 s	−07° 50' 47"
1237	Eri	**	...	...	03 h 10 m 08.9 s	−08° 41' 32"
1238	Eri	Gxy	1.6' × 1.2'	13.3	03 h 10 m 52.7 s	−10° 44' 53"
1239	Eri	Gxy	1.1' × 0.6'	13.6	03 h 10 m 53.7 s	−02° 33' 11"
1241	Eri	Gxy	2.8' × 1.7'	12	03 h 11 m 14.7 s	−08° 55' 20"
1242	Eri	Gxy	1.2' × 0.7'	13.7	03 h 11 m 19.2 s	−08° 54' 07"

(continued)

Table 10.2 (continued)

NGC #	Const	Type	Size	V Mag	RA (2000)	Dec (2000)
1243	Eri	**	...	...	03 h 11 m 25.5 s	−08° 56' 43"
1247	Eri	Gxy	3.4'×0.5'	12.7	03 h 12 m 14.1 s	−10° 28' 50"
1248	Eri	Gxy	1.1'×1.0'	12.5	03 h 12 m 48.5 s	−05° 13' 29"
1253	Eri	Gxy	5.2'×2.3'	11.7	03 h 14 m 09.2 s	−02° 49' 21"
1256	Eri	Gxy	1.1'×0.4'	13.5	03 h 13 m 58.1 s	−21° 59' 12"
1258	Eri	Gxy	1.3'×0.9'	13.5	03 h 14 m 05.5 s	−21° 46' 30"
1262	Eri	Gxy	0.8'×0.7'	14.3	03 h 15 m 33.6 s	−15° 52' 47"
1263	Eri	Gxy	0.6'×0.5'	14.1	03 h 15 m 39.5 s	−15° 05' 55"
1266	Eri	Gxy	1.6'×1.0	13	03 h 16 m 00.7 s	−02° 25' 38"
1269	Eri	Gxy	11.1'×10.1'	8.5	03 h 17 m 18.3 s	−41° 06' 26"
1284	Eri	Gxy	1.7'×1.5'	13.8	03 h 17 m 45.5 s	−10° 17' 20"
1285	Eri	Gxy	1.5'×1.1'	12.8	03 h 17 m 53.4 s	−07° 17' 55"
1286	Eri	Gxy	0.7'×0.5'	13.8	03 h 17 m 48.5 s	−07° 37' 01"
1287	Eri	Gxy	0.8'×0.7'	13.9	03 h 18 m 33.3 s	−02° 43' 52"
1289	Eri	Gxy	1.8'×1.1'	12.6	03 h 18 m 49.7 s	−01° 58' 25"
1290	Eri	Gxy	0.8'×0.5'	14.8	03 h 19 m 25.2 s	−13° 59' 23"
1291	Eri	Gxy	9.8'×8.1'	8.6	03 h 17 m 18.3 s	−41° 06' 26"
1295	Eri	Gxy	0.9'×0.5'	14.3	03 h 20 m 03.2 s	−13° 59' 54"
1296	Eri	Gxy	1.1'×0.9'	14.2	03 h 18 m 49.7 s	−13° 03' 46"
1297	Eri	Gxy	2.2'×1.9'	11.9	03 h 19 m 14.3 s	−19° 06' 01"
1298	Eri	Gxy	1.6'×1.3'	12.8	03 h 20 m 13.0 s	−02° 06' 51"
1299	Eri	Gxy	1.1'×0.6'	13	03 h 20 m 09.7 s	−06° 15' 44"
1300	Eri	Gxy	6.2'×4.1'	10.5	03 h 19 m 40.9 s	−19° 24' 40"
1301	Eri	Gxy	2.2'×0.4'	13.5	03 h 20 m 35.3 s	−18° 42' 56"
1303	Eri	Gxy	0.5'×0.4'	13.8	03 h 20 m 40.8 s	−07° 23' 40"
1304	Eri	Gxy	1.3'×0.8'	13.5	03 h 21 m 12.7 s	−04° 35' 03"
1305	Eri	Gxy	1.4'×0.9'	13.4	03 h 21 m 23.1 s	−02° 19' 00"
1307	Eri	Gxy	1.4'×0.8'	13.5	03 h 21 m 12.7 s	−04° 35' 03"
1308	Eri	Gxy	1.2'×0.9'	13.7	03 h 22 m 28.6 s	−02° 45' 26"
1309	Eri	Gxy	2.2'×2.0'	11.3	03 h 22 m 06.5 s	−15° 24' 02"
1314	Eri	Gxy	1.5'×1.4'	14.2	03 h 22 m 41.2 s	−04° 11' 12"
1315	Eri	Gxy	1.6'×1.4'	12.5	03 h 23 m 06.5 s	−21° 22' 30"
1319	Eri	Gxy	1.3'×0.7'	13	03 h 23 m 56.5 s	−21° 31' 39"
1320	Eri	Gxy	1.9'×0.6'	12.5	03 h 24 m 48.7 s	−03° 02' 32"
1321	Eri	Gxy	0.9'×0.5'	13.6	03 h 24 m 48.8 s	−03° 00' 57"
1322	Eri	Gxy	1.0'×0.8'	13.4	03 h 24 m 54.6 s	−02° 55' 09"
1323	Eri	Gxy	0.9'×0.3'	15.1	03 h 24 m 55.9 s	−02° 49' 19"
1324	Eri	Gxy	2.1'×0.8'	12.7	03 h 25 m 01.8 s	−05° 44' 45"
1325	Eri	Gxy	4.7'×1.6'	11.7	03 h 24 m 25.2 s	−21° 32' 41"
1328	Eri	Gxy	1.0'×0.6'	14	03 h 25 m 39.1 s	−04° 07' 30"
1329	Eri	Gxy	1.4'×1.1'	12.7	03 h 26 m 02.6 s	−17° 35' 30"
1331	Eri	Gxy	0.9'×0.9'	13.3	03 h 26 m 28.2 s	−21° 21' 20"
1332	Eri	Gxy	4.7'×1.4'	10.4	03 h 26 m 17.1 s	−21° 20' 04"
1337	Eri	Gxy	5.8'×1.5'	11.6	03 h 28 m 05.9 s	−08° 23' 21"
1338	Eri	Gxy	1.4'×1.3'	12.9	03 h 28 m 54.4 s	−12° 09' 13"

(continued)

Table 10.2 (continued)

NGC #	Const	Type	Size	V Mag	RA (2000)	Dec (2000)
1345	Eri	Gxy	1.5'×1.1'	13.2	03 h 29 m 31.6 s	−17° 46' 44"
1346	Eri	Gxy	1.0'×0.7'	13.6	03 h 30 m 13.1 s	−05° 32' 35"
1347	Eri	Gxy	1.5'×1.3'	13.2	03 h 29 m 41.8 s	−22° 16' 47"
1352	Eri	Gxy	1.0'×0.7'	13.2	03 h 31 m 32.9 s	−19° 16' 42"
1353	Eri	Gxy	3.4'×1.4'	11.7	03 h 32 m 03.1 s	−20° 49' 09"
1354	Eri	Gxy	2.2'×0.8'	12.3	03 h 32 m 29.3 s	−15° 13' 20"
1355	Eri	Gxy	1.4'×0.4'	13.3	03 h 33 m 23.6 s	−04° 59' 54"
1357	Eri	Gxy	2.8'×1.9'	11.8	03 h 33 m 17.2 s	−13° 39' 53"
1358	Eri	Gxy	2.6'×2.0'	11.9	03 h 33 m 39.7 s	−05° 05' 22"
1359	Eri	Gxy	2.4'×1.7'	12.2	03 h 33 m 47.3 s	−19° 29' 23"
1361	Eri	Gxy	1.6'×1.4'	13.9	03 h 34 m 17.7 s	−06° 15' 55"
1362	Eri	Gxy	1.2'×1.1'	12.6	03 h 33 m 52.9 s	−20° 16' 58"
1363	Eri	Gxy	0.9'×0.7'	13.1	03 h 34 m 49.5 s	−09° 50' 34"
1364	Eri	Gxy	0.5'×0.5'	14.7	03 h 34 m 58.9 s	−09° 50' 20"
1368	Eri	Gxy	0.7'×0.4'	14.3	03 h 34 m 58.9 s	−15° 39' 22"
1370	Eri	Gxy	1.5'×1.0'	12.5	03 h 35 m 14.5 s	−20° 22' 26"
1372	Eri	Gxy	0.5'×0.5'	14.3	03 h 36 m 59.7 s	−15° 52' 53"
1376	Eri	Gxy	2.0'×1.7'	12.3	03 h 37 m 05.9 s	−05° 02' 35"
1377	Eri	Gxy	1.8'×0.9'	12.5	03 h 36 m 39.0 s	−20° 54' 06"
1383	Eri	Gxy	1.9'×0.9'	12.5	03 h 37 m 39.1 s	−18° 20' 23"
1386	Eri	Gxy	3.4'×1.3'	11.4	03 h 36 m 47.1 s	−35° 59' 57"
1388	Eri	Gxy	0.7'×0.7'	13.8	03 h 38 m 12.0 s	−15° 53' 58"
1389	Eri	Gxy	2.3'×1.4'	11.6	03 h 37 m 11.7 s	−35° 44' 40"
1390	Eri	Gxy	1.4'×0.5'	13.9	03 h 37 m 52.1 s	−19° 00' 32"
1391	Eri	Gxy	1.1'×0.5'	13.3	03 h 38 m 52.9 s	−18° 21' 15"
1392	Eri	NF	...	...	03 h 37 m 31.6 s	−37° 07' 13"
1393	Eri	Gxy	1.9'×1.3'	12.1	03 h 38 m 38.5 s	−18° 25' 41"
1394	Eri	Gxy	1.3'×0.4'	12.7	03 h 39 m 06.9 s	−18° 17' 33"
1395	Eri	Gxy	5.9'×4.5'	10	03 h 38 m 29.7 s	−23° 01' 40"
1396	Eri	Gxy	1.0'×0.9'	13.8	03 h 38 m 06.5 s	−35° 26' 25"
1397	Eri	Gxy	1.6'×1.3'	13.7	03 h 39 m 47.1 s	−04° 40' 12"
1400	Eri	Gxy	2.3'×2.0'	11.3	03 h 39 m 30.8 s	−18° 41' 18"
1401	Eri	Gxy	2.4'×0.6'	12.4	03 h 39 m 21.7 s	−22° 43' 30"
1402	Eri	Gxy	0.8'×0.6'	13	03 h 39 m 30.6 s	−18° 31' 37"
1403	Eri	Gxy	1.3'×1.0'	12.9	03 h 39 m 10.8 s	−22° 23' 19"
1405	Eri	Gxy	1.5'×0.5'	14.1	03 h 40 m 18.9 s	−15° 31' 49"
1407	Eri	Gxy	4.6'×4.3'	10.1	03 h 40 m 11.8 s	−18° 34' 48"
1413	Eri	Gxy	0.6'×0.5'	14.3	03 h 40 m 11.5 s	−15° 36' 40"
1414	Eri	Gxy	1.4'×0.3'	14	03 h 40 m 56.9 s	−21° 42' 50"
1415	Eri	Gxy	3.5'×1.8'	11.6	03 h 40 m 56.9 s	−22° 33' 53"
1416	Eri	Gxy	1.3'×1.3'	13	03 h 41 m 02.9 s	−22° 43' 09"
1417	Eri	Gxy	2.7'×1.7'	12	03 h 41 m 57.1 s	−04° 42' 21"
1418	Eri	Gxy	1.3'×0.9'	13.6	03 h 42 m 16.2 s	−04° 43' 56"
1419	Eri	Gxy	1.1'×1.1'	12.6	03 h 40 m 42.0 s	−37° 30' 41"
1420	Eri	***	...	...	03 h 42 m 39.8 s	−05° 51' 09"

(continued)

Table 10.2 (continued)

NGC #	Const	Type	Size	V Mag	RA (2000)	Dec (2000)
1421	Eri	Gxy	3.5'×0.9'	11.6	03 h 42 m 29.3 s	−13° 29' 18"
1422	Eri	Gxy	2.2'×0.5'	13.4	03 h 41 m 31.3 s	−21° 40' 53"
1423	Eri	Gxy	0.9'×0.5'	14.1	03 h 42 m 40.1 s	−06° 22' 56"
1424	Eri	Gxy	1.7'×0.6'	13.8	03 h 43 m 13.9 s	−04° 43' 48"
1426	Eri	Gxy	2.6'×1.7'	11.5	03 h 42 m 49.3 s	−22° 06' 29"
1429	Eri	NF	...	...	03 h 44 m 04.1 s	−04° 43' 05"
1430	Eri	*	...	...	03 h 43 m 25.1 s	−18° 13' 30"
1434	Eri	Gxy	0.8'×0.5'	14.3	03 h 46 m 12.8 s	−09° 40' 56"
1436	Eri	Gxy	2.9'×2.1'	11.7	03 h 43 m 37.0 s	−35° 51' 17"
1437	Eri	Gxy	3.0'×2.0'	11.9	03 h 43 m 37.0 s	−35° 51' 17"
1438	Eri	Gxy	2.0'×0.9'	12.5	03 h 45 m 17.1 s	−23° 00' 09"
1439	Eri	Gxy	2.5'×2.3'	11.4	03 h 44 m 49.7 s	−21° 55' 15"
1440	Eri	Gxy	2.1'×1.6'	11.7	03 h 45 m 02.9 s	−18° 15' 58"
1441	Eri	Gxy	1.6'×0.6'	13.1	03 h 45 m 43.0 s	−04° 05' 31"
1442	Eri	Gxy	3.3'×1.9'	11.6	03 h 45 m 02.9 s	−18° 15' 58"
1443	Eri	*	...	...	03 h 45 m 53.0 s	−04° 03' 10"
1445	Eri	Gxy	0.9'×0.5'	14	03 h 44 m 56.1 s	−09° 51' 20"
1446	Eri	*	...	...	03 h 45 m 57.5 s	−04° 06' 44"
1447	Eri	Gxy	1.0'×0.5'	14.7	03 h 45 m 47.1 s	−09° 01' 08"
1449	Eri	Gxy	0.7'×0.5'	13.5	03 h 46 m 03.0 s	−04° 08' 17"
1450	Eri	Gxy	0.8'×0.6'	14.1	03 h 45 m 36.5 s	−09° 14' 04"
1451	Eri	Gxy	0.7'×0.4'	13.4	03 h 46 m 07.1 s	−04° 04' 10"
1452	Eri	Gxy	2.2'×1.5'	12.3	03 h 45 m 22.3 s	−18° 38' 00"
1453	Eri	Gxy	2.4'×1.9'	11.7	03 h 46 m 27.2 s	−03° 58' 09"
1454	Eri	*	...	...	03 h 45 m 59.3 s	−20° 39' 09"
1455	Eri	Gxy	3.8'×2.3'	11.9	03 h 45 m 22.3 s	−18° 38' 00"
1458	Eri	Gxy	3.3'×1.9'	11.6	03 h 45 m 02.9 s	−18° 15' 58"
1460	Eri	Gxy	1.7'×1.4'	12.6	03 h 46 m 13.6 s	−36° 41' 49"
1461	Eri	Gxy	3.0'×0.9'	11.7	03 h 48 m 27.1 s	−16° 23' 34"
1464	Eri	Gxy	0.8'×0.6'	14.6	03 h 51 m 24.5 s	−15° 24' 09"
1467	Eri	Gxy	1.1'×0.7'	14.1	03 h 51 m 52.7 s	−08° 50' 17"
1468	Eri	Gxy	1.2'×0.8'	14.2	03 h 52 m 12.5 s	−06° 20' 56"
1470	Eri	Gxy	1.3'×0.3'	14.2	03 h 52 m 09.7 s	−08° 59' 57"
1471	Eri	Gxy	0.8'×0.6'	14.6	03 h 51 m 24.5 s	−15° 24' 09"
1472	Eri	Gxy	0.7'×0.6'	14.4	03 h 53 m 47.3 s	−08° 34' 07"
1474	Eri	Gxy	1.1'×1.0'	13.9	03 h 54 m 30.3 s	+10° 42' 24"
1475	Eri	Gxy	0.6'×0.5'	...	03 h 53 m 49.7 s	−08° 08' 15"
1477	Eri	Gxy	0.6'×0.5'	14.2	03 h 54 m 02.9 s	−08° 34' 30"
1478	Eri	Gxy	0.6'×0.3'	14.8	03 h 54 m 07.3 s	−08° 33' 19"
1479	Eri	NF	...	...	03 h 54 m 20.4 s	−10° 12' 31"
1480	Eri	NF	...	...	03 h 54 m 32.3 s	−10° 15' 32"
1481	Eri	Gxy	1.0'×0.7'	13.5	03 h 54 m 29.0 s	−20° 25' 38"
1482	Eri	Gxy	2.5'×1.4'	12.3	03 h 54 m 38.9 s	−20° 30' 14"
1484	Eri	Gxy	2.5'×0.6'	13.1	03 h 54 m 17.4 s	−36° 58' 17"
1486	Eri	Gxy	0.9'×0.5'	14.6	03 h 56 m 18.7 s	−21° 49' 17"

(continued)

Table 10.2 (continued)

NGC #	Const	Type	Size	V Mag	RA (2000)	Dec (2000)
1487	Eri	Gxy	2.0'×1.4'	11.9	03 h 55 m 45.3 s	−42° 22' 05"
1489	Eri	Gxy	1.4'×0.6'	14	03 h 57 m 38.1 s	−19° 13' 02"
1492	Eri	Gxy	1.0'×0.7'	13.6	03 h 58 m 13.1 s	−35° 26' 47"
1498	Eri	***	...	...	04 h 00 m 19.3 s	−12° 01' 11"
1504	Eri	Gxy	0.7'×0.5'	14.4	04 h 02 m 29.7 s	−09° 20' 08"
1505	Eri	Gxy	0.9'×0.7'	13.7	04 h 02 m 36.5 s	−09° 19' 21"
1507	Eri	Gxy	3.6'×0.9'	12.4	04 h 04 m 27.1 s	−02° 11' 20"
1509	Eri	Gxy	0.4'×0.3'	13.7	04 h 03 m 55.1 s	−11° 10' 44"
1516	Eri	Gxy	1.5'×1.2'	11.8	04 h 08 m 07.9 s	−08° 49' 51"
1518	Eri	Gxy	3.0'×1.3'	11.7	04 h 06 m 49.0 s	−21° 10' 45"
1519	Eri	Gxy	2.1'×0.5'	13.1	04 h 08 m 07.3 s	−17° 11' 36"
1521	Eri	Gxy	2.8'×1.7'	11.5	04 h 08 m 18.7 s	−21° 03' 07"
1524	Eri	Gxy	0.4'×0.3'	11.8	04 h 08 m 07.5 s	−08° 49' 47"
1525	Eri	Gxy	0.5'×0.5'	13.8	04 h 08 m 08.3 s	−08° 50' 07"
1531	Eri	Gxy	1.3'×0.9'	12.3	04 h 11 m 59.1 s	−32° 51' 05"
1532	Eri	Gxy	11.1'×3.2'	10.6	04 h 12 m 04.2 s	−32° 52' 30"
1535	Eri	PN	48"×42"	9.6	04 h 14 m 15.9 s	−12° 44' 21"
1537	Eri	Gxy	3.9'×2.6'	10.6	04 h 13 m 40.7 s	−31° 38' 46"
1538	Eri	Gxy	0.4'×0.4'	15.1	04 h 14 m 56.2 s	−13° 11' 30"
1540	Eri	Gxy	0.5'×0.3'	14.5	04 h 15 m 10.6 s	−28° 29' 20"
1547	Eri	Gxy	1.3'×0.6'	13.8	04 h 17 m 12.3 s	−17° 51' 29"
1552	Eri	Gxy	1.8'×1.2'	13	04 h 20 m 17.5 s	−00° 41' 36"
1561	Eri	Gxy	0.6'×0.5'	13.9	04 h 23 m 01.1 s	−15° 50' 45"
1562	Eri	Gxy	0.8'×0.4'	14.3	04 h 21 m 47.7 s	−15° 45' 20"
1563	Eri	Gxy	0.7'×0.3'	15	04 h 22 m 53.9 s	−15° 43' 57"
1564	Eri	Gxy	0.7'×0.5'	14.6	04 h 23 m 00.9 s	−15° 44' 21"
1565	Eri	Gxy	0.9'×0.8'	14	04 h 23 m 23.5 s	−15° 44' 40"
1568	Eri	Gxy	1.7'×1.3'	14.6	04 h 24 m 25.4 s	−00° 44' 47"
1575	Eri	Gxy	1.5'×1.3'	12.7	04 h 26 m 20.7 s	−10° 05' 56"
1576	Eri	Gxy	1.1'×0.6'	13.3	04 h 26 m 18.8 s	−03° 37' 19"
1577	Eri	Gxy	1.5'×1.3'	12.7	04 h 26 m 20.7 s	−10° 05' 56"
1580	Eri	Gxy	0.9'×0.8'	13.7	04 h 28 m 18.5 s	−05° 10' 46"
1583	Eri	Gxy	0.9'×0.8'	14.2	04 h 28 m 20.7 s	−17° 35' 43"
1584	Eri	Gxy	0.8'×0.7'	13.9	04 h 28 m 10.2 s	−17° 31' 24"
1586	Eri	Gxy	1.7'×0.9'	13.4	04 h 30 m 38.3 s	−00° 18' 18"
1591	Eri	Gxy	1.2'×0.8'	13	04 h 29 m 30.7 s	−26° 42' 47"
1592	Eri	Gxy	1.4'×0.7'	14.1	04 h 29 m 40.7 s	−27° 24' 31"
1594	Eri	Gxy	1.8'×1.3'	13	04 h 30 m 51.6 s	−05° 47' 54"
1597	Eri	Gxy	0.9'×0.7'	14	04 h 31 m 13.4 s	−11° 17' 26"
1599	Eri	Gxy	0.9'×0.8'	13.7	04 h 31 m 38.6 s	−04° 35' 18"
1600	Eri	Gxy	2.5'×1.7'	11.5	04 h 31 m 39.9 s	−05° 05' 16"
1601	Eri	Gxy	0.6'×0.3'	13.8	04 h 31 m 41.7 s	−05° 03' 37"
1603	Eri	Gxy	0.8'×0.5'	13.8	04 h 31 m 49.9 s	−05° 05' 40"
1604	Eri	Gxy	0.7'×0.5'	13.7	04 h 31 m 58.6 s	−05° 22' 11"
1606	Eri	Gxy	0.4'×0.4'	14.9	04 h 32 m 03.3 s	−05° 01' 57"

(continued)

Table 10.2 (continued)

NGC #	Const	Type	Size	V Mag	RA (2000)	Dec (2000)
1607	Eri	Gxy	1.0'×0.4'	13.3	04 h 32 m 03.1 s	−04° 27' 37"
1609	Eri	Gxy	1.1'×0.7'	13.5	04 h 32 m 45.1 s	−04° 22' 21"
1610	Eri	NF	...	...	04 h 32 m 44.7 s	−04° 34' 55"
1611	Eri	Gxy	1.9'×0.6'	13.4	04 h 33 m 05.9 s	−04° 17' 49"
1612	Eri	Gxy	1.2'×1.0'	13.4	04 h 33 m 13.1 s	−04° 10' 20"
1613	Eri	Gxy	1.0'×0.8'	13.7	04 h 33 m 25.3 s	−04° 15' 55"
1614	Eri	Gxy	1.3'×1.1'	12.9	04 h 34 m 00.1 s	−08° 34' 45"
1618	Eri	Gxy	2.3'×0.8'	12.7	04 h 36 m 06.7 s	−03° 08' 57"
1619	Eri	NF	...	...	04 h 36 m 11.4 s	−04° 49' 57"
1620	Eri	Gxy	2.9'×1.0'	12.8	04 h 36 m 37.3 s	−00° 08' 35"
1621	Eri	Gxy	1.3'×0.8'	13.6	04 h 36 m 25.0 s	−04° 59' 14"
1622	Eri	Gxy	3.6'×0.7'	13.1	04 h 36 m 36.6 s	−03° 11' 20"
1623	Eri	Gxy	0.8'×0.5'	15.6	04 h 35 m 32.3 s	−13° 33' 22"
1625	Eri	Gxy	2.1'×0.5'	12.6	04 h 37 m 06.3 s	−03° 18' 16"
1626	Eri	Gxy	0.5'×0.4'	13.6	04 h 36 m 25.0 s	−04° 59' 14"
1627	Eri	Gxy	1.6'×1.5'	13	04 h 37 m 37.9 s	−04° 53' 20"
1628	Eri	Gxy	1.8'×0.4'	13.4	04 h 37 m 36.3 s	−04° 43' 01"
1630	Eri	Gxy	0.7'×0.5'	14.3	04 h 37 m 15.5 s	−18° 54' 05"
1631	Eri	Gxy	1.4'×0.9'	13.6	04 h 38 m 24.3 s	−20° 39' 01"
1632	Eri	Gxy	1.0'×0.6'	14.4	04 h 39 m 58.5 s	−09° 27' 22"
1635	Eri	Gxy	1.4'×1.3'	12.6	04 h 40 m 07.8 s	−00° 32' 51"
1636	Eri	Gxy	1.2'×0.8'	13	04 h 40 m 40.2 s	−08° 36' 28"
1637	Eri	Gxy	4.0'×3.2'	11.1	04 h 41 m 28.0 s	−02° 51' 29"
1638	Eri	Gxy	2.0'×1.5'	12	04 h 41 m 36.3 s	−01° 48' 33"
1639	Eri	***	...	...	04 h 40 m 53.5 s	−16° 59' 40"
1640	Eri	Gxy	2.6'×2.0'	11.9	04 h 42 m 14.5 s	−20° 26' 07"
1643	Eri	Gxy	1.1'×1.1'	13.3	04 h 43 m 44.0 s	−05° 19' 11"
1645	Eri	Gxy	2.3'×1.0'	13	04 h 44 m 06.4 s	−05° 27' 56"
1646	Eri	Gxy	0.5'×0.4'	13	04 h 44 m 23.4 s	−08° 31' 54"
1648	Eri	Gxy	0.4'×0.3'	14.6	04 h 44 m 34.8 s	−08° 28' 44"
1650	Eri	Gxy	2.3'×1.3'	12.7	04 h 45 m 11.5 s	−15° 52' 12"
1653	Eri	Gxy	1.5'×1.5'	11.8	04 h 45 m 47.3 s	−02° 23' 36"
1654	Eri	Gxy	0.7'×0.7'	13.6	04 h 45 m 48.4 s	−02° 05' 02"
1656	Eri	Gxy	1.5'×1.0'	12.9	04 h 45 m 53.4 s	−05° 08' 13"
1657	Eri	Gxy	1.2'×0.8'	14	04 h 46 m 07.3 s	−02° 04' 37"
1659	Eri	Gxy	1.6'×1.1'	12.5	04 h 46 m 29.8 s	−04° 47' 18"
1665	Eri	Gxy	1.7'×1.1'	12.8	04 h 48 m 17.1 s	−05° 25' 39"
1666	Eri	Gxy	1.4'×1.1'	12.7	04 h 48 m 32.9 s	−06° 34' 11"
1667	Eri	Gxy	1.8'×1.4'	12.6	04 h 48 m 37.2 s	−06° 19' 12"
1677	Eri	Gxy	1.7'×1.3'	12.5	04 h 46 m 29.8 s	−04° 47' 18"
1681	Eri	Gxy	1.3'×1.1'	12.9	04 h 51 m 50.1 s	−05° 48' 13"
1686	Eri	Gxy	1.7'×0.3'	13.7	04 h 52 m 54.5 s	−15° 20' 53"
1689	Eri	Gxy	1.5'×1.3'	12.1	04 h 48 m 37.2 s	−06° 19' 12"
1692	Eri	Gxy	1.3'×1.2'	13.3	04 h 55 m 23.7 s	−20° 34' 16"
1694	Eri	Gxy	0.8'×0.4'	14.3	04 h 55 m 16.7 s	−04° 39' 10"

(continued)

Table 10.2 (continued)

NGC #	Const	Type	Size	V Mag	RA (2000)	Dec (2000)
1699	Eri	Gxy	0.9'×0.5'	13.9	04 h 56 m 59.6 s	−04° 45' 27"
1700	Eri	Gxy	3.3'×2.1'	11.1	04 h 56 m 56.4 s	−04° 51' 53"
1720	Eri	Gxy	1.6'×1.0'	13.1	04 h 59 m 20.7 s	−07° 51' 35"
1721	Eri	Gxy	2.1'×1.0'	12.8	04 h 59 m 17.3 s	−11° 07' 08"
1723	Eri	Gxy	3.2'×2.0'	11.7	04 h 59 m 25.9 s	−10° 58' 51"
1725	Eri	Gxy	1.9'×1.2'	12.8	04 h 59 m 22.8 s	−11° 07' 56"
1726	Eri	Gxy	1.5'×1.1'	12.3	04 h 59 m 41.9 s	−07° 45' 19"
1728	Eri	Gxy	2.0'×0.7'	13.9	04 h 59 m 27.7 s	−11° 07' 21"
1741	Eri	Gxy	1.4'×0.7'	14.4	05 h 01 m 38.7 s	−04° 15' 34"
1752	Eri	Gxy	2.6'×0.8'	12.6	05 h 02 m 09.4 s	−08° 14' 27"
1757	Eri	NF	...	...	05 h 02 m 39.3 s	−04° 43' 23"
1779	Eri	Gxy	2.3'×1.3'	12.1	05 h 05 m 18.1 s	−09° 08' 51"
1797	Eri	Gxy	1.1'×0.8'	13.5	05 h 07 m 44.7 s	−08° 01' 10"
1799	Eri	Gxy	1.1'×0.6'	13.7	05 h 07 m 44.7 s	−07° 58' 09"
686	For	Gxy	1.8'×1.4'	12.5	01 h 48 m 56.3 s	−23° 47' 55"
689	For	Gxy	1.0'×0.6'	13.8	01 h 49 m 51.8 s	−27° 27' 56"
696	For	Gxy	1.7'×0.6'	13.5	01 h 49 m 31.1 s	−34° 54' 20"
698	For	Gxy	0.9'×0.8'	14.1	01 h 49 m 43.5 s	−34° 49' 53"
727	For	Gxy	1.1'×0.6'	14.2	01 h 53 m 49.3 s	−35° 51' 23"
729	For	Gxy	1.6'×1.1'	14.1	01 h 53 m 49.3 s	−35° 51' 23"
749	For	Gxy	1.9'×1.4'	12.7	01 h 55 m 41.1 s	−29° 55' 20"
775	For	Gxy	1.7'×1.2'	12.9	01 h 58 m 32.5 s	−26° 17' 32"
823	For	Gxy	1.8'×1.3'	12.7	02 h 07 m 21.0 s	−25° 26' 34"
824	For	Gxy	1.4'×1.2'	13.5	02 h 06 m 52.9 s	−36° 27' 09"
854	For	Gxy	1.8'×0.6'	13.3	02 h 11 m 30.6 s	−35° 50' 14"
857	For	Gxy	1.5'×1.3'	12.5	02 h 12 m 37.0 s	−31° 56' 40"
897	For	Gxy	2.0'×1.2'	12.4	02 h 21 m 06.0 s	−33° 43' 19"
922	For	Gxy	1.9'×1.8'	12	02 h 25 m 04.7 s	−24° 47' 25"
964	For	Gxy	2.0'×0.5'	12.8	02 h 31 m 05.7 s	−36° 02' 07"
986	For	Gxy	3.9'×3.0'	11	02 h 33 m 34.1 s	−39° 02' 47"
1049	For	GC	0.8'	12.6	02 h 39 m 48.1 s	−34° 15' 30"
1079	For	Gxy	3.5'×2.1'	11.5	02 h 43 m 44.4 s	−29° 00' 11"
1097	For	Gxy	9.3'×6.3'	9.4	02 h 46 m 19.0 s	−30° 16' 28"
1124	For	Gxy	1.0'×0.8'	14	02 h 51 m 36.0 s	−25° 42' 07"
1165	For	Gxy	2.5'×1.0'	12.9	02 h 58 m 47.7 s	−32° 05' 59"
1201	For	Gxy	3.6'×2.1'	10.8	03 h 04 m 07.9 s	−26° 04' 10"
1210	For	Gxy	2.0'×1.8'	12.8	03 h 06 m 45.3 s	−25° 43' 01"
1217	For	Gxy	1.8'×1.3'	12.6	03 h 06 m 06.0 s	−39° 02' 10"
1255	For	Gxy	4.2'×2.6'	11.2	03 h 13 m 31.9 s	−25° 43' 30"
1288	For	Gxy	2.3'×1.9'	12.2	03 h 17 m 13.1 s	−32° 34' 37"
1292	For	Gxy	3.0'×1.3'	12.2	03 h 18 m 14.7 s	−27° 36' 39"
1302	For	Gxy	3.9'×3.7'	10.6	03 h 19 m 51.1 s	−26° 03' 40"
1306	For	Gxy	1.1'×0.8'	13	03 h 21 m 03.0 s	−25° 30' 45"
1310	For	Gxy	2.0'×1.5'	12.5	03 h 21 m 03.5 s	−37° 06' 07"
1316	For	Gxy	12.0'×8.5'	8.5	03 h 22 m 41.7 s	−37° 12' 30"

(continued)

Table 10.2 (continued)

NGC #	Const	Type	Size	V Mag	RA (2000)	Dec (2000)
1317	For	Gxy	2.8'×2.4'	11.2	03 h 22 m 44.4 s	−37° 06' 12"
1318	For	Gxy	3.2'×2.8'	11.1	03 h 22 m 44.4 s	−37° 06' 12"
1326	For	Gxy	3.9'×2.9'	10.5	03 h 23 m 56.3 s	−36° 27' 53"
1327	For	Gxy	1.0'×0.3'	14.9	03 h 25 m 23.1 s	−25° 40' 47"
1336	For	Gxy	2.1'×1.5'	12.2	03 h 26 m 32.0 s	−35° 42' 50"
1339	For	Gxy	1.9'×1.4'	11.6	03 h 28 m 06.5 s	−32° 17' 10"
1340	For	Gxy	4.6'×2.9'	10.5	03 h 28 m 19.7 s	−31° 04' 07"
1341	For	Gxy	1.5'×1.3'	12.4	03 h 27 m 58.3 s	−37° 09' 00"
1344	For	Gxy	6.0'×3.5'	10.3	03 h 28 m 19.7 s	−31° 04' 07"
1350	For	Gxy	5.2'×2.8'	10.7	03 h 31 m 08.1 s	−33° 37' 44"
1351	For	Gxy	2.8'×1.7'	11.7	03 h 30 m 35.0 s	−34° 51' 15"
1360	For	PN	11.0'×7.5'	9.4	03 h 33 m 14.6 s	−25° 52' 18"
1365	For	Gxy	11.2'×6.2'	9.6	03 h 33 m 36.3 s	−36° 08' 25"
1366	For	Gxy	2.1'×0.9'	12	03 h 33 m 53.6 s	−31° 11' 38"
1367	For	Gxy	9.0'×6.0'	10.8	03 h 35 m 01.2 s	−24° 56' 02"
1369	For	Gxy	1.5'×1.4'	12.9	03 h 36 m 45.2 s	−36° 15' 22"
1371	For	Gxy	9.0'×6.0'	10.8	03 h 35 m 01.2 s	−24° 56' 02"
1373	For	Gxy	1.1'×0.9'	13.1	03 h 34 m 59.0 s	−35° 10' 18"
1374	For	Gxy	2.5'×2.3'	11.2	03 h 35 m 16.8 s	−35° 13' 34"
1375	For	Gxy	2.2'×0.9'	12.1	03 h 35 m 16.8 s	−35° 15' 56"
1378	For	**	...	...	03 h 35 m 58.2 s	−35° 12' 40"
1379	For	Gxy	2.4'×2.3'	11.2	03 h 36 m 03.8 s	−35° 26' 29"
1380	For	Gxy	4.8'×2.3'	10.2	03 h 36 m 27.1 s	−34° 58' 34"
1381	For	Gxy	2.7'×0.7'	11.7	03 h 36 m 31.8 s	−35° 17' 44"
1382	For	Gxy	1.5'×1.3'	12.9	03 h 37 m 08.9 s	−35° 11' 46"
1385	For	Gxy	3.4'×2.0'	11.1	03 h 37 m 28.8 s	−24° 30' 07"
1387	For	Gxy	2.8'×2.8'	10.9	03 h 36 m 57.1 s	−35° 30' 23"
1398	For	Gxy	7.1'×5.4'	9.8	03 h 38 m 52.0 s	−26° 20' 12"
1399	For	Gxy	6.9'×6.5'	9.4	03 h 38 m 29.0 s	−35° 27' 04"
1404	For	Gxy	3.3'×3.0'	10.2	03 h 38 m 51.9 s	−35° 35' 35"
1406	For	Gxy	3.8'×0.8'	11.9	03 h 39 m 23.0 s	−31° 19' 21"
1408	For	NF	...	...	03 h 39 m 21.1 s	−35° 31' 31"
1412	For	Gxy	1.9'×0.8'	12.6	03 h 40 m 29.2 s	−26° 51' 45"
1425	For	Gxy	5.8'×2.6'	11	03 h 42 m 11.4 s	−29° 53' 36"
1427	For	Gxy	3.6'×2.5'	11.1	03 h 42 m 19.1 s	−35° 23' 36"
1428	For	Gxy	1.6'×0.8'	12.8	03 h 42 m 22.9 s	−35° 09' 15"
1459	For	Gxy	1.7'×1.1'	13	03 h 46 m 57.9 s	−25° 31' 18"
2129	Gem	OC	6'	6.7	06 h 01 m 06.5 s	+23° 19' 20"
2158	Gem	OC	5'	8.6	06 h 07 m 25.6 s	+24° 05' 46"
2168	Gem	OC	28'	5.1	06 h 08 m 55.9 s	+24° 21' 28"
2218	Gem	Ast	1.5'	...	06 h 24 m 41.5 s	+19° 20' 29"
2224	Gem	Ast	...	...	06 h 27 m 28.6 s	+12° 35' 36"
2234	Gem	OC	25'	...	06 h 29 m 21.7 s	+16° 43' 22"
2248	Gem	Ast	...	...	06 h 34 m 35.7 s	+26° 18' 16"
2265	Gem	Ast	10.0'×5.0'	...	06 h 41 m 41.6 s	+11° 54' 17"

(continued)

Table 10.2 (continued)

NGC #	Const	Type	Size	V Mag	RA (2000)	Dec (2000)
2266	Gem	OC	6'	9.5	06 h 43 m 19.2 s	+26° 58' 10"
2274	Gem	Gxy	1.7'×1.7'	12.2	06 h 47 m 17.3 s	+33° 34' 02"
2275	Gem	Gxy	1.3'×1.0'	13.3	06 h 47 m 17.9 s	+33° 35' 58"
2277	Gem	Ast	…	…	06 h 47 m 47.0 s	+33° 27' 05"
2278	Gem	**	…	…	06 h 48 m 16.3 s	+33° 23' 35"
2279	Gem	***	…	…	06 h 48 m 24.7 s	+33° 24' 44"
2284	Gem	Ast	…	…	06 h 49 m 09.5 s	+33° 11' 38"
2285	Gem	**	…	…	06 h 49 m 36.0 s	+33° 21' 53"
2288	Gem	Gxy	0.2'×0.2'	14.4	06 h 50 m 52.0 s	+33° 27' 45"
2289	Gem	Gxy	1.1'×0.7'	13.4	06 h 50 m 53.7 s	+33° 28' 43"
2290	Gem	Gxy	1.3'×0.7'	13.6	06 h 50 m 56.9 s	+33° 26' 16"
2291	Gem	Gxy	1.0'×1.0'	13.8	06 h 50 m 58.6 s	+33° 31' 30"
2294	Gem	Gxy	1.7'×0.6'	13.6	06 h 51 m 11.3 s	+33° 31' 37"
2304	Gem	OC	5'	10	06 h 55 m 11.9 s	+17° 59' 19"
2331	Gem	OC	18.0'	8.5	07 h 06 m 59.8 s	+27° 15' 42"
2333	Gem	Gxy	1.0'×0.7'	13.5	07 h 08 m 21.3 s	+35° 10' 11"
2339	Gem	Gxy	2.7'×2.0'	11.4	07 h 08 m 20.5 s	+18° 46' 49"
2341	Gem	Gxy	0.8'×0.8'	13.4	07 h 09 m 12.0 s	+20° 36' 09"
2342	Gem	Gxy	1.4'×1.3'	12.5	07 h 09 m 18.5 s	+20° 38' 12"
2355	Gem	OC	9'	9.7	07 h 16 m 59.3 s	+13° 44' 59"
2356	Gem	OC	9'	9.7	07 h 16 m 59.3 s	+13° 44' 59"
2357	Gem	Gxy	3.7'×0.5'	13.5	07 h 17 m 41.0 s	+23° 21' 22"
2365	Gem	Gxy	2.4'×1.3'	12.6	07 h 22 m 22.6 s	+22° 05' 00"
2370	Gem	Gxy	0.9'×0.5'	13.7	07 h 25 m 01.7 s	+23° 47' 01"
2371	Gem	PN	74"×54"	11.2	07 h 25 m 33.8 s	+29° 29' 18"
2372	Gem	PN	2.2'×0.9'	11.2	07 h 25 m 35.8 s	+29° 29' 30"
2373	Gem	Gxy	0.6'×0.5'	14	07 h 26 m 37.0 s	+33° 49' 26"
2375	Gem	Gxy	1.3'×1.0'	13.8	07 h 27 m 09.5 s	+33° 49' 55"
2376	Gem	Gxy	0.5'×0.5'	14.4	07 h 26 m 35.9 s	+23° 04' 22"
2378	Gem	**	…	…	07 h 27 m 24.1 s	+33° 49' 54"
2379	Gem	Gxy	1.1'×1.0'	13.9	07 h 27 m 26.3 s	+33° 48' 40"
2385	Gem	Gxy	0.6'×0.3'	14.3	07 h 28 m 28.2 s	+33° 50' 15"
2386	Gem	***	…	…	07 h 28 m 38.1 s	+33° 46' 29"
2388	Gem	Gxy	1.0'×0.6'	13.9	07 h 28 m 53.4 s	+33° 49' 08"
2389	Gem	Gxy	2.0'×1.4'	12.8	07 h 29 m 04.7 s	+33° 51' 39"
2390	Gem	*	…	…	07 h 29 m 04.3 s	+33° 50' 13"
2391	Gem	*	…	…	07 h 29 m 07.4 s	+33° 49' 34"
2392	Gem	PN	47"×43"	9.1	07 h 29 m 10.8 s	+20° 54' 42"
2393	Gem	Gxy	1.2'×0.8'	14.2	07 h 30 m 04.7 s	+34° 01' 39"
2395	Gem	OC	12'	8	07 h 27 m 12.9 s	+13° 36' 29"
2398	Gem	Gxy	0.4'×0.3'	14.4	07 h 30 m 16.3 s	+24° 29' 15"
2405	Gem	Gxy	0.6'×0.4'	14.1	07 h 32 m 13.9 s	+25° 54' 23"
2406	Gem	Gxy	0.7'×0.6'	14.2	07 h 31 m 47.8 s	+18° 17' 15"
2407	Gem	Gxy	1.0'×0.8'	13.5	07 h 31 m 56.7 s	+18° 19' 58"
2410	Gem	Gxy	2.5'×0.7'	13.2	07 h 35 m 02.3 s	+32° 49' 18"

(continued)

Table 10.2 (continued)

NGC #	Const	Type	Size	V Mag	RA (2000)	Dec (2000)
2411	Gem	Gxy	1.0'×0.6'	13.6	07 h 34 m 36.3 s	+18° 16' 53"
2418	Gem	Gxy	1.8'×1.8'	12.3	07 h 36 m 37.5 s	+17° 53' 02"
2420	Gem	OC	10'	8.3	07 h 38 m 23.9 s	+21° 34' 27"
2435	Gem	Gxy	2.1'×0.5'	13	07 h 44 m 13.5 s	+31° 39' 02"
2449	Gem	Gxy	1.3'×0.6'	13.6	07 h 47 m 20.4 s	+26° 55' 49"
2450	Gem	Gxy	0.8'×0.2'	14.7	07 h 47 m 32.3 s	+27° 01' 10"
2454	Gem	Gxy	1.1'×0.6'	14	07 h 50 m 35.0 s	+16° 22' 07"
2480	Gem	Gxy	1.3'×0.7'	14.3	07 h 57 m 10.5 s	+23° 46' 45"
2481	Gem	Gxy	1.4'×0.5'	12.7	07 h 57 m 13.7 s	+23° 46' 04"
2486	Gem	Gxy	1.7'×0.9'	13.3	07 h 57 m 55.9 s	+25° 09' 38"
2487	Gem	Gxy	2.6'×2.1'	12.5	07 h 58 m 20.5 s	+25° 08' 57"
2490	Gem	Gxy	0.4'×0.3'	14.6	07 h 59 m 17.9 s	+27° 04' 39"
2492	Gem	Gxy	1.0'×1.0'	13.3	07 h 59 m 29.7 s	+27° 01' 35"
2498	Gem	Gxy	1.1'×0.8'	13.6	07 h 59 m 38.8 s	+24° 58' 57"
7070	Gru	Gxy	2.3'×1.8'	12.1	21 h 30 m 25.2 s	−43° 05' 14"
7072	Gru	Gxy	0.9'×0.7'	13.8	21 h 30 m 36.9 s	−43° 09' 10"
7075	Gru	Gxy	1.2'×0.9'	12.9	21 h 31 m 32.9 s	−38° 37' 03"
7079	Gru	Gxy	2.1'×1.3'	11.5	21 h 32 m 34.9 s	−44° 04' 04"
7087	Gru	Gxy	1.3'×0.8'	13.2	21 h 34 m 36.3 s	−40° 49' 08"
7091	Gru	Gxy	2.1'×1.7'	12.9	21 h 34 m 07.7 s	−36° 39' 15"
7097	Gru	Gxy	1.9'×1.3'	11.6	21 h 40 m 12.7 s	−42° 32' 22"
7107	Gru	Gxy	2.0'×1.5'	12.7	21 h 42 m 26.7 s	−44° 47' 33"
7117	Gru	Gxy	1.4'×0.9'	12.7	21 h 45 m 47.0 s	−48° 25' 18"
7118	Gru	Gxy	1.5'×1.2'	12.6	21 h 46 m 09.7 s	−48° 21' 14"
7119	Gru	Gxy	1.1'×0.6'	12.8	21 h 46 m 16.1 s	−46° 30' 58"
7144	Gru	Gxy	3.7'×3.6'	10.8	21 h 52 m 42.5 s	−48° 15' 19"
7145	Gru	Gxy	2.5'×2.4'	11.3	21 h 53 m 20.5 s	−47° 52' 59"
7162	Gru	Gxy	2.8'×1.0'	12.7	21 h 59 m 38.7 s	−43° 18' 28"
7166	Gru	Gxy	2.5'×0.9'	11.7	22 h 00 m 32.7 s	−43° 23' 26"
7169	Gru	Gxy	1.0'×0.5'	13.6	22 h 02 m 48.7 s	−47° 41' 52"
7213	Gru	Gxy	3.1'×2.8'	10.8	22 h 09 m 16.2 s	−47° 10' 00"
7232	Gru	Gxy	2.6'×0.9'	12.1	22 h 15 m 37.5 s	−45° 51' 03"
7233	Gru	Gxy	1.7'×1.3'	12.1	22 h 15 m 48.9 s	−45° 50' 49"
7249	Gru	Gxy	1.1'×0.8'	13.5	22 h 20 m 30.7 s	−55° 07' 30"
7297	Gru	Gxy	0.9'×0.6'	13.9	22 h 31 m 10.1 s	−37° 49' 39"
7299	Gru	Gxy	0.7'×0.6'	14.2	22 h 31 m 32.9 s	−37° 48' 37"
7307	Gru	Gxy	3.5'×0.9'	12.4	22 h 33 m 52.5 s	−40° 56' 01"
7322	Gru	Gxy	0.9'×0.6'	13.7	22 h 37 m 51.5 s	−37° 13' 55"
7334	Gru	Gxy	0.9'×0.6'	13.6	22 h 37 m 51.5 s	−37° 13' 55"
7355	Gru	Gxy	1.0'×0.4'	14.6	22 h 43 m 30.5 s	−36° 51' 53"
7368	Gru	Gxy	3.0'×0.5'	12.6	22 h 45 m 31.8 s	−39° 20' 29"
7382	Gru	Gxy	1.2'×0.3'	13.8	22 h 50 m 23.8 s	−36° 51' 26"
7400	Gru	Gxy	2.6'×0.5'	13	22 h 54 m 20.7 s	−45° 20' 51"
7404	Gru	Gxy	1.5'×0.9'	12.7	22 h 54 m 18.4 s	−39° 18' 55"
7410	Gru	Gxy	5.5'×2.0'	10.6	22 h 54 m 59.3 s	−39° 39' 45"

(continued)

Table 10.2 (continued)

NGC #	Const	Type	Size	V Mag	RA (2000)	Dec (2000)
7412	Gru	Gxy	3.9'×2.9'	11.2	22 h 55 m 46.1 s	−42° 38' 32"
7418	Gru	Gxy	3.5'×2.6'	11.2	22 h 56 m 36.3 s	−37° 01' 54"
7421	Gru	Gxy	2.0'×1.8'	12	22 h 56 m 54.1 s	−37° 20' 54"
7424	Gru	Gxy	9.5'×8.1'	10.4	22 h 57 m 18.4 s	−41° 04' 15"
7456	Gru	Gxy	5.5'×1.6'	11.7	23 h 02 m 10.2 s	−39° 34' 11"
7462	Gru	Gxy	4.2'×0.7'	11.9	23 h 02 m 46.7 s	−40° 50' 05"
7470	Gru	Gxy	1.4'×0.9'	13.5	23 h 05 m 13.8 s	−50° 06' 44"
7476	Gru	Gxy	1.3'×0.5'	12.9	23 h 05 m 11.9 s	−43° 05' 56"
7496	Gru	Gxy	3.3'×3.0'	11.3	23 h 09 m 46.9 s	−43° 25' 37"
7531	Gru	Gxy	4.5'×1.8'	11.6	23 h 14 m 48.3 s	−43° 36' 00"
7545	Gru	Gxy	1.1'×0.7'	13.2	23 h 15 m 32.3 s	−38° 32' 08"
7552	Gru	Gxy	3.4'×2.7'	10.7	23 h 16 m 10.7 s	−42° 35' 05"
7582	Gru	Gxy	5.0'×2.1'	10.6	23 h 18 m 23.7 s	−42° 22' 15"
7590	Gru	Gxy	2.7'×1.0'	11.6	23 h 18 m 54.7 s	−42° 14' 21"
7599	Gru	Gxy	4.4'×1.3'	11.4	23 h 19 m 21.1 s	−42° 15' 26"
7632	Gru	Gxy	2.2'×1.1'	12.2	23 h 22 m 00.8 s	−42° 28' 49"
7658	Gru	Gxy	0.8'×0.3'	14.3	23 h 26 m 24.9 s	−39° 13' 21"
6013	Her	Gxy	1.5'×0.8'	13.8	15 h 52 m 52.7 s	+40° 38' 47"
6028	Her	Gxy	1.3'×1.2'	13.5	16 h 01 m 29.1 s	+19° 21' 33"
6030	Her	Gxy	1.1'×0.8'	13.2	16 h 01 m 51.4 s	+17° 57' 27"
6032	Her	Gxy	1.6'×0.7'	13.9	16 h 03 m 01.2 s	+20° 57' 20"
6034	Her	Gxy	1.1'×0.8'	13.7	16 h 03 m 32.2 s	+17° 11' 53"
6035	Her	Gxy	1.1'×0.9'	14	16 h 03 m 24.1 s	+20° 53' 27"
6039	Her	Gxy	0.6'×0.6'	13.9	16 h 04 m 39.6 s	+17° 42' 02"
6040	Her	Gxy	1.3'×0.5'	13.9	16 h 04 m 26.8 s	+17° 45' 00"
6041	Her	Gxy	1.2'×1.1'	13.4	16 h 04 m 35.8 s	+17° 43' 17"
6042	Her	Gxy	0.9'×0.7'	14.1	16 h 04 m 39.6 s	+17° 42' 02"
6043	Her	Gxy	0.7'×0.4'	14.4	16 h 05 m 01.4 s	+17° 46' 32"
6044	Her	Gxy	0.6'×0.6'	14.3	16 h 04 m 59.7 s	+17° 52' 12"
6045	Her	Gxy	1.3'×0.3'	14.4	16 h 05 m 07.8 s	+17° 45' 26"
6046	Her	Gxy	1.3'×1.1'	13.5	16 h 01 m 29.1 s	+19° 21' 33"
6047	Her	Gxy	1.1'×0.8'	13.8	16 h 05 m 08.9 s	+17° 43' 45"
6050	Her	Gxy	0.9'×0.6'	14.1	16 h 05 m 23.4 s	+17° 45' 26"
6052	Her	Gxy	0.9'×0.7'	13.6	16 h 05 m 13.0 s	+20° 32' 32"
6053	Her	Gxy	0.6'×0.5'	14.7	16 h 05 m 39.6 s	+18° 09' 51"
6054	Her	Gxy	0.8'×0.4'	14.3	16 h 05 m 38.0 s	+17° 46' 02"
6055	Her	Gxy	1.0'×0.6'	14	16 h 05 m 32.7 s	+18° 09' 34"
6056	Her	Gxy	0.9'×0.5'	14.1	16 h 05 m 31.3 s	+17° 57' 49"
6057	Her	Gxy	0.6'×0.5'	14.5	16 h 05 m 39.6 s	+18° 09' 51"
6058	Her	PN	25"×20"	12.9	16 h 04 m 26.5 s	+40° 40' 59"
6060	Her	Gxy	2.0'×1.1'	13.3	16 h 05 m 52.1 s	+21° 29' 03"
6061	Her	Gxy	1.0'×0.8'	13.6	16 h 06 m 16.1 s	+18° 14' 59"
6062	Her	Gxy	1.2'×0.9'	13.6	16 h 06 m 22.9 s	+19° 46' 39"
6064	Her	Gxy	0.8'×0.6'	13.8	16 h 05 m 13.0 s	+20° 32' 32"
6073	Her	Gxy	1.3'×0.7'	13.7	16 h 10 m 10.9 s	+16° 41' 58"

(continued)

Table 10.2 (continued)

NGC #	Const	Type	Size	V Mag	RA (2000)	Dec (2000)
6074	Her	Gxy	0.3'×0.2'	14.5	16 h 11 m 17.2 s	+14° 15' 32"
6075	Her	Gxy	0.8'×0.6'	14.2	16 h 11 m 22.6 s	+23° 57' 53"
6078	Her	Gxy	0.3'×0.3'	15.2	16 h 12 m 05.4 s	+14° 12' 32"
6081	Her	Gxy	1.8'×0.6'	13.2	16 h 12 m 56.8 s	+09° 52' 02"
6083	Her	Gxy	0.6'×0.4'	14.6	16 h 13 m 12.6 s	+14° 11' 07"
6084	Her	Gxy	1.0'×0.5'	14.3	16 h 14 m 16.6 s	+17° 45' 26"
6098	Her	Gxy	1.6'×0.9'	13.3	16 h 15 m 34.0 s	+19° 27' 41"
6099	Her	Gxy	1.3'×1.3'	14.2	16 h 15 m 35.4 s	+19° 27' 12"
6106	Her	Gxy	2.5'×1.4'	12.4	16 h 18 m 47.1 s	+07° 24' 39"
6113	Her	Gxy	0.9'×0.4'	14.1	16 h 19 m 10.7 s	+14° 08' 01"
6132	Her	Gxy	1.5'×0.5'	13.8	16 h 23 m 38.7 s	+11° 47' 09"
6138	Her	Gxy	1.1'×0.9'	14.1	17 h 22 m 40.0 s	+41° 06' 05"
6141	Her	Gxy	0.4'×0.3'	14.9	16 h 23 m 06.3 s	+40° 51' 30"
6145	Her	Gxy	0.8'×0.4'	14.4	16 h 25 m 02.4 s	+40° 56' 47"
6146	Her	Gxy	1.3'×1.0'	12.5	16 h 25 m 10.2 s	+40° 53' 35"
6147	Her	Gxy	0.4'×0.3'	15.2	16 h 25 m 05.8 s	+40° 55' 43"
6148	Her	Gxy	0.6'×0.3'	16.1	16 h 27 m 04.0 s	+24° 05' 31"
6149	Her	Gxy	1.1'×0.8'	13.6	16 h 27 m 24.2 s	+19° 35' 50"
6150	Her	Gxy	...'×...'	14	16 h 25 m 50.0 s	+40° 29' 19"
6154	Her	Gxy	2.1'×2.0'	12.9	16 h 25 m 30.5 s	+49° 50' 25"
6155	Her	Gxy	1.3'×0.9'	12.4	16 h 26 m 08.3 s	+48° 22' 01"
6158	Her	Gxy	...'×...'	13.8	16 h 27 m 40.9 s	+39° 22' 59"
6159	Her	Gxy	1.1'×1.0'	14.3	16 h 27 m 25.1 s	+42° 40' 47"
6160	Her	Gxy	1.8'×1.5'	13.3	16 h 27 m 41.0 s	+40° 55' 37"
6161	Her	Gxy	0.9'×0.3'	14.9	16 h 28 m 20.6 s	+32° 48' 39"
6162	Her	Gxy	0.9'×0.7'	14	16 h 28 m 22.5 s	+32° 50' 57"
6163	Her	Gxy	0.6'×0.3'	14.4	16 h 28 m 27.9 s	+32° 50' 46"
6166	Her	Gxy	1.9'×1.4'	12.4	16 h 28 m 38.5 s	+39° 33' 06"
6168	Her	Gxy	1.4'×0.3'	14.3	16 h 31 m 21.3 s	+20° 11' 06"
6173	Her	Gxy	1.9'×1.4'	12.7	16 h 29 m 44.8 s	+40° 48' 41"
6174	Her	Gxy	0.4'×0.4'	...	16 h 29 m 23.8 s	+40° 52' 28"
6175	Her	Gxy	1.2'×0.7'	13.7	16 h 29 m 57.6 s	+40° 37' 48"
6177	Her	Gxy	1.7'×1.2'	13.8	16 h 30 m 38.9 s	+35° 03' 22"
6179	Her	Gxy	0.2'×0.2'	15.3	16 h 30 m 47.0 s	+35° 06' 08"
6180	Her	Gxy	0.4'×0.2'	14.1	16 h 30 m 33.9 s	+40° 32' 21"
6181	Her	Gxy	2.5'×1.1'	12.2	16 h 32 m 20.9 s	+19° 49' 34"
6184	Her	Gxy	0.8'×0.5'	14.3	16 h 31 m 34.5 s	+40° 33' 55"
6185	Her	Gxy	1.2'×0.9'	13.5	16 h 33 m 17.9 s	+35° 20' 31"
6186	Her	Gxy	1.5'×1.2'	13.1	16 h 34 m 25.4 s	+21° 32' 26"
6194	Her	Gxy	0.4'×0.3'	13.8	16 h 36 m 37.1 s	+36° 12' 01"
6195	Her	Gxy	1.6'×1.1'	13.8	16 h 36 m 32.3 s	+39° 01' 40"
6196	Her	Gxy	2.0'×1.2'	12.9	16 h 37 m 53.8 s	+36° 04' 22"
6197	Her	Gxy	0.6'×0.3'	14.6	16 h 37 m 59.9 s	+35° 59' 43"
6199	Her	*	...	...	16 h 39 m 28.9 s	+36° 03' 31"
6201	Her	Gxy	0.4'×0.3'	14.5	16 h 40 m 14.5 s	+23° 45' 55"

(continued)

Table 10.2 (continued)

NGC #	Const	Type	Size	V Mag	RA (2000)	Dec (2000)
6203	Her	Gxy	0.7'×0.7'	14.4	16 h 40 m 27.3 s	+23° 46' 30"
6205	Her	GC	17'	5.8	16 h 41 m 41.4 s	+36° 27' 36"
6207	Her	Gxy	3.0'×1.3'	11.7	16 h 43 m 03.9 s	+36° 49' 56"
6210	Her	PN	48"×8"	8.8	16 h 44 m 29.4 s	+23° 48' 00"
6212	Her	Gxy	0.6'×0.4'	14.2	16 h 43 m 23.1 s	+39° 48' 24"
6219	Her	Gxy	0.6'×0.6'	14.2	16 h 46 m 22.6 s	+09° 02' 18"
6224	Her	Gxy	1.1'×1.1'	13.6	16 h 48 m 18.5 s	+06° 18' 43"
6225	Her	Gxy	1.1'×0.8'	13.9	16 h 48 m 21.5 s	+06° 13' 22"
6228	Her	Gxy	1.1'×0.6'	14.3	16 h 48 m 02.7 s	+26° 12' 48"
6229	Her	GC	4.5'	9.4	16 h 46 m 58.8 s	+47° 31' 40"
6230	Her	Gxy	1.0'×0.9'	14.5	16 h 50 m 46.7 s	+04° 36' 17"
6233	Her	Gxy	1.4'×1.0'	13.4	16 h 50 m 15.7 s	+23° 34' 46"
6239	Her	Gxy	2.6'×1.1'	12.3	16 h 50 m 05.2 s	+42° 44' 22"
6241	Her	Gxy	1.0'×0.8'	15.2	16 h 50 m 10.9 s	+45° 25' 14"
6243	Her	Gxy	1.3'×0.5'	14.3	16 h 52 m 26.3 s	+23° 19' 57"
6255	Her	Gxy	3.6'×1.5'	12.9	16 h 54 m 47.1 s	+36° 30' 06"
6257	Her	Gxy	0.8'×0.2'	14.8	16 h 56 m 03.5 s	+39° 38' 42"
6261	Her	Gxy	1.2'×0.5'	14.2	16 h 56 m 30.6 s	+27° 58' 39"
6263	Her	Gxy	0.9'×0.9'	13.8	16 h 56 m 43.2 s	+27° 49' 19"
6264	Her	Gxy	0.7'×0.4'	14.7	16 h 57 m 16.1 s	+27° 50' 58"
6265	Her	Gxy	0.8'×0.5'	14.3	16 h 57 m 29.0 s	+27° 50' 39"
6267	Her	Gxy	1.3'×1.0'	13.3	16 h 58 m 08.6 s	+22° 59' 05"
6269	Her	Gxy	2.0'×1.6'	12.6	16 h 57 m 58.1 s	+27° 51' 15"
6270	Her	Gxy	0.7'×0.4'	13.3	16 h 58 m 44.2 s	+27° 51' 33"
6271	Her	Gxy	0.8'×0.5'	14.6	16 h 58 m 50.5 s	+27° 57' 54"
6272	Her	Gxy	0.5'×0.3'	14.8	16 h 58 m 58.3 s	+27° 55' 50"
6274	Her	Gxy	0.6'×0.4'	13.8	16 h 59 m 20.4 s	+29° 56' 47"
6276	Her	Gxy	0.4'×0.3'	14.6	17 h 00 m 45.0 s	+23° 02' 38"
6277	Her	*	...	...	17 h 00 m 48.9 s	+23° 02' 22"
6278	Her	Gxy	2.0'×1.2'	12.5	17 h 00 m 50.3 s	+23° 00' 40"
6279	Her	Gxy	1.1'×1.0'	13.7	16 h 59 m 01.3 s	+47° 14' 14"
6282	Her	Gxy	0.7'×0.5'	14.4	17 h 00 m 47.2 s	+29° 49' 15"
6283	Her	Gxy	1.2'×1.2'	12.9	16 h 59 m 26.5 s	+49° 55' 19"
6301	Her	Gxy	2.3'×1.4'	13.6	17 h 08 m 32.6 s	+42° 20' 20"
6308	Her	Gxy	1.2'×1.1'	13.6	17 h 11 m 59.7 s	+23° 22' 48"
6311	Her	Gxy	1.1'×1.1'	13.6	17 h 10 m 43.6 s	+41° 39' 04"
6312	Her	Gxy	0.6'×0.6'	14.4	17 h 10 m 48.1 s	+42° 17' 15"
6313	Her	Gxy	1.3'×0.4'	14.1	17 h 10 m 20.9 s	+48° 19' 54"
6314	Her	Gxy	1.4'×0.7'	13.5	17 h 12 m 38.7 s	+23° 16' 13"
6315	Her	Gxy	1.0'×0.9'	14.3	17 h 12 m 46.1 s	+23° 13' 24"
6320	Her	Gxy	1.2'×0.9'	14.1	17 h 12 m 55.6 s	+40° 15' 59"
6321	Her	Gxy	1.1'×1.0'	13.6	17 h 14 m 24.3 s	+20° 18' 49"
6323	Her	Gxy	1.1'×0.4'	14.2	17 h 13 m 18.0 s	+43° 46' 56"
6327	Her	Gxy	0.4'×0.4'	15	17 h 14 m 02.1 s	+43° 38' 57"
6329	Her	Gxy	1.8'×1.8'	12.9	17 h 14 m 15.1 s	+43° 41' 04"

(continued)

Table 10.2 (continued)

NGC #	Const	Type	Size	V Mag	RA (2000)	Dec (2000)
6330	Her	Gxy	1.4'×0.5'	14.2	17 h 15 m 44.3 s	+29° 24' 14"
6332	Her	Gxy	1.2'×0.8'	13.8	17 h 15 m 02.9 s	+43° 39' 35"
6336	Her	Gxy	1.0'×0.7'	13.8	17 h 16 m 16.5 s	+43° 49' 15"
6339	Her	Gxy	2.9'×1.7'	12.8	17 h 17 m 06.4 s	+40° 50' 42"
6341	Her	GC	14'	6.5	17 h 17 m 07.5 s	+43° 08' 11"
6343	Her	Gxy	1.0'×1.0'	13.8	17 h 17 m 16.3 s	+41° 03' 10"
6344	Her	**	…	…	17 h 17 m 18.1 s	+42° 26' 03"
6347	Her	Gxy	1.2'×0.7'	13.9	17 h 19 m 54.6 s	+16° 39' 38"
6348	Her	Gxy	0.6'×0.5'	14.4	17 h 18 m 21.2 s	+41° 38' 51"
6349	Her	Gxy	0.7'×0.2'	14.3	17 h 19 m 06.5 s	+36° 03' 38"
6350	Her	Gxy	1.0'×1.0'	13.3	17 h 18 m 42.1 s	+41° 41' 38"
6351	Her	Gxy	0.2'×0.2'	15.2	17 h 19 m 11.3 s	+36° 03' 38"
6353	Her	Ast	…	…	17 h 21 m 12.4 s	+15° 41' 18"
6363	Her	Gxy	1.1'×0.9'	13.4	17 h 22 m 40.0 s	+41° 06' 05"
6364	Her	Gxy	1.5'×1.2'	13	17 h 24 m 27.3 s	+29° 23' 23"
6367	Her	Gxy	0.8'×0.6'	14.3	17 h 25 m 08.8 s	+37° 45' 35"
6371	Her	Gxy	0.7'×0.4'	14.3	17 h 27 m 20.5 s	+26° 30' 18"
6372	Her	Gxy	1.7'×1.1'	13.1	17 h 27 m 31.9 s	+26° 28' 30"
6375	Her	Gxy	1.6'×1.6'	13.3	17 h 29 m 21.9 s	+16° 12' 24"
6379	Her	Gxy	1.2'×1.1'	13.8	17 h 30 m 34.7 s	+16° 17' 18"
6389	Her	Gxy	2.8'×1.9'	12.3	17 h 32 m 39.8 s	+16° 24' 07"
6406	Her	**	…	…	17 h 38 m 18.9 s	+18° 49' 59"
6408	Her	Gxy	1.6'×1.4'	12.9	17 h 38 m 47.4 s	+18° 52' 40"
6417	Her	Gxy	1.4'×1.2'	13.3	17 h 41 m 47.8 s	+23° 40' 19"
6427	Her	Gxy	1.5'×0.6'	13.4	17 h 43 m 38.5 s	+25° 29' 37"
6428	Her	**	…	…	17 h 43 m 52.5 s	+25° 33' 16"
6429	Her	Gxy	1.9'×0.8'	13.3	17 h 44 m 05.3 s	+25° 21' 01"
6430	Her	Gxy	1.9'×0.6'	13.8	17 h 45 m 14.3 s	+18° 08' 21"
6431	Her	Gxy	1.5'×0.6'	13.4	17 h 43 m 38.5 s	+25° 29' 37"
6433	Her	Gxy	2.0'×0.5'	13.5	17 h 43 m 56.3 s	+36° 47' 59"
6442	Her	Gxy	1.9'×1.6'	12.7	17 h 46 m 51.3 s	+20° 45' 39"
6443	Her	Gxy	1.2'×0.5'	14	17 h 44 m 33.7 s	+48° 06' 52"
6446	Her	Gxy	0.6'×0.3'	15.2	17 h 46 m 07.5 s	+35° 34' 10"
6447	Her	Gxy	1.6'×0.9'	12.9	17 h 46 m 17.2 s	+35° 34' 18"
6450	Her	NF	…	…	17 h 47 m 32.3 s	+18° 34' 31"
6452	Her	Gxy	0.5'×0.5'	14.4	17 h 47 m 58.4 s	+20° 50' 15"
6458	Her	Gxy	1.1'×0.8'	13.5	17 h 49 m 10.9 s	+20° 48' 14"
6460	Her	Gxy	1.9'×1.1'	13.3	17 h 49 m 30.3 s	+20° 45' 47"
6467	Her	Gxy	2.6'×1.7'	13.3	17 h 50 m 40.0 s	+17° 32' 16"
6468	Her	Gxy	2.3'×1.7'	13.3	17 h 50 m 40.0 s	+17° 32' 16"
6482	Her	Gxy	2.0'×1.7'	11.5	17 h 51 m 48.9 s	+23° 04' 19"
6484	Her	Gxy	1.9'×1.9'	12.5	17 h 51 m 47.1 s	+24° 29' 02"
6485	Her	Gxy	1.5'×1.4'	13.1	17 h 51 m 52.7 s	+31° 27' 42"
6486	Her	Gxy	0.8'×0.8'	14.3	17 h 52 m 35.3 s	+29° 49' 04"
6487	Her	Gxy	1.9'×1.7'	12.4	17 h 52 m 41.8 s	+29° 50' 19"

(continued)

Table 10.2 (continued)

NGC #	Const	Type	Size	V Mag	RA (2000)	Dec (2000)
6490	Her	Gxy	1.0'×0.8'	13.6	17 h 54 m 30.4 s	+18° 22' 32"
6495	Her	Gxy	2.0'×1.8'	12.3	17 h 54 m 50.7 s	+18° 19' 37"
6499	Her	**	...	...	17 h 55 m 19.9 s	+18° 21' 35"
6500	Her	Gxy	2.2'×1.6'	12.4	17 h 55 m 59.8 s	+18° 20' 18"
6501	Her	Gxy	2.0'×1.8'	12.2	17 h 56 m 03.7 s	+18° 22' 24"
6504	Her	Gxy	2.2'×0.5'	12.6	17 h 56 m 05.6 s	+33° 12' 31"
6513	Her	Gxy	1.2'×0.9'	13.4	17 h 59 m 34.2 s	+24° 53' 13"
6518	Her	Gxy	0.4'×0.4'	13.9	17 h 59 m 43.7 s	+28° 52' 00"
6524	Her	Gxy	1.3'×1.0'	12.9	17 h 59 m 14.7 s	+45° 53' 15"
6527	Her	Gxy	1.4'×1.0'	13.6	18 h 01 m 46.3 s	+19° 43' 42"
6547	Her	Gxy	1.3'×0.4'	13.7	18 h 05 m 09.9 s	+25° 13' 57"
6548	Her	Gxy	3.0'×2.8'	11.5	18 h 05 m 59.3 s	+18° 35' 13"
6549	Her	Gxy	1.4'×0.5'	14	18 h 05 m 49.4 s	+18° 32' 16"
6550	Her	Gxy	1.4'×0.5'	14	18 h 05 m 49.4 s	+18° 32' 16"
6555	Her	Gxy	2.0'×1.5'	12.3	18 h 07 m 49.1 s	+17° 36' 15"
6560	Her	Gxy	1.2'×0.8'	13.4	18 h 05 m 13.9 s	+46° 52' 54"
6564	Her	***	...	...	18 h 09 m 02.3 s	+17° 23' 41"
6571	Her	Gxy	0.3'×0.3'	14.4	18 h 10 m 49.4 s	+21° 14' 19"
6574	Her	Gxy	1.4'×1.1'	12.3	18 h 11 m 51.1 s	+14° 58' 54"
6575	Her	Gxy	1.8'×1.3'	12.8	18 h 10 m 57.5 s	+31° 06' 58"
6576	Her	Gxy	0.7'×0.5'	14.6	18 h 11 m 47.9 s	+21° 25' 43"
6577	Her	Gxy	1.5'×1.3'	13	18 h 12 m 01.1 s	+21° 27' 49"
6579	Her	Gxy	0.4'×0.4'	13.7	18 h 12 m 31.7 s	+21° 25' 14"
6580	Her	Gxy	1.3'×0.7'	13.1	18 h 12 m 33.7 s	+21° 25' 34"
6581	Her	Gxy	0.4'×0.4'	14.3	18 h 12 m 18.4 s	+25° 39' 43"
6582	Her	Gxy	1.8'×1.2'	13.9	18 h 11 m 01.7 s	+49° 54' 43"
6585	Her	Gxy	1.9'×0.4'	12.9	18 h 12 m 21.7 s	+39° 37' 58"
6586	Her	Gxy	1.0'×0.5'	13.8	18 h 13 m 38.4 s	+21° 05' 25"
6587	Her	Gxy	1.3'×1.1'	13	18 h 13 m 50.8 s	+18° 49' 30"
6591	Her	Gxy	0.3'×0.1'	15.5	18 h 14 m 03.5 s	+21° 03' 48"
6593	Her	Gxy	0.8'×0.4'	14.3	18 h 14 m 03.5 s	+22° 17' 02"
6599	Her	Gxy	1.3'×1.2'	12.7	18 h 15 m 43.0 s	+24° 54' 44"
6600	Her	Gxy	1.3'×1.2'	12.8	18 h 15 m 43.0 s	+24° 54' 44"
6602	Her	Gxy	1.0'×0.7'	13.8	18 h 16 m 34.3 s	+25° 02' 38"
6610	Her	Gxy	1.4'×1.1'	12	18 h 11 m 51.1 s	+14° 58' 54"
6616	Her	Gxy	1.5'×0.6'	14	18 h 17 m 41.1 s	+22° 14' 17"
6619	Her	Gxy	1.2'×1.1'	13.1	18 h 18 m 54.3 s	+23° 39' 20"
6623	Her	Gxy	1.3'×1.2'	13.3	18 h 19 m 42.7 s	+23° 42' 34"
6627	Her	Gxy	1.3'×1.1'	13.2	18 h 22 m 38.9 s	+15° 41' 52"
6628	Her	Gxy	1.7'×1.1'	13	18 h 22 m 21.9 s	+23° 28' 40"
6632	Her	Gxy	3.0'×1.4'	12.3	18 h 25 m 03.1 s	+27° 32' 06"
6635	Her	Gxy	1.0'×1.0'	13.5	18 h 27 m 37.1 s	+14° 49' 09"
6641	Her	Gxy	0.9'×0.7'	13.5	18 h 28 m 57.3 s	+22° 54' 11"
6658	Her	Gxy	1.7'×0.4'	13.6	18 h 33 m 55.6 s	+22° 53' 17"
6659	Her	OC	7'	...	18 h 33 m 59.9 s	+23° 35' 42"

(continued)

Table 10.2 (continued)

NGC #	Const	Type	Size	V Mag	RA (2000)	Dec (2000)
6660	Her	Gxy	1.9'×1.2'	12.3	18 h 34 m 36.6 s	+22° 54' 34"
6661	Her	Gxy	1.7'×1.1'	12.7	18 h 34 m 36.6 s	+22° 54' 34"
6669	Her	Ast	...	...	18 h 37 m 15.0 s	+22° 11' 45"
6674	Her	Gxy	4.0'×2.2'	11.9	18 h 38 m 33.7 s	+25° 22' 30"
6680	Her	Gxy	0.7'×0.5'	14.6	18 h 39 m 43.8 s	+22° 18' 57"
6697	Her	Gxy	1.2'×1.0'	12.8	18 h 45 m 15.1 s	+25° 30' 45"
888	Hor	Gxy	1.1'×0.9'	13.5	02 h 17 m 27.7 s	−59° 51' 47"
1025	Hor	Gxy	0.9'×0.5'	14	02 h 36 m 19.9 s	−54° 51' 51"
1031	Hor	Gxy	1.9'×1.1'	12.9	02 h 36 m 38.6 s	−54° 51' 34"
1096	Hor	Gxy	1.9'×1.8'	13.2	02 h 43 m 49.3 s	−59° 54' 52"
1135	Hor	Gxy	1.4'×1.2'	13.1	02 h 50 m 53.6 s	−54° 58' 36"
1136	Hor	Gxy	1.4'×1.2'	13.1	02 h 50 m 53.6 s	−54° 58' 36"
1244	Hor	Gxy	1.9'×0.4'	13.3	03 h 06 m 31.1 s	−66° 46' 35"
1246	Hor	Gxy	1.2'×0.8'	13.2	03 h 07 m 02.3 s	−66° 56' 19"
1249	Hor	Gxy	5.4'×3.1'	11.5	03 h 10 m 01.3 s	−53° 20' 09"
1252	Hor	OC	2.0'	...	03 h 10 m 44.3 s	−57° 45' 31"
1261	Hor	GC	6.9'	8.3	03 h 12 m 15.3 s	−55° 13' 01"
1311	Hor	Gxy	3.0'×0.8'	12.8	03 h 20 m 06.7 s	−52° 11' 16"
1356	Hor	Gxy	1.4'×1.1'	13.5	03 h 30 m 40.7 s	−50° 18' 35"
1411	Hor	Gxy	2.3'×1.7'	11	03 h 38 m 44.9 s	−44° 06' 04"
1433	Hor	Gxy	6.5'×5.9'	10.1	03 h 42 m 01.4 s	−47° 13' 20"
1448	Hor	Gxy	7.6'×1.7'	11	03 h 44 m 32.2 s	−44° 38' 44"
1457	Hor	Gxy	8.1'×1.8'	10.7	03 h 44 m 32.2 s	−44° 38' 44"
1476	Hor	Gxy	1.4'×0.6'	13.3	03 h 52 m 08.5 s	−44° 31' 57"
1483	Hor	Gxy	1.6'×1.3'	12.7	03 h 52 m 47.7 s	−47° 28' 44"
1493	Hor	Gxy	3.5'×3.2'	11.4	03 h 57 m 27.5 s	−46° 12' 43"
1494	Hor	Gxy	3.2'×1.9'	11.8	03 h 57 m 42.5 s	−48° 54' 35"
1495	Hor	Gxy	3.0'×0.5'	12.8	03 h 58 m 21.3 s	−44° 27' 59"
1510	Hor	Gxy	1.3'×0.7'	12.6	04 h 03 m 32.5 s	−43° 24' 01"
1512	Hor	Gxy	8.9'×5.6'	10.6	04 h 03 m 54.2 s	−43° 20' 57"
1527	Hor	Gxy	3.7'×1.4'	11	04 h 08 m 24.1 s	−47° 53' 50"
1558	Hor	Gxy	2.5'×1.0'	12.7	04 h 20 m 15.8 s	−45° 01' 53"
2548	Hya	OC	54'	5.8	08 h 13 m 43.1 s	−05° 45' 02"
2555	Hya	Gxy	1.9'×1.4'	12.4	08 h 17 m 56.3 s	+00° 44' 44"
2561	Hya	Gxy	1.1'×0.6'	13.4	08 h 19 m 36.9 s	+04° 39' 25"
2574	Hya	Gxy	2.2'×1.2'	12.9	08 h 20 m 48.2 s	−08° 55' 07"
2583	Hya	Gxy	0.8'×0.8'	13.4	08 h 23 m 07.9 s	−05° 00' 09"
2584	Hya	Gxy	1.1'×0.6'	13.8	08 h 23 m 15.3 s	−04° 58' 13"
2585	Hya	Gxy	1.8'×0.8'	13.7	08 h 23 m 26.2 s	−04° 54' 56"
2586	Hya	***	...	...	08 h 23 m 31.3 s	−04° 57' 07"
2589	Hya	NF	...	...	08 h 24 m 29.5 s	−08° 46' 05"
2590	Hya	Gxy	2.2'×0.7'	13.1	08 h 25 m 01.9 s	−00° 35' 30"
2610	Hya	PN	50"×47"	12.7	08 h 33 m 23.4 s	−16° 08' 57"
2612	Hya	Gxy	2.7'×0.6'	12.7	08 h 33 m 50.1 s	−13° 10' 29"
2615	Hya	Gxy	1.9'×1.0'	12.7	08 h 34 m 33.3 s	−02° 32' 47"

(continued)

Table 10.2 (continued)

NGC #	Const	Type	Size	V Mag	RA (2000)	Dec (2000)
2616	Hya	Gxy	1.6'×1.3'	13.2	08 h 35 m 34.0 s	−01° 51' 00"
2617	Hya	Gxy	1.1'×0.8'	13.8	08 h 35 m 38.9 s	−04° 05' 21"
2618	Hya	Gxy	2.4'×1.9'	12.3	08 h 35 m 53.5 s	+00° 42' 24"
2642	Hya	Gxy	2.0'×1.9'	11.9	08 h 40 m 44.5 s	−04° 07' 19"
2644	Hya	Gxy	2.1'×0.8'	12.7	08 h 41 m 31.9 s	+04° 58' 49"
2662	Hya	Gxy	0.5'×0.5'	12.8	08 h 45 m 32.0 s	−15° 07' 17"
2665	Hya	Gxy	2.0'×1.5'	12.6	08 h 46 m 01.0 s	−19° 18' 09"
2674	Hya	Gxy	0.6'×0.3'	14.9	08 h 49 m 13.1 s	−14° 17' 39"
2690	Hya	Gxy	1.9'×0.5'	13.3	08 h 52 m 38.0 s	−02° 36' 12"
2695	Hya	Gxy	1.7'×1.2'	12.3	08 h 54 m 27.1 s	−03° 04' 00"
2696	Hya	Gxy	1.3'×1.3'	13.3	08 h 50 m 42.1 s	−05° 00' 34"
2697	Hya	Gxy	1.8'×1.1'	12.4	08 h 54 m 59.4 s	−02° 59' 14"
2698	Hya	Gxy	1.4'×0.6'	12.7	08 h 55 m 36.5 s	−03° 11' 03"
2699	Hya	Gxy	1.0'×1.0'	12.7	08 h 55 m 48.7 s	−03° 07' 39"
2700	Hya	*	…	…	08 h 55 m 50.6 s	−03° 06' 59"
2702	Hya	*	…	…	08 h 55 m 54.6 s	−03° 03' 55"
2703	Hya	**	…	…	08 h 55 m 47.1 s	−03° 18' 25"
2705	Hya	*	…	…	08 h 56 m 00.1 s	−03° 00' 53"
2706	Hya	Gxy	1.8'×0.6'	13.2	08 h 56 m 12.5 s	−02° 33' 46"
2707	Hya	*	…	…	08 h 56 m 05.6 s	−03° 03' 59"
2708	Hya	Gxy	2.6'×1.3'	12.4	08 h 56 m 07.9 s	−03° 21' 37"
2709	Hya	Gxy	0.8'×0.6'	13.8	08 h 56 m 12.8 s	−03° 14' 35"
2713	Hya	Gxy	3.6'×1.5'	11.7	08 h 57 m 20.5 s	+02° 55' 16"
2716	Hya	Gxy	1.3'×1.0'	12.7	08 h 57 m 35.9 s	+03° 05' 26"
2718	Hya	Gxy	2.1'×2.1'	12	08 h 58 m 50.4 s	+06° 17' 34"
2721	Hya	Gxy	2.3'×1.6'	12.2	08 h 58 m 56.5 s	−04° 54' 08"
2722	Hya	Gxy	2.0'×1.3'	12.7	08 h 58 m 46.1 s	−03° 42' 37"
2723	Hya	Gxy	0.9'×0.9'	13.3	09 h 00 m 14.4 s	+03° 10' 40"
2727	Hya	Gxy	2.5'×1.2'	12	08 h 56 m 07.9 s	−03° 21' 37"
2729	Hya	Gxy	0.8'×0.5'	13.5	09 h 01 m 28.6 s	+03° 43' 15"
2733	Hya	Gxy	2.0'×1.3'	12.7	08 h 58 m 46.1 s	−03° 42' 37"
2754	Hya	Gxy	0.8'×0.5'	14.3	09 h 05 m 11.1 s	−19° 05' 05"
2757	Hya	***	…	…	09 h 05 m 25.7 s	−19° 02' 52"
2758	Hya	Gxy	1.9'×0.5'	13.4	09 h 05 m 31.2 s	−19° 02' 33"
2763	Hya	Gxy	2.3'×2.0'	12	09 h 06 m 49.1 s	−15° 30' 04"
2765	Hya	Gxy	2.1'×1.1'	12.2	09 h 07 m 36.7 s	+03° 23' 33"
2781	Hya	Gxy	3.0'×1.5'	11.3	09 h 11 m 27.5 s	−14° 49' 01"
2784	Hya	Gxy	5.5'×2.2'	10.3	09 h 12 m 19.3 s	−24° 10' 18"
2811	Hya	Gxy	2.5'×0.9'	12	09 h 16 m 11.2 s	−16° 18' 48"
2815	Hya	Gxy	3.5'×1.1'	12.1	09 h 16 m 19.5 s	−23° 38' 00"
2817	Hya	Gxy	2.0'×1.7'	13	09 h 17 m 10.5 s	−04° 45' 12"
2835	Hya	Gxy	6.6'×4.4'	10.6	09 h 17 m 52.8 s	−22° 21' 17"
2837	Hya	**	…	…	09 h 18 m 23.3 s	−16° 28' 54"
2846	Hya	**	…	…	09 h 19 m 40.3 s	−14° 40' 34"
2847	Hya	HIIRgn	…	…	09 h 20 m 08.5 s	−16° 31' 06"

(continued)

Table 10.2 (continued)

NGC #	Const	Type	Size	V Mag	RA (2000)	Dec (2000)
2848	Hya	Gxy	2.7'×1.7'	12.1	09 h 20 m 09.6 s	−16° 31' 34"
2850	Hya	Gxy	0.9'×0.7'	14.2	09 h 20 m 56.9 s	−04° 56' 24"
2851	Hya	Gxy	1.2'×0.5'	13.4	09 h 20 m 30.7 s	−16° 29' 45"
2855	Hya	Gxy	2.5'×2.2'	11.4	09 h 21 m 27.5 s	−11° 54' 34"
2858	Hya	Gxy	1.7'×0.9'	12.8	09 h 22 m 54.9 s	+03° 09' 25"
2861	Hya	Gxy	1.5'×1.4'	12.9	09 h 23 m 36.6 s	+02° 08' 11"
2863	Hya	Gxy	1.0'×0.9'	12.9	09 h 23 m 36.5 s	−10° 26' 01"
2864	Hya	Gxy	1.0'×0.7'	14.6	09 h 24 m 15.4 s	+05° 56' 27"
2865	Hya	Gxy	2.5'×1.8'	11.5	09 h 23 m 30.3 s	−23° 09' 37"
2868	Hya	Gxy	0.6'×0.5'	14.3	09 h 23 m 27.2 s	−10° 25' 46"
2869	Hya	Gxy	1.0'×0.9'	12.9	09 h 23 m 36.5 s	−10° 26' 01"
2876	Hya	Gxy	1.7'×1.1'	13.1	09 h 25 m 13.8 s	−06° 43' 02"
2877	Hya	Gxy	0.5'×0.5'	14.7	09 h 25 m 46.9 s	+02° 13' 44"
2878	Hya	Gxy	0.8'×0.3'	14.4	09 h 25 m 47.5 s	+02° 05' 23"
2879	Hya	Ast	...	...	09 h 25 m 22.5 s	−11° 39' 05"
2881	Hya	Gxy	1.1'×0.9'	13.3	09 h 25 m 54.7 s	−11° 59' 47"
2884	Hya	Gxy	2.0'×1.1'	12.6	09 h 26 m 24.5 s	−11° 33' 21"
2886	Hya	Ast	...	...	09 h 26 m 38.7 s	−21° 44' 16"
2889	Hya	Gxy	2.2'×1.9'	12	09 h 27 m 12.5 s	−11° 38' 37"
2890	Hya	Gxy	0.8'×0.5'	14.5	09 h 26 m 29.8 s	−14° 31' 46"
2897	Hya	Gxy	0.7'×0.4'	14.5	09 h 29 m 45.7 s	+02° 12' 25"
2898	Hya	Gxy	1.0'×0.8'	13.5	09 h 29 m 46.3 s	+02° 03' 51"
2900	Hya	Gxy	1.7'×1.4'	13.2	09 h 30 m 15.2 s	+04° 08' 39"
2902	Hya	Gxy	1.4'×1.2'	12.4	09 h 30 m 52.9 s	−14° 44' 10"
2907	Hya	Gxy	1.8'×1.1'	12.3	09 h 31 m 36.5 s	−16° 44' 07"
2917	Hya	Gxy	1.3'×0.4'	13.7	09 h 34 m 27.0 s	−02° 30' 15"
2920	Hya	Gxy	0.9'×0.6'	13.2	09 h 34 m 12.1 s	−20° 51' 34"
2921	Hya	Gxy	2.8'×1.0'	12.2	09 h 34 m 31.5 s	−20° 55' 14"
2924	Hya	Gxy	1.4'×1.3'	12.1	09 h 35 m 10.9 s	−16° 23' 55"
2935	Hya	Gxy	3.6'×2.8'	11.3	09 h 36 m 44.9 s	−21° 07' 46"
2936	Hya	Gxy	1.3'×1.1'	12.9	09 h 37 m 44.3 s	+02° 45' 40"
2937	Hya	Gxy	2.1'×0.7'	13.6	09 h 37 m 45.0 s	+02° 44' 51"
2945	Hya	Gxy	1.6'×1.2'	12.3	09 h 37 m 41.1 s	−22° 02' 03"
2947	Hya	Gxy	1.5'×1.3'	12.4	09 h 36 m 06.0 s	−12° 26' 14"
2951	Hya	Gxy	0.5'×0.3'	14.7	09 h 39 m 40.7 s	−00° 14' 08"
2952	Hya	Gxy	0.8'×0.4'	14.5	09 h 37 m 36.6 s	−10° 10' 58"
2956	Hya	Gxy	0.9'×0.3'	14.6	09 h 39 m 17.0 s	−19° 06' 04"
2960	Hya	Gxy	1.8'×1.2'	12.6	09 h 40 m 36.4 s	+03° 34' 37"
2962	Hya	Gxy	2.6'×1.9'	11.4	09 h 40 m 53.9 s	+05° 09' 56"
2975	Hya	Gxy	0.8'×0.7'	14.8	09 h 41 m 16.1 s	−16° 40' 27"
2983	Hya	Gxy	2.5'×1.5'	11.8	09 h 43 m 41.1 s	−20° 28' 40"
2986	Hya	Gxy	3.2'×2.8'	10.9	09 h 44 m 16.2 s	−21° 16' 45"
2989	Hya	Gxy	1.7'×0.9'	13	09 h 45 m 25.3 s	−18° 22' 25"
2992	Hya	Gxy	3.5'×1.1'	11.9	09 h 45 m 41.9 s	−14° 19' 34"
2993	Hya	Gxy	1.3'×0.9'	12.5	09 h 45 m 48.3 s	−14° 22' 08"

(continued)

Table 10.2 (continued)

NGC #	Const	Type	Size	V Mag	RA (2000)	Dec (2000)
2996	Hya	Gxy	1.5' × 1.3'	12.6	09 h 46 m 30.5 s	−21° 34' 22"
3025	Hya	Gxy	1.5' × 1.2'	13	09 h 49 m 29.3 s	−21° 44' 37"
3028	Hya	Gxy	1.3' × 1.0'	12.9	09 h 49 m 54.2 s	−19° 11' 04"
3030	Hya	Gxy	1.0' × 0.9'	13.3	09 h 50 m 10.5 s	−12° 13' 35"
3037	Hya	Gxy	1.2' × 1.1'	13.2	09 h 51 m 24.1 s	−27° 00' 40"
3045	Hya	Gxy	1.4' × 0.6'	13.3	09 h 53 m 17.8 s	−18° 38' 44"
3052	Hya	Gxy	2.0' × 1.3'	12.4	09 h 54 m 28.0 s	−18° 38' 24"
3054	Hya	Gxy	3.8' × 2.3'	12	09 h 54 m 28.7 s	−25° 42' 14"
3058	Hya	Gxy	1.2' × 0.6'	13.5	09 h 53 m 35.7 s	−12° 28' 58"
3072	Hya	Gxy	1.9' × 0.6'	12.9	09 h 57 m 23.9 s	−19° 21' 17"
3076	Hya	Gxy	1.0' × 0.9'	13.3	09 h 57 m 37.7 s	−18° 10' 43"
3078	Hya	Gxy	2.5' × 2.1'	11.3	09 h 58 m 24.3 s	−26° 55' 35"
3081	Hya	Gxy	2.1' × 1.6'	12	09 h 59 m 29.5 s	−22° 49' 34"
3085	Hya	Gxy	1.2' × 0.4'	13.1	09 h 59 m 29.2 s	−19° 29' 32"
3091	Hya	Gxy	3.0' × 1.9'	11.3	10 h 00 m 14.1 s	−19° 38' 11"
3096	Hya	Gxy	1.0' × 0.8'	13.2	10 h 00 m 33.1 s	−19° 39' 44"
3109	Hya	Gxy	19.1' × 3.7'	9.9	10 h 03 m 06.7 s	−26° 09' 31"
3112	Hya	Gxy	1.1' × 0.3'	15.3	10 h 03 m 59.0 s	−20° 46' 56"
3124	Hya	Gxy	3.0' × 2.5'	12	10 h 06 m 39.9 s	−19° 13' 20"
3127	Hya	Gxy	1.2' × 0.2'	13.8	10 h 06 m 24.7 s	−16° 07' 36"
3128	Hya	Gxy	1.6' × 0.7'	13.5	10 h 06 m 01.4 s	−16° 07' 23"
3133	Hya	Gxy	0.7' × 0.4'	14.5	10 h 07 m 12.8 s	−11° 57' 54"
3138	Hya	Gxy	1.1' × 0.3'	14.8	10 h 09 m 16.6 s	−11° 57' 24"
3139	Hya	Gxy	1.4' × 1.2'	13.5	10 h 10 m 05.3 s	−11° 46' 42"
3140	Hya	Gxy	0.9' × 0.5'	14.2	10 h 09 m 27.7 s	−16° 37' 42"
3141	Hya	Gxy	0.4' × 0.2'	15.4	10 h 09 m 19.8 s	−16° 39' 12"
3143	Hya	Gxy	0.8' × 0.6'	14.3	10 h 10 m 03.9 s	−12° 34' 54"
3145	Hya	Gxy	3.1' × 1.6'	11.7	10 h 10 m 09.9 s	−12° 26' 07"
3146	Hya	Gxy	1.0' × 0.9'	13.3	10 h 11 m 09.9 s	−20° 52' 15"
3171	Hya	Gxy	1.7' × 1.1'	12.9	10 h 15 m 36.8 s	−20° 38' 51"
3176	Hya	Gxy	0.6' × 0.5'	14.4	10 h 14 m 52.1 s	−20° 00' 45"
3178	Hya	Gxy	1.3' × 0.7'	12.7	10 h 16 m 09.1 s	−15° 47' 30"
3200	Hya	Gxy	4.2' × 1.3'	12.1	10 h 18 m 36.4 s	−17° 58' 57"
3203	Hya	Gxy	2.9' × 0.6'	12.2	10 h 19 m 33.8 s	−26° 41' 54"
3208	Hya	Gxy	1.8' × 1.6'	13.2	10 h 19 m 41.3 s	−25° 48' 54"
3233	Hya	Gxy	1.7' × 0.9'	12.7	10 h 21 m 57.5 s	−22° 16' 06"
3240	Hya	Gxy	1.1' × 0.9'	13.4	10 h 24 m 30.7 s	−21° 47' 31"
3242	Hya	PN	45" × 36"	7.7	10 h 24 m 46.2 s	−18° 38' 34"
3280	Hya	Gxy	0.7' × 0.5'	14.7	10 h 32 m 45.5 s	−12° 38' 14"
3282	Hya	Gxy	1.9' × 0.6'	13.1	10 h 32 m 21.7 s	−22° 18' 08"
3285	Hya	Gxy	2.6' × 1.5'	12.3	10 h 33 m 35.9 s	−27° 27' 15"
3290	Hya	Gxy	1.0' × 0.5'	13.5	10 h 35 m 17.5 s	−17° 16' 37"
3295	Hya	Gxy	0.7' × 0.5'	14.7	10 h 32 m 45.5 s	−12° 38' 14"
3296	Hya	Gxy	0.7' × 0.7'	13.9	10 h 32 m 45.4 s	−12° 43' 02"
3297	Hya	Gxy	0.8' × 0.5'	14.5	10 h 33 m 11.7 s	−12° 40' 18"

(continued)

Table 10.2 (continued)

NGC #	Const	Type	Size	V Mag	RA (2000)	Dec (2000)
3305	Hya	Gxy	1.1'×1.1'	13	10 h 36 m 11.7 s	−27° 09' 45"
3307	Hya	Gxy	0.9'×0.3'	14.5	10 h 36 m 17.2 s	−27° 31' 45"
3308	Hya	Gxy	1.7'×1.3'	12.4	10 h 36 m 22.5 s	−27° 26' 17"
3309	Hya	Gxy	1.9'×1.6'	11.3	10 h 36 m 35.8 s	−27° 31' 08"
3311	Hya	Gxy	3.5'×2.9'	11	10 h 36 m 42.9 s	−27° 31' 43"
3312	Hya	Gxy	3.3'×1.3'	12.1	10 h 37 m 02.5 s	−27° 33' 56"
3313	Hya	Gxy	3.9'×3.2'	11.8	10 h 37 m 25.5 s	−25° 19' 09"
3314	Hya	Gxy	1.6'×0.7'	13	10 h 37 m 12.5 s	−27° 41' 05"
3315	Hya	Gxy	1.1'×1.0'	12.9	10 h 37 m 19.5 s	−27° 11' 28"
3316	Hya	Gxy	1.3'×1.1'	12.7	10 h 37 m 37.2 s	−27° 35' 35"
3317	Hya	***	...	...	10 h 37 m 43.1 s	−27° 31' 11"
3331	Hya	Gxy	1.2'×0.9'	13.4	10 h 40 m 09.0 s	−23° 49' 15"
3335	Hya	Gxy	1.1'×0.9'	13.1	10 h 39 m 34.1 s	−23° 55' 20"
3336	Hya	Gxy	1.9'×1.5'	12.5	10 h 40 m 16.7 s	−27° 46' 39"
3355	Hya	Gxy	3.2'×0.7'	13	10 h 42 m 37.8 s	−23° 56' 08"
3369	Hya	Gxy	1.4'×0.8'	13.7	10 h 46 m 44.7 s	−25° 14' 38"
3383	Hya	Gxy	1.4'×1.1'	12.9	10 h 47 m 19.1 s	−24° 26' 18"
3390	Hya	Gxy	3.5'×0.6'	12.6	10 h 48 m 04.4 s	−31° 32' 02"
3393	Hya	Gxy	2.2'×2.0'	12.4	10 h 48 m 23.5 s	−25° 09' 43"
3402	Hya	Gxy	2.1'×2.1'	11.9	10 h 50 m 26.1 s	−12° 50' 44"
3404	Hya	Gxy	2.1'×0.5'	13.2	10 h 50 m 18.1 s	−12° 06' 31"
3409	Hya	Gxy	1.2'×0.2'	14.1	10 h 50 m 20.3 s	−17° 02' 38"
3411	Hya	Gxy	2.1'×2.1'	11.9	10 h 50 m 26.1 s	−12° 50' 44"
3420	Hya	Gxy	1.3'×1.1'	13.8	10 h 50 m 09.7 s	−17° 14' 33"
3421	Hya	Gxy	2.0'×1.6'	13.7	10 h 50 m 57.6 s	−12° 26' 54"
3422	Hya	Gxy	1.3'×0.3'	14.1	10 h 51 m 17.3 s	−12° 24' 10"
3450	Hya	Gxy	2.5'×2.2'	12.1	10 h 48 m 03.7 s	−20° 50' 58"
3453	Hya	Gxy	1.1'×0.6'	13.2	10 h 53 m 40.4 s	−21° 47' 35"
3463	Hya	Gxy	1.5'×0.7'	13.2	10 h 55 m 13.4 s	−26° 08' 25"
3464	Hya	Gxy	2.6'×1.7'	12.6	10 h 54 m 40.1 s	−21° 04' 01"
3483	Hya	Gxy	1.7'×1.3'	12.3	10 h 59 m 00.1 s	−28° 28' 42"
3585	Hya	Gxy	4.7'×2.6'	10.2	11 h 13 m 17.1 s	−26° 45' 18"
3606	Hya	Gxy	1.5'×1.3'	12.5	11 h 16 m 15.6 s	−33° 49' 38"
3617	Hya	Gxy	1.8'×1.3'	12.5	11 h 17 m 50.9 s	−26° 08' 02"
3621	Hya	Gxy	12.3'×7.1'	9.2	11 h 18 m 16.5 s	−32° 48' 50"
3673	Hya	Gxy	3.6'×2.4'	11.7	11 h 25 m 12.7 s	−26° 44' 12"
3717	Hya	Gxy	6.0'×1.1'	11.6	11 h 31 m 32.1 s	−30° 18' 25"
3885	Hya	Gxy	2.4'×1.0'	12.2	11 h 46 m 46.4 s	−27° 55' 17"
3904	Hya	Gxy	2.7'×1.9'	11	11 h 49 m 13.2 s	−29° 16' 35"
3923	Hya	Gxy	5.9'×3.9'	10	11 h 51 m 01.8 s	−28° 48' 21"
3936	Hya	Gxy	3.9'×0.6'	12.2	11 h 52 m 20.5 s	−26° 54' 14"
4087	Hya	Gxy	2.1'×1.7'	12.2	12 h 05 m 35.1 s	−26° 31' 21"
4105	Hya	Gxy	2.7'×2.0'	10.7	12 h 06 m 40.7 s	−29° 45' 38"
4106	Hya	Gxy	1.6'×1.3'	11.2	12 h 06 m 45.3 s	−29° 46' 06"
4304	Hya	Gxy	2.6'×2.5'	11.8	12 h 22 m 12.7 s	−33° 29' 05"

(continued)

Table 10.2 (continued)

NGC #	Const	Type	Size	V Mag	RA (2000)	Dec (2000)
4456	Hya	Gxy	1.2' × 0.6'	13.5	12 h 27 m 52.2 s	−30° 05' 54"
4590	Hya	GC	12'	7.3	12 h 39 m 28.0 s	−26° 44' 34"
4806	Hya	Gxy	1.2' × 1.0'	12.9	12 h 56 m 12.7 s	−29° 30' 14"
4831	Hya	Gxy	1.7' × 0.9'	12.5	12 h 57 m 36.7 s	−27° 17' 30"
4955	Hya	Gxy	1.8' × 1.3'	12	13 h 06 m 04.8 s	−29° 45' 18"
4965	Hya	Gxy	2.6' × 2.0'	12.3	13 h 07 m 09.2 s	−28° 13' 43"
4968	Hya	Gxy	1.9' × 0.9'	12.9	13 h 07 m 05.9 s	−23° 40' 38"
4970	Hya	Gxy	1.8' × 1.0'	12.3	13 h 07 m 33.8 s	−24° 00' 34"
4980	Hya	Gxy	1.7' × 0.7'	12.8	13 h 09 m 10.1 s	−28° 38' 30"
4993	Hya	Gxy	1.3' × 1.1'	12.4	13 h 09 m 47.6 s	−23° 23' 03"
4994	Hya	Gxy	1.2' × 1.0'	12.4	13 h 09 m 47.6 s	−23° 23' 03"
5042	Hya	Gxy	4.2' × 2.2'	12	13 h 15 m 30.5 s	−23° 59' 00"
5048	Hya	Gxy	1.5' × 0.8'	12.8	13 h 16 m 07.7 s	−28° 24' 37"
5051	Hya	Gxy	1.3' × 0.6'	13.4	13 h 16 m 19.9 s	−28° 17' 11"
5061	Hya	Gxy	3.5' × 3.0'	10.5	13 h 18 m 05.2 s	−26° 50' 16"
5078	Hya	Gxy	4.0' × 1.9'	10.8	13 h 19 m 49.8 s	−27° 24' 36"
5085	Hya	Gxy	3.4' × 3.0'	11.5	13 h 20 m 17.4 s	−24° 26' 26"
5101	Hya	Gxy	5.4' × 4.6'	10.6	13 h 21 m 46.1 s	−27° 25' 47"
5135	Hya	Gxy	2.6' × 1.8'	12.1	13 h 25 m 44.2 s	−29° 50' 01"
5150	Hya	Gxy	1.3' × 1.0'	12.7	13 h 27 m 36.4 s	−29° 33' 43"
5152	Hya	Gxy	2.0' × 0.6'	12.7	13 h 27 m 51.5 s	−29° 37' 11"
5153	Hya	Gxy	2.1' × 1.4'	12.4	13 h 27 m 54.3 s	−29° 37' 06"
5182	Hya	Gxy	1.9' × 1.3'	12.6	13 h 30 m 40.9 s	−28° 08' 59"
5236	Hya	Gxy	15.4' × 13.1'	7.8	13 h 37 m 00.2 s	−29° 52' 04"
5260	Hya	Gxy	1.6' × 1.4'	13.1	13 h 40 m 19.8 s	−23° 51' 29"
5264	Hya	Gxy	2.5' × 1.5'	12.2	13 h 41 m 36.5 s	−29° 54' 44"
5328	Hya	Gxy	1.7' × 1.3'	11.9	13 h 52 m 53.0 s	−28° 29' 23"
5330	Hya	Gxy	0.4' × 0.3'	13.9	13 h 52 m 59.2 s	−28° 28' 16"
5393	Hya	Gxy	1.2' × 1.0'	13.3	14 h 00 m 31.9 s	−28° 52' 28"
5464	Hya	Gxy	1.3' × 0.8'	12.9	14 h 07 m 04.3 s	−30° 01' 03"
5495	Hya	Gxy	1.4' × 1.1'	13	14 h 12 m 23.1 s	−27° 06' 27"
5556	Hya	Gxy	4.0' × 3.2'	11.9	14 h 20 m 34.2 s	−29° 14' 32"
5592	Hya	Gxy	1.5' × 0.8'	12.9	14 h 23 m 55.1 s	−28° 41' 16"
5626	Hya	Gxy	1.1' × 0.9'	13.1	14 h 29 m 48.7 s	−29° 44' 56"
5694	Hya	GC	3.6'	10.2	14 h 39 m 36.5 s	−26° 32' 18"

Table 10.3 These NGC objects are found in the constellations of Hydrus through Reticulum (Copyright © The NGC/IC Project, LLC – All rights reserved. Used with permission)

NGC #	Const	Type	Size	V Mag	RA (2000)	Dec (2000)
602	Hyi	OC + Neb	3.4'	...	01 h 29 m 26.4 s	−73° 33' 26"
643	Hyi	OC	...	...	01 h 35 m 00.8 s	−75° 33' 26"
646	Hyi	Gxy	1.3'×1.0'	13.7	01 h 37 m 22.1 s	−64° 53' 45"
796	Hyi	OC + Neb	3.4'	...	01 h 56 m 44.5 s	−74° 13' 09"
802	Hyi	Gxy	0.9'×0.6'	13.3	01 h 59 m 07.0 s	−67° 52' 11"
813	Hyi	Gxy	1.3'×0.9'	13	02 h 01 m 37.0 s	−68° 26' 28"
1466	Hyi	OC	2.3'	11.4	03 h 44 m 33.4 s	−71° 40' 18"
1473	Hyi	Gxy	1.5'×0.8'	13.3	03 h 47 m 26.3 s	−68° 13' 16"
1511	Hyi	Gxy	5.0'×2.0'	11.4	03 h 59 m 36.9 s	−67° 38' 07"
1557	Hyi	Ast	10.0'	...	04 h 13 m 14.9 s	−70° 26' 19"
1629	Hyi	OC	2.0'	12.7	04 h 29 m 37.0 s	−71° 50' 18"
6918	Ind	Gxy	0.9'×0.7'	13.7	20 h 30 m 46.9 s	−47° 28' 28"
6935	Ind	Gxy	2.0'×1.7'	12.1	20 h 38 m 19.9 s	−52° 06' 39"
6937	Ind	Gxy	2.3'×2.0'	12.9	20 h 38 m 45.7 s	−52° 08' 40"
6942	Ind	Gxy	2.0'×1.5'	12	20 h 40 m 37.8 s	−54° 18' 10"
6948	Ind	Gxy	2.2'×1.0'	13.1	20 h 43 m 28.6 s	−53° 21' 27"
6970	Ind	Gxy	1.4'×0.9'	12.5	20 h 52 m 09.3 s	−48° 46' 43"
6982	Ind	Gxy	1.1'×0.6'	13.5	20 h 57 m 18.2 s	−51° 51' 47"
6984	Ind	Gxy	1.8'×1.2'	12.7	20 h 57 m 53.9 s	−51° 52' 15"
6987	Ind	Gxy	1.4'×1.2'	12.7	20 h 58 m 10.3 s	−48° 37' 49"
6990	Ind	Gxy	1.1'×0.5'	13.3	20 h 59 m 56.9 s	−55° 33' 44"
7002	Ind	Gxy	1.5'×1.2'	12.4	21 h 03 m 45.0 s	−49° 01' 47"
7004	Ind	Gxy	1.3'×0.6'	13.8	21 h 04 m 01.9 s	−49° 06' 52"
7007	Ind	Gxy	1.9'×1.1'	12	21 h 05 m 27.7 s	−52° 33' 09"
7014	Ind	Gxy	1.9'×1.5'	12	21 h 07 m 51.9 s	−47° 10' 44"
7022	Ind	Gxy	1.5'×1.1'	13.1	21 h 09 m 35.1 s	−49° 18' 14"
7029	Ind	Gxy	2.6'×1.4'	11.7	21 h 11 m 51.5 s	−49° 17' 03"
7038	Ind	Gxy	3.2'×1.6'	12.1	21 h 15 m 07.6 s	−47° 13' 14"
7041	Ind	Gxy	3.6'×1.5'	11.1	21 h 16 m 32.1 s	−48° 21' 49"
7049	Ind	Gxy	4.2'×3.2'	10.6	21 h 19 m 00.2 s	−48° 33' 43"
7061	Ind	Gxy	1.2'×0.7'	13.2	21 h 27 m 26.6 s	−49° 03' 49"
7064	Ind	Gxy	3.8'×0.6'	12.3	21 h 29 m 03.2 s	−52° 46' 04"
7083	Ind	Gxy	3.9'×2.3'	11.4	21 h 35 m 44.5 s	−63° 54' 14"
7090	Ind	Gxy	7.4'×1.3'	11	21 h 36 m 28.5 s	−54° 33' 24"
7096	Ind	Gxy	1.9'×1.6'	12	21 h 41 m 18.9 s	−63° 54' 30"
7106	Ind	Gxy	1.7'×1.0'	13.2	21 h 42 m 36.6 s	−52° 41' 57"
7123	Ind	Gxy	3.0'×1.4'	12.6	21 h 50 m 46.4 s	−70° 20' 04"
7124	Ind	Gxy	3.0'×1.3'	12.5	21 h 48 m 05.3 s	−50° 33' 58"
7125	Ind	Gxy	3.1'×2.1'	12.2	21 h 49 m 16.1 s	−60° 42' 47"
7126	Ind	Gxy	2.8'×1.3'	12.6	21 h 49 m 18.3 s	−60° 36' 36"
7140	Ind	Gxy	6.0'×4.0'	12.1	21 h 52 m 15.1 s	−55° 34' 12"
7141	Ind	Gxy	4.2'×3.0'	12.1	21 h 52 m 15.1 s	−55° 34' 12"
7151	Ind	Gxy	3.0'×1.2'	13	21 h 55 m 03.9 s	−50° 39' 31"
7155	Ind	Gxy	2.2'×1.8'	12.1	21 h 56 m 09.7 s	−49° 31' 21"

(continued)

Table 10.3 (continued)

NGC #	Const	Type	Size	V Mag	RA (2000)	Dec (2000)
7168	Ind	Gxy	2.0'×1.5'	11.8	22 h 02 m 07.2 s	−51° 44' 38"
7179	Ind	Gxy	2.0'×0.8'	12.8	22 h 04 m 49.1 s	−64° 02' 51"
7191	Ind	Gxy	1.6'×0.6'	13.3	22 h 06 m 51.2 s	−64° 38' 03"
7192	Ind	Gxy	2.0'×1.9'	11.4	22 h 06 m 50.3 s	−64° 18' 57"
7196	Ind	Gxy	2.5'×1.8'	11.5	22 h 05 m 54.9 s	−50° 07' 11"
7199	Ind	Gxy	1.1'×0.9'	13.1	22 h 08 m 30.0 s	−64° 42' 24"
7200	Ind	Gxy	1.6'×1.2'	12.8	22 h 07 m 09.5 s	−49° 59' 46"
7205	Ind	Gxy	4.1'×2.0'	11.1	22 h 08 m 33.1 s	−57° 26' 34"
7216	Ind	Gxy	1.7'×1.2'	12.5	22 h 12 m 35.9 s	−68° 39' 45"
7633	Ind	Gxy	2.1'×1.2'	12.6	23 h 23 m 03.1 s	−67° 39' 15"
7655	Ind	Gxy	0.7'×0.6'	13.3	23 h 26 m 45.9 s	−68° 01' 41"
7197	Lac	Gxy	1.6'×0.8'	13	22 h 02 m 58.0 s	+41° 03' 34"
7209	Lac	OC	25'	7.7	22 h 05 m 07.8 s	+46° 29' 01"
7223	Lac	Gxy	1.7'×1.2'	12.4	22 h 10 m 09.0 s	+41° 00' 59"
7227	Lac	Gxy	1.3'×0.6'	13.6	22 h 11 m 31.3 s	+38° 43' 16"
7228	Lac	Gxy	1.3'×1.3'	13.7	22 h 11 m 48.5 s	+38° 41' 55"
7231	Lac	Gxy	1.9'×0.7'	13.2	22 h 12 m 30.5 s	+45° 19' 43"
7240	Lac	Gxy	0.7'×0.7'	14.3	22 h 15 m 22.6 s	+37° 16' 51"
7242	Lac	Gxy	1.2'×1.0'	12.9	22 h 15 m 39.4 s	+37° 17' 55"
7243	Lac	OC	21'	6.4	22 h 15 m 08.5 s	+49° 53' 51"
7245	Lac	OC	5'	9.2	22 h 15 m 11.5 s	+54° 20' 33"
7248	Lac	Gxy	1.7'×0.9'	12.3	22 h 16 m 52.7 s	+40° 30' 16"
7250	Lac	Gxy	1.7'×0.8'	12.4	22 h 18 m 17.7 s	+40° 33' 46"
7263	Lac	Gxy	0.7'×0.5'	14.7	22 h 21 m 45.0 s	+36° 21' 01"
7264	Lac	Gxy	2.2'×0.3'	14	22 h 22 m 13.8 s	+36° 23' 12"
7265	Lac	Gxy	2.4'×1.9'	12.4	22 h 22 m 27.4 s	+36° 12' 35"
7273	Lac	Gxy	0.8'×0.5'	13.8	22 h 24 m 09.2 s	+36° 11' 59"
7274	Lac	Gxy	1.5'×1.5'	12.9	22 h 24 m 11.1 s	+36° 07' 33"
7276	Lac	Gxy	0.9'×0.9'	13.9	22 h 24 m 14.4 s	+36° 05' 16"
7282	Lac	Gxy	2.5'×1.0'	13.9	22 h 25 m 53.9 s	+40° 18' 55"
7295	Lac	OC	4.0'	...	22 h 28 m 02.8 s	+52° 17' 21"
7296	Lac	OC	4.0'	...	22 h 28 m 02.8 s	+52° 17' 21"
7330	Lac	Gxy	1.8'×1.7'	12.3	22 h 36 m 56.5 s	+38° 32' 49"
7379	Lac	Gxy	1.1'×0.9'	13.6	22 h 47 m 32.8 s	+40° 14' 20"
7394	Lac	OC	10'×3'	...	22 h 50 m 23.8 s	+52° 08' 07"
7395	Lac	Gxy	1.2'×1.1'	13.9	22 h 51 m 03.0 s	+37° 05' 16"
7426	Lac	Gxy	1.7'×1.4'	12.4	22 h 56 m 02.9 s	+36° 21' 40"
2862	Leo	Gxy	2.5'×0.5'	13	09 h 24 m 55.1 s	+26° 46' 30"
2871	Leo	*	...	...	09 h 25 m 39.4 s	+11° 26' 36"
2872	Leo	Gxy	2.1'×1.8'	11.7	09 h 25 m 42.6 s	+11° 25' 54"
2873	Leo	Gxy	0.8'×0.2'	15.4	09 h 25 m 48.5 s	+11° 27' 15"
2874	Leo	Gxy	2.4'×0.7'	12.8	09 h 25 m 47.3 s	+11° 25' 28"
2875	Leo	GxyCld	...	...	09 h 25 m 48.8 s	+11° 25' 54"
2882	Leo	Gxy	1.5'×0.8'	12.6	09 h 26 m 36.2 s	+07° 57' 15"
2885	Leo	Gxy	0.7'×0.4'	13.9	09 h 27 m 18.5 s	+23° 01' 12"

(continued)

Table 10.3 (continued)

NGC #	Const	Type	Size	V Mag	RA (2000)	Dec (2000)
2893	Leo	Gxy	1.1'×1.0'	13	09 h 30 m 16.9 s	+29° 32' 23"
2894	Leo	Gxy	1.9'×1.0'	12.6	09 h 29 m 30.4 s	+07° 43' 07"
2896	Leo	Gxy	0.8'×0.8'	14	09 h 30 m 16.9 s	+23° 39' 45"
2901	Leo	NF	...	...	09 h 32 m 34.2 s	+31° 06' 42"
2903	Leo	Gxy	12.6'×6.0'	9	09 h 32 m 09.9 s	+21° 30' 07"
2905	Leo	GxyCld	...	...	09 h 32 m 11.9 s	+21° 31' 05"
2906	Leo	Gxy	1.4'×0.9'	12.9	09 h 32 m 06.2 s	+08° 26' 30"
2911	Leo	Gxy	4.1'×3.2'	11.3	09 h 33 m 46.1 s	+10° 09' 09"
2912	Leo	*	...	...	09 h 33 m 56.9 s	+10° 11' 33"
2913	Leo	Gxy	1.1'×0.7'	13.4	09 h 34 m 02.7 s	+09° 28' 45"
2914	Leo	Gxy	1.0'×0.7'	13.3	09 h 34 m 02.7 s	+10° 06' 31"
2916	Leo	Gxy	2.5'×1.7'	11.8	09 h 34 m 57.9 s	+21° 42' 29"
2918	Leo	Gxy	1.4'×1.0'	12.7	09 h 35 m 44.1 s	+31° 42' 20"
2919	Leo	Gxy	1.7'×0.6'	13.1	09 h 34 m 47.5 s	+10° 17' 01"
2923	Leo	Gxy	0.5'×0.4'	14.2	09 h 36 m 03.9 s	+16° 45' 37"
2926	Leo	Gxy	0.8'×0.8'	13.6	09 h 37 m 31.0 s	+32° 50' 29"
2927	Leo	Gxy	1.9'×1.1'	13.1	09 h 37 m 15.2 s	+23° 35' 26"
2928	Leo	Gxy	1.1'×0.6'	14.7	09 h 37 m 10.2 s	+16° 58' 39"
2929	Leo	Gxy	1.2'×0.3'	13.8	09 h 37 m 30.0 s	+23° 09' 38"
2930	Leo	Gxy	0.7'×0.6'	14.2	09 h 37 m 32.6 s	+23° 12' 11"
2931	Leo	Gxy	0.7'×0.5'	14.2	09 h 37 m 37.7 s	+23° 14' 27"
2933	Leo	Gxy	1.0'×0.4'	14.6	09 h 37 m 55.0 s	+17° 00' 53"
2934	Leo	Gxy	0.2'×0.2'	16	09 h 37 m 55.1 s	+17° 03' 16"
2939	Leo	Gxy	2.5'×0.9'	12.7	09 h 38 m 08.1 s	+09° 31' 23"
2940	Leo	Gxy	0.9'×0.7'	13.6	09 h 38 m 05.2 s	+09° 36' 59"
2941	Leo	Gxy	0.8'×0.4'	15	09 h 38 m 24.4 s	+17° 02' 41"
2943	Leo	Gxy	2.2'×1.2'	12.5	09 h 38 m 32.8 s	+17° 01' 51"
2944	Leo	Gxy	1.1'×0.4'	14.2	09 h 39 m 16.7 s	+32° 18' 39"
2946	Leo	Gxy	1.2'×0.4'	14	09 h 39 m 01.6 s	+17° 01' 30"
2948	Leo	Gxy	1.4'×0.9'	13.1	09 h 38 m 59.1 s	+06° 57' 19"
2949	Leo	Gxy	0.7'×0.5'	14.4	09 h 39 m 55.1 s	+16° 47' 06"
2953	Leo	*	...	...	09 h 40 m 27.7 s	+14° 50' 37"
2954	Leo	Gxy	1.7'×1.1'	12.5	09 h 40 m 24.1 s	+14° 55' 20"
2958	Leo	Gxy	1.0'×0.8'	13.4	09 h 40 m 41.7 s	+11° 53' 18"
2964	Leo	Gxy	2.9'×1.6'	11.4	09 h 42 m 54.3 s	+31° 50' 49"
2968	Leo	Gxy	2.3'×1.6'	12.5	09 h 43 m 12.0 s	+31° 55' 44"
2970	Leo	Gxy	0.6'×0.5'	14	09 h 43 m 31.1 s	+31° 58' 37"
2981	Leo	Gxy	1.2'×1.0'	13.8	09 h 44 m 56.6 s	+31° 05' 51"
2984	Leo	Gxy	0.7'×0.7'	13.4	09 h 43 m 40.4 s	+11° 03' 38"
2988	Leo	Gxy	1.4'×1.1'	15	09 h 46 m 47.9 s	+22° 00' 45"
2991	Leo	Gxy	1.4'×1.1'	12.5	09 h 46 m 50.0 s	+22° 00' 49"
2994	Leo	Gxy	1.3'×1.0'	13.2	09 h 47 m 16.1 s	+22° 05' 23"
3011	Leo	Gxy	0.9'×0.8'	13.4	09 h 49 m 41.3 s	+32° 13' 16"
3016	Leo	Gxy	1.2'×0.9'	13.2	09 h 49 m 50.7 s	+12° 41' 43"
3019	Leo	Gxy	1.0'×0.7'	15.2	09 h 50 m 07.2 s	+12° 44' 46"

(continued)

Table 10.3 (continued)

NGC #	Const	Type	Size	V Mag	RA (2000)	Dec (2000)
3020	Leo	Gxy	3.2'×1.6'	12.1	09 h 50 m 06.6 s	+12° 48' 50"
3024	Leo	Gxy	2.1'×0.5'	13.3	09 h 50 m 27.5 s	+12° 45' 54"
3026	Leo	Gxy	2.6'×0.7'	13.1	09 h 50 m 55.4 s	+28° 33' 04"
3032	Leo	Gxy	2.0'×1.8'	11.7	09 h 52 m 08.1 s	+29° 14' 10"
3040	Leo	Gxy	1.6'×0.6'	13.4	09 h 53 m 05.1 s	+19° 25' 56"
3041	Leo	Gxy	3.7'×2.4'	12	09 h 53 m 07.1 s	+16° 40' 38"
3048	Leo	Gxy	0.6'×0.3'	14.1	09 h 54 m 57.4 s	+16° 27' 24"
3049	Leo	Gxy	2.2'×1.4'	12.3	09 h 54 m 49.7 s	+09° 16' 17"
3053	Leo	Gxy	1.8'×0.9'	12.9	09 h 55 m 33.5 s	+16° 25' 57"
3060	Leo	Gxy	2.2'×0.6'	13.2	09 h 56 m 19.2 s	+16° 49' 52"
3067	Leo	Gxy	2.5'×0.9'	12.2	09 h 58 m 21.5 s	+32° 22' 11"
3068	Leo	Gxy	1.1'×0.9'	14.4	09 h 58 m 40.1 s	+28° 52' 39"
3069	Leo	Gxy	0.6'×0.3'	14.2	09 h 57 m 56.7 s	+10° 25' 57"
3070	Leo	Gxy	1.4'×1.4'	12.1	09 h 58 m 06.9 s	+10° 21' 35"
3071	Leo	Gxy	0.5'×0.3'	14.5	09 h 58 m 53.0 s	+31° 37' 13"
3075	Leo	Gxy	1.2'×0.8'	13.8	09 h 58 m 56.2 s	+14° 25' 07"
3080	Leo	Gxy	0.9'×0.8'	13.6	09 h 59 m 55.9 s	+13° 02' 36"
3088	Leo	Gxy	1.4'×0.7'	13.7	10 h 01 m 08.3 s	+22° 24' 21"
3094	Leo	Gxy	2.0'×1.4'	12.5	10 h 01 m 26.0 s	+15° 46' 13"
3098	Leo	Gxy	2.3'×0.6'	12.1	10 h 02 m 16.7 s	+24° 42' 39"
3107	Leo	Gxy	0.7'×0.6'	13.6	10 h 04 m 22.5 s	+13° 37' 17"
3119	Leo	Gxy	1.1'×0.8'	12.9	10 h 06 m 51.9 s	+14° 22' 26"
3121	Leo	Gxy	1.7'×1.4'	12.7	10 h 06 m 51.9 s	+14° 22' 26"
3129	Leo	**	...	...	10 h 08 m 19.3 s	+18° 25' 50"
3130	Leo	Gxy	1.0'×0.6'	13.6	10 h 08 m 12.3 s	+09° 58' 36"
3131	Leo	Gxy	2.4'×0.7'	13.2	10 h 08 m 36.5 s	+18° 13' 52"
3134	Leo	Gxy	0.8'×0.2'	13.7	10 h 12 m 29.2 s	+12° 22' 38"
3153	Leo	Gxy	2.1'×0.9'	12.9	10 h 12 m 50.5 s	+12° 39' 58"
3154	Leo	Gxy	1.0'×0.5'	13.5	10 h 13 m 01.3 s	+17° 02' 03"
3162	Leo	Gxy	3.0'×2.5'	11.4	10 h 13 m 31.5 s	+22° 44' 17"
3177	Leo	Gxy	1.4'×1.2'	12.3	10 h 16 m 34.1 s	+21° 07' 23"
3185	Leo	Gxy	2.3'×1.6'	12	10 h 17 m 38.7 s	+21° 41' 18"
3186	Leo	Gxy	0.4'×0.4'	14.2	10 h 15 m 53.3 s	+06° 57' 49"
3187	Leo	Gxy	3.0'×1.3'	12.7	10 h 17 m 48.1 s	+21° 52' 23"
3189	Leo	GxyCld	...	...	10 h 18 m 04.3 s	+21° 49' 54"
3190	Leo	Gxy	4.4'×1.5'	11.1	10 h 18 m 05.9 s	+21° 49' 59"
3193	Leo	Gxy	3.0'×2.7'	11.2	10 h 18 m 24.9 s	+21° 53' 37"
3196	Leo	Gxy	0.3'×0.2'	14.9	10 h 18 m 49.0 s	+27° 40' 07"
3204	Leo	Gxy	1.3'×0.9'	13.7	10 h 20 m 11.1 s	+27° 49' 01"
3209	Leo	Gxy	1.3'×1.1'	12.8	10 h 20 m 38.5 s	+25° 30' 17"
3213	Leo	Gxy	1.1'×0.9'	13.7	10 h 21 m 17.3 s	+19° 39' 07"
3216	Leo	Gxy	1.4'×1.2'	13.5	10 h 21 m 41.2 s	+23° 55' 22"
3217	Leo	Gxy	0.5'×0.4'	14.6	10 h 23 m 32.6 s	+10° 57' 35"
3221	Leo	Gxy	3.2'×0.7'	13.3	10 h 22 m 20.1 s	+21° 34' 07"
3222	Leo	Gxy	1.3'×1.1'	13.2	10 h 22 m 34.5 s	+19° 53' 12"

(continued)

Table 10.3 (continued)

NGC #	Const	Type	Size	V Mag	RA (2000)	Dec (2000)
3226	Leo	Gxy	3.2'×2.8'	11.7	10 h 23 m 27.0 s	+19° 53' 54"
3227	Leo	Gxy	5.4'×3.6'	10.9	10 h 23 m 30.6 s	+19° 51' 55"
3230	Leo	Gxy	2.3'×1.1'	12.9	10 h 23 m 43.9 s	+12° 34' 03"
3239	Leo	Gxy	5.0'×3.3'	11.9	10 h 25 m 05.8 s	+17° 09' 39"
3248	Leo	Gxy	2.5'×1.1'	12.5	10 h 27 m 45.4 s	+22° 50' 48"
3251	Leo	Gxy	2.0'×0.4'	13.5	10 h 29 m 16.8 s	+26° 05' 58"
3253	Leo	Gxy	1.2'×1.1'	13.4	10 h 28 m 27.5 s	+12° 42' 14"
3270	Leo	Gxy	3.2'×0.8'	13	10 h 31 m 29.9 s	+24° 52' 10"
3274	Leo	Gxy	2.1'×1.0'	12.6	10 h 32 m 17.1 s	+27° 40' 07"
3279	Leo	Gxy	2.9'×0.3'	13.5	10 h 34 m 42.7 s	+11° 11' 49"
3287	Leo	Gxy	2.1'×1.0'	12.4	10 h 34 m 47.3 s	+21° 38' 51"
3299	Leo	Gxy	2.2'×1.7'	12.8	10 h 36 m 23.9 s	+12° 42' 26"
3300	Leo	Gxy	1.9'×1.0'	12.2	10 h 36 m 38.5 s	+14° 10' 15"
3301	Leo	Gxy	3.5'×1.0'	11.5	10 h 36 m 56.0 s	+21° 52' 55"
3303	Leo	Gxy	3.0'×2.1'	13.5	10 h 36 m 59.5 s	+18° 08' 13"
3306	Leo	Gxy	1.3'×0.5'	13.6	10 h 37 m 10.2 s	+12° 39' 09"
3328	Leo	**	...	...	10 h 39 m 40.3 s	+09° 12' 50"
3332	Leo	Gxy	1.4'×1.4'	12.4	10 h 40 m 28.5 s	+09° 10' 57"
3338	Leo	Gxy	5.9'×3.6'	10.9	10 h 42 m 07.5 s	+13° 44' 48"
3342	Leo	Gxy	1.4'×1.4'	12.5	10 h 40 m 28.5 s	+09° 10' 57"
3345	Leo	**	...	...	10 h 43 m 32.0 s	+11° 59' 07"
3346	Leo	Gxy	2.9'×2.5'	11.9	10 h 43 m 38.9 s	+14° 52' 18"
3349	Leo	Gxy	0.6'×0.5'	14.5	10 h 43 m 50.7 s	+06° 45' 46"
3351	Leo	Gxy	7.4'×5.0'	10	10 h 43 m 57.7 s	+11° 42' 13"
3352	Leo	Gxy	1.6'×1.2'	12.8	10 h 44 m 14.9 s	+22° 22' 15"
3356	Leo	Gxy	1.7'×0.8'	12.8	10 h 44 m 12.2 s	+06° 45' 33"
3357	Leo	Gxy	1.4'×1.3'	12.8	10 h 44 m 20.9 s	+14° 05' 02"
3362	Leo	Gxy	1.4'×1.1'	13	10 h 44 m 51.8 s	+06° 35' 48"
3363	Leo	Gxy	1.3'×0.8'	13.8	10 h 45 m 09.5 s	+22° 04' 42"
3367	Leo	Gxy	2.5'×2.2'	11.6	10 h 46 m 34.8 s	+13° 45' 02"
3368	Leo	Gxy	7.6'×5.2'	9.3	10 h 46 m 45.9 s	+11° 49' 25"
3370	Leo	Gxy	3.2'×1.8'	11.8	10 h 47 m 04.0 s	+17° 16' 24"
3371	Leo	Gxy	5.4'×2.7'	9.9	10 h 48 m 16.9 s	+12° 37' 46"
3373	Leo	Gxy	2.7'×1.5'	11.8	10 h 48 m 28.0 s	+12° 31' 58"
3377	Leo	Gxy	5.2'×3.0'	10.1	10 h 47 m 42.2 s	+13° 59' 08"
3379	Leo	Gxy	5.4'×4.8'	9.4	10 h 47 m 49.7 s	+12° 34' 53"
3384	Leo	Gxy	5.5'×2.5'	9.7	10 h 48 m 16.9 s	+12° 37' 46"
3388	Leo	Gxy	0.9'×0.9'	13.3	10 h 51 m 25.5 s	+08° 34' 00"
3389	Leo	Gxy	2.8'×1.3'	11.8	10 h 48 m 28.0 s	+12° 31' 58"
3391	Leo	Gxy	1.0'×0.7'	13	10 h 48 m 56.3 s	+14° 13' 11"
3399	Leo	Gxy	1.3'×1.3'	13.3	10 h 49 m 27.6 s	+16° 13' 06"
3405	Leo	Gxy	0.9'×0.9'	14.4	10 h 49 m 43.2 s	+16° 14' 18"
3412	Leo	Gxy	3.6'×2.0'	10.5	10 h 50 m 53.3 s	+13° 24' 43"
3417	Leo	Gxy	0.6'×0.3'	14.8	10 h 51 m 01.7 s	+08° 28' 24"
3419	Leo	Gxy	1.2'×1.0'	12.6	10 h 51 m 17.7 s	+13° 56' 45"

(continued)

Table 10.3 (continued)

NGC #	Const	Type	Size	V Mag	RA (2000)	Dec (2000)
3425	Leo	Gxy	1.0' × 1.0'	13.2	10 h 51 m 25.5 s	+08° 34' 00"
3426	Leo	Gxy	1.1' × 0.8'	13.2	10 h 51 m 41.7 s	+18° 28' 51"
3427	Leo	Gxy	1.1' × 0.5'	13.4	10 h 51 m 26.3 s	+08° 17' 54"
3428	Leo	Gxy	1.6' × 0.7'	13.4	10 h 51 m 29.5 s	+09° 16' 46"
3429	Leo	Gxy	1.6' × 0.7'	13.4	10 h 51 m 29.5 s	+09° 16' 46"
3433	Leo	Gxy	3.5' × 3.2'	11.8	10 h 52 m 03.9 s	+10° 08' 53"
3434	Leo	Gxy	2.1' × 1.9'	12.3	10 h 51 m 58.1 s	+03° 47' 31"
3436	Leo	Gxy	0.7' × 0.5'	13.9	10 h 52 m 27.5 s	+08° 05' 39"
3437	Leo	Gxy	2.5' × 0.8'	12.3	10 h 52 m 35.7 s	+22° 56' 04"
3438	Leo	Gxy	0.8' × 0.8'	13.4	10 h 52 m 26.0 s	+10° 32' 50"
3439	Leo	Gxy	0.4' × 0.3'	13.9	10 h 52 m 25.7 s	+08° 33' 30"
3441	Leo	Gxy	0.7' × 0.4'	13.5	10 h 52 m 31.1 s	+07° 13' 30"
3443	Leo	Gxy	2.6' × 1.3'	13.2	10 h 53 m 00.2 s	+17° 34' 25"
3444	Leo	Gxy	1.1' × 0.2'	14.9	10 h 52 m 59.5 s	+10° 12' 37"
3447	Leo	Gxy	3.7' × 2.1'	12.7	10 h 53 m 23.9 s	+16° 46' 21"
3454	Leo	Gxy	2.1' × 0.4'	13.7	10 h 54 m 29.4 s	+17° 20' 37"
3455	Leo	Gxy	2.5' × 1.5'	12.2	10 h 54 m 31.0 s	+17° 17' 03"
3457	Leo	Gxy	0.9' × 0.9'	12.3	10 h 54 m 48.7 s	+17° 37' 16"
3460	Leo	Gxy	1.3' × 1.3'	12.3	10 h 54 m 48.7 s	+17° 37' 16"
3461	Leo	Gxy	0.5' × 0.4'	14.5	10 h 54 m 55.3 s	+17° 42' 29"
3462	Leo	Gxy	1.7' × 1.2'	12.3	10 h 55 m 21.1 s	+07° 41' 48"
3466	Leo	Gxy	1.2' × 0.7'	13.8	10 h 56 m 15.5 s	+09° 45' 16"
3467	Leo	Gxy	0.9' × 0.8'	13.2	10 h 56 m 44.0 s	+09° 45' 34"
3473	Leo	Gxy	1.1' × 0.9'	13.8	10 h 58 m 05.2 s	+17° 07' 27"
3474	Leo	Gxy	0.7' × 0.6'	13.8	10 h 58 m 08.7 s	+17° 05' 44"
3475	Leo	Gxy	1.7' × 1.1'	13.3	10 h 58 m 25.2 s	+24° 13' 34"
3476	Leo	Gxy	0.8' × 0.7'	13.9	10 h 58 m 07.5 s	+09° 16' 34"
3477	Leo	Gxy	0.7' × 0.3'	14.8	10 h 58 m 12.5 s	+09° 13' 03"
3480	Leo	Gxy	0.8' × 0.7'	13.9	10 h 58 m 07.5 s	+09° 16' 34"
3485	Leo	Gxy	2.3' × 2.0'	12	11 h 00 m 02.3 s	+14° 50' 28"
3487	Leo	Gxy	1.0' × 0.4'	13.9	11 h 00 m 46.7 s	+17° 35' 14"
3489	Leo	Gxy	3.5' × 2.0'	10.4	11 h 00 m 18.9 s	+13° 54' 07"
3490	Leo	Gxy	...' × ...'	13.9	10 h 59 m 54.4 s	+09° 21' 42"
3491	Leo	Gxy	0.9' × 0.9'	13.1	11 h 00 m 35.3 s	+12° 09' 42"
3492	Leo	Gxy	1.1' × 0.7'	13.2	11 h 00 m 57.3 s	+10° 30' 20"
3494	Leo	**	...	...	11 h 01 m 10.9 s	+03° 46' 28"
3495	Leo	Gxy	4.9' × 1.2'	12	11 h 01 m 16.3 s	+03° 37' 41"
3498	Leo	***	...	...	11 h 01 m 41.7 s	+14° 20' 57"
3501	Leo	Gxy	3.9' × 0.5'	13.1	11 h 02 m 47.3 s	+17° 59' 20"
3506	Leo	Gxy	1.2' × 1.1'	12.7	11 h 03 m 12.9 s	+11° 04' 37"
3507	Leo	Gxy	3.4' × 2.9'	11.9	11 h 03 m 25.5 s	+18° 08' 09"
3509	Leo	Gxy	2.1' × 1.0'	12.9	11 h 04 m 23.5 s	+04° 49' 42"
3521	Leo	Gxy	11.0' × 5.1'	9.3	11 h 05 m 48.9 s	−00° 02' 06"
3522	Leo	Gxy	1.2' × 0.7'	13.2	11 h 06 m 40.4 s	+20° 05' 07"
3524	Leo	Gxy	1.6' × 0.5'	13	11 h 06 m 32.0 s	+11° 23' 07"

(continued)

Table 10.3 (continued)

NGC #	Const	Type	Size	V Mag	RA (2000)	Dec (2000)
3526	Leo	Gxy	1.9'×0.4'	13.4	11 h 06 m 56.7 s	+07° 10' 28"
3531	Leo	Gxy	2.0'×0.5'	13.1	11 h 06 m 56.7 s	+07° 10' 28"
3534	Leo	Gxy	1.4'×0.4'	14.5	11 h 08 m 55.7 s	+26° 36' 38"
3535	Leo	Gxy	1.1'×0.5'	13.7	11 h 08 m 34.0 s	+04° 49' 54"
3547	Leo	Gxy	1.9'×0.9'	12.3	11 h 09 m 55.9 s	+10° 43' 15"
3551	Leo	Gxy	1.8'×1.7'	13.5	11 h 09 m 44.4 s	+21° 45' 32"
3555	Leo	Gxy	0.4'×0.3'	14.7	11 h 09 m 50.3 s	+21° 48' 36"
3559	Leo	Gxy	1.3'×0.9'	12.8	11 h 10 m 45.2 s	+12° 00' 58"
3560	Leo	Gxy	1.3'×0.9'	12.8	11 h 10 m 45.2 s	+12° 00' 58"
3563	Leo	Gxy	1.1'×0.8'	13.5	11 h 11 m 25.2 s	+26° 57' 47"
3567	Leo	Gxy	1.2'×0.9'	13.4	11 h 11 m 18.7 s	+05° 50' 10"
3570	Leo	Gxy	1.0'×1.0'	13.6	11 h 12 m 03.4 s	+27° 35' 23"
3574	Leo	Gxy	0.4'×0.4'	15.2	11 h 12 m 12.1 s	+27° 37' 29"
3575	Leo	Gxy	3.1'×2.7'	11.4	10 h 13 m 31.5 s	+22° 44' 17"
3580	Leo	Gxy	0.8'×0.3'	14.1	11 h 13 m 15.9 s	+03° 39' 26"
3588	Leo	Gxy	1.1'×0.9'	14.4	11 h 14 m 02.5 s	+20° 23' 15"
3592	Leo	Gxy	1.8'×0.6'	13.9	11 h 14 m 27.5 s	+17° 15' 34"
3593	Leo	Gxy	5.2'×1.9'	10.8	11 h 14 m 37.0 s	+12° 49' 03"
3596	Leo	Gxy	4.0'×3.8'	11	11 h 15 m 06.1 s	+14° 47' 12"
3598	Leo	Gxy	1.9'×1.3'	12.4	11 h 15 m 11.7 s	+17° 15' 44"
3599	Leo	Gxy	2.7'×2.1'	11.7	11 h 15 m 27.1 s	+18° 06' 37"
3601	Leo	Gxy	0.5'×0.4'	14	11 h 15 m 33.2 s	+05° 06' 56"
3602	Leo	Gxy	0.7'×0.2'	15	11 h 15 m 48.3 s	+17° 24' 57"
3604	Leo	Gxy	2.4'×2.0'	11.9	11 h 17 m 30.0 s	+04° 33' 19"
3605	Leo	Gxy	1.5'×1.0'	12	11 h 16 m 46.5 s	+18° 01' 01"
3607	Leo	Gxy	4.9'×2.5'	9.9	11 h 16 m 54.7 s	+18° 03' 05"
3608	Leo	Gxy	3.2'×2.6'	10.8	11 h 16 m 59.0 s	+18° 08' 54"
3609	Leo	Gxy	1.2'×1.0'	13.3	11 h 17 m 50.5 s	+26° 37' 32"
3611	Leo	Gxy	2.1'×1.7'	11.6	11 h 17 m 30.0 s	+04° 33' 19"
3612	Leo	Gxy	1.0'×0.8'	14.2	11 h 18 m 14.6 s	+26° 37' 13"
3615	Leo	Gxy	1.4'×0.9'	12.9	11 h 18 m 06.7 s	+23° 23' 50"
3616	Leo	**	...	...	11 h 18 m 23.0 s	+14° 45' 39"
3618	Leo	Gxy	0.9'×0.8'	13.8	11 h 18 m 32.5 s	+23° 28' 07"
3623	Leo	Gxy	9.8'×2.9'	9.3	11 h 18 m 55.9 s	+13° 05' 37"
3624	Leo	Gxy	0.9'×0.6'	14.1	11 h 18 m 51.0 s	+07° 31' 17"
3626	Leo	Gxy	2.7'×1.9'	10.6	11 h 20 m 03.9 s	+18° 21' 23"
3627	Leo	Gxy	9.1'×4.2'	9	11 h 20 m 15.1 s	+12° 59' 28"
3628	Leo	Gxy	14.8'×3.0'	9.8	11 h 20 m 16.9 s	+13° 35' 20"
3629	Leo	Gxy	2.3'×1.6'	12.3	11 h 20 m 31.8 s	+26° 57' 48"
3630	Leo	Gxy	4.6'×3.0'	12	11 h 20 m 17.0 s	+02° 57' 52"
3632	Leo	Gxy	3.1'×2.2'	10.9	11 h 20 m 03.9 s	+18° 21' 23"
3633	Leo	Gxy	1.2'×0.4'	13.8	11 h 20 m 26.2 s	+03° 35' 09"
3639	Leo	Gxy	0.6'×0.5'	13.3	11 h 21 m 35.7 s	+18° 27' 30"
3640	Leo	Gxy	4.0'×3.2'	10.4	11 h 21 m 06.9 s	+03° 14' 05"
3641	Leo	Gxy	1.1'×1.1'	13.1	11 h 21 m 08.8 s	+03° 11' 40"

(continued)

Table 10.3 (continued)

NGC #	Const	Type	Size	V Mag	RA (2000)	Dec (2000)
3643	Leo	Gxy	0.8'×0.3'	14.2	11 h 21 m 24.9 s	+03° 00' 50"
3644	Leo	Gxy	1.5'×0.7'	13.9	11 h 21 m 32.9 s	+02° 48' 38"
3645	Leo	Gxy	2.3'×0.9'	12.1	11 h 20 m 17.0 s	+02° 57' 52"
3646	Leo	Gxy	3.9'×2.2'	11	11 h 21 m 43.1 s	+20° 10' 11"
3647	Leo	Gxy	0.4'×0.4'	14.3	11 h 21 m 38.5 s	+02° 53' 30"
3649	Leo	Gxy	1.2'×0.6'	13.9	11 h 22 m 14.7 s	+20° 12' 29"
3650	Leo	Gxy	1.7'×0.3'	14.1	11 h 22 m 35.4 s	+20° 42' 14"
3651	Leo	Gxy	1.1'×1.1'	13.3	11 h 22 m 26.3 s	+24° 17' 55"
3653	Leo	Gxy	0.9'×0.6'	13.7	11 h 22 m 30.1 s	+24° 16' 45"
3655	Leo	Gxy	1.5'×1.0'	12	11 h 22 m 54.7 s	+16° 35' 24"
3659	Leo	Gxy	2.1'×1.1'	12.4	11 h 23 m 45.3 s	+17° 49' 04"
3662	Leo	Gxy	1.4'×0.9'	13.1	11 h 23 m 46.4 s	−01° 06' 18"
3664	Leo	Gxy	2.0'×1.9'	12.5	11 h 24 m 25.1 s	+03° 19' 41"
3666	Leo	Gxy	4.4'×1.2'	11.8	11 h 24 m 26.1 s	+11° 20' 31"
3670	Leo	Gxy	1.3'×0.7'	13.7	11 h 24 m 49.7 s	+23° 56' 42"
3678	Leo	Gxy	0.8'×0.8'	13.8	11 h 26 m 15.7 s	+27° 52' 00"
3679	Leo	Gxy	1.0'×0.8'	13.6	11 h 21 m 48.0 s	−05° 45' 28"
3681	Leo	Gxy	2.5'×2.0'	11.7	11 h 26 m 29.8 s	+16° 51' 47"
3684	Leo	Gxy	3.1'×2.1'	11.5	11 h 27 m 11.2 s	+17° 01' 48"
3685	Leo	Gxy	0.7'×0.4'	14.1	11 h 28 m 16.2 s	+04° 19' 40"
3686	Leo	Gxy	3.2'×2.5'	11.1	11 h 27 m 44.1 s	+17° 13' 25"
3689	Leo	Gxy	1.7'×1.1'	12.5	11 h 28 m 11.0 s	+25° 39' 40"
3691	Leo	Gxy	1.3'×1.0'	12.7	11 h 28 m 09.4 s	+16° 55' 13"
3692	Leo	Gxy	3.2'×0.7'	12.3	11 h 28 m 23.9 s	+09° 24' 26"
3697	Leo	Gxy	2.3'×0.7'	13.2	11 h 28 m 50.3 s	+20° 47' 43"
3701	Leo	Gxy	1.9'×0.9'	13.2	11 h 29 m 28.9 s	+24° 05' 35"
3705	Leo	Gxy	4.9'×2.0'	11.1	11 h 30 m 07.3 s	+09° 16' 36"
3708	Leo	NF	…	…	11 h 30 m 39.3 s	−03° 13' 21"
3709	Leo	NF	…	…	11 h 30 m 39.3 s	−03° 15' 21"
3710	Leo	Gxy	1.0'×0.8'	13.4	11 h 31 m 06.9 s	+22° 46' 04"
3713	Leo	Gxy	1.2'×0.8'	13.3	11 h 31 m 41.9 s	+28° 09' 12"
3716	Leo	Gxy	0.7'×0.6'	13.6	11 h 31 m 41.2 s	+03° 29' 16"
3719	Leo	Gxy	1.8'×1.3'	12.8	11 h 32 m 13.3 s	+00° 49' 10"
3720	Leo	Gxy	1.0'×0.9'	12.9	11 h 32 m 21.7 s	+00° 48' 15"
3728	Leo	Gxy	2.5'×1.6'	13.2	11 h 33 m 15.7 s	+24° 26' 48"
3731	Leo	Gxy	1.0'×0.9'	13.1	11 h 34 m 11.7 s	+12° 30' 44"
3739	Leo	Gxy	1.3'×0.3'	14.7	11 h 35 m 37.6 s	+25° 05' 18"
3743	Leo	Gxy	0.5'×0.5'	14.4	11 h 35 m 57.4 s	+21° 43' 21"
3744	Leo	Gxy	0.6'×0.2'	14.4	11 h 35 m 57.9 s	+23° 00' 41"
3745	Leo	Gxy	0.4'×0.2'	15.2	11 h 37 m 44.7 s	+22° 01' 15"
3746	Leo	Gxy	1.1'×0.5'	14.4	11 h 37 m 43.5 s	+22° 00' 34"
3748	Leo	Gxy	0.7'×0.4'	14.6	11 h 37 m 49.1 s	+22° 01' 33"
3750	Leo	Gxy	0.8'×0.7'	14	11 h 37 m 51.7 s	+21° 58' 26"
3751	Leo	Gxy	0.8'×0.5'	14.3	11 h 37 m 54.1 s	+21° 56' 10"
3753	Leo	Gxy	1.7'×0.5'	13.7	11 h 37 m 53.7 s	+21° 58' 51"

(continued)

Table 10.3 (continued)

NGC #	Const	Type	Size	V Mag	RA (2000)	Dec (2000)
3754	Leo	Gxy	0.4'×0.3'	14.3	11 h 37 m 55.0 s	+21° 59' 07"
3758	Leo	Gxy	0.5'×0.5'	14.3	11 h 36 m 29.1 s	+21° 35' 46"
3760	Leo	Gxy	3.6'×1.2'	11.4	10 h 36 m 56.0 s	+21° 52' 55"
3761	Leo	Gxy	0.9'×0.8'	14.3	11 h 36 m 44.2 s	+22° 59' 30"
3764	Leo	Gxy	1.0'×0.8'	14.7	11 h 36 m 54.6 s	+17° 53' 18"
3765	Leo	Gxy	0.8'×0.6'	14.3	11 h 37 m 04.1 s	+24° 05' 46"
3767	Leo	Gxy	1.0'×0.9'	13.5	11 h 37 m 15.5 s	+16° 52' 37"
3768	Leo	Gxy	1.8'×1.2'	12.5	11 h 37 m 14.4 s	+17° 50' 23"
3772	Leo	Gxy	1.1'×0.6'	13.7	11 h 37 m 48.5 s	+22° 41' 28"
3773	Leo	Gxy	1.2'×1.0'	12.1	11 h 38 m 13.0 s	+12° 06' 44"
3781	Leo	Gxy	...'×...'	13.8	11 h 39 m 03.7 s	+26° 21' 42"
3784	Leo	Gxy	0.9'×0.3'	14.4	11 h 39 m 29.7 s	+26° 18' 35"
3785	Leo	Gxy	1.0'×0.4'	14.3	11 h 39 m 32.8 s	+26° 18' 08"
3787	Leo	Gxy	0.4'×0.3'	13.9	11 h 39 m 37.9 s	+20° 27' 16"
3790	Leo	Gxy	1.1'×0.3'	14.1	11 h 39 m 47.2 s	+17° 42' 43"
3798	Leo	Gxy	2.5'×1.8'	12.2	11 h 40 m 13.9 s	+24° 41' 48"
3799	Leo	Gxy	0.8'×0.5'	14.1	11 h 40 m 09.3 s	+15° 19' 38"
3800	Leo	Gxy	2.0'×0.6'	12.9	11 h 40 m 13.3 s	+15° 20' 31"
3801	Leo	Gxy	3.5'×2.1'	11.7	11 h 40 m 16.7 s	+17° 43' 41"
3802	Leo	Gxy	1.1'×0.3'	13.6	11 h 40 m 18.7 s	+17° 45' 56"
3803	Leo	Gxy	0.4'×0.4'	15.5	11 h 40 m 17.2 s	+17° 48' 06"
3805	Leo	Gxy	1.4'×1.1'	12.8	11 h 40 m 41.6 s	+20° 20' 36"
3806	Leo	Gxy	1.6'×1.5'	13.4	11 h 40 m 46.7 s	+17° 47' 45"
3807	Leo	*	...	...	11 h 41 m 52.5 s	+17° 49' 12"
3808	Leo	Gxy	1.7'×0.9'	13.4	11 h 40 m 44.2 s	+22° 25' 47"
3810	Leo	Gxy	4.3'×3.0'	10.8	11 h 40 m 58.7 s	+11° 28' 17"
3812	Leo	Gxy	1.7'×1.6'	12.5	11 h 41 m 07.7 s	+24° 49' 18"
3814	Leo	Gxy	0.9'×0.3'	14.7	11 h 41 m 27.7 s	+24° 48' 19"
3815	Leo	Gxy	1.7'×0.9'	13.2	11 h 41 m 39.4 s	+24° 48' 00"
3816	Leo	Gxy	1.9'×1.1'	12.5	11 h 41 m 48.0 s	+20° 06' 13"
3817	Leo	Gxy	1.0'×0.9'	13.4	11 h 41 m 52.9 s	+10° 18' 15"
3819	Leo	Gxy	0.8'×0.7'	14	11 h 42 m 05.9 s	+10° 21' 03"
3820	Leo	Gxy	0.7'×0.4'	14.5	11 h 42 m 04.9 s	+10° 23' 02"
3821	Leo	Gxy	1.4'×1.3'	12.9	11 h 42 m 09.1 s	+20° 18' 56"
3826	Leo	Gxy	0.9'×0.7'	13.5	11 h 42 m 32.9 s	+26° 29' 20"
3827	Leo	Gxy	0.9'×0.7'	13.2	11 h 42 m 36.3 s	+18° 50' 43"
3828	Leo	Gxy	0.8'×0.5'	14.8	11 h 42 m 58.5 s	+16° 29' 15"
3830	Leo	Gxy	0.9'×0.7'	13.3	11 h 42 m 32.9 s	+26° 29' 20"
3832	Leo	Gxy	2.2'×1.8'	12.8	11 h 43 m 31.5 s	+22° 43' 32"
3834	Leo	Gxy	1.3'×0.9'	14	11 h 43 m 37.7 s	+19° 05' 26"
3837	Leo	Gxy	0.6'×0.6'	13.7	11 h 43 m 56.4 s	+19° 53' 41"
3839	Leo	Gxy	1.0'×0.5'	13.6	11 h 43 m 54.3 s	+10° 47' 06"
3840	Leo	Gxy	1.2'×0.8'	13.8	11 h 43 m 58.9 s	+20° 04' 38"
3841	Leo	Gxy	0.9'×0.4'	13.8	11 h 44 m 02.1 s	+19° 58' 18"
3842	Leo	Gxy	1.4'×1.0'	12.5	11 h 44 m 02.1 s	+19° 56' 59"

(continued)

Table 10.3 (continued)

NGC #	Const	Type	Size	V Mag	RA (2000)	Dec (2000)
3844	Leo	Gxy	1.2'×0.2'	14.4	11 h 44 m 00.8 s	+20° 01' 46"
3845	Leo	Gxy	1.3'×0.5'	14.2	11 h 44 m 05.4 s	+19° 59' 46"
3851	Leo	Gxy	0.3'×0.3'	14.7	11 h 44 m 20.3 s	+19° 58' 50"
3853	Leo	Gxy	1.7'×1.0'	12.5	11 h 44 m 28.4 s	+16° 33' 27"
3857	Leo	Gxy	1.0'×0.6'	14.1	11 h 44 m 50.1 s	+19° 31' 58"
3859	Leo	Gxy	1.2'×0.3'	14.1	11 h 44 m 52.2 s	+19° 27' 15"
3860	Leo	Gxy	1.0'×0.5'	13.4	11 h 44 m 49.1 s	+19° 47' 42"
3861	Leo	Gxy	2.3'×1.3'	...	11 h 45 m 03.9 s	+19° 58' 24"
3862	Leo	Gxy	1.5'×1.5'	12.6	11 h 45 m 04.9 s	+19° 36' 22"
3864	Leo	Gxy	0.9'×0.7'	14.8	11 h 45 m 15.7 s	+19° 23' 31"
3867	Leo	Gxy	1.5'×0.6'	13.3	11 h 45 m 29.5 s	+19° 24' 00"
3868	Leo	Gxy	0.8'×0.4'	14.4	11 h 45 m 29.9 s	+19° 26' 40"
3869	Leo	Gxy	1.9'×0.5'	13	11 h 45 m 45.6 s	+10° 49' 28"
3872	Leo	Gxy	1.9'×1.2'	11.7	11 h 45 m 49.1 s	+13° 46' 00"
3873	Leo	Gxy	1.5'×1.3'	13.1	11 h 45 m 46.1 s	+19° 46' 24"
3875	Leo	Gxy	1.3'×0.3'	14.1	11 h 45 m 49.5 s	+19° 46' 02"
3883	Leo	Gxy	3.0'×2.4'	12.5	11 h 46 m 47.1 s	+20° 40' 31"
3884	Leo	Gxy	2.1'×1.3'	12.7	11 h 46 m 12.3 s	+20° 23' 31"
3886	Leo	Gxy	1.2'×0.9'	13.2	11 h 47 m 05.5 s	+19° 50' 14"
3899	Leo	Gxy	1.7'×1.0'	12.6	11 h 50 m 04.5 s	+26° 28' 46"
3900	Leo	Gxy	3.2'×1.7'	11.4	11 h 49 m 09.3 s	+27° 01' 18"
3902	Leo	Gxy	1.8'×1.5'	13	11 h 49 m 18.7 s	+26° 07' 17"
3908	Leo	Gxy	0.4'×0.4'	15.4	11 h 49 m 52.7 s	+12° 11' 09"
3910	Leo	Gxy	1.6'×1.2'	12.9	11 h 49 m 59.3 s	+21° 20' 00"
3911	Leo	Gxy	1.4'×1.3'	14.5	11 h 50 m 06.0 s	+24° 55' 12"
3912	Leo	Gxy	1.5'×0.9'	12.6	11 h 50 m 04.5 s	+26° 28' 46"
3919	Leo	Gxy	0.9'×0.9'	13.4	11 h 50 m 41.5 s	+20° 00' 54"
3920	Leo	Gxy	1.3'×1.0'	13.3	11 h 49 m 22.1 s	+24° 56' 19"
3925	Leo	Gxy	0.7'×0.5'	14.5	11 h 51 m 20.9 s	+21° 53' 21"
3926	Leo	Gxy	1.0'×0.8'	14.5	11 h 51 m 26.5 s	+22° 01' 40"
3927	Leo	NF	...	...	11 h 51 m 32.3 s	+28° 08' 25"
3929	Leo	Gxy	0.5'×0.4'	13.9	11 h 51 m 42.5 s	+21° 00' 10"
3933	Leo	Gxy	1.1'×0.6'	13.4	11 h 52 m 01.9 s	+16° 48' 36"
3934	Leo	Gxy	1.1'×1.0'	14	11 h 52 m 12.5 s	+16° 51' 07"
3937	Leo	Gxy	1.8'×1.6'	12.5	11 h 52 m 42.6 s	+20° 37' 52"
3940	Leo	Gxy	1.7'×1.6'	12.7	11 h 52 m 46.5 s	+20° 59' 19"
3943	Leo	Gxy	1.1'×1.1'	13.7	11 h 52 m 56.5 s	+20° 28' 44"
3944	Leo	Gxy	1.4'×1.1'	13	11 h 53 m 05.1 s	+26° 12' 24"
3946	Leo	Gxy	0.5'×0.4'	14.6	11 h 53 m 20.5 s	+21° 01' 16"
3947	Leo	Gxy	1.4'×1.2'	13.2	11 h 53 m 20.3 s	+20° 45' 05"
3948	Leo	*	...	...	11 h 53 m 36.7 s	+20° 57' 03"
3951	Leo	Gxy	1.1'×0.5'	13.4	11 h 53 m 41.2 s	+23° 22' 56"
3954	Leo	Gxy	0.6'×0.6'	13.4	11 h 53 m 41.7 s	+20° 52' 56"
3964	Leo	Gxy	0.8'×0.6'	14.1	11 h 54 m 53.5 s	+28° 15' 44"
3968	Leo	Gxy	2.7'×1.9'	12	11 h 55 m 28.7 s	+11° 58' 05"

(continued)

Table 10.3 (continued)

NGC #	Const	Type	Size	V Mag	RA (2000)	Dec (2000)
3973	Leo	Gxy	0.4'×0.2'	15	11 h 55 m 37.0 s	+11° 59' 50"
3983	Leo	Gxy	1.2'×0.3'	14.2	11 h 56 m 23.6 s	+23° 52' 04"
3987	Leo	Gxy	2.2'×0.4'	13.7	11 h 57 m 20.9 s	+25° 11' 42"
3988	Leo	Gxy	...'×...'	13.6	11 h 57 m 24.1 s	+27° 52' 38"
3989	Leo	Gxy	0.6'×0.3'	14.9	11 h 57 m 26.7 s	+25° 13' 58"
3993	Leo	Gxy	1.8'×0.5'	13.9	11 h 57 m 37.7 s	+25° 14' 25"
3996	Leo	Gxy	0.9'×0.7'	13.9	11 h 57 m 46.1 s	+14° 17' 52"
3997	Leo	Gxy	1.7'×0.6'	13.6	11 h 57 m 48.3 s	+25° 16' 14"
3999	Leo	Gxy	0.4'×0.2'	14.7	11 h 57 m 56.5 s	+25° 04' 05"
4000	Leo	Gxy	1.0'×0.2'	14.5	11 h 57 m 57.0 s	+25° 08' 39"
4002	Leo	Gxy	0.6'×0.3'	14.1	11 h 57 m 59.3 s	+23° 12' 07"
4003	Leo	Gxy	1.5'×1.1'	13.4	11 h 57 m 59.0 s	+23° 07' 29"
4004	Leo	Gxy	1.8'×0.6'	13.6	11 h 58 m 05.3 s	+27° 52' 46"
4005	Leo	Gxy	1.2'×0.7'	13	11 h 58 m 10.1 s	+25° 07' 20"
4007	Leo	Gxy	1.2'×0.7'	13	11 h 58 m 10.1 s	+25° 07' 20"
4008	Leo	Gxy	2.5'×1.3'	11.8	11 h 58 m 17.0 s	+28° 11' 34"
4009	Leo	*	...	...	11 h 58 m 15.1 s	+25° 11' 24"
4011	Leo	Gxy	0.4'×0.2'	14.7	11 h 58 m 25.3 s	+25° 05' 51"
4016	Leo	Gxy	1.5'×0.8'	13.7	11 h 58 m 28.9 s	+27° 31' 44"
1710	Lep	Gxy	1.3'×1.0'	13	04 h 57 m 17.0 s	−15° 17' 20"
1716	Lep	Gxy	1.4'×1.1'	13.4	04 h 58 m 13.3 s	−20° 21' 51"
1730	Lep	Gxy	2.2'×1.0'	12.3	04 h 59 m 31.8 s	−15° 49' 26"
1738	Lep	Gxy	1.3'×0.7'	13.1	05 h 01 m 46.7 s	−18° 09' 28"
1739	Lep	Gxy	1.4'×0.7'	13.7	05 h 01 m 47.4 s	−18° 10' 02"
1744	Lep	Gxy	8.1'×4.4'	11.5	04 h 59 m 57.6 s	−26° 01' 19"
1780	Lep	Gxy	0.9'×0.5'	13.8	05 h 06 m 20.7 s	−19° 28' 03"
1781	Lep	Gxy	1.4'×1.3'	12.7	05 h 07 m 55.1 s	−18° 11' 25"
1784	Lep	Gxy	4.0'×2.5'	11.7	05 h 05 m 27.1 s	−11° 52' 18"
1794	Lep	Gxy	1.3'×1.1'	12.8	05 h 07 m 55.1 s	−18° 11' 25"
1821	Lep	Gxy	1.1'×0.8'	13.9	05 h 11 m 46.1 s	−15° 08' 07"
1832	Lep	Gxy	2.6'×1.7'	11.8	05 h 12 m 03.3 s	−15° 41' 16"
1886	Lep	Gxy	3.1'×0.4'	13	05 h 21 m 48.7 s	−23° 48' 45"
1888	Lep	Gxy	3.0'×0.8'	12.1	05 h 22 m 34.7 s	−11° 29' 58"
1889	Lep	Gxy	0.7'×0.5'	13.1	05 h 22 m 35.3 s	−11° 29' 49"
1904	Lep	GC	8.7'	7.7	05 h 24 m 10.6 s	−24° 31' 27"
1906	Lep	Gxy	0.9'×0.5'	14.1	05 h 24 m 47.1 s	−15° 56' 34"
1954	Lep	Gxy	4.2'×2.2'	11.8	05 h 32 m 48.3 s	−14° 03' 45"
1957	Lep	Gxy	1.1'×0.8'	13.9	05 h 32 m 55.2 s	−14° 08' 00"
1964	Lep	Gxy	5.6'×2.1'	10.9	05 h 33 m 21.6 s	−21° 56' 48"
1979	Lep	Gxy	2.2'×1.8'	11.9	05 h 39 m 39.9 s	−69° 46' 26"
1993	Lep	Gxy	1.5'×1.4'	12.5	05 h 35 m 25.6 s	−17° 48' 56"
2017	Lep	Ast	10.0'	...	05 h 39 m 15.3 s	−17° 51' 01"
2073	Lep	Gxy	1.5'×1.4'	12.5	05 h 45 m 53.8 s	−21° 59' 59"
2076	Lep	Gxy	2.2'×1.3'	13	05 h 46 m 47.2 s	−16° 46' 56"
2089	Lep	Gxy	1.9'×1.1'	12	05 h 47 m 51.5 s	−17° 36' 10"

(continued)

Table 10.3 (continued)

NGC #	Const	Type	Size	V Mag	RA (2000)	Dec (2000)
2106	Lep	Gxy	2.7'×1.3'	12.2	05 h 50 m 46.3 s	−21° 34' 03"
2124	Lep	Gxy	2.7'×0.9'	12.8	05 h 57 m 52.2 s	−20° 05' 06"
2131	Lep	Gxy	1.1'×0.5'	14.2	05 h 58 m 47.2 s	−26° 39' 12"
2139	Lep	Gxy	2.6'×1.9'	11.7	06 h 01 m 07.8 s	−23° 40' 20"
2179	Lep	Gxy	1.7'×1.2'	12.5	06 h 08 m 02.1 s	−21° 44' 50"
2196	Lep	Gxy	2.8'×2.2'	11.5	06 h 12 m 09.5 s	−21° 48' 24"
5595	Lib	Gxy	2.2'×1.2'	12.2	14 h 24 m 13.2 s	−16° 43' 21"
5597	Lib	Gxy	2.1'×1.7'	12.1	14 h 24 m 27.4 s	−16° 45' 46"
5605	Lib	Gxy	1.6'×1.3'	12.6	14 h 25 m 07.5 s	−13° 09' 48"
5663	Lib	Gxy	1.4'×1.3'	14.3	14 h 33 m 56.3 s	−16° 34' 51"
5664	Lib	Gxy	0.8'×0.3'	14.1	14 h 33 m 43.4 s	−14° 37' 09"
5716	Lib	Gxy	1.8'×1.3'	13	14 h 41 m 05.5 s	−17° 28' 35"
5726	Lib	Gxy	1.3'×1.0'	12.9	14 h 42 m 56.0 s	−18° 26' 41"
5728	Lib	Gxy	3.1'×1.8'	11.7	14 h 42 m 24.0 s	−17° 15' 12"
5729	Lib	Gxy	2.6'×0.7'	12.6	14 h 42 m 06.7 s	−09° 00' 31"
5734	Lib	Gxy	1.5'×1.0'	12.8	14 h 45 m 09.1 s	−20° 52' 15"
5741	Lib	Gxy	0.4'×0.4'	13.6	14 h 45 m 51.7 s	−11° 54' 51"
5742	Lib	Gxy	1.3'×0.5'	13.5	14 h 45 m 36.8 s	−11° 48' 36"
5743	Lib	Gxy	1.3'×0.5'	13.2	14 h 45 m 10.9 s	−20° 54' 48"
5744	Lib	Gxy	1.1'×0.9'	13.4	14 h 46 m 38.4 s	−18° 30' 48"
5745	Lib	Gxy	1.7'×1.1'	13.4	14 h 45 m 01.8 s	−13° 56' 49"
5756	Lib	Gxy	2.1'×0.9'	12.6	14 h 47 m 33.7 s	−14° 51' 14"
5757	Lib	Gxy	2.0'×1.6'	12.1	14 h 47 m 46.2 s	−19° 04' 39"
5761	Lib	Gxy	1.3'×1.0'	12.7	14 h 49 m 08.5 s	−20° 22' 34"
5766	Lib	Gxy	1.0'×0.7'	13.6	14 h 53 m 09.5 s	−21° 23' 40"
5768	Lib	Gxy	1.9'×1.4'	12.7	14 h 52 m 08.1 s	−02° 31' 47"
5781	Lib	Gxy	1.4'×0.8'	13.7	14 h 56 m 41.2 s	−17° 14' 37"
5791	Lib	Gxy	2.6'×1.3'	11.8	14 h 58 m 46.1 s	−19° 16' 02"
5792	Lib	Gxy	6.9'×1.7'	11.6	14 h 58 m 22.7 s	−01° 05' 27"
5793	Lib	Gxy	1.7'×0.6'	13.5	14 h 59 m 24.7 s	−16° 41' 37"
5796	Lib	Gxy	2.5'×1.8'	11.7	14 h 59 m 24.0 s	−16° 37' 27"
5801	Lib	Gxy	0.7'×0.5'	14.7	15 h 00 m 25.9 s	−13° 54' 16"
5802	Lib	Gxy	0.9'×0.5'	14.2	15 h 00 m 29.9 s	−13° 55' 08"
5803	Lib	Gxy	0.6'×0.5'	14.8	15 h 00 m 34.5 s	−13° 53' 40"
5809	Lib	Gxy	1.4'×0.7'	13.6	15 h 00 m 52.2 s	−14° 09' 57"
5810	Lib	Gxy	1.1'×0.9'	13.4	15 h 02 m 42.5 s	−17° 52' 06"
5812	Lib	Gxy	2.1'×1.9'	11.3	15 h 00 m 55.7 s	−07° 27' 29"
5815	Lib	Gxy	0.8'×0.4'	14.3	15 h 00 m 29.1 s	−16° 50' 03"
5816	Lib	Gxy	0.5'×0.3'	14.8	15 h 00 m 04.9 s	−16° 05' 37"
5817	Lib	Gxy	0.9'×0.7'	14.2	14 h 59 m 40.9 s	−16° 10' 47"
5849	Lib	Gxy	0.9'×0.8'	13.9	15 h 06 m 50.6 s	−14° 34' 19"
5858	Lib	Gxy	1.5'×0.7'	12.8	15 h 08 m 49.0 s	−11° 12' 28"
5861	Lib	Gxy	3.0'×1.7'	11.8	15 h 09 m 15.9 s	−11° 19' 18"
5863	Lib	Gxy	1.4'×1.2'	13	15 h 10 m 48.4 s	−18° 25' 54"
5872	Lib	Gxy	1.5'×0.9'	13.8	15 h 10 m 55.5 s	−11° 28' 49"

(continued)

Table 10.3 (continued)

NGC #	Const	Type	Size	V Mag	RA (2000)	Dec (2000)
5877	Lib	***	...	...	15 h 12 m 53.1 s	−04° 55' 38"
5878	Lib	Gxy	3.5'×1.4'	12	15 h 13 m 45.7 s	−14° 16' 14"
5880	Lib	Gxy	0.7'×0.6'	14.3	15 h 15 m 01.1 s	−14° 34' 44"
5883	Lib	Gxy	0.9'×0.5'	13.6	15 h 15 m 10.1 s	−14° 37' 01"
5885	Lib	Gxy	3.5'×3.1'	11.4	15 h 15 m 04.0 s	−10° 05' 11"
5890	Lib	Gxy	1.7'×1.1'	12.8	15 h 17 m 51.1 s	−17° 35' 21"
5891	Lib	Gxy	0.8'×0.5'	14.2	15 h 16 m 13.3 s	−11° 29' 38"
5892	Lib	Gxy	3.5'×2.8'	11.8	15 h 13 m 48.2 s	−15° 27' 50"
5897	Lib	GC	13'	8.4	15 h 17 m 24.4 s	−21° 00' 37"
5898	Lib	Gxy	2.2'×2.0'	11.5	15 h 18 m 13.7 s	−24° 05' 54"
5903	Lib	Gxy	2.7'×2.1'	11.5	15 h 18 m 36.9 s	−24° 03' 59"
5915	Lib	Gxy	1.7'×1.3'	12.2	15 h 21 m 33.0 s	−13° 05' 32"
5916	Lib	Gxy	2.8'×0.9'	13.2	15 h 21 m 37.7 s	−13° 10' 11"
5917	Lib	Gxy	1.5'×0.9'	13.2	15 h 21 m 32.5 s	−07° 22' 42"
5959	Lib	Gxy	1.9'×1.6'	13.6	15 h 37 m 22.3 s	−16° 35' 45"
5973	Lib	Gxy	0.9'×0.3'	14.4	15 h 40 m 15.6 s	−08° 36' 08"
5978	Lib	Gxy	0.7'×0.6'	14	15 h 42 m 27.1 s	−13° 14' 04"
5995	Lib	Gxy	0.9'×0.7'	13.6	15 h 48 m 24.9 s	−13° 45' 27"
2859	LMi	Gxy	4.3'×3.8'	10.6	09 h 24 m 18.5 s	+34° 30' 47"
2922	LMi	Gxy	1.1'×0.5'	14.2	09 h 36 m 52.4 s	+37° 41' 40"
2942	LMi	Gxy	2.2'×1.8'	12.7	09 h 39 m 08.0 s	+34° 00' 22"
2955	LMi	Gxy	1.7'×0.9'	13	09 h 41 m 16.6 s	+35° 52' 56"
2965	LMi	Gxy	1.2'×0.9'	13.5	09 h 43 m 19.1 s	+36° 14' 52"
2971	LMi	Gxy	1.1'×0.8'	14.2	09 h 43 m 46.1 s	+36° 10' 45"
3003	LMi	Gxy	5.8'×1.3'	11.5	09 h 48 m 36.5 s	+33° 25' 22"
3012	LMi	Gxy	1.1'×1.0'	13.6	09 h 49 m 52.1 s	+34° 42' 51"
3013	LMi	Gxy	0.5'×0.4'	15.2	09 h 50 m 09.3 s	+33° 34' 09"
3021	LMi	Gxy	1.6'×0.9'	12.3	09 h 50 m 57.3 s	+33° 33' 14"
3074	LMi	Gxy	2.3'×2.1'	13.7	09 h 59 m 41.1 s	+35° 23' 34"
3099	LMi	Gxy	0.4'×0.4'	13.5	10 h 02 m 36.5 s	+32° 42' 24"
3104	LMi	Gxy	3.3'×2.2'	13.2	10 h 03 m 57.3 s	+40° 45' 25"
3106	LMi	Gxy	1.8'×1.8'	12.5	10 h 04 m 05.3 s	+31° 11' 07"
3116	LMi	Gxy	0.3'×0.3'	14.5	10 h 06 m 45.1 s	+31° 05' 52"
3118	LMi	Gxy	2.5'×0.4'	13.7	10 h 07 m 11.6 s	+33° 01' 40"
3126	LMi	Gxy	2.8'×0.5'	13	10 h 08 m 20.7 s	+31° 51' 46"
3150	LMi	Gxy	0.5'×0.4'	14.6	10 h 13 m 26.3 s	+38° 39' 27"
3151	LMi	Gxy	0.6'×0.3'	13.9	10 h 13 m 28.9 s	+38° 37' 12"
3152	LMi	Gxy	1.0'×0.6'	14.3	10 h 13 m 34.1 s	+38° 50' 36"
3158	LMi	Gxy	2.0'×1.8'	11.8	10 h 13 m 50.5 s	+38° 45' 54"
3159	LMi	Gxy	2.1'×2.1'	13.2	10 h 13 m 52.8 s	+38° 39' 15"
3160	LMi	Gxy	1.4'×0.3'	14.3	10 h 13 m 55.1 s	+38° 50' 33"
3161	LMi	Gxy	2.1'×1.6'	13.6	10 h 13 m 59.2 s	+38° 39' 26"
3163	LMi	Gxy	1.5'×1.5'	13.1	10 h 14 m 07.1 s	+38° 39' 09"
3167	LMi	NF	...	...	10 h 14 m 35.9 s	+29° 35' 47"
3219	LMi	Gxy	0.2'×0.2'	13.2	10 h 22 m 37.4 s	+38° 34' 46"

(continued)

Table 10.3 (continued)

NGC #	Const	Type	Size	V Mag	RA (2000)	Dec (2000)
3232	LMi	Gxy	0.7'×0.7'	14.4	10 h 24 m 24.3 s	+28° 01' 40"
3234	LMi	Gxy	1.2'×0.9'	13.4	10 h 24 m 59.3 s	+28° 01' 25"
3235	LMi	Gxy	1.2'×0.9'	13.4	10 h 24 m 59.3 s	+28° 01' 25"
3245	LMi	Gxy	3.2'×1.8'	10.8	10 h 27 m 18.4 s	+28° 30' 26"
3254	LMi	Gxy	5.0'×1.6'	11.6	10 h 29 m 19.9 s	+29° 29' 31"
3265	LMi	Gxy	1.3'×1.0'	13	10 h 31 m 06.7 s	+28° 47' 48"
3272	LMi	**	...	...	10 h 31 m 48.1 s	+28° 28' 07"
3277	LMi	Gxy	1.9'×1.7'	11.7	10 h 32 m 55.5 s	+28° 30' 41"
3291	LMi	*	...	...	10 h 36 m 06.2 s	+37° 16' 28"
3294	LMi	Gxy	3.5'×1.8'	11.2	10 h 36 m 16.5 s	+37° 19' 28"
3304	LMi	Gxy	1.7'×0.6'	13.6	10 h 37 m 38.1 s	+37° 27' 20"
3323	LMi	Gxy	1.3'×0.7'	13.4	10 h 39 m 39.0 s	+25° 19' 21"
3327	LMi	Gxy	1.1'×0.9'	13.6	10 h 39 m 57.9 s	+24° 05' 28"
3334	LMi	Gxy	1.1'×1.0'	13.1	10 h 41 m 31.2 s	+37° 18' 45"
3344	LMi	Gxy	7.1'×6.5'	10.2	10 h 43 m 30.7 s	+24° 55' 19"
3350	LMi	Gxy	0.5'×0.5'	14.5	10 h 44 m 22.9 s	+30° 43' 30"
3380	LMi	Gxy	1.7'×1.3'	12.7	10 h 48 m 12.2 s	+28° 36' 06"
3381	LMi	Gxy	2.0'×1.9'	11.8	10 h 48 m 24.5 s	+34° 42' 42"
3382	LMi	**	...	...	10 h 48 m 25.6 s	+36° 43' 28"
3395	LMi	Gxy	2.1'×1.2'	12	10 h 49 m 50.1 s	+32° 58' 58"
3396	LMi	Gxy	3.1'×1.2'	12.2	10 h 49 m 55.3 s	+32° 59' 26"
3400	LMi	Gxy	1.3'×0.8'	13.4	10 h 50 m 45.5 s	+28° 28' 08"
3413	LMi	Gxy	2.2'×0.9'	12.2	10 h 51 m 20.7 s	+32° 45' 57"
3414	LMi	Gxy	3.5'×2.6'	10.9	10 h 51 m 16.3 s	+27° 58' 30"
3418	LMi	Gxy	1.4'×1.1'	13.4	10 h 51 m 24.0 s	+28° 06' 42"
3424	LMi	Gxy	2.8'×0.8'	12.6	10 h 51 m 46.5 s	+32° 54' 00"
3430	LMi	Gxy	4.0'×2.2'	11.6	10 h 52 m 11.7 s	+32° 56' 58"
3432	LMi	Gxy	6.8'×1.5'	11.4	10 h 52 m 31.2 s	+36° 37' 10"
3442	LMi	Gxy	0.6'×0.5'	13	10 h 53 m 08.1 s	+33° 54' 36"
3451	LMi	Gxy	1.7'×0.8'	13.1	10 h 54 m 20.9 s	+27° 14' 21"
3486	LMi	Gxy	7.1'×5.2'	10.3	11 h 00 m 23.9 s	+28° 58' 29"
3493	LMi	Gxy	1.1'×0.4'	14.6	11 h 01 m 27.8 s	+27° 43' 11"
3504	LMi	Gxy	2.7'×2.1'	11	11 h 03 m 11.2 s	+27° 58' 20"
3510	LMi	Gxy	4.0'×0.8'	12.3	11 h 03 m 43.5 s	+28° 53' 05"
3512	LMi	Gxy	1.6'×1.5'	12.4	11 h 04 m 02.9 s	+28° 02' 11"
3515	LMi	Gxy	1.0'×0.8'	14.1	11 h 04 m 37.3 s	+28° 13' 40"
5530	Lup	Gxy	4.2'×1.9'	11.3	14 h 18 m 27.1 s	−43° 23' 19"
5593	Lup	OC	7.0'	...	14 h 25 m 39.1 s	−54° 47' 55"
5643	Lup	Gxy	4.6'×4.0'	10.5	14 h 32 m 41.3 s	−44° 10' 24"
5670	Lup	Gxy	2.2'×1.0'	12.1	14 h 35 m 36.3 s	−45° 57' 55"
5688	Lup	Gxy	3.8'×2.5'	12.1	14 h 39 m 35.1 s	−45° 01' 12"
5749	Lup	OC	7'	8.8	14 h 48 m 53.9 s	−54° 29' 51"
5764	Lup	OC	3.0'	12.6	14 h 53 m 32.2 s	−52° 40' 14"
5800	Lup	OC	5.0'	...	15 h 01 m 47.8 s	−51° 55' 07"
5822	Lup	OC	39.0'	6.5	15 h 04 m 21.2 s	−54° 23' 47"

(continued)

Table 10.3 (continued)

NGC #	Const	Type	Size	V Mag	RA (2000)	Dec (2000)
5824	Lup	GC	6.2'	9.1	15 h 03 m 58.5 s	−33° 04' 04"
5834	Lup	GC	6.2'	9.1	15 h 03 m 58.5 s	−33° 04' 04"
5843	Lup	Gxy	1.9'×1.1'	12.5	15 h 07 m 28.2 s	−36° 19' 42"
5873	Lup	PN	0.22'	11	15 h 12 m 50.7 s	−38° 07' 33"
5882	Lup	PN	0.33'	9.4	15 h 16 m 49.9 s	−45° 38' 57"
5927	Lup	GC	6.0'	8	15 h 28 m 00.4 s	−50° 40' 22"
5968	Lup	Gxy	2.1'×1.9'	12.4	15 h 39 m 57.0 s	−30° 33' 08"
5986	Lup	GC	9.8'	7.6	15 h 46 m 03.4 s	−37° 47' 10"
6026	Lup	PN	54"×36"	12.9	16 h 01 m 21.1 s	−34° 32' 37"
2273	Lyn	Gxy	3.2'×2.5'	11.6	06 h 50 m 08.8 s	+60° 50' 45"
2308	Lyn	Gxy	1.8'×1.2'	13.4	06 h 58 m 37.6 s	+45° 12' 37"
2315	Lyn	Gxy	1.3'×0.4'	13.8	07 h 02 m 33.1 s	+50° 35' 26"
2320	Lyn	Gxy	1.4'×0.8'	12.9	07 h 05 m 42.0 s	+50° 34' 51"
2321	Lyn	Gxy	1.4'×1.1'	13.8	07 h 05 m 59.0 s	+50° 45' 22"
2322	Lyn	Gxy	1.3'×0.4'	14	07 h 06 m 00.3 s	+50° 30' 36"
2326	Lyn	Gxy	1.9'×1.8'	13.3	07 h 08 m 11.0 s	+50° 40' 54"
2329	Lyn	Gxy	1.3'×1.1'	12.7	07 h 09 m 08.0 s	+48° 36' 55"
2330	Lyn	Gxy	0.4'×0.3'	14.8	07 h 09 m 28.4 s	+50° 09' 09"
2332	Lyn	Gxy	1.5'×1.0'	12.9	07 h 09 m 34.1 s	+50° 10' 55"
2334	Lyn	Gxy	0.9'×0.7'	13.7	07 h 11 m 33.5 s	+50° 14' 52"
2337	Lyn	Gxy	2.2'×1.7'	12.6	07 h 10 m 13.5 s	+44° 27' 25"
2340	Lyn	Gxy	1.8'×1.2'	12.7	07 h 11 m 10.9 s	+50° 10' 28"
2344	Lyn	Gxy	1.7'×1.7'	12.6	07 h 12 m 28.6 s	+47° 09' 59"
2415	Lyn	Gxy	0.9'×0.9'	12.5	07 h 36 m 56.7 s	+35° 14' 31"
2419	Lyn	GC	4.1'	10.3	07 h 38 m 08.4 s	+38° 52' 53"
2424	Lyn	Gxy	3.8'×0.6'	13.1	07 h 40 m 39.4 s	+39° 13' 58"
2426	Lyn	Gxy	1.1'×1.1'	13.2	07 h 43 m 18.5 s	+52° 19' 06"
2429	Lyn	Gxy	1.8'×0.4'	13.8	07 h 43 m 47.6 s	+52° 21' 26"
2431	Lyn	Gxy	0.9'×0.9'	13.6	07 h 45 m 13.4 s	+53° 04' 30"
2436	Lyn	Gxy	0.9'×0.8'	13.5	07 h 45 m 13.4 s	+53° 04' 30"
2444	Lyn	Gxy	1.2'×0.8'	12.9	07 h 46 m 53.0 s	+39° 01' 55"
2445	Lyn	Gxy	1.4'×1.1'	12.9	07 h 46 m 55.1 s	+39° 00' 56"
2446	Lyn	Gxy	1.9'×1.0'	13.1	07 h 48 m 39.1 s	+54° 36' 41"
2456	Lyn	Gxy	1.1'×0.8'	13.2	07 h 54 m 10.6 s	+55° 29' 43"
2457	Lyn	Gxy	0.5'×0.2'	15.4	07 h 54 m 45.7 s	+55° 32' 48"
2458	Lyn	Gxy	0.4'×0.3'	14.5	07 h 55 m 51.4 s	+56° 42' 38"
2461	Lyn	*	...	...	07 h 56 m 26.3 s	+56° 40' 24"
2462	Lyn	Gxy	0.4'×0.3'	13.3	07 h 56 m 32.1 s	+56° 41' 14"
2463	Lyn	Gxy	0.4'×0.4'	14.2	07 h 57 m 12.5 s	+56° 40' 35"
2464	Lyn	***	0.4'	...	07 h 57 m 32.7 s	+56° 41' 26"
2465	Lyn	*	...	...	07 h 57 m 26.1 s	+56° 49' 20"
2468	Lyn	Gxy	1.3'×0.6'	14	07 h 58 m 02.4 s	+56° 21' 33"
2469	Lyn	Gxy	1.1'×0.8'	12.9	07 h 58 m 03.5 s	+56° 40' 49"
2471	Lyn	**	...	...	07 h 58 m 32.9 s	+56° 46' 34"
2472	Lyn	Gxy	0.5'×0.5'	15.3	07 h 58 m 41.7 s	+56° 42' 04"

(continued)

Table 10.3 (continued)

NGC #	Const	Type	Size	V Mag	RA (2000)	Dec (2000)
2473	Lyn	Gxy	0.5'×0.3'	15.2	07 h 55 m 34.8 s	+56° 44' 09"
2474	Lyn	Gxy	0.6'×0.6'	13.1	07 h 57 m 58.9 s	+52° 51' 27"
2475	Lyn	Gxy	0.8'×0.8'	13.3	07 h 58 m 00.5 s	+52° 51' 44"
2476	Lyn	Gxy	1.4'×0.8'	12.5	07 h 56 m 45.1 s	+39° 55' 41"
2484	Lyn	Gxy	0.9'×0.8'	13.8	07 h 58 m 28.1 s	+37° 47' 11"
2488	Lyn	Gxy	1.4'×0.8'	13.2	08 h 01 m 45.9 s	+56° 33' 13"
2493	Lyn	Gxy	1.9'×1.9'	12	08 h 00 m 23.7 s	+39° 49' 50"
2495	Lyn	Gxy	0.4'×0.2'	15.2	08 h 00 m 33.1 s	+39° 50' 23"
2497	Lyn	Gxy	1.4'×1.2'	13.2	08 h 02 m 11.1 s	+56° 56' 32"
2500	Lyn	Gxy	2.9'×2.6'	11.7	08 h 01 m 53.1 s	+50° 44' 15"
2505	Lyn	Gxy	1.2'×0.6'	13.4	08 h 04 m 06.9 s	+53° 32' 57"
2518	Lyn	Gxy	1.2'×1.0'	13.1	08 h 07 m 20.1 s	+51° 07' 53"
2519	Lyn	*	...	...	08 h 07 m 58.7 s	+51° 07' 42"
2521	Lyn	Gxy	1.2'×0.7'	13.1	08 h 08 m 49.3 s	+57° 46' 10"
2524	Lyn	Gxy	1.4'×1.0'	12.9	08 h 08 m 09.6 s	+39° 09' 25"
2528	Lyn	Gxy	1.5'×1.5'	12.8	08 h 07 m 24.9 s	+39° 11' 38"
2532	Lyn	Gxy	2.2'×1.8'	12.3	08 h 10 m 15.4 s	+33° 57' 25"
2534	Lyn	Gxy	1.4'×1.2'	12.7	08 h 12 m 54.1 s	+55° 40' 18"
2537	Lyn	Gxy	1.7'×1.5'	11.7	08 h 13 m 14.5 s	+45° 59' 31"
2541	Lyn	Gxy	6.3'×3.2'	11.5	08 h 14 m 40.1 s	+49° 03' 41"
2543	Lyn	Gxy	2.3'×1.3'	12.1	08 h 12 m 57.9 s	+36° 15' 14"
2549	Lyn	Gxy	3.9'×1.3'	11.2	08 h 18 m 58.3 s	+57° 48' 10"
2552	Lyn	Gxy	3.5'×2.3'	12.3	08 h 19 m 19.7 s	+50° 00' 26"
2638	Lyn	Gxy	1.7'×0.6'	13	08 h 42 m 25.8 s	+37° 13' 14"
2649	Lyn	Gxy	1.6'×1.5'	12.5	08 h 44 m 08.3 s	+34° 43' 01"
2668	Lyn	Gxy	1.2'×0.6'	14	08 h 49 m 22.6 s	+36° 42' 36"
2683	Lyn	Gxy	9.3'×2.2'	9.6	08 h 52 m 41.8 s	+33° 25' 20"
2691	Lyn	Gxy	1.2'×0.8'	13.2	08 h 54 m 46.4 s	+39° 32' 20"
2704	Lyn	Gxy	1.0'×1.0'	13.6	08 h 56 m 47.7 s	+39° 22' 56"
2712	Lyn	Gxy	2.9'×1.6'	11.8	08 h 59 m 30.6 s	+44° 54' 50"
2719	Lyn	Gxy	1.3'×0.3'	13.3	09 h 00 m 15.5 s	+35° 43' 39"
2724	Lyn	Gxy	1.8'×1.6'	13.8	09 h 01 m 01.7 s	+35° 45' 44"
2746	Lyn	Gxy	1.6'×1.5'	13.3	09 h 05 m 59.5 s	+35° 22' 38"
2755	Lyn	Gxy	1.2'×0.8'	13.4	09 h 07 m 58.3 s	+41° 42' 31"
2759	Lyn	Gxy	1.0'×0.7'	13.3	09 h 08 m 37.3 s	+37° 37' 16"
2770	Lyn	Gxy	3.8'×1.1'	11.7	09 h 09 m 33.7 s	+33° 07' 25"
2776	Lyn	Gxy	3.0'×2.7'	11.5	09 h 12 m 14.5 s	+44° 57' 18"
2778	Lyn	Gxy	1.4'×1.0'	12.2	09 h 12 m 24.3 s	+35° 01' 39"
2779	Lyn	Gxy	0.4'×0.3'	15	09 h 12 m 28.3 s	+35° 03' 12"
2780	Lyn	Gxy	0.9'×0.7'	13.4	09 h 12 m 44.3 s	+34° 55' 33"
2782	Lyn	Gxy	3.5'×2.6'	11.3	09 h 14 m 05.2 s	+40° 06' 48"
2785	Lyn	Gxy	1.5'×0.5'	14.3	09 h 15 m 15.3 s	+40° 55' 03"
2793	Lyn	Gxy	1.3'×1.1'	13.2	09 h 16 m 47.2 s	+34° 25' 48"
2798	Lyn	Gxy	2.6'×1.0'	12.2	09 h 17 m 22.8 s	+41° 59' 59"
2799	Lyn	Gxy	1.9'×0.5'	13.9	09 h 17 m 31.1 s	+41° 59' 38"

(continued)

Table 10.3 (continued)

NGC #	Const	Type	Size	V Mag	RA (2000)	Dec (2000)
2823	Lyn	Gxy	0.9'×0.5'	14.8	09 h 19 m 17.5 s	+34° 00' 31"
2825	Lyn	Gxy	1.0'×0.3'	14.6	09 h 19 m 22.5 s	+33° 44' 33"
2826	Lyn	Gxy	1.5'×0.3'	13.9	09 h 19 m 24.2 s	+33° 37' 25"
2827	Lyn	Gxy	0.8'×0.3'	14.8	09 h 19 m 18.9 s	+33° 52' 51"
2828	Lyn	Gxy	0.4'×0.2'	14.7	09 h 19 m 34.8 s	+33° 53' 19"
2829	Lyn	Gxy	0.3'×0.2'	16	09 h 19 m 52.3 s	+33° 39' 00"
2830	Lyn	Gxy	1.3'×0.3'	14.5	09 h 19 m 41.4 s	+33° 44' 17"
2831	Lyn	Gxy	1.4'×1.4'	13.4	09 h 19 m 45.5 s	+33° 44' 42"
2832	Lyn	Gxy	2.3'×1.9'	11.3	09 h 19 m 46.8 s	+33° 44' 58"
2833	Lyn	Gxy	0.9'×0.3'	14.8	09 h 19 m 57.9 s	+33° 55' 41"
2834	Lyn	Gxy	0.2'×0.2'	14.5	09 h 20 m 02.5 s	+33° 42' 37"
2838	Lyn	Gxy	0.4'×0.3'	13.8	09 h 20 m 43.0 s	+39° 18' 57"
2839	Lyn	Gxy	...'×...'	14.2	09 h 20 m 36.3 s	+33° 39' 03"
2840	Lyn	Gxy	1.0'×0.9'	14	09 h 20 m 52.7 s	+35° 22' 05"
2844	Lyn	Gxy	1.5'×0.8'	12.7	09 h 21 m 47.9 s	+40° 09' 05"
2852	Lyn	Gxy	0.9'×0.8'	13.4	09 h 23 m 14.6 s	+40° 09' 50"
2853	Lyn	Gxy	1.7'×0.9'	13.4	09 h 23 m 17.3 s	+40° 11' 59"
2860	Lyn	Gxy	1.4'×0.6'	13.9	09 h 24 m 53.3 s	+41° 03' 36"
6606	Lyr	Gxy	1.0'×0.8'	13.6	18 h 14 m 41.5 s	+43° 16' 07"
6612	Lyr	Gxy	0.6'×0.6'	14.5	18 h 16 m 10.9 s	+36° 04' 43"
6640	Lyr	Gxy	1.1'×0.8'	13.7	18 h 28 m 08.3 s	+34° 18' 10"
6646	Lyr	Gxy	1.2'×1.0'	13	18 h 29 m 38.7 s	+39° 51' 55"
6657	Lyr	Gxy	0.8'×0.4'	13.7	18 h 33 m 01.5 s	+34° 03' 38"
6662	Lyr	Gxy	1.6'×0.5'	13.9	18 h 34 m 11.2 s	+32° 03' 52"
6663	Lyr	Gxy	1.0'×0.9'	14.1	18 h 33 m 33.6 s	+40° 02' 56"
6665	Lyr	Gxy	1.1'×0.6'	13.9	18 h 34 m 29.9 s	+30° 43' 13"
6666	Lyr	NF	...	...	18 h 34 m 44.9 s	+33° 35' 15"
6671	Lyr	Gxy	1.5'×1.3'	12.8	18 h 37 m 26.2 s	+26° 25' 02"
6672	Lyr	***	...	...	18 h 36 m 14.3 s	+42° 56' 47"
6675	Lyr	Gxy	1.7'×1.3'	12.7	18 h 37 m 26.4 s	+40° 03' 27"
6685	Lyr	Gxy	1.1'×0.9'	13.6	18 h 39 m 58.5 s	+39° 58' 55"
6686	Lyr	Gxy	0.8'×0.7'	14.7	18 h 40 m 07.0 s	+40° 08' 14"
6688	Lyr	Gxy	1.7'×1.7'	12.7	18 h 40 m 40.2 s	+36° 17' 22"
6692	Lyr	Gxy	1.1'×0.7'	13.2	18 h 41 m 41.5 s	+34° 50' 36"
6693	Lyr	NF	...	...	18 h 41 m 32.3 s	+36° 54' 57"
6695	Lyr	Gxy	1.1'×0.7'	13.7	18 h 42 m 42.7 s	+40° 22' 00"
6700	Lyr	Gxy	1.4'×1.0'	13.3	18 h 46 m 04.5 s	+32° 16' 45"
6702	Lyr	Gxy	1.8'×1.3'	12.6	18 h 46 m 57.5 s	+45° 42' 21"
6703	Lyr	Gxy	2.7'×2.5'	11.3	18 h 47 m 18.9 s	+45° 33' 02"
6710	Lyr	Gxy	1.7'×1.0'	12.8	18 h 50 m 34.2 s	+26° 50' 16"
6713	Lyr	Gxy	0.4'×0.3'	13.4	18 h 50 m 44.1 s	+33° 57' 39"
6720	Lyr	PN	86"×62"	8.8	18 h 53 m 34.9 s	+33° 01' 43"
6731	Lyr	**	...	...	18 h 57 m 13.4 s	+43° 04' 36"
6740	Lyr	Gxy	1.0'×0.8'	14.2	19 h 00 m 50.5 s	+28° 46' 17"
6743	Lyr	OC	...	...	19 h 01 m 20.6 s	+29° 16' 39"

(continued)

Table 10.3 (continued)

NGC #	Const	Type	Size	V Mag	RA (2000)	Dec (2000)
6745	Lyr	Gxy	1.4'×0.7'	13.9	19 h 01 m 41.5 s	+40° 44' 43"
6765	Lyr	PN	38"	12.9	19 h 11 m 06.4 s	+30° 32' 43"
6767	Lyr	**	…	…	19 h 11 m 33.8 s	+37° 43' 31"
6779	Lyr	GC	7.1'	8.4	19 h 16 m 35.4 s	+30° 11' 05"
6791	Lyr	OC	10'	9.5	19 h 20 m 53.1 s	+37° 46' 19"
6792	Lyr	Gxy	2.2'×1.3'	12.3	19 h 20 m 57.3 s	+43° 07' 56"
1520	Men	OC	5'	…	03 h 57 m 31.1 s	−76° 50' 02"
1651	Men	OC	3.0'	12.3	04 h 37 m 32.7 s	−70° 35' 08"
1673	Men	OC	1.0'	14.1	04 h 42 m 39.7 s	−69° 49' 17"
1702	Men	OC	0.9'	12.5	04 h 49 m 27.7 s	−69° 51' 03"
1711	Men	OC	3.4'	10.1	04 h 50 m 36.2 s	−69° 59' 08"
1751	Men	OC	1.6'	14.5	04 h 54 m 12.1 s	−69° 48' 23"
1754	Men	OC	1.0'×1.0'	12	04 h 54 m 17.9 s	−70° 26' 30"
1766	Men	OC	0.7'	12.2	04 h 55 m 57.7 s	−70° 13' 30"
1775	Men	OC	0.7'	12.6	04 h 56 m 53.4 s	−70° 25' 46"
1777	Men	OC	2.1'	12.8	04 h 55 m 47.7 s	−74° 17' 08"
1789	Men	OC	1.5'×1.0'	13.1	04 h 57 m 51.3 s	−71° 54' 03"
1791	Men	OC	1.3'	13.1	04 h 59 m 06.5 s	−70° 10' 08"
1813	Men	OC	0.7'	12.8	05 h 02 m 40.3 s	−70° 19' 05"
1815	Men	OC	0.3'	12.4	05 h 02 m 27.3 s	−70° 37' 16"
1823	Men	OC	0.9'	12.1	05 h 03 m 25.0 s	−70° 20' 08"
1833	Men	OC+Neb	2.0'	…	05 h 04 m 21.8 s	−70° 43' 54"
1837	Men	OC+Neb	1.4'	10.6	05 h 04 m 55.9 s	−70° 42' 51"
1840	Men	OC	0.6'	…	05 h 05 m 19.2 s	−71° 45' 47"
1841	Men	GC	…	14.1	04 h 45 m 23.1 s	−83° 59' 49"
1845	Men	OC	21'	10.2	05 h 06 m 22.0 s	−70° 28' 45"
1848	Men	OC	3.0'	9.7	05 h 07 m 27.2 s	−71° 11' 44"
1861	Men	OC	1.2'	13.2	05 h 10 m 22.2 s	−70° 46' 38"
1878	Men	OC	1.1'	12.9	05 h 12 m 51.0 s	−70° 28' 18"
1890	Men	OC	1.1'	12.8	05 h 13 m 45.9 s	−72° 04' 41"
1914	Men	OC+Neb	2.1'	12	05 h 17 m 39.9 s	−71° 15' 21"
1938	Men	GC	0.6'	11.8	05 h 21 m 24.8 s	−69° 56' 22"
1939	Men	GC	0.5'	12.9	05 h 21 m 26.7 s	−69° 56' 59"
1943	Men	OC	…	…	05 h 22 m 28.8 s	−70° 09' 18"
1944	Men	OC	3.5'	11.8	05 h 21 m 57.4 s	−72° 29' 40"
1950	Men	OC	1.0'	13.2	05 h 24 m 33.1 s	−69° 54' 09"
1956	Men	Gxy	2.5'×1.1'	13.3	05 h 19 m 35.7 s	−77° 43' 49"
1959	Men	OC	0.5'	12.2	05 h 25 m 36.7 s	−69° 55' 37"
1986	Men	GC	0.7'	11.1	05 h 27 m 39.0 s	−69° 58' 22"
1987	Men	OC	1.7'	…	05 h 27 m 17.2 s	−70° 44' 15"
2000	Men	OC	1.7'	11.9	05 h 27 m 29.3 s	−71° 52' 46"
2010	Men	OC	2.2'	11.7	05 h 30 m 35.0 s	−70° 49' 11"
2012	Men	Gxy	1.2'×0.7'	13.6	05 h 22 m 35.9 s	−79° 51' 03"
2016	Men	OC	0.3'	10.4	05 h 31 m 38.9 s	−69° 56' 45"
2018	Men	OC+Neb	25'×18'	10.9	05 h 30 m 45.3 s	−71° 04' 06"

(continued)

Table 10.3 (continued)

NGC #	Const	Type	Size	V Mag	RA (2000)	Dec (2000)
2019	Men	GC	1.0'	10.9	05 h 31 m 56.7 s	−70° 09' 35"
2025	Men	OC	1.9'	10.9	05 h 32 m 33.7 s	−71° 42' 56"
2028	Men	OC	0.5'	12.9	05 h 33 m 48.7 s	−69° 57' 07"
2031	Men	OC	3.5'	10.8	05 h 33 m 41.9 s	−70° 59' 13"
2036	Men	OC	0.8'	12.8	05 h 34 m 31.5 s	−70° 03' 52"
2038	Men	OC	1.7'	11.9	05 h 34 m 42.3 s	−70° 33' 47"
2043	Men	OC	...	...	05 h 35 m 57.3 s	−70° 04' 28"
2046	Men	OC	1.3'	12.6	05 h 35 m 37.6 s	−70° 14' 27"
2047	Men	OC	0.8'	13.2	05 h 35 m 53.1 s	−70° 11' 34"
2051	Men	OC	1.7'	11.7	05 h 36 m 07.4 s	−71° 00' 41"
2056	Men	OC	1.7'	12.3	05 h 36 m 34.1 s	−70° 40' 19"
2057	Men	OC	1.8'	12.2	05 h 36 m 55.2 s	−70° 16' 08"
2058	Men	OC	1.8'	11.9	05 h 36 m 54.3 s	−70° 09' 44"
2059	Men	OC	1.0'	12.9	05 h 37 m 00.6 s	−70° 07' 45"
2065	Men	OC	2.5'	11.2	05 h 37 m 38.5 s	−70° 14' 12"
2066	Men	OC	0.7'	13.1	05 h 37 m 43.2 s	−70° 10' 00"
2072	Men	OC	1.3'	13.2	05 h 38 m 24.5 s	−70° 14' 03"
2075	Men	OC + Neb	2.0'	11.5	05 h 38 m 20.7 s	−70° 41' 09"
2103	Men	Neb	3.0'×4.0'	...	05 h 41 m 40.4 s	−71° 20' 00"
2107	Men	OC	1.0'×1.0'	11.5	05 h 43 m 12.9 s	−70° 38' 24"
2111	Men	OC	0.8'×0.7'	12.5	05 h 44 m 33.1 s	−70° 59' 36"
2121	Men	OC	1.8'	12.4	05 h 48 m 12.4 s	−71° 28' 52"
2122	Men	OC + Neb	6.0'×5.0'	10.4	05 h 48 m 52.6 s	−70° 04' 12"
2133	Men	OC	0.7'×0.7'	...	05 h 51 m 28.8 s	−71° 10' 30"
2134	Men	OC	1.7'×1.7'	11.1	05 h 51 m 56.7 s	−71° 05' 50"
2144	Men	Gxy	1.4'×1.1'	13.2	05 h 40 m 56.5 s	−82° 07' 10"
2145	Men	OC	0.8'×0.4'	12.1	05 h 54 m 22.7 s	−70° 54' 07"
2161	Men	OC	1.0'×1.0'	13	05 h 55 m 43.1 s	−74° 21' 14"
2171	Men	OC	...	...	05 h 44 m 13.9 s	−70° 40' 09"
2173	Men	OC	1.5'×1.5'	11.9	05 h 57 m 58.5 s	−72° 58' 41"
2190	Men	OC	1.0'×1.0'	12.9	06 h 01 m 04.1 s	−74° 43' 32"
2199	Men	Gxy	1.9'×0.8'	13	06 h 04 m 44.9 s	−73° 24' 01"
2203	Men	OC	3.4'	12	06 h 04 m 42.7 s	−75° 26' 18"
2209	Men	OC	2.7'	13.2	06 h 08 m 33.5 s	−73° 50' 16"
2213	Men	OC	2.0'	12.4	06 h 10 m 42.0 s	−71° 31' 42"
6919	Mic	Gxy	1.7'×1.2'	13.2	20 h 31 m 38.1 s	−44° 12' 58"
6923	Mic	Gxy	2.6'×1.3'	12.2	20 h 31 m 39.2 s	−30° 49' 54"
6925	Mic	Gxy	4.5'×1.2'	11.5	20 h 34 m 20.8 s	−31° 58' 50"
6947	Mic	Gxy	1.7'×1.0'	13.5	20 h 41 m 15.0 s	−32° 29' 13"
6958	Mic	Gxy	2.1'×1.7'	11.3	20 h 48 m 42.4 s	−37° 59' 50"
6983	Mic	Gxy	0.8'×0.6'	13.5	20 h 56 m 43.5 s	−43° 59' 12"
6998	Mic	Gxy	0.8'×0.7'	14.3	21 h 01 m 37.7 s	−28° 01' 54"
6999	Mic	Gxy	0.9'×0.7'	14.1	21 h 01 m 59.5 s	−28° 03' 34"
7012	Mic	Gxy	2.5'×1.4'	12.8	21 h 06 m 45.5 s	−44° 48' 54"
7057	Mic	Gxy	1.4'×1.0'	12.7	21 h 24 m 58.5 s	−42° 27' 39"

(continued)

Table 10.3 (continued)

NGC #	Const	Type	Size	V Mag	RA (2000)	Dec (2000)
7060	Mic	Gxy	1.7'×0.9'	12.9	21 h 25 m 53.4 s	−42° 24' 40"
2142	Mon	*	...	...	06 h 01 m 50.4 s	−10° 35' 53"
2149	Mon	Neb	3'×2'	...	06 h 03 m 30.8 s	−09° 43' 50"
2167	Mon	Neb	4'	...	06 h 06 m 58.5 s	−06° 12' 09"
2170	Mon	Neb	10'	...	06 h 07 m 31.3 s	−06° 23' 53"
2182	Mon	Neb	3.0'×2.0'	...	06 h 09 m 30.9 s	−06° 19' 35"
2183	Mon	Neb	2'×2'	...	06 h 10 m 47.0 s	−06° 12' 43"
2185	Mon	Neb	4'×2'	...	06 h 11 m 00.5 s	−06° 13' 37"
2215	Mon	OC	11'	8.4	06 h 20 m 49.2 s	−07° 17' 02"
2219	Mon	OC	8.0'×8.0'	...	06 h 23 m 44.3 s	−04° 40' 38"
2225	Mon	OC	1.5'	...	06 h 26 m 34.5 s	−09° 37' 51"
2226	Mon	OC	1.5'	...	06 h 26 m 37.6 s	−09° 38' 34"
2232	Mon	OC	29'	4.2	06 h 28 m 01.1 s	−04° 50' 51"
2236	Mon	OC	6'	8.5	06 h 29 m 39.7 s	+06° 49' 50"
2237	Mon	Neb	90'×90'	...	06 h 30 m 54.6 s	+05° 02' 57"
2238	Mon	Neb	80'×60'	...	06 h 30 m 40.3 s	+05° 00' 47"
2239	Mon	OC	24'	...	06 h 31 m 55.6 s	+04° 56' 35"
2244	Mon	OC	24'	...	06 h 31 m 55.6 s	+04° 56' 35"
2245	Mon	Neb	2'×2'	...	06 h 32 m 41.2 s	+10° 09' 24"
2246	Mon	Neb	80'×60'	...	06 h 32 m 33.7 s	+05° 07' 42"
2247	Mon	Neb	2.0'×2.0'	...	06 h 33 m 05.1 s	+10° 19' 18"
2250	Mon	OC	7'	...	06 h 33 m 49.9 s	−05° 05' 04"
2251	Mon	OC	10'	7.3	06 h 34 m 38.5 s	+08° 21' 59"
2252	Mon	OC	20'	...	06 h 34 m 42.9 s	+05° 21' 59"
2254	Mon	OC	4.0'	9.1	06 h 35 m 49.7 s	+07° 40' 24"
2259	Mon	OC	4.5'	10.8	06 h 38 m 21.4 s	+10° 53' 01"
2260	Mon	OC	10'	...	06 h 38 m 03.1 s	−01° 28' 22"
2261	Mon	Neb	2'×1'	...	06 h 39 m 09.5 s	+08° 44' 39"
2262	Mon	OC	3.5'	11.3	06 h 39 m 38.1 s	+01° 08' 37"
2264	Mon	OC+Neb	20'	4.1	06 h 40 m 58.3 s	+09° 53' 44"
2269	Mon	OC	4.0'	10	06 h 43 m 17.1 s	+04° 37' 28"
2270	Mon	OC	...	...	06 h 43 m 57.7 s	+03° 28' 43"
2282	Mon	Neb	3'×3'	...	06 h 46 m 51.3 s	+01° 18' 54"
2286	Mon	OC	14'	7.5	06 h 47 m 40.1 s	−03° 08' 52"
2299	Mon	OC	2.50'	...	06 h 51 m 56.7 s	−07° 05' 04"
2301	Mon	OC	12'	6	06 h 51 m 45.3 s	+00° 27' 33"
2302	Mon	OC	2.50'	...	06 h 51 m 56.7 s	−07° 05' 04"
2306	Mon	OC	20'×10'	...	06 h 54 m 29.5 s	−07° 12' 15"
2309	Mon	OC	3.0'	10.5	06 h 56 m 03.6 s	−07° 10' 28"
2311	Mon	OC	6'	9.6	06 h 57 m 47.5 s	−04° 36' 41"
2312	Mon	OC	5.0'×2.5'	...	06 h 58 m 47.5 s	+10° 17' 40"
2313	Mon	Neb	0.5'×0.3'	...	06 h 58 m 02.8 s	−07° 56' 42"
2316	Mon	Neb	...	...	06 h 59 m 40.8 s	−07° 46' 40"
2317	Mon	Neb	...	...	06 h 59 m 41.5 s	−07° 46' 29"
2319	Mon	Ast	...	...	07 h 00 m 32.2 s	+03° 02' 32"

(continued)

Table 10.3 (continued)

NGC #	Const	Type	Size	V Mag	RA (2000)	Dec (2000)
2323	Mon	OC	16'	5.9	07 h 02 m 42.3 s	−08° 23' 26"
2324	Mon	OC	7'	8.4	07 h 04 m 07.9 s	+01° 02' 41"
2335	Mon	OC	12'	7.2	07 h 06 m 49.5 s	−10° 01' 43"
2338	Mon	OC	6.0'×5.0'	...	07 h 07 m 47.3 s	−05° 43' 11"
2343	Mon	OC	6'	6.7	07 h 08 m 06.8 s	−10° 37' 01"
2346	Mon	PN	60"×50"	11.6	07 h 09 m 22.5 s	−00° 48' 24"
2349	Mon	OC	5.0'	...	07 h 10 m 48.1 s	−08° 35' 36"
2351	Mon	OC	5'	...	07 h 13 m 32.0 s	−10° 29' 29"
2353	Mon	OC	20'	7.1	07 h 14 m 30.3 s	−10° 15' 57"
2364	Mon	OC	...	...	07 h 20 m 46.5 s	−07° 32' 59"
2368	Mon	OC	5'	11.8	07 h 21 m 06.3 s	−10° 22' 18"
2377	Mon	Gxy	1.7'×1.3'	12.8	07 h 24 m 56.7 s	−09° 39' 37"
2494	Mon	Gxy	0.9'×0.7'	13.3	07 h 59 m 06.9 s	−00° 38' 17"
2506	Mon	OC	8'	7.6	08 h 00 m 01.7 s	−10° 46' 11"
4071	Mus	PN	1.3'	13	12 h 04 m 15.3 s	−67° 18' 35"
4372	Mus	GC	18.6'	7.2	12 h 25 m 45.4 s	−72° 39' 33"
4463	Mus	OC	5'	7.2	12 h 29 m 55.2 s	−64° 47' 23"
4815	Mus	OC	3.0'	8.6	12 h 57 m 58.3 s	−64° 57' 42"
4833	Mus	GC	13.5'	8.4	12 h 59 m 34.9 s	−70° 52' 28"
5189	Mus	PN	140"	...	13 h 33 m 32.9 s	−65° 58' 26"
5925	Nor	OC	20'	8.4	15 h 27 m 26.7 s	−54° 31' 43"
5946	Nor	GC	3.0'	8.4	15 h 35 m 28.5 s	−50° 39' 34"
5999	Nor	OC	3'	9	15 h 52 m 08.6 s	−56° 28' 22"
6005	Nor	OC	3.0'	10.7	15 h 55 m 48.7 s	−57° 26' 14"
6031	Nor	OC	2.0'	8.5	16 h 07 m 35.0 s	−54° 00' 54"
6067	Nor	OC	12'	5.6	16 h 13 m 11.0 s	−54° 13' 06"
6087	Nor	OC	12'	5.4	16 h 18 m 50.5 s	−57° 56' 04"
6115	Nor	OC	3.4'	9.8	16 h 24 m 26.3 s	−51° 56' 54"
6134	Nor	OC	6'	7.2	16 h 27 m 46.5 s	−49° 09' 04"
6152	Nor	OC	29.0'	8.1	16 h 32 m 45.5 s	−52° 38' 38"
6164	Nor	Neb	1.0'×0.3'	...	16 h 33 m 41.6 s	−48° 04' 46"
6165	Nor	Neb	2.5'×0.5'	...	16 h 34 m 02.8 s	−48° 09' 09"
6167	Nor	OC	7'	6.7	16 h 34 m 34.9 s	−49° 46' 19"
6169	Nor	OC	6'	6.6	16 h 34 m 04.6 s	−44° 02' 44"
2573	Oct	Gxy	2.0'×0.8'	13.6	01 h 41 m 42.8 s	−89° 20' 04"
6438	Oct	Gxy	1.2'×1.1'	12.5	18 h 22 m 17.4 s	−85° 24' 06"
6557	Oct	Gxy	1.6'×1.1'	12.9	18 h 21 m 25.2 s	−76° 35' 00"
6920	Oct	Gxy	1.8'×1.5'	12.2	20 h 43 m 57.0 s	−80° 00' 02"
7095	Oct	Gxy	2.8'×2.7'	11.7	21 h 52 m 24.7 s	−81° 31' 55"
7098	Oct	Gxy	4.1'×2.6'	11.6	21 h 44 m 15.9 s	−75° 06' 41"
7637	Oct	Gxy	2.1'×1.9'	12.7	23 h 26 m 28.5 s	−81° 54' 42"
6059	Oph	NF	...	...	16 h 07 m 13.0 s	−06° 24' 48"
6171	Oph	GC	10'	7.8	16 h 32 m 31.9 s	−13° 03' 10"
6218	Oph	GC	15'	6.1	16 h 47 m 14.4 s	−01° 56' 52"
6220	Oph	Gxy	1.6'×0.9'	13.9	16 h 47 m 13.3 s	−00° 16' 31"

(continued)

Table 10.3 (continued)

NGC #	Const	Type	Size	V Mag	RA (2000)	Dec (2000)
6234	Oph	Gxy	0.3'×0.3'	14.4	16 h 51 m 57.4 s	+04° 23' 01"
6235	Oph	GC	5.0'	8.9	16 h 53 m 25.3 s	–22° 10' 38"
6240	Oph	Gxy	2.1'×1.1'	12.8	16 h 52 m 58.9 s	+02° 24' 07"
6254	Oph	GC	15'	6.6	16 h 57 m 09.0 s	–04° 05' 58"
6266	Oph	GC	14'	6.4	17 h 01 m 12.5 s	–30° 06' 44"
6273	Oph	GC	14'	6.8	17 h 02 m 37.7 s	–26° 16' 05"
6280	Oph	Gxy	0.4'×0.2'	14.6	17 h 01 m 57.4 s	+06° 39' 58"
6284	Oph	GC	5.6'	8.9	17 h 04 m 28.5 s	–24° 45' 49"
6287	Oph	GC	5.1'	9.3	17 h 05 m 09.3 s	–22° 42' 29"
6293	Oph	GC	7.9'	8.3	17 h 10 m 10.3 s	–26° 34' 57"
6294	Oph	**	…	…	17 h 10 m 16.1 s	–26° 34' 28"
6296	Oph	Gxy	1.0'×0.8'	13.6	17 h 08 m 44.5 s	+03° 53' 38"
6304	Oph	GC	6.8'	8.3	17 h 14 m 32.5 s	–29° 27' 44"
6309	Oph	PN	20"	11.5	17 h 14 m 04.3 s	–12° 54' 39"
6316	Oph	GC	4.9'	8.1	17 h 16 m 37.3 s	–28° 08' 24"
6325	Oph	GC	4.3'	10.2	17 h 17 m 59.2 s	–23° 45' 57"
6333	Oph	GC	9.3'	7.8	17 h 19 m 11.7 s	–18° 30' 59"
6342	Oph	GC	4.4'	9.5	17 h 21 m 10.1 s	–19° 35' 14"
6355	Oph	GC	5.0'	8.6	17 h 23 m 58.5 s	–26° 21' 13"
6356	Oph	GC	7.2'	8.2	17 h 23 m 34.9 s	–17° 48' 47"
6360	Oph	Ast	…	…	17 h 24 m 27.5 s	–29° 52' 18"
6366	Oph	GC	8.3'	9.5	17 h 27 m 44.3 s	–05° 04' 36"
6368	Oph	Gxy	3.8'×1.0'	12.5	17 h 27 m 11.5 s	+11° 32' 36"
6369	Oph	PN	58"×34"	11.4	17 h 29 m 20.4 s	–23° 45' 34"
6378	Oph	Gxy	1.4'×1.1'	14	17 h 30 m 41.8 s	+06° 16' 56"
6384	Oph	Gxy	6.2'×4.1'	11.1	17 h 32 m 24.3 s	+07° 03' 37"
6401	Oph	GC	5.6'	7.4	17 h 38 m 36.9 s	–23° 54' 32"
6402	Oph	GC	12'	7.6	17 h 37 m 36.1 s	–03° 14' 46"
6413	Oph	Ast	…	…	17 h 40 m 40.7 s	+12° 37' 27"
6426	Oph	GC	3.2'	10.9	17 h 44 m 54.6 s	+03° 10' 13"
6481	Oph	Ast	…	…	17 h 52 m 48.9 s	+04° 10' 04"
6509	Oph	Gxy	1.6'×1.2'	12.6	17 h 59 m 25.1 s	+06° 17' 12"
6517	Oph	GC	4.3'	10.1	18 h 01 m 50.6 s	–08° 57' 32"
6525	Oph	OC	10'	…	18 h 02 m 04.7 s	+11° 02' 18"
6570	Oph	Gxy	1.8'×1.1'	12.6	18 h 11 m 07.3 s	+14° 05' 35"
6572	Oph	PN	16"×13"	8.1	18 h 12 m 06.3 s	+06° 51' 13"
6615	Oph	Gxy	0.9'×0.9'	13.2	18 h 18 m 33.3 s	+13° 15' 53"
6633	Oph	OC	20'	4.6	18 h 27 m 15.2 s	+06° 30' 30"
1661	Ori	Gxy	1.4'×0.9'	13.4	04 h 47 m 07.7 s	–02° 03' 18"
1662	Ori	OC	20.0'	6.4	04 h 48 m 28.9 s	+10° 55' 49"
1663	Ori	OC	…	…	04 h 49 m 19.5 s	+13° 08' 52"
1670	Ori	Gxy	2.1'×1.0'	12.8	04 h 49 m 42.5 s	–02° 45' 37"
1671	Ori	Gxy	1.1'×0.9'	12.9	04 h 49 m 34.1 s	+00° 15' 11"
1678	Ori	Gxy	1.1'×0.8'	13.3	04 h 51 m 35.5 s	–02° 37' 25"
1682	Ori	Gxy	0.3'×0.3'	13.5	04 h 52 m 19.8 s	–03° 06' 21"

(continued)

Table 10.3 (continued)

NGC #	Const	Type	Size	V Mag	RA (2000)	Dec (2000)
1683	Ori	Gxy	1.0'×0.4'	14.7	04 h 52 m 17.6 s	−03° 01' 29"
1684	Ori	Gxy	2.5'×1.7'	12	04 h 52 m 31.1 s	−03° 06' 21"
1685	Ori	Gxy	1.3'×0.9'	14.1	04 h 52 m 34.2 s	−02° 56' 57"
1690	Ori	Gxy	1.1'×1.1'	14.3	04 h 54 m 19.3 s	+01° 38' 25"
1691	Ori	Gxy	1.7'×1.5'	12.2	04 h 54 m 38.3 s	+03° 16' 02"
1707	Ori	Ast	...	...	04 h 58 m 20.6 s	+08° 14' 18"
1709	Ori	Gxy	0.9'×0.7'	14.3	04 h 58 m 43.9 s	−00° 28' 41"
1713	Ori	Gxy	1.4'×1.2'	12.7	04 h 58 m 54.6 s	−00° 29' 22"
1717	Ori	*	...	...	04 h 59 m 30.1 s	−00° 14' 18"
1719	Ori	Gxy	1.1'×0.3'	13.8	04 h 59 m 34.6 s	−00° 15' 37"
1729	Ori	Gxy	1.6'×1.3'	12.9	05 h 00 m 15.7 s	−03° 21' 09"
1740	Ori	Gxy	1.5'×1.2'	12.9	05 h 01 m 54.7 s	−03° 17' 46"
1742	Ori	*	...	...	05 h 02 m 00.4 s	−03° 17' 15"
1753	Ori	Gxy	1.4'×0.6'	14.6	05 h 02 m 32.3 s	−03° 20' 40"
1762	Ori	Gxy	1.7'×1.1'	12.8	05 h 03 m 36.9 s	+01° 34' 25"
1788	Ori	Neb	5'×3'	...	05 h 06 m 53.1 s	−03° 20' 27"
1819	Ori	Gxy	1.7'×1.2'	12.5	05 h 11 m 46.1 s	+05° 12' 01"
1843	Ori	Gxy	2.0'×1.6'	12.7	05 h 14 m 06.1 s	−10° 37' 38"
1875	Ori	Gxy	1.6'×0.4'	13.7	05 h 21 m 45.9 s	+06° 41' 19"
1908	Ori	NF	...	...	05 h 25 m 53.8 s	−02° 31' 44"
1909	Ori	Neb	180'×60'	...	05 h 04 m 55.4 s	−07° 15' 56"
1924	Ori	Gxy	1.6'×1.2'	12.5	05 h 28 m 01.9 s	−05° 18' 38"
1927	Ori	NF	...	...	05 h 28 m 42.9 s	−08° 22' 38"
1973	Ori	Neb	5.0'×5.0'	...	05 h 35 m 04.8 s	−04° 43' 55"
1975	Ori	Neb	10'×5'	...	05 h 35 m 18.3 s	−04° 41' 04"
1976	Ori	OC+Neb	90'×60'	...	05 h 35 m 17.2 s	−05° 23' 27"
1977	Ori	OC+Neb	40'×25'	...	05 h 35 m 15.8 s	−04° 50' 40"
1980	Ori	OC+Neb	13'	...	05 h 35 m 25.9 s	−05° 54' 35"
1981	Ori	OC	25'	4.2	05 h 35 m 09.6 s	−04° 25' 30"
1982	Ori	Neb	20.0'×15.0'	...	05 h 35 m 31.3 s	−05° 16' 03"
1990	Ori	Neb?	50.0'×10.0'	...	05 h 36 m 12.7 s	−01° 12' 07"
1999	Ori	Neb	2'×2'	...	05 h 36 m 25.3 s	−06° 42' 57"
2022	Ori	PN	28"×27"	11.6	05 h 42 m 06.1 s	+09° 05' 10"
2023	Ori	Neb	10'×8'	...	05 h 41 m 38.3 s	−02° 15' 33"
2024	Ori	Neb	30'×30'	...	05 h 41 m 42.5 s	−01° 51' 23"
2039	Ori	OC	...	...	05 h 44 m 00.9 s	+08° 41' 28"
2054	Ori	Ast	...	...	05 h 45 m 15.3 s	−10° 04' 59"
2063	Ori	OC	20'	...	05 h 46 m 43.0 s	+08° 46' 52"
2064	Ori	Neb	10'×10'	...	05 h 46 m 18.5 s	+00° 00' 21"
2067	Ori	Neb	8'×3'	...	05 h 46 m 31.9 s	+00° 07' 52"
2068	Ori	Neb	8'×6'	...	05 h 46 m 45.8 s	+00° 04' 45"
2071	Ori	Neb	7'×5'	...	05 h 47 m 07.3 s	+00° 17' 39"
2110	Ori	Gxy	1.7'×1.3'	12.5	05 h 52 m 11.4 s	−07° 27' 24"
2112	Ori	OC	11'	9.1	05 h 53 m 45.2 s	+00° 24' 39"
2119	Ori	Gxy	1.2'×1.0'	13.6	05 h 57 m 26.9 s	+11° 56' 56"

(continued)

Table 10.3 (continued)

NGC #	Const	Type	Size	V Mag	RA (2000)	Dec (2000)
2141	Ori	OC	10'	9.4	06 h 02 m 55.1 s	+10° 26' 47"
2143	Ori	OC	12'×10'	...	06 h 02 m 53.4 s	+05° 42' 53"
2163	Ori	Neb	2.0'×0.5'	...	06 h 07 m 49.5 s	+18° 39' 27"
2169	Ori	OC	6'	5.9	06 h 08 m 24.3 s	+13° 57' 53"
2174	Ori	Neb	2.0'×3.0'	...	06 h 09 m 23.7 s	+20° 39' 34"
2175	Ori	OC+Neb	40'×30'	...	06 h 09 m 39.6 s	+20° 29' 15"
2180	Ori	OC	6'	...	06 h 09 m 36.3 s	+04° 42' 42"
2184	Ori	OC	15'	...	06 h 10 m 59.7 s	−03° 29' 43"
2186	Ori	OC	4'	8.7	06 h 12 m 07.1 s	+05° 27' 31"
2189	Ori	OC	...	...	06 h 12 m 02.0 s	+01° 04' 13"
2194	Ori	OC	10'	8.5	06 h 13 m 45.9 s	+12° 48' 24"
2195	Ori	**	...	...	06 h 14 m 33.9 s	+17° 38' 22"
2198	Ori	NF	...	...	06 h 13 m 54.9 s	+00° 59' 41"
2202	Ori	OC	...	...	06 h 16 m 50.7 s	+05° 59' 46"
6398	Pav	Gxy	2.0'×1.7'	12.7	17 h 42 m 43.8 s	−61° 41' 40"
6403	Pav	Gxy	1.1'×0.9'	13.4	17 h 43 m 23.3 s	−61° 40' 53"
6407	Pav	Gxy	2.1'×1.6'	12	17 h 44 m 57.6 s	−60° 44' 22"
6483	Pav	Gxy	1.0'×0.5'	12.3	17 h 59 m 31.0 s	−63° 40' 09"
6492	Pav	Gxy	2.5'×1.2'	12.3	18 h 02 m 47.9 s	−66° 25' 47"
6502	Pav	Gxy	1.3'×1.1'	12.7	18 h 04 m 13.9 s	−65° 24' 36"
6545	Pav	Gxy	1.0'×0.9'	13.3	18 h 12 m 15.3 s	−63° 46' 35"
6588	Pav	Ast	4.0'×1.5'	...	18 h 20 m 58.7 s	−63° 48' 36"
6614	Pav	Gxy	1.4'×1.1'	12.9	18 h 25 m 07.7 s	−63° 14' 51"
6630	Pav	Gxy	0.8'×0.7'	13.7	18 h 32 m 34.4 s	−63° 17' 32"
6653	Pav	Gxy	1.7'×1.5'	12.5	18 h 44 m 38.5 s	−73° 15' 48"
6673	Pav	Gxy	2.2'×0.9'	11.8	18 h 45 m 06.2 s	−62° 17' 51"
6684	Pav	Gxy	4.0'×2.6'	10.5	18 h 48 m 57.8 s	−65° 10' 24"
6699	Pav	Gxy	1.5'×1.5'	12.2	18 h 52 m 01.8 s	−57° 19' 15"
6706	Pav	Gxy	1.5'×0.7'	13	18 h 56 m 50.9 s	−63° 09' 58"
6718	Pav	Gxy	1.4'×0.8'	13.5	19 h 01 m 28.9 s	−66° 06' 39"
6719	Pav	Gxy	1.7'×0.8'	13	19 h 03 m 07.5 s	−68° 35' 17"
6721	Pav	Gxy	1.7'×1.4'	12.1	19 h 00 m 50.9 s	−57° 45' 35"
6722	Pav	Gxy	2.9'×0.4'	12.9	19 h 03 m 40.1 s	−64° 53' 43"
6730	Pav	Gxy	1.7'×1.5'	12.2	19 h 07 m 33.7 s	−68° 54' 46"
6733	Pav	Gxy	1.8'×1.2'	12.5	19 h 06 m 10.7 s	−62° 11' 49"
6734	Pav	Gxy	1.3'×1.1'	12.8	19 h 07 m 14.3 s	−65° 27' 40"
6736	Pav	Gxy	1.1'×0.9'	13.4	19 h 07 m 29.2 s	−65° 25' 39"
6739	Pav	Gxy	2.4'×0.9'	12.2	19 h 07 m 48.5 s	−61° 22' 04"
6744	Pav	Gxy	20.0'×12.9'	8.8	19 h 09 m 46.1 s	−63° 51' 25"
6746	Pav	Gxy	1.4'×0.9'	12.8	19 h 10 m 22.2 s	−61° 58' 08"
6752	Pav	GC	20.4'	5.3	19 h 10 m 51.7 s	−59° 58' 55"
6753	Pav	Gxy	2.5'×2.1'	11.2	19 h 11 m 23.6 s	−57° 02' 58"
6769	Pav	Gxy	2.3'×1.5'	11.9	19 h 18 m 23.1 s	−60° 30' 05"
6770	Pav	Gxy	2.3'×1.7'	11.8	19 h 18 m 37.0 s	−60° 29' 46"
6771	Pav	Gxy	2.3'×0.5'	12.5	19 h 18 m 39.4 s	−60° 32' 45"

(continued)

Table 10.3 (continued)

NGC #	Const	Type	Size	V Mag	RA (2000)	Dec (2000)
6776	Pav	Gxy	1.7'×1.4'	12	19 h 25 m 19.1 s	−63° 51' 35"
6777	Pav	**	...	...	19 h 26 m 32.2 s	−71° 27' 54"
6782	Pav	Gxy	2.2'×1.4'	11.9	19 h 23 m 57.9 s	−59° 55' 22"
6784	Pav	Gxy	0.8'×0.5'	14.1	19 h 26 m 33.7 s	−65° 37' 24"
6808	Pav	Gxy	1.5'×0.8'	12.6	19 h 43 m 54.3 s	−70° 37' 58"
6810	Pav	Gxy	3.2'×0.9'	11.7	19 h 43 m 34.2 s	−58° 39' 24"
6844	Pav	Gxy	1.4'×1.1'	12.9	20 h 02 m 50.0 s	−65° 13' 46"
6860	Pav	Gxy	1.3'×0.8'	13.1	20 h 08 m 46.9 s	−61° 06' 01"
6872	Pav	Gxy	6.0'×1.7'	11.4	20 h 16 m 56.6 s	−70° 46' 03"
6876	Pav	Gxy	2.4'×2.0'	11.2	20 h 18 m 18.9 s	−70° 51' 31"
6877	Pav	Gxy	1.8'×0.8'	12.3	20 h 18 m 36.1 s	−70° 51' 10"
6880	Pav	Gxy	2.0'×0.9'	12.3	20 h 19 m 29.6 s	−70° 51' 36"
6932	Pav	Gxy	2.1'×1.5'	12.4	20 h 42 m 08.5 s	−73° 37' 10"
6943	Pav	Gxy	4.0'×2.0'	11.5	20 h 44 m 33.6 s	−68° 44' 54"
7020	Pav	Gxy	3.7'×1.8'	11.7	21 h 11 m 20.0 s	−64° 01' 33"
7021	Pav	Gxy	4.2'×1.8'	11.8	21 h 11 m 20.0 s	−64° 01' 33"
7032	Pav	Gxy	1.1'×1.0'	13.1	21 h 15 m 22.9 s	−68° 17' 17"
7059	Pav	Gxy	3.5'×1.7'	12.1	21 h 27 m 21.6 s	−60° 00' 55"
1	Peg	Gxy	1.6'×1.2'	12.7	00 h 07 m 15.9 s	+27° 42' 32"
2	Peg	Gxy	1.0'×0.6'	14	00 h 07 m 17.2 s	+27° 40' 46"
8	Peg	**	...	...	00 h 08 m 45.7 s	+23° 50' 16"
9	Peg	Gxy	1.3'×0.7'	13.8	00 h 08 m 54.7 s	+23° 49' 04"
14	Peg	Gxy	2.8'×2.1'	12.5	00 h 08 m 46.3 s	+15° 48' 56"
15	Peg	Gxy	1.2'×0.7'	14	00 h 09 m 02.4 s	+21° 37' 28"
16	Peg	Gxy	1.8'×1.0'	11.8	00 h 09 m 04.6 s	+27° 43' 50"
18	Peg	**	...	...	00 h 09 m 23.0 s	+27° 43' 56"
22	Peg	Gxy	1.7'×1.3'	13.8	00 h 09 m 48.1 s	+27° 49' 57"
23	Peg	Gxy	2.1'×1.3'	11.9	00 h 09 m 53.4 s	+25° 55' 25"
26	Peg	Gxy	1.9'×1.4'	12.9	00 h 10 m 25.9 s	+25° 49' 51"
30	Peg	**	...	...	00 h 10 m 50.7 s	+21° 58' 37"
32	Peg	*	...	...	00 h 10 m 53.5 s	+18° 47' 45"
41	Peg	Gxy	0.9'×0.5'	13.9	00 h 12 m 47.9 s	+22° 01' 25"
42	Peg	Gxy	1.1'×0.6'	13.9	00 h 12 m 56.3 s	+22° 06' 02"
52	Peg	Gxy	2.6'×0.5'	13.7	00 h 14 m 40.2 s	+18° 34' 48"
7033	Peg	Gxy	0.7'×0.4'	14.2	21 h 09 m 36.2 s	+15° 07' 25"
7034	Peg	Gxy	1.0'×0.6'	13.9	21 h 09 m 38.1 s	+15° 08' 59"
7036	Peg	OC	8'×5'	...	21 h 10 m 12.3 s	+15° 22' 34"
7042	Peg	Gxy	2.0'×1.8'	12.2	21 h 13 m 45.9 s	+13° 34' 28"
7043	Peg	Gxy	1.1'×0.8'	13.9	21 h 14 m 04.1 s	+13° 37' 33"
7053	Peg	Gxy	1.4'×1.3'	13.1	21 h 21 m 07.5 s	+23° 05' 06"
7056	Peg	Gxy	1.0'×0.9'	13.2	21 h 22 m 07.5 s	+18° 39' 56"
7066	Peg	Gxy	1.1'×1.1'	14.2	21 h 26 m 13.8 s	+14° 10' 58"
7068	Peg	Gxy	0.7'×0.3'	14	21 h 26 m 32.4 s	+12° 11' 04"
7074	Peg	Gxy	0.7'×0.3'	14.5	21 h 29 m 39.0 s	+06° 40' 56"
7078	Peg	GC	12'	6.3	21 h 29 m 58.4 s	+12° 10' 00"

(continued)

Table 10.3 (continued)

NGC #	Const	Type	Size	V Mag	RA (2000)	Dec (2000)
7084	Peg	OC	20'	...	21 h 32 m 33.1 s	+17° 30' 31"
7085	Peg	Gxy	1.0'×0.4'	14.5	21 h 32 m 25.1 s	+06° 34' 52"
7094	Peg	PN	99"×91"	13.4	21 h 36 m 52.9 s	+12° 47' 16"
7100	Peg	*	...	...	21 h 39 m 06.9 s	+08° 57' 06"
7101	Peg	Gxy	0.5'×0.5'	13.5	21 h 39 m 34.5 s	+08° 52' 37"
7102	Peg	Gxy	1.7'×1.1'	12.9	21 h 39 m 44.5 s	+06° 17' 09"
7112	Peg	Gxy	0.9'×0.9'	14.2	21 h 42 m 26.7 s	+12° 34' 10"
7113	Peg	Gxy	0.9'×0.9'	14.2	21 h 42 m 26.7 s	+12° 34' 10"
7132	Peg	Gxy	1.0'×0.6'	14.2	21 h 47 m 16.5 s	+10° 14' 28"
7137	Peg	Gxy	1.6'×1.4'	12.7	21 h 48 m 12.9 s	+22° 09' 34"
7138	Peg	Gxy	1.1'×0.5'	14.4	21 h 49 m 01.0 s	+12° 30' 50"
7146	Peg	Gxy	0.9'×0.6'	14.5	21 h 51 m 47.3 s	+03° 01' 01"
7147	Peg	Gxy	1.0'×0.9'	13.6	21 h 51 m 58.4 s	+03° 04' 18"
7148	Peg	**	...	...	21 h 52 m 08.5 s	+03° 20' 29"
7149	Peg	Gxy	1.3'×0.9'	13.3	21 h 52 m 11.7 s	+03° 18' 03"
7156	Peg	Gxy	1.6'×1.4'	12.8	21 h 54 m 33.7 s	+02° 56' 35"
7159	Peg	Gxy	0.6'×0.5'	14.2	21 h 56 m 25.7 s	+13° 33' 45"
7161	Peg	**	...	...	21 h 56 m 57.1 s	+02° 55' 39"
7177	Peg	Gxy	3.1'×2.0'	11.4	22 h 00 m 41.2 s	+17° 44' 16"
7186	Peg	Ast	...	...	22 h 01 m 05.2 s	+35° 04' 44"
7190	Peg	Gxy	0.9'×0.5'	13.9	22 h 03 m 06.7 s	+11° 11' 56"
7193	Peg	OC	...	...	22 h 03 m 01.5 s	+10° 48' 14"
7194	Peg	Gxy	1.1'×0.9'	13.2	22 h 03 m 30.9 s	+12° 38' 12"
7195	Peg	Gxy	0.5'×0.4'	14.7	22 h 03 m 30.2 s	+12° 39' 39"
7206	Peg	Gxy	1.0'×0.7'	13.4	22 h 05 m 40.9 s	+16° 47' 07"
7207	Peg	Gxy	0.5'×0.2'	14.8	22 h 05 m 45.7 s	+16° 46' 05"
7210	Peg	NF	...	...	22 h 06 m 22.4 s	+27° 06' 33"
7212	Peg	Gxy	1.2'×0.6'	14.1	22 h 07 m 01.9 s	+10° 14' 01"
7217	Peg	Gxy	3.9'×3.2'	10.5	22 h 07 m 52.5 s	+31° 21' 32"
7224	Peg	Gxy	1.6'×1.0'	13.3	22 h 11 m 35.3 s	+25° 51' 52"
7236	Peg	Gxy	0.7'×0.7'	13.6	22 h 14 m 44.9 s	+13° 50' 47"
7237	Peg	Gxy	0.7'×0.7'	13.9	22 h 14 m 46.9 s	+13° 50' 27"
7238	Peg	NF	...	...	22 h 15 m 20.5 s	+22° 31' 09"
7241	Peg	Gxy	3.4'×1.1'	12.8	22 h 15 m 49.7 s	+19° 13' 54"
7244	Peg	Gxy	0.7'×0.3'	13.9	22 h 16 m 26.8 s	+16° 28' 16"
7253	Peg	Gxy	1.8'×0.8	13.7	22 h 19 m 27.1 s	+29° 23' 47"
7270	Peg	Gxy	1.0'×0.6'	14.2	22 h 23 m 47.7 s	+32° 24' 10"
7271	Peg	Gxy	0.7'×0.3'	15.6	22 h 23 m 57.6 s	+32° 22' 02"
7272	Peg	Gxy	1.0'×0.9'	13.8	22 h 24 m 31.7 s	+16° 35' 16"
7275	Peg	Gxy	0.9'×0.2'	14.5	22 h 24 m 17.1 s	+32° 26' 46"
7280	Peg	Gxy	2.2'×1.5'	12.3	22 h 26 m 27.6 s	+16° 08' 53"
7283	Peg	Gxy	0.8'×0.4'	14.5	22 h 28 m 32.6 s	+17° 28' 12"
7286	Peg	Gxy	1.7'×0.7'	12.7	22 h 27 m 50.5 s	+29° 05' 44"
7290	Peg	Gxy	1.6'×1.0'	13	22 h 28 m 26.5 s	+17° 08' 50"
7291	Peg	Gxy	1.8'×1.7'	13.2	22 h 28 m 29.5 s	+16° 46' 58"

(continued)

Table 10.3 (continued)

NGC #	Const	Type	Size	V Mag	RA (2000)	Dec (2000)
7292	Peg	Gxy	2.1'×1.7'	12.5	22 h 28 m 25.8 s	+30° 17' 32"
7303	Peg	Gxy	1.5'×1.2'	12.8	22 h 31 m 32.7 s	+30° 57' 22"
7304	Peg	***	...	...	22 h 31 m 44.4 s	+30° 58' 47"
7305	Peg	Gxy	0.7'×0.6'	14.1	22 h 32 m 13.9 s	+11° 42' 44"
7311	Peg	Gxy	1.6'×0.8'	12.7	22 h 34 m 06.7 s	+05° 34' 13"
7312	Peg	Gxy	1.4'×0.8'	13.6	22 h 34 m 34.7 s	+05° 49' 01"
7315	Peg	Gxy	1.6'×1.6'	12.6	22 h 35 m 31.7 s	+34° 48' 12"
7316	Peg	Gxy	1.1'×0.9'	12.9	22 h 35 m 56.3 s	+20° 19' 21"
7317	Peg	Gxy	1.1'×1.1'	13.9	22 h 35 m 51.9 s	+33° 56' 41"
7318	Peg	Gxy	1.2'×1.0'	13.4	22 h 35 m 58.4 s	+33° 57' 57"
7319	Peg	Gxy	1.7'×1.3'	13.8	22 h 36 m 03.5 s	+33° 58' 33"
7320	Peg	Gxy	2.2'×1.1'	13.1	22 h 36 m 03.4 s	+33° 56' 54"
7321	Peg	Gxy	1.6'×1.1'	13.1	22 h 36 m 28.0 s	+21° 37' 18"
7323	Peg	Gxy	1.4'×1.1'	13.1	22 h 36 m 53.6 s	+19° 08' 37"
7324	Peg	Gxy	0.8'×0.4'	14	22 h 37 m 00.9 s	+19° 08' 45"
7325	Peg	**	...	...	22 h 36 m 48.5 s	+34° 22' 05"
7326	Peg	**	...	...	22 h 36 m 52.1 s	+34° 25' 23"
7327	Peg	Gxy	0.5'×0.4'	...	22 h 36 m 33.5 s	+34° 30' 07"
7328	Peg	Gxy	2.0'×0.7'	13.3	22 h 37 m 29.3 s	+10° 31' 52"
7331	Peg	Gxy	10.5'×3.7'	9.7	22 h 37 m 04.5 s	+34° 25' 01"
7332	Peg	Gxy	4.1'×1.1'	11.2	22 h 37 m 24.5 s	+23° 47' 54"
7333	Peg	*	...	...	22 h 37 m 11.6 s	+34° 26' 17"
7335	Peg	Gxy	1.3'×0.6'	13.6	22 h 37 m 19.3 s	+34° 26' 52"
7336	Peg	Gxy	1.0'×0.4'	14.6	22 h 37 m 21.9 s	+34° 28' 56"
7337	Peg	Gxy	1.1'×0.9'	14.6	22 h 37 m 26.6 s	+34° 22' 28"
7338	Peg	**	...	...	22 h 37 m 31.3 s	+34° 24' 52"
7339	Peg	Gxy	3.0'×0.7'	12.5	22 h 37 m 47.5 s	+23° 47' 11"
7340	Peg	Gxy	0.9'×0.6'	13.9	22 h 37 m 44.1 s	+34° 24' 35"
7342	Peg	Gxy	1.3'×1.3'	14.1	22 h 38 m 13.1 s	+35° 29' 55"
7343	Peg	Gxy	1.1'×0.9'	13.6	22 h 38 m 38.0 s	+34° 04' 17"
7345	Peg	Gxy	1.2'×0.2'	14.5	22 h 38 m 44.8 s	+35° 32' 25"
7346	Peg	Gxy	0.4'×0.3'	14.7	22 h 39 m 35.4 s	+11° 05' 00"
7347	Peg	Gxy	1.5'×0.3'	13.9	22 h 39 m 56.1 s	+11° 01' 40"
7348	Peg	Gxy	1.1'×0.7'	14	22 h 40 m 36.3 s	+11° 54' 23"
7350	Peg	**	...	...	22 h 40 m 48.5 s	+12° 00' 23"
7353	Peg	Gxy	0.6'×0.5'	15.1	22 h 42 m 12.5 s	+11° 52' 38"
7356	Peg	Gxy	1.1'×0.5'	14.3	22 h 42 m 02.3 s	+30° 42' 31"
7357	Peg	Gxy	1.5'×0.7'	14.2	22 h 42 m 23.9 s	+30° 10' 17"
7360	Peg	Gxy	0.7'×0.3'	13.8	22 h 43 m 33.9 s	+04° 09' 04"
7362	Peg	Gxy	1.1'×0.8'	13.6	22 h 43 m 49.0 s	+08° 42' 20"
7363	Peg	Gxy	1.1'×1.0'	13.9	22 h 43 m 18.4 s	+33° 59' 57"
7366	Peg	Gxy	0.3'×0.3'	14.4	22 h 44 m 26.6 s	+10° 46' 53"
7367	Peg	Gxy	1.5'×0.4'	14	22 h 44 m 34.6 s	+03° 38' 44"
7369	Peg	Gxy	0.9'×0.9'	14	22 h 44 m 12.1 s	+34° 21' 04"
7370	Peg	Gxy	0.5'×0.2'	15.3	22 h 45 m 37.1 s	+11° 03' 28"

(continued)

Table 10.3 (continued)

NGC #	Const	Type	Size	V Mag	RA (2000)	Dec (2000)
7372	Peg	Gxy	1.0'×0.9'	13.7	22 h 45 m 45.9 s	+11° 07' 50"
7373	Peg	Gxy	1.2'×0.4'	13.7	22 h 46 m 19.4 s	+03° 12' 35"
7374	Peg	Gxy	0.9'×0.7'	14	22 h 46 m 00.9 s	+10° 51' 13"
7375	Peg	Gxy	0.7'×0.5'	13.7	22 h 46 m 32.1 s	+21° 05' 00"
7376	Peg	Gxy	0.7'×0.4'	14.5	22 h 47 m 17.4 s	+03° 38' 44"
7383	Peg	Gxy	0.8'×0.7'	14.1	22 h 49 m 35.7 s	+11° 33' 23"
7384	Peg	*	...	...	22 h 49 m 42.5 s	+11° 29' 15"
7385	Peg	Gxy	1.5'×1.3'	13	22 h 49 m 54.6 s	+11° 36' 30"
7386	Peg	Gxy	1.4'×1.2'	13.2	22 h 50 m 02.1 s	+11° 41' 54"
7387	Peg	Gxy	0.7'×0.4'	14.3	22 h 50 m 17.7 s	+11° 38' 12"
7388	Peg	*	...	...	22 h 50 m 20.9 s	+11° 42' 40"
7389	Peg	Gxy	1.4'×0.9'	13.9	22 h 50 m 15.9 s	+11° 33' 58"
7390	Peg	Gxy	0.8'×0.6'	14.3	22 h 50 m 19.5 s	+11° 31' 51"
7405	Peg	NF	...	...	22 h 53 m 36.5 s	+12° 28' 35"
7407	Peg	Gxy	2.0'×0.9'	13.4	22 h 53 m 21.0 s	+32° 07' 45"
7409	Peg	Gxy	0.5'×0.4'	14.4	22 h 53 m 48.1 s	+20° 12' 37"
7411	Peg	Gxy	1.0'×0.9'	13.5	22 h 54 m 34.9 s	+20° 14' 09"
7413	Peg	Gxy	1.0'×0.7'	14.2	22 h 55 m 03.1 s	+13° 13' 14"
7414	Peg	Gxy	0.4'×0.2'	15.8	22 h 55 m 24.4 s	+13° 14' 54"
7415	Peg	Gxy	1.1'×0.3'	14.7	22 h 54 m 53.6 s	+20° 15' 45"
7420	Peg	Gxy	0.8'×0.6'	14.4	22 h 55 m 32.0 s	+29° 48' 17"
7427	Peg	Gxy	0.6'×0.6'	15.1	22 h 57 m 10.0 s	+08° 30' 20"
7430	Peg	Gxy	0.4'×0.2'	14.5	22 h 57 m 29.7 s	+08° 47' 38"
7431	Peg	Gxy	0.4'×0.2'	15	22 h 57 m 38.7 s	+26° 09' 52"
7432	Peg	Gxy	1.5'×1.0'	13.5	22 h 58 m 01.9 s	+13° 08' 03"
7433	Peg	Gxy	0.7'×0.2'	14.9	22 h 57 m 51.7 s	+26° 09' 43"
7435	Peg	Gxy	1.5'×0.7'	14.4	22 h 57 m 54.5 s	+26° 08' 19"
7436	Peg	Gxy	2.0'×2.0'	12.9	22 h 57 m 57.6 s	+26° 08' 58"
7437	Peg	Gxy	1.8'×1.8'	13.4	22 h 58 m 10.1 s	+14° 18' 31"
7439	Peg	Gxy	1.3'×0.7'	14.1	22 h 58 m 09.9 s	+29° 13' 42"
7442	Peg	Gxy	1.1'×1.1'	13.5	22 h 59 m 26.5 s	+15° 32' 54"
7448	Peg	Gxy	2.7'×1.2'	11.5	23 h 00 m 03.7 s	+15° 58' 50"
7451	Peg	Gxy	1.0'×0.5'	14.2	23 h 00 m 40.9 s	+08° 28' 02"
7454	Peg	Gxy	2.2'×1.6'	12.2	23 h 01 m 06.5 s	+16° 23' 19"
7457	Peg	Gxy	4.3'×2.3'	11	23 h 01 m 00.0 s	+30° 08' 41"
7461	Peg	Gxy	0.9'×0.7'	13.4	23 h 01 m 48.3 s	+15° 34' 57"
7463	Peg	Gxy	2.9'×0.7'	12.7	23 h 01 m 51.9 s	+15° 58' 55"
7464	Peg	Gxy	0.8'×0.8'	13.4	23 h 01 m 53.6 s	+15° 58' 26"
7465	Peg	Gxy	1.2'×0.7'	12.4	23 h 02 m 00.9 s	+15° 57' 53"
7466	Peg	Gxy	1.5'×0.5'	13.8	23 h 02 m 03.5 s	+27° 03' 09"
7467	Peg	Gxy	0.4'×0.3'	14.5	23 h 02 m 28.0 s	+15° 33' 14"
7468	Peg	Gxy	0.9'×0.6'	13	23 h 02 m 59.0 s	+16° 36' 15"
7469	Peg	Gxy	1.5'×1.1'	12.3	23 h 03 m 15.5 s	+08° 52' 24"
7473	Peg	Gxy	1.1'×0.5'	13.8	23 h 03 m 57.1 s	+30° 09' 36"
7474	Peg	Gxy	0.5'×0.5'	14.5	23 h 04 m 04.3 s	+20° 04' 00"

(continued)

Table 10.3 (continued)

NGC #	Const	Type	Size	V Mag	RA (2000)	Dec (2000)
7475	Peg	Gxy	1.5'×0.9'	14.1	23 h 04 m 10.8 s	+20° 04' 54"
7479	Peg	Gxy	4.1'×3.1'	11	23 h 04 m 56.4 s	+12° 19' 00"
7485	Peg	Gxy	1.2'×0.6'	13.2	23 h 06 m 04.8 s	+34° 06' 27"
7486	Peg	Ast	...	...	23 h 06 m 12.9 s	+34° 06' 06"
7487	Peg	Gxy	2.0'×1.8'	13.8	23 h 06 m 50.5 s	+28° 10' 43"
7489	Peg	Gxy	2.1'×1.1'	13.5	23 h 07 m 32.6 s	+22° 59' 54"
7490	Peg	Gxy	2.8'×2.6'	12.5	23 h 07 m 25.1 s	+32° 22' 30"
7495	Peg	Gxy	2.0'×1.8'	13.6	23 h 08 m 57.1 s	+12° 02' 52"
7497	Peg	Gxy	4.9'×1.1'	12.4	23 h 09 m 03.3 s	+18° 10' 35"
7500	Peg	Gxy	2.1'×1.1'	13.4	23 h 10 m 29.9 s	+11° 00' 44"
7504	Peg	*	...	...	23 h 10 m 41.1 s	+14° 23' 10"
7505	Peg	Gxy	0.8'×0.2'	14.5	23 h 11 m 00.7 s	+13° 37' 53"
7508	Peg	Gxy	1.1'×0.3'	14.8	23 h 11 m 49.2 s	+12° 56' 25"
7509	Peg	Gxy	1.0'×1.0'	13.8	23 h 12 m 21.5 s	+14° 36' 33"
7511	Peg	Gxy	1.1'×0.5'	14	23 h 12 m 26.3 s	+13° 43' 35"
7512	Peg	Gxy	1.8'×1.5'	12.7	23 h 12 m 21.0 s	+31° 07' 32"
7514	Peg	Gxy	1.4'×0.9'	12.8	23 h 12 m 25.8 s	+34° 52' 53"
7515	Peg	Gxy	1.7'×1.6'	12.5	23 h 12 m 48.6 s	+12° 40' 46"
7516	Peg	Gxy	1.1'×0.9'	13.4	23 h 12 m 51.9 s	+20° 14' 54"
7519	Peg	Gxy	1.3'×1.1'	14	23 h 13 m 11.2 s	+10° 46' 19"
7523	Peg	Gxy	1.0'×0.2'	14.9	23 h 13 m 34.7 s	+13° 59' 11"
7525	Peg	Gxy	0.5'×0.3'	14.1	23 h 13 m 40.5 s	+14° 01' 18"
7527	Peg	Gxy	1.3'×0.9'	13.4	23 h 13 m 41.8 s	+24° 54' 08"
7528	Peg	Gxy	0.4'×0.3'	15.1	23 h 14 m 20.3 s	+10° 13' 54"
7529	Peg	Gxy	0.9'×0.8'	13.8	23 h 14 m 03.2 s	+08° 59' 31"
7535	Peg	Gxy	1.5'×1.5'	14.3	23 h 14 m 12.7 s	+13° 34' 55"
7536	Peg	Gxy	2.2'×0.8'	13.8	23 h 14 m 13.3 s	+13° 25' 33"
7539	Peg	Gxy	1.5'×1.2'	12.6	23 h 14 m 29.4 s	+23° 41' 05"
7540	Peg	Gxy	0.5'×0.3'	14.7	23 h 14 m 36.2 s	+15° 57' 03"
7542	Peg	Gxy	0.8'×0.5'	14.6	23 h 14 m 41.6 s	+10° 38' 37"
7543	Peg	Gxy	1.1'×0.9'	13.2	23 h 14 m 34.6 s	+28° 19' 38"
7547	Peg	Gxy	1.1'×0.5'	13.9	23 h 15 m 03.4 s	+18° 58' 23"
7548	Peg	Gxy	1.0'×0.9'	13.4	23 h 15 m 11.2 s	+25° 16' 55"
7549	Peg	Gxy	2.8'×0.7'	13.2	23 h 15 m 17.2 s	+19° 02' 29"
7550	Peg	Gxy	1.4'×1.2'	12.7	23 h 15 m 16.1 s	+18° 57' 40"
7551	Peg	Gxy	0.3'×0.2'	16.2	23 h 15 m 21.9 s	+15° 56' 27"
7553	Peg	Gxy	0.4'×0.4'	14.4	23 h 15 m 32.9 s	+19° 03' 01"
7555	Peg	NF	...	...	23 h 15 m 30.8 s	+12° 34' 22"
7558	Peg	Gxy	0.4'×0.4'	14.9	23 h 15 m 38.1 s	+18° 55' 11"
7559	Peg	Gxy	1.0'×0.8'	13.7	23 h 15 m 46.5 s	+13° 17' 25"
7563	Peg	Gxy	1.9'×1.0'	13.4	23 h 15 m 55.9 s	+13° 11' 46"
7567	Peg	Gxy	1.0'×0.3'	14.8	23 h 16 m 10.8 s	+15° 51' 03"
7568	Peg	Gxy	0.9'×0.5'	13.6	23 h 16 m 24.9 s	+24° 29' 48"
7569	Peg	Gxy	1.0'×0.7'	13.3	23 h 16 m 44.5 s	+08° 54' 21"
7570	Peg	Gxy	1.5'×0.9'	13.3	23 h 16 m 44.7 s	+13° 28' 58"

(continued)

Table 10.3 (continued)

NGC #	Const	Type	Size	V Mag	RA (2000)	Dec (2000)
7571	Peg	Gxy	1.0'×1.0'	14.1	23 h 18 m 30.1 s	+18° 41' 18"
7572	Peg	Gxy	0.7'×0.2'	14.4	23 h 16 m 50.3 s	+18° 28' 58"
7574	Peg	Gxy	1.0'×0.6'	13.6	23 h 16 m 24.9 s	+24° 29' 48"
7578	Peg	Gxy	0.8'×0.8'	13.9	23 h 17 m 13.5 s	+18° 42' 29"
7579	Peg	Gxy	0.5'×0.4'	14.1	23 h 17 m 38.9 s	+09° 25' 59"
7580	Peg	Gxy	0.8'×0.6'	13.4	23 h 17 m 36.5 s	+14° 00' 02"
7584	Peg	Gxy	0.4'×0.4'	14.4	23 h 17 m 52.9 s	+09° 25' 59"
7586	Peg	Gxy	0.4'×0.3'	15.5	23 h 17 m 55.5 s	+08° 35' 04"
7587	Peg	Gxy	1.3'×0.3'	14.1	23 h 17 m 59.1 s	+09° 40' 47"
7588	Peg	Gxy	0.3'×0.2'	14.9	23 h 17 m 57.7 s	+18° 45' 07"
7593	Peg	Gxy	1.0'×0.6'	13.5	23 h 17 m 57.1 s	+11° 20' 56"
7594	Peg	Gxy	1.4'×0.9'	13.8	23 h 18 m 14.0 s	+10° 17' 53"
7595	Peg	Gxy	0.5'×0.5'	14.9	23 h 18 m 30.1 s	+09° 55' 57"
7597	Peg	Gxy	0.9'×0.9'	14.1	23 h 18 m 30.1 s	+18° 41' 18"
7598	Peg	Gxy	0.3'×0.2'	14.9	23 h 18 m 33.3 s	+18° 45' 00"
7601	Peg	Gxy	1.2'×0.9'	13.9	23 h 18 m 47.1 s	+09° 14' 01"
7602	Peg	Gxy	0.5'×0.5'	14.4	23 h 18 m 43.3 s	+18° 41' 52"
7607	Peg	**	...	...	23 h 18 m 59.1 s	+11° 20' 30"
7608	Peg	Gxy	1.6'×0.5'	14.2	23 h 19 m 15.3 s	+08° 21' 01"
7609	Peg	Gxy	1.3'×1.1'	14.2	23 h 19 m 29.9 s	+09° 30' 29"
7610	Peg	Gxy	2.5'×1.9'	13.6	23 h 19 m 41.4 s	+10° 11' 07"
7612	Peg	Gxy	1.6'×0.8'	13.1	23 h 19 m 44.1 s	+08° 34' 35"
7615	Peg	Gxy	1.0'×0.5'	14.4	23 h 19 m 54.3 s	+08° 23' 57"
7616	Peg	Gxy	2.7'×2.2'	12.7	23 h 19 m 41.4 s	+10° 11' 07"
7619	Peg	Gxy	2.5'×2.3'	11.3	23 h 20 m 14.5 s	+08° 12' 21"
7620	Peg	Gxy	1.1'×1.1'	13	23 h 20 m 05.7 s	+24° 13' 15"
7621	Peg	Gxy	0.8'×0.2'	14.9	23 h 20 m 24.5 s	+08° 21' 58"
7623	Peg	Gxy	1.2'×0.9'	12.8	23 h 20 m 30.0 s	+08° 23' 43"
7624	Peg	Gxy	1.0'×0.7'	13.3	23 h 20 m 22.7 s	+27° 18' 54"
7625	Peg	Gxy	1.6'×1.4'	12.1	23 h 20 m 30.1 s	+17° 13' 32"
7626	Peg	Gxy	2.6'×2.3'	11.5	23 h 20 m 42.6 s	+08° 13' 01"
7627	Peg	Gxy	1.7'×0.5'	14	23 h 22 m 30.9 s	+11° 53' 33"
7628	Peg	Gxy	1.1'×0.9'	12.8	23 h 20 m 54.9 s	+25° 53' 55"
7630	Peg	Gxy	1.1'×0.4'	14.4	23 h 21 m 16.2 s	+11° 23' 50"
7631	Peg	Gxy	1.8'×0.7'	13.1	23 h 21 m 26.7 s	+08° 13' 02"
7634	Peg	Gxy	1.2'×0.9'	12.7	23 h 21 m 41.7 s	+08° 53' 12"
7638	Peg	Gxy	0.6'×0.6'	14.9	23 h 22 m 33.0 s	+11° 19' 45"
7639	Peg	Gxy	0.5'×0.4'	14.6	23 h 22 m 48.3 s	+11° 22' 25"
7641	Peg	Gxy	1.7'×0.5'	14.1	23 h 22 m 30.9 s	+11° 53' 33"
7643	Peg	Gxy	1.4'×0.8'	13.2	23 h 22 m 50.4 s	+11° 59' 19"
7644	Peg	Gxy	1.4'×0.8'	13.2	23 h 22 m 50.4 s	+11° 59' 19"
7647	Peg	Gxy	1.4'×1.0'	13.7	23 h 23 m 57.3 s	+16° 46' 38"
7648	Peg	Gxy	1.6'×1.0'	12.4	23 h 23 m 54.1 s	+09° 40' 02"
7649	Peg	Gxy	1.2'×0.8'	14.1	23 h 24 m 20.0 s	+14° 38' 49"
7651	Peg	Gxy	0.7'×0.5'	14.8	23 h 24 m 26.0 s	+13° 58' 19"

(continued)

Table 10.3 (continued)

NGC #	Const	Type	Size	V Mag	RA (2000)	Dec (2000)
7653	Peg	Gxy	1.6′×1.4′	12.9	23 h 24 m 49.4 s	+15° 16′ 31″
7659	Peg	Gxy	0.8′×0.3′	14.2	23 h 25 m 55.7 s	+14° 12′ 34″
7660	Peg	Gxy	1.4′×1.1′	12.5	23 h 25 m 48.7 s	+27° 01′ 46″
7664	Peg	Gxy	2.6′×1.5′	12.6	23 h 26 m 39.7 s	+25° 04′ 49″
7671	Peg	Gxy	1.4′×0.8′	13.1	23 h 27 m 19.4 s	+12° 28′ 02″
7672	Peg	Gxy	1.2′×0.9′	13.8	23 h 27 m 31.5 s	+12° 23′ 06″
7673	Peg	Gxy	1.3′×1.2′	12.4	23 h 27 m 41.1 s	+23° 35′ 21″
7674	Peg	Gxy	1.1′×1.0′	12.9	23 h 27 m 56.7 s	+08° 46′ 43″
7675	Peg	Gxy	0.7′×0.4′	15.2	23 h 28 m 05.9 s	+08° 46′ 06″
7677	Peg	Gxy	1.6′×1.0′	13.2	23 h 28 m 06.3 s	+23° 31′ 53″
7678	Peg	Gxy	2.3′×1.7′	12.2	23 h 28 m 27.9 s	+22° 25′ 16″
7680	Peg	Gxy	1.9′×1.9′	12.3	23 h 28 m 35.1 s	+32° 24′ 56″
7681	Peg	Gxy	1.8′×1.5′	13.8	23 h 28 m 54.8 s	+17° 18′ 33″
7683	Peg	Gxy	1.9′×1.0′	13.2	23 h 29 m 03.7 s	+11° 26′ 42″
7688	Peg	Gxy	0.4′×0.3′	14.5	23 h 31 m 05.5 s	+21° 24′ 41″
7691	Peg	Gxy	2.1′×1.6′	13.1	23 h 32 m 24.3 s	+15° 50′ 49″
7698	Peg	Gxy	1.0′×0.9′	13.4	23 h 34 m 01.5 s	+24° 56′ 40″
7703	Peg	Gxy	2.3′×0.4′	13.5	23 h 34 m 46.9 s	+16° 04′ 33″
7711	Peg	Gxy	2.6′×1.3′	12.6	23 h 35 m 39.3 s	+15° 18′ 07″
7712	Peg	Gxy	0.9′×0.8′	12.8	23 h 35 m 51.5 s	+23° 37′ 07″
7718	Peg	Gxy	0.9′×0.6′	14.2	23 h 38 m 05.0 s	+25° 43′ 11″
7720	Peg	Gxy	1.6′×1.3′	12.4	23 h 38 m 29.1 s	+27° 01′ 57″
7722	Peg	Gxy	1.7′×1.4′	12.6	23 h 38 m 41.0 s	+15° 57′ 13″
7726	Peg	Gxy	1.5′×0.5′	14.2	23 h 39 m 11.9 s	+27° 06′ 56″
7728	Peg	Gxy	1.0′×0.8′	13.2	23 h 40 m 00.8 s	+27° 08′ 00″
7729	Peg	Gxy	1.9′×0.6′	13.7	23 h 40 m 33.7 s	+29° 11′ 17″
7735	Peg	Gxy	1.2′×1.0′	13.7	23 h 42 m 17.2 s	+26° 13′ 55″
7737	Peg	Gxy	1.1′×0.5′	14	23 h 42 m 46.3 s	+27° 03′ 10″
7740	Peg	Gxy	0.8′×0.5′	13.9	23 h 43 m 32.3 s	+27° 18′ 43″
7741	Peg	Gxy	4.4′×3.0′	11.2	23 h 43 m 54.3 s	+26° 04′ 31″
7742	Peg	Gxy	1.7′×1.7′	11.9	23 h 44 m 15.8 s	+10° 46′ 00″
7743	Peg	Gxy	3.0′×2.6′	11.5	23 h 44 m 21.1 s	+09° 56′ 03″
7745	Peg	Gxy	0.7′×0.7′	14.2	23 h 44 m 45.7 s	+25° 54′ 33″
7747	Peg	Gxy	1.5′×0.5′	13.8	23 h 45 m 32.3 s	+27° 21′ 38″
7752	Peg	Gxy	0.8′×0.5′	14	23 h 46 m 58.5 s	+29° 27′ 32″
7753	Peg	Gxy	3.3′×2.1′	12.2	23 h 47 m 04.9 s	+29° 29′ 00″
7760	Peg	Gxy	1.0′×1.0′	13.6	23 h 49 m 11.8 s	+30° 58′ 58″
7765	Peg	Gxy	0.7′×0.7′	14.7	23 h 50 m 52.3 s	+27° 09′ 56″
7766	Peg	Gxy	0.5′×0.2′	15.5	23 h 50 m 55.9 s	+27° 07′ 34″
7767	Peg	Gxy	1.0′×0.2′	13.7	23 h 50 m 56.4 s	+27° 05′ 12″
7768	Peg	Gxy	1.6′×1.3′	12.8	23 h 50 m 58.6 s	+27° 08′ 48″
7769	Peg	Gxy	1.7′×1.6′	12.2	23 h 51 m 03.9 s	+20° 08′ 59″
7770	Peg	Gxy	0.8′×0.7′	13.6	23 h 51 m 22.6 s	+20° 05′ 46″
7771	Peg	Gxy	2.5′×1.0′	12.2	23 h 51 m 24.6 s	+20° 06′ 41″
7772	Peg	OC	5.0′	…	23 h 51 m 45.9 s	+16° 14′ 53″

(continued)

Table 10.3 (continued)

NGC #	Const	Type	Size	V Mag	RA (2000)	Dec (2000)
7773	Peg	Gxy	1.2'×1.2'	13.6	23 h 52 m 09.9 s	+31° 16' 35"
7774	Peg	Gxy	1.0'×0.6'	13.2	23 h 52 m 11.0 s	+11° 28' 12"
7775	Peg	Gxy	1.0'×0.8'	13.5	23 h 52 m 24.3 s	+28° 46' 21"
7777	Peg	Gxy	1.2'×0.8'	13.4	23 h 53 m 12.5 s	+28° 17' 00"
7784	Peg	Gxy	0.5'×0.5'	14.5	23 h 55 m 13.6 s	+21° 45' 43"
7786	Peg	Gxy	0.6'×0.4'	13.3	23 h 55 m 21.7 s	+21° 35' 15"
7791	Peg	**	...	...	23 h 57 m 57.3 s	+10° 45' 57"
7792	Peg	Gxy	1.0'×0.8'	14.1	23 h 58 m 03.7 s	+16° 30' 07"
7794	Peg	Gxy	1.5'×1.3'	12.8	23 h 58 m 34.1 s	+10° 43' 42"
7798	Peg	Gxy	1.4'×1.3'	12.3	23 h 59 m 25.5 s	+20° 44' 57"
7799	Peg	*	...	...	23 h 59 m 31.5 s	+31° 17' 44"
7800	Peg	Gxy	2.3'×1.6'	12.7	23 h 59 m 36.7 s	+14° 48' 24"
7803	Peg	Gxy	1.0'×0.6'	13.1	00 h 01 m 20.0 s	+13° 06' 39"
7805	Peg	Gxy	1.2'×0.9'	13.2	00 h 01 m 26.7 s	+31° 25' 58"
7806	Peg	Gxy	1.1'×0.8'	13.7	00 h 01 m 30.0 s	+31° 26' 28"
7810	Peg	Gxy	1.2'×0.9'	13.1	00 h 02 m 19.1 s	+12° 58' 15"
7814	Peg	Gxy	5.5'×2.3'	11	00 h 03 m 14.8 s	+16° 08' 43"
7815	Peg	*	...	...	00 h 03 m 25.3 s	+20° 42' 17"
7817	Peg	Gxy	3.5'×0.9'	12	00 h 03 m 58.4 s	+20° 44' 59"
7819	Peg	Gxy	1.5'×1.2'	13.4	00 h 04 m 24.5 s	+31° 28' 19"
7833	Peg	Ast	...	...	00 h 06 m 31.5 s	+27° 38' 30"
7839	Peg	**	...	...	00 h 07 m 00.6 s	+27° 38' 08"
650	Per	PN	2.7'×1.8'	10.1	01 h 42 m 18.1 s	+51° 34' 16"
651	Per	PN	163"×107"	10.1	01 h 42 m 21.9 s	+51° 34' 49"
744	Per	OC	11'	7.9	01 h 58 m 29.9 s	+55° 28' 29"
869	Per	OC	29'	5.3	02 h 19 m 03.8 s	+57° 08' 06"
884	Per	OC	29'	6.1	02 h 22 m 32.1 s	+57° 08' 39"
957	Per	OC	11'	7.6	02 h 33 m 19.0 s	+57° 34' 11"
1001	Per	Gxy	0.7'×0.3'	14.8	02 h 39 m 12.6 s	+41° 40' 18"
1003	Per	Gxy	5.5'×1.9'	11.4	02 h 39 m 16.4 s	+40° 52' 22"
1005	Per	Gxy	0.3'×0.3'	13.8	02 h 39 m 27.5 s	+41° 29' 36"
1023	Per	Gxy	8.7'×3.0'	9.5	02 h 40 m 23.9 s	+39° 03' 48"
1039	Per	OC	35'	5.2	02 h 42 m 07.4 s	+42° 44' 46"
1040	Per	Gxy	1.7'×0.8'	12.9	02 h 43 m 12.4 s	+41° 30' 03"
1050	Per	Gxy	1.4'×1.1'	12.8	02 h 42 m 35.7 s	+34° 45' 49"
1053	Per	Gxy	1.7'×0.8'	13	02 h 43 m 12.4 s	+41° 30' 03"
1058	Per	Gxy	3.0'×2.8'	11.4	02 h 43 m 29.9 s	+37° 20' 29"
1077	Per	Gxy	1.3'×0.9'	13.7	02 h 46 m 00.5 s	+40° 05' 23"
1086	Per	Gxy	1.5'×1.0'	13	02 h 47 m 56.3 s	+41° 14' 47"
1106	Per	Gxy	1.8'×1.8'	12.4	02 h 50 m 40.5 s	+41° 40' 19"
1122	Per	Gxy	1.7'×1.3'	12.3	02 h 52 m 51.0 s	+42° 12' 19"
1123	Per	Gxy	2.2'×1.7'	12.2	02 h 52 m 51.0 s	+42° 12' 19"
1129	Per	Gxy	4.0'×3.1'	12.6	02 h 54 m 27.4 s	+41° 34' 47"
1130	Per	Gxy	0.4'×0.4'	15.1	02 h 54 m 24.4 s	+41° 36' 20"
1131	Per	Gxy	0.4'×0.4'	14.6	02 h 54 m 33.9 s	+41° 33' 32"

(continued)

Table 10.3 (continued)

NGC #	Const	Type	Size	V Mag	RA (2000)	Dec (2000)
1138	Per	Gxy	1.1'×0.9'	12.9	02 h 56 m 36.5 s	+43° 02' 51"
1146	Per	Ast	0.5'×0.5'	…	02 h 57 m 36.9 s	+46° 26' 16"
1159	Per	Gxy	0.5'×0.5'	13.4	03 h 00 m 46.5 s	+43° 09' 45"
1160	Per	Gxy	1.9'×0.9'	12.1	03 h 01 m 13.3 s	+44° 57' 15"
1161	Per	Gxy	2.8'×2.0'	11.1	03 h 01 m 14.1 s	+44° 53' 49"
1164	Per	Gxy	1.3'×1.0'	13.3	03 h 01 m 59.9 s	+42° 35' 06"
1167	Per	Gxy	2.8'×2.3'	12.5	03 h 01 m 42.2 s	+35° 12' 20"
1169	Per	Gxy	4.2'×2.8'	11.6	03 h 03 m 34.7 s	+46° 23' 08"
1171	Per	Gxy	2.6'×1.1'	12.5	03 h 03 m 58.9 s	+43° 23' 53"
1173	Per	NF	…	…	03 h 03 m 57.7 s	+42° 23' 01"
1174	Per	Gxy	3.2'×1.2'	11.7	03 h 05 m 30.9 s	+42° 50' 07"
1175	Per	Gxy	1.9'×0.6'	12.3	03 h 04 m 32.4 s	+42° 20' 22"
1176	Per	*	…	…	03 h 04 m 35.0 s	+42° 23' 37"
1177	Per	Gxy	0.4'×0.4'	14.6	03 h 04 m 37.1 s	+42° 21' 46"
1178	Per	*	…	…	03 h 04 m 38.7 s	+42° 18' 48"
1183	Per	*	…	…	03 h 04 m 46.1 s	+42° 22' 08"
1186	Per	Gxy	3.2'×1.2'	11.6	03 h 05 m 30.9 s	+42° 50' 07"
1193	Per	OC	1.5'	12.6	03 h 05 m 55.7 s	+44° 22' 59"
1197	Per	NF	…	…	03 h 06 m 09.7 s	+44° 03' 49"
1198	Per	Gxy	1.9'×1.1'	12.6	03 h 06 m 13.3 s	+41° 50' 55"
1207	Per	Gxy	2.3'×1.7'	12.8	03 h 08 m 15.3 s	+38° 22' 57"
1212	Per	Gxy	0.3'×0.3'	14.6	03 h 09 m 42.3 s	+40° 53' 34"
1213	Per	Gxy	1.8'×1.4'	14.5	03 h 09 m 17.3 s	+38° 39' 00"
1220	Per	OC	2.0'	11.8	03 h 11 m 40.7 s	+53° 20' 54"
1224	Per	Gxy	1.4'×1.2'	13.8	03 h 11 m 13.6 s	+41° 21' 49"
1226	Per	Gxy	2.1'×1.9'	13	03 h 11 m 05.3 s	+35° 23' 12"
1227	Per	Gxy	1.0'×0.9'	14.3	03 h 11 m 07.7 s	+35° 19' 28"
1233	Per	Gxy	1.8'×0.6'	13.2	03 h 12 m 33.1 s	+39° 19' 09"
1235	Per	Gxy	2.3'×0.9'	13.1	03 h 12 m 33.1 s	+39° 19' 09"
1245	Per	OC	10'	8.4	03 h 14 m 41.5 s	+47° 14' 20"
1250	Per	Gxy	2.1'×0.9'	12.8	03 h 15 m 21.1 s	+41° 21' 20"
1257	Per	**	…	…	03 h 16 m 59.5 s	+41° 31' 45"
1259	Per	Gxy	0.6'×0.6'	14.7	03 h 17 m 17.2 s	+41° 23' 06"
1260	Per	Gxy	1.1'×0.5'	13.2	03 h 17 m 27.3 s	+41° 24' 19"
1264	Per	Gxy	1.0'×0.5'	14.3	03 h 17 m 59.5 s	+41° 31' 13"
1265	Per	Gxy	1.8'×1.6'	12.2	03 h 18 m 15.7 s	+41° 51' 29"
1267	Per	Gxy	1.1'×0.9'	13.1	03 h 18 m 44.9 s	+41° 28' 04"
1268	Per	Gxy	1.0'×0.6'	13.6	03 h 18 m 45.1 s	+41° 29' 20"
1270	Per	Gxy	1.5'×1.2'	12.8	03 h 18 m 58.0 s	+41° 28' 13"
1271	Per	Gxy	…'×…'	14.5	03 h 19 m 11.3 s	+41° 21' 11"
1272	Per	Gxy	2.0'×1.9'	12.7	03 h 19 m 21.3 s	+41° 29' 27"
1273	Per	Gxy	1.1'×1.1'	13.2	03 h 19 m 26.7 s	+41° 32' 26"
1274	Per	Gxy	0.5'×0.4'	13.9	03 h 19 m 40.5 s	+41° 32' 57"
1275	Per	Gxy	2.2'×1.7'	12	03 h 19 m 48.2 s	+41° 30' 42"
1276	Per	**	…	…	03 h 19 m 51.2 s	+41° 38' 31"

(continued)

Table 10.3 (continued)

NGC #	Const	Type	Size	V Mag	RA (2000)	Dec (2000)
1277	Per	Gxy	1.0'×0.4'	13.7	03 h 19 m 51.4 s	+41° 34' 25"
1278	Per	Gxy	1.5'×1.3'	12.7	03 h 19 m 54.1 s	+41° 33' 48"
1279	Per	Gxy	0.6'×0.4'	15	03 h 19 m 58.8 s	+41° 28' 47"
1281	Per	Gxy	1.0'×0.7'	13.5	03 h 20 m 06.1 s	+41° 37' 48"
1282	Per	Gxy	1.4'×1.1'	12.8	03 h 20 m 12.1 s	+41° 22' 02"
1283	Per	Gxy	0.7'×0.6'	14.1	03 h 20 m 15.5 s	+41° 23' 55"
1293	Per	Gxy	1.0'×1.0'	13.4	03 h 21 m 36.4 s	+41° 23' 35"
1294	Per	Gxy	1.3'×1.1'	13.4	03 h 21 m 39.9 s	+41° 21' 38"
1330	Per	Ast	...	...	03 h 29 m 04.4 s	+41° 40' 31"
1333	Per	Neb	9'×7'	...	03 h 29 m 15.2 s	+31° 23' 05"
1334	Per	Gxy	1.5'×0.7'	13.8	03 h 30 m 01.8 s	+41° 49' 58"
1335	Per	Gxy	1.1'×0.6'	13.9	03 h 30 m 19.4 s	+41° 34' 24"
1342	Per	OC	14'	6.7	03 h 31 m 40.1 s	+37° 22' 28"
1348	Per	OC	5.0'	...	03 h 34 m 08.5 s	+51° 25' 14"
1444	Per	OC	4.0'	6.6	03 h 49 m 22.9 s	+52° 39' 44"
1465	Per	Gxy	1.9'×0.5'	13.9	03 h 53 m 31.9 s	+32° 29' 31"
1491	Per	Neb	9'×6'	...	04 h 03 m 13.5 s	+51° 18' 58"
1496	Per	OC	6.0'	9.6	04 h 04 m 31.9 s	+52° 39' 41"
1499	Per	Neb	160'×40'	...	04 h 03 m 14.4 s	+36° 22' 03"
1513	Per	OC	9.0'	8.4	04 h 09 m 54.7 s	+49° 31' 02"
1528	Per	OC	23.0'	6.4	04 h 15 m 18.9 s	+51° 12' 41"
1545	Per	OC	18.0'	6.2	04 h 20 m 56.3 s	+50° 15' 19"
1548	Per	OC	...	...	04 h 21 m 19.9 s	+36° 55' 00"
1579	Per	Neb	12'×8'	...	04 h 30 m 14.2 s	+35° 16' 47"
1582	Per	OC	37.0'	7	04 h 32 m 00.0 s	+43° 50' 53"
1605	Per	OC	5.0'	10.7	04 h 34 m 54.3 s	+45° 16' 11"
1624	Per	OC+Neb	5'×5'	11.8	04 h 40 m 36.5 s	+50° 27' 42"
25	Phe	Gxy	1.4'×0.8'	13.1	00 h 09 m 59.2 s	−57° 01' 16"
28	Phe	Gxy	1.5'×1.3'	13.8	00 h 10 m 25.3 s	−56° 59' 21"
31	Phe	Gxy	1.1'×0.6'	14.1	00 h 10 m 38.5 s	−56° 59' 11"
37	Phe	Gxy	1.1'×0.7'	13.8	00 h 11 m 22.9 s	−56° 57' 26"
87	Phe	Gxy	0.9'×0.7'	14.3	00 h 21 m 14.2 s	−48° 37' 42"
88	Phe	Gxy	0.8'×0.5'	14.3	00 h 21 m 21.8 s	−48° 38' 25"
89	Phe	Gxy	1.2'×0.6'	13.5	00 h 21 m 24.3 s	−48° 39' 57"
92	Phe	Gxy	1.9'×0.9'	13	00 h 21 m 32.0 s	−48° 37' 34"
98	Phe	Gxy	1.7'×1.3'	13.1	00 h 22 m 49.3 s	−45° 16' 10"
119	Phe	Gxy	1.0'×1.0'	13.2	00 h 26 m 57.1 s	−56° 58' 42"
159	Phe	Gxy	1.4'×0.4'	14	00 h 34 m 36.3 s	−55° 47' 22"
212	Phe	Gxy	1.3'×1.0'	13.5	00 h 40 m 12.7 s	−56° 09' 08"
215	Phe	Gxy	1.1'×0.9'	13.2	00 h 40 m 48.5 s	−56° 12' 51"
238	Phe	Gxy	1.9'×1.6'	12.7	00 h 43 m 25.9 s	−50° 11' 05"
312	Phe	Gxy	1.4'×1.1'	12.5	00 h 56 m 16.7 s	−52° 46' 59"
319	Phe	Gxy	1.0'×0.8'	13.5	00 h 56 m 57.2 s	−43° 50' 17"
322	Phe	Gxy	1.1'×0.6'	13.4	00 h 57 m 10.2 s	−43° 43' 36"
323	Phe	Gxy	1.0'×1.0'	12.6	00 h 56 m 41.3 s	−52° 58' 35"

(continued)

Table 10.3 (continued)

NGC #	Const	Type	Size	V Mag	RA (2000)	Dec (2000)
324	Phe	Gxy	1.4'×0.5'	13.1	00 h 57 m 15.6 s	−40° 57' 36"
328	Phe	Gxy	2.7'×0.5'	13.5	00 h 56 m 57.3 s	−52° 55' 23"
348	Phe	Gxy	0.8'×0.7'	13.9	01 h 00 m 51.5 s	−53° 14' 40"
368	Phe	Gxy	0.7'×0.7'	13.8	01 h 04 m 21.5 s	−43° 16' 32"
405	Phe	**	...	...	01 h 08 m 34.1 s	−46° 40' 07"
454	Phe	Gxy	1.8'×1.8'	12.7	01 h 14 m 23.1 s	−55° 23' 51"
482	Phe	Gxy	2.2'×0.5'	13.9	01 h 20 m 20.0 s	−40° 57' 35"
576	Phe	Gxy	1.0'×0.8'	13.6	01 h 28 m 57.0 s	−51° 35' 55"
625	Phe	Gxy	5.8'×1.9'	11.4	01 h 35 m 05.1 s	−41° 26' 11"
641	Phe	Gxy	1.4'×1.3'	12.4	01 h 38 m 39.3 s	−42° 31' 35"
644	Phe	Gxy	1.3'×0.6'	14.2	01 h 38 m 53.2 s	−42° 35' 06"
692	Phe	Gxy	2.1'×1.8'	12.5	01 h 48 m 42.2 s	−48° 38' 54"
822	Phe	Gxy	1.1'×0.6'	13.3	02 h 06 m 39.1 s	−41° 09' 27"
862	Phe	Gxy	0.9'×0.8'	12.9	02 h 13 m 02.8 s	−42° 02' 00"
889	Phe	Gxy	1.0'×0.9'	13.4	02 h 19 m 06.8 s	−41° 44' 56"
893	Phe	Gxy	1.3'×1.0'	13.1	02 h 19 m 59.0 s	−41° 24' 16"
7689	Phe	Gxy	2.9'×1.9'	11.7	23 h 33 m 16.3 s	−54° 05' 43"
7690	Phe	Gxy	2.2'×0.9'	12.3	23 h 33 m 02.3 s	−51° 41' 55"
7702	Phe	Gxy	2.2'×1.2'	12.2	23 h 35 m 28.7 s	−56° 00' 45"
7744	Phe	Gxy	2.2'×1.7'	11.6	23 h 44 m 59.1 s	−42° 54' 40"
7764	Phe	Gxy	1.9'×1.3'	12.5	23 h 50 m 53.5 s	−40° 43' 46"
7796	Phe	Gxy	2.2'×1.9'	11.5	23 h 58 m 59.7 s	−55° 27' 31"
1680	Pic	Gxy	1.2'×0.5'	13.8	04 h 48 m 33.7 s	−47° 49' 02"
1705	Pic	Gxy	1.9'×1.4'	11.9	04 h 54 m 13.9 s	−53° 21' 40"
1803	Pic	Gxy	1.3'×0.8'	12.9	05 h 05 m 26.6 s	−49° 34' 05"
1930	Pic	Gxy	1.9'×1.2'	12.5	05 h 25 m 56.6 s	−46° 43' 44"
1995	Pic	**	...	...	05 h 33 m 03.3 s	−48° 40' 31"
1998	Pic	Gxy	0.9'×0.4'	14.4	05 h 33 m 15.6 s	−48° 41' 46"
2007	Pic	Gxy	1.7'×0.6'	14.3	05 h 34 m 58.9 s	−50° 55' 20"
2008	Pic	Gxy	1.5'×0.7'	14.1	05 h 35 m 03.8 s	−50° 58' 00"
2087	Pic	Gxy	0.8'×0.6'	14	05 h 44 m 16.1 s	−55° 32' 00"
2101	Pic	Gxy	1.9'×1.3'	13.3	05 h 46 m 22.4 s	−52° 05' 26"
2104	Pic	Gxy	2.0'×0.9'	13.1	05 h 47 m 04.2 s	−51° 33' 13"
2115	Pic	Gxy	1.2'×1.1'	13.5	05 h 51 m 19.9 s	−50° 34' 59"
2132	Pic	Ast	4.0'×4.0'	...	05 h 55 m 18.0 s	−59° 55' 00"
2148	Pic	Gxy	1.1'×0.8'	14	05 h 58 m 45.7 s	−59° 07' 37"
2152	Pic	Gxy	1.1'×0.8'	14	06 h 00 m 55.0 s	−50° 44' 28"
2178	Pic	Gxy	1.1'×1.0'	12.7	06 h 02 m 47.5 s	−63° 45' 52"
2205	Pic	Gxy	1.3'×0.9'	12.8	06 h 10 m 32.8 s	−62° 32' 20"
2221	Pic	Gxy	1.9'×0.4'	13.1	06 h 20 m 15.7 s	−57° 34' 45"
2222	Pic	Gxy	1.2'×0.3'	13.5	06 h 20 m 16.9 s	−57° 32' 02"
2297	Pic	Gxy	1.4'×1.3'	13.1	06 h 44 m 24.5 s	−63° 43' 04"
7109	PsA	Gxy	0.8'×0.8'	13.5	21 h 41 m 58.5 s	−34° 26' 48"
7110	PsA	Gxy	1.3'×0.6'	13.6	21 h 42 m 11.9 s	−34° 09' 47"
7115	PsA	Gxy	1.7'×0.4'	12.7	21 h 43 m 39.1 s	−25° 21' 01"

(continued)

Table 10.3 (continued)

NGC #	Const	Type	Size	V Mag	RA (2000)	Dec (2000)
7130	PsA	Gxy	1.5'×1.4'	12.3	21 h 48 m 19.3 s	−34° 57' 11"
7135	PsA	Gxy	3.0'×2.1'	11.5	21 h 49 m 45.9 s	−34° 52' 36"
7152	PsA	Gxy	1.2'×0.6'	13.8	21 h 53 m 58.9 s	−29° 17' 23"
7153	PsA	Gxy	1.9'×0.3'	13.6	21 h 54 m 35.3 s	−29° 03' 51"
7154	PsA	Gxy	2.1'×1.6'	12.6	21 h 55 m 21.0 s	−34° 48' 52"
7157	PsA	Gxy	1.1'×0.5'	14.3	21 h 56 m 56.7 s	−25° 21' 05"
7163	PsA	Gxy	1.9'×1.1'	12.9	21 h 59 m 20.2 s	−31° 53' 03"
7172	PsA	Gxy	2.5'×1.4'	12	22 h 02 m 02.1 s	−31° 52' 13"
7173	PsA	Gxy	1.2'×0.9'	11.2	22 h 02 m 03.4 s	−31° 58' 29"
7174	PsA	Gxy	2.3'×1.2'	11.6	22 h 02 m 06.2 s	−31° 59' 40"
7176	PsA	Gxy	1.0'×0.8'	11.2	22 h 02 m 08.7 s	−31° 59' 27"
7178	PsA	Gxy	1.1'×0.5'	14.2	22 h 02 m 25.0 s	−35° 47' 29"
7187	PsA	Gxy	1.4'×1.3'	12.5	22 h 02 m 44.3 s	−32° 48' 13"
7201	PsA	Gxy	1.6'×0.5'	13	22 h 06 m 32.1 s	−31° 15' 47"
7202	PsA	*	...	...	22 h 06 m 43.2 s	−31° 13' 05"
7203	PsA	Gxy	1.6'×0.9'	12.8	22 h 06 m 43.8 s	−31° 09' 46"
7204	PsA	Gxy	1.2'×0.5'	13.7	22 h 06 m 54.1 s	−31° 03' 02"
7208	PsA	Gxy	0.9'×0.5'	12.9	22 h 08 m 24.7 s	−29° 03' 06"
7214	PsA	Gxy	2.2'×1.4'	12	22 h 09 m 07.6 s	−27° 48' 36"
7221	PsA	Gxy	2.0'×1.6'	12.5	22 h 11 m 15.2 s	−30° 33' 48"
7225	PsA	Gxy	2.0'×1.0'	12.4	22 h 13 m 07.7 s	−26° 08' 56"
7229	PsA	Gxy	1.8'×1.5'	12.8	22 h 14 m 03.2 s	−29° 23' 02"
7258	PsA	Gxy	1.4'×0.6'	13.3	22 h 22 m 58.0 s	−28° 20' 45"
7259	PsA	Gxy	1.1'×0.9'	13.3	22 h 23 m 05.6 s	−28° 57' 16"
7262	PsA	Gxy	0.7'×0.7'	13.9	22 h 23 m 28.3 s	−32° 21' 55"
7267	PsA	Gxy	1.6'×1.3'	12.3	22 h 24 m 21.6 s	−33° 41' 38"
7268	PsA	Gxy	1.2'×1.0'	13.5	22 h 25 m 40.6 s	−31° 12' 04"
7277	PsA	Gxy	1.5'×0.6'	13.5	22 h 26 m 10.9 s	−31° 08' 47"
7279	PsA	Gxy	1.2'×0.8'	13.7	22 h 27 m 12.5 s	−35° 08' 25"
7289	PsA	Gxy	1.4'×1.1'	13.3	22 h 29 m 20.1 s	−35° 28' 21"
7294	PsA	Gxy	1.9'×1.2'	12.6	22 h 32 m 08.1 s	−25° 23' 53"
7306	PsA	Gxy	1.7'×0.7'	12.9	22 h 33 m 16.3 s	−27° 14' 47"
7313	PsA	Gxy	0.7'×0.5'	14.4	22 h 35 m 32.4 s	−26° 06' 08"
7314	PsA	Gxy	4.6'×2.1'	11	22 h 35 m 45.9 s	−26° 03' 03"
7361	PsA	Gxy	3.8'×1.0'	12.4	22 h 42 m 17.9 s	−30° 03' 31"
3	Psc	Gxy	1.1'×0.6'	13.5	00 h 07 m 16.7 s	+08° 18' 05"
4	Psc	Gxy	0.6'×0.3'	15.9	00 h 07 m 24.4 s	+08° 22' 22"
12	Psc	Gxy	1.7'×1.5'	13.5	00 h 08 m 44.8 s	+04° 36' 47"
33	Psc	**	...	...	00 h 10 m 56.6 s	+03° 40' 33"
36	Psc	Gxy	2.2'×1.3'	13.4	00 h 11 m 22.3 s	+06° 23' 21"
38	Psc	Gxy	1.3'×1.3'	13.3	00 h 11 m 47.0 s	−05° 35' 13"
46	Psc	*	...	...	00 h 14 m 09.9 s	+05° 59' 16"
56	Psc	NF	...	...	00 h 15 m 20.6 s	+12° 26' 40"
57	Psc	Gxy	2.2'×1.9'	12.2	00 h 15 m 30.7 s	+17° 19' 38"
60	Psc	Gxy	1.3'×1.2'	14.3	00 h 15 m 58.3 s	−00° 18' 12"

(continued)

Table 10.3 (continued)

NGC #	Const	Type	Size	V Mag	RA (2000)	Dec (2000)
61	Psc	Gxy	1.0'×0.6'	13.4	00 h 16 m 24.5 s	−06° 19' 21"
63	Psc	Gxy	1.7'×1.1'	11.8	00 h 17 m 45.5 s	+11° 26' 58"
75	Psc	Gxy	1.9'×1.9'	13.3	00 h 19 m 25.7 s	+06° 26' 53"
78	Psc	Gxy	1.1'×0.7'	13.3	00 h 20 m 27.5 s	+00° 50' 01"
95	Psc	Gxy	1.9'×1.1'	12.6	00 h 22 m 14.1 s	+10° 29' 28"
99	Psc	Gxy	1.4'×1.3'	13.2	00 h 23 m 59.6 s	+15° 46' 10"
100	Psc	Gxy	5.5'×0.7'	13.5	00 h 24 m 02.5 s	+16° 29' 09"
105	Psc	Gxy	1.1'×0.7'	13.1	00 h 25 m 16.9 s	+12° 53' 07"
106	Psc	Gxy	0.9'×0.7'	13.7	00 h 24 m 43.7 s	−05° 08' 57"
125	Psc	Gxy	1.7'×1.5'	12.9	00 h 28 m 50.3 s	+02° 50' 19"
126	Psc	Gxy	0.9'×0.4'	14.3	00 h 29 m 08.1 s	+02° 48' 40"
127	Psc	Gxy	0.8'×0.6'	14.8	00 h 29 m 12.3 s	+02° 52' 21"
128	Psc	Gxy	3.0'×0.9'	11.8	00 h 29 m 15.1 s	+02° 51' 51"
130	Psc	Gxy	0.7'×0.4'	14.4	00 h 29 m 18.5 s	+02° 52' 14"
137	Psc	Gxy	1.3'×1.3'	12.9	00 h 30 m 57.4 s	+10° 12' 26"
138	Psc	Gxy	1.4'×0.7'	13.9	00 h 30 m 59.2 s	+05° 09' 36"
139	Psc	Gxy	0.8'×0.5'	14.6	00 h 31 m 06.3 s	+05° 04' 44"
141	Psc	Gxy	0.9'×0.6'	14.4	00 h 31 m 17.5 s	+05° 10' 47"
164	Psc	Gxy	0.4'×0.4'	15.8	00 h 36 m 33.0 s	+02° 45' 03"
180	Psc	Gxy	2.4'×1.9'	13.1	00 h 37 m 57.6 s	+08° 38' 05"
182	Psc	Gxy	2.0'×1.7'	12.7	00 h 38 m 12.4 s	+02° 43' 42"
186	Psc	Gxy	1.4'×0.8'	13.6	00 h 38 m 25.3 s	+03° 09' 59"
190	Psc	Gxy	1.1'×0.9'	14	00 h 38 m 54.7 s	+07° 03' 45"
193	Psc	Gxy	1.4'×1.2'	13	00 h 39 m 18.5 s	+03° 19' 52"
194	Psc	Gxy	1.5'×1.4'	12.6	00 h 39 m 18.3 s	+03° 02' 14"
198	Psc	Gxy	1.2'×1.2'	13.4	00 h 39 m 23.0 s	+02° 47' 52"
199	Psc	Gxy	1.2'×0.7'	13.7	00 h 39 m 33.1 s	+03° 08' 19"
200	Psc	Gxy	1.9'×1.0'	13.1	00 h 39 m 34.9 s	+02° 53' 17"
202	Psc	Gxy	0.9'×0.3'	14.4	00 h 39 m 39.7 s	+03° 32' 08"
203	Psc	Gxy	0.9'×0.3'	14.1	00 h 39 m 39.5 s	+03° 26' 34"
204	Psc	Gxy	1.2'×1.1'	13.3	00 h 39 m 44.1 s	+03° 17' 58"
208	Psc	Gxy	0.7'×0.7'	14.5	00 h 40 m 17.5 s	+02° 45' 23"
211	Psc	Gxy	0.8'×0.3'	14.1	00 h 39 m 39.5 s	+03° 26' 34"
213	Psc	Gxy	1.8'×1.7'	13.5	00 h 41 m 10.1 s	+16° 28' 08"
234	Psc	Gxy	1.6'×1.6'	12.7	00 h 43 m 32.4 s	+14° 20' 35"
236	Psc	Gxy	1.1'×1.0'	13.7	00 h 43 m 27.5 s	+02° 57' 31"
240	Psc	Gxy	1.6'×1.6'	13.7	00 h 45 m 01.9 s	+06° 06' 47"
250	Psc	Gxy	1.4'×0.8'	13.8	00 h 47 m 16.1 s	+07° 54' 36"
251	Psc	Gxy	2.4'×1.9'	13.4	00 h 47 m 53.9 s	+19° 35' 48"
257	Psc	Gxy	1.9'×1.3'	12.9	00 h 48 m 01.5 s	+08° 17' 48"
266	Psc	Gxy	3.0'×2.9'	11.6	00 h 49 m 48.1 s	+32° 16' 43"
282	Psc	Gxy	1.0'×0.9'	13.7	00 h 52 m 42.2 s	+30° 38' 21"
287	Psc	Gxy	0.8'×0.4'	13.9	00 h 53 m 28.3 s	+32° 28' 56"
295	Psc	Gxy	2.2'×1.0'	12.8	00 h 55 m 05.2 s	+31° 31' 51"
296	Psc	Gxy	1.2'×0.3'	14.8	00 h 55 m 07.6 s	+31° 32' 32"

(continued)

Table 10.3 (continued)

NGC #	Const	Type	Size	V Mag	RA (2000)	Dec (2000)
305	Psc	Ast	...	...	00 h 56 m 20.9 s	+12° 03' 54"
311	Psc	Gxy	1.5'×0.8'	13	00 h 57 m 32.7 s	+30° 16' 47"
313	Psc	***	3.6'	...	00 h 57 m 45.6 s	+30° 21' 58"
315	Psc	Gxy	3.2'×2.0'	11.3	00 h 57 m 48.8 s	+30° 21' 09"
316	Psc	*	3.6'	...	00 h 57 m 52.4 s	+30° 21' 16"
318	Psc	Gxy	0.5'×0.3'	14.4	00 h 58 m 05.1 s	+30° 25' 32"
326	Psc	Gxy	1.2'×1.2'	13.3	00 h 58 m 22.7 s	+26° 51' 56"
332	Psc	Gxy	1.3'×1.1'	13.5	00 h 58 m 49.1 s	+07° 06' 41"
338	Psc	Gxy	1.9'×0.6'	13.2	01 h 00 m 36.1 s	+30° 40' 07"
354	Psc	Gxy	0.8'×0.4'	13.5	01 h 03 m 16.8 s	+22° 20' 32"
370	Psc	***	...	...	01 h 06 m 44.5 s	+32° 25' 44"
372	Psc	***	...	...	01 h 06 m 44.5 s	+32° 25' 44"
373	Psc	Gxy	0.3'×0.3	14.9	01 h 06 m 58.8 s	+32° 18' 32"
374	Psc	Gxy	1.1'×0.5'	13.6	01 h 07 m 05.5 s	+32° 47' 47"
375	Psc	Gxy	1.4'×1.4'	14.5	01 h 07 m 05.9 s	+32° 20' 53"
379	Psc	Gxy	1.4'×0.8'	13	01 h 07 m 15.6 s	+32° 31' 13"
380	Psc	Gxy	1.4'×1.2'	12.7	01 h 07 m 17.5 s	+32° 28' 59"
382	Psc	Gxy	0.7'×0.7'	13.3	01 h 07 m 23.9 s	+32° 24' 15"
383	Psc	Gxy	1.6'×1.4'	12.5	01 h 07 m 24.9 s	+32° 24' 45"
384	Psc	Gxy	1.1'×0.9'	13.2	01 h 07 m 24.9 s	+32° 17' 34"
385	Psc	Gxy	1.1'×1.0'	13.2	01 h 07 m 27.1 s	+32° 19' 12"
386	Psc	Gxy	0.9'×0.8'	14.1	01 h 07 m 31.3 s	+32° 21' 43"
387	Psc	Gxy	0.3'×0.3'	15.5	01 h 07 m 33.0 s	+32° 23' 28"
388	Psc	Gxy	0.9'×0.8'	14.2	01 h 07 m 47.1 s	+32° 18' 35"
390	Psc	*	...	...	01 h 07 m 54.4 s	+32° 25' 58"
392	Psc	Gxy	1.2'×0.9'	12.9	01 h 08 m 23.3 s	+33° 08' 00"
394	Psc	Gxy	0.4'×0.2'	14	01 h 08 m 26.1 s	+33° 08' 53"
396	Psc	Gxy	0.3'×0.1'	15.7	01 h 08 m 08.4 s	+04° 31' 51"
397	Psc	Gxy	0.3'×0.2'	14.8	01 h 08 m 31.2 s	+33° 06' 33"
398	Psc	Gxy	0.3'×0.2'	14.5	01 h 08 m 53.7 s	+32° 30' 52"
399	Psc	Gxy	0.9'×0.7'	13.8	01 h 08 m 59.3 s	+32° 38' 03"
400	Psc	*	...	...	01 h 09 m 02.7 s	+32° 43' 58"
401	Psc	*	...	...	01 h 09 m 07.7 s	+32° 45' 37"
402	Psc	*	...	...	01 h 09 m 13.3 s	+32° 48' 23"
403	Psc	Gxy	1.9'×0.6'	12.7	01 h 09 m 14.1 s	+32° 45' 08"
407	Psc	Gxy	1.7'×0.4'	13.6	01 h 10 m 36.5 s	+33° 07' 35"
408	Psc	*	...	...	01 h 10 m 51.1 s	+33° 09' 05"
410	Psc	Gxy	2.4'×1.3'	11.6	01 h 10 m 58.9 s	+33° 09' 06"
414	Psc	Gxy	0.8'×0.4'	13.5	01 h 11 m 17.6 s	+33° 06' 48"
420	Psc	Gxy	2.0'×2.0'	12.2	01 h 12 m 09.6 s	+32° 07' 24"
421	Psc	NF	...	...	01 h 12 m 14.4 s	+32° 07' 25"
437	Psc	Gxy	1.3'×1.0'	13	01 h 14 m 21.9 s	+05° 55' 33"
443	Psc	Gxy	0.8'×0.7'	13.7	01 h 15 m 06.8 s	+33° 22' 41"
444	Psc	Gxy	1.9'×0.4'	14	01 h 15 m 49.5 s	+31° 04' 50"
446	Psc	Gxy	2.0'×1.6'	12.8	01 h 16 m 03.5 s	+04° 17' 38"

(continued)

Table 10.3 (continued)

NGC #	Const	Type	Size	V Mag	RA (2000)	Dec (2000)
447	Psc	Gxy	2.2'×2.2'	12.7	01 h 15 m 37.6 s	+33° 04' 04"
449	Psc	Gxy	0.8'×0.5'	14.2	01 h 16 m 07.2 s	+33° 05' 21"
451	Psc	Gxy	0.7'×0.5'	14.2	01 h 16 m 12.4 s	+33° 03' 51"
452	Psc	Gxy	2.5'×0.8'	13.1	01 h 16 m 14.8 s	+31° 02' 01"
453	Psc	Ast	...	...	01 h 16 m 17.4 s	+33° 01' 51"
455	Psc	Gxy	1.9'×1.2'	12.6	01 h 15 m 57.2 s	+05° 10' 40"
459	Psc	Gxy	1.0'×0.9'	14.6	01 h 18 m 08.1 s	+17° 33' 45"
462	Psc	Gxy	0.4'×0.4'	14.7	01 h 18 m 11.0 s	+04° 13' 33"
463	Psc	Gxy	1.2'×0.4'	14.2	01 h 18 m 58.2 s	+16° 19' 34"
467	Psc	Gxy	1.7'×1.7'	12.1	01 h 19 m 10.1 s	+03° 18. 02"
468	Psc	Gxy	0.7'×0.4'	14.4	01 h 19 m 48.4 s	+32° 46' 03"
469	Psc	Gxy	0.6'×0.4'	14.3	01 h 19 m 32.9 s	+14° 52' 18"
470	Psc	Gxy	2.8'×1.7'	11.6	01 h 19 m 44.9 s	+03° 24' 35"
471	Psc	Gxy	1.0'×0.7'	13.2	01 h 19 m 59.6 s	+14° 47' 11"
472	Psc	Gxy	1.3'×1.1'	13.1	01 h 20 m 28.7 s	+32° 42' 32"
473	Psc	Gxy	1.7'×1.1'	12.2	01 h 19 m 54.7 s	+16° 32' 42"
474	Psc	Gxy	7.1'×6.3'	10.4	01 h 20 m 06.6 s	+03° 24' 55"
475	Psc	Gxy	0.3'×0.3'	15	01 h 20 m 02.1 s	+14° 51' 40"
476	Psc	Gxy	0.5'×0.4'	14.4	01 h 20 m 19.9 s	+16° 01' 12"
479	Psc	Gxy	1.1'×0.9'	14.1	01 h 21 m 15.7 s	+03° 51' 44"
483	Psc	Gxy	0.7'×0.7'	13.2	01 h 21 m 56.3 s	+33° 31' 17"
485	Psc	Gxy	1.7'×0.6'	13.2	01 h 21 m 27.7 s	+07° 01' 07"
486	Psc	Gxy	0.2'×0.2'	...	01 h 21 m 43.0 s	+05° 20' 45"
488	Psc	Gxy	5.2'×3.9'	10.6	01 h 21 m 46.8 s	+05° 15' 25"
489	Psc	Gxy	1.7'×0.4'	12.6	01 h 21 m 53.8 s	+09° 12' 21"
490	Psc	Gxy	0.7'×0.6'	14.4	01 h 22 m 02.9 s	+05° 22' 02"
492	Psc	Gxy	0.6'×0.5'	15.6	01 h 22 m 13.6 s	+05° 25' 01"
494	Psc	Gxy	2.0'×0.8'	13	01 h 22 m 55.4 s	+33° 10' 27"
495	Psc	Gxy	1.3'×0.8'	13.1	01 h 22 m 56.0 s	+33° 28' 18"
496	Psc	Gxy	1.6'×0.9'	13.5	01 h 23 m 11.7 s	+33° 31' 48"
498	Psc	Gxy	0.3'×0.3'	14.3	01 h 23 m 11.3 s	+33° 29' 22"
499	Psc	Gxy	1.6'×1.3'	12	01 h 23 m 11.5 s	+33° 27' 37"
500	Psc	Gxy	0.6'×0.5'	14.2	01 h 22 m 39.3 s	+05° 23' 15"
501	Psc	Gxy	...'×...'	14.5	01 h 23 m 22.3 s	+33° 25' 59"
502	Psc	Gxy	1.1'×1.0'	12.7	01 h 22 m 55.6 s	+09° 02' 56"
503	Psc	Gxy	...'×...'	14.1	01 h 23 m 28.5 s	+33° 19' 56"
504	Psc	Gxy	1.7'×0.4'	13.1	01 h 23 m 27.9 s	+33° 12' 17"
505	Psc	Gxy	1.1'×0.7'	13.9	01 h 22 m 57.0 s	+09° 28' 08"
506	Psc	*	...	...	01 h 23 m 35.3 s	+33° 14' 42"
507	Psc	Gxy	3.1'×3.1'	11.7	01 h 23 m 40.0 s	+33° 15' 22"
508	Psc	Gxy	1.3'×1.3'	13.2	01 h 23 m 40.5 s	+33° 16' 50"
509	Psc	Gxy	1.6'×0.6'	13.5	01 h 23 m 24.1 s	+09° 26' 01"
510	Psc	**	...	...	01 h 23 m 55.5 s	+33° 29' 49"
511	Psc	Gxy	1.2'×1.2'	13.8	01 h 23 m 30.7 s	+11° 17' 27"
514	Psc	Gxy	3.5'×2.8'	11.7	01 h 24 m 03.9 s	+12° 55' 00"

(continued)

Table 10.3 (continued)

NGC #	Const	Type	Size	V Mag	RA (2000)	Dec (2000)
515	Psc	Gxy	1.4'×1.1'	13.1	01 h 24 m 38.5 s	+33° 28' 23"
516	Psc	Gxy	1.4'×0.5'	13.2	01 h 24 m 08.3 s	+09° 33' 06"
517	Psc	Gxy	2.0'×1.0'	12.5	01 h 24 m 43.9 s	+33° 25' 48"
518	Psc	Gxy	1.7'×0.6'	13.5	01 h 24 m 17.6 s	+09° 19' 55"
520	Psc	Gxy	4.5'×1.8'	11.6	01 h 24 m 34.3 s	+03° 47' 42"
522	Psc	Gxy	2.7'×0.4'	13.4	01 h 24 m 45.7 s	+09° 59' 40"
524	Psc	Gxy	2.8'×2.8'	10.5	01 h 24 m 47.8 s	+09° 32' 21"
525	Psc	Gxy	1.5'×0.7'	13.3	01 h 24 m 52.4 s	+09° 42' 08"
532	Psc	Gxy	2.5'×0.8'	12.7	01 h 25 m 17.1 s	+09° 15' 54"
552	Psc	*	...	...	01 h 26 m 10.1 s	+33° 24' 21"
553	Psc	Gxy	0.5'×0.3'	14.1	01 h 26 m 12.5 s	+33° 24' 19"
566	Psc	Gxy	1.6'×0.4'	13.6	01 h 29 m 02.9 s	+32° 19' 57"
569	Psc	Gxy	1.0'×0.5'	13.9	01 h 29 m 07.3 s	+11° 07' 52"
571	Psc	Gxy	1.3'×1.3'	13.8	01 h 29 m 55.9 s	+32° 30' 05"
575	Psc	Gxy	1.7'×1.6'	13	01 h 30 m 46.5 s	+21° 26' 20"
606	Psc	Gxy	1.4'×1.2'	13.6	01 h 34 m 50.4 s	+21° 25' 00"
628	Psc	Gxy	10.5'×9.5'	9.5	01 h 36 m 41.6 s	+15° 47' 03"
631	Psc	Gxy	1.7'×1.5'	13.4	01 h 36 m 47.0 s	+05° 50' 07"
632	Psc	Gxy	1.5'×1.2'	12.4	01 h 37 m 17.5 s	+05° 52' 39"
638	Psc	Gxy	0.8'×0.5'	13.7	01 h 39 m 37.8 s	+07° 14' 16"
645	Psc	Gxy	2.6'×1.2'	12.8	01 h 40 m 08.7 s	+05° 43' 40"
652	Psc	Gxy	1.0'×0.6'	13.8	01 h 40 m 43.2 s	+07° 58' 57"
656	Psc	Gxy	1.5'×1.3'	12.5	01 h 42 m 27.6 s	+26° 08' 36"
658	Psc	Gxy	3.0'×1.6'	12.5	01 h 42 m 09.8 s	+12° 36' 09"
660	Psc	Gxy	8.3'×3.2'	11.4	01 h 43 m 01.7 s	+13° 38' 34"
664	Psc	Gxy	1.5'×1.3'	13	01 h 43 m 45.9 s	+04° 13' 17"
665	Psc	Gxy	2.4'×1.6'	12.2	01 h 44 m 56.0 s	+10° 25' 22"
676	Psc	Gxy	4.0'×1.2'	11.9	01 h 48 m 57.4 s	+05° 54' 27"
693	Psc	Gxy	2.1'×1.0'	12.5	01 h 50 m 30.9 s	+06° 08' 42"
706	Psc	Gxy	1.9'×1.4'	12.4	01 h 51 m 50.3 s	+06° 17' 44"
718	Psc	Gxy	2.3'×2.0'	11.6	01 h 53 m 12.5 s	+04° 11' 48"
728	Psc	***	...	...	01 h 55 m 01.4 s	+04° 13' 21"
730	Psc	*	...	...	01 h 55 m 18.5 s	+05° 38' 11"
741	Psc	Gxy	3.0'×2.9'	11.5	01 h 56 m 20.9 s	+05° 37' 44"
742	Psc	Gxy	0.2'×0.2'	14.4	01 h 56 m 24.1 s	+05° 37' 36"
766	Psc	Gxy	2.0'×2.0'	12.8	01 h 58 m 41.9 s	+08° 20' 54"
791	Psc	Gxy	1.6'×1.6'	13.2	02 h 01 m 44.1 s	+08° 29' 59"
7396	Psc	Gxy	1.9'×1.1'	13.1	22 h 52 m 22.7 s	+01° 05' 33"
7397	Psc	Gxy	0.7'×0.5'	14.4	22 h 52 m 46.7 s	+01° 07' 58"
7398	Psc	Gxy	1.2'×0.8'	13.8	22 h 52 m 49.1 s	+01° 12' 04"
7401	Psc	Gxy	1.0'×0.6'	14.9	22 h 52 m 58.4 s	+01° 08' 33"
7402	Psc	Gxy	0.7'×0.5'	14.5	22 h 53 m 04.5 s	+01° 08' 39"
7403	Psc	*	...	...	22 h 53 m 06.3 s	+01° 28' 57"
7422	Psc	Gxy	0.9'×0.8'	13.6	22 h 56 m 12.4 s	+03° 55' 34"
7428	Psc	Gxy	2.4'×1.3'	12.7	22 h 57 m 19.5 s	−01° 02' 57"

(continued)

Table 10.3 (continued)

NGC #	Const	Type	Size	V Mag	RA (2000)	Dec (2000)
7434	Psc	Gxy	0.4'×0.2'	15.1	22 h 58 m 21.4 s	−01° 11' 02"
7452	Psc	Gxy	0.4'×0.2'	15.5	23 h 00 m 47.4 s	+06° 44' 45"
7455	Psc	Gxy	...'×...'	14.4	23 h 00 m 40.9 s	+07° 18' 10"
7458	Psc	Gxy	1.4'×1.2'	12.6	23 h 01 m 28.5 s	+01° 45' 12"
7459	Psc	Gxy	0.8'×0.4'	15.2	23 h 01 m 00.3 s	+06° 45' 04"
7460	Psc	Gxy	1.4'×1.0'	13.3	23 h 01 m 42.9 s	+02° 15' 49"
7472	Psc	Gxy	0.4'×0.3'	13.7	23 h 05 m 38.5 s	+03° 03' 33"
7477	Psc	Gxy	0.3'×0.2'	15.7	23 h 04 m 40.5 s	+03° 07' 05"
7478	Psc	Gxy	0.4'×0.3'	15.1	23 h 04 m 56.5 s	+02° 34' 40"
7480	Psc	Gxy	1.3'×0.3'	14.2	23 h 05 m 13.6 s	+02° 32' 56"
7482	Psc	Gxy	0.6'×0.4'	13.7	23 h 05 m 38.5 s	+03° 03' 33"
7483	Psc	Gxy	1.6'×1.1'	13.2	23 h 05 m 48.2 s	+03° 32' 40"
7488	Psc	Gxy	0.6'×0.5'	13.9	23 h 07 m 48.9 s	+00° 56' 25"
7493	Psc	*	...	...	23 h 08 m 31.6 s	+00° 54' 38"
7499	Psc	Gxy	1.1'×0.7'	13.8	23 h 10 m 22.4 s	+07° 34' 51"
7501	Psc	Gxy	1.0'×1.0'	13.8	23 h 10 m 30.5 s	+07° 35' 21"
7503	Psc	Gxy	1.1'×1.1'	13.5	23 h 10 m 42.2 s	+07° 34' 04"
7506	Psc	Gxy	1.7'×1.1'	13	23 h 11 m 41.0 s	−02° 09' 36"
7517	Psc	Gxy	0.6'×0.3'	14.5	23 h 13 m 13.7 s	−02° 06' 02"
7518	Psc	Gxy	1.4'×1.0'	13.4	23 h 13 m 12.7 s	+06° 19' 17"
7521	Psc	Gxy	0.6'×0.5'	14	23 h 13 m 35.3 s	−01° 43' 52"
7524	Psc	Gxy	0.7'×0.2'	15.1	23 h 13 m 46.5 s	−01° 43' 49"
7530	Psc	Gxy	0.9'×0.5'	14.6	23 h 14 m 11.8 s	−02° 46' 46"
7532	Psc	Gxy	1.4'×0.7'	13.8	23 h 14 m 22.2 s	−02° 43' 40"
7533	Psc	Gxy	0.7'×0.3'	14.9	23 h 14 m 22.1 s	−02° 02' 02"
7534	Psc	Gxy	1.0'×0.6'	13.9	23 h 14 m 26.6 s	−02° 41' 57"
7537	Psc	Gxy	2.2'×0.6'	13.1	23 h 14 m 34.5 s	+04° 29' 54"
7541	Psc	Gxy	3.5'×1.2'	11.9	23 h 14 m 43.7 s	+04° 32' 03"
7544	Psc	Gxy	0.8'×0.2'	15.1	23 h 14 m 57.0 s	−02° 11' 57"
7546	Psc	Gxy	1.1'×0.8'	15.1	23 h 15 m 05.5 s	−02° 19' 29"
7554	Psc	Gxy	0.4'×0.4'	14.9	23 h 15 m 41.3 s	−02° 22' 44"
7556	Psc	Gxy	2.5'×1.6'	12.7	23 h 15 m 44.5 s	−02° 22' 54"
7557	Psc	Gxy	0.6'×0.6'	14.4	23 h 15 m 39.8 s	+06° 42' 29"
7560	Psc	**	...	...	23 h 15 m 53.7 s	+04° 29' 46"
7561	Psc	*	...	...	23 h 15 m 57.5 s	+04° 31' 22"
7562	Psc	Gxy	2.2'×1.5'	11.8	23 h 15 m 57.4 s	+06° 41' 15"
7564	Psc	*	...	...	23 h 16 m 01.2 s	+07° 20' 53"
7565	Psc	NF	...	...	23 h 16 m 19.8 s	−00° 03' 31"
7566	Psc	Gxy	1.3'×0.7'	13.6	23 h 16 m 37.3 s	−02° 19' 53"
7575	Psc	Gxy	0.5'×0.3'	13.8	23 h 17 m 20.7 s	+05° 39' 39"
7577	Psc	Gxy	0.3'×0.2'	15.7	23 h 17 m 17.1 s	+07° 21' 58"
7581	Psc	Gxy	3.5'×1.4'	11.7	23 h 14 m 43.7 s	+04° 32' 03"
7583	Psc	Gxy	0.6'×0.6'	14.3	23 h 17 m 52.7 s	+07° 22' 47"
7589	Psc	Gxy	1.1'×0.7'	14.3	23 h 18 m 15.7 s	+00° 15' 41"
7591	Psc	Gxy	1.9'×0.8'	13	23 h 18 m 16.4 s	+06° 35' 09"

(continued)

Table 10.3 (continued)

NGC #	Const	Type	Size	V Mag	RA (2000)	Dec (2000)
7603	Psc	Gxy	1.5'×1.0'	13.4	23 h 18 m 56.6 s	+00° 14' 37"
7604	Psc	Gxy	0.3'×0.2'	14.5	23 h 17 m 51.9 s	+07° 25' 49"
7605	Psc	Gxy	0.6'×0.6'	14.3	23 h 17 m 52.7 s	+07° 22' 47"
7611	Psc	Gxy	1.5'×0.6'	13	23 h 19 m 36.6 s	+08° 03' 48"
7613	Psc	NF	...	...	23 h 19 m 51.7 s	+00° 11' 56"
7614	Psc	NF	...	...	23 h 19 m 51.7 s	+00° 11' 56"
7617	Psc	Gxy	0.9'×0.7'	13.9	23 h 20 m 09.0 s	+08° 09' 55"
7629	Psc	Gxy	1.0'×0.8'	13.9	23 h 21 m 19.4 s	+01° 24' 10"
7642	Psc	Gxy	0.5'×0.5'	13.9	23 h 22 m 53.3 s	+01° 26' 33"
7667	Psc	Gxy	1.6'×1.1'	14.1	23 h 24 m 23.1 s	−00° 06' 29"
7668	Psc	NF	...	...	23 h 27 m 21.8 s	−00° 11' 29"
7669	Psc	NF	...	...	23 h 27 m 21.8 s	−00° 11' 29"
7670	Psc	NF	...	...	23 h 27 m 21.8 s	−00° 11' 29"
7679	Psc	Gxy	1.3'×0.9'	12.1	23 h 28 m 46.7 s	+03° 30' 41"
7682	Psc	Gxy	1.2'×1.1'	13.3	23 h 29 m 03.8 s	+03° 31' 55"
7684	Psc	Gxy	1.4'×0.4'	13.8	23 h 30 m 31.9 s	+00° 04' 50"
7685	Psc	Gxy	1.9'×1.4'	13.9	23 h 30 m 33.5 s	+03° 54' 05"
7687	Psc	Gxy	1.3'×1.0'	13.5	23 h 30 m 54.5 s	+03° 32' 47"
7693	Psc	Gxy	1.3'×0.9'	13.5	23 h 33 m 10.5 s	−01° 17' 30"
7694	Psc	Gxy	1.6'×0.9'	13.4	23 h 33 m 16.1 s	−02° 42' 09"
7695	Psc	Gxy	0.6'×0.3'	15.1	23 h 33 m 14.9 s	−02° 43' 12"
7696	Psc	Gxy	1.3'×0.6'	13.9	23 h 33 m 50.1 s	+04° 52' 14"
7699	Psc	Gxy	0.6'×0.3'	15	23 h 34 m 26.9 s	−02° 53' 57"
7700	Psc	Gxy	2.0'×0.4'	13.8	23 h 34 m 30.1 s	−02° 57' 14"
7701	Psc	Gxy	1.6'×0.4'	13.8	23 h 34 m 31.4 s	−02° 51' 15"
7704	Psc	Gxy	1.1'×0.9'	13.5	23 h 35 m 01.1 s	+04° 53' 51"
7705	Psc	Gxy	0.6'×0.5'	14.2	23 h 35 m 02.5 s	+04° 48' 14"
7706	Psc	Gxy	1.2'×1.0'	13.3	23 h 35 m 10.5 s	+04° 57' 51"
7710	Psc	Gxy	1.2'×0.4'	13.9	23 h 35 m 46.0 s	−02° 52' 53"
7714	Psc	Gxy	1.9'×1.4'	12.3	23 h 36 m 14.2 s	+02° 09' 15"
7715	Psc	Gxy	2.6'×0.5'	14.1	23 h 36 m 21.7 s	+02° 09' 22"
7716	Psc	Gxy	2.1'×1.8'	12.1	23 h 36 m 31.5 s	+00° 17' 48"
7731	Psc	Gxy	1.4'×1.1'	13.3	23 h 41 m 29.1 s	+03° 44' 24"
7732	Psc	Gxy	1.9'×0.6'	13.7	23 h 41 m 33.9 s	+03° 43' 29"
7738	Psc	Gxy	2.1'×1.5'	13.2	23 h 44 m 02.1 s	+00° 30' 59"
7739	Psc	Gxy	1.1'×0.9'	14.1	23 h 44 m 30.0 s	+00° 19' 12"
7746	Psc	Gxy	1.4'×1.1'	13.2	23 h 45 m 19.9 s	−01° 41' 05"
7750	Psc	Gxy	1.6'×0.8'	13.1	23 h 46 m 37.9 s	+03° 47' 58"
7751	Psc	Gxy	1.0'×1.0'	13.1	23 h 46 m 58.4 s	+06° 51' 41"
7756	Psc	*	...	...	23 h 48 m 28.5 s	+04° 07' 31"
7757	Psc	Gxy	2.5'×1.8'	12.9	23 h 48 m 45.5 s	+04° 10' 16"
7778	Psc	Gxy	1.0'×1.0'	12.7	23 h 53 m 19.6 s	+07° 52' 15"
7779	Psc	Gxy	1.4'×1.1'	12.6	23 h 53 m 26.7 s	+07° 52' 32"
7780	Psc	Gxy	1.2'×0.7'	14	23 h 53 m 32.2 s	+08° 07' 05"
7781	Psc	Gxy	0.9'×0.3'	14.3	23 h 53 m 45.9 s	+07° 51' 38"

(continued)

Table 10.3 (continued)

NGC #	Const	Type	Size	V Mag	RA (2000)	Dec (2000)
7782	Psc	Gxy	2.4'×1.3'	12.4	23 h 53 m 53.9 s	+07° 58' 14"
7783	Psc	Gxy	1.3'×0.6	13	23 h 54 m 10.0 s	+00° 22' 57"
7785	Psc	Gxy	2.5'×1.3'	11.8	23 h 55 m 18.9 s	+05° 54' 56"
7787	Psc	Gxy	1.8'×0.5'	14.4	23 h 56 m 07.7 s	+00° 32' 59"
7797	Psc	Gxy	1.0'×0.9'	13.9	23 h 58 m 58.8 s	+03° 38' 04"
7802	Psc	Gxy	1.1'×0.6'	13.6	00 h 01 m 00.5 s	+06° 14' 31"
7804	Psc	**	…	…	00 h 01 m 18.7 s	+07° 44' 52"
7809	Psc	Gxy	0.5'×0.4'	14.7	00 h 02 m 09.5 s	+02° 56' 26"
7811	Psc	Gxy	0.4'×0.4'	14.7	00 h 02 m 26.5 s	+03° 21' 07"
7816	Psc	Gxy	1.7'×1.5'	13	00 h 03 m 48.9 s	+07° 28' 39"
7818	Psc	Gxy	1.1'×1.1'	14.2	00 h 04 m 08.9 s	+07° 22' 42"
7820	Psc	Gxy	1.3'×0.6'	13.1	00 h 04 m 30.9 s	+05° 11' 56"
7824	Psc	Gxy	1.6'×1.2'	13.4	00 h 05 m 06.1 s	+06° 55' 12"
7825	Psc	Gxy	1.3'×0.4'	13.6	00 h 05 m 06.6 s	+05° 12' 08"
7827	Psc	Gxy	1.4'×1.3'	14	00 h 05 m 27.6 s	+05° 13' 20"
7830	Psc	*	…	…	00 h 06 m 01.8 s	+08° 20' 34"
7832	Psc	Gxy	1.9'×1.0'	13.2	00 h 06 m 28.4 s	−03° 42' 57"
7834	Psc	Gxy	1.1'×0.8'	14.5	00 h 06 m 37.9 s	+08° 22' 04"
7835	Psc	Gxy	0.4'×0.2'	14.6	00 h 06 m 46.8 s	+08° 25' 30"
7837	Psc	Gxy	0.4'×0.3'	15.8	00 h 06 m 51.4 s	+08° 21' 05"
7838	Psc	Gxy	0.9'×0.6'	14.6	00 h 06 m 53.9 s	+08° 21' 04"
7840	Psc	Gxy	0.4'×0.3'	15.2	00 h 07 m 08.9 s	+08° 22' 55"
2200	Pup	Gxy	1.0'×0.9'	14.3	06 h 13 m 17.3 s	−43° 39' 49"
2201	Pup	Gxy	1.4'×1.0'	13.6	06 h 13 m 31.4 s	−43° 42' 20"
2220	Pup	Ast	5.0'	…	06 h 21 m 11.9 s	−44° 45' 29"
2298	Pup	GC	6.8'	9.3	06 h 48 m 59.1 s	−36° 00' 19"
2310	Pup	Gxy	4.4'×0.8'	11.5	06 h 53 m 53.8 s	−40° 51' 50"
2328	Pup	Gxy	1.6'×1.3'	12.1	07 h 02 m 36.1 s	−42° 04' 07"
2396	Pup	OC	10'	7.4	07 h 28 m 02.9 s	−11° 43' 11"
2401	Pup	OC	2.0'	12.6	07 h 29 m 24.4 s	−13° 57' 58"
2409	Pup	OC	2.3'	…	07 h 31 m 36.7 s	−17° 11' 26"
2413	Pup	OC	10'	…	07 h 33 m 16.5 s	−13° 05' 44"
2414	Pup	OC	4.0'	7.9	07 h 33 m 12.8 s	−15° 27' 14"
2421	Pup	OC	10'	8.3	07 h 36 m 11.8 s	−20° 36' 44"
2422	Pup	OC	29'	4.4	07 h 36 m 35.0 s	−14° 28' 57"
2423	Pup	OC	19'	6.7	07 h 37 m 06.7 s	−13° 52' 17"
2425	Pup	OC	3.3'	…	07 h 38 m 17.6 s	−14° 52' 40"
2427	Pup	Gxy	5.2'×2.2'	11.5	07 h 36 m 27.9 s	−47° 38' 11"
2428	Pup	OC	…	…	07 h 39 m 21.7 s	−16° 31' 45"
2430	Pup	Ast	…	…	07 h 39 m 41.1 s	−16° 17' 46"
2432	Pup	OC	7'	10.2	07 h 40 m 53.8 s	−19° 05' 09"
2437	Pup	OC	27'	6.1	07 h 41 m 46.8 s	−14° 48' 36"
2438	Pup	PN	73"×68"	10.8	07 h 41 m 50.4 s	−14° 44' 09"
2439	Pup	OC	10'	6.9	07 h 40 m 45.4 s	−31° 41' 33"
2440	Pup	PN	74"×42"	9.4	07 h 41 m 55.3 s	−18° 12' 32"

(continued)

Table 10.3 (continued)

NGC #	Const	Type	Size	V Mag	RA (2000)	Dec (2000)
2447	Pup	OC	22'	6.2	07 h 44 m 29.2 s	−23° 51' 11"
2448	Pup	Ast	5'	...	07 h 44 m 36.0 s	−24° 41' 00"
2451	Pup	OC	45'	2.8	07 h 45 m 15.0 s	−37° 58' 03"
2452	Pup	PN	31"×24"	12	07 h 47 m 26.1 s	−27° 20' 08"
2453	Pup	OC	5.0'	8.3	07 h 47 m 34.1 s	−27° 11' 41"
2455	Pup	OC	7'	10.2	07 h 48 m 58.6 s	−21° 17' 53"
2467	Pup	OC+Neb	15'	7.1	07 h 52 m 29.5 s	−26° 25' 48"
2477	Pup	OC	27'	5.8	07 h 52 m 09.8 s	−38° 32' 00"
2478	Pup	OC	29'	4.4	07 h 36 m 35.0 s	−14° 28' 57"
2479	Pup	OC	7'	9.6	07 h 55 m 06.1 s	−17° 42' 28"
2482	Pup	OC	12'	7.3	07 h 55 m 10.3 s	−24° 15' 17"
2483	Pup	OC	10'	7.6	07 h 55 m 38.8 s	−27° 53' 13"
2489	Pup	OC	8'	7.9	07 h 56 m 15.9 s	−30° 03' 51"
2501	Pup	Gxy	1.3'×0.9'	13.5	07 h 58 m 30.1 s	−14° 21' 19"
2509	Pup	OC	8'	9.3	08 h 00 m 47.8 s	−19° 03' 02"
2517	Pup	Gxy	1.5'×1.1'	11.8	08 h 02 m 47.0 s	−12° 19' 04"
2520	Pup	OC	22'	6.5	08 h 04 m 58.2 s	−28° 08' 48"
2525	Pup	Gxy	2.9'×1.9'	11.6	08 h 05 m 38.1 s	−11° 25' 41"
2527	Pup	OC	22'	6.5	08 h 04 m 58.2 s	−28° 08' 48"
2533	Pup	OC	3.5'	7.6	08 h 07 m 04.1 s	−29° 53' 02"
2539	Pup	OC	21'	6.5	08 h 10 m 37.9 s	−12° 49' 09"
2542	Pup	*	...	...	08 h 11 m 16.3 s	−12° 55' 38"
2546	Pup	OC	40'	6.3	08 h 12 m 15.6 s	−37° 35' 39"
2559	Pup	Gxy	3.7'×1.7'	11.1	08 h 17 m 06.2 s	−27° 27' 27"
2564	Pup	Gxy	1.2'×0.8'	12.5	08 h 18 m 30.0 s	−21° 48' 58"
2566	Pup	Gxy	3.4'×2.3'	11.2	08 h 18 m 45.5 s	−25° 30' 04"
2567	Pup	OC	10'	7.4	08 h 18 m 29.1 s	−30° 38' 44"
2568	Pup	OC	2.0'	10.7	08 h 18 m 18.1 s	−37° 06' 19"
2571	Pup	OC	13'	7	08 h 18 m 56.3 s	−29° 44' 57"
2578	Pup	Gxy	2.0'×1.2'	12.6	08 h 21 m 24.3 s	−13° 19' 08"
2579	Pup	OC+Neb	10'	7.5	08 h 20 m 53.0 s	−36° 13' 02"
2580	Pup	OC	7'	9.7	08 h 21 m 27.9 s	−30° 17' 36"
2587	Pup	OC	9'	9.2	08 h 23 m 24.1 s	−29° 30' 31"
2588	Pup	OC	2.0'	11.8	08 h 23 m 09.5 s	−32° 58' 31"
2613	Pyx	Gxy	7.2'×1.8'	10.6	08 h 33 m 23.0 s	−22° 58' 23"
2627	Pyx	OC	11'	8.4	08 h 37 m 14.9 s	−29° 57' 01"
2635	Pyx	OC	3.0'	11.2	08 h 38 m 26.0 s	−34° 46' 18"
2658	Pyx	OC	12'	9.2	08 h 43 m 27.3 s	−32° 39' 22"
2663	Pyx	Gxy	3.5'×2.4'	11	08 h 45 m 08.3 s	−33° 47' 43"
2717	Pyx	Gxy	2.1'×1.5'	12.3	08 h 57 m 01.0 s	−24° 40' 27"
2772	Pyx	Gxy	1.5'×0.9'	13.6	09 h 07 m 41.7 s	−23° 37' 15"
2818	Pyx	PN	35"×35"	11.6	09 h 16 m 01.5 s	−36° 37' 37"
2821	Pyx	Gxy	2.0'×0.4'	13.2	09 h 16 m 47.8 s	−26° 49' 03"
2883	Pyx	Gxy	2.8'×0.9'	14.4	09 h 25 m 18.4 s	−34° 06' 12"
2888	Pyx	Gxy	1.4'×1.0'	12.6	09 h 26 m 19.6 s	−28° 02' 08"

(continued)

Table 10.3 (continued)

NGC #	Const	Type	Size	V Mag	RA (2000)	Dec (2000)
2891	Pyx	Gxy	1.5'×1.4'	12.4	09 h 26 m 56.6 s	−24° 46' 58"
1313	Ret	Gxy	9.1'×6.9'	9.2	03 h 18 m 16.0 s	−66° 29' 55"
1463	Ret	Gxy	1.4'×1.2'	13.4	03 h 46 m 15.5 s	−59° 48' 39"
1490	Ret	Gxy	1.3'×1.1'	12.5	03 h 53 m 34.3 s	−66° 01' 07"
1503	Ret	Gxy	0.9'×0.7'	13.5	03 h 56 m 33.4 s	−66° 02' 28"
1526	Ret	Gxy	0.8'×0.5'	14	04 h 05 m 12.3 s	−65° 50' 26"
1529	Ret	Gxy	1.2'×0.3'	13.5	04 h 07 m 19.9 s	−62° 54' 00"
1534	Ret	Gxy	1.7'×0.8'	13	04 h 08 m 46.2 s	−62° 47' 52"
1536	Ret	Gxy	2.0'×1.4'	12.7	04 h 11 m 00.2 s	−56° 28' 59"
1543	Ret	Gxy	4.9'×2.8'	10.3	04 h 12 m 42.9 s	−57° 44' 17"
1559	Ret	Gxy	3.5'×2.0'	10.7	04 h 17 m 36.7 s	−62° 47' 04"
1574	Ret	Gxy	3.4'×3.1'	10.7	04 h 21 m 58.6 s	−56° 58' 29"

Table 10.4 These NGC objects are found in the constellations of Sculptor through Vulpecula (Copyright © The NGC/IC Project, LLC – All rights reserved. Used with permission)

NGC #	Const	Type	Size	V Mag	RA (2000)	Dec (2000)
7	Scl	Gxy	2.2'×0.5'	13.4	00 h 08 m 20.8 s	−29° 54' 55"
10	Scl	Gxy	2.4'×1.2'	12.5	00 h 08 m 34.6 s	−33° 51' 20"
24	Scl	Gxy	5.8'×1.3'	11.5	00 h 09 m 56.3 s	−24° 57' 49"
55	Scl	Gxy	32.4'×5.6'	8.3	00 h 15 m 08.4 s	−39° 13' 14"
101	Scl	Gxy	2.2'×2.0'	12.8	00 h 23 m 54.5 s	−32° 32' 11"
115	Scl	Gxy	1.9'×0.9'	13.1	00 h 26 m 46.7 s	−33° 40' 36"
131	Scl	Gxy	1.9'×0.6'	13.1	00 h 29 m 38.3 s	−33° 15' 31"
134	Scl	Gxy	8.5'×2.0'	10.5	00 h 30 m 21.4 s	−33° 14' 50"
148	Scl	Gxy	2.0'×0.8'	12.1	00 h 34 m 15.3 s	−31° 47' 08"
150	Scl	Gxy	3.9'×1.9'	11.4	00 h 34 m 16.1 s	−27° 48' 16"
174	Scl	Gxy	1.4'×0.6'	13	00 h 36 m 58.8 s	−29° 28' 42"
253	Scl	Gxy	27.5'×6.8'	7.8	00 h 47 m 33.1 s	−25° 17' 17"
254	Scl	Gxy	2.5'×1.5'	12	00 h 47 m 27.7 s	−31° 25' 16"
264	Scl	Gxy	1.0'×0.3'	13.5	00 h 48 m 20.9 s	−38° 14' 03"
288	Scl	GC	14'	8.1	00 h 52 m 47.4 s	−26° 35' 24"
289	Scl	Gxy	5.1'×3.6'	11	00 h 52 m 41.7 s	−31° 12' 28"
300	Scl	Gxy	21.9'×15.5'	8.3	00 h 54 m 53.4 s	−37° 41' 00"
314	Scl	Gxy	1.0'×0.8'	13.3	00 h 56 m 52.7 s	−31° 57' 47"
334	Scl	Gxy	1.2'×0.6'	14	00 h 58 m 50.0 s	−35° 06' 55"
365	Scl	Gxy	1.0'×0.6'	13.7	01 h 04 m 18.9 s	−35° 07' 14"
378	Scl	Gxy	1.5'×1.1'	13.3	01 h 06 m 11.8 s	−30° 10' 40"
409	Scl	Gxy	1.3'×1.1'	13.1	01 h 09 m 32.5 s	−35° 48' 20"
415	Scl	Gxy	1.4'×0.8'	13.7	01 h 10 m 05.5 s	−35° 29' 27"
418	Scl	Gxy	2.0'×1.7'	12.7	01 h 10 m 36.1 s	−30° 13' 15"

(continued)

Table 10.4 (continued)

NGC #	Const	Type	Size	V Mag	RA (2000)	Dec (2000)
423	Scl	Gxy	1.0'×0.4'	13.6	01 h 11 m 22.4 s	−29° 14' 04"
424	Scl	Gxy	1.8'×0.8'	13	01 h 11 m 27.9 s	−38° 04' 59"
427	Scl	Gxy	1.0'×0.7'	14.3	01 h 12 m 19.9 s	−32° 03' 42"
438	Scl	Gxy	1.4'×1.1'	13	01 h 13 m 33.5 s	−37° 54' 07"
439	Scl	Gxy	2.5'×1.5'	11.6	01 h 13 m 47.2 s	−31° 44' 51"
441	Scl	Gxy	1.4'×1.1'	12.9	01 h 13 m 51.7 s	−31° 47' 14"
461	Scl	Gxy	1.2'×0.9'	13.6	01 h 17 m 20.1 s	−33° 50' 25"
491	Scl	Gxy	1.4'×1.0'	12.6	01 h 21 m 20.1 s	−34° 03' 48"
526	Scl	Gxy	0.8'×0.6'	14.1	01 h 23 m 54.1 s	−35° 03' 56"
527	Scl	Gxy	1.7'×0.4'	13.2	01 h 23 m 58.0 s	−35° 06' 55"
534	Scl	Gxy	1.1'×1.0'	13.5	01 h 24 m 44.7 s	−38° 07' 45"
544	Scl	Gxy	1.2'×0.9'	13.5	01 h 25 m 12.1 s	−38° 05' 41"
546	Scl	Gxy	1.4'×0.5'	13.8	01 h 25 m 12.7 s	−38° 04' 09"
549	Scl	Gxy	0.5'×0.4'	13.7	01 h 25 m 07.0 s	−38° 00' 28"
568	Scl	Gxy	2.2'×1.4'	12.5	01 h 27 m 56.9 s	−35° 43' 05"
572	Scl	Gxy	0.8'×0.7'	14.2	01 h 28 m 36.3 s	−39° 18' 24"
574	Scl	Gxy	1.1'×0.7'	13.5	01 h 29 m 02.7 s	−35° 35' 55"
597	Scl	Gxy	1.4'×1.3'	13.4	01 h 32 m 14.5 s	−33° 29' 54"
612	Scl	Gxy	1.4'×0.9'	13	01 h 33 m 57.6 s	−36° 29' 36"
613	Scl	Gxy	5.5'×4.2'	10.1	01 h 34 m 17.5 s	−29° 24' 58"
619	Scl	Gxy	1.5'×1.0'	13.7	01 h 34 m 52.0 s	−36° 29' 16"
623	Scl	Gxy	2.0'×1.5'	12.7	01 h 35 m 05.9 s	−36° 29' 23"
626	Scl	Gxy	1.9'×1.9'	12.9	01 h 35 m 11.9 s	−39° 08' 53"
630	Scl	Gxy	1.6'×1.4'	12.7	01 h 35 m 36.7 s	−39° 21' 36"
633	Scl	Gxy	1.3'×1.1'	12.9	01 h 36 m 24.1 s	−37° 19' 25"
639	Scl	Gxy	1.0'×0.2'	13.9	01 h 38 m 58.7 s	−29° 55' 30"
642	Scl	Gxy	2.0'×1.1'	13.1	01 h 39 m 06.7 s	−29° 54' 54"
7484	Scl	Gxy	1.8'×1.7'	11.9	23 h 07 m 04.9 s	−36° 16' 24"
7507	Scl	Gxy	2.8'×2.7'	10.8	23 h 12 m 07.7 s	−28° 32' 26"
7513	Scl	Gxy	3.2'×2.1'	12.1	23 h 13 m 13.9 s	−28° 21' 31"
7636	Scl	Gxy	0.9'×0.6'	13.8	23 h 22 m 33.1 s	−29° 16' 53"
7645	Scl	Gxy	1.4'×1.2'	13.3	23 h 23 m 47.3 s	−29° 23' 19"
7713	Scl	Gxy	4.6'×2.2'	11.2	23 h 36 m 14.9 s	−37° 56' 20"
7749	Scl	Gxy	1.6'×1.1'	12.9	23 h 45 m 47.5 s	−29° 31' 06"
7755	Scl	Gxy	3.8'×2.9'	11.6	23 h 47 m 51.9 s	−30° 31' 26"
7793	Scl	Gxy	9.3'×6.3'	9.2	23 h 57 m 49.7 s	−32° 35' 30"
7812	Scl	Gxy	1.1'×0.7'	13.3	00 h 02 m 54.2 s	−34° 14' 11"
5998	Sco	OC	2.0'×1.0'	...	15 h 49 m 38.1 s	−28° 34' 41"
6000	Sco	Gxy	1.9'×1.6'	12.4	15 h 49 m 49.4 s	−29° 23' 15"
6072	Sco	PN	70"×70"	11.7	16 h 12 m 58.1 s	−36° 13' 48"
6082	Sco	Gxy	1.2'×1.0'	...	16 h 17 m 39.6 s	−34° 21' 59"
6093	Sco	GC	8.9'	7.3	16 h 17 m 03.1 s	−22° 58' 30"
6121	Sco	GC	26'	5.4	16 h 23 m 35.4 s	−26° 31' 31"
6124	Sco	OC	40'	5.8	16 h 25 m 20.0 s	−40° 39' 13"
6139	Sco	GC	5.5'	9.1	16 h 27 m 40.4 s	−38° 50' 56"

(continued)

Table 10.4 (continued)

NGC #	Const	Type	Size	V Mag	RA (2000)	Dec (2000)
6144	Sco	GC	9.3'	9	16 h 27 m 14.1 s	−26° 01' 29"
6153	Sco	PN	28"×21"	10.9	16 h 31 m 30.5 s	−40° 15' 13"
6178	Sco	OC	4.0'	7.2	16 h 35 m 47.2 s	−45° 38' 37"
6192	Sco	OC	7'	8.5	16 h 40 m 23.8 s	−43° 22' 00"
6216	Sco	OC	4.0'	10.1	16 h 49 m 23.5 s	−44° 43' 53"
6222	Sco	OC	4.0'	10.1	16 h 49 m 23.5 s	−44° 43' 53"
6227	Sco	MWSC	18'	…	16 h 51 m 37.5 s	−41° 13' 26"
6231	Sco	OC	14'	2.6	16 h 54 m 10.9 s	−41° 49' 27"
6242	Sco	OC	9'	6.4	16 h 55 m 33.4 s	−39° 27' 39"
6249	Sco	OC	6'	8.2	16 h 57 m 41.5 s	−44° 48' 43"
6256	Sco	GC	6.6'	11.3	16 h 59 m 32.6 s	−37° 07' 17"
6259	Sco	OC	10'	8	17 h 00 m 45.3 s	−44° 39' 18"
6268	Sco	OC	6'	9.5	17 h 02 m 10.3 s	−39° 43' 42"
6281	Sco	OC	8'	5.4	17 h 04 m 41.2 s	−37° 59' 07"
6302	Sco	PN	83"×24"	9.6	17 h 13 m 44.6 s	−37° 06' 12"
6318	Sco	OC	5.0'	11.8	17 h 16 m 11.5 s	−39° 25' 30"
6322	Sco	OC	10'	6	17 h 18 m 25.7 s	−42° 56' 02"
6334	Sco	OC+Neb	35'×20'	…	17 h 20 m 49.7 s	−36° 06' 10"
6335	Sco	Ast	…	…	17 h 19 m 31.9 s	−30° 09' 51"
6337	Sco	PN	49"×45"	12.3	17 h 22 m 15.6 s	−38° 29' 02"
6354	Sco	Ast	…	…	17 h 24 m 33.7 s	−38° 32' 32"
6357	Sco	OC+Neb	35'?	…	17 h 24 m 43.5 s	−34° 12' 05"
6374	Sco	OC	2.5'	…	17 h 34 m 42.4 s	−32° 34' 53"
6380	Sco	GC	3.9'	11.5	17 h 34 m 28.3 s	−39° 04' 11"
6383	Sco	OC	2.5'	…	17 h 34 m 42.4 s	−32° 34' 53"
6388	Sco	GC	8.7'	6.8	17 h 36 m 16.9 s	−44° 44' 05"
6396	Sco	OC	3.0'	8.5	17 h 37 m 36.3 s	−35° 01' 33"
6400	Sco	OC	12'	8.8	17 h 40 m 12.7 s	−36° 56' 52"
6404	Sco	OC	5'	10.6	17 h 39 m 37.3 s	−33° 14' 48"
6405	Sco	OC	33'	4.2	17 h 40 m 20.7 s	−32° 15' 15"
6415	Sco	MWSC	…	…	17 h 44 m 20.5 s	−35° 04' 16"
6416	Sco	OC	15'	5.7	17 h 44 m 19.9 s	−32° 21' 40"
6421	Sco	MWSC	45.0'	…	17 h 45 m 44.1 s	−33° 41' 34"
6425	Sco	OC	15'	7.2	17 h 47 m 01.6 s	−31° 31' 46"
6437	Sco	MWSC	…	…	17 h 48 m 21.1 s	−35° 21' 58"
6441	Sco	GC	7.8'	7.2	17 h 50 m 12.9 s	−37° 03' 04"
6444	Sco	OC	12'	…	17 h 49 m 35.1 s	−34° 49' 11"
6451	Sco	OC	7'	8.2	17 h 50 m 40.6 s	−30° 12' 42"
6453	Sco	GC	3.5'	10.2	17 h 50 m 51.6 s	−34° 35' 55"
6455	Sco	MWSC	…	…	17 h 51 m 08.1 s	−35° 20' 16"
6475	Sco	OC	80'	3.3	17 h 53 m 51.1 s	−34° 47' 34"
6480	Sco	MWSC	…	…	17 h 54 m 26.0 s	−30° 27' 07"
6496	Sco	GC	6.9'	8.6	17 h 59 m 03.7 s	−44° 15' 59"
6625	Sct	OC	39'	9	18 h 23 m 01.9 s	−12° 01' 26"
6631	Sct	OC	7'	11.7	18 h 27 m 11.3 s	−12° 01' 52"

(continued)

Table 10.4 (continued)

NGC #	Const	Type	Size	V Mag	RA (2000)	Dec (2000)
6639	Sct	OC	7'	…	18 h 30 m 59.3 s	−13° 09' 21"
6649	Sct	OC	5'	8.9	18 h 33 m 27.9 s	−10° 24' 10"
6655	Sct	**	…	…	18 h 34 m 30.8 s	−05° 55' 15"
6664	Sct	OC	16'	7.8	18 h 36 m 33.3 s	−08° 13' 15"
6682	Sct	OC	18.0'	…	18 h 39 m 37.3 s	−04° 48' 49"
6683	Sct	OC	3.0'	9.4	18 h 42 m 13.9 s	−06° 12' 44"
6694	Sct	OC	14'	8	18 h 45 m 18.6 s	−09° 23' 01"
6704	Sct	OC	5'	9.2	18 h 50 m 45.7 s	−05° 12' 20"
6705	Sct	OC	13'	5.8	18 h 51 m 05.9 s	−06° 16' 12"
6712	Sct	GC	7.2'	8.1	18 h 53 m 04.3 s	−08° 42' 22"
6728	Sct	Ast	…	…	18 h 58 m 44.9 s	−08° 56' 58"
6535	SerCd	GC	3.6'	9.3	18 h 03 m 50.7 s	−00° 17' 49"
6539	SerCd	GC	6.9'	8.9	18 h 04 m 50.2 s	−07° 35' 09"
6604	SerCd	OC+Neb	4.0'	6.5	18 h 18 m 02.9 s	−12° 14' 35"
6605	SerCd	OC	29'	6	18 h 18 m 21.6 s	−15° 00' 46"
6611	SerCd	OC+Neb	120'×25'	6	18 h 18 m 48.1 s	−13° 48' 26"
5887	SerCp	Gxy	1.2'×1.0'	13.6	15 h 14 m 43.9 s	+01° 09' 14"
5904	SerCp	GC	23'	5.7	15 h 18 m 33.7 s	+02° 04' 58"
5910	SerCp	Gxy	0.9'×0.8'	13.6	15 h 19 m 24.6 s	+20° 53' 46"
5911	SerCp	Gxy	1.0'×0.7'	14	15 h 20 m 18.1 s	+03° 31' 06"
5913	SerCp	Gxy	1.6'×0.7'	13.4	15 h 20 m 55.3 s	−02° 34' 40"
5919	SerCp	Gxy	0.3'×0.3'	15.9	15 h 21 m 36.9 s	+07° 43' 08"
5920	SerCp	Gxy	1.1'×0.9'	13.7	15 h 21 m 51.9 s	+07° 42' 32"
5921	SerCp	Gxy	4.9'×4.0'	11.1	15 h 21 m 56.5 s	+05° 04' 13"
5926	SerCp	Gxy	0.9'×0.9'	13.8	15 h 23 m 24.8 s	+12° 42' 54"
5928	SerCp	Gxy	2.2'×1.6'	12.3	15 h 26 m 02.9 s	+18° 04' 24"
5931	SerCp	Gxy	1.0'×0.5'	14	15 h 29 m 29.6 s	+07° 34' 22"
5936	SerCp	Gxy	1.4'×1.3'	12.4	15 h 30 m 00.7 s	+12° 59' 21"
5937	SerCp	Gxy	1.9'×1.1'	12.5	15 h 30 m 46.1 s	02° 49' 45"
5940	SerCp	Gxy	0.8'×0.8'	13.6	15 h 31 m 18.0 s	+07° 27' 27"
5941	SerCp	Gxy	0.4'×0.3'	13.9	15 h 31 m 40.2 s	+07° 20' 20"
5942	SerCp	Gxy	0.4'×0.4'	14.3	15 h 31 m 36.7 s	+07° 18' 42"
5944	SerCp	Gxy	0.7'×0.2'	15	15 h 31 m 47.6 s	+07° 18' 29"
5948	SerCp	**	…	…	15 h 32 m 58.7 s	+03° 58' 58"
5951	SerCp	Gxy	3.5'×0.8'	12.9	15 h 33 m 43.1 s	+15° 00' 27"
5952	SerCp	Gxy	0.4'×0.3'	14.2	15 h 34 m 56.4 s	+04° 57' 32"
5953	SerCp	Gxy	1.6'×1.3'	12	15 h 34 m 32.4 s	+15° 11' 39"
5954	SerCp	Gxy	1.3'×0.6'	12.2	15 h 34 m 34.9 s	+15° 12' 00"
5955	SerCp	Gxy	0.8'×0.5'	14.9	15 h 35 m 12.5 s	+05° 03' 47"
5956	SerCp	Gxy	1.6'×1.6'	12.5	15 h 34 m 58.5 s	+11° 45' 00"
5957	SerCp	Gxy	2.8'×2.6'	11.9	15 h 35 m 23.2 s	+12° 02' 51"
5960	SerCp	Gxy	0.7'×0.6'	14.3	15 h 36 m 18.3 s	+05° 39' 56"
5962	SerCp	Gxy	3.0'×2.1'	11.7	15 h 36 m 31.7 s	+16° 36' 29"
5964	SerCp	Gxy	4.2'×3.2'	12.1	15 h 37 m 36.3 s	+05° 58' 25"
5970	SerCp	Gxy	2.9'×1.9'	11.6	15 h 38 m 29.9 s	+12° 11' 12"

(continued)

Table 10.4 (continued)

NGC #	Const	Type	Size	V Mag	RA (2000)	Dec (2000)
5972	SerCp	Gxy	1.1'×0.7'	13.8	15 h 38 m 54.2 s	+17° 01' 33"
5975	SerCp	Gxy	1.0'×0.3'	14.1	15 h 39 m 58.0 s	+21° 28' 13"
5977	SerCp	Gxy	1.1'×0.9'	13.5	15 h 40 m 33.3 s	+17° 07' 40"
5980	SerCp	Gxy	1.9'×0.7'	12.7	15 h 41 m 30.5 s	+15° 47' 14"
5983	SerCp	Gxy	1.0'×1.0'	13.5	15 h 42 m 45.5 s	+08° 14' 28"
5984	SerCp	Gxy	2.9'×0.8'	12.6	15 h 42 m 53.3 s	+14° 13' 51"
5988	SerCp	Gxy	1.2'×1.0'	14	15 h 44 m 33.8 s	+10° 17' 35"
5990	SerCp	Gxy	1.5'×0.9'	12.6	15 h 46 m 16.4 s	+02° 24' 55"
5991	SerCp	Gxy	1.0'×0.9'	14	15 h 45 m 16.7 s	+24° 37' 49"
5994	SerCp	Gxy	0.4'×0.2'	14.8	15 h 46 m 53.3 s	+17° 52' 21"
5996	SerCp	Gxy	1.7'×0.9'	12.5	15 h 46 m 58.7 s	+17° 53' 05"
5997	SerCp	Gxy	0.5'×0.4'	14.2	15 h 47 m 27.5 s	+08° 19' 16"
6003	SerCp	Gxy	0.9'×0.9'	13.4	15 h 49 m 25.6 s	+19° 01' 55"
6004	SerCp	Gxy	1.9'×1.7'	12.5	15 h 50 m 22.7 s	+18° 56' 21"
6006	SerCp	Gxy	0.7'×0.4'	14.2	15 h 53 m 02.5 s	+12° 00' 19"
6007	SerCp	Gxy	1.7'×1.2'	13	15 h 53 m 23.1 s	+11° 57' 32"
6008	SerCp	Gxy	1.4'×1.3'	13.2	15 h 52 m 56.0 s	+21° 06' 01"
6009	SerCp	Gxy	0.5'×0.2'	14.5	15 h 53 m 24.3 s	+12° 03' 30"
6010	SerCp	Gxy	1.9'×0.5'	12.8	15 h 54 m 19.1 s	+00° 32' 34"
6012	SerCp	Gxy	2.1'×1.5'	12.3	15 h 54 m 13.9 s	+14° 36' 04"
6014	SerCp	Gxy	1.7'×1.6'	12.3	15 h 55 m 57.5 s	+05° 55' 55"
6017	SerCp	Gxy	0.8'×0.7'	13.2	15 h 57 m 15.4 s	+05° 59' 55"
6018	SerCp	Gxy	1.4'×0.7'	13.5	15 h 57 m 29.9 s	+15° 52' 23"
6020	SerCp	Gxy	1.4'×1.0'	13.1	15 h 57 m 08.2 s	+22° 24' 15"
6021	SerCp	Gxy	1.4'×0.8'	13.1	15 h 57 m 30.7 s	+15° 57' 21"
6022	SerCp	Gxy	0.7'×0.5'	14.8	15 h 57 m 47.8 s	+16° 16' 55"
6023	SerCp	Gxy	1.1'×0.8'	13.4	15 h 57 m 49.6 s	+16° 18' 35"
6027	SerCp	Gxy	0.4'×0.2'	13.8	15 h 59 m 12.6 s	+20° 45' 48"
6029	SerCp	Gxy	1.2'×0.7'	14.5	16 h 01 m 58.9 s	+12° 34' 28"
6033	SerCp	Gxy	1.0'×0.9'	13.9	16 h 04 m 27.9 s	−02° 07' 16"
6036	SerCp	Gxy	1.1'×0.4'	13.6	16 h 04 m 30.6 s	+03° 52' 07"
6037	SerCp	Gxy	0.7'×0.7'	14.2	16 h 04 m 29.9 s	+03° 48' 53"
6049	SerCp	*	…	…	16 h 05 m 37.9 s	+08° 05' 46"
6051	SerCp	Gxy	1.6'×1.0'	13.4	16 h 04 m 56.7 s	+23° 55' 54"
6063	SerCp	Gxy	1.7'×0.9'	13.3	16 h 07 m 13.1 s	+07° 58' 43"
6065	SerCp	Gxy	0.6'×0.5'	14	16 h 07 m 23.0 s	+13° 53' 16"
6066	SerCp	Gxy	0.8'×0.7'	14.1	16 h 07 m 35.4 s	+13° 56' 37"
6070	SerCp	Gxy	3.5'×1.9'	11.8	16 h 09 m 58.7 s	+00° 42' 31"
6080	SerCp	Gxy	1.1'×1.0'	13.4	16 h 12 m 58.6 s	+02° 10' 38"
6100	SerCp	Gxy	1.9'×1.1'	13.2	16 h 16 m 52.4 s	+00° 50' 27"
6118	SerCp	Gxy	4.7'×2.0'	11.6	16 h 21 m 48.5 s	−02° 17' 01"
6172	SerCp	Gxy	1.0'×1.0'	12.9	16 h 22 m 10.3 s	−01° 30' 53"
2652	Sex	Gxy	3.4'×2.1'	10.9	09 h 42 m 32.9 s	−03° 41' 58"
2966	Sex	Gxy	2.2'×0.9'	13	09 h 42 m 11.7 s	+04° 40' 24"
2967	Sex	Gxy	3.0'×2.8'	11.4	09 h 42 m 03.5 s	+00° 20' 09"

(continued)

Table 10.4 (continued)

NGC #	Const	Type	Size	V Mag	RA (2000)	Dec (2000)
2969	Sex	Gxy	1.3' × 1.2'	13.1	09 h 41 m 54.5 s	−08° 36' 16"
2974	Sex	Gxy	3.5' × 2.0'	11	09 h 42 m 32.9 s	−03° 41' 58"
2978	Sex	Gxy	1.0' × 0.9'	12.8	09 h 43 m 16.8 s	−09° 44' 45"
2979	Sex	Gxy	1.5' × 0.9'	12.7	09 h 43 m 08.7 s	−10° 23' 01"
2980	Sex	Gxy	1.6' × 0.9'	13	09 h 43 m 12.0 s	−09° 36' 47"
2987	Sex	Gxy	1.5' × 0.7'	13.2	09 h 45 m 41.4 s	+04° 56' 29"
2990	Sex	Gxy	1.3' × 0.7'	12.9	09 h 46 m 17.3 s	+05° 42' 32"
3007	Sex	Gxy	1.3' × 0.5'	13.6	09 h 47 m 45.5 s	−06° 26' 18"
3014	Sex	Gxy	0.9' × 0.8'	14	09 h 49 m 07.9 s	−04° 44' 36"
3015	Sex	Gxy	0.5' × 0.4'	14	09 h 49 m 22.8 s	+01° 08' 42"
3017	Sex	Gxy	1.0' × 1.0'	13.2	09 h 49 m 03.0 s	−02° 49' 19"
3018	Sex	Gxy	1.2' × 0.7'	13.5	09 h 49 m 41.5 s	+00° 37' 20"
3022	Sex	Gxy	1.6' × 1.6'	13.2	09 h 49 m 39.2 s	−05° 10' 01"
3023	Sex	Gxy	2.9' × 1.4'	12.3	09 h 49 m 52.5 s	+00° 37' 07"
3029	Sex	Gxy	1.4' × 1.0'	14	09 h 48 m 54.1 s	−08° 03' 04"
3035	Sex	Gxy	1.6' × 1.4'	12.7	09 h 51 m 54.9 s	−06° 49' 24"
3039	Sex	Gxy	1.3' × 0.6'	13.6	09 h 52 m 29.7 s	+02° 09' 15"
3042	Sex	Gxy	1.2' × 0.8'	13	09 h 53 m 20.0 s	+00° 41' 52"
3044	Sex	Gxy	4.8' × 0.9'	11.9	09 h 53 m 41.1 s	+01° 34' 41"
3047	Sex	Gxy	0.9' × 0.5'	14.1	09 h 54 m 32.1 s	−01° 17' 30"
3050	Sex	Gxy	1.5' × 1.0'	12.7	09 h 43 m 08.7 s	−10° 23' 01"
3055	Sex	Gxy	2.1' × 1.3'	12	09 h 55 m 18.1 s	+04° 16' 11"
3062	Sex	Gxy	0.6' × 0.3'	14.9	09 h 56 m 35.8 s	+01° 25' 42"
3064	Sex	Gxy	1.1' × 0.3'	14.1	09 h 55 m 41.5 s	−06° 21' 50"
3083	Sex	Gxy	1.0' × 0.4'	13.9	09 h 59 m 49.7 s	−02° 52' 39"
3086	Sex	Gxy	1.1' × 0.4'	14.1	10 h 00 m 11.0 s	−02° 58' 35"
3090	Sex	Gxy	1.7' × 1.4'	12.6	10 h 00 m 30.3 s	−02° 58' 06"
3092	Sex	Gxy	1.2' × 0.6'	13.4	10 h 00 m 47.4 s	−03° 00' 45"
3093	Sex	Gxy	0.7' × 0.3'	14.3	10 h 00 m 53.5 s	−02° 58' 20"
3101	Sex	Gxy	1.2' × 0.3'	14.6	10 h 01 m 35.3 s	−02° 59' 41"
3110	Sex	Gxy	1.5' × 0.7'	12.7	10 h 04 m 01.9 s	−06° 28' 28"
3115	Sex	Gxy	7.2' × 2.5'	8.6	10 h 05 m 14.1 s	−07° 43' 09"
3117	Sex	Gxy	0.9' × 0.9'	13.4	10 h 06 m 10.5 s	+02° 54' 46"
3122	Sex	Gxy	1.5' × 0.7'	12.7	10 h 04 m 01.9 s	−06° 28' 28"
3123	Sex	*	…	…	10 h 07 m 05.8 s	+00° 03' 35"
3142	Sex	Gxy	0.6' × 0.5'	13.8	10 h 10 m 06.3 s	−08° 28' 49"
3156	Sex	Gxy	1.9' × 1.1'	12	10 h 12 m 41.3 s	+03° 07' 45"
3165	Sex	Gxy	1.6' × 0.8'	13.6	10 h 13 m 31.3 s	+03° 22' 33"
3166	Sex	Gxy	4.8' × 2.3'	10.4	10 h 13 m 45.6 s	+03° 25' 31"
3169	Sex	Gxy	4.4' × 2.8'	10.7	10 h 14 m 14.8 s	+03° 28' 00"
3229	Sex	***	…	…	10 h 23 m 24.4 s	+00° 03' 55"
3243	Sex	Gxy	1.4' × 1.1'	12.8	10 h 26 m 21.5 s	−02° 37' 20"
3246	Sex	Gxy	2.4' × 1.3'	12.7	10 h 26 m 41.8 s	+03° 51' 43"
3292	Sex	Gxy	0.6' × 0.4'	14.1	10 h 35 m 34.4 s	−06° 10' 45"
3321	Sex	Gxy	2.5' × 1.2'	13.5	10 h 38 m 50.7 s	−11° 38' 54"

(continued)

Table 10.4 (continued)

NGC #	Const	Type	Size	V Mag	RA (2000)	Dec (2000)
3322	Sex	Gxy	2.5'×1.2'	13.5	10 h 38 m 50.7 s	−11° 38' 54"
3325	Sex	Gxy	1.3'×1.1'	12.8	10 h 39 m 20.4 s	−00° 12' 03"
3326	Sex	Gxy	0.6'×0.5'	13.9	10 h 39 m 31.9 s	+05° 06' 27"
3337	Sex	Gxy	0.6'×0.3'	14	10 h 41 m 47.5 s	+04° 59' 17"
3339	Sex	*	...	...	10 h 42 m 10.0 s	−00° 22' 08"
3340	Sex	Gxy	1.0'×0.9'	13.2	10 h 42 m 17.9 s	−00° 22' 36"
3341	Sex	Gxy	1.4'×0.4'	13.8	10 h 42 m 31.5 s	+05° 02' 37"
3360	Sex	Gxy	1.2'×0.9'	13.7	10 h 44 m 16.2 s	−11° 14' 35"
3361	Sex	Gxy	2.0'×1.2'	12.8	10 h 44 m 29.0 s	−11° 12' 29"
3365	Sex	Gxy	4.5'×0.8'	12.5	10 h 46 m 12.5 s	+01° 48' 47"
3375	Sex	Gxy	0.5'×0.4'	12.6	10 h 47 m 00.8 s	−09° 56' 29"
3376	Sex	Gxy	0.8'×0.4'	13.4	10 h 47 m 26.6 s	+06° 02' 52"
3385	Sex	Gxy	1.5'×0.9'	12.7	10 h 48 m 11.6 s	+04° 55' 39"
3386	Sex	Gxy	0.4'×0.4'	13.8	10 h 48 m 11.9 s	+04° 59' 54"
3387	Sex	Gxy	0.25'×0.2'	15	10 h 48 m 16.7 s	+04° 58' 00"
3401	Sex	NF	...	...	10 h 50 m 19.9 s	+05° 48' 41"
3423	Sex	Gxy	3.8'×3.2'	11	10 h 51 m 14.2 s	+05° 50' 23"
3518	Sex	Gxy	1.5'×0.7'	12.7	10 h 04 m 01.9 s	−06° 28' 28"
6838	Sge	GC	7.2'	8.4	19 h 53 m 46.1 s	+18° 46' 42"
6839	Sge	Ast	...	...	19 h 54 m 33.7 s	+17° 56' 20"
6873	Sge	OC	...	...	20 h 07 m 13.8 s	+21° 06' 08"
6879	Sge	PN	4.7"×4.1"	12.5	20 h 10 m 26.5 s	+16° 55' 22"
6886	Sge	PN	9"×9"	11.4	20 h 12 m 42.7 s	+19° 59' 22"
6892	Sge	Ast	...	...	20 h 16 m 57.2 s	+18° 01' 11"
6432	Sgr	Ast	...	...	17 h 47 m 22.4 s	−24° 53' 15"
6439	Sgr	PN	6.1"×5.1"	12.6	17 h 48 m 20.32 s	−16° 27' 33.8"
6440	Sgr	GC	5.4'	9.3	17 h 48 m 52.5 s	−20° 21' 34"
6445	Sgr	PN	2.8'×0.9'	11.2	17 h 49 m 15.0 s	−20° 00' 34"
6465	Sgr	Ast	...	...	17 h 52 m 55.5 s	−25° 23' 52"
6469	Sgr	OC	8'	8.2	17 h 53 m 12.1 s	−22° 16' 30"
6476	Sgr	MWSC	...	...	17 h 54 m 01.9 s	−29° 08' 39"
6494	Sgr	OC	30'	5.5	17 h 57 m 04.7 s	−18° 59' 07"
6506	Sgr	OC	4'	...	17 h 59 m 53.5 s	−24° 41' 07"
6507	Sgr	OC	6'	...	17 h 59 m 50.7 s	−17° 27' 01"
6514	Sgr	OC+Neb	20'×20'	...	18 h 02 m 20.9 s	−23° 01' 38"
6519	Sgr	**	...	...	18 h 03 m 20.1 s	−29° 48' 15"
6520	Sgr	OC	6'	7.6	18 h 03 m 25.1 s	−27° 53' 28"
6522	Sgr	GC	5.6'	9.9	18 h 03 m 35.1 s	−30° 02' 06"
6523	Sgr	Neb	45'×30'	...	18 h 03 m 41.2 s	−24° 22' 49"
6526	Sgr	Neb	40'	...	18 h 04 m 06.1 s	−24° 26' 31"
6528	Sgr	GC	3.7'	9.6	18 h 04 m 49.6 s	−30° 03' 21"
6529	Sgr	NF	...	...	18 h 05 m 28.8 s	−36° 17' 43"
6530	Sgr	OC+Neb	14'	4.6	18 h 04 m 31.0 s	−24° 21' 29"
6531	Sgr	OC	15'	5.9	18 h 04 m 13.4 s	−22° 29' 24"
6533	Sgr	OC+Neb	90'×40'	...	18 h 04 m 04.0 s	−24° 23' 49"

(continued)

Table 10.4 (continued)

NGC #	Const	Type	Size	V Mag	RA (2000)	Dec (2000)
6537	Sgr	PN	5"	11.6	18 h 05 m 13.0 s	−19° 50' 35"
6540	Sgr	GC	1.5'	14.6	18 h 06 m 08.5 s	−27° 45' 55"
6544	Sgr	GC	8.9'	7.5	18 h 07 m 20.5 s	−24° 59' 51"
6546	Sgr	OC	15'	8	18 h 07 m 22.5 s	−23° 17' 46"
6551	Sgr	GC	5.6'	...	18 h 08 m 59.6 s	−29° 33' 28"
6553	Sgr	GC	9.2'	8.3	18 h 09 m 17.5 s	−25° 54' 28"
6554	Sgr	OC	20'	...	18 h 08 m 58.0 s	−18° 26' 15"
6556	Sgr	OC	15'×12'	...	18 h 09 m 57.5 s	−27° 31' 29"
6558	Sgr	GC	3.7'	8.6	18 h 10 m 18.3 s	−31° 45' 49"
6559	Sgr	Neb	15'×10'	...	18 h 09 m 56.8 s	−24° 06' 23"
6561	Sgr	OC	5'	...	18 h 10 m 30.8 s	−16° 43' 32"
6563	Sgr	PN	54"×41"	11	18 h 12 m 02.6 s	−33° 52' 07"
6565	Sgr	PN	10"×8"	11.6	18 h 11 m 52.4 s	−28° 10' 42"
6567	Sgr	PN	11"×7"	11	18 h 13 m 45.1 s	−19° 04' 34"
6568	Sgr	OC	12'	8.6	18 h 12 m 45.1 s	−21° 34' 59"
6569	Sgr	GC	5.8'	8.4	18 h 13 m 38.6 s	−31° 49' 35"
6573	Sgr	Ast	...	...	18 h 13 m 49.9 s	−22° 08' 54"
6578	Sgr	PN	8.5"	12.9	18 h 16 m 16.4 s	−20° 27' 02"
6583	Sgr	OC	5'	10	18 h 15 m 49.9 s	−22° 08' 09"
6589	Sgr	Neb	5'×3'	...	18 h 16 m 55.3 s	−19° 46' 38"
6590	Sgr	OC+Neb	4'×3'	...	18 h 17 m 04.9 s	−19° 51' 58"
6595	Sgr	OC+Neb	4'×3'	...	18 h 17 m 04.9 s	−19° 51' 58"
6596	Sgr	OC	10'	...	18 h 17 m 33.7 s	−16° 39' 02"
6603	Sgr	OC	5.0'	11.1	18 h 18 m 26.9 s	−18° 24' 22"
6613	Sgr	OC	9'	6.9	18 h 19 m 58.4 s	−17° 06' 07"
6618	Sgr	OC+Neb	20'×15'	...	18 h 20 m 47.1 s	−16° 10' 18"
6620	Sgr	PN	5.3"×3.8"	12.7	18 h 22 m 54.1 s	−26° 49' 18"
6624	Sgr	GC	8.8'	7.6	18 h 23 m 40.6 s	−30° 21' 39"
6626	Sgr	GC	11'	6.9	18 h 24 m 32.8 s	−24° 52' 12"
6629	Sgr	PN	16"×14"	11.3	18 h 25 m 42.3 s	−23° 12' 10"
6634	Sgr	GC	7.1'	...	18 h 31 m 23.2 s	−32° 20' 53"
6637	Sgr	GC	7.1'	...	18 h 31 m 23.2 s	−32° 20' 53"
6638	Sgr	GC	7.0'	9.2	18 h 30 m 56.1 s	−25° 29' 47"
6642	Sgr	GC	4.5'	8.9	18 h 31 m 54.3 s	−23° 28' 35"
6644	Sgr	PN	2.5"	10.7	18 h 32 m 34.6 s	−25° 07' 44"
6645	Sgr	OC	15'	8.5	18 h 32 m 37.9 s	−16° 53' 02"
6647	Sgr	OC	...	...	18 h 32 m 49.3 s	−17° 13' 43"
6652	Sgr	GC	3.5'	8.5	18 h 35 m 45.7 s	−32° 59' 25"
6656	Sgr	GC	24'	5.2	18 h 36 m 24.1 s	−23° 54' 12"
6681	Sgr	GC	7.8'	7.8	18 h 43 m 12.5 s	−32° 17' 31"
6698	Sgr	MWSC	...	...	18 h 48 m 04.9 s	−25° 28' 38"
6715	Sgr	GC	9.1'	7.7	18 h 55 m 03.3 s	−30° 28' 42"
6716	Sgr	OC	6'	7.5	18 h 54 m 34.3 s	−19° 54' 04"
6717	Sgr	GC	3.9'	8.4	18 h 55 m 05.9 s	−22° 42' 06"
6723	Sgr	GC	11'	6.8	18 h 59 m 33.1 s	−36° 37' 54"

(continued)

Table 10.4 (continued)

NGC #	Const	Type	Size	V Mag	RA (2000)	Dec (2000)
6737	Sgr	Ast	…	…	19 h 02 m 17.4 s	−18° 32' 49"
6774	Sgr	OC	25.0'	…	19 h 16 m 16.3 s	−16° 15' 39"
6794	Sgr	Gxy	1.7'×1.5'	13.3	19 h 28 m 03.8 s	−38° 55' 08"
6797	Sgr	***	…	…	19 h 29 m 00.7 s	−25° 40' 00"
6805	Sgr	Gxy	1.1'×0.9'	13.4	19 h 36 m 45.7 s	−37° 33' 16"
6806	Sgr	Gxy	1.6'×1.0'	13.3	19 h 37 m 04.7 s	−42° 17' 47"
6809	Sgr	GC	19'	6.3	19 h 39 m 59.3 s	−30° 57' 44"
6816	Sgr	Gxy	1.6'×1.1'	12.8	19 h 44 m 02.3 s	−28° 24' 05"
6818	Sgr	PN	22"×15"	9.3	19 h 43 m 57.7 s	−14° 09' 11"
6822	Sgr	Gxy	15.5'×13.5'	9.9	19 h 44 m 56.3 s	−14° 48' 37"
6835	Sgr	Gxy	2.3'×0.5'	12.4	19 h 54 m 32.3 s	−12° 33' 50"
6836	Sgr	Gxy	1.5'×1.3'	12.9	19 h 54 m 40.3 s	−12° 41' 17"
6841	Sgr	Gxy	1.5'×1.4'	12.5	19 h 57 m 49.1 s	−31° 48' 40"
6849	Sgr	Gxy	1.9'×1.1'	12	20 h 06 m 15.6 s	−40° 11' 56"
6864	Sgr	GC	6.0'	8.6	20 h 06 m 04.7 s	−21° 55' 17"
6878	Sgr	Gxy	1.6'×1.2'	12.9	20 h 13 m 53.1 s	−44° 31' 33"
6890	Sgr	Gxy	1.5'×1.2'	12.5	20 h 18 m 18.0 s	−44° 48' 24"
6902	Sgr	Gxy	5.6'×3.9'	11.5	20 h 24 m 27.9 s	−43° 39' 12"
1312	Tau	**	…	…	03 h 23 m 41.4 s	+01° 11' 05"
1349	Tau	Gxy	0.7'×0.7'	13.9	03 h 31 m 27.5 s	+04° 22' 51"
1384	Tau	Gxy	0.8'×0.4'	14.8	03 h 39 m 13.5 s	+15° 49' 08"
1409	Tau	Gxy	1.0'×0.8'	13.7	03 h 41 m 10.4 s	−01° 18' 09"
1410	Tau	Gxy	1.2'×1.2'	14	03 h 41 m 10.7 s	−01° 17' 55"
1431	Tau	Gxy	1.0'×0.8'	14.1	03 h 44 m 40.8 s	+02° 50' 06"
1432	Tau	Neb	60'×40'	…	03 h 45 m 49.5 s	+24° 22' 06"
1435	Tau	Neb	30'×30'	…	03 h 46 m 10.1 s	+23° 45' 54"
1456	Tau	**	…	…	03 h 48 m 08.2 s	+22° 33' 31"
1462	Tau	Gxy	0.9'×0.5'	14.4	03 h 50 m 23.5 s	+06° 58' 21"
1488	Tau	**	…	…	04 h 00 m 04.3 s	+18° 34' 02"
1497	Tau	Gxy	1.8'×1.2'	13.2	04 h 02 m 06.8 s	+23° 07' 59"
1508	Tau	Gxy	0.4'×0.3'	14.6	04 h 05 m 47.6 s	+25° 24' 31"
1514	Tau	PN	2.3'×2.0'	10.9	04 h 09 m 16.9 s	+30° 46' 34"
1517	Tau	Gxy	1.1'×1.0'	13.6	04 h 09 m 11.9 s	+08° 38' 53"
1539	Tau	Gxy	0.5'×0.5'	14.6	04 h 19 m 02.1 s	+26° 49' 38"
1541	Tau	Gxy	1.3'×0.6'	13.9	04 h 17 m 00.3 s	+00° 50' 04"
1542	Tau	Gxy	1.4'×0.6'	14.1	04 h 17 m 14.1 s	+04° 46' 55"
1550	Tau	Gxy	2.2'×1.9'	12.5	04 h 19 m 37.9 s	+02° 24' 35"
1551	Tau	Gxy	1.7'×1.7'	12.1	04 h 19 m 37.9 s	+02° 24' 35"
1554	Tau	Neb	Var'	…	04 h 21 m 43.5 s	+19° 31' 15"
1555	Tau	Neb	0.5'	…	04 h 21 m 56.7 s	+19° 32' 04"
1587	Tau	Gxy	1.7'×1.5'	12.1	04 h 30 m 40.1 s	+00° 39' 40"
1588	Tau	Gxy	1.4'×0.8'	12.9	04 h 30 m 43.8 s	+00° 39' 53"
1589	Tau	Gxy	3.2'×1.0'	12.8	04 h 30 m 45.7 s	+00° 51' 48"
1590	Tau	Gxy	0.9'×0.7'	14.1	04 h 31 m 10.2 s	+07° 37' 52"
1593	Tau	Gxy	1.7'×0.6'	13.8	04 h 32 m 06.1 s	+00° 34' 02"

(continued)

Table 10.4 (continued)

NGC #	Const	Type	Size	V Mag	RA (2000)	Dec (2000)
1608	Tau	Gxy	1.6'×0.6'	13.5	04 h 32 m 06.1 s	+00° 34' 02"
1615	Tau	Gxy	1.2'×0.7'	13.7	04 h 36 m 01.9 s	+19° 57' 00"
1633	Tau	Gxy	1.0'×0.9'	13.7	04 h 40 m 09.0 s	+07° 20' 57"
1634	Tau	Gxy	0.4'×0.3'	14.1	04 h 40 m 09.9 s	+07° 20' 19"
1642	Tau	Gxy	1.8'×1.6'	12.8	04 h 42 m 55.1 s	+00° 37' 07"
1647	Tau	OC	45'	6.4	04 h 45 m 55.6 s	+19° 06' 42"
1655	Tau	NF	…	…	04 h 47 m 11.9 s	+20° 55' 25"
1674	Tau	NF	…	…	04 h 52 m 24.9 s	+23° 54' 28"
1675	Tau	NF	…	…	04 h 52 m 24.9 s	+23° 54' 28"
1746	Tau	OC	42'	6.1	05 h 03 m 50.2 s	+23° 46' 04"
1750	Tau	OC	…	…	05 h 04 m 00.0 s	+23° 38' 45"
1758	Tau	OC	39'	…	05 h 04 m 29.3 s	+23° 48' 49"
1802	Tau	OC	…	…	05 h 10 m 16.8 s	+24° 05' 30"
1807	Tau	OC	17'	7	05 h 10 m 41.1 s	+16° 31' 52"
1817	Tau	OC	16'	7.7	05 h 12 m 26.3 s	+16° 41' 03"
1896	Tau	Ast	7.0'	…	05 h 25 m 34.7 s	+29° 15' 37"
1952	Tau	Neb	6'×4'	…	05 h 34 m 31.9 s	+22° 00' 52"
1988	Tau	*	…	…	05 h 37 m 26.4 s	+21° 13' 06"
1996	Tau	Ast	…	…	05 h 38 m 18.9 s	+25° 49' 23"
2026	Tau	OC	10'×7'	…	05 h 43 m 10.1 s	+20° 08' 20"
2045	Tau	*	…	…	05 h 45 m 01.3 s	+12° 53' 18"
6584	Tel	GC	7.9'	7.9	18 h 18 m 37.7 s	−52° 12' 54"
6707	Tel	Gxy	2.1'×0.9'	12.7	18 h 55 m 21.9 s	−53° 49' 11"
6708	Tel	Gxy	1.1'×0.9'	12.7	18 h 55 m 35.5 s	−53° 43' 26"
6725	Tel	Gxy	2.2'×0.5'	12.5	19 h 01 m 56.3 s	−53° 51' 53"
6754	Tel	Gxy	1.9'×0.9'	12.4	19 h 11 m 24.9 s	−50° 38' 33"
6758	Tel	Gxy	2.2'×1.7'	11.6	19 h 13 m 52.4 s	−56° 18' 35"
6761	Tel	Gxy	1.6'×1.2'	13.2	19 h 15 m 04.7 s	−50° 39' 27"
6780	Tel	Gxy	1.9'×1.6'	12.8	19 h 22 m 50.7 s	−55° 46' 35"
6788	Tel	Gxy	2.9'×0.9'	12.2	19 h 26 m 49.6 s	−54° 57' 04"
6799	Tel	Gxy	1.6'×1.2'	12.5	19 h 32 m 16.7 s	−55° 54' 26"
6812	Tel	Gxy	1.6'×1.1'	12.7	19 h 45 m 24.2 s	−55° 20' 49"
6845	Tel	Gxy	1.1'×0.6'	15.3	20 h 00 m 58.5 s	−47° 04' 11"
6848	Tel	Gxy	2.5'×1.0'	12.7	20 h 02 m 46.9 s	−56° 05' 27"
6850	Tel	Gxy	2.1'×1.1'	12.6	20 h 03 m 29.7 s	−54° 50' 45"
6851	Tel	Gxy	2.0'×1.5'	11.9	20 h 03 m 34.3 s	−48° 17' 05"
6854	Tel	Gxy	2.0'×1.3'	12.2	20 h 05 m 38.7 s	−54° 22' 32"
6855	Tel	Gxy	1.5'×1.2'	12.9	20 h 06 m 49.9 s	−56° 23' 22"
6861	Tel	Gxy	2.8'×1.8'	11.2	20 h 07 m 19.3 s	−48° 22' 12"
6862	Tel	Gxy	1.6'×1.1'	12.9	20 h 08 m 54.4 s	−56° 23' 32"
6867	Tel	Gxy	2.0'×0.7'	13.3	20 h 10 m 29.9 s	−54° 47' 05"
6868	Tel	Gxy	3.5'×2.8'	10.8	20 h 09 m 53.9 s	−48° 22' 47"
6870	Tel	Gxy	2.6'×1.2'	12.5	20 h 10 m 10.7 s	−48° 17' 12"
6875	Tel	Gxy	2.3'×1.3'	11.7	20 h 13 m 12.3 s	−46° 09' 42"
6887	Tel	Gxy	3.4'×1.3'	12.2	20 h 17 m 16.9 s	−52° 47' 52"

(continued)

Table 10.4 (continued)

NGC #	Const	Type	Size	V Mag	RA (2000)	Dec (2000)
6889	Tel	Gxy	1.7'×1.1'	13.1	20 h 18 m 53.3 s	−53° 57' 26"
6893	Tel	Gxy	2.6'×1.7'	11.6	20 h 20 m 49.6 s	−48° 14' 24"
6899	Tel	Gxy	1.7'×1.0'	13	20 h 24 m 22.3 s	−50° 26' 03"
6909	Tel	Gxy	2.2'×1.1'	11.7	20 h 27 m 38.8 s	−47° 01' 39"
5844	TrA	PN	2.0'×1.2'	…	15 h 10 m 40.7 s	−64° 40' 24"
5938	TrA	Gxy	2.8'×2.5'	10.9	15 h 36 m 26.0 s	−66° 51' 35"
5979	TrA	PN	0.3'	11.5	15 h 47 m 41.1 s	−61° 13' 04"
6025	TrA	OC	12'	5.1	16 h 03 m 17.0 s	−60° 25' 54"
6156	TrA	Gxy	1.6'×1.4'	11.8	16 h 34 m 52.5 s	−60° 37' 11"
6183	TrA	Gxy	1.7'×0.5'	12.2	16 h 41 m 41.7 s	−69° 22' 23"
579	Tri	Gxy	1.1'×1.0'	13.1	01 h 31 m 46.7 s	+33° 36' 54"
582	Tri	Gxy	2.2'×0.6'	13.3	01 h 31 m 58.1 s	+33° 28' 36"
587	Tri	Gxy	2.2'×0.8'	13	01 h 32 m 33.3 s	+35° 21' 32"
588	Tri	GxyCld	…	…	01 h 32 m 45.5 s	+30° 38' 57"
592	Tri	GxyCld	…	…	01 h 33 m 12.2 s	+30° 38' 46"
595	Tri	GxyCld	…	…	01 h 33 m 34.1 s	+30° 41' 32"
598	Tri	Gxy	70.8'×41.7'	5.8	01 h 33 m 50.8 s	+30° 39' 37"
603	Tri	***	…	…	01 h 34 m 44.0 s	+30° 13' 58"
604	Tri	GxyCld	…	…	01 h 34 m 32.8 s	+30° 47' 03"
608	Tri	Gxy	0.8'×0.5'	13.4	01 h 35 m 28.3 s	+33° 39' 24"
614	Tri	Gxy	1.4'×1.4'	12.8	01 h 35 m 52.3 s	+33° 40' 54"
616	Tri	**	…	…	01 h 36 m 04.3 s	+33° 46' 12"
618	Tri	Gxy	1.6'×1.5'	12.7	01 h 35 m 28.3 s	+33° 39' 24"
621	Tri	Gxy	1.9'×1.8'	12.8	01 h 36 m 48.9 s	+35° 30' 45"
627	Tri	Gxy	1.6'×1.5'	12.7	01 h 35 m 52.3 s	+33° 40' 54"
634	Tri	Gxy	2.1'×0.6'	13.2	01 h 38 m 18.1 s	+35° 21' 55"
661	Tri	Gxy	1.7'×1.4'	12	01 h 44 m 14.5 s	+28° 42' 20"
666	Tri	Gxy	0.7'×0.5'	13.3	01 h 46 m 06.0 s	+34° 22' 27"
669	Tri	Gxy	3.2'×0.6'	12.3	01 h 47 m 16.1 s	+35° 33' 48"
670	Tri	Gxy	2.0'×1.0'	12	01 h 47 m 24.9 s	+27° 53' 08"
672	Tri	Gxy	7.2'×2.6'	10.8	01 h 47 m 54.3 s	+27° 25' 59"
684	Tri	Gxy	3.2'×0.6'	12.6	01 h 50 m 14.3 s	+27° 38' 52"
688	Tri	Gxy	2.5'×1.5'	12.5	01 h 50 m 44.4 s	+35° 17' 05"
733	Tri	*	…	…	01 h 56 m 33.8 s	+33° 03' 19"
735	Tri	Gxy	1.8'×0.9'	13.2	01 h 56 m 37.7 s	+34° 10' 38"
736	Tri	Gxy	1.5'×1.5'	12.5	01 h 56 m 40.9 s	+33° 02' 37"
737	Tri	Ast	…	…	01 h 56 m 40.8 s	+33° 03' 00"
738	Tri	Gxy	0.4'×0.2'	15	01 h 56 m 45.7 s	+33° 03' 30"
739	Tri	Gxy	0.5'×0.5'	14.1	01 h 56 m 54.7 s	+33° 16' 00"
740	Tri	Gxy	1.6'×0.4'	14.2	01 h 56 m 54.9 s	+33° 00' 56"
750	Tri	Gxy	1.7'×1.3'	12	01 h 57 m 32.5 s	+33° 12' 36"
751	Tri	Gxy	1.4'×1.4'	12.2	01 h 57 m 32.9 s	+33° 12' 16"
760	Tri	**	…	…	01 h 57 m 47.4 s	+33° 21' 20"
761	Tri	Gxy	1.5'×0.5'	13.7	01 h 57 m 49.6 s	+33° 22' 37"
769	Tri	Gxy	0.8'×0.5'	12.9	01 h 59 m 36.1 s	+30° 54' 36"

(continued)

Table 10.4 (continued)

NGC #	Const	Type	Size	V Mag	RA (2000)	Dec (2000)
777	Tri	Gxy	2.5'×2.0'	11.4	02 h 00 m 15.0 s	+31° 25' 47"
778	Tri	Gxy	1.1'×0.5'	13.3	02 h 00 m 19.4 s	+31° 18' 49"
780	Tri	Gxy	1.8'×1.0'	14	02 h 00 m 35.3 s	+28° 13' 32"
783	Tri	Gxy	1.6'×1.4'	12.3	02 h 01 m 06.5 s	+31° 52' 56"
784	Tri	Gxy	6.6'×1.5'	11.4	02 h 01 m 17.0 s	+28° 50' 15"
785	Tri	Gxy	1.5'×1.1'	12.8	02 h 01 m 40.0 s	+31° 49' 37"
789	Tri	Gxy	0.6'×0.4'	13.6	02 h 02 m 26.0 s	+32° 04' 20"
793	Tri	**	...	...	02 h 02 m 54.5 s	+31° 58' 51"
798	Tri	Gxy	1.2'×0.5'	13.6	02 h 03 m 19.7 s	+32° 04' 40"
804	Tri	Gxy	1.4'×0.3'	13.8	02 h 04 m 02.1 s	+30° 49' 58"
805	Tri	Gxy	1.1'×0.8'	13.6	02 h 04 m 29.5 s	+28° 48' 45"
807	Tri	Gxy	1.8'×1.3'	12.6	02 h 04 m 55.7 s	+28° 59' 18"
816	Tri	Gxy	0.4'×0.4'	14.2	02 h 08 m 08.9 s	+29° 15' 24"
819	Tri	Gxy	0.6'×0.4'	13.4	02 h 08 m 34.4 s	+29° 14' 03"
826	Tri	Gxy	0.5'×0.1'	14.6	02 h 09 m 25.2 s	+30° 44' 22"
832	Tri	**	...	...	02 h 11 m 00.8 s	+35° 32' 28"
843	Tri	***	...	...	02 h 11 m 08.1 s	+32° 05' 51"
855	Tri	Gxy	2.6'×1.0'	12	02 h 14 m 03.3 s	+27° 52' 35"
860	Tri	Gxy	0.5'×0.3'	14.2	02 h 15 m 00.2 s	+30° 46' 43"
861	Tri	Gxy	1.5'×0.5'	14	02 h 15 m 51.2 s	+35° 54' 48"
865	Tri	Gxy	1.5'×0.4'	13.2	02 h 16 m 15.1 s	+28° 36' 02"
890	Tri	Gxy	2.5'×1.7'	11.4	02 h 22 m 01.1 s	+33° 15' 47"
917	Tri	Gxy	1.6'×1.0'	13.7	02 h 26 m 07.7 s	+31° 54' 44"
925	Tri	Gxy	10.5'×5.9'	9.7	02 h 27 m 17.1 s	+33° 34' 42"
931	Tri	Gxy	3.9'×0.8'	13	02 h 28 m 14.5 s	+31° 18' 41"
940	Tri	Gxy	1.2'×1.0'	12.5	02 h 29 m 27.7 s	+31° 38' 27"
949	Tri	Gxy	2.4'×1.3'	11.7	02 h 30 m 48.8 s	+37° 08' 09"
952	Tri	NF	...	...	02 h 31 m 18.8 s	+34° 44' 52"
953	Tri	Gxy	...'×...'	13.6	02 h 31 m 09.5 s	+29° 35' 18"
959	Tri	Gxy	2.3'×1.4'	12	02 h 32 m 23.9 s	+35° 29' 42"
968	Tri	Gxy	3.6'×1.9'	12.3	02 h 34 m 06.3 s	+34° 28' 47"
969	Tri	Gxy	1.7'×1.6'	12.4	02 h 34 m 08.0 s	+32° 56' 51"
970	Tri	Gxy	0.6'×0.2'	14.9	02 h 34 m 11.8 s	+32° 58' 39"
971	Tri	*	...	...	02 h 34 m 16.0 s	+32° 58' 44"
973	Tri	Gxy	3.7'×0.5'	13	02 h 34 m 20.1 s	+32° 30' 21"
974	Tri	Gxy	2.5'×1.9'	12.9	02 h 34 m 25.9 s	+32° 57' 16"
978	Tri	Gxy	2.0'×1.7'	12.2	02 h 34 m 46.9 s	+32° 50' 46"
983	Tri	Gxy	1.6'×1.0'	13.2	02 h 38 m 55.7 s	+34° 37' 21"
987	Tri	Gxy	1.3'×1.1'	12.6	02 h 36 m 49.7 s	+33° 19' 39"
1002	Tri	Gxy	1.2'×0.9'	13.3	02 h 38 m 55.7 s	+34° 37' 21"
1057	Tri	Gxy	1.1'×0.8'	14.3	02 h 43 m 03.0 s	+32° 29' 27"
1060	Tri	Gxy	2.3'×1.7'	12.2	02 h 43 m 15.0 s	+32° 25' 30"
1061	Tri	Gxy	0.9'×0.6'	14.5	02 h 43 m 15.8 s	+32° 28' 00"
1062	Tri	*	...	...	02 h 43 m 24.8 s	+32° 27' 42"
1066	Tri	Gxy	1.7'×1.6'	13.4	02 h 43 m 50.0 s	+32° 28' 29"

(continued)

Table 10.4 (continued)

NGC #	Const	Type	Size	V Mag	RA (2000)	Dec (2000)
1067	Tri	Gxy	1.0' × 1.0'	13.9	02 h 43 m 50.5 s	+32° 30' 42"
1093	Tri	Gxy	1.8' × 1.1'	13.3	02 h 48 m 16.2 s	+34° 25' 12"
53	Tuc	Gxy	2.0' × 1.4'	13.1	00 h 14 m 41.8 s	−60° 19' 43"
104	Tuc	GC	50'	4	00 h 24 m 05.1 s	−72° 04' 51"
121	Tuc	GC	1.5'	...	00 h 26 m 47.1 s	−71° 32' 12"
152	Tuc	OC	3.0'	...	00 h 32 m 55.5 s	−73° 06' 59"
176	Tuc	OC	11.3'	12.7	00 h 35 m 58.6 s	−73° 09' 57"
220	Tuc	OC	11.3'	12.4	00 h 40 m 30.6 s	−73° 24' 11"
222	Tuc	OC	5.7'	12.2	00 h 40 m 44.5 s	−73° 23' 03"
231	Tuc	OC	9.1'	12.7	00 h 41 m 06.4 s	−73° 21' 08"
241	Tuc	OC	...	...	00 h 43 m 33.8 s	−73° 26' 47"
242	Tuc	OC	...	...	00 h 43 m 33.8 s	−73° 26' 47"
248	Tuc	Neb	...	...	00 h 45 m 24.2 s	−73° 22' 49"
249	Tuc	Neb	...	...	00 h 45 m 32.9 s	−73° 04' 39"
256	Tuc	OC + Neb	...	12.7	00 h 45 m 53.4 s	−73° 30' 28"
261	Tuc	Neb	...	...	00 h 46 m 27.9 s	−73° 06' 13"
265	Tuc	OC	...	12.2	00 h 47 m 11.3 s	−73° 28' 43"
267	Tuc	OC + Neb	...	...	00 h 48 m 02.9 s	−73° 16' 27"
269	Tuc	OC	...	13.3	00 h 48 m 21.1 s	−73° 31' 54"
290	Tuc	OC	...	12	00 h 51 m 14.8 s	−73° 09' 45"
292	Tuc	Gxy	316' × 186	2.2	00 h 52 m 38.0 s	−72° 48' 01"
294	Tuc	OC	...	12.7	00 h 53 m 05.8 s	−73° 22' 52"
299	Tuc	OC + Neb	...	...	00 h 53 m 24.8 s	−72° 11' 47"
306	Tuc	OC + Neb	...	...	00 h 54 m 14.8 s	−72° 14' 28"
330	Tuc	GC	1.9'	9.6	00 h 56 m 19.8 s	−72° 27' 44"
339	Tuc	GC	2.2'	...	00 h 57 m 45.4 s	−74° 28' 21
346	Tuc	OC + Neb	14' × 11'	10.3	00 h 59 m 05.1 s	−72° 10' 38"
360	Tuc	Gxy	3.5' × 0.5'	12.8	01 h 02 m 51.1 s	−65° 36' 36"
361	Tuc	OC	1.5'	...	01 h 02 m 11.3 s	−71° 36' 25"
362	Tuc	GC	12.9'	6.8	01 h 03 m 14.2 s	−70° 50' 54"
371	Tuc	OC + Neb	7.5'	...	01 h 03 m 29.5 s	−72° 03' 25"
376	Tuc	OC	...	10.9	01 h 03 m 53.6 s	−72° 49' 31"
395	Tuc	OC + Neb	...	...	01 h 05 m 08.1 s	−71° 59' 49"
406	Tuc	Gxy	3.3' × 1.3'	12.3	01 h 07 m 24.1 s	−69° 52' 48"
411	Tuc	OC	...	...	01 h 07 m 55.7 s	−71° 46' 09"
416	Tuc	GC	1.1'	12.6	01 h 07 m 58.4 s	−72° 21' 25"
419	Tuc	GC	2.6'	11.2	01 h 08 m 17.5 s	−72° 53' 01"
422	Tuc	OC	...	13.4	01 h 09 m 25.2 s	−71° 45' 58"
432	Tuc	Gxy	1.3' × 1.2'	13	01 h 11 m 46.1 s	−61° 31' 35"
434	Tuc	Gxy	2.1' × 1.2'	12.3	01 h 12 m 13.5 s	−58° 14' 48"
440	Tuc	Gxy	1.1' × 0.7'	13.1	01 h 12 m 48.2 s	−58° 16' 55"
456	Tuc	OC + Neb	15' × 15'	...	01 h 13 m 44.4 s	−73° 17' 26"
458	Tuc	OC	...	...	01 h 14 m 53.7 s	−71° 32' 58"
460	Tuc	OC + Neb	...	12.5	01 h 14 m 37.6 s	−73° 17' 18"
465	Tuc	OC	...	11.5	01 h 15 m 42.7 s	−73° 19' 27"

(continued)

Table 10.4 (continued)

NGC #	Const	Type	Size	V Mag	RA (2000)	Dec (2000)
466	Tuc	Gxy	1.8'×1.5'	12.8	01 h 17 m 13.3 s	−58° 54' 37"
484	Tuc	Gxy	1.9'×1.4'	12.2	01 h 19 m 34.8 s	−58° 31' 28"
7219	Tuc	Gxy	1.7'×1.0'	12.5	22 h 13 m 09.1 s	−64° 50' 55"
7278	Tuc	Gxy	0.8'×0.5'	14.6	22 h 28 m 22.5 s	−60° 10' 12"
7329	Tuc	Gxy	3.7'×2.7'	11.8	22 h 40 m 24.6 s	−66° 28' 43"
7358	Tuc	Gxy	1.9'×0.5'	12.7	22 h 45 m 36.3 s	−65° 07' 19"
7408	Tuc	Gxy	1.5'×1.2'	12.8	22 h 55 m 56.7 s	−63° 41' 39"
7417	Tuc	Gxy	1.9'×1.3'	12.5	22 h 57 m 49.1 s	−65° 02' 19"
7622	Tuc	Gxy	1.2'×0.4'	13.5	23 h 21 m 38.4 s	−62° 07' 05"
7650	Tuc	Gxy	1.4'×1.1'	12.9	23 h 25 m 21.6 s	−57° 47' 28"
7652	Tuc	Gxy	1.2'×0.7'	13.8	23 h 25 m 37.6 s	−57° 53' 17"
7657	Tuc	Gxy	1.4'×0.4'	14.2	23 h 26 m 47.1 s	−57° 48' 21"
7661	Tuc	Gxy	1.8'×1.2'	13.6	23 h 27 m 14.2 s	−65° 16' 18"
7676	Tuc	Gxy	1.7'×0.9'	12.5	23 h 29 m 01.6 s	−59° 43' 02"
7697	Tuc	Gxy	1.6'×0.3'	13.7	23 h 34 m 52.7 s	−65° 23' 44"
7733	Tuc	Gxy	1.3'×0.8'	13.9	23 h 42 m 33.0 s	−65° 57' 24"
7734	Tuc	Gxy	1.4'×1.1'	13.3	23 h 42 m 43.1 s	−65° 56' 41"
7823	Tuc	Gxy	1.2'×1.0'	12.8	00 h 04 m 45.8 s	−62° 03' 42"
2600	UMa	Gxy	1.3'×0.4'	14.4	08 h 34 m 45.1 s	+52° 42' 56"
2602	UMa	Gxy	0.3'×0.2'	15.2	08 h 34 m 31.3 s	+52° 50' 24"
2603	UMa	Gxy	0.6'×0.3'	15.2	08 h 34 m 53.5 s	+52° 48' 14"
2605	UMa	Gxy	0.3'×0.2'	14.9	08 h 35 m 04.3 s	+52° 49' 54"
2606	UMa	Gxy	0.7'×0.3'	14.3	08 h 35 m 34.3 s	+52° 47' 20"
2614	UMa	Gxy	2.5'×2.0'	13.1	08 h 42 m 48.0 s	+72° 58' 35"
2629	UMa	Gxy	2.3'×1.8'	11.9	08 h 47 m 15.8 s	+72° 59' 07"
2630	UMa	NF	…	…	08 h 47 m 07.3 s	+73° 00' 01"
2631	UMa	NF	…	…	08 h 47 m 07.3 s	+73° 00' 01"
2639	UMa	Gxy	1.8'×1.1'	11.9	08 h 43 m 37.9 s	+50° 12' 20"
2641	UMa	Gxy	1.3'×1.0'	13.7	08 h 47 m 57.3 s	+72° 53' 44"
2650	UMa	Gxy	1.8'×1.3'	13.5	08 h 49 m 58.7 s	+70° 17' 57"
2654	UMa	Gxy	4.3'×0.8'	12	08 h 49 m 11.9 s	+60° 13' 15"
2656	UMa	Gxy	1.0'×1.0'	13.9	08 h 47 m 53.1 s	+53° 52' 35"
2666	UMa	OC	…	…	08 h 49 m 47.3 s	+44° 42' 16"
2675	UMa	Gxy	1.5'×1.1'	13.1	08 h 52 m 04.7 s	+53° 37' 02"
2676	UMa	Gxy	1.2'×1.1'	13.2	08 h 51 m 35.6 s	+47° 33' 27"
2681	UMa	Gxy	3.6'×3.3'	10.1	08 h 53 m 32.8 s	+51° 18' 49"
2684	UMa	Gxy	0.9'×0.8'	13.1	08 h 54 m 54.1 s	+49° 09' 38"
2685	UMa	Gxy	4.5'×2.3'	10.9	08 h 55 m 34.6 s	+58° 44' 02"
2686	UMa	Gxy	0.5'×0.3'	15.2	08 h 54 m 58.9 s	+49° 08' 33"
2687	UMa	Gxy	0.3'×0.2'	17.1	08 h 55 m 05.9 s	+49° 09' 22"
2688	UMa	Gxy	0.3'×0.2'	15.8	08 h 55 m 11.7 s	+49° 07' 21"
2689	UMa	Gxy	0.3'×0.2'	16.3	08 h 55 m 25.4 s	+49° 06' 55"
2692	UMa	Gxy	1.3'×0.5'	13.5	08 h 56 m 58.0 s	+52° 03' 57"
2693	UMa	Gxy	2.6'×1.8'	11.7	08 h 56 m 59.3 s	+51° 20' 49"
2694	UMa	Gxy	1.2'×1.2'	13.8	08 h 56 m 59.3 s	+51° 19' 55"

(continued)

Table 10.4 (continued)

NGC #	Const	Type	Size	V Mag	RA (2000)	Dec (2000)
2701	UMa	Gxy	2.2'×1.6'	12	08 h 59 m 05.4 s	+53° 46' 14"
2710	UMa	Gxy	2.0'×1.0'	13.1	08 h 59 m 48.4 s	+55° 42' 23"
2726	UMa	Gxy	1.6'×0.5'	12.7	09 h 04 m 56.8 s	+59° 55' 57"
2739	UMa	Gxy	0.8'×0.2'	14.7	09 h 06 m 03.0 s	+51° 44' 40"
2740	UMa	Gxy	1.0'×0.9'	14.2	09 h 06 m 05.0 s	+51° 44' 07"
2742	UMa	Gxy	3.0'×1.5'	11.7	09 h 07 m 33.3 s	+60° 28' 45"
2756	UMa	Gxy	1.7'×1.1'	12.6	09 h 09 m 00.9 s	+53° 50' 57"
2762	UMa	Gxy	0.3'×0.2'	15.7	09 h 09 m 54.5 s	+50° 25' 06"
2767	UMa	Gxy	0.6'×0.5'	13.8	09 h 10 m 11.9 s	+50° 24' 05"
2768	UMa	Gxy	8.1'×4.3'	10.1	09 h 11 m 37.5 s	+60° 02' 14"
2769	UMa	Gxy	1.7'×0.4'	13.2	09 h 10 m 32.1 s	+50° 26' 00"
2771	UMa	Gxy	2.2'×1.9'	12.9	09 h 10 m 39.7 s	+50° 22' 47"
2787	UMa	Gxy	3.2'×2.0'	10.9	09 h 19 m 18.5 s	+69° 12' 11"
2800	UMa	Gxy	1.4'×0.9'	12.9	09 h 18 m 35.2 s	+52° 30' 52"
2805	UMa	Gxy	6.3'×4.8'	11	09 h 20 m 20.4 s	+64° 06' 09"
2810	UMa	Gxy	1.7'×1.7'	12.3	09 h 22 m 04.5 s	+71° 50' 38"
2814	UMa	Gxy	1.4'×0.4'	13.7	09 h 21 m 11.5 s	+64° 15' 06"
2816	UMa	Gxy	3.1'×1.7'	11.4	09 h 07 m 33.3 s	+60° 28' 45"
2820	UMa	Gxy	4.3'×0.7'	12.7	09 h 21 m 49.9 s	+64° 15' 41"
2841	UMa	Gxy	8.1'×3.5'	9.5	09 h 22 m 02.3 s	+50° 58' 44"
2854	UMa	Gxy	1.7'×0.6'	13.2	09 h 24 m 03.1 s	+49° 12' 15"
2856	UMa	Gxy	1.1'×0.5'	13.2	09 h 24 m 16.0 s	+49° 14' 56"
2857	UMa	Gxy	2.2'×2.0'	13.3	09 h 24 m 37.8 s	+49° 21' 25"
2870	UMa	Gxy	2.5'×0.6'	13.2	09 h 27 m 53.5 s	+57° 22' 33"
2880	UMa	Gxy	2.0'×1.2'	11.7	09 h 29 m 34.5 s	+62° 29' 26"
2892	UMa	Gxy	1.4'×1.4'	13.2	09 h 32 m 53.0 s	+67° 37' 02"
2895	UMa	Gxy	0.8'×0.7'	14	09 h 32 m 25.1 s	+57° 28' 57"
2909	UMa	**	...	...	09 h 36 m 59.9 s	+65° 56' 26"
2950	UMa	Gxy	3.2'×2.1'	11	09 h 42 m 35.1 s	+58° 51' 04"
2959	UMa	Gxy	1.3'×1.3'	13	09 h 45 m 09.2 s	+68° 35' 40"
2961	UMa	Gxy	1.1'×0.3'	14.9	09 h 45 m 22.3 s	+68° 36' 30"
2976	UMa	Gxy	5.9'×2.7'	10.6	09 h 47 m 15.3 s	+67° 54' 59"
2985	UMa	Gxy	4.6'×3.6'	10.5	09 h 50 m 21.8 s	+72° 16' 47"
2998	UMa	Gxy	2.9'×1.3'	12.4	09 h 48 m 43.6 s	+44° 04' 51"
3000	UMa	**	...	...	09 h 48 m 51.3 s	+44° 07' 48"
3002	UMa	*	...	...	09 h 48 m 57.3 s	+44° 03' 26"
3004	UMa	*	...	...	09 h 49 m 02.2 s	+44° 06' 40"
3005	UMa	Gxy	1.1'×0.2'	14.4	09 h 49 m 14.9 s	+44° 07' 49"
3006	UMa	Gxy	0.7'×0.2'	14.9	09 h 49 m 17.3 s	+44° 01' 32"
3008	UMa	Gxy	0.5'×0.3'	14.6	09 h 49 m 34.3 s	+44° 06' 08"
3009	UMa	Gxy	0.8'×0.7'	12.7	09 h 50 m 11.1 s	+44° 17' 41"
3010	UMa	Gxy	1.9'×0.6'	14.6	09 h 50 m 33.9 s	+44° 19' 07"
3027	UMa	Gxy	4.3'×2.0'	11.7	09 h 55 m 40.4 s	+72° 12' 13"
3031	UMa	Gxy	26.9'×14.1'	7.3	09 h 55 m 32.9 s	+69° 03' 55"
3034	UMa	Gxy	11'×4.6'	8.9	09 h 55 m 50.7 s	+69° 40' 43"

(continued)

Table 10.4 (continued)

NGC #	Const	Type	Size	V Mag	RA (2000)	Dec (2000)
3043	UMa	Gxy	1.7'×0.6'	12.8	09 h 56 m 14.6 s	+59° 18' 24"
3063	UMa	**	...	...	10 h 01 m 41.8 s	+72° 07' 04"
3065	UMa	Gxy	2.0'×1.9'	11.9	10 h 01 m 55.3 s	+72° 10' 13"
3066	UMa	Gxy	1.1'×1.0'	12.5	10 h 02 m 11.0 s	+72° 07' 31"
3073	UMa	Gxy	1.3'×1.2'	12.8	10 h 00 m 52.1 s	+55° 37' 06"
3077	UMa	Gxy	5.4'×4.5'	10.5	10 h 03 m 20.4 s	+68° 44' 02"
3079	UMa	Gxy	7.9'×1.4'	10.8	10 h 01 m 57.3 s	+55° 40' 54"
3097	UMa	NF	...	...	10 h 04 m 16.1 s	+60° 07' 33"
3102	UMa	Gxy	0.9'×0.9'	13.4	10 h 04 m 31.7 s	+60° 06' 29"
3111	UMa	Gxy	0.9'×0.8'	13.1	10 h 06 m 07.5 s	+47° 15' 45"
3135	UMa	Gxy	0.9'×0.6'	13.6	10 h 10 m 54.5 s	+45° 57' 00"
3148	UMa	*	...	...	10 h 13 m 43.9 s	+50° 29' 47"
3164	UMa	Gxy	0.9'×0.7'	13.6	10 h 15 m 11.4 s	+56° 40' 20"
3168	UMa	Gxy	1.0'×0.9'	13.5	10 h 16 m 23.0 s	+60° 14' 06"
3170	UMa	**	...	...	10 h 16 m 14.5 s	+46° 36' 45"
3179	UMa	Gxy	1.9'×0.5'	13.2	10 h 17 m 57.2 s	+41° 06' 50"
3180	UMa	HIIRgn	...	...	10 h 18 m 09.3 s	+41° 26' 41"
3181	UMa	HIIRgn	...	...	10 h 18 m 11.5 s	+41° 24' 46"
3182	UMa	Gxy	1.8'×1.5'	12.3	10 h 19 m 33.0 s	+58° 12' 21"
3184	UMa	Gxy	7.4'×6.9'	9.9	10 h 18 m 17.0 s	+41° 25' 27"
3188	UMa	Gxy	0.9'×0.9'	13.9	10 h 19 m 42.8 s	+57° 25' 23"
3191	UMa	Gxy	0.8'×0.6'	13.5	10 h 19 m 05.1 s	+46° 27' 15"
3192	UMa	Gxy	0.7'×0.5'	13.2	10 h 19 m 05.1 s	+46° 27' 15"
3198	UMa	Gxy	8.5'×3.3'	10.2	10 h 19 m 54.9 s	+45° 33' 00"
3202	UMa	Gxy	1.2'×0.8'	13.4	10 h 20 m 31.7 s	+43° 01' 16"
3205	UMa	Gxy	1.4'×1.1'	13.2	10 h 20 m 49.9 s	+42° 58' 17"
3206	UMa	Gxy	3.0'×1.9'	12.1	10 h 21 m 47.8 s	+56° 55' 48"
3207	UMa	Gxy	1.4'×1.0'	13.7	10 h 21 m 00.5 s	+42° 59' 07"
3214	UMa	Gxy	1.5'×0.8'	14.1	10 h 23 m 08.9 s	+57° 02' 20"
3220	UMa	Gxy	1.7'×0.6'	13.2	10 h 23 m 45.1 s	+57° 01' 37"
3225	UMa	Gxy	2.0'×1.0'	12.8	10 h 25 m 09.9 s	+58° 08' 59"
3231	UMa	OC	4.0'	...	10 h 26 m 57.9 s	+66° 48' 54"
3236	UMa	Gxy	0.6'×0.4'	14.4	10 h 26 m 48.4 s	+61° 16' 22"
3237	UMa	Gxy	1.3'×1.3'	13.1	10 h 25 m 43.3 s	+39° 38' 46"
3238	UMa	Gxy	1.4'×1.3'	13	10 h 26 m 42.9 s	+57° 13' 34"
3259	UMa	Gxy	2.2'×1.2'	12.3	10 h 32 m 34.8 s	+65° 02' 28"
3264	UMa	Gxy	2.9'×1.2'	13.4	10 h 32 m 19.7 s	+56° 05' 06"
3266	UMa	Gxy	1.5'×1.3'	12.5	10 h 33 m 17.5 s	+64° 44' 57"
3284	UMa	Gxy	1.4'×1.1'	13.6	10 h 36 m 21.2 s	+58° 37' 12"
3286	UMa	Gxy	1.4'×1.1'	13.6	10 h 36 m 21.2 s	+58° 37' 12"
3288	UMa	Gxy	1.1'×0.9'	14.2	10 h 36 m 25.8 s	+58° 33' 22"
3298	UMa	Gxy	0.3'×0.2'	14	10 h 37 m 12.3 s	+50° 07' 13"
3310	UMa	Gxy	3.1'×2.4'	10.4	10 h 38 m 45.9 s	+53° 30' 10"
3319	UMa	Gxy	6.2'×3.4'	11.1	10 h 39 m 09.4 s	+41° 41' 14"
3320	UMa	Gxy	2.2'×1.0'	12.5	10 h 39 m 36.5 s	+47° 23' 52"

(continued)

Table 10.4 (continued)

NGC #	Const	Type	Size	V Mag	RA (2000)	Dec (2000)
3348	UMa	Gxy	2.0'×2.0'	11.1	10 h 47 m 10.1 s	+72° 50' 23"
3353	UMa	Gxy	1.3'×1.0'	12.5	10 h 45 m 22.3 s	+55° 57' 37"
3359	UMa	Gxy	6.8'×4.3'	10.5	10 h 46 m 36.7 s	+63° 13' 26"
3364	UMa	Gxy	1.9'×1.8'	13.2	10 h 48 m 29.8 s	+72° 25' 30"
3374	UMa	Gxy	1.3'×0.9'	13.9	10 h 48 m 01.2 s	+43° 11' 11"
3392	UMa	Gxy	0.5'×0.3'	14	10 h 51 m 03.0 s	+65° 46' 53"
3394	UMa	Gxy	1.9'×1.4'	12.6	10 h 50 m 39.9 s	+65° 43' 37"
3398	UMa	Gxy	1.5'×1.2'	13.6	10 h 51 m 31.5 s	+55° 23' 24"
3406	UMa	Gxy	1.1'×1.0'	12.7	10 h 51 m 44.0 s	+51° 01' 23"
3407	UMa	Gxy	1.4'×0.7'	13.7	10 h 52 m 17.9 s	+61° 22' 45"
3408	UMa	Gxy	1.0'×0.9'	13.7	10 h 52 m 11.7 s	+58° 26' 17"
3410	UMa	Gxy	0.4'×0.3'	14.2	10 h 51 m 53.5 s	+51° 00' 24"
3415	UMa	Gxy	2.1'×1.3'	12.2	10 h 51 m 42.7 s	+43° 42' 44"
3416	UMa	Gxy	0.6'×0.2'	14.9	10 h 51 m 48.1 s	+43° 45' 49"
3435	UMa	Gxy	1.9'×1.2'	13.4	10 h 54 m 48.3 s	+61° 17' 22"
3440	UMa	Gxy	2.1'×0.5'	13.4	10 h 53 m 49.6 s	+57° 07' 07"
3445	UMa	Gxy	1.6'×1.5'	12.5	10 h 54 m 35.7 s	+56° 59' 25"
3448	UMa	Gxy	5.6'×1.8'	11.7	10 h 54 m 39.0 s	+54° 18' 18"
3458	UMa	Gxy	1.4'×0.9'	12.3	10 h 56 m 01.5 s	+57° 07' 01"
3468	UMa	Gxy	1.6'×1.0'	13.1	10 h 57 m 31.1 s	+40° 56' 45"
3470	UMa	Gxy	1.4'×1.2'	13.4	10 h 58 m 44.9 s	+59° 30' 38"
3471	UMa	Gxy	1.7'×0.8'	12.5	10 h 59 m 09.1 s	+61° 31' 51"
3478	UMa	Gxy	2.6'×1.2'	12.6	10 h 59 m 27.3 s	+46° 07' 20"
3488	UMa	Gxy	1.9'×1.3'	13.1	11 h 01 m 23.5 s	+57° 40' 38"
3499	UMa	Gxy	0.8'×0.7'	14	11 h 03 m 11.0 s	+56° 13' 17"
3516	UMa	Gxy	1.7'×1.3'	11.4	11 h 06 m 47.5 s	+72° 34' 06"
3517	UMa	Gxy	1.1'×0.9'	13	11 h 05 m 36.9 s	+56° 31' 28"
3527	UMa	Gxy	1.0'×0.9'	13.9	11 h 07 m 18.2 s	+28° 31' 39"
3530	UMa	Gxy	0.6'×0.3'	13.8	11 h 08 m 40.3 s	+57° 13' 48"
3536	UMa	Gxy	1.1'×0.8'	14.1	11 h 08 m 51.3 s	+28° 28' 31"
3539	UMa	Gxy	1.0'×0.2'	14.6	11 h 09 m 08.9 s	+28° 40' 20"
3540	UMa	Gxy	1.4'×1.2'	13.4	11 h 09 m 16.0 s	+36° 01' 16"
3542	UMa	Gxy	0.8'×0.3'	14.3	11 h 09 m 55.5 s	+36° 56' 46"
3543	UMa	Gxy	1.4'×0.3'	14.2	11 h 10 m 56.5 s	+61° 20' 48"
3545	UMa	Gxy	0.2'×0.2'	13.9	11 h 10 m 13.3 s	+36° 57' 59"
3548	UMa	Gxy	1.4'×1.3'	13.8	11 h 09 m 16.0 s	+36° 01' 16"
3549	UMa	Gxy	3.2'×1.2'	12.2	11 h 10 m 56.7 s	+53° 23' 16"
3550	UMa	Gxy	1.0'×1.0'	13.7	11 h 10 m 38.6 s	+28° 46' 03"
3552	UMa	Gxy	0.6'×0.5'	14.5	11 h 10 m 42.9 s	+28° 41' 34"
3553	UMa	Gxy	0.5'×0.4'	16	11 h 10 m 40.4 s	+28° 41' 06"
3554	UMa	Gxy	0.4'×0.4'	14.2	11 h 10 m 47.8 s	+28° 39' 35"
3556	UMa	Gxy	8.7'×2.2'	10.2	11 h 11 m 31.2 s	+55° 40' 24"
3558	UMa	Gxy	0.9'×0.8'	13.8	11 h 10 m 55.9 s	+28° 32' 37"
3561	UMa	Gxy	5.3'×1.0'	13.8	11 h 11 m 13.2 s	+28° 41' 46"
3569	UMa	Gxy	1.1'×1.0'	13.4	11 h 12 m 08.1 s	+35° 27' 08"

(continued)

Table 10.4 (continued)

NGC #	Const	Type	Size	V Mag	RA (2000)	Dec (2000)
3577	UMa	Gxy	1.4'×1.4'	13.6	11 h 13 m 44.9 s	+48° 16' 21"
3583	UMa	Gxy	2.8'×1.8'	11.3	11 h 14 m 10.7 s	+48° 19' 06"
3587	UMa	PN	3.4'×3.3'	9.9	11 h 14 m 47.7 s	+55° 01' 07"
3589	UMa	Gxy	1.5'×0.8'	13.9	11 h 15 m 13.6 s	+60° 42' 02"
3594	UMa	Gxy	1.3'×1.1'	13.8	11 h 16 m 14.0 s	+55° 42' 14"
3595	UMa	Gxy	1.6'×0.7'	12.1	11 h 15 m 25.5 s	+47° 26' 49"
3600	UMa	Gxy	4.1'×0.9'	11.9	11 h 15 m 52.0 s	+41° 35' 27"
3610	UMa	Gxy	2.7'×2.3'	10.5	11 h 18 m 25.3 s	+58° 47' 10"
3613	UMa	Gxy	3.9'×1.9'	10.7	11 h 18 m 36.1 s	+57° 59' 59"
3614	UMa	Gxy	4.6'×2.6'	11.8	11 h 18 m 21.3 s	+45° 44' 52"
3619	UMa	Gxy	2.7'×2.3'	11.5	11 h 19 m 21.6 s	+57° 45' 29"
3622	UMa	Gxy	1.2'×0.5'	12.8	11 h 20 m 12.5 s	+67° 14' 29"
3625	UMa	Gxy	2.0'×0.6'	13.3	11 h 20 m 31.3 s	+57° 46' 54"
3631	UMa	Gxy	5.0'×4.8'	10.6	11 h 21 m 02.9 s	+53° 10' 09"
3642	UMa	Gxy	5.4'×4.5'	10.9	11 h 22 m 18.0 s	+59° 04' 27"
3648	UMa	Gxy	1.3'×0.8'	12.7	11 h 22 m 31.5 s	+39° 52' 36"
3652	UMa	Gxy	2.0'×0.7'	12.4	11 h 22 m 39.0 s	+37° 45' 53"
3654	UMa	Gxy	1.2'×0.6'	12.7	11 h 24 m 10.9 s	+69° 24' 47"
3656	UMa	Gxy	1.6'×1.6'	12.9	11 h 23 m 38.5 s	+53° 50' 31"
3657	UMa	Gxy	1.0'×0.8'	12.6	11 h 23 m 55.6 s	+52° 55' 15"
3658	UMa	Gxy	1.6'×1.5'	12.3	11 h 23 m 58.2 s	+38° 33' 44"
3665	UMa	Gxy	2.5'×2.0'	10.8	11 h 24 m 43.7 s	+38° 45' 47"
3668	UMa	Gxy	2.1'×1.3'	12.5	11 h 25 m 30.3 s	+63° 26' 46"
3669	UMa	Gxy	2.2'×0.5'	12.6	11 h 25 m 26.6 s	+57° 43' 16"
3671	UMa	Gxy	0.5'×0.4'	14.8	11 h 25 m 52.6 s	+60° 28' 46"
3674	UMa	Gxy	1.9'×0.6'	12.3	11 h 26 m 26.5 s	+57° 02' 54"
3675	UMa	Gxy	5.9'×3.1'	10.2	11 h 26 m 07.9 s	+43° 35' 10"
3677	UMa	Gxy	1.9'×1.7'	12.5	11 h 26 m 17.7 s	+46° 58' 26"
3683	UMa	Gxy	1.9'×0.7'	12.7	11 h 27 m 31.9 s	+56° 52' 37"
3687	UMa	Gxy	2.0'×2.0'	12.2	11 h 28 m 00.5 s	+29° 30' 39"
3690	UMa	Gxy	2.4'×1.9'	11.6	11 h 28 m 31.9 s	+58° 33' 45"
3694	UMa	Gxy	0.7'×0.6'	12.7	11 h 28 m 54.1 s	+35° 24' 50"
3695	UMa	Gxy	1.1'×0.7'	14	11 h 29 m 17.3 s	+35° 34' 32"
3698	UMa	Gxy	1.1'×0.7'	14	11 h 29 m 17.3 s	+35° 34' 32"
3700	UMa	Gxy	1.0'×0.7'	14.2	11 h 29 m 38.5 s	+35° 30' 52"
3712	UMa	Gxy	1.7'×0.6'	14.6	11 h 31 m 09.3 s	+28° 33' 59"
3714	UMa	Gxy	0.5'×0.4'	13.3	11 h 31 m 53.6 s	+28° 21' 29"
3718	UMa	Gxy	8.1'×4.0'	10.6	11 h 32 m 35.0 s	+53° 04' 06"
3725	UMa	Gxy	1.2'×0.9'	13.2	11 h 33 m 40.9 s	+61° 53' 15"
3726	UMa	Gxy	6.2'×4.3'	10.5	11 h 33 m 21.3 s	+47° 01' 44"
3729	UMa	Gxy	2.8'×1.9'	11.6	11 h 33 m 49.3 s	+53° 07' 32"
3733	UMa	Gxy	4.8'×2.2'	12.3	11 h 35 m 01.8 s	+54° 51' 02"
3737	UMa	Gxy	1.0'×1.0'	13	11 h 35 m 36.3 s	+54° 56' 55"
3738	UMa	Gxy	2.5'×1.9'	11.4	11 h 35 m 48.5 s	+54° 31' 27"
3740	UMa	Gxy	1.0'×0.4'	14.1	11 h 36 m 12.3 s	+59° 58' 34"

(continued)

Table 10.4 (continued)

NGC #	Const	Type	Size	V Mag	RA (2000)	Dec (2000)
3741	UMa	Gxy	2.0'×1.1'	13.5	11 h 36 m 06.1 s	+45° 17' 04"
3755	UMa	Gxy	3.2'×1.4'	13	11 h 36 m 33.3 s	+36° 24' 38"
3756	UMa	Gxy	4.2'×2.1'	11.4	11 h 36 m 48.1 s	+54° 17' 39"
3757	UMa	Gxy	1.1'×1.1'	12.7	11 h 37 m 02.9 s	+58° 24' 55"
3759	UMa	Gxy	1.1'×1.1'	13.2	11 h 36 m 54.1 s	+54° 49' 23"
3762	UMa	Gxy	1.9'×0.5'	12.8	11 h 37 m 23.9 s	+61° 45' 32"
3769	UMa	Gxy	3.1'×1.0'	11.7	11 h 37 m 44.1 s	+47° 53' 34"
3770	UMa	Gxy	1.0'×0.7'	13.1	11 h 37 m 58.8 s	+59° 37' 00"
3780	UMa	Gxy	3.1'×2.6'	11.7	11 h 39 m 22.3 s	+56° 16' 14"
3782	UMa	Gxy	1.7'×1.1'	12.6	11 h 39 m 20.8 s	+46° 30' 48"
3786	UMa	Gxy	2.2'×1.3'	12.5	11 h 39 m 42.3 s	+31° 54' 32"
3788	UMa	Gxy	2.1'×0.7'	12.8	11 h 39 m 44.7 s	+31° 55' 50"
3793	UMa	*	...	...	11 h 40 m 02.0 s	+31° 52' 39"
3794	UMa	Gxy	2.4'×1.7'	12.9	11 h 40 m 54.1 s	+56° 12' 05"
3795	UMa	Gxy	2.1'×0.5'	13.2	11 h 40 m 06.7 s	+58° 36' 46"
3796	UMa	Gxy	1.3'×0.9'	12.6	11 h 40 m 31.0 s	+60° 17' 55"
3797	UMa	*	...	...	11 h 40 m 13.3 s	+31° 54' 23"
3804	UMa	Gxy	2.2'×1.4'	13	11 h 40 m 54.1 s	+56° 12' 05"
3809	UMa	Gxy	1.0'×0.8'	12.8	11 h 41 m 16.1 s	+59° 53' 09"
3811	UMa	Gxy	2.2'×1.7'	12.4	11 h 41 m 16.8 s	+47° 41' 28"
3813	UMa	Gxy	2.2'×1.1'	12.1	11 h 41 m 18.7 s	+36° 32' 47"
3824	UMa	Gxy	1.3'×0.7'	13.8	11 h 42 m 44.9 s	+52° 46' 46"
3829	UMa	Gxy	1.1'×0.7'	14.2	11 h 43 m 27.3 s	+52° 42' 39"
3835	UMa	Gxy	1.9'×0.8'	12.6	11 h 44 m 04.8 s	+60° 07' 11"
3838	UMa	Gxy	1.5'×0.6'	12.3	11 h 44 m 13.7 s	+57° 56' 54"
3846	UMa	Gxy	1.1'×0.8'	14.1	11 h 44 m 29.1 s	+55° 39' 07"
3847	UMa	Gxy	1.1'×1.1'	13.4	11 h 44 m 14.0 s	+33° 30' 52"
3850	UMa	Gxy	2.2'×1.0'	13.5	11 h 45 m 35.5 s	+55° 53' 10"
3855	UMa	Gxy	1.2'×0.9'	14	11 h 44 m 25.8 s	+33° 21' 17"
3856	UMa	Gxy	1.1'×1.1'	13.3	11 h 44 m 14.0 s	+33° 30' 52"
3870	UMa	Gxy	1.0'×0.9'	12.5	11 h 45 m 56.5 s	+50° 11' 59"
3871	UMa	Gxy	1.0'×0.2'	14.7	11 h 46 m 10.2 s	+33° 06' 31"
3877	UMa	Gxy	5.5'×1.3'	11.3	11 h 46 m 07.7 s	+47° 29' 41"
3878	UMa	Gxy	0.4'×0.4'	13.3	11 h 46 m 17.8 s	+33° 12' 15"
3880	UMa	Gxy	0.7'×0.7'	13.9	11 h 46 m 22.3 s	+33° 09' 41"
3881	UMa	Gxy	0.5'×0.5'	14.5	11 h 46 m 34.5 s	+33° 06' 22"
3888	UMa	Gxy	1.7'×1.3'	12.3	11 h 47 m 34.4 s	+55° 58' 01"
3889	UMa	Gxy	0.7'×0.4'	14.8	11 h 47 m 48.8 s	+56° 01' 01"
3891	UMa	Gxy	2.0'×1.7'	12.6	11 h 48 m 03.4 s	+30° 21' 33"
3893	UMa	Gxy	4.5'×2.8'	10.4	11 h 48 m 38.1 s	+48° 42' 38"
3894	UMa	Gxy	2.0'×1.4'	11.8	11 h 48 m 50.3 s	+59° 24' 55"
3895	UMa	Gxy	1.3'×1.0'	13.3	11 h 49 m 04.0 s	+59° 25' 56"
3896	UMa	Gxy	1.4'×1.0'	13.1	11 h 48 m 56.4 s	+48° 40' 28"
3897	UMa	Gxy	1.9'×1.9'	13.2	11 h 48 m 59.5 s	+35° 00' 57"
3898	UMa	Gxy	4.4'×2.6'	11	11 h 49 m 15.1 s	+56° 05' 04"

(continued)

Table 10.4 (continued)

NGC #	Const	Type	Size	V Mag	RA (2000)	Dec (2000)
3906	UMa	Gxy	1.9' × 1.7'	13.3	11 h 49 m 40.2 s	+48° 25' 35"
3913	UMa	Gxy	2.6' × 2.6'	13.1	11 h 50 m 38.9 s	+55° 21' 12"
3916	UMa	Gxy	1.6' × 0.4'	14.1	11 h 50 m 51.0 s	+55° 08' 35"
3917	UMa	Gxy	5.1' × 1.3'	11.9	11 h 50 m 45.3 s	+51° 49' 28"
3921	UMa	Gxy	2.1' × 1.3'	12.5	11 h 51 m 06.8 s	+55° 04' 42"
3922	UMa	Gxy	1.7' × 0.8'	13	11 h 51 m 13.3 s	+50° 09' 25"
3924	UMa	Gxy	1.9' × 0.9'	12.9	11 h 51 m 13.3 s	+50° 09' 25"
3928	UMa	Gxy	1.5' × 1.5'	12.4	11 h 51 m 47.5 s	+48° 40' 59"
3930	UMa	Gxy	3.2' × 2.4'	12.6	11 h 51 m 45.9 s	+38° 00' 54"
3931	UMa	Gxy	1.1' × 0.9'	13.5	11 h 51 m 13.5 s	+52° 00' 02"
3932	UMa	Gxy	1.1' × 0.5'	14.5	11 h 52 m 29.2 s	+48° 27' 31"
3935	UMa	Gxy	1.0' × 0.5'	13.3	11 h 52 m 24.1 s	+32° 24' 14"
3938	UMa	Gxy	5.4' × 4.9'	10.4	11 h 52 m 49.3 s	+44° 07' 13"
3941	UMa	Gxy	3.5' × 2.3'	10.4	11 h 52 m 55.3 s	+36° 59' 10"
3945	UMa	Gxy	5.5' × 3.6'	10.5	11 h 53 m 13.5 s	+60° 40' 31"
3949	UMa	Gxy	2.9' × 1.7'	10.9	11 h 53 m 41.7 s	+47° 51' 30"
3950	UMa	Gxy	0.2' × 0.2'	15.7	11 h 53 m 41.3 s	+47° 53' 05"
3953	UMa	Gxy	6.9' × 3.5'	10.2	11 h 53 m 48.9 s	+52° 19' 35"
3958	UMa	Gxy	1.5' × 0.7'	12.7	11 h 54 m 33.7 s	+58° 22' 00"
3963	UMa	Gxy	2.8' × 2.5'	11.6	11 h 54 m 58.6 s	+58° 29' 36"
3966	UMa	Gxy	2.5' × 0.5'	12.7	11 h 56 m 44.1 s	+32° 01' 17"
3971	UMa	Gxy	1.4' × 1.2'	12.8	11 h 55 m 36.4 s	+29° 59' 44"
3972	UMa	Gxy	3.9' × 1.1'	12.2	11 h 55 m 45.2 s	+55° 19' 12"
3975	UMa	Gxy	1.0' × 0.6'	15.3	11 h 55 m 53.7 s	+60° 31' 45"
3977	UMa	Gxy	1.7' × 1.6'	13.6	11 h 56 m 07.2 s	+55° 23' 25"
3978	UMa	Gxy	1.6' × 1.5'	12.6	11 h 56 m 10.3 s	+60° 31' 20"
3980	UMa	Gxy	1.7' × 1.6'	13.6	11 h 56 m 07.2 s	+55° 23' 25"
3982	UMa	Gxy	2.3' × 2.0'	11.2	11 h 56 m 28.0 s	+55° 07' 29"
3984	UMa	Gxy	1.4' × 1.2'	12.8	11 h 55 m 36.4 s	+29° 59' 44"
3985	UMa	Gxy	1.3' × 0.8'	12.8	11 h 56 m 42.1 s	+48° 20' 02"
3986	UMa	Gxy	3.1' × 0.7'	12.7	11 h 56 m 44.1 s	+32° 01' 17"
3990	UMa	Gxy	1.4' × 0.8'	12.7	11 h 57 m 35.7 s	+55° 27' 30"
3991	UMa	Gxy	1.4' × 0.4'	13.6	11 h 57 m 30.8 s	+32° 20' 13"
3992	UMa	Gxy	7.6' × 4.7'	9.9	11 h 57 m 36.0 s	+53° 22' 28"
3994	UMa	Gxy	1.0' × 0.6'	13.4	11 h 57 m 36.9 s	+32° 16' 38"
3995	UMa	Gxy	2.8' × 1.0'	12.4	11 h 57 m 44.1 s	+32° 17' 39"
3998	UMa	Gxy	2.7' × 2.2'	10.5	11 h 57 m 56.1 s	+55° 27' 12"
4001	UMa	Gxy	0.6' × 0.3'	15.3	11 h 58 m 06.7 s	+47° 20' 05"
4010	UMa	Gxy	4.3' × 0.8'	12.5	11 h 58 m 37.5 s	+47° 15' 39"
4013	UMa	Gxy	5.2' × 1.0'	11.7	11 h 58 m 31.1 s	+43° 56' 51"
4020	UMa	Gxy	2.1' × 0.9'	12.8	11 h 58 m 56.5 s	+30° 24' 42"
4025	UMa	Gxy	2.8' × 1.6'	13.8	11 h 59 m 10.0 s	+37° 47' 32"
4026	UMa	Gxy	5.2' × 1.3'	10.7	11 h 59 m 25.1 s	+50° 57' 42"
4031	UMa	Gxy	0.4' × 0.4'	14.4	12 h 00 m 31.3 s	+31° 56' 51"
4036	UMa	Gxy	4.3' × 1.7'	10.6	12 h 01 m 26.7 s	+61° 53' 44"

(continued)

Table 10.4 (continued)

NGC #	Const	Type	Size	V Mag	RA (2000)	Dec (2000)
4041	UMa	Gxy	2.7'×2.5'	11.1	12 h 02 m 12.1 s	+62° 08' 14"
4047	UMa	Gxy	1.6'×1.3'	12.4	12 h 02 m 50.7 s	+48° 38' 11"
4051	UMa	Gxy	5.2'×3.9'	10.7	12 h 03 m 09.6 s	+44° 31' 53"
4054	UMa	Gxy	0.5'×0.3'	14.4	12 h 03 m 12.4 s	+57° 53' 36"
4062	UMa	Gxy	4.1'×1.7'	11.3	12 h 04 m 03.7 s	+31° 53' 44"
4068	UMa	Gxy	3.2'×1.8'	12.6	12 h 04 m 00.8 s	+52° 35' 17"
4081	UMa	Gxy	1.5'×0.6'	13.1	12 h 04 m 33.8 s	+64° 26' 13"
4085	UMa	Gxy	2.8'×0.8'	12.4	12 h 05 m 22.7 s	+50° 21' 11"
4088	UMa	Gxy	5.8'×2.2'	11.1	12 h 05 m 34.1 s	+50° 32' 23"
4096	UMa	Gxy	6.6'×1.8'	11.1	12 h 06 m 01.0 s	+47° 28' 41"
4097	UMa	Gxy	1.1'×0.7'	13.5	12 h 06 m 02.5 s	+36° 51' 48"
4100	UMa	Gxy	5.4'×1.8'	11	12 h 06 m 08.3 s	+49° 34' 58"
4102	UMa	Gxy	3.0'×1.7'	11.3	12 h 06 m 22.9 s	+52° 42' 40"
4141	UMa	Gxy	1.3'×0.9'	13.9	12 h 09 m 47.5 s	+58° 50' 55"
4142	UMa	Gxy	2.2'×1.2'	13.5	12 h 09 m 30.5 s	+53° 06' 13"
4144	UMa	Gxy	6.0'×1.3'	11.6	12 h 09 m 59.3 s	+46° 27' 27"
4149	UMa	Gxy	1.3'×0.3'	13.2	12 h 10 m 32.9 s	+58° 18' 14"
4154	UMa	Gxy	1.3'×0.3'	13.2	12 h 10 m 32.9 s	+58° 18' 14"
4157	UMa	Gxy	6.8'×1.3'	11.1	12 h 11 m 05.0 s	+50° 29' 08"
4161	UMa	Gxy	1.1'×0.7'	13.1	12 h 11 m 33.5 s	+57° 44' 13"
4172	UMa	Gxy	1.3'×1.1'	13.3	12 h 12 m 15.0 s	+56° 10' 38"
4194	UMa	Gxy	1.8'×1.1'	12.7	12 h 14 m 09.6 s	+54° 31' 35"
4195	UMa	Gxy	1.6'×1.4'	14.4	12 h 14 m 18.0 s	+59° 36' 55"
4198	UMa	Gxy	1.1'×0.6'	13.8	12 h 14 m 22.1 s	+56° 00' 40"
4199	UMa	Gxy	1.1'×0.6'	14.9	12 h 14 m 48.6 s	+59° 54' 21"
4271	UMa	Gxy	1.5'×1.3'	12.7	12 h 19 m 32.6 s	+56° 44' 11"
4284	UMa	Gxy	2.5'×1.2'	13.7	12 h 20 m 12.6 s	+58° 05' 34"
4290	UMa	Gxy	2.3'×1.6'	12	12 h 20 m 47.5 s	+58° 05' 33"
4335	UMa	Gxy	1.9'×1.5'	12.5	12 h 23 m 01.9 s	+58° 26' 40"
4358	UMa	Gxy	1.3'×1.2'	13.2	12 h 24 m 02.1 s	+58° 23' 06"
4362	UMa	Gxy	0.7'×0.3'	14.4	12 h 24 m 11.3 s	+58° 21' 38"
4364	UMa	Gxy	0.7'×0.3'	14.4	12 h 24 m 11.3 s	+58° 21' 38"
4384	UMa	Gxy	1.3'×1.0'	12.8	12 h 25 m 11.9 s	+54° 30' 22"
4500	UMa	Gxy	1.6'×1.0'	12.6	12 h 31 m 22.1 s	+57° 57' 52"
4511	UMa	Gxy	0.7'×0.4'	14.1	12 h 32 m 08.1 s	+56° 28' 15"
4547	UMa	Gxy	0.7'×0.3'	14.5	12 h 34 m 51.8 s	+58° 55' 00"
4549	UMa	Gxy	0.3'×0.2'	15.2	12 h 35 m 21.1 s	+58° 56' 58"
4566	UMa	Gxy	1.2'×0.8'	13.1	12 h 36 m 00.1 s	+54° 13' 14"
4605	UMa	Gxy	5.5'×2.3'	10.4	12 h 40 m 00.0 s	+61° 36' 28"
4644	UMa	Gxy	1.7'×0.7'	13.8	12 h 42 m 42.6 s	+55° 08' 43"
4646	UMa	Gxy	0.6'×0.3'	13.4	12 h 42 m 52.2 s	+54° 51' 21"
4652	UMa	Gxy	1.0'×0.2'	14.8	12 h 43 m 19.7 s	+58° 57' 53"
4669	UMa	Gxy	1.7'×0.4'	13.7	12 h 44 m 46.8 s	+54° 52' 32"
4675	UMa	Gxy	1.6'×0.5'	14.6	12 h 45 m 31.9 s	+54° 44' 14"
4686	UMa	Gxy	2.0'×0.6'	13	12 h 46 m 39.7 s	+54° 32' 02"

(continued)

Table 10.4 (continued)

NGC #	Const	Type	Size	V Mag	RA (2000)	Dec (2000)
4695	UMa	Gxy	1.1'×0.7'	13.5	12 h 47 m 32.1 s	+54° 22' 28"
4732	UMa	Gxy	1.2'×0.6'	14	12 h 50 m 07.2 s	+52° 51' 00"
4801	UMa	Gxy	0.7'×0.5'	14.2	12 h 54 m 37.9 s	+53° 05' 24"
4814	UMa	Gxy	3.1'×2.3'	11.6	12 h 55 m 21.9 s	+58° 20' 38"
4964	UMa	Gxy	1.3'×0.6'	13.1	13 h 05 m 24.8 s	+56° 19' 21"
4967	UMa	Gxy	0.5'×0.4'	14.2	13 h 05 m 36.3 s	+53° 33' 51"
4973	UMa	Gxy	0.7'×0.7'	13.9	13 h 05 m 32.1 s	+53° 41' 06"
4974	UMa	Gxy	1.4'×1.2'	13.3	13 h 05 m 56.0 s	+53° 39' 34"
4977	UMa	Gxy	1.5'×1.5'	13.5	13 h 06 m 04.5 s	+55° 39' 21"
5001	UMa	Gxy	1.3'×0.4'	13.8	13 h 09 m 33.2 s	+53° 29' 39"
5007	UMa	Gxy	0.9'×0.6'	13.4	13 h 09 m 14.2 s	+62° 10' 29"
5109	UMa	Gxy	1.7'×0.5'	12.9	13 h 20 m 52.7 s	+57° 38' 32"
5113	UMa	Gxy	1.8'×0.5'	12.9	13 h 20 m 52.7 s	+57° 38' 32"
5163	UMa	Gxy	1.1'×0.7'	13.7	13 h 26 m 54.2 s	+52° 45' 12"
5164	UMa	Gxy	1.0'×0.9'	13.9	13 h 27 m 11.9 s	+55° 29' 14"
5201	UMa	Gxy	1.7'×1.0'	13.2	13 h 29 m 16.4 s	+53° 04' 54"
5204	UMa	Gxy	5.0'×3.0'	11.3	13 h 29 m 36.5 s	+58° 25' 08"
5205	UMa	Gxy	3.2'×1.8'	12.4	13 h 30 m 03.5 s	+62° 30' 42"
5216	UMa	Gxy	2.5'×1.5'	12.7	13 h 32 m 06.9 s	+62° 42' 01"
5218	UMa	Gxy	1.8'×1.3'	12.5	13 h 32 m 10.3 s	+62° 46' 03"
5250	UMa	Gxy	1.0'×0.9'	13.1	13 h 36 m 07.2 s	+51° 14' 09"
5255	UMa	Gxy	0.7'×0.2'	14.5	13 h 37 m 18.1 s	+57° 06' 31"
5256	UMa	Gxy	1.2'×1.1'	13.6	13 h 38 m 17.6 s	+48° 16' 35"
5278	UMa	Gxy	1.3'×1.0'	12.9	13 h 41 m 39.7 s	+55° 40' 14"
5279	UMa	Gxy	0.6'×0.4'	14	13 h 41 m 43.7 s	+55° 40' 24"
5294	UMa	Gxy	0.6'×0.5'	14.2	13 h 45 m 18.1 s	+55° 17' 26"
5308	UMa	Gxy	3.7'×0.7'	11.7	13 h 47 m 00.5 s	+60° 58' 22"
5322	UMa	Gxy	5.9'×3.9'	10.1	13 h 49 m 15.4 s	+60° 11' 27"
5342	UMa	Gxy	1.1'×0.4'	13.6	13 h 51 m 25.8 s	+59° 51' 49"
5368	UMa	Gxy	0.9'×0.7'	13.2	13 h 54 m 29.2 s	+54° 19' 49"
5370	UMa	Gxy	1.1'×1.1'	13.3	13 h 54 m 09.4 s	+60° 40' 40"
5372	UMa	Gxy	0.6'×0.4'	13	13 h 54 m 45.8 s	+58° 40' 00"
5376	UMa	Gxy	2.1'×1.3'	12.3	13 h 55 m 15.9 s	+59° 30' 23"
5379	UMa	Gxy	2.2'×1.0'	13	13 h 55 m 34.4 s	+59° 44' 34"
5389	UMa	Gxy	3.5'×1.0'	12.3	13 h 56 m 06.5 s	+59° 44' 30"
5402	UMa	Gxy	1.2'×0.3'	13.8	13 h 58 m 16.4 s	+59° 48' 56"
5422	UMa	Gxy	3.9'×0.7'	11.9	14 h 00 m 42.0 s	+55° 09' 52"
5425	UMa	Gxy	1.9'×0.5'	13.7	14 h 00 m 47.7 s	+48° 26' 38"
5430	UMa	Gxy	2.2'×1.1'	12.1	14 h 00 m 45.7 s	+59° 19' 43"
5443	UMa	Gxy	2.7'×1.0'	12.5	14 h 02 m 11.5 s	+55° 48' 52"
5447	UMa	GxyCld	...	...	14 h 02 m 28.3 s	+54° 16' 32"
5448	UMa	Gxy	4.0'×1.8'	11.6	14 h 02 m 49.9 s	+49° 10' 21"
5449	UMa	GxyCld	...	...	14 h 02 m 27.2 s	+54° 19' 48"
5450	UMa	GxyCld	...	...	14 h 02 m 29.5 s	+54° 16' 15"
5451	UMa	GxyCld	...	...	14 h 02 m 37.1 s	+54° 21' 45"

(continued)

Table 10.4 (continued)

NGC #	Const	Type	Size	V Mag	RA (2000)	Dec (2000)
5453	UMa	GxyCld	...	...	14 h 02 m 56.4 s	+54° 18' 29"
5455	UMa	GxyCld	...	...	14 h 03 m 01.2 s	+54° 14' 27"
5457	UMa	Gxy	28.8'×26.9'	7.5	14 h 03 m 12.4 s	+54° 20' 55"
5458	UMa	GxyCld	...	...	14 h 03 m 12.4 s	+54° 17' 55"
5461	UMa	GxyCld	...	...	14 h 03 m 40.9 s	+54° 19' 02"
5462	UMa	GxyCld	...	...	14 h 03 m 52.9 s	+54° 21' 54"
5471	UMa	GxyCld	...	...	14 h 04 m 29.0 s	+54° 23' 48"
5473	UMa	Gxy	2.3'×1.7'	11.5	14 h 04 m 43.2 s	+54° 53' 32"
5474	UMa	Gxy	4.8'×4.3'	11	14 h 05 m 01.5 s	+53° 39' 44"
5475	UMa	Gxy	2.0'×0.5'	12.8	14 h 05 m 12.3 s	+55° 44' 31"
5477	UMa	Gxy	1.7'×1.3'	13.8	14 h 05 m 32.9 s	+54° 27' 38"
5480	UMa	Gxy	1.7'×1.1'	12.3	14 h 06 m 21.4 s	+50° 43' 31"
5481	UMa	Gxy	1.8'×1.5'	12.4	14 h 06 m 41.2 s	+50° 43' 23"
5484	UMa	Gxy	...'×...'	14.8	14 h 06 m 48.1 s	+55° 01' 47"
5485	UMa	Gxy	2.3'×1.9'	11.5	14 h 07 m 11.3 s	+55° 00' 05"
5486	UMa	Gxy	1.9'×1.2'	13.4	14 h 07 m 24.9 s	+55° 06' 11"
5502	UMa	Gxy	0.3'×0.1'	15.3	14 h 09 m 33.9 s	+60° 24' 34"
5503	UMa	Gxy	0.3'×0.1'	15.3	14 h 09 m 33.9 s	+60° 24' 34"
5526	UMa	Gxy	1.8'×0.2'	13.5	14 h 13 m 53.3 s	+57° 46' 18"
5540	UMa	Gxy	0.5'×0.3'	13.9	14 h 14 m 54.3 s	+60° 00' 40"
5561	UMa	Gxy	0.5'×0.5'	14.5	14 h 17 m 22.7 s	+58° 45' 02"
5585	UMa	Gxy	5.8'×3.7'	11	14 h 19 m 47.8 s	+56° 43' 46"
5631	UMa	Gxy	1.7'×1.7'	11.5	14 h 26 m 33.1 s	+56° 34' 57"
3172	UMi	Gxy	1.0'×0.7'	14.4	11 h 47 m 15.1 s	+89° 05' 34"
5034	UMi	Gxy	0.9'×0.7'	13.5	13 h 12 m 19.1 s	+70° 38' 58"
5144	UMi	Gxy	1.2'×0.9'	12.9	13 h 22 m 54.2 s	+70° 30' 51"
5262	UMi	Gxy	1.2'×0.7'	13.9	13 h 35 m 38.5 s	+75° 02' 22"
5314	UMi	Gxy	...'×...'	13.9	13 h 46 m 11.4 s	+70° 20' 21"
5323	UMi	Gxy	1.5'×0.5'	13.7	13 h 45 m 36.4 s	+76° 49' 40"
5340	UMi	Gxy	0.8'×0.5'	14.7	13 h 49 m 00.0 s	+72° 39' 13"
5344	UMi	Gxy	0.6'×0.4'	14.7	13 h 50 m 12.0 s	+73° 57' 10"
5385	UMi	Ast	6.0'	...	13 h 52 m 31.9 s	+76° 09' 46"
5412	UMi	Gxy	1.2'×1.0'	13.5	13 h 57 m 13.4 s	+73° 36' 59"
5415	UMi	Gxy	1.0'×0.6'	14.3	13 h 56 m 56.9 s	+70° 45' 15"
5452	UMi	Gxy	2.0'×1.5'	13.4	13 h 54 m 24.9 s	+78° 13' 13"
5479	UMi	Gxy	0.5'×0.3'	14.3	14 h 05 m 57.4 s	+65° 41' 26"
5547	UMi	Gxy	1.0'×0.4'	13.9	14 h 09 m 45.1 s	+78° 36' 04"
5607	UMi	Gxy	0.9'×0.8'	13.7	14 h 19 m 26.7 s	+71° 35' 17"
5620	UMi	Gxy	0.5'×0.5'	14.2	14 h 22 m 40.2 s	+69° 35' 41"
5671	UMi	Gxy	1.7'×1.2'	13.5	14 h 27 m 41.9 s	+69° 41' 38"
5712	UMi	Gxy	0.7'×0.7'	14.5	14 h 29 m 41.8 s	+78° 51' 51"
5808	UMi	Gxy	0.9'×0.9'	13.5	14 h 54 m 02.9 s	+73° 07' 53"
5819	UMi	Gxy	0.9'×0.9'	13.7	14 h 54 m 02.9 s	+73° 07' 53"
5832	UMi	Gxy	3.7'×2.2'	12.3	14 h 57 m 45.7 s	+71° 40' 50"
5836	UMi	Gxy	1.3'×1.2'	14.1	14 h 59 m 31.4 s	+73° 53' 34"

(continued)

Table 10.4 (continued)

NGC #	Const	Type	Size	V Mag	RA (2000)	Dec (2000)
5909	UMi	Gxy	1.1'×0.5'	14	15 h 11 m 27.2 s	+75° 23' 01"
5912	UMi	Gxy	1.2'×1.1'	13.8	15 h 11 m 41.2 s	+75° 23' 05"
5939	UMi	Gxy	0.9'×0.5'	13.1	15 h 24 m 46.1 s	+68° 43' 50"
6011	UMi	Gxy	2.0'×0.7'	13.7	15 h 46 m 32.8 s	+72° 10' 08"
6048	UMi	Gxy	2.2'×1.7'	12.4	15 h 57 m 30.5 s	+70° 41' 21"
6068	UMi	Gxy	1.1'×0.7'	13	15 h 55 m 26.4 s	+78° 59' 48"
6071	UMi	Gxy	0.8'×0.8'	14	16 h 02 m 06.9 s	+70° 25' 01"
6091	UMi	Gxy	0.8'×0.5'	14.2	16 h 07 m 52.8 s	+69° 54' 17"
6094	UMi	Gxy	1.8'×1.4'	13.3	16 h 06 m 33.8 s	+72° 29' 39"
6217	UMi	Gxy	3.0'×2.5'	11.5	16 h 32 m 39.6 s	+78° 11' 51"
6251	UMi	Gxy	1.8'×1.5'	12.7	16 h 32 m 31.5 s	+82° 32' 16"
6252	UMi	Gxy	0.7'×0.3'	14.8	16 h 32 m 41.1 s	+82° 34' 34"
6324	UMi	Gxy	0.9'×0.5'	12.9	17 h 05 m 25.3 s	+75° 24' 25"
6331	UMi	Gxy	0.6'×0.4'	14.4	17 h 03 m 34.0 s	+78° 37' 47"
2547	Vel	OC	20'	4.7	08 h 10 m 10.5 s	−49° 13' 32"
2626	Vel	Neb	5'×5'	...	08 h 35 m 31.0 s	−40° 40' 20"
2645	Vel	OC	3.0'	7	08 h 39 m 03.1 s	−46° 13' 38"
2659	Vel	OC	2.7'	8.6	08 h 42 m 33.0 s	−45° 00' 02"
2660	Vel	OC	4.0'	8.8	08 h 42 m 38.0 s	−47° 12' 02"
2669	Vel	OC	12'	6.1	08 h 46 m 22.6 s	−52° 56' 51"
2670	Vel	OC	9'	7.8	08 h 45 m 29.5 s	−48° 47' 30"
2671	Vel	OC	4.0'	11.6	08 h 46 m 11.9 s	−41° 52' 38"
2736	Vel	Neb	30.0'×7.0'	...	09 h 00 m 16.9 s	−45° 56' 53"
2792	Vel	PN	13"×13"	11.6	09 h 12 m 26.5 s	−42° 25' 41"
2845	Vel	Gxy	2.0'×1.0'	13	09 h 18 m 36.9 s	−38° 00' 41"
2849	Vel	OC	2.3'	12.5	09 h 19 m 22.9 s	−40° 31' 13"
2866	Vel	OC	2.0'	...	09 h 22 m 05.1 s	−51° 06' 09"
2899	Vel	PN	120"	11.8	09 h 27 m 02.9 s	−56° 06' 21"
2910	Vel	OC	5'	7.2	09 h 30 m 29.0 s	−52° 54' 50"
2925	Vel	OC	12'	8.3	09 h 33 m 10.9 s	−53° 23' 45"
2932	Vel	GxyCld	60'×60'	...	09 h 35 m 27.7 s	−46° 49' 15"
2972	Vel	OC	4.0'	9.9	09 h 40 m 11.5 s	−50° 19' 15"
2982	Vel	OC	12'	...	09 h 42 m 00.1 s	−44° 01' 38"
2995	Vel	Ast	...	...	09 h 43 m 59.1 s	−54° 35' 49"
2999	Vel	OC	4.0'	9.9	09 h 40 m 11.5 s	−50° 19' 15"
3033	Vel	OC	5'	8.8	09 h 48 m 39.1 s	−56° 24' 42"
3105	Vel	OC	2.0'	9.7	10 h 00 m 39.5 s	−54° 47' 15"
3132	Vel	PN	84"×53"	9.2	10 h 07 m 01.4 s	−40° 26' 11"
3201	Vel	GC	18'	6.9	10 h 17 m 36.7 s	−46° 24' 40"
3228	Vel	OC	5'	6	10 h 21 m 22.2 s	−51° 43' 57"
3256	Vel	Gxy	3.8'×2.1'	11.5	10 h 27 m 51.4 s	−43° 54' 20"
3261	Vel	Gxy	3.7'×2.8'	11.4	10 h 29 m 01.4 s	−44° 39' 29"
3262	Vel	Gxy	1.1'×0.9'	13.3	10 h 29 m 06.1 s	−44° 09' 36"
3263	Vel	Gxy	5.1'×1.4'	11	10 h 29 m 13.3 s	−44° 07' 23"
3283	Vel	Gxy	2.7'×1.5'	...	10 h 31 m 11.5 s	−46° 15' 04"

(continued)

Table 10.4 (continued)

NGC #	Const	Type	Size	V Mag	RA (2000)	Dec (2000)
3318	Vel	Gxy	2.4'×1.3'	12.1	10 h 37 m 15.0 s	−41° 37' 39"
3330	Vel	OC	6'	7.4	10 h 38 m 47.5 s	−54° 06' 56"
3366	Vel	Gxy	2.2'×1.1'	12.2	10 h 35 m 08.5 s	−43° 41' 33"
3446	Vel	OC	7'	...	10 h 52 m 06.9 s	−45° 08' 21"
3482	Vel	Gxy	1.9'×1.4'	12.8	10 h 58 m 34.2 s	−46° 35' 04"
3776	Vir	Gxy	0.3'×0.2'	14.7	11 h 38 m 17.9 s	−03° 21' 16"
3792	Vir	**	...	...	11 h 39 m 38.5 s	+05° 05' 58"
3818	Vir	Gxy	2.0'×1.2'	11.6	11 h 41 m 57.3 s	−06° 09' 22"
3822	Vir	Gxy	1.4'×0.8'	12.8	11 h 42 m 11.1 s	+10° 16' 39"
3825	Vir	Gxy	1.3'×1.0'	13	11 h 42 m 23.7 s	+10° 15' 50"
3833	Vir	Gxy	1.5'×0.7'	13.7	11 h 43 m 28.9 s	+10° 09' 42"
3843	Vir	Gxy	0.9'×0.4'	13.7	11 h 43 m 54.7 s	+07° 55' 33"
3848	Vir	Gxy	1.3'×0.7'	12.8	11 h 42 m 11.1 s	+10° 16' 39"
3849	Vir	Gxy	0.8'×0.5'	13.7	11 h 45 m 35.2 s	+03° 13' 53"
3852	Vir	Gxy	1.4'×1.1'	12.8	11 h 42 m 23.7 s	+10° 15' 50"
3863	Vir	Gxy	2.8'×0.6'	13.1	11 h 45 m 05.6 s	+08° 28' 10"
3874	Vir	**	...	...	11 h 45 m 37.7 s	+08° 34' 26"
3876	Vir	Gxy	1.1'×0.7'	13	11 h 45 m 26.7 s	+09° 09' 39"
3907	Vir	Gxy	1.2'×0.9'	13.2	11 h 49 m 30.1 s	−01° 05' 12"
3914	Vir	Gxy	1.1'×0.6'	13.4	11 h 50 m 32.7 s	+06° 34' 06"
3915	Vir	Gxy	1.5'×0.4'	13.9	11 h 49 m 24.5 s	−05° 07' 07"
3952	Vir	Gxy	1.6'×0.7'	13.1	11 h 53 m 40.5 s	−03° 59' 46"
3976	Vir	Gxy	3.8'×1.2'	11.7	11 h 55 m 57.3 s	+06° 44' 56"
3979	Vir	Gxy	1.4'×1.2'	12.7	11 h 56 m 01.1 s	−02° 43' 14"
4006	Vir	Gxy	1.7'×1.2'	12.7	11 h 58 m 05.7 s	−02° 07' 13"
4012	Vir	Gxy	2.1'×0.5'	13.7	11 h 58 m 27.6 s	+10° 01' 17"
4029	Vir	Gxy	1.2'×0.7'	13.7	12 h 00 m 03.2 s	+08° 10' 54"
4030	Vir	Gxy	4.2'×3.0'	10.8	12 h 00 m 23.5 s	−01° 06' 00"
4043	Vir	Gxy	0.6'×0.5'	13.7	12 h 02 m 22.9 s	+04° 19' 48"
4044	Vir	Gxy	1.2'×1.1'	13.1	12 h 02 m 29.5 s	−00° 12' 44"
4045	Vir	Gxy	2.7'×1.9'	11.9	12 h 02 m 42.3 s	+01° 58' 38"
4046	Vir	Gxy	2.7'×1.9'	11.9	12 h 02 m 42.3 s	+01° 58' 38"
4058	Vir	Gxy	1.2'×0.6'	13.2	12 h 03 m 49.1 s	+03° 32' 54"
4063	Vir	Gxy	1.2'×0.3'	13.9	12 h 04 m 05.9 s	+01° 50' 49"
4067	Vir	Gxy	1.4'×1.0'	12.7	12 h 04 m 11.5 s	+10° 51' 15"
4073	Vir	Gxy	3.2'×2.3'	11.8	12 h 04 m 27.1 s	+01° 53' 45"
4075	Vir	Gxy	1.2'×0.6'	13.7	12 h 04 m 37.7 s	+02° 04' 22"
4077	Vir	Gxy	1.3'×0.9'	13.2	12 h 04 m 38.0 s	+01° 47' 16"
4078	Vir	Gxy	1.3'×0.5'	13.3	12 h 04 m 47.7 s	+10° 35' 45"
4079	Vir	Gxy	2.2'×1.6'	12.6	12 h 04 m 49.9 s	−02° 23' 00"
4082	Vir	Gxy	0.8'×0.3'	14.8	12 h 05 m 11.5 s	+10° 40' 14"
4083	Vir	Gxy	0.9'×0.7'	14.6	12 h 05 m 14.1 s	+10° 36' 47"
4107	Vir	Gxy	1.3'×0.4'	13.2	12 h 04 m 47.7 s	+10° 35' 45"
4116	Vir	Gxy	3.8'×2.2'	11.6	12 h 07 m 37.1 s	+02° 41' 29"
4119	Vir	Gxy	3.9'×1.8'	11.3	12 h 08 m 09.6 s	+10° 22' 42"

(continued)

Table 10.4 (continued)

NGC #	Const	Type	Size	V Mag	RA (2000)	Dec (2000)
4123	Vir	Gxy	4.4'×3.2'	11.3	12 h 08 m 11.1 s	+02° 52' 41"
4124	Vir	Gxy	4.3'×1.4'	11.1	12 h 08 m 09.6 s	+10° 22' 42"
4129	Vir	Gxy	2.3'×0.6'	12.4	12 h 08 m 53.2 s	−09° 02' 14"
4130	Vir	Gxy	2.3'×0.6'	12.5	12 h 08 m 53.2 s	−09° 02' 14"
4139	Vir	Gxy	1.0'×0.5'	13.8	12 h 04 m 34.0 s	+01° 48' 05"
4140	Vir	Gxy	1.0'×0.7'	13.3	12 h 04 m 38.0 s	+01° 47' 16"
4164	Vir	Gxy	0.3'×0.3'	14.4	12 h 12 m 05.4 s	+13° 12' 20"
4165	Vir	Gxy	1.3'×0.9'	13.7	12 h 12 m 11.8 s	+13° 14' 47"
4168	Vir	Gxy	2.8'×2.3'	11.2	12 h 12 m 17.1 s	+13° 12' 19"
4176	Vir	Gxy	0.7'×0.5'	14.5	12 h 12 m 36.9 s	−09° 09' 39"
4178	Vir	Gxy	5.1'×1.8'	11.4	12 h 12 m 46.5 s	+10° 52' 05"
4179	Vir	Gxy	4.0'×1.1'	11.2	12 h 12 m 52.4 s	+01° 17' 57"
4180	Vir	Gxy	1.6'×0.6'	12.8	12 h 13 m 03.0 s	+07° 02' 20"
4182	Vir	Gxy	1.7'×0.5'	12.7	12 h 13 m 03.0 s	+07° 02' 20"
4191	Vir	Gxy	1.1'×0.8'	12.9	12 h 13 m 50.4 s	+07° 12' 02"
4193	Vir	Gxy	2.0'×1.0'	12.6	12 h 13 m 53.6 s	+13° 10' 21"
4197	Vir	Gxy	3.4'×0.6'	13	12 h 14 m 38.5 s	+05° 48' 21"
4200	Vir	Gxy	1.5'×0.8'	13.1	12 h 14 m 44.2 s	+12° 10' 50"
4201	Vir	Gxy	0.9'×0.8'	13.6	12 h 14 m 41.9 s	−11° 35' 00"
4202	Vir	Gxy	1.2'×0.7'	13.8	12 h 18 m 08.5 s	−01° 03' 55"
4206	Vir	Gxy	6.2'×1.2'	12.2	12 h 15 m 16.7 s	+13° 01' 26"
4207	Vir	Gxy	1.6'×0.8'	12.9	12 h 15 m 30.4 s	+09° 35' 08"
4215	Vir	Gxy	1.9'×0.7'	12.2	12 h 15 m 54.5 s	+06° 24' 04"
4216	Vir	Gxy	8.1'×1.8'	10.2	12 h 15 m 54.1 s	+13° 08' 59"
4223	Vir	Gxy	2.5'×1.4'	12.1	12 h 17 m 25.7 s	+06° 41' 24"
4224	Vir	Gxy	2.6'×1.0'	12.3	12 h 16 m 33.7 s	+07° 27' 42"
4233	Vir	Gxy	2.4'×1.1'	12	12 h 17 m 07.6 s	+07° 37' 26"
4234	Vir	Gxy	1.3'×1.3'	12.9	12 h 17 m 09.1 s	+03° 40' 58"
4235	Vir	Gxy	4.2'×0.9'	11.9	12 h 17 m 09.9 s	+07° 11' 28"
4240	Vir	Gxy	1.2'×1.2'	12.7	12 h 17 m 24.3 s	−09° 57' 08"
4241	Vir	Gxy	2.6'×1.3'	13.1	12 h 17 m 59.9 s	+06° 39' 13"
4243	Vir	Gxy	0.9'×0.9'	12.7	12 h 17 m 24.3 s	−09° 57' 08"
4246	Vir	Gxy	2.4'×1.3'	12.7	12 h 17 m 58.1 s	+07° 11' 08"
4247	Vir	Gxy	1.1'×0.9'	13.7	12 h 17 m 57.9 s	+07° 16' 25"
4249	Vir	Gxy	0.4'×0.4'	14	12 h 17 m 59.4 s	+05° 35' 54"
4252	Vir	Gxy	1.5'×0.4'	14.2	12 h 18 m 30.8 s	+05° 33' 34"
4255	Vir	Gxy	1.3'×0.5'	12.9	12 h 18 m 56.1 s	+04° 47' 10"
4257	Vir	Gxy	1.3'×0.4'	14.2	12 h 19 m 06.5 s	+05° 43' 33"
4259	Vir	Gxy	1.2'×0.4'	13.7	12 h 19 m 22.2 s	+05° 22' 35"
4260	Vir	Gxy	2.7'×1.3'	11.9	12 h 19 m 22.2 s	+06° 05' 55"
4261	Vir	Gxy	4.1'×3.6'	10.5	12 h 19 m 23.2 s	+05° 49' 30"
4264	Vir	Gxy	1.0'×0.8'	13.1	12 h 19 m 35.7 s	+05° 50' 48"
4266	Vir	Gxy	2.0'×0.4'	13.9	12 h 19 m 42.3 s	+05° 32' 17"
4267	Vir	Gxy	3.2'×3.0'	10.8	12 h 19 m 45.3 s	+12° 47' 53"
4268	Vir	Gxy	1.6'×0.6'	13	12 h 19 m 47.2 s	+05° 17' 00"

(continued)

Table 10.4 (continued)

NGC #	Const	Type	Size	V Mag	RA (2000)	Dec (2000)
4269	Vir	Gxy	1.1'×0.8'	13	12 h 19 m 49.3 s	+06° 00' 52"
4270	Vir	Gxy	2.0'×0.9'	12.1	12 h 19 m 49.5 s	+05° 27' 47"
4273	Vir	Gxy	2.3'×1.5'	11.8	12 h 19 m 56.1 s	+05° 20' 36"
4276	Vir	Gxy	1.6'×1.4'	12.7	12 h 20 m 07.5 s	+07° 41' 29"
4277	Vir	Gxy	1.0'×0.9'	13.8	12 h 20 m 03.7 s	+05° 20' 28"
4279	Vir	Gxy	1.1'×0.7'	13.6	12 h 20 m 24.9 s	−11° 40' 00"
4280	Vir	Ast	…	…	12 h 20 m 31.9 s	−11° 39' 09"
4281	Vir	Gxy	3.0'×1.5'	11.3	12 h 20 m 21.5 s	+05° 23' 10"
4282	Vir	Gxy	0.8'×0.5'	14.1	12 h 20 m 24.3 s	+05° 34' 21"
4285	Vir	Gxy	0.8'×0.4'	14.1	12 h 20 m 39.9 s	−11° 38' 31"
4287	Vir	Gxy	1.0'×0.2'	14.3	12 h 20 m 48.5 s	+05° 38' 23"
4289	Vir	Gxy	3.9'×0.4'	13.8	12 h 21 m 02.2 s	+03° 43' 19"
4292	Vir	Gxy	1.7'×1.1'	12.7	12 h 21 m 16.5 s	+04° 35' 44"
4294	Vir	Gxy	3.2'×1.2'	12	12 h 21 m 17.7 s	+11° 30' 37"
4296	Vir	Gxy	1.3'×0.9'	12.8	12 h 21 m 28.3 s	+06° 39' 12"
4297	Vir	Gxy	0.4'×0.2'	14.6	12 h 21 m 27.4 s	+06° 40' 16"
4299	Vir	Gxy	1.7'×1.6'	12.2	12 h 21 m 40.7 s	+11° 30' 06"
4300	Vir	Gxy	1.5'×0.6'	13.2	12 h 21 m 41.5 s	+05° 23' 05"
4301	Vir	Gxy	1.6'×1.3'	12.9	12 h 22 m 27.1 s	+04° 33' 57"
4303	Vir	Gxy	6.5'×5.8'	9.6	12 h 21 m 54.9 s	+04° 28' 24"
4305	Vir	Gxy	2.2'×1.2'	12.6	12 h 22 m 03.5 s	+12° 44' 28"
4306	Vir	Gxy	1.6'×1.3'	12.7	12 h 22 m 04.1 s	+12° 47' 15"
4307	Vir	Gxy	3.6'×0.8'	12.2	12 h 22 m 05.7 s	+09° 02' 38"
4309	Vir	Gxy	1.9'×1.1'	12.8	12 h 22 m 12.4 s	+07° 08' 39"
4313	Vir	Gxy	4.0'×1.0'	12	12 h 22 m 38.6 s	+11° 48' 04"
4315	Vir	*	…	…	12 h 22 m 45.3 s	+09° 18' 20"
4316	Vir	Gxy	2.5'×0.5'	13.3	12 h 22 m 42.3 s	+09° 19' 56"
4318	Vir	Gxy	0.7'×0.5'	13.8	12 h 22 m 43.3 s	+08° 11' 53"
4320	Vir	Gxy	1.0'×0.6'	14	12 h 22 m 57.8 s	+10° 32' 55"
4323	Vir	Gxy	1.1'×0.8'	13.8	12 h 23 m 01.8 s	+15° 54' 20"
4324	Vir	Gxy	2.8'×1.2'	11.7	12 h 23 m 06.1 s	+05° 15' 00"
4325	Vir	Gxy	1.0'×0.6'	13.4	12 h 23 m 06.6 s	+10° 37' 16"
4326	Vir	Gxy	1.5'×1.1'	13.5	12 h 23 m 11.6 s	+06° 04' 18"
4330	Vir	Gxy	4.5'×0.9'	12.7	12 h 23 m 16.9 s	+11° 22' 05"
4333	Vir	Gxy	0.9'×0.7'	13.8	12 h 23 m 22.2 s	+06° 02' 25"
4334	Vir	Gxy	2.3'×1.1'	13.2	12 h 23 m 23.9 s	+07° 28' 22"
4339	Vir	Gxy	2.4'×2.3'	11.5	12 h 23 m 34.9 s	+06° 04' 53"
4341	Vir	Gxy	1.6'×0.5'	13.2	12 h 23 m 53.5 s	+07° 06' 25"
4342	Vir	Gxy	1.3'×0.2'	12.2	12 h 23 m 39.1 s	+07° 03' 14"
4343	Vir	Gxy	2.5'×0.7'	12.6	12 h 23 m 38.9 s	+06° 57' 15"
4347	Vir	*	…	…	12 h 23 m 52.3 s	−03° 14' 24"
4348	Vir	Gxy	3.2'×0.7'	12.4	12 h 23 m 53.9 s	−03° 26' 34"
4351	Vir	Gxy	2.0'×1.3'	12.3	12 h 24 m 01.5 s	+12° 12' 16"
4352	Vir	Gxy	2.1'×1.0'	12.6	12 h 24 m 05.0 s	+11° 13' 06"
4353	Vir	Gxy	1.1'×0.7'	13.8	12 h 24 m 00.2 s	+07° 47' 05"

(continued)

Table 10.4 (continued)

NGC #	Const	Type	Size	V Mag	RA (2000)	Dec (2000)
4354	Vir	Gxy	2.0'×1.3'	12.2	12 h 24 m 01.5 s	+12° 12' 16"
4355	Vir	Gxy	1.4'×0.7'	13.2	12 h 26 m 54.6 s	−00° 52' 40"
4356	Vir	Gxy	2.8'×0.5'	13.5	12 h 24 m 14.8 s	+08° 32' 14"
4360	Vir	Gxy	1.4'×1.2'	12.6	12 h 24 m 21.8 s	+09° 17' 34"
4365	Vir	Gxy	6.9'×5.0'	9.6	12 h 24 m 28.1 s	+07° 19' 03"
4366	Vir	Gxy	0.8'×0.4'	14.3	12 h 24 m 46.9 s	+07° 21' 09"
4367	Vir	**	...	...	12 h 24 m 35.2 s	+12° 10' 56"
4368	Vir	Gxy	1.0'×0.6'	13.4	12 h 23 m 06.6 s	+10° 37' 16"
4370	Vir	Gxy	1.4'×0.7'	13.2	12 h 24 m 55.0 s	+07° 26' 39"
4371	Vir	Gxy	4.0'×2.2'	10.7	12 h 24 m 55.5 s	+11° 42' 14"
4374	Vir	Gxy	6.5'×5.6'	9.4	12 h 25 m 04.7 s	+12° 53' 13"
4376	Vir	Gxy	1.4'×0.9'	13.4	12 h 25 m 18.2 s	+05° 44' 28"
4378	Vir	Gxy	2.9'×2.7'	11.4	12 h 25 m 18.1 s	+04° 55' 29"
4380	Vir	Gxy	3.5'×1.9'	11.7	12 h 25 m 22.1 s	+10° 00' 59"
4385	Vir	Gxy	2.2'×1.4'	12	12 h 25 m 42.8 s	+00° 34' 20"
4387	Vir	Gxy	1.8'×1.1'	12.1	12 h 25 m 41.7 s	+12° 48' 37"
4388	Vir	Gxy	5.6'×1.3'	11.3	12 h 25 m 46.8 s	+12° 39' 43"
4390	Vir	Gxy	1.7'×1.3'	12.8	12 h 25 m 50.7 s	+10° 27' 32"
4398	Vir	*	...	...	12 h 26 m 07.5 s	+10° 41' 11"
4402	Vir	Gxy	3.9'×1.1'	12.2	12 h 26 m 07.6 s	+13° 06' 48"
4403	Vir	Gxy	1.9'×0.6'	13	12 h 26 m 12.7 s	−07° 41' 04"
4404	Vir	Gxy	1.2'×0.9'	13	12 h 26 m 16.1 s	−07° 40' 51"
4406	Vir	Gxy	8.9'×5.8'	9	12 h 26 m 11.9 s	+12° 56' 47"
4407	Vir	Gxy	2.5'×1.7'	11.9	12 h 26 m 32.2 s	+12° 36' 39"
4409	Vir	Gxy	2.2'×1.2'	12	12 h 26 m 58.5 s	+02° 29' 39"
4410	Vir	Gxy	1.0'×0.6'	13.8	12 h 26 m 29.5 s	+09° 01' 09"
4411	Vir	Gxy	2.2'×2.0'	12.9	12 h 26 m 30.0 s	+08° 52' 20"
4412	Vir	Gxy	1.4'×1.3'	12.5	12 h 26 m 36.1 s	+03° 57' 53"
4413	Vir	Gxy	2.3'×1.5'	12.3	12 h 26 m 32.2 s	+12° 36' 39"
4415	Vir	Gxy	1.3'×1.2'	13	12 h 26 m 40.5 s	+08° 26' 08"
4416	Vir	Gxy	1.7'×1.5'	12.4	12 h 26 m 46.8 s	+07° 55' 08"
4417	Vir	Gxy	3.4'×1.3'	11.2	12 h 26 m 50.6 s	+09° 35' 03"
4418	Vir	Gxy	1.4'×0.7'	13.3	12 h 26 m 54.6 s	−00° 52' 40"
4420	Vir	Gxy	2.0'×1.0'	12.1	12 h 26 m 58.5 s	+02° 29' 39"
4422	Vir	Gxy	1.2'×1.2'	13.8	12 h 27 m 12.2 s	−05° 49' 52"
4423	Vir	Gxy	2.3'×0.4'	13.8	12 h 27 m 08.9 s	+05° 52' 46"
4424	Vir	Gxy	3.6'×1.8'	11.5	12 h 27 m 11.7 s	+09° 25' 12"
4425	Vir	Gxy	3.0'×1.0'	11.8	12 h 27 m 13.4 s	+12° 44' 04"
4428	Vir	Gxy	1.9'×0.8'	12.8	12 h 27 m 28.3 s	−08° 10' 04"
4429	Vir	Gxy	5.6'×2.6'	10.2	12 h 27 m 26.7 s	+11° 06' 26"
4430	Vir	Gxy	2.3'×2.0'	12	12 h 27 m 26.1 s	+06° 15' 44"
4431	Vir	Gxy	1.7'×1.1'	12.9	12 h 27 m 27.3 s	+12° 17' 24"
4432	Vir	Gxy	0.9'×0.7'	14.2	12 h 27 m 33.0 s	+06° 13' 59"
4433	Vir	Gxy	2.2'×1.0'	12.3	12 h 27 m 38.7 s	−08° 16' 46"
4434	Vir	Gxy	1.4'×1.4'	12.1	12 h 27 m 36.7 s	+08° 09' 15"

(continued)

Table 10.4 (continued)

NGC #	Const	Type	Size	V Mag	RA (2000)	Dec (2000)
4435	Vir	Gxy	2.8'×2.0'	10.8	12 h 27 m 40.5 s	+13° 04' 44"
4436	Vir	Gxy	1.5'×0.7'	13.2	12 h 27 m 41.2 s	+12° 18' 56"
4437	Vir	Gxy	10.8'×1.5'	10.5	12 h 32 m 45.5 s	+00° 06' 48"
4438	Vir	Gxy	8.5'×3.2'	10	12 h 27 m 45.6 s	+13° 00' 31"
4440	Vir	Gxy	1.9'×1.5'	12.1	12 h 27 m 53.5 s	+12° 17' 35"
4442	Vir	Gxy	4.6'×1.8'	10.4	12 h 28 m 03.8 s	+09° 48' 14"
4443	Vir	Gxy	3.7'×1.5'	…	12 h 29 m 03.0 s	+13° 11' 01"
4445	Vir	Gxy	2.6'×0.5'	13	12 h 28 m 16.0 s	+09° 26' 11"
4451	Vir	Gxy	1.5'×1.0'	12.8	12 h 28 m 40.5 s	+09° 15' 33"
4452	Vir	Gxy	2.8'×0.6'	12.3	12 h 28 m 43.3 s	+11° 45' 18"
4453	Vir	Gxy	0.5'×0.2'	14.9	12 h 28 m 46.9 s	+06° 30' 43"
4454	Vir	Gxy	2.0'×1.7'	12	12 h 28 m 50.8 s	−01° 56' 21"
4457	Vir	Gxy	2.7'×2.3'	10.8	12 h 28 m 59.1 s	+03° 34' 13"
4458	Vir	Gxy	1.7'×1.6'	12.1	12 h 28 m 57.6 s	+13° 14' 30"
4461	Vir	Gxy	3.5'×1.4'	11.1	12 h 29 m 03.0 s	+13° 11' 01"
4464	Vir	Gxy	1.1'×0.8'	12.8	12 h 29 m 21.3 s	+08° 09' 22"
4465	Vir	Gxy	0.3'×0.2'	14.9	12 h 29 m 23.5 s	+08° 01' 33"
4466	Vir	Gxy	1.3'×0.4'	13.9	12 h 29 m 30.6 s	+07° 41' 46"
4467	Vir	Gxy	1.2'×0.9'	13.4	12 h 29 m 30.1 s	+07° 59' 33"
4469	Vir	Gxy	3.8'×1.3'	11.4	12 h 29 m 28.1 s	+08° 45' 01"
4470	Vir	Gxy	1.3'×0.9'	12.5	12 h 29 m 37.9 s	+07° 49' 25"
4471	Vir	*	…	…	12 h 29 m 41.9 s	+07° 53' 46"
4472	Vir	Gxy	10.2'×8.3'	8.4	12 h 29 m 46.7 s	+08° 00' 00"
4476	Vir	Gxy	1.7'×1.2'	12.1	12 h 29 m 59.1 s	+12° 20' 56"
4478	Vir	Gxy	1.9'×1.6'	11.3	12 h 30 m 17.3 s	+12° 19' 43"
4480	Vir	Gxy	2.3'×1.2'	12.5	12 h 30 m 26.8 s	+04° 14' 48"
4482	Vir	Gxy	1.7'×1.0'	12.8	12 h 30 m 10.4 s	+10° 46' 46"
4483	Vir	Gxy	1.6'×0.9'	12.3	12 h 30 m 40.7 s	+09° 00' 55"
4484	Vir	Gxy	1.5'×1.5'	13.6	12 h 28 m 52.7 s	−11° 39' 08"
4486	Vir	Gxy	8.3'×6.6'	8.8	12 h 30 m 49.3 s	+12° 23' 26"
4487	Vir	Gxy	4.2'×2.8'	11.1	12 h 31 m 04.3 s	−08° 03' 15"
4488	Vir	Gxy	3.7'×1.5'	12	12 h 30 m 51.3 s	+08° 21' 35"
4491	Vir	Gxy	1.9'×1.0'	12.8	12 h 30 m 58.1 s	+11° 29' 00"
4492	Vir	Gxy	1.7'×1.6'	12.6	12 h 30 m 59.7 s	+08° 04' 39"
4493	Vir	Gxy	1.2'×1.0'	14.1	12 h 31 m 08.4 s	+00° 36' 48"
4496	Vir	Gxy	3.9'×3.1'	11.2	12 h 31 m 39.6 s	+03° 56' 27"
4497	Vir	Gxy	2.0'×0.9'	12.6	12 h 31 m 32.5 s	+11° 37' 28"
4503	Vir	Gxy	3.5'×1.7'	11.2	12 h 32 m 06.2 s	+11° 10' 35"
4504	Vir	Gxy	4.4'×2.7'	11.4	12 h 32 m 17.3 s	−07° 33' 50"
4505	Vir	Gxy	3.9'×3.1'	11.2	12 h 31 m 39.6 s	+03° 56' 27"
4508	Vir	**	…	…	12 h 32 m 17.4 s	+05° 49' 12"
4517	Vir	Gxy	10.5'×1.5'	10.8	12 h 32 m 45.5 s	+00° 06' 48"
4518	Vir	Gxy	1.0'×0.4'	13.8	12 h 33 m 11.7 s	+07° 51' 05"
4519	Vir	Gxy	3.2'×2.5'	11.7	12 h 33 m 30.3 s	+08° 39' 17"
4520	Vir	Gxy	3.5'×2.2'	14	12 h 33 m 49.8 s	−07° 22' 32"

(continued)

Table 10.4 (continued)

NGC #	Const	Type	Size	V Mag	RA (2000)	Dec (2000)
4522	Vir	Gxy	3.7'×1.0'	12.3	12 h 33 m 39.5 s	+09° 10' 25"
4526	Vir	Gxy	7.2'×2.4'	9.5	12 h 34 m 03.1 s	+07° 41' 58"
4527	Vir	Gxy	6.2'×2.1'	10.8	12 h 34 m 08.5 s	+02° 39' 09"
4528	Vir	Gxy	1.7'×1.0'	12.2	12 h 34 m 06.1 s	+11° 19' 16"
4531	Vir	Gxy	3.1'×2.0'	11.6	12 h 34 m 15.9 s	+13° 04' 30"
4532	Vir	Gxy	2.8'×1.1'	11.7	12 h 34 m 19.3 s	+06° 28' 07"
4533	Vir	Gxy	2.1'×0.4'	13.9	12 h 34 m 22.1 s	+02° 19' 32"
4535	Vir	Gxy	7.1'×5.0'	10	12 h 34 m 20.3 s	+08° 11' 52"
4536	Vir	Gxy	7.6'×3.2'	10.4	12 h 34 m 27.1 s	+02° 11' 16"
4538	Vir	Gxy	0.7'×0.3'	14.4	12 h 34 m 40.8 s	+03° 19' 24"
4541	Vir	Gxy	1.6'×0.7'	13.2	12 h 35 m 10.7 s	−00° 13' 17"
4543	Vir	Gxy	0.4'×0.3'	13.6	12 h 35 m 20.3 s	+06° 06' 54"
4544	Vir	Gxy	2.0'×0.6'	13.2	12 h 35 m 36.6 s	+03° 02' 04"
4546	Vir	Gxy	3.3'×1.4'	10.5	12 h 35 m 29.4 s	−03° 47' 38"
4550	Vir	Gxy	3.3'×0.9'	11.6	12 h 35 m 30.5 s	+12° 13' 14"
4551	Vir	Gxy	1.8'×1.4'	11.9	12 h 35 m 37.9 s	+12° 15' 49"
4552	Vir	Gxy	5.1'×4.7'	9.8	12 h 35 m 39.9 s	+12° 33' 23"
4554	Vir	NF	...	...	12 h 35 m 42.0 s	+11° 11' 00"
4560	Vir	Gxy	7.2'×2.3'	9.6	12 h 34 m 03.1 s	+07° 41' 58"
4564	Vir	Gxy	3.5'×1.5'	11.2	12 h 36 m 26.9 s	+11° 26' 21"
4567	Vir	Gxy	3.0'×2.0'	11.5	12 h 36 m 32.8 s	+11° 15' 28"
4568	Vir	Gxy	4.6'×2.0'	11.2	12 h 36 m 34.3 s	+11° 14' 17"
4569	Vir	Gxy	9.5'×4.4'	9.6	12 h 36 m 49.9 s	+13° 09' 45"
4570	Vir	Gxy	3.8'×1.1'	10.9	12 h 36 m 53.4 s	+07° 14' 47"
4576	Vir	Gxy	1.2'×0.8'	13.8	12 h 37 m 33.5 s	+04° 22' 03"
4577	Vir	Gxy	1.8'×0.9'	13	12 h 39 m 12.3 s	+06° 00' 43"
4578	Vir	Gxy	3.3'×2.5'	11	12 h 37 m 30.5 s	+09° 33' 17"
4579	Vir	Gxy	5.9'×4.7'	10.1	12 h 37 m 43.5 s	+11° 49' 05"
4580	Vir	Gxy	2.1'×1.6'	11.9	12 h 37 m 48.4 s	+05° 22' 05"
4581	Vir	Gxy	1.9'×1.1'	12.2	12 h 38 m 05.1 s	+01° 28' 39"
4582	Vir	*	...	...	12 h 38 m 10.1 s	+00° 10' 58"
4584	Vir	Gxy	1.4'×1.0'	13.1	12 h 38 m 17.9 s	+13° 06' 35"
4586	Vir	Gxy	4.0'×1.3'	11.9	12 h 38 m 28.3 s	+04° 19' 08"
4587	Vir	Gxy	1.1'×0.5'	13.6	12 h 38 m 35.5 s	+02° 39' 27"
4588	Vir	Gxy	1.3'×0.4'	14.5	12 h 38 m 45.4 s	+06° 46' 05"
4591	Vir	Gxy	1.6'×0.8'	13	12 h 39 m 12.3 s	+06° 00' 43"
4592	Vir	Gxy	5.8'×1.5'	11.7	12 h 39 m 18.9 s	−00° 31' 54"
4593	Vir	Gxy	3.9'×2.9'	11.1	12 h 39 m 39.4 s	−05° 20' 39"
4594	Vir	Gxy	8.7'×3.5'	7.9	12 h 39 m 59.4 s	−11° 37' 23"
4596	Vir	Gxy	4.0'×3.0'	10.7	12 h 39 m 55.9 s	+10° 10' 33"
4597	Vir	Gxy	4.1'×1.9'	12.2	12 h 40 m 12.8 s	−05° 48' 00"
4598	Vir	Gxy	1.4'×1.1'	12.8	12 h 40 m 11.8 s	+08° 23' 01"
4599	Vir	Gxy	1.7'×0.8'	12.8	12 h 40 m 27.1 s	+01° 11' 47"
4600	Vir	Gxy	1.2'×0.8'	12.8	12 h 40 m 23.0 s	+03° 07' 03"
4602	Vir	Gxy	3.4'×1.2'	11.7	12 h 40 m 36.7 s	−05° 07' 55"

(continued)

Table 10.4 (continued)

NGC #	Const	Type	Size	V Mag	RA (2000)	Dec (2000)
4604	Vir	Gxy	1.0'×0.4'	13.8	12 h 40 m 45.3 s	−05° 18' 13"
4606	Vir	Gxy	3.2'×1.6'	11.8	12 h 40 m 57.5 s	+11° 54' 42"
4607	Vir	Gxy	2.9'×0.7'	13.3	12 h 41 m 12.3 s	+11° 53' 07"
4608	Vir	Gxy	3.2'×2.7'	11.1	12 h 41 m 13.2 s	+10° 09' 19"
4610	Vir	Gxy	1.5'×1.1'	12.2	12 h 29 m 37.9 s	+07° 49' 25"
4612	Vir	Gxy	2.5'×1.9'	11.7	12 h 41 m 32.7 s	+07° 18' 52"
4620	Vir	Gxy	1.8'×1.5'	12.3	12 h 41 m 59.3 s	+12° 56' 34"
4621	Vir	Gxy	5.4'×3.7'	9.8	12 h 42 m 02.4 s	+11° 38' 48"
4623	Vir	Gxy	2.2'×0.7'	12.4	12 h 42 m 10.7 s	+07° 40' 37"
4624	Vir	Gxy	3.5'×3.5'	10.3	12 h 45 m 06.2 s	+03° 03' 27"
4626	Vir	Gxy	1.3'×0.4'	13.4	12 h 42 m 25.4 s	−07° 02' 39"
4628	Vir	Gxy	1.7'×0.5'	13.5	12 h 42 m 25.2 s	−06° 58' 17"
4629	Vir	Gxy	1.4'×1.2'	13.4	12 h 42 m 32.7 s	−01° 21' 04"
4630	Vir	Gxy	1.8'×1.3'	12.8	12 h 42 m 31.1 s	+03° 57' 30"
4632	Vir	Gxy	3.1'×1.2'	11.9	12 h 42 m 32.8 s	−00° 04' 47"
4636	Vir	Gxy	6.0'×4.7'	9.7	12 h 42 m 49.9 s	+02° 41' 16"
4637	Vir	Gxy	1.2'×0.5'	13.8	12 h 42 m 54.1 s	+11° 26' 16"
4638	Vir	Gxy	2.2'×1.4'	11.2	12 h 42 m 47.4 s	+11° 26' 32"
4639	Vir	Gxy	2.8'×1.9'	11.5	12 h 42 m 52.3 s	+13° 15' 25"
4640	Vir	Gxy	1.4'×0.8'	13.4	12 h 42 m 57.8 s	+12° 17' 12"
4641	Vir	Gxy	1.2'×1.0'	13.3	12 h 43 m 07.7 s	+12° 03' 02"
4642	Vir	Gxy	1.9'×0.6'	13.2	12 h 43 m 17.6 s	−00° 38' 40"
4643	Vir	Gxy	3.1'×2.3'	10.8	12 h 43 m 20.1 s	+01° 58' 42"
4647	Vir	Gxy	2.9'×2.3'	11.6	12 h 43 m 32.0 s	+11° 34' 55"
4649	Vir	Gxy	7.4'×6.0'	8.8	12 h 43 m 39.7 s	+11° 33' 07"
4653	Vir	Gxy	3.1'×2.7'	12.2	12 h 43 m 50.8 s	−00° 33' 40"
4654	Vir	Gxy	4.9'×2.8'	10.7	12 h 43 m 56.5 s	+13° 07' 32"
4658	Vir	Gxy	2.1'×0.9'	12.1	12 h 44 m 37.7 s	−10° 05' 02"
4660	Vir	Gxy	2.2'×1.6'	11	12 h 44 m 32.0 s	+11° 11' 26"
4663	Vir	Gxy	1.1'×0.9'	13.5	12 h 44 m 47.1 s	−10° 11' 52"
4664	Vir	Gxy	3.5'×3.5'	10.3	12 h 45 m 06.2 s	+03° 03' 27"
4665	Vir	Gxy	3.8'×3.2'	10.7	12 h 45 m 06.2 s	+03° 03' 27"
4666	Vir	Gxy	4.6'×1.3'	11.2	12 h 45 m 08.0 s	−00° 27' 44"
4667	Vir	Gxy	2.8'×1.6'	11.1	12 h 42 m 47.4 s	+11° 26' 32"
4668	Vir	Gxy	1.4'×0.8'	13	12 h 45 m 31.8 s	−00° 32' 09"
4671	Vir	Gxy	1.3'×1.1'	13	12 h 45 m 47.6 s	−07° 04' 09"
4674	Vir	Gxy	1.7'×0.6'	13.2	12 h 46 m 03.4 s	−08° 39' 21"
4678	Vir	Gxy	1.0'×0.5'	14.3	12 h 49 m 42.2 s	−04° 34' 47"
4680	Vir	Gxy	1.4'×1.2'	12.8	12 h 46 m 54.7 s	−11° 38' 10"
4682	Vir	Gxy	2.6'×1.3'	12.4	12 h 47 m 15.5 s	−10° 03' 47"
4684	Vir	Gxy	2.9'×1.0'	11.5	12 h 47 m 17.4 s	−02° 43' 37"
4688	Vir	Gxy	3.2'×2.8'	12.1	12 h 47 m 46.5 s	+04° 20' 08"
4690	Vir	Gxy	1.2'×0.9'	13	12 h 47 m 55.5 s	−01° 39' 23"
4691	Vir	Gxy	2.8'×2.3'	10.9	12 h 48 m 13.0 s	−03° 20' 01"
4694	Vir	Gxy	3.2'×1.5'	11.2	12 h 48 m 15.1 s	+10° 59' 01"

(continued)

Table 10.4 (continued)

NGC #	Const	Type	Size	V Mag	RA (2000)	Dec (2000)
4697	Vir	Gxy	7.2′×4.7′	9.2	12 h 48 m 35.7 s	−05° 48′ 03″
4698	Vir	Gxy	4.0′×2.5′	10.9	12 h 48 m 22.9 s	+08° 29′ 16″
4699	Vir	Gxy	3.8′×2.6′	10.2	12 h 49 m 02.2 s	−08° 39′ 52″
4700	Vir	Gxy	3.0′×0.6′	12.1	12 h 49 m 07.7 s	−11° 24′ 39″
4701	Vir	Gxy	2.8′×2.1′	11.9	12 h 49 m 11.6 s	+03° 23′ 19″
4703	Vir	Gxy	3.0′×0.6′	13.7	12 h 49 m 18.9 s	−09° 06′ 34″
4705	Vir	Gxy	3.0′×1.0′	12.8	12 h 49 m 24.9 s	−05° 11′ 46″
4708	Vir	Gxy	1.3′×0.9′	13.8	12 h 49 m 41.5 s	−11° 05′ 35″
4713	Vir	Gxy	2.7′×1.7′	11.6	12 h 49 m 57.8 s	+05° 18′ 38″
4716	Vir	Gxy	1.0′×0.7′	13.3	12 h 50 m 33.1 s	−09° 27′ 04″
4717	Vir	Gxy	1.4′×0.5′	14.2	12 h 50 m 34.3 s	−09° 27′ 47″
4718	Vir	Gxy	1.8′×0.6′	13.4	12 h 50 m 32.5 s	−05° 16′ 55″
4720	Vir	Gxy	0.9′×0.5′	13.3	12 h 50 m 42.7 s	−04° 09′ 22″
4731	Vir	Gxy	6.6′×3.2′	11	12 h 51 m 01.2 s	−06° 23′ 34″
4733	Vir	Gxy	2.0′×0.7′	12.3	12 h 51 m 06.9 s	+10° 54′ 43″
4734	Vir	Gxy	1.0′×0.9′	13.7	12 h 51 m 12.9 s	+04° 51′ 31″
4739	Vir	Gxy	1.6′×1.4′	12.5	12 h 51 m 37.1 s	−08° 24′ 38″
4742	Vir	Gxy	2.6′×1.5′	11	12 h 51 m 48.0 s	−10° 27′ 19″
4746	Vir	Gxy	3.8′×1.1′	12.1	12 h 51 m 55.3 s	+12° 04′ 58″
4753	Vir	Gxy	6.0′×2.8′	10.5	12 h 52 m 22.0 s	−01° 12′ 16″
4754	Vir	Gxy	4.6′×2.5′	10.3	12 h 52 m 17.5 s	+11° 18′ 49″
4757	Vir	Gxy	1.5′×0.4′	14.1	12 h 52 m 50.0 s	−10° 18′ 37″
4759	Vir	Gxy	0.4′×0.3′	13	12 h 53 m 05.1 s	−09° 12′ 06″
4760	Vir	Gxy	2.0′×1.9′	11.5	12 h 53 m 07.3 s	−10° 29′ 41″
4761	Vir	Gxy	0.4′×0.3′	13.9	12 h 53 m 09.7 s	−09° 11′ 52″
4762	Vir	Gxy	8.7′×1.7′	10	12 h 52 m 56.0 s	+11° 13′ 52″
4764	Vir	Gxy	…′×…′	15.1	12 h 53 m 06.7 s	−09° 15′ 28″
4765	Vir	Gxy	1.1′×0.8′	12.3	12 h 53 m 14.6 s	+04° 27′ 47″
4766	Vir	Gxy	1.1′×0.3′	14.4	12 h 53 m 08.1 s	−10° 22′ 42″
4768	Vir	*	…	…	12 h 53 m 17.2 s	−09° 31′ 54″
4769	Vir	**	…	…	12 h 53 m 18.0 s	−09° 32′ 10″
4770	Vir	Gxy	1.4′×0.7′	12.9	12 h 53 m 32.1 s	−09° 32′ 30″
4771	Vir	Gxy	3.9′×0.8′	12.3	12 h 53 m 21.1 s	+01° 16′ 10″
4772	Vir	Gxy	3.4′×1.7′	11.7	12 h 53 m 29.1 s	+02° 10′ 06″
4773	Vir	Gxy	1.3′×1.0′	13.9	12 h 53 m 35.9 s	−08° 38′ 35″
4775	Vir	Gxy	2.1′×2.0′	11.2	12 h 53 m 45.9 s	−06° 37′ 21″
4776	Vir	Gxy	…′×…′	13	12 h 53 m 04.4 s	−09° 11′ 56″
4777	Vir	Gxy	1.9′×0.9′	13.6	12 h 53 m 58.6 s	−08° 46′ 32″
4778	Vir	Gxy	…′×…′	12.7	12 h 53 m 05.9 s	−09° 12′ 17″
4779	Vir	Gxy	2.1′×1.8′	12.3	12 h 53 m 50.9 s	+09° 42′ 35″
4780	Vir	Gxy	1.9′×1.3′	13.4	12 h 54 m 05.1 s	−08° 37′ 16″
4781	Vir	Gxy	3.5′×1.5′	11.2	12 h 54 m 23.3 s	−10° 32′ 09″
4784	Vir	Gxy	1.9′×0.4′	14	12 h 54 m 37.0 s	−10° 36′ 46″
4786	Vir	Gxy	1.6′×1.3′	11.6	12 h 54 m 32.5 s	−06° 51′ 31″
4790	Vir	Gxy	1.7′×1.1′	12.3	12 h 54 m 51.9 s	−10° 14′ 52″

(continued)

Table 10.4 (continued)

NGC #	Const	Type	Size	V Mag	RA (2000)	Dec (2000)
4791	Vir	Gxy	1.6'×1.1'	14.1	12 h 54 m 44.1 s	+08° 03' 12"
4795	Vir	Gxy	1.9'×1.6'	12.4	12 h 55 m 02.9 s	+08° 03' 56"
4796	Vir	Gxy	0.4'×0.3'	14.5	12 h 55 m 04.7 s	+08° 03' 58"
4799	Vir	Gxy	1.3'×0.6'	13.6	12 h 55 m 15.5 s	+02° 53' 46"
4803	Vir	Gxy	0.6'×0.4'	14.4	12 h 55 m 33.6 s	+08° 14' 25"
4808	Vir	Gxy	2.8'×1.1'	11.9	12 h 55 m 49.1 s	+04° 18' 15"
4809	Vir	Gxy	1.7'×0.7'	14	12 h 54 m 51.1 s	+02° 39' 13"
4810	Vir	Gxy	0.8'×0.5'	14.5	12 h 54 m 51.2 s	+02° 38' 24"
4813	Vir	Gxy	1.2'×0.5'	13.5	12 h 56 m 36.3 s	−06° 49' 01"
4818	Vir	Gxy	4.3'×1.5'	11.2	12 h 56 m 48.9 s	−08° 31' 32"
4820	Vir	Gxy	1.0'×0.2'	13.9	12 h 57 m 00.5 s	−13° 43' 10"
4822	Vir	Gxy	1.2'×0.7'	13.5	12 h 57 m 03.6 s	−10° 45' 44"
4823	Vir	Gxy	0.7'×0.2'	14.7	12 h 57 m 25.5 s	−13° 41' 55"
4825	Vir	Gxy	1.9'×1.2'	11.8	12 h 57 m 12.2 s	−13° 39' 55"
4829	Vir	Gxy	0.4'×0.3'	14.8	12 h 57 m 24.4 s	−13° 44' 15"
4830	Vir	Gxy	1.9'×1.2'	11.8	12 h 57 m 27.9 s	−19° 41' 29"
4836	Vir	Gxy	1.4'×1.1'	13.4	12 h 57 m 34.3 s	−12° 44' 39"
4838	Vir	Gxy	1.6'×1.4'	13	12 h 57 m 56.3 s	−13° 03' 35"
4843	Vir	Gxy	2.1'×0.5'	13.2	12 h 58 m 00.7 s	−03° 37' 18"
4844	Vir	*	…	…	12 h 58 m 08.2 s	−13° 04' 48"
4845	Vir	Gxy	5.0'×1.3'	11.5	12 h 58 m 01.3 s	+01° 34' 32"
4847	Vir	Gxy	0.6'×0.5'	14.3	12 h 58 m 29.0 s	−13° 08' 26"
4855	Vir	Gxy	1.8'×1.3'	12.9	12 h 59 m 18.4 s	−13° 13' 52"
4856	Vir	Gxy	4.3'×1.2'	10.7	12 h 59 m 21.3 s	−15° 02' 32"
4862	Vir	Gxy	1.0'×0.8'	14.2	12 h 59 m 30.9 s	−14° 07' 55"
4863	Vir	Gxy	1.8'×0.4'	13.7	12 h 59 m 42.4 s	−14° 01' 47"
4866	Vir	Gxy	6.3'×1.3'	10.6	12 h 59 m 27.2 s	+14° 10' 17"
4877	Vir	Gxy	2.2'×1.0'	12.4	13 h 00 m 26.3 s	−15° 17' 00"
4878	Vir	Gxy	1.3'×1.1'	13.6	13 h 00 m 20.1 s	−06° 06' 14"
4879	Vir	*	…	…	13 h 00 m 25.6 s	−06° 06' 40"
4880	Vir	Gxy	3.2'×2.5'	11.5	13 h 00 m 10.5 s	+12° 29' 00"
4885	Vir	Gxy	0.5'×0.4'	14	13 h 00 m 33.8 s	−06° 51' 12"
4887	Vir	Gxy	1.1'×0.6'	13.6	13 h 00 m 39.1 s	−14° 39' 59"
4888	Vir	Gxy	0.8'×0.3'	13.4	13 h 00 m 36.5 s	−06° 04' 30"
4890	Vir	Gxy	1.1'×0.8'	13.2	13 h 00 m 39.0 s	−04° 36' 15"
4891	Vir	*	…	…	13 h 00 m 47.0 s	−13° 25' 34"
4897	Vir	Gxy	2.6'×2.3'	12	13 h 00 m 53.0 s	−13° 27' 00"
4899	Vir	Gxy	2.6'×1.4'	12.1	13 h 00 m 56.5 s	−13° 56' 41"
4900	Vir	Gxy	2.2'×2.1'	11.8	13 h 00 m 39.0 s	+02° 29' 59"
4902	Vir	Gxy	3.0'×2.7'	10.9	13 h 00 m 59.6 s	−14° 30' 49"
4904	Vir	Gxy	2.2'×1.4'	12.3	13 h 00 m 58.9 s	−00° 01' 45"
4910	Vir	Gxy	5.0'×1.6'	11.3	12 h 58 m 01.3 s	+01° 34' 32"
4915	Vir	Gxy	1.6'×1.3'	11.9	13 h 01 m 28.1 s	−04° 32' 48"
4918	Vir	Gxy	1.0'×0.5'	14.4	13 h 01 m 50.7 s	−04° 30' 05"
4920	Vir	Gxy	1.0'×0.7'	13.5	13 h 02 m 04.1 s	−11° 22' 42"

(continued)

Table 10.4 (continued)

NGC #	Const	Type	Size	V Mag	RA (2000)	Dec (2000)
4924	Vir	Gxy	1.2'×1.1'	12.8	13 h 02 m 12.9 s	−14° 58' 13"
4925	Vir	Gxy	0.6'×0.5'	13.1	13 h 02 m 07.5 s	−07° 42' 38"
4928	Vir	Gxy	1.3'×1.0'	12.7	13 h 03 m 00.4 s	−08° 05' 07"
4933	Vir	Gxy	2.0'×1.1'	12.1	13 h 03 m 56.9 s	−11° 29' 52"
4939	Vir	Gxy	5.5'×2.8'	11	13 h 04 m 14.3 s	−10° 20' 25"
4941	Vir	Gxy	3.6'×1.9'	11.4	13 h 04 m 13.1 s	−05° 33' 06"
4942	Vir	Gxy	1.9'×1.3'	13	13 h 04 m 19.1 s	−07° 39' 00"
4948	Vir	Gxy	2.0'×0.7'	14.6	13 h 04 m 55.7 s	−07° 56' 47"
4951	Vir	Gxy	3.3'×1.2'	12.1	13 h 05 m 07.7 s	−06° 29' 41"
4958	Vir	Gxy	4.1'×1.3'	10.7	13 h 05 m 48.9 s	−08° 01' 15"
4969	Vir	Gxy	0.7'×0.6'	13.9	13 h 07 m 02.8 s	+13° 38' 14"
4975	Vir	Gxy	1.0'×0.6'	13.9	13 h 07 m 50.0 s	−05° 01' 04"
4981	Vir	Gxy	2.8'×2.0'	11.5	13 h 08 m 48.7 s	−06° 46' 44"
4982	Vir	Ast	...	...	13 h 08 m 46.1 s	−10° 35' 19"
4984	Vir	Gxy	2.8'×2.2'	10.9	13 h 08 m 57.3 s	−15° 31' 01"
4989	Vir	Gxy	1.5'×0.9'	13.3	13 h 09 m 15.9 s	−05° 23' 49"
4990	Vir	Gxy	1.0'×0.8'	13.9	13 h 09 m 17.3 s	−05° 16' 23"
4991	Vir	Gxy	0.6'×0.4'	14.7	13 h 09 m 15.1 s	+02° 20' 51"
4992	Vir	Gxy	1.2'×0.7'	13.6	13 h 09 m 05.6 s	+11° 38' 02"
4995	Vir	Gxy	2.5'×1.6'	11.3	13 h 09 m 40.6 s	−07° 49' 59"
4996	Vir	Gxy	1.8'×1.4'	12.8	13 h 09 m 31.9 s	+00° 51' 25"
4997	Vir	Gxy	1.4'×1.0'	12.9	13 h 09 m 51.7 s	−16° 31' 00"
4999	Vir	Gxy	2.5'×1.9'	12	13 h 09 m 33.3 s	+01° 40' 23"
5006	Vir	Gxy	2.0'×1.7'	12.5	13 h 11 m 45.7 s	−19° 15' 41"
5010	Vir	Gxy	1.3'×0.6'	13.3	13 h 12 m 26.2 s	−15° 47' 51"
5013	Vir	Gxy	0.7'×0.4'	15.1	13 h 12 m 07.3 s	+03° 11' 57"
5015	Vir	Gxy	1.8'×1.5'	12.3	13 h 12 m 22.6 s	−04° 20' 17"
5017	Vir	Gxy	1.8'×1.3'	12.2	13 h 12 m 54.4 s	−16° 46' 01"
5018	Vir	Gxy	3.3'×2.5'	10.7	13 h 13 m 00.9 s	−19° 31' 06"
5019	Vir	Gxy	0.8'×0.7'	13.7	13 h 12 m 42.4 s	+04° 43' 47"
5020	Vir	Gxy	3.2'×2.7'	11.9	13 h 12 m 39.9 s	+12° 35' 58"
5022	Vir	Gxy	2.4'×0.4'	13.1	13 h 13 m 30.7 s	−19° 32' 47"
5027	Vir	Gxy	1.5'×1.3'	13.6	13 h 13 m 20.9 s	+06° 03' 41"
5028	Vir	Gxy	1.8'×0.9'	12.3	13 h 13 m 45.8 s	−13° 02' 33"
5030	Vir	Gxy	1.8'×1.3'	12.7	13 h 13 m 54.1 s	−16° 29' 29"
5031	Vir	Gxy	1.6'×0.4'	13.5	13 h 14 m 03.3 s	−16° 07' 24"
5035	Vir	Gxy	1.4'×1.1'	12.8	13 h 14 m 49.3 s	−16° 29' 33"
5036	Vir	Gxy	0.7'×0.5'	14.6	13 h 14 m 42.8 s	−04° 10' 42"
5037	Vir	Gxy	2.2'×0.7'	12.5	13 h 14 m 59.3 s	−16° 35' 28"
5038	Vir	Gxy	1.4'×0.3'	13.5	13 h 15 m 02.3 s	−15° 57' 07"
5039	Vir	Gxy	0.7'×0.3'	15.4	13 h 14 m 51.9 s	−04° 09' 29"
5044	Vir	Gxy	3.0'×3.0'	10.9	13 h 15 m 23.9 s	−16° 23' 09"
5046	Vir	Gxy	0.8'×0.7'	13.5	13 h 15 m 45.1 s	−16° 19' 37"
5047	Vir	Gxy	2.8'×0.6'	12.6	13 h 15 m 48.4 s	−16° 31' 10"
5049	Vir	Gxy	1.9'×0.6'	13	13 h 15 m 59.2 s	−16° 23' 52"

(continued)

Table 10.4 (continued)

NGC #	Const	Type	Size	V Mag	RA (2000)	Dec (2000)
5050	Vir	Gxy	1.1'×0.4'	13.9	13 h 15 m 41.7 s	+02° 52' 44"
5054	Vir	Gxy	5.1'×3.0'	10.8	13 h 16 m 58.3 s	−16° 38' 07"
5058	Vir	Gxy	0.8'×0.7'	13.8	13 h 16 m 52.3 s	+12° 32' 54"
5059	Vir	Gxy	1.0'×0.2'	14.9	13 h 16 m 58.5 s	+07° 50' 40"
5060	Vir	Gxy	1.1'×0.8'	13.5	13 h 17 m 16.3 s	+06° 02' 15"
5066	Vir	Gxy	0.5'×0.4'	13.5	13 h 18 m 28.4 s	−10° 14' 02"
5067	Vir	**	...	...	13 h 18 m 27.7 s	−10° 08' 43"
5068	Vir	Gxy	7.2'×6.3'	9.8	13 h 18 m 54.6 s	−21° 02' 19"
5069	Vir	Gxy	0.5'×0.4'	13.5	13 h 18 m 28.4 s	−10° 14' 02"
5070	Vir	Gxy	0.7'×0.2'	13.5	13 h 19 m 12.4 s	−12° 32' 23"
5071	Vir	Gxy	0.4'×0.2'	14.5	13 h 18 m 37.2 s	+07° 56' 07"
5072	Vir	Gxy	1.0'×0.7'	13.5	13 h 19 m 12.4 s	−12° 32' 23"
5073	Vir	Gxy	3.4'×0.7'	12.6	13 h 19 m 20.9 s	−14° 50' 41"
5075	Vir	Gxy	1.0'×0.7'	13.9	13 h 19 m 06.2 s	+07° 49' 51"
5076	Vir	Gxy	1.3'×0.7'	13.1	13 h 19 m 30.5 s	−12° 44' 28"
5077	Vir	Gxy	1.9'×1.5'	11.3	13 h 19 m 31.6 s	−12° 39' 23"
5079	Vir	Gxy	1.6'×1.0'	12	13 h 19 m 37.9 s	−12° 41' 55"
5080	Vir	Gxy	1.2'×1.1'	13.6	13 h 19 m 19.1 s	+08° 25' 44"
5084	Vir	Gxy	9.3'×1.7'	10.8	13 h 20 m 16.8 s	−21° 49' 40"
5087	Vir	Gxy	2.3'×1.7'	11.2	13 h 20 m 24.9 s	−20° 36' 40"
5088	Vir	Gxy	2.6'×0.8'	12.6	13 h 20 m 20.1 s	−12° 34' 18"
5094	Vir	Gxy	0.8'×0.6'	13.2	13 h 20 m 46.7 s	−14° 04' 50"
5095	Vir	Gxy	1.1'×0.4'	13.8	13 h 20 m 36.8 s	−02° 17' 22"
5097	Vir	Gxy	0.5'×0.3'	14.6	13 h 20 m 59.6 s	−12° 28' 16"
5099	Vir	Gxy	0.6'×0.6'	14.3	13 h 21 m 19.5 s	−13° 02' 33"
5100	Vir	Gxy	1.3'×0.9'	14.3	13 h 20 m 59.5 s	+08° 58' 42"
5104	Vir	Gxy	1.2'×0.4'	13.9	13 h 21 m 23.0 s	+00° 20' 31"
5105	Vir	Gxy	2.0'×1.5'	12.1	13 h 21 m 49.1 s	−13° 12' 24"
5106	Vir	Gxy	1.3'×0.9'	14.3	13 h 20 m 59.5 s	+08° 58' 42"
5110	Vir	Gxy	2.0'×1.7'	13.3	13 h 22 m 56.4 s	−12° 57' 53"
5111	Vir	Gxy	2.0'×1.7'	13.3	13 h 22 m 56.4 s	−12° 57' 53"
5115	Vir	Gxy	1.4'×0.7'	13.9	13 h 23 m 00.4 s	+13° 57' 03"
5118	Vir	Gxy	0.8'×0.7'	13.9	13 h 23 m 27.5 s	+06° 23' 33"
5119	Vir	Gxy	1.2'×0.3'	13	13 h 24 m 00.1 s	−12° 16' 36"
5122	Vir	Gxy	0.8'×0.3'	13.4	13 h 24 m 14.9 s	−10° 39' 17"
5125	Vir	Gxy	1.9'×1.4'	12.6	13 h 24 m 00.6 s	+09° 42' 36"
5129	Vir	Gxy	1.7'×1.4'	12.1	13 h 24 m 10.0 s	+13° 58' 35"
5130	Vir	Gxy	0.8'×0.4'	13.5	13 h 24 m 27.1 s	−10° 12' 38"
5132	Vir	Gxy	1.3'×0.9'	13	13 h 24 m 28.9 s	+14° 05' 32"
5133	Vir	Gxy	1.0'×0.7'	13.7	13 h 24 m 52.9 s	−04° 04' 56"
5134	Vir	Gxy	2.8'×1.7'	11.6	13 h 25 m 18.5 s	−21° 08' 04"
5136	Vir	Gxy	0.7'×0.6'	14	13 h 24 m 51.3 s	+13° 44' 16"
5137	Vir	Gxy	0.6'×0.3'	14.8	13 h 24 m 52.5 s	+14° 04' 35"
5146	Vir	Gxy	0.6'×0.4'	12.6	13 h 26 m 37.5 s	−12° 19' 26"
5147	Vir	Gxy	1.9'×1.5'	12.1	13 h 26 m 19.8 s	+02° 06' 02"

(continued)

Table 10.4 (continued)

NGC #	Const	Type	Size	V Mag	RA (2000)	Dec (2000)
5148	Vir	Gxy	0.9'×0.8'	14.3	13 h 26 m 38.7 s	+02° 18' 47"
5159	Vir	Gxy	1.3'×0.4'	14.3	13 h 28 m 16.1 s	+02° 58' 59"
5160	Vir	**	...	...	13 h 28 m 21.6 s	+05° 59' 44"
5162	Vir	Gxy	3.2'×1.7'	11.6	13 h 29 m 25.9 s	+11° 00' 28"
5165	Vir	Gxy	0.6'×0.4'	13.9	13 h 28 m 39.2 s	+11° 23' 14"
5167	Vir	Gxy	0.9'×0.9'	13.9	13 h 28 m 40.2 s	+12° 42' 41"
5170	Vir	Gxy	8.3'×1.0'	11.6	13 h 29 m 48.7 s	−17° 57' 57"
5171	Vir	Gxy	1.1'×0.8'	13.4	13 h 29 m 21.5 s	+11° 44' 05"
5174	Vir	Gxy	3.2'×1.7'	11.6	13 h 29 m 25.9 s	+11° 00' 28"
5175	Vir	*	...	...	13 h 29 m 26.3 s	+10° 59' 42"
5176	Vir	Gxy	0.4'×0.4'	14.4	13 h 29 m 25.0 s	+11° 46' 53"
5177	Vir	Gxy	0.6'×0.2'	14.6	13 h 29 m 24.2 s	+11° 47' 49"
5178	Vir	Gxy	1.1'×0.7'	14	13 h 29 m 29.3 s	+11° 37' 28"
5179	Vir	Gxy	0.6'×0.4'	14.2	13 h 29 m 30.9 s	+11° 44' 44"
5181	Vir	Gxy	0.4'×0.3'	13.9	13 h 29 m 41.9 s	+13° 18' 14"
5183	Vir	Gxy	1.9'×0.8'	12.9	13 h 30 m 06.0 s	−01° 43' 16"
5184	Vir	Gxy	1.9'×1.1'	12.8	13 h 30 m 11.5 s	−01° 39' 48"
5185	Vir	Gxy	1.9'×0.7'	13.5	13 h 30 m 02.3 s	+13° 24' 58"
5186	Vir	Gxy	0.4'×0.3'	14.8	13 h 30 m 03.8 s	+12° 10' 28"
5191	Vir	Gxy	0.6'×0.4'	14.1	13 h 30 m 47.3 s	+11° 12' 02"
5192	Vir	Gxy	0.6'×0.3'	14.8	13 h 30 m 51.7 s	−01° 46' 43"
5196	Vir	Gxy	0.8'×0.7'	14.1	13 h 31 m 19.6 s	−01° 36' 55"
5197	Vir	Gxy	0.6'×0.3'	14.3	13 h 31 m 25.0 s	−01° 41' 35"
5200	Vir	**	...	...	13 h 31 m 42.3 s	−00° 01' 48"
5202	Vir	Gxy	1.2'×0.3'	14.7	13 h 32 m 00.4 s	−01° 41' 57"
5203	Vir	Gxy	1.9'×1.1'	12.6	13 h 32 m 13.4 s	−08° 47' 12"
5207	Vir	Gxy	1.7'×1.0'	13.4	13 h 32 m 14.0 s	+13° 53' 32"
5208	Vir	Gxy	1.7'×0.6'	13.2	13 h 32 m 27.9 s	+07° 18' 58"
5209	Vir	Gxy	1.2'×1.1'	13.1	13 h 32 m 42.5 s	+07° 19' 37"
5210	Vir	Gxy	1.3'×1.3'	13.2	13 h 32 m 49.3 s	+07° 10' 13"
5211	Vir	Gxy	2.1'×1.6'	12.5	13 h 33 m 05.3 s	−01° 02' 10"
5212	Vir	Gxy	0.4'×0.3'	14.3	13 h 32 m 56.1 s	+07° 17' 16"
5213	Vir	Gxy	1.0'×0.9'	13.9	13 h 34 m 39.2 s	+04° 07' 48"
5221	Vir	Gxy	2.4'×0.8'	13.2	13 h 34 m 55.9 s	+13° 49' 56"
5222	Vir	Gxy	1.6'×1.2'	13.1	13 h 34 m 55.9 s	+13° 44' 31"
5224	Vir	Gxy	0.6'×0.6'	14	13 h 35 m 08.9 s	+06° 28' 52"
5226	Vir	Gxy	0.3'×0.1'	15.8	13 h 35 m 03.6 s	+13° 55' 21"
5227	Vir	Gxy	1.8'×1.5'	13.1	13 h 35 m 24.5 s	+01° 24' 37"
5230	Vir	Gxy	2.2'×1.9'	12.3	13 h 35 m 31.8 s	+13° 40' 33"
5231	Vir	Gxy	1.2'×1.1'	13.6	13 h 35 m 48.3 s	+02° 59' 57"
5232	Vir	Gxy	1.5'×1.4'	13	13 h 36 m 08.2 s	−08° 29' 52"
5235	Vir	Gxy	1.2'×0.5'	14	13 h 36 m 01.3 s	+06° 35' 06"
5241	Vir	Gxy	1.1'×0.5'	14.2	13 h 36 m 39.9 s	−08° 24' 08"
5242	Vir	NF	...	...	13 h 37 m 07.3 s	+02° 46' 14"
5245	Vir	Gxy	0.8'×0.3'	14.2	13 h 37 m 23.3 s	+03° 53' 51"

(continued)

Table 10.4 (continued)

NGC #	Const	Type	Size	V Mag	RA (2000)	Dec (2000)
5246	Vir	Gxy	1.0'×0.8'	13.9	13 h 37 m 29.5 s	+04° 06' 17"
5247	Vir	Gxy	5.6'×4.9'	10.4	13 h 38 m 03.0 s	−17° 53' 03"
5252	Vir	Gxy	1.4'×0.9'	13.1	13 h 38 m 15.9 s	+04° 32' 32"
5254	Vir	Gxy	3.2'×1.5'	12.4	13 h 39 m 37.9 s	−11° 29' 39"
5257	Vir	Gxy	1.8'×0.9'	12.9	13 h 39 m 52.9 s	+00° 50' 24"
5258	Vir	Gxy	1.7'×1.1'	12.9	13 h 39 m 57.7 s	+00° 49' 52"
5261	Vir	Gxy	0.8'×0.4'	14.2	13 h 40 m 16.1 s	+05° 04' 35"
5268	Vir	*	...	...	13 h 42 m 12.5 s	−13° 51' 35"
5270	Vir	Gxy	1.2'×0.9'	13.7	13 h 42 m 10.9 s	+04° 15' 44"
5285	Vir	Gxy	0.7'×0.7'	14.1	13 h 44 m 25.8 s	+02° 06' 36"
5300	Vir	Gxy	3.9'×2.6'	11.6	13 h 48 m 15.9 s	+03° 57' 02"
5306	Vir	Gxy	1.4'×1.1'	12.2	13 h 49 m 11.2 s	−07° 13' 26"
5309	Vir	Gxy	0.8'×0.7'	...	13 h 50 m 10.9 s	−15° 37' 05"
5310	Vir	*	...	...	13 h 49 m 47.8 s	+00° 04' 08"
5317	Vir	Gxy	6.8'×4.4'	10.4	13 h 56 m 11.9 s	+05° 00' 52"
5324	Vir	Gxy	2.3'×2.1'	11.9	13 h 52 m 05.9 s	−06° 03' 30"
5327	Vir	Gxy	1.9'×1.6'	12.8	13 h 52 m 04.1 s	−02° 12' 23"
5329	Vir	Gxy	1.3'×1.3'	12.4	13 h 52 m 10.0 s	+02° 19' 29"
5331	Vir	Gxy	0.6'×0.4'	14.1	13 h 52 m 16.3 s	+02° 06' 28"
5334	Vir	Gxy	4.2'×3.0'	11.5	13 h 52 m 54.4 s	−01° 06' 52"
5335	Vir	Gxy	2.1'×1.4'	13	13 h 52 m 56.4 s	+02° 48' 50"
5338	Vir	Gxy	2.5'×1.4'	12.5	13 h 53 m 26.6 s	+05° 12' 26"
5339	Vir	Gxy	1.7'×1.5'	12.1	13 h 54 m 00.3 s	−07° 55' 51"
5343	Vir	Gxy	1.5'×1.2'	12.9	13 h 54 m 11.7 s	−07° 35' 20"
5345	Vir	Gxy	1.6'×1.5'	12.6	13 h 54 m 14.2 s	−01° 26' 12"
5348	Vir	Gxy	3.5'×0.5'	13.3	13 h 54 m 11.3 s	+05° 13' 38"
5356	Vir	Gxy	3.1'×0.9'	12.8	13 h 54 m 58.5 s	+05° 20' 01"
5360	Vir	Gxy	2.2'×0.8'	13.7	13 h 55 m 38.2 s	+04° 59' 02"
5363	Vir	Gxy	4.1'×2.6'	10.8	13 h 56 m 07.1 s	+05° 15' 19"
5364	Vir	Gxy	6.8'×4.4'	10.5	13 h 56 m 11.9 s	+05° 00' 52"
5366	Vir	Gxy	0.7'×0.6'	13.8	13 h 56 m 24.9 s	−00° 14' 49"
5369	Vir	Gxy	0.9'×0.8'	13.4	13 h 56 m 37.5 s	−05° 28' 14"
5373	Vir	Gxy	0.5'×0.4'	14	13 h 57 m 07.5 s	+05° 15' 06"
5374	Vir	Gxy	1.7'×1.5'	12.7	13 h 57 m 29.6 s	+06° 05' 49"
5382	Vir	Gxy	1.5'×1.1'	12.7	13 h 58 m 14.9 s	+06° 15' 31"
5384	Vir	Gxy	1.5'×0.8'	12.9	13 h 58 m 12.9 s	+06° 31' 04"
5386	Vir	Gxy	1.0'×0.4'	13.4	13 h 58 m 22.2 s	+06° 20' 21"
5387	Vir	Gxy	1.8'×0.3'	14.1	13 h 58 m 24.8 s	+06° 04' 17"
5388	Vir	NF	...	...	13 h 58 m 57.9 s	−14° 09' 03"
5392	Vir	Gxy	1.2'×0.8'	13.6	13 h 59 m 24.9 s	−03° 12' 31"
5400	Vir	Gxy	1.5'×0.9'	13	14 h 00 m 37.1 s	−02° 51' 27"
5404	Vir	**	...	...	14 h 01 m 07.5 s	+00° 05' 15"
5420	Vir	Gxy	1.4'×0.6'	13.1	14 h 03 m 59.8 s	−14° 37' 00"
5426	Vir	Gxy	3.0'×1.6'	12.1	14 h 03 m 25.1 s	−06° 04' 08"
5427	Vir	Gxy	2.8'×2.4'	11.5	14 h 03 m 26.1 s	−06° 01' 51"

(continued)

Table 10.4 (continued)

NGC #	Const	Type	Size	V Mag	RA (2000)	Dec (2000)
5428	Vir	**	...	...	14 h 03 m 28.0 s	−05° 59' 04"
5429	Vir	**	...	...	14 h 03 m 33.3 s	−06° 02' 18"
5432	Vir	***	...	...	14 h 03 m 40.6 s	−05° 58' 32"
5435	Vir	**	...	...	14 h 04 m 00.0 s	−05° 55' 54"
5442	Vir	Gxy	1.2'×0.5'	13.7	14 h 04 m 43.3 s	−09° 42' 47"
5465	Vir	*	...	...	14 h 06 m 27.3 s	−05° 30' 24"
5467	Vir	*	...	...	14 h 06 m 29.5 s	−05° 28' 53"
5468	Vir	Gxy	2.6'×2.4'	11.7	14 h 06 m 34.9 s	−05° 27' 11"
5470	Vir	Gxy	2.5'×0.4'	13.6	14 h 06 m 31.9 s	+06° 01' 46"
5472	Vir	Gxy	1.3'×0.4'	14.3	14 h 06 m 54.9 s	−05° 27' 38"
5476	Vir	Gxy	1.4'×1.1'	12.8	14 h 08 m 08.5 s	−06° 05' 34"
5478	Vir	Gxy	1.1'×0.8'	13.8	14 h 08 m 08.5 s	−01° 42' 09"
5491	Vir	Gxy	1.4'×0.8'	13.1	14 h 10 m 57.4 s	+06° 21' 53"
5493	Vir	Gxy	1.6'×1.3'	11.6	14 h 11 m 29.3 s	−05° 02' 39"
5496	Vir	Gxy	4.7'×0.9'	12.2	14 h 11 m 37.9 s	−01° 09' 27"
5501	Vir	Gxy	1.1'×0.8'	13.6	14 h 12 m 20.1 s	+01° 16' 19"
5506	Vir	Gxy	2.8'×0.9'	12.5	14 h 13 m 14.7 s	−03° 12' 27"
5507	Vir	Gxy	1.7'×0.9'	12.6	14 h 13 m 19.7 s	−03° 08' 55"
5510	Vir	Gxy	1.4'×1.1'	13	14 h 13 m 37.2 s	−17° 59' 02"
5521	Vir	Gxy	0.6'×0.5'	13.6	14 h 15 m 23.7 s	+04° 24' 30"
5534	Vir	Gxy	1.4'×0.8'	12.8	14 h 17 m 40.3 s	−07° 25' 02"
5537	Vir	Gxy	1.0'×0.5'	14.3	14 h 17 m 37.0 s	+07° 03' 16"
5549	Vir	Gxy	1.6'×0.8'	13	14 h 18 m 38.7 s	+07° 22' 37"
5551	Vir	Gxy	0.4'×0.3'	14.2	14 h 18 m 54.9 s	+05° 27' 04"
5552	Vir	Gxy	0.8'×0.3'	14.1	14 h 19 m 04.0 s	+07° 01' 54"
5554	Vir	Gxy	0.7'×0.5'	14.4	14 h 19 m 15.1 s	+07° 01' 14"
5555	Vir	Gxy	1.0'×0.5'	14.9	14 h 18 m 48.1 s	−19° 08' 20"
5558	Vir	Gxy	0.8'×0.3'	14.1	14 h 19 m 04.0 s	+07° 01' 54"
5560	Vir	Gxy	3.7'×0.7'	12.7	14 h 20 m 04.4 s	+03° 59' 32"
5563	Vir	Gxy	0.7'×0.4'	14.5	14 h 20 m 13.0 s	+07° 03' 19"
5564	Vir	Gxy	0.7'×0.5'	14.4	14 h 19 m 15.1 s	+07° 01' 14"
5565	Vir	*	...	...	14 h 19 m 18.5 s	+06° 59' 42"
5566	Vir	Gxy	6.6'×2.2'	10.7	14 h 20 m 19.9 s	+03° 56' 02"
5569	Vir	Gxy	1.7'×1.4'	13.4	14 h 20 m 32.2 s	+03° 58' 59"
5573	Vir	Gxy	1.2'×0.2'	14.5	14 h 20 m 41.5 s	+06° 54' 27"
5574	Vir	Gxy	1.6'×1.0'	12.4	14 h 20 m 56.1 s	+03° 14' 17"
5575	Vir	Gxy	0.9'×0.9'	13.4	14 h 20 m 59.3 s	+06° 12' 09"
5576	Vir	Gxy	3.5'×2.2'	10.9	14 h 21 m 03.7 s	+03° 16' 15"
5577	Vir	Gxy	3.4'×1.0'	12.5	14 h 21 m 13.1 s	+03° 26' 10"
5578	Vir	Gxy	1.0'×0.9'	13.4	14 h 20 m 59.3 s	+06° 12' 09"
5584	Vir	Gxy	3.4'×2.5'	11.6	14 h 22 m 23.7 s	−00° 23' 09"
5599	Vir	Gxy	1.4'×0.5'	13.8	14 h 23 m 50.5 s	+06° 34' 34"
5604	Vir	Gxy	1.8'×1.0'	12.8	14 h 24 m 42.7 s	−03° 12' 43"
5618	Vir	Gxy	1.6'×1.2'	13.5	14 h 27 m 11.8 s	−02° 15' 46"
5619	Vir	Gxy	2.2'×1.2'	12.8	14 h 27 m 18.1 s	+04° 48' 08"

(continued)

Table 10.4 (continued)

NGC #	Const	Type	Size	V Mag	RA (2000)	Dec (2000)
5632	Vir	*	...	...	14 h 29 m 19.5 s	−00° 26' 51"
5634	Vir	GC	4.9'	9.5	14 h 29 m 37.3 s	−05° 58' 36"
5636	Vir	Gxy	1.9'×1.4'	12.8	14 h 29 m 38.9 s	+03° 15' 59"
5638	Vir	Gxy	2.7'×2.4'	11.2	14 h 29 m 40.4 s	+03° 14' 00"
5645	Vir	Gxy	2.4'×1.5'	12.1	14 h 30 m 39.5 s	+07° 16' 28"
5650	Vir	Gxy	2.1'×1.5'	12.8	14 h 31 m 01.1 s	+05° 58' 41"
5651	Vir	*	...	...	14 h 31 m 12.8 s	−00° 19' 20"
5652	Vir	Gxy	2.0'×1.4'	12.7	14 h 31 m 01.1 s	+05° 58' 41"
5658	Vir	*	...	...	14 h 31 m 55.2 s	−00° 22' 01"
5661	Vir	Gxy	1.5'×0.6'	13.5	14 h 31 m 57.4 s	+06° 15' 01"
5668	Vir	Gxy	3.3'×3.0'	11.5	14 h 33 m 24.3 s	+04° 27' 02"
5674	Vir	Gxy	1.1'×1.0'	13.2	14 h 33 m 52.3 s	+05° 27' 31"
5679	Vir	Gxy	1.1'×0.6'	13	14 h 35 m 08.8 s	+05° 21' 31"
5680	Vir	Gxy	0.9'×0.9'	13.8	14 h 35 m 44.5 s	−00° 00' 48"
5690	Vir	Gxy	3.4'×1.0'	12.2	14 h 37 m 40.9 s	+02° 17' 27"
5691	Vir	Gxy	1.9'×1.4'	12.1	14 h 37 m 53.3 s	−00° 23' 52"
5692	Vir	Gxy	0.9'×0.6'	12.7	14 h 38 m 18.1 s	+03° 24' 37"
5701	Vir	Gxy	4.3'×4.1'	11	14 h 39 m 11.1 s	+05° 21' 48"
5705	Vir	Gxy	2.9'×1.7'	12.8	14 h 39 m 49.6 s	−00° 43' 08"
5713	Vir	Gxy	2.8'×2.5'	11.1	14 h 40 m 11.3 s	−00° 17' 27"
5718	Vir	Gxy	1.5'×1.1'	13	14 h 40 m 42.9 s	+03° 27' 55"
5719	Vir	Gxy	3.2'×1.2'	12.4	14 h 40 m 56.4 s	−00° 19' 05"
5725	Vir	Gxy	1.1'×0.7'	13.8	14 h 40 m 58.2 s	+02° 11' 09"
5733	Vir	Gxy	1.1'×0.4'	14.1	14 h 42 m 45.8 s	−00° 21' 07"
5738	Vir	Gxy	1.0'×0.3'	14	14 h 43 m 56.3 s	+01° 36' 15"
5740	Vir	Gxy	3.0'×1.5'	12	14 h 44 m 24.3 s	+01° 40' 47"
5746	Vir	Gxy	7.4'×1.3'	10.8	14 h 44 m 55.9 s	+01° 57' 21"
5750	Vir	Gxy	3.0'×1.6'	11.7	14 h 46 m 11.0 s	−00° 13' 25"
5765	Vir	Gxy	0.9'×0.4'	14.6	14 h 50 m 51.1 s	+05° 07' 03"
5770	Vir	Gxy	1.7'×1.3'	12.2	14 h 53 m 15.0 s	+03° 57' 35"
5774	Vir	Gxy	3.0'×2.5'	12.3	14 h 53 m 42.6 s	+03° 34' 58"
5775	Vir	Gxy	4.2'×1.0'	12	14 h 53 m 57.4 s	+03° 32' 41"
5776	Vir	Gxy	0.9'×0.7'	14.1	14 h 54 m 32.7 s	+02° 58' 00"
5806	Vir	Gxy	3.1'×1.7'	11.8	15 h 00 m 00.2 s	+01° 53' 28"
5811	Vir	Gxy	0.9'×0.8'	14.1	15 h 00 m 27.1 s	+01° 37' 25"
5813	Vir	Gxy	4.2'×3.0'	10.9	15 h 01 m 11.2 s	+01° 42' 06"
5814	Vir	Gxy	0.9'×0.5'	14	15 h 01 m 21.1 s	+01° 38' 13"
5831	Vir	Gxy	2.0'×1.9'	11.6	15 h 04 m 06.9 s	+01° 13' 10"
5838	Vir	Gxy	4.2'×1.5'	10.8	15 h 05 m 26.1 s	+02° 05' 57"
5839	Vir	Gxy	1.3'×1.2'	12.7	15 h 05 m 27.5 s	+01° 38' 04"
5841	Vir	Gxy	1.1'×0.4'	13.9	15 h 06 m 34.9 s	+02° 00' 18"
5845	Vir	Gxy	0.8'×0.5'	13.3	15 h 06 m 00.8 s	+01° 38' 01"
5846	Vir	Gxy	4.1'×3.8'	10.3	15 h 06 m 29.3 s	+01° 36' 22"
5847	Vir	Gxy	0.7'×0.4'	14.9	15 h 06 m 22.3 s	+06° 22' 47"
5848	Vir	Gxy	1.0'×0.4'	13.9	15 h 06 m 34.9 s	+02° 00' 18"

(continued)

Table 10.4   (continued)

NGC #	Const	Type	Size	V Mag	RA (2000)	Dec (2000)
5850	Vir	Gxy	4.3'×3.7'	11.4	15 h 07 m 07.6 s	+01° 32' 39"
5854	Vir	Gxy	2.8'×0.8'	12	15 h 07 m 47.7 s	+02° 34' 07"
5855	Vir	Gxy	0.4'×0.4'	14.2	15 h 07 m 49.1 s	+03° 59' 03"
5864	Vir	Gxy	2.8'×0.9'	11.9	15 h 09 m 33.6 s	+03° 03' 10"
5865	Vir	Gxy	1.1'×1.0'	14.1	15 h 09 m 49.2 s	+00° 31' 46"
5868	Vir	Gxy	1.1'×1.0'	13.5	15 h 09 m 49.2 s	+00° 31' 46"
5869	Vir	Gxy	2.3'×1.7'	12	15 h 09 m 49.4 s	+00° 28' 11"
5871	Vir	*	...	...	15 h 10 m 04.7 s	+00° 29' 53"
2305	Vol	Gxy	2.0'×1.5'	11.9	06 h 48 m 37.3 s	−64° 16' 23"
2307	Vol	Gxy	1.7'×1.6'	12.8	06 h 48 m 50.9 s	−64° 20' 05"
2348	Vol	OC	18.0'×10.0'	...	07 h 03 m 02.6 s	−67° 23' 38"
2397	Vol	Gxy	2.7'×1.3'	12.3	07 h 21 m 19.5 s	−69° 00' 07"
2434	Vol	Gxy	2.5'×2.3'	11.5	07 h 34 m 51.5 s	−69° 17' 05"
2442	Vol	Gxy	6.0'×5.5'	10.5	07 h 36 m 19.7 s	−69° 32' 31"
2443	Vol	Gxy	6.0'×5.5'	10.4	07 h 36 m 31.0 s	−69° 31' 08"
2466	Vol	Gxy	1.5'×1.4'	12.9	07 h 45 m 15.6 s	−71° 24' 40"
2601	Vol	Gxy	1.6'×1.1'	12.7	08 h 25 m 30.4 s	−68° 07' 06"
6748	Vul	NF	...	...	19 h 03 m 50.0 s	+21° 36' 32"
6793	Vul	OC	6'	...	19 h 23 m 14.3 s	+22° 08' 28"
6800	Vul	OC	15.0'	...	19 h 27 m 07.6 s	+25° 08' 26"
6802	Vul	OC	3.2'	8.8	19 h 30 m 35.0 s	+20° 15' 39"
6813	Vul	Neb	1.0'	...	19 h 40 m 22.3 s	+27° 18' 33"
6815	Vul	OC	...	...	19 h 40 m 44.3 s	+26° 45' 32"
6820	Vul	Neb	0.5'×0.5'	...	19 h 42 m 27.9 s	+23° 05' 17"
6823	Vul	OC+Neb	12'	7.1	19 h 43 m 09.8 s	+23° 18' 00"
6827	Vul	OC	4.0'	...	19 h 48 m 53.4 s	+21° 12' 54"
6830	Vul	OC	12'	7.9	19 h 50 m 59.5 s	+23° 06' 00"
6842	Vul	PN	53"×48"	13.1	19 h 55 m 02.3 s	+29° 17' 20"
6853	Vul	PN	8.0'×5.7'	7.4	19 h 59 m 36.3 s	+22° 43' 17"
6882	Vul	OC	20'	8.1	20 h 11 m 55.8 s	+26° 29' 20"
6885	Vul	OC	20'	8.1	20 h 11 m 55.8 s	+26° 29' 20"
6904	Vul	OC	3'	...	20 h 21 m 48.1 s	+25° 44' 30"
6921	Vul	Gxy	0.9'×0.2'	13.5	20 h 28 m 28.9 s	+25° 43' 24"
6938	Vul	OC	18'×8'	...	20 h 34 m 32.4 s	+22° 12' 52"
6940	Vul	OC	31'	6.3	20 h 34 m 26.6 s	+28° 16' 58"
7052	Vul	Gxy	2.5'×1.4'	12.5	21 h 18 m 33.1 s	+26° 26' 49"
7080	Vul	Gxy	1.8'×1.7'	13.1	21 h 30 m 01.9 s	+26° 43' 03"

Chapter 11

Halton Arp and
the Arp Catalog

Halton Arp is a cum laude graduate of Harvard and California Institute of Technology. He has won many awards, including the Newcomb Cleveland research award, the Alexander von Humboldt Senior Scientist Award, and the Helen B. Warner prize. Early in his career he conducted observations at the Mt. Wilson Observatory and the Palomar Observatory. By 1957 he became a full time staff member at the Palomar Observatory and continued to work there for 29 years.

A major area of research during this time was the classification of galaxies (galaxy morphology). Many astronomers were creating mathematical models of galaxies to explore the processes that shape them. Halton Arp was not interested in the perfectly formed galaxies used in some of these models. The process of perfectly formed galaxies was pretty well understood. The misshapen and malformed galaxies did not easily fit into the models. Halton Arp created his catalog so the galaxy modelers would have a large variety of special cases to examine. These special cases would eventually lead to today's more comprehensive understanding of galaxies.

Halton Arp was not the only astronomer that investigated these peculiar galaxies. In 1959 the Russian Astronomer B. A. Vorontsoc-Velyaminov published a catalog of 355 interacting galaxies he collected from the National Geographic Society Palomar Sky Survey. Arp also was inspired by a fellow astronomer at the Palomar Observatory, Fritz Zwicky (1898–1974). Zwicky developed a method for investigating the relationships between all forms of galaxies. In 1969 Zwicky published the book *Discovery, Invention, Research Through the Morphological Approach*, documenting his findings.

Another early attempt at cataloging these peculiar galaxies was made by Gerard de Vaucouleurs. In the 1950s de Vaucouleurs created a catalog of over 500 images

J.D. Cavin, *The Amateur Astronomer's Guide to the Deep-Sky Catalogs*,
Patrick Moore's Practical Astronomy Series, DOI 10.1007/978-1-4614-0656-3_11,
© Springer Science+Business Media, LLC 2012

of galaxies to help in developing an improved Hubble classification system. His illustrated catalog *The de Vaucouleurs Atlas of Galaxies* still provides an up to date reference of galaxy morphological processes and is still used by professionals and amateurs to aid in galaxy classifications.

The Arp Catalog contains 338 galaxies that exhibit irregular structures. When Halton Arp created the *Arp Catalog of Peculiar Galaxies* astronomers were trying to understand the processes that produced their misshapen form. The catalog provided his fellow astronomers with 338 examples to study every effect in great detail. The *Arp Catalog of Peculiar Galaxies* is divided into the following object types:

- Objects 1–101: Individual spiral galaxies and spiral galaxies with a companion.
- Objects 102–145: Elliptical and elliptical-like galaxies.
- Objects 146–268: Individual and groups of galaxies.
- Objects 269–327: Double galaxies.
- Objects 332–338: Do not fit into any category.

Tables 11.1–11.5 of the Arp catalog are provided with the permission of Dennis Web.

Table 11.1 The Arp catalog of spiral galaxies (Printed with the permission of Dennis Web)

Arp #	Name	FOV	Or	RA (2000)	Dec (2000)
Low surface brightness					
1	NGC 2857	5.2'	E	09 h 24.63 m	48° 21.4 min
2	UGC 10310	3.5'	E	16 h 16.30 m	47° 02.8 min
3	MCG-01-57-016	5.2'	N	22 h 36.57 m	−02° 54.3 min
4	MCG-02-05-050+A	3.5'	E	01 h 48.43 m	−12° 22.9 min
5	NGC 3664	2.6'	N	11 h 24.41 m	03° 19.6 min
6	NGC 2537	2.6'	E	08 h 13.24 m	45° 59.5 min
Split arm					
7	MCG-03-23-009	2.6'	N	08 h 50.29 m	−16° 34.6 min
8	NGC 0497	3.5'	S	01 h 22.39 m	−00° 52.5 min
9	NGC 2523	3.5'	E	08 h 14.99 m	73° 34.8 min
10	UGC 01775	2.6'	E	02 h 18.44 m	05° 39.2 min
11	UGC 00717	5.2'	W	01 h 09.39 m	14° 20.2 min
12	NGC 2608	3.5'	S	08 h 35.29 m	28° 28.4 min
Detached segments					
13	NGC 7448	2.6'	S	23 h 00.04 m	15° 59.4 min
14	NGC 7314	5.2'	N	22 h 35.76 m	−26° 03.0 min
15	NGC 7393	2.6'	E	22 h 51.66 m	−05° 33.5 min
16	Messier 66	10.4'	S	11 h 20.25 m	12° 59.3 min
17	UGC 03972	2.1'	W	07 h 44.55 m	73° 49.8 min
18	NGC 4088	5.2'	E	12 h 05.59 m	50° 32.5 min

(continued)

Table 11.1 (continued)

Arp #	Name	FOV	Or	RA (2000)	Dec (2000)
Three armed					
19	NGC 0145	2.6'	E	00 h 31.75 m	−05° 09.2 min
20	UGC 03014	2.1'	W	04 h 19.90 m	02° 05.7 min
21	CGCG 155-056	2.1'	S	11 h 04.98 m	30° 01.6 min
One armed					
22	NGC 4027	10.4'	N	11 h 59.51 m	−19° 15.8 min
23	NGC 4618	20.8'	W	12 h 41.54 m	41° 09.0 min
24	NGC 3445	3.5'	E	10 h 54.61 m	56° 59.4 min
One heavy arm					
25	NGC 2276	5.2'	N	07 h 27.22 m	85° 45.3 min
26	Messier 101	32.4'	N	14 h 03.21 m	54° 21.0 min
27	NGC 3631	5.2'	E	11 h 21.04 m	53° 10.3 min
28	NGC 7678	2.6'	W	23 h 28.46 m	22° 25.3 min
29	NGC 6946	20.8'	E	20 h 34.87 m	60° 09.2 min
30	NGC 6365A + B	2.6'	N	17 h 22.72 m	62° 10.4 min
Integral sign					
31	IC 0167	3.5	E	01 h 51.14 m	21° 54.8 min
32	UGC 10770	2.1	N	17 h 13.13 m	59° 19.4 min
33	UGC 08613	3.5	E	13 h 37.40 m	06° 26.2 min
34	NGC 4615 + 13 + 14	5.2	W	12 h 41.62 m	26° 04.3 min
35	UGC 00212 + comp	5.2	S	00 h 22.38 m	−01° 18.2 min
36	UGC 08548	2.6	N	13 h 34.25 m	31° 25.5 min

Table 11.2 The Arp catalog of spiral galaxies with companions on arms (Printed with the permission of Dennis Web)

Arp #	Name	FOV	Or	RA (2000)	Dec (2000)
Low surface brightness companions					
37	Messier 77	3.5	E	02 h 42.68 m	00° 00.8 min
38	NGC 6412	3.5	N	17 h 29.60 m	75° 42.3 min
39	NGC 1347	3.5	E	03 h 29.70 m	−22° 16.7 min
40	IC 4271	2.1	E	13 h 29.34 m	37° 24.5 min
41	NGC 1232 + A	10.4	E	03 h 09.76 m	−20° 34.7 min
42	NGC 5829 + IC 4526	3.5	N	15 h 02.70 m	23° 19.9 min
43	IC 0607	2.6	W	10 h 24.26 m	16° 44.6 min
44	IC 0609	3.5	E	10 h 25.55 m	−02° 13.3 min
45	UGC 09178	2.6	S	14 h 19.85 m	51° 54.3 min
46	UGC 12665	2.1	N	23 h 33.69 m	30° 02.6 min
47	MCG + 03-38-014 + comp	2.6	N	14 h 47.21 m	18° 51.5 min
48	CGCG 436-026	2.1	E	01 h 19.94 m	12° 20.7 min

(continued)

Table 11.2 (continued)

Arp #	Name	FOV	Or	RA (2000)	Dec (2000)
Small, high surface brightness companions					
49	NGC 5665	2.6	E	14 h 32.43 m	08° 04.8 min
50	IC 1520	2.1	E	23 h 57.97 m	−14° 01.8 min
51	PGC 475	2.1	N	00 h 06.28 m	−13° 26.9 min
52	CGCG 421-027	2.1	N	05 h 19.73 m	03° 43.0 min
53	NGC 3290	2.6	W	10 h 35.29 m	−17° 16.5 min
54	MCG-01-07-007+comp	2.6	N	02 h 24.03 m	−04° 41.6 min
55	UGC 4881 (Grasshopper)	2.6	S	09 h 15.93 m	44° 19.9 min
56	UGC 01432	2.1	W	01 h 57.43 m	17° 13.0 min
57	MCG+03-34-012/-013	2.6	E	13 h 16.67 m	14° 26.2 min
58	UGC 04457	2.6	S	08 h 31.97 m	19° 12.8 min
59	NGC 0341A+B	2.6	S	01 h 00.76 m	−09° 11.1 min
60	ARP 060	2.1	W	13 h 14.79 m	26° 05.1 min
61	UGC 03104	2.1	W	04 h 36.70 m	−02° 17.2 min
62	UGC 06865	2.1	S	11 h 53.60 m	43° 27.3 min
63	NGC 2944+comp	2.1	W	09 h 39.28 m	32° 18.6 min
64	UGC 09503	2.6	W	14 h 45.43 m	19° 27.9 min
65	NGC 0091	5.2	E	00 h 21.86 m	22° 24.0 min
66	UGC 10396	2.6	N	16 h 26.88 m	51° 33.3 min
67	UGC 00892	2.6	S	01 h 21.28 m	−00° 32.7 min
68	NGC 7757+56	3.5	W	23 h 48.75 m	04° 10.4 min
69	NGC 5579+80	2.6	S	14 h 20.44 m	35° 11.3 min
70	UGC 00934	2.6	S	01 h 23.47 m	30° 47.1 min
71	NGC 6045+comp	2.1	E	16 h 05.13 m	17° 45.4 min
72	NGC 5994+96	3.5	W	15 h 46.89 m	17° 52.3 min
73	IC 1222	2.6	E	16 h 35.15 m	46° 12.8 min
74	UGC 01626	2.6	S	02 h 08.36 m	41° 28.8 min
75	NGC 0702	3.5	E	01 h 51.31 m	−04° 03.3 min
76	Messier 90+IC 3583	10.4	N	12 h 36.83 m	13° 09.8 min
77	NGC 1097+A	10.4	W	02 h 46.32 m	−30° 16.5 min
78	NGC 0772+0770	10.4	W	01 h 59.34 m	19° 00.4 min
Large, high surface brightness companions					
79	NGC 5490 C	2.1	E	14 h 10.12 m	17° 36.9 min
80	NGC 2633	3.5	S	08 h 48.11 m	74° 06.0 min
81	NGC 6621+22	2.6	S	18 h 12.91 m	68° 21.8 min
82	NGC 2535+36	5.2	S	08 h 11.23 m	25° 12.4 min
83	NGC 3800+3799	3.5	W	11 h 40.22 m	15° 20.6 min
84	NGC 5395+94	5.2	N	13 h 58.64 m	37° 25.5 min
85	Messier 51	21.8	N	13 h 29.87 m	47° 11.9 min
86	NGC 7753+52	5.2	W	23 h 47.08 m	29° 29.0 min
87	NGC 3808+A	3.5	N	11 h 40.74 m	22° 25.7 min
88	VV 445	2.1	E	01 h 19.09 m	12° 28.7 min
89	NGC 2648+comp	5.2	S	08 h 42.67 m	14° 17.1 min
90	NGC 5930+29	2.6	S	15 h 26.13 m	41° 40.5 min
91	NGC 5953+54	2.6	N	15 h 34.54 m	15° 11.7 min

(continued)

Table 11.2 (continued)

Arp #	Name	FOV	Or	RA (2000)	Dec (2000)
Elliptical galaxy companions					
92	NGC 7603+comp	2.6	N	23 h 18.94 m	00° 14.6 min
93	NGC 7284+85	5.2	N	22 h 28.60 m	−24° 50.6 min
94	NGC 3226+27	10.4	N	10 h 23.45 m	19° 53.9 min
95	IC 4461+62	2.6	N	14 h 35.03 m	26° 32.6 min
96	UGC 03528A	3.5	E	07 h 03.01 m	86° 36.2 min
97	UGC 07085A	2.6	N	12 h 05.76 m	31° 03.6 min
98	UGC 01095+comp	2.6	W	01 h 32.28 m	32° 05.4 min
99	NGC 7550+49	10.4	S	23 h 15.27 m	18° 57.7 min
100	IC 0018+19	10.4	S	00 h 28.64 m	−11° 34.4 min
101	UGC 10169+164	5.2	S	16 h 04.53 m	14° 49.1 min
Elliptical and elliptical-like galaxies					
102	UGC 10814	10.4	S	17 h 19.24 m	48° 58.8 min
103	Zwicky's Triplet	3.5	N	16 h 49.43 m	45° 27.5 min
104	Keenan's System	104	S	13 h 32.18 m	62° 46.0 min
105	NGC 3561A+B	5.2	S	11 h 11.22 m	28° 41.7 min
106	NGC 4211+A	2.6	S	12 h 15.60 m	28° 10.6 min
107	UGC 05984	3.5	E	10 h 52.31 m	30° 04.0 min
108	ESO 547-G002+G003	2.6	E	03 h 03.09 m	−22° 13.0 min
Repelling spiral arms					
109	UGC 10053	3.5	E	15 h 48.13 m	69° 27.5 min
110	MCG-03-58-011	2.1	W	22 h 54.15 m	−15° 14.2 min
111	NGC 5421+comp	2.6	S	14 h 01.68 m	33° 49.7 min
112	NGC 7805+06	3.5	S	00 h 01.45 m	31° 26.0 min
Close to and perturbing spirals					
113	NGC 0071+70+68	5.2	N	00 h 18.37 m	30° 04.8 min
114	NGC 2300+2276	10.4	W	07 h 32.33 m	85° 42.5 min
115	UGC 06678	2.6	S	11 h 43.05 m	26° 16.6 min
116	Messier 60	10.4	W	12 h 43.66 m	11° 33.2 min
117	IC 0982+983	5.2	S	14 h 09.98 m	17° 41.8 min
118	NGC 1144+43	2.1	E	02 h 55.19 m	00° 10.6 min
119	UGC 00849	2.6	S	01 h 19.40 m	12° 26.8 min
120	NGC 4438+35	10.4	S	12 h 27.76 m	13° 00.5 min
121	MCG-01-03-051/-052	2.6	S	00 h 59.39 m	−04° 48.2 min
122	NGC 6040A+B	2.6	W	16 h 04.45 m	17° 45.0 min
123	NGC 1888+89	3.5	S	05 h 22.54 m	−11° 29.7 min
124	NGC 6361+comp	3.5	E	17 h 18.68 m	60° 36.4 min
125	UGC 10491	2.1	E	16 h 38.23 m	41° 55.8 min
126	UGC 01449	2.6	E	01 h 58.09 m	03° 06.0 min
127	NGC 0191+IC 1563	2.6	E	00 h 38.99 m	−09° 00.1 min
128	UGC 00827	2.1	W	01 h 17.41 m	14° 41.8 min
129	UGC 05146	2.1	W	09 h 39.41 m	32° 21.7 min
130	IC 5378	2.6	N	00 h 02.63 m	16° 38.6 min
131	MCG-03-08-025/-026	2.6	N	02 h 47.30 m	−14° 48.1 min
132	CGCG 011-053	2.1	S	11 h 19.35 m	−03° 05.4 min

(continued)

Table 11.2 (continued)

Arp #	Name	FOV	Or	RA (2000)	Dec (2000)
With nearby fragments					
133	NGC 0541	2.6	E	01 h 25.74 m	−01° 22.8 min
134	Messier 49	10.4	S	12 h 29.77 m	07° 59.8 min
135	NGC 1023	10.4	E	02 h 40.40 m	39° 03.8 min
136	NGC 5820	3.5	S	14 h 58.67 m	53° 53.2 min
Material emanating from the elliptical galaxies					
137	NGC 2914	2.1	N	09 h 34.05 m	10° 06.5 min
138	NGC 4015	2.1	N	11 h 58.71 m	25° 02.2 min
139	MCG + 05-31-135	2.1	W	13 h 07.40 m	26° 43.1 min
140	NGC 0275 + 74	2.6	E	00 h 51.08 m	−07° 03.8 min
141	UGC 03730	2.6	S	07 h 14.35 m	73° 28.8 min
142	NGC 2936 + 37	3.5	N	09 h 37.74 m	02° 45.6 min
143	NGC 2444 + 45	3.5	S	07 h 46.88 m	39° 01.9 min
144	NGC 7828	2.6	W	00 h 06.45 m	−13° 25.0 min
145	ARP 145	2.6	N	02 h 23.13 m	41° 22.2 min

Table 11.3 The Arp catalog of galaxies (not classifiable as S or E) (Printed with the permission of Dennis Web)

Arp #	Name	FOV	Or	RA (2000)	Dec (2000)
With associated rings					
146	VV 790	2.1	N	00 h 06.74 m	−06° 38.2 min
147	IC 0298	2.1	N	03 h 11.31 m	01° 18.8 min
148	Mayall's Object	2.1	W	11 h 04.00 m	40° 50.8 min
With Jets					
149	IC 0803	2.1	E	12 h 39.71 m	16° 35.5 min
150	Hickson 95	2.1	E	23 h 19.50 m	09° 30.5 min
151	Markarian 0040	2.1	N	11 h 25.60 m	54° 22.9 min
152	Messier 87	2.6	N	12 h 30.82 m	12° 23.5 min
Disturbed with interior absorption					
153	NGC 5128 (Centaurus A)	21.8	W	13 h 25.46 m	−43° 01.1 min
154	NGC 1316	5.2	E	03 h 22.69 m	−37° 12.6 min
155	NGC 3656	2.6	E	11 h 23.64 m	53° 50.5 min
156	UGC 05814	2.6	E	10 h 42.63 m	77° 29.7 min
157	NGC 0520	10.4	N	01 h 24.59 m	03° 47.6 min
158	NGC 0523	3.5	E	01 h 25.33 m	34° 01.5 min
159	NGC 4747	10.4	E	12 h 51.76 m	25° 46.5 min
160	NGC 4194	3.5	N	12 h 14.17 m	54° 31.7 min
With diffuse elements					
161	UGC 06665	2.6	S	11 h 42.21 m	00° 20.0 min
162	NGC 3414	3.5	N	10 h 51.27 m	27° 58.5 min
163	NGC 4670	2.1	W	12 h 45.28 m	27° 07.5 min
164	NGC 0455	3.5	E	01 h 15.95 m	05° 10.7 min

(continued)

Table 11.3 (continued)

Arp #	Name	FOV	Or	RA (2000)	Dec (2000)
165	NGC 2418	3.5	S	07 h 36.63 m	17° 53.1 min
166	NGC 0750+751	3.5	S	01 h 57.54 m	33° 12.6 min
Diffuse counter-tails					
167	NGC 2672+73	2.6	E	08 h 49.37 m	19° 04.5 min
168	Messier 32	16.2	E	00 h 42.70 m	40° 51.9 min
169	NGC 7236+37	3.5	E	22 h 14.75 m	13° 50.8 min
170	NGC 7578A+B	3.5	N	23 h 17.20 m	18° 42.0 min
171	NGC 5718	2.6	E	14 h 40.72 m	03° 27.2 min
172	IC 1178	3.5	S	16 h 05.55 m	17° 36.1 min
Narrow counter-tails					
173	UGC 09561	3.5	S	14 h 51.44 m	09° 19.7 min
174	NGC 3068A	5.2	S	09 h 58.67 m	28° 52.6 min
175	IC 3481+A+83	10.4	E	12 h 32.87 m	11° 24.2 min
176	NGC 4933A+B	3.5	W	13 h 03.91 m	−11° 30.4 min
177	MCG+04-35-017	2.1	W	14 h 55.81 m	24° 35.9 min
178	NGC 5613+14+15	3.5	N	14 h 24.10 m	34° 53.5 min
Narrow filaments					
179	NPM1G − 04.0134	2.1	W	03 h 01.68 m	−04° 40.3 min
180	MCG-01-13-034	2.6	S	04 h 53.37 m	−04° 48.1 min
181	NGC 3212	3.5	W	10 h 28.28 m	79° 49.4 min
182	Hickson 96	3.5	E	23 h 27.95 m	08° 46.7 min
183	UGC 08560	3.5	E	13 h 34.92 m	31° 23.5 min
184	NGC 1961	10.4	W	05 h 42.07 m	69° 22.8 min
185	NGC 6217	5.2	W	16 h 32.67 m	78° 11.9 min
186	NGC 1614	2.6	S	04 h 34.00 m	−08° 34.7 min
187	MCG-02-13-040A	2.1	S	05 h 04.88 m	−10° 14.9 min
188	UGC 10214	5.2	E	16 h 06.06 m	5° 25.3 min
189	NGC 4651	10.4	E	12 h 43.71 m	16° 23.7 min
190	UGC 02320	5.2	N	02 h 50.33 m	12° 53.3 min
191	UGC 06175	2.6	E	11 h 07.34 m	18° 25.9 min
192	NGC 3303	3.5	N	10 h 36.99 m	18° 08.2 min
193	IC 0883	2.6	S	13 h 20.59 m	34° 08.3 min
Material ejected from nuclei					
194	UGC 06945	2.6	N	11 h 57.88 m	36° 23.3 min
195	UGC 04653	2.6	W	08 h 53.84 m	35° 08.6 min
196	Herzog 21	2.1	N	13 h 14.63 m	26° 07.4 min
197	IC 0701	2.6	E	11 h 31.01 m	20° 28.2 min
198	UGC 06073	2.6	W	10 h 59.76 m	17° 38.9 min
199	NGC 5544+45	2.1	E	14 h 17.04 m	36° 34.3 min
200	NGC 1134	3.5	S	02 h 53.69 m	13° 00.9 min
201	UGC 00224	2.1	N	00 h 23.56 m	−00° 29.4 min
202	NGC 2719+A	3.5	W	09 h 00.26 m	35° 43.6 min
203	NGC 3712	3.5	N	11 h 31.15 m	28° 34.0 min
204	UGC 08454	2.6	E	13 h 22.83 m	84° 30.4 min
205	NGC 3448+comp	10.4	E	10 h 54.65 m	54° 18.4 min

(continued)

Table 11.3 (continued)

Arp #	Name	FOV	Or	RA (2000)	Dec (2000)
206	NGC 3432	10.4	N	10 h 52.52 m	36° 37.2 min
207	UGC 05050	2.1	S	09 h 31.11 m	76° 27.9 min
208	MCG+08-31-010	2.6	E	16 h 51.05 m	47° 13.2 min
Irregularities, absorption and resolution					
209	NGC 6052	2.1	W	16 h 05.21 m	20° 32.5 min
210	NGC 1569	3.5	W	04 h 30.82 m	64° 50.9 min
211	MCG+07-26-034	2.1	E	12 h 37.32 m	38° 43.5 min
212	NGC 7625	2.6		23 h 20.50 m	17° 13.5 min
213	IC 0356	5.2	W	04 h 07.79 m	69° 48.8 min
214	NGC 3718	10.4	S	11 h 32.59 m	53° 04.0 min
Adjacent loops					
215	NGC 2782	5.2	E	09 h 14.09 m	40° 06.8 min
216	NGC 7679+82	10.4	E	23 h 28.78 m	03° 30.7 min
217	NGC 3310	10.4	W	10 h 38.77 m	53° 30.1 min
218	CGCG 107-052	2.1	N	15 h 53.54 m	18° 36.2 min
219	UGC 02812	2.1	S	03 h 39.70 m	−02° 06.8 min
220	IC 4553	2.6	S	15 h 34.95 m	23° 30.3 min
Amorphous spiral arms					
221	MCG-02-25-006	2.6	N	09 h 36.40 m	−11° 19.7 min
222	NGC 7727	10.4	N	23 h 39.90 m	−12° 17.5 min
223	NGC 7585	3.5	E	23 h 18.04 m	−04° 38.9 min
224	NGC 3921	5.2	S	11 h 51.11 m	55° 04.7 min
225	NGC 2655	10.4	W	08 h 55.65 m	78° 13.5 min
226	NGC 7252	10.4	N	22 h 20.75 m	−24° 40.7 min
Concentric rings					
227	NGC 0474+70	10.4	E	01 h 20.11 m	03° 25.0 min
228	IC 0162	2.6	N	01 h 48.89 m	10° 31.3 min
229	NGC 0508+07	10.4	E	01 h 23.68 m	33° 16.8 min
230	IC 0051	3.5	N	00 h 46.40 m	−13° 26.5 min
231	IC 1575	5.2	S	00 h 43.56 m	−04° 07.1 min
232	NGC 2911	2.1	W	09 h 33.78 m	10° 09.1 min
Appearance of fission					
233	UGC 05720	2.6	W	10 h32.53 m	54° 24.0 min
234	NGC 3738	3.5	E	11 h 35.81 m	54° 31.5 min
235	NGC 0014	3.5	N	00 h 08.77 m	15° 48.9 min
236	IC 1623	5.2	E	01 h 07.79 m	−17° 30.4 min
237	UGC 05044	2.6	S	09 h 27.73 m	12° 17.2 min
238	UGC 08335	2.6	W	13 h 15.49 m	62° 07.7 min
239	NGC 5278+79	4.2	E	13 h 41.66 m	55° 40.1 min
240	NGC 5257+58	4.2	E	13 h 39.88 m	00° 50.4 min
241	UGC 09425	2.1	N	14 h 37.85 m	30° 28.9 min
242	NGC 4676 (The minice)	5.2	N	12 h 46.17 m	30° 44.0 min
243	NGC 2623	2.6	W	08 h 38.40 m	25° 45.3 min
244	NGC 4038/9 (Antennae)	20.8	E	12 h 01.88 m	−18° 51.9 min
245	NGC 2992+93	10.4	N	09 h 45.70 m	−14° 19.6 min

(continued)

Table 11.3 (continued)

Arp #	Name	FOV	Or	RA (2000)	Dec (2000)
246	NGC 7838+37	2.1	E	00 h 06.90 m	08° 21.0 min
247	IC 2338+39	2.6	S	08 h 23.54 m	21° 20.3 min
248	Wild's Triplet	5.2	W	11 h 46.75 m	–03° 50.8 min
249	UGC 12891	2.1	N	00 h 00.32 m	22° 59.4 min
250	IRAS F07327+3529	2.1	E	07 h 36.00 m	35° 22 5 min
251	VV 674	2.1	S	00 h 53.78 m	–13° 51.7 min
252	ESO 566-IG007	2.6	S	09 h 44.97 m	–19° 43.5 min
253	UGCA 173+174	3.5	E	09 h 43.41 m	–05° 16.8 min
254	NGC 5917	10.4	S	15 h 21.54 m	–07° 22.6 min
255	UGC 05304	2.6	N	09 h 53.15 m	07° 52.8 min
256	MCG-02-01-052/-051	2.6	W	00 h 18.83 m	–10° 21.7 min
Irregular clumps					
257	UGC 04638	2.6	W	08 h 51.64 m	–02° 22.0 min
258	Hickson 18	3.5	S	02 h 39.08 m	18° 23.6 min
259	Hickson 31	3.5	W	05 h 01.63 m	–04° 15.5 min
260	UGC 07230	2.6	N	12 h 13.65 m	16° 07.3 min
261	MCG-02-38-016	10.4	N	14 h 49.51 m	–10° 10.4 min
262	UGC 12856	2.6	N	23 h 56.75 m	16° 48.7 min
263	NGC 3239	5.2	S	10 h 25.09 m	17° 09.6 min
264	NGC 3104	3.5	E	10 h 03.96 m	40° 45.4 min
265	IC 3862	2.1	N	12 h 53.89 m	36° 05.2 min
266	NGC 4861	3.5	N	12 h 59.03 m	34° 51.6 min
267	UGC 05764	2.6	S	10 h 36.71 m	31° 32.8 min
268	Holmberg II	10.4	S	08 h 19.10 m	70° 42.8 min

Table 11.4 The Arp catalog of galaxies with group characteristics (Printed with the permission of Dennis Web)

Arp #	Name	FOV	Or	RA (2000)	Dec (2000)
Connected arms					
269	NGC 4490+85	10.4	N	12 h 30.61 m	41° 38.4 min
270	NGC 3395+96	3.5	E	10 h 49.83 m	32° 58.8 min
271	NGC 5426+27	5.2	S	14 h 03.42 m	–06° 04.2 min
272	NGC 6050	2.1	W	16 h 05.40 m	17° 45.5 min
273	UGC 01810+13	3.5	E	02 h 21.48 m	39° 22.5 min
274	NGC 5679A+B+C	2.6	E	14 h 35.10 m	05° 20.9 min
Interacting					
275	NGC 2881	2.1	W	09 h 25.92 m	–11° 59.0 min
276	NGC 0935+IC 1801	2.6	S	02 h 28.18 m	19° 36.0 min
277	NGC 4809+10	2.1	E	12 h 54.85 m	02° 39.2 min
278	NGC 7253A+B	2.6	E	22 h 19.44 m	29° 23.9 min
279	NGC 1253+A	10.4	E	03 h 14.15 m	–02° 49.4 min
280	NGC 3769+A	3.5	S	11 h 37.74 m	47° 53.5 min

(continued)

Table 11.4 (continued)

Arp #	Name	FOV	Or	RA (2000)	Dec (2000)
Infall and attraction					
281	NGC 4631+27	21.8	E	12 h 42.10 m	32° 32.4 min
282	NGC 0169+IC 1559	3.5	W	00 h 36.87 m	23° 59.5 min
283	NGC 2798+99	3.5	S	09 h 17.38 m	41° 60.0 min
284	NGC 7714+15	5.2	E	23 h 36.24 m	02° 09.3 min
285	NGC 2854+56	5.2	W	09 h 24.05 m	49° 12.2 min
286	NGC 5566+60+69	10.4	W	14 h 20.34 m	03° 56.0 min
Wind effects					
287	NGC 2735+A	2.6	N	09 h 02.64 m	25° 56.1 min
288	NGC 5221+22	10.4	N	13 h 34.94 m	13° 49.9 min
289	NGC 3981	10.4	N	11 h 56.12 m	−19° 53.7 min
290	IC 0195+196	3.5	N	02 h 03.74 m	14° 42.5 min
291	UGC 05832	2.6	E	10 h 42.81 m	13° 27.6 min
292	IC 0575	2.6	S	09 h 54.55 m	−06° 51.4 min
293	NGC 6286+85	3.5	N	16 h 58.52 m	58° 56.2 min
Long filaments					
294	NGC 3786+88	10.4	W	11 h 39.71 m	31° 54.5 min
295	MCG-01-60-021+comp	10.4	N	23 h 41.80 m	−03° 40.6 min
296	MCG+10-17-005+comp	3.5	S	11 h 28.33 m	58° 32.5 min
297	NGC 5754+55	10.4	W	14 h 45.33 m	38° 43.9 min
Uncertain category, probably continuation of filaments					
298	NGC 7469+IC 5283	3.5	N	23 h 03.26 m	08° 52.4 min
299	NGC 3690	3.5	S	11 h 28.56 m	58° 33.8 min
300	UGC 05029	3.5	W	09 h 28.04 m	68° 25.3 min
301	UGC 06204	2.6	E	11 h 09.86 m	24° 15.7 min
302	UGC 09618	2.6	N	14 h 57.02 m	24° 37.0 min
303	IC 0563+64	3.5	S	09 h 46.34 m	03° 02.7 min
304	NGC 1241+42	3.5	E	03 h 11.25 m	−08° 55.3 min
305	NGC 4016+17	10.4	N	11 h 58.49 m	27° 31.7 min
306	UGC 01102	5.2	E	01 h 32.49 m	04° 35.7 min
307	NGC 2872+73+74	3.5	N	09 h 25.71 m	11° 25.9 min
308	NGC 0545+47	2.6	W	01 h 25.99 m	−01° 20.4 min
309	NGC 0942+43	3.5	N	02 h 29.17 m	−10° 50.2 min
310	IC 1259	2.1	W	17 h 27.42 m	58° 31.0 min
Groups of galaxies					
311	IC 1259 Field	5.2	E	17 h 27.29 m	58° 29.1 min
312	MCG+08-31-004	2.6	W	16 h 49.81 m	46° 43.1 min
313	NGC 3994+95	10.4	W	11 h 57.61 m	32° 16.6 min
314	MCG-03-58-009/10/11	10.4	E	22 h 58.02 m	−03° 46.1 min
315	NGC 2832+30+31	2.6	W	09 h 19.78 m	33° 45.0 min
316	Hickson 44	16.2	W	10 h 18.42 m	21° 53.6 min
317	Messier 65+66	62.5	S	11 h 18.92 m	13° 05.6 min
318	Hickson 16	10.4	E	02 h 09.34 m	−10° 08.0 min
319	Stephan's Quintet	10.4	E	22 h 35.95 m	33° 57.9 min

(continued)

Table 11.4 (continued)

Arp #	Name	FOV	Or	RA (2000)	Dec (2000)
320	Copeland's Septet	5.2	N	11 h 37.86 m	21° 58.4 min
321	Hickson 40	2.6	N	09 h 38.89 m	−04° 51.6 min
Chains of galaxies					
322	Hickson 56 (part)	2.6	E	11 h 32.67 m	52° 57.0 min
323	Hickson 98	3.5	S	23 h 54.17 m	00° 23.0 min
324	UGC 10143	12.8	S	16 h 02.21 m	15° 54.4 min
325	ESO 601-G018	2.1	N	22 h 06.40 m	−21° 04.7 min
326	UGC 08610+08613	10.4	N	13 h 37.31 m	06° 28.7 min
327	Hickson 34	2.6	E	05 h 21.77 m	06° 41.3 min
328	Hickson 72	5.2	N	14 h 47.93 m	19° 03.4 min
329	Hickson 55	2.6	N	11 h 32.12 m	70° 48.9 min
330	I Zw 167	10.4	N	16 h 49.13 m	53° 25.9 min
331	Pisces Cloud	20.8	N	01 h 07.42 m	32° 24.7 min
333	IC 1892+comps	20.8	N	03 h 08.47 m	−23° 03.1 min

Table 11.5 The Arp catalog of miscellaneous galaxies (Printed with the permission of Dennis Web)

Arp #	Name	FOV	Or	RA (2000)	Dec (2000)
333	NGC 1024	5.2	S	02 h 39.20 m	10° 50.8 min
334	UGC 08498	3.5	S	13 h 30.43 m	31° 37.2 min
335	NGC 3509	2.6	E	11 h 04.40 m	04° 49.8 min
336	NGC 2685	6.9	S	08 h 55.59 m	58° 44.0 min
337	Messier 82	10.4	E	09 h 55.86 m	69° 40.8 min
338	Arp 338	2.1	S	10 h 10.99 m	−07° 54.9 min

Chapter 12

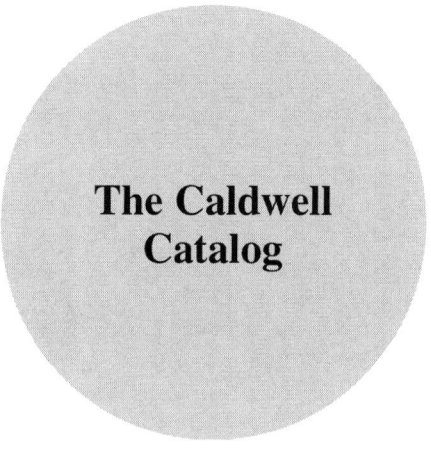

The Caldwell Catalog

Sir Patrick Alfred Caldwell-Moore (1923–) is a British amateur astronomer who has gained worldwide admiration and respect as an authority on astronomy. His interest in astronomy began at the age of 6 when his mother gave him a copy of *The Story of the Solar System* (1898) by G. F. Chambers. He is the author of over 70 books and has hosted the popular BBC series *Sky At Night* since 1957, making it the longest running television show ever. Sir Patrick Moore is the former president of the British Astronomical Association and is a co-founder and former president of the Society of Popular Astronomy. His dedication and his popularization of astronomy to the British public have gained him a long list of notable awards and honors.

The Caldwell Catalogue was created by Sir Patrick Moore when he noted that the Messier Catalog did not include some of the skies brightest objects and that the objects included were only found in the northern hemisphere. To complement the Messier Catalog he compiled a catalog of 109 interesting deep-sky objects containing star clusters, nebulae, and galaxies found in both the northern and southern hemispheres. The Caldwell Catalog was first published by *Sky & Telescope* magazine in December 1995. Since its publication it has become very popular among amateur astronomers. It is seen as a perfect progression for amateurs after they have viewed the Messier Catalog and before they move on to the Herschel catalogs (Table 12.1).

J.D. Cavin, *The Amateur Astronomer's Guide to the Deep-Sky Catalogs*,
Patrick Moore's Practical Astronomy Series, DOI 10.1007/978-1-4614-0656-3_12,
© Springer Science+Business Media, LLC 2012

Table 12.1   The Caldwell catalogue (With permission from Tony Flanders, Associate Editor, Sky & Telescope Magazine)

Caldwell number	NGC number	Common name(s)	Object type	Const	RA (2000)	Dec (2000)	V Mag
C1	NGC 188		Open Cluster	Cep	00 h 47 m 29.7 s	+85° 14′ 29″	8.1
C2	NGC 40	Bow-Tie Nebula	Planetary Nebula	Cep	00 h 13 m 00.9 s	+72° 31′20″	11
C3	NGC 4236		Galaxy	Dra	12 h 16 m 42.1 s	+69° 27′ 45″	9.7
C4	NGC 7023	Iris Nebula	Open Cluster and Nebula	Cep	21 h 01 m 35.5 s	+68° 10′ 11″	7
C5	IC 342		Galaxy	Cam	03 h 46 m 48.0 s	+68° 6′ 00″	9
C6	NGC 6543	Cat's Eye Nebula	Planetary Nebula	Dra	17 h 58 m 33.37 s	+66° 37′59.1″	9
C7	NGC 2403		Galaxy	Cam	07 h 36 m 51.8 s	+65° 36′ 13″	8.4
C8	NGC 559		Open Cluster	Cas	01 h 29 m 29.1 s	+63° 18′ 30″	9.5
C9	Sh2-155	Cave Nebula	Nebula	Cep	22 h 57 m 17.14 s	+62° 28′ 33.4″	–
C10	NGC 663		Open Cluster	Cas	01 h 46 m 16.0 s	+61° 13′ 06″	7.1
C11	NGC 7635	Bubble Nebula	Nebula	Cas	23 h 20 m 45.5 s	+61° 12′ 45″	–
C12	NGC 6946		Galaxy	Cep	20 h 34 m 52.7 s	+60° 09′ 11″	8.9
C13	NGC 457	Owl Cluster, E.T. Cluster	Open Cluster	Cas	01 h 19 m 32.6 s	+58° 17′ 27″	6.4
C14	NGC 869 and NGC 884	Double Cluster, H and χ Persei	Open Cluster	Per	02 h 19 m 03.8 s 02 h 22 m 32.1 s	+57° 08′ 06″ +57° 08′ 39″	5.3 4
C15	NGC 6826	Blinking Planetary	Planetary Nebula	Cyg	19 h 44 m 48.17 s	+50° 31′ 30.4″	10
C16	NGC 7243		Open Cluster	Lac	22 h 15 m 08.5 s	+49° 53′ 51″	6.4
C17	NGC 147		Galaxy	Cas	00 h 33 m 12.0 s	+48° 30′ 31″	9.3
C18	NGC 185		Galaxy	Cas	00 h 38 m 58.1 s	+48° 20′ 27″	9.2
C19	IC 5146	Cocoon Nebula	Open Cluster and Nebula	Cyg	21 h 53 m 24.0 s	+47° 16′ 00″	7.2
C20	NGC 7000	North America Nebula	Nebula	Cyg	21 h 01 m 48.0 s	+44° 12′ 00″	–
C21	NGC 4449		Galaxy	CVn	12 h 28 m 11.1 s	+44° 05′ 36″	9.4
C22	NGC 7662	Blue Snowball	Planetary Nebula	And	23 h 25 m 53.93 s	+42° 32′ 06.1″	9
C23	NGC 891		Galaxy	And	02 h 22 m 33.5 s	+42° 21′ 03″	10
C24	NGC 1275	Per A	Galaxy	Per	03 h 19 m 48.2 s	+41° 30′ 42″	11.6
C25	NGC 2419		Globular Cluster	Lyn	07 h 38 m 08.4 s	+38° 52′ 53″	10.4

C26	NGC 4244		Galaxy	CVn	12 h 17 m 29.7 s	+37° 48' 24"	10.2
C27	NGC 6888	Crescent Nebula	Nebula	Cyg	20 h 12 m 06.4 s	+38° 21' 18"	7.4
C28	NGC 752		Open Cluster	And	01 h 57 m 47.9 s	+37° 51' 00"	5.7
C29	NGC 5005		Galaxy	CVn	13 h 10 m 56.6 s	+37° 03' 33"	9.8
C30	NGC 7331		Galaxy	Peg	22 h 37 m 04.5 s	+34° 25' 01"	9.5
C31	IC 405	Flaming Star Nebula	Nebula	Aur	05 h 16 m 12.0 s	+34° 16' 00"	–
C32	NGC 4631	Whale Galaxy	Galaxy	CVn	12 h 42 m 08.1 s	+32° 32' 26"	9.3
C33	NGC 6992	Network Nebula	Supernova Remnant	Cyg	20 h 56 m 19.0 s	+31° 44' 34"	–
C34	NGC 6960	Veil Nebula	Supernova Remnant	Cyg	20 h 45 m 58.1 s	+30° 35' 43"	–
C35	NGC 4889		Galaxy	Com	13 h 00 m 08.1 s	+27° 58' 37"	11.4
C36	NGC 4559		Galaxy	Com	12 h 35 m 57.7 s	+27° 57' 35"	9.9
C37	NGC 6885		Open Cluster	Vul	20 h 11 m 55.8 s	+26° 29' 20"	6
C38	NGC 4565	Needle Galaxy	Galaxy	Com	12 h 36 m 20.7 s	+25° 59' 19"	9.6
C39	NGC 2392	Eskimo Nebula/Clown Face Nebula	Planetary Nebula	Gem	07 h 29 m 10.8 s	+20° 54' 42"	10
C40	NGC 3626		Galaxy	Leo	11 h 20 m 03.9 s	+18° 21' 23"	10.9
C41	Mel25	Hyades	Open Cluster	Tau			0.5
C42	NGC 7006		Globular Cluster	Del	21 h 01 m 29.3 s	+16° 11' 16"	10.6
C43	NGC 7814		Galaxy	Peg	00 h 03 m 14.8 s	+16° 08' 43"	10.5
C44	NGC 7479		Galaxy	Peg	23 h 04 m 56.4 s	+12° 19' 00"	11
C45	NGC 5248		Galaxy	Boö	13 h 37 m 32.0 s	+08° 53' 08"	10.2
C46	NGC 2261	Hubble's Variable Nebula	Nebula	Mon	06 h 39 m 09.5 s	+08° 44' 39"	–
C47	NGC 6934		Globular Cluster	Del	20 h 34 m 11.5 s	+07° 24' 16"	8.9
C48	NGC 2775		Galaxy	Cnc	09 h 10 m 20.3 s	+07° 02' 16"	10.3
C49	NGC 2237	Western half of the Rosette Nebula	Nebula	Mon	06 h 30 m 54.6 s	+05° 02' 57"	9.0
C50	NGC 2244	Center of the Rosette Nebula	Open Cluster	Mon	06 h 31 m 55.6 s	+04° 56' 35"	4.8
C51	IC 1613		Galaxy	Cet	01 h 04 m 54.0 s	+2° 8' 00"	9.3

(continued)

Table 12.1 (continued)

Caldwell number	NGC number	Common name(s)	Object type	Const	RA (2000)	Dec (2000)	V Mag
C52	NGC 4697		Galaxy	Vir	12 h 48 m 35.7 s	−05° 48' 03"	9.3
C53	NGC 3115	Spindle Galaxy	Galaxy	Sex	10 h 05 m 14.1 s	−07° 43' 09"	9.2
C54	NGC 2506		Open Cluster	Mon	08 h 00 m 01.7 s	−10° 46' 11"	7.6
C55	NGC 7009	Saturn Nebula	Planetary Nebula	Aqr	21 h 04 m 10.7 s	−11° 21' 49"	8
C56	NGC 246		Planetary Nebula	Cet	00 h 47 m 03.2 s	−11° 52' 20"	8
C57	NGC 6822	Barnard's Galaxy	Galaxy	Sgr	19 h 44 m 56.3 s	−14° 48' 37"	9
C58	NGC 2360		Open Cluster	CMa	07 h 17 m 43.1 s	−15° 38' 29"	7.2
C59	NGC 3242	Ghost of Jupiter	Planetary Nebula	Hya	10 h 24 m 46.2 s	−18° 38' 34"	9
C60	NGC 4038	Antennae Galaxies	Galaxy	Crv	12 h 01 m 53.0 s	−18° 52' 07"	10.7
C61	NGC 4039	Antennae Galaxies	Galaxy	Crv	12 h 01 m 53.6 s	−18° 53' 10"	13
C62	NGC 247		Galaxy	Cetus	00 h 47 m 08.4 s	−20° 45' 36"	8.9
C63	NGC 7293	Helix Nebula	Planetary Nebula	Aqr	22 h 29 m 38.35 s	−20° 50' 13.2"	7.3
C64	NGC 2362		Open Cluster and Nebula	CMa	07 h 18 m 41.5 s	−24° 57' 15"	4.1
C65	NGC 253	Sculptor Galaxy/Silver Coin Galaxy	Galaxy	Scl	00 h 47 m 33.1 s	−25° 17' 17"	7.1
C66	NGC 5694		Globular Cluster	Hya	14 h 39 m 36.5 s	−26° 32' 18"	10.2
C67	NGC 1097		Galaxy	For	02 h 46 m 19.0 s	−30° 16' 28"	9.3
C68	NGC 6729		Nebula	CrA	19 h 01 m 54.1 s	−36° 57' 12"	–
C69	NGC 6302	Bug Nebula	Planetary Nebula	Sco	17 h 13 m 44.6 s	−37° 06' 12"	13
C70	NGC 300		Galaxy	Scl	00 h 54 m 53.4 s	−37° 41' 00"	9
C71	NGC 2477		Open Cluster	Pup	07 h 52 m 09.8 s	−38° 32' 00"	5.8
C72	NGC 55		Galaxy	Scl	00 h 15 m 08.4 s	−39° 13' 14	8
C73	NGC 1851		Globular Cluster	Col	05 h 14 m 06.3 s	−40° 02' 50"	7.3
C74	NGC 3132	Eight Burst Nebula	Planetary Nebula	Vela	10 h 07 m 01.77 s	−40° 26' 11.7"	8
C75	NGC 6124		Open Cluster	Sco	16 h 25 m 20.0 s	−40° 39' 13"	5.8
C76	NGC 6231		Open Cluster and Nebula	Sco	16 h 54 m 10.9 s	−41° 49' 27"	2.6

C77	NGC 5128	Cen A	Galaxy	Cen	13 h 25 m 29.0 s	−43° 01' 00"	7
C78	NGC 6541		Globular Cluster	CrA	18 h 08 m 02.3 s	−43° 42' 57"	6.6
C79	NGC 3201		Globular Cluster	Vela	10 h 17 m 36.7 s	−46° 24' 40"	6.8
C80	NGC 5139	Omega Centauri	Globular Cluster	Cen	13 h 26 m 45.9 s	−47° 28' 37"	3.7
C81	NGC 6352		Globular Cluster	Ara	17 h 25 m 29.1 s	−48° 25' 22"	8.2
C82	NGC 6193		Open Cluster	Ara	16 h 41 m 20.2 s	−48° 45' 45"	5.2
C83	NGC 4945		Galaxy	Cen	13 h 05 m 27.1 s	−49° 28' 03"	9
C84	NGC 5286		Globular Cluster	Cen	13 h 46 m 26.5 s	−51° 22' 24"	7.6
C85	IC 2391	Omicron Velorum Cluster	Open Cluster	Vela	08 h 40 m 12.0 s	−53° 4' 00"	2.5
C86	NGC 6397		Globular Cluster	Ara	17 h 40 m 41.2 s	−53° 40' 25"	5.7
C87	NGC 1261		Globular Cluster	Hor	03 h 12 m 15.3 s	−55° 13' 01"	8.4
C88	NGC 5823		Open Cluster	Cir	15 h 05 m 30.6 s	−55° 36' 13"	7.9
C89	NGC 6087	S Nor Cluster	Open Cluster	Nor	16 h 18 m 50.5 s	−57° 56' 04"	5.4
C90	NGC 2867		Planetary Nebula	Car	09 h 21 m 25.35 s	−58° 18' 41.5"	10
C91	NGC 3532		Open Cluster	Car	11 h 05 m 47.5 s	−58° 46' 13"	3
C92	NGC 3372	Eta Carina Nebula	Nebula	Car	10 h 45 m 08.5 s	−59° 52' 04"	—
C93	NGC 6752		Globular Cluster	Pav	19 h 10 m 51.7 s	−59° 58' 55"	5.4
C94	NGC 4755	Jewel Box	Open Cluster	Cru	12 h 53 m 37.1 s	−60° 21' 22"	4.2
C95	NGC 6025		Open Cluster	TrA	16 h 03 m 17.0 s	−60° 25' 54"	5.1
C96	NGC 2516	The Running Man Cluster	Open Cluster	Car	07 h 58 m 07.1 s	−60° 45' 12"	3.8
C97	NGC 3766		Open Cluster	Cen	11 h 36 m 14.4 s	−61° 36' 36"	5.3
C98	NGC 4609		Open Cluster	Cru	12 h 42 m 19.9 s	−62° 59' 38"	6.9
C99	—	Coalsack Nebula	Dark Nebula	Cru	12 h 50 m	−62° 30'	—
C100	IC 2944	Lambda Centauri Nebula	Open Cluster and Nebula	Cen	11 h 36 m 36.0 s	−63° 2' 00"	4.5
C101	NGC 6744		Galaxy	Pav	19 h 09 m 46.1 s	−63° 51' 25"	9
C102	IC 2602	Theta Car Cluster	Open Cluster	Car	10 h 43 m 12.0 s	−64° 24' 00"	1.9
C103	NGC 2070	Tarantula Nebula	Open Cluster and Nebula	Dor	05 h 38 m 38.4 s	−69° 05' 39"	8.2
C104	NGC 362		Globular Cluster	Tuc	01 h 03 m 14.2 s	−70° 50' 54"	6.6
C105	NGC 4833		Globular Cluster	Mus	12 h 59 m 34.9 s	−70° 52' 28"	7.4

(continued)

Table 12.1 (continued)

Caldwell number	NGC number	Common name(s)	Object type	Const	RA (2000)	Dec (2000)	V Mag
C106	NGC 104	47 Tucanae	Globular Cluster	Tuc	00 h 24 m 05.1 s	−72° 04′ 51″	4
C107	NGC 6101		Globular Cluster	Aps	16 h 25 m 48.5 s	−72° 12′ 06″	9.3
C108	NGC 4372		Globular Cluster	Mus	12 h 25 m 45.4 s	−72° 39′ 33″	7.8
C109	NGC 3195		Planetary Nebula	Cha	10 h 09 m 21.07 s	−80° 51′ 30.8″	–

Appendix A

Constellation name	Abbr	Location	Meaning	Proposer
Andromeda	And	Northern	Andromeda	Ptolemy
Antlia	Ant	Southern	The air pump	La Caille
Apus	Aps	Southern	The bird of paradise	Bayer
Aquarius	Aqr	Zodiac	The water carrier	Ptolemy
Aquila	Aql	Northern	The eagle	Ptolemy
Ara	Ara	Southern	The altar	Ptolemy
Aries	Ari	Zodiac	The ram	Ptolemy
Auriga	Aur	Northern	The charioteer	Ptolemy
Boötes	Boo	Northern	The herdsman	Ptolemy
Caelum	Cae	Southern	The chisel (or Sculptor's tool)	La Caille
Camelopardalis	Cam	Northern	The giraffe	Hevelius or Bartschius
Cancer	Cnc	Zodiac	The crab	Ptolemy
Canes Venatici	CVn	Northern	The hunting dogs	Hevelius
Canis Major	CMa	Northern	The greater dog	Ptolemy
Canis Minor	CMi	Northern	The lesser dog	Ptolemy
Capricornus	Cap	Zodiac	The goat-fish	Ptolemy
Carina	Car	Southern	The keel	La Caille
Cassiopeia	Cas	Northern		Ptolemy
Centaurus	Cen	Southern	The centaur	Ptolemy

(continued)

J.D. Cavin, *The Amateur Astronomer's Guide to the Deep-Sky Catalogs*,
Patrick Moore's Practical Astronomy Series, DOI 10.1007/978-1-4614-0656-3,
© Springer Science+Business Media, LLC 2012

(continued)

Constellation name	Abbr	Location	Meaning	Proposer
Cepheus	Cep	Northern		Ptolemy
Cetus	Cet	Southern	The whale	Ptolemy
Chameleon	Cha	Southern	The chameleon	Bayer
Circinus	Cir	Southern	The compass	La Caille
Columba	Col	Southern	The dove	Royer or Bartschius
Coma Berenices	Com	Northern	Berenice's hair	Tycho Brahe or Eratosthenes
Corona Australis	CrA	Southern	The southern crown	Ptolemy
Corona Borealis	CrB	Northern	The northern crown	Ptolemy
Corvus	Crv	Southern	The crow (or Raven)	Ptolemy
Crater	Crt	Southern	The cup	Ptolemy
Crux	Cru	Southern	The cross	Royer
Cygnus	Cyg	Northern	The swan	Ptolemy
Delphinus	Del	Northern	The dolphin	Ptolemy
Dorado	Dor	Southern	The swordfish	Bayer
Draco	Dra	Northern	The dragon	Ptolemy
Equuleus	Equ	Northern	The little horse	Ptolemy
Eridanus	Eri	Southern	The river	Ptolemy
Fornax	For	Southern	The furnace	La Caille
Gemini	Gem	Zodiac	The twins	Ptolemy
Grus	Gru	Southern	The crane	Bayer
Hercules	Her	Northern	Hercules	Ptolemy
Horologium	Hor	Southern	The pendulum clock	La Caille
Hydra	Hya	Northern	The snake	Ptolemy
Hydrus	Hyi	Southern	The water snake	Bayer
Indus	Ind	Southern	The Indian	Bayer
Lacerta	Lac	Northern	The lizard	Hevelius
Leo	Leo	Zodiac	The lion	Ptolemy
Leo Minor	LMi	Northern	The lesser lion	Hevelius
Lepus	Lep	Southern	The hare	Ptolemy
Libra	Lib	Zodiac	The scales	Ptolemy
Lupus	Lup	Southern	The wolf	Ptolemy
Lynx	Lyn	Northern	The lynx	Hevelius
Lyra	Lyr	Northern	The lyre (harp)	Ptolemy
Mensa	Men	Southern	The table	La Caille
Microscopium	Mic	Southern	The microscope	La Caille
Monoceros	Mon	Southern	The unicorn	La Caille
Musca	Mus	Southern	The southern fly	Bayer
Norma	Nor	Southern	The rule (or straightedge)	La Caille
Octans	Oct	Southern	The octant	Ptolemy
Ophiuchus	Oph	Northern	Serpent-bearer	Ptolemy
Orion	Ori	Southern	The hunter	Ptolemy
Pavo	Pav	Southern	The peacock	Bayer
Pegasus	Peg	Northern	The winged horse	Ptolemy
Perseus	Per	Northern	Perseus	Ptolemy
Phoenix	Phe	Southern	The phoenix	Bayer

(continued)

(continued)

Constellation name	Abbr	Location	Meaning	Proposer
Pictor	Pic	Southern	The painter (or easel)	La Caille
Pisces	Psc	Zodiac	The pishes	Ptolemy
Pisces Austrinus	PsA	Southern	The southern fish	Ptolemy
Puppis	Pup	Puppis	Poop deck of the Argo	La Caille
Pyxis	Pyx	Southern	The Mariner's compass	La Caille
Reticulum	Ret	Southern	The reticle	La Caille
Sagitta	Sge	Northern	The arrow	Ptolemy
Sagittarius	Sgr	Zodiac	The archer	Ptolemy
Scorpius	Sco	Zodiac	The scorpion	Ptolemy
Sculptor	Scl	Southern	The sculptor	La Caille
Scutum	Sct	Northern	The shield	Hevelius
Serpens	Ser	Northern	The serpent	Ptolemy
Sextans	Sex	Southern	The sextant	Ptolemy
Taurus	Tau	Zodiac	The bull	Ptolemy
Telescopium	Tel	Southern	The telescope	La Caille
Triangulum	Tri	Northern	The triangle	Ptolemy
Triangulum Australe	TrA	Southern	Southern triangle	Bayer
Tucana	Tuc	Southern	The toucan	Bayer
Ursa Major	UMa	Northern	The great bear	Ptolemy
Ursa Minor	UMi	Northern	The lesser bear	Ptolemy
Vela	Vel	Southern	The sails	La Caille
Virgo	Vir	Zodiac	The virgin	Ptolemy
Volans	Vol	Southern	The flying fish	Bayer
Vulpecula	Vul	Northern	The fox	Hevelius

Appendix B

Upper case	Lower case	Greek name
A	α	Alpha
B	β	Beta
Γ	γ	Gamma
Δ	δ	Delta
E	ε	Epsilon
Z	ζ	Zeta
H	η	Eta
Θ	θ	Theta
I	ι	Iota
K	κ	Kappa
Λ	λ	Lambda
M	μ	Mu
N	ν	Nu
Ξ	ξ	Xi
O	o	Omicron
Π	π	Pi
P	ρ	Rho
Σ	σ	Sigma
T	τ	Tau
Y	υ	Upsilon
Φ	φ	Phi
X	χ	Chi
Ψ	ψ	Psi
Ω	ω	Omega

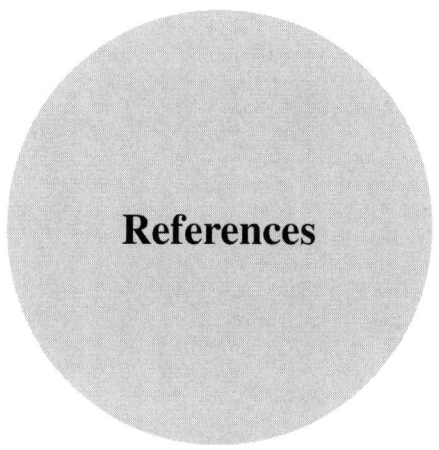

References

Brewster, David. *The Martyrs of Science, or, The Lives of Galileo, Tycho Brahe, and Kepler*. G S Tullis Printers, London (1841).

Couper, Heather, and Nigel Henbest. *The History of Astronomy*, Firefly Books, Buffalo, New York (2007).

Forbes, George. *The History of Astronomy*, G P Putnam's Sons, New York (1909).

Freely, John. *Aladdin's Lamp: How Greek Science Came to Europe Through the Islamic World*, Vintage Books, a division of Random House, New York (2009)

Grasshoff, Gerd. *The History of Ptolemy's Star Catalog*, Springer-Verlag, New York (1990).

Horowitz, Wayne. *Mesopotamian Cosmic Geography*, Eisenbrauns; illustrated edition (2008).

Jones, Kenneth Glyn. *Messier's Nebulae and Star Clusters*, American Elsevier Publishing Company, New York (1968).

Kanipe, Jeff, and Dennis Webb. *The Arp Atlas of Peculiar Galaxies, A Chronicle and Observer's Guide*, William-Bell, Inc. (2006).

Knoble, Edward Ball. *Ulugh Beg's Catalogue of Stars*, Gibson's Brothers Press, Washington (1917).

Machholtz, Don. *The Observing Guide to the Messier Marathon*, Cambridge University Press, New York (2002).

Mullaney, James, *The Herschel Objects and How to Observe Them*, Springer Books, New York (2007).

Mullaney, James, and W. Tirion. *The Cambridge Atlas of Herschel Objects*, Cambridge University Press, New York (2011).

Newcomb, Simon. *The Stars: A Study of the Universe*, G P Putnam's Sons, New York (1901).

North, John. *Cosmos, An Illustrated History of Astronomy and Cosmology*, the University of Chicago Press (2008),Chicago .

Prince, Charles Leeson, ed. *The Illustrated Account given by Hevelius in his "Machina Celestas" of the Method of Mounting His Telescope and Erecting an Observatory* (1882).

Ratledge, David. *Observing the Caldwell Objects*, Springer Books, New York (2000).

Rogers, John H. "Origins of Ancient Constellations: I. Mesopotamian Traditions," *Journal of the British Astronomical Association,* vol.108, no.1, pp. 9–28.

Steinicke, Wolfgang. *Observing and Cataloguing Nebulae and Star Clusters, From Herschel to Dreyer's New General Catalogue*, Cambridge University Press, New York (2010).

White, Gavin. *Babylonian Star Lore*, Solaria Publications, London (2008).

Index